Electromagnetic Fields and Waves

Magdy F. Iskander

University of Hawaii at Manoa

WAVELAND

PRESS, INC.

Long Grove, Illinois

To the memory of my parents

For information about this book, contact:
Waveland Press, Inc.
4180 IL Route 83, Suite 101
Long Grove, IL 60047-9580
(847) 634-0081
info@waveland.com
www.waveland.com

CONTENTS

CHAPTER **2**

MAXWELL'S EQUATIONS IN DIFFERENTIAL FORM   99

CHAPTER **3**

MAXWELL'S EQUATIONS AND PLANE WAVE
PROPAGATION IN MATERIALS   179

CHAPTER 4

STATIC ELECTRIC AND MAGNETIC FIELDS   273

CHAPTER 5

NORMAL-INCIDENCE PLANE WAVE REFLECTION
AND TRANSMISSION AT PLANE BOUNDARIES   371

CHAPTER **6**

OBLIQUE INCIDENCE PLANE WAVE REFLECTION AND TRANSMISSION

CHAPTER **7**

TRANSMISSION LINES

CHAPTER **8**

WAVE GUIDES   591

CHAPTER **9**

ANTENNAS   637

APPENDIXES

INDEX   752

PREFACE

Electromagnetic energy has highly diversified applications in communications, medicine, processing, and characterization of materials, biology, atmospheric sciences, radar systems, and in high-speed electronics and integrated circuits. Students in their junior or senior year of electrical engineering are expected to have either academically or in practice encountered applications involving electromagnetic fields, waves, and energy. For example, students should be familiar academically with electromagnetics in their introductory physics courses. Practical applications based on electromagnetics technology such as electric power lines, antennas, microwave ovens, and broadcast stations are encountered in our daily activities. Therefore, when students take electromagnetics courses they are expected to be excited and prepared to gain in-depth knowledge of this important subject. Instead, however, they quickly get bogged down with equations and mathematical relations involving vector quantities and soon lose sight of the interesting subject and exciting applications of electromagnetics.

It is true that the mathematical formulation of electromagnetics concepts is essential in quantifying the relationship between the electromagnetic fields and their sources. Integral and differential equations involving vector quantities are important in describing the characteristics and behavior of electromagnetic fields under a wide variety of propagation and interaction conditions. It is unfortunate, however, that the overall emphasis of the subject may be placed on these mathematical relations and their clever manipulation. Instead, the physical and exciting phenomena associated with electromagnetic radiation should be foremost, and mathematics should always be approached as a way to quantify and characterize electromagnetic fields, their radiation, propagation, and interactions. It is with this in mind that I have approached the development of this junior-level electrical engineering book on electromagnetic fields and waves.

There are several ways of organizing an introductory book on electromagnetics. One way is to start with the electrostatic and magnetostatic concepts, and continue to work toward the development of time-varying fields and dynamic electromagnetics. This has been the traditional procedure adopted in many textbooks. The other approach involves describing the mathematical relations between the time-varying electromagnetic fields and their sources by first introducing Maxwell's equations in integral forms. This allows a quick move toward the introduction of the propagation characteristics of plane waves. It is generally agreed that the second approach provides a faster pace toward the development of more exciting and dynamic aspects of electromagnetics, the subject matter that maintains high levels of enthusiasm for students and helps them carry on their otherwise difficult mathematical tasks.

I found the second method of organization to be helpful because students at the junior level usually have previous exposure to static fields. Also, the delay in discussing Maxwell's equations toward the end of the course does not help in consolidating and comprehending these important concepts and ideas. A few introductory textbooks adopt this approach. Although I used some of these books as texts when I initially taught the electromagnetics course series, I found it to be more constructive to include a concise description of the properties of the static electric and magnetic fields in terms of their charge and current sources before introducing Maxwell's equations. In addition, I have tried in this text to show how Maxwell's equations actually evolved from experimental observations made by Coulomb, Biot and Savart, Faraday and Ampere.

This brief introduction of the properties of electromagnetic fields and the experiments by pioneers in this field provides students with insight into the physical properties of these fields and help in developing a smoother transition from experimental observations to the mathematical relations that quantify them. In a sense, therefore, we may consider the adopted approach in this book to be a combination and a middle ground of the traditional approach of introducing the subject of electromagnetics in terms of static fields and the fast-paced approach of promptly introducing Maxwell's equations.

Additional features of this text are the inclusion of many examples in each chapter to help emphasize key concepts, detailed description of the subject of "reflection and refraction of plane waves of oblique incidence on a dielectric interface," including some of its applications in optics, and a detailed introduction to antennas including physical mechanisms of radiation and practical design of antenna arrays. The treatment of the subject of transmission lines was comprehensive and included a detailed treatment of transients and sinusoidal steady-state analysis of propagation on two conductor lines. Another important feature of this text is the introductory section on "numerical techniques" included in chapter 4. At this time and age, many solutions are handled by computers and, with the availability of this technology, solutions to more realistic and exciting engineering problems may be included in homework assignments and even simulated and demonstrated in classrooms. It is essential, however, that students be familiar with the commonly used computational procedures such as the finite difference method and the method of moments, learn of the various approximations involved, and be aware of the limitations of such methods. Recently, some focused efforts* have

---

* NSF/IEEE Center on Computer Applications in Electromagnetic Education (CAEME), University of Utah, Salt Lake City, UT 84112.

attempted to stimulate, accelerate, and encourage the use of computers and software tools to help electromagnetic education. Many educational software packages are now available to educators, and it is imperative that students be aware of the capabilities, accuracies, and limitations of some of these software tools—particularly those that use computational techniques and numerical methods. It is with this in mind that we prepared the introductory material on computational methods in chapter 4. Furthermore, educators and students are encouraged to use available software from CAEME* to help comprehend concepts, visualize the dynamic-field phenomena, and solve interesting practical applications.

I would like to conclude by expressing my sincere thanks and appreciation to my students who, during the years, provided me with valuable feedback on the manuscript. Comments and suggestions by Professor Robert S. Elliott of University of California, Los Angeles, were deeply appreciated. I would also like to express my sincere appreciation to Ruth Eichers and Holly Cox for their expert efforts in typing and preparing the manuscript. My gratitude, sincere thanks, deep appreciation, and love are also expressed to my family for patience, sacrifice, and understanding during the completion of this endeavor.

Magdy F. Iskander

* NSF/IEEE Center on Computer Applications in Electromagnetic Education (CAEME), University of Utah, Salt Lake City, UT 84112.

CHAPTER 1

VECTOR ANALYSIS AND MAXWELL'S EQUATIONS IN INTEGRAL FORM

1.1 INTRODUCTION

In this chapter we will first review some simple rules of vector algebra. These basic vector operations are first defined independent of any coordinate system and then specifically applied to the Cartesian, cylindrical, and spherical coordinate systems. Transformation of vector representation from one coordinate system to another will also be described. Scalar and vector fields will then be defined, with emphasis on understanding the concepts of electric and magnetic fields because they constitute the basic elements of electromagnetics. Vector integration will be introduced to pave the way for the introduction of Maxwell's equations in integral form. Maxwell's equations are simply the mathematical relations that govern the relationships between the electric and magnetic fields, and their associated charge and current distribution sources. These relations include the following:

1. Gauss's law for the electric field.

2. Gauss's law for the magnetic field.

3. Faraday's law.

4. Ampere's circuital law.

A brief description of the experimental evidence that led to Maxwell's hypothesis will also be given.

1.2 VECTOR ALGEBRA

Familiarity with some of the mathematical rules of the vector calculus certainly helps in simplifying the development of the electromagnetic fields theory. This is simply because the electric and magnetic fields, which are the bases of our study, are vector quantities, the matter that makes it useful for us to start with reviewing our vector algebra. Let us first distinguish between scalar and vector quantities.

Scalar: Is a physical quantity completely specified by a single number describing the magnitude of the quantity (e.g., temperature, size of a class, mass, humidity, etc.).

Vector: Is a physical quantity that can only be specified if both magnitude and direction of the quantity are given. This class of physical quantities *cannot* be described by *one number* only (e.g., force field, velocity of a car or a tornado, etc.).

Graphically, a vector is represented as shown in Figure 1.1 by a straight line with an arrowhead pointing in the direction of the vector and of length proportional to the magnitude of the vector.

Unit Vector: A unit vector in a given direction is a vector along the described direction with magnitude equal to unity.

In Figure 1.2, **A** is a vector along the x axis, and $\mathbf{a}_x$ is a unit vector along the x axis.

$$\mathbf{a}_x = \frac{\mathbf{A}}{|\mathbf{A}|}$$

Hence, any vector can be represented as a product of a unit vector in the direction of the vector with the magnitude of the vector

$$\mathbf{A} = |\mathbf{A}|\,\mathbf{a}_x$$

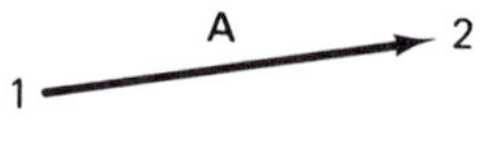

Figure 1.1 Vector representation by an arrow. The length of the arrow is proportional to the magnitude of the vector, and the direction of the vector is indicated by the direction of the arrow.

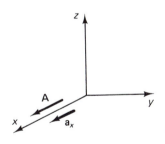

Figure 1.2   A unit vector $\mathbf{a}_x$ along the direction of the vector **A**.

1.2.1 Vector Addition and Subtraction

Four possible types of vector algebraic operations exist. This includes vector additions, subtractions, scalar, and vector products. In the following two sections, we will discuss these operations in more detail. Let us start with the process of adding and subtracting vector quantities.

The displacement of a point for a certain distance along a straight line is a good illustration of a physical vector quantity. For example, the displacement of a point from location 1 to location 2 in Figure 1.1 represents a vector quantity where its magnitude equals the distance between the end points 1 and 2, and the vector direction is along the straight line connecting 1 to 2. The addition of two vectors, therefore, can be described as the net displacement that results from two consecutive displacements. In Figure 1.3, vector **A** represents the vector displacement between 1 and 2, whereas the vector **B** represents the vector displacement between 2 and 3. The total displacement between 1 and 3 is described by the vector **C**, which is the sum of the individual displacements **A** and **B**. Hence,

$$\mathbf{C} = \mathbf{A} + \mathbf{B}$$

Based on similar reasoning, it is fairly simple to show that

$$(\mathbf{A} + \mathbf{B}) + \mathbf{D} = \mathbf{A} + (\mathbf{B} + \mathbf{D})$$

Because the negative of a vector is defined as a vector with the same magnitude but opposite direction, *vector subtraction* can be easily defined in terms of vector addition. In other words, the subtraction of two vectors can be thought of as the summation of one vector and the negative of the other,

$$\mathbf{A} - \mathbf{B} = \mathbf{A} + (-\mathbf{B})$$

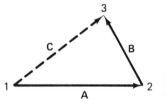

Figure 1.3   The vector addition of two displacements.

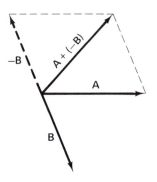

Figure 1.4   Vector subtraction performed as the addition of one vector to the negative of the other.

Figure 1.4 illustrates the process of vector subtraction where it is shown that a vector $-\mathbf{B}$ was first obtained and then added to the vector $\mathbf{A}$ to provide the resultant vector $\mathbf{A} + (-\mathbf{B})$.

1.2.2 Vector Multiplication

The process of vector multiplication is more involved than the simple multiplication of scalar quantities. The directions of the vectors are involved in the multiplication process, which further complicates the procedure. Two kinds of multiplications are commonly encountered in physical problems and hence are given special shorthand notations. These are the scalar (dot) product and the vector (cross) product. In the following sections, these two vector product procedures will be explained in more detail.

     Scalar (dot) product of two vectors.   The name "scalar product" emerged from the fact that the result of this multiplication process is a scalar quantity. To appreciate the physical reasoning behind the scalar product of two vectors, let us assume an object of mass m placed on a rough surface s. To move this object from location 1 to location 2, a vector force $\mathbf{F}$, which makes an angle α with respect to the displacement vector $\mathbf{r}$, is applied as shown in Figure 1.5. It is required to calculate the work done in moving m from location 1 to 2. This work W is actually equal to the component of the force along the direction of motion multiplied by the distance between 1 and 2, hence

$$W = |\mathbf{F}| \cos \alpha \, |\mathbf{r}|$$

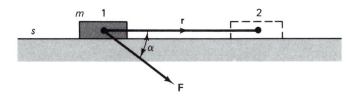

Figure 1.5   Explanation of the scalar product in terms of a physical problem. Force $\mathbf{F}$ is applied to move the mass m from location 1 to 2. The scalar product of $\mathbf{F}$ and $\mathbf{r}$ is related to the work required to achieve this motion.

Scalar quantities, such as the work W, which are calculated by multiplying the magnitudes of two vectors and the cosine of the angle between them, are encountered in many other physical problems, which led to identifying them by the shorthand notation of the dot product. For example, the desired work in Figure 1.5 may be expressed in the form $W = \mathbf{F} \cdot \mathbf{r}$.

The scalar or dot product of two vectors $\mathbf{A}$ and $\mathbf{B}$ is therefore equal to the product of the magnitudes of $\mathbf{A}$ and $\mathbf{B}$, and the cosine of the angle between them. It is represented by a dot between $\mathbf{A}$ and $\mathbf{B}$. Thus,

$$\mathbf{A} \cdot \mathbf{B} = |\mathbf{A}||\mathbf{B}| \cos \alpha = AB \cos \alpha$$

where α is the angle between $\mathbf{A}$ and $\mathbf{B}$.

The dot product operation can also be interpreted as the multiplication of the magnitude of one vector by the scalar obtained by projecting the second vector onto the first vector as shown in Figure 1.6.

The dot product can therefore be expressed as $\mathbf{A} \cdot \mathbf{B} = |\mathbf{A}||\mathbf{B}| \cos \alpha = |\mathbf{A}|$ multiplied by the projection of $\mathbf{B}$ along $\mathbf{A}$ (i.e., $|\mathbf{B}| \cos \alpha$ as shown in Figure 1.6b) $= |\mathbf{B}|$ multiplied by the projection of $\mathbf{A}$ along $\mathbf{B}$ (i.e., $|\mathbf{A}| \cos \alpha$ as shown in Figure 1.6a). Based on this interpretation, it may be emphasized that the dot product of two perpendicular vectors is zero. This can be seen by simply noting that the projection of one vector along the other that is perpendicular to it is zero. Such an observation is usually more useful than going through the mathematical substitution and recognizing that the angle α between the two perpendicular vectors is $\pi/2$ and that $\cos \pi/2 = 0$.

The distributive property for the dot product of the sum of two vectors with a third vector is:

$$\mathbf{A} \cdot (\mathbf{B} + \mathbf{C}) = \mathbf{A} \cdot \mathbf{B} + \mathbf{A} \cdot \mathbf{C}$$

Figure 1.7 illustrates that the projection of $\mathbf{B} + \mathbf{C}$ onto $\mathbf{A}$ is equal to the sum of the individual projections of $\mathbf{B}$ and $\mathbf{C}$ onto $\mathbf{A}$.

Vector (cross) product of two vectors.   The vector or cross product of two vectors $\mathbf{A}$ and $\mathbf{B}$ is a *vector*, perpendicular to $\mathbf{A}$ and $\mathbf{B}$ or equivalently perpendicular to the plane containing $\mathbf{A}$ and $\mathbf{B}$. The direction of the vector product is obtained by the right-hand rule rotating the first vector $\mathbf{A}$ to coincide with the second vector $\mathbf{B}$ in the

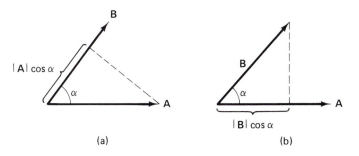

(a)   (b)

Figure 1.6   Dot product of two vectors.

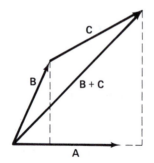

Figure 1.7 The distributive property for the dot product.

shortest way (through the angle α of Figure 1.8a). The magnitude of the cross product of two vectors is obtained by multiplying the magnitudes of the two individual vectors and sine of the angle between them. Figure 1.8 shows the magnitude and direction of vector **C**, which resulted from the cross product of **A** and **B**

$$\mathbf{C} = \mathbf{A} \times \mathbf{B} = AB \sin \alpha \, \mathbf{a}_c$$

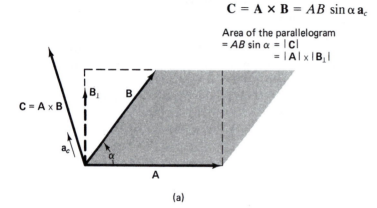

Area of the parallelogram
$= AB \sin \alpha = |\mathbf{C}|$
$= |\mathbf{A}| \times |\mathbf{B}_\perp|$

(a)

(b)

Figure 1.8 The cross product of two vectors **A** and **B**. The magnitude of the resultant vector **C** is $|\mathbf{C}| = |\mathbf{A}||\mathbf{B}| \sin \alpha$. The direction of **C** is obtained according to the right-hand rule shown in b.

where $\mathbf{a}_c$ is a unit vector perpendicular to $\mathbf{A}$ and $\mathbf{B}$ and in the direction indicated by the right-hand rule shown in Figure 1.8b.

To illustrate the importance of the cross product in physical problems, let us consider the lever ℓ that is free to rotate around a pivot O. A force $\mathbf{F}$ is applied to the lever at point a as shown in Figure 1.9. It is required to calculate the moment $\mathbf{M}$ of the force $\mathbf{F}$ around the pivot O. From Figure 1.9, it is clear that the moment $\mathbf{M}$ is actually related to the component of $\mathbf{F}$ perpendicular to $\mathbf{r}$—that is, $|\mathbf{F}| \sin \alpha$. The other component of $\mathbf{F}$ in the direction of $\mathbf{r}$ does not contribute to the rotation of the lever around O. The magnitude of the moment $|\mathbf{M}|$ is therefore given by

$$|\mathbf{M}| = |\mathbf{F}| \sin \alpha \, |\mathbf{r}|$$

Figure 1.9 shows that in certain physical problems parameters of interest, such as the moment in our case, are obtained by multiplying the magnitudes of two vectors by the sine of the angle between them. The magnitude of the moment, however, does not provide a *complete description* of the amount and direction of rotation of the lever. An indication of the direction of the moment is still required. To obtain the direction of the moment, it may be seen from Figure 1.9 that for the indicated direction of the force $\mathbf{F}$ the rotation of the lever will be in the counterclockwise direction. Therefore, if we imagine the presence of a screw at O, it can be seen that such a screw will proceed in the direction out of the plane of the paper as a result of the rotation. The direction to which a screw proceeds as a result of the rotation is taken to be the direction of the moment $\mathbf{M}$. From Figure 1.9, it may be seen that such a direction is the same as that obtained according to the right-hand rule when applied to the vectors $\mathbf{r}$ and $\mathbf{F}$ in the sequence from $\mathbf{r}$ to $\mathbf{F}$. Hence, a complete description of the moment $\mathbf{M}$ (i.e., magnitude and direction) is given by

$$\mathbf{M} = \mathbf{r} \times \mathbf{F}$$

in which case the magnitude of $\mathbf{M}$ is obtained by multiplying the magnitudes of $\mathbf{r}$ and $\mathbf{F}$ by the sine of the angle α, and the direction of $\mathbf{M}$ is indicated by the right-hand rule from $\mathbf{r}$ to $\mathbf{F}$ as explained earlier.

Therefore, the shorthand notation of the cross product of two vectors $\mathbf{A}$ and $\mathbf{B}$ is simply a vector with its magnitude equal to $|\mathbf{A}||\mathbf{B}| \sin \alpha$, where α is the angle between $\mathbf{A}$ and $\mathbf{B}$, and the direction of the resultant vector is obtained according to the right-hand rule shown in Figure 1.8b.

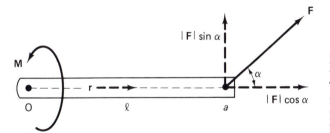

Figure 1.9 Physical illustration of the cross product of two vectors. The magnitude and direction of the moment $\mathbf{M}$ is related to the cross product of the force vector $\mathbf{F}$ and the distance vector $\mathbf{r}$, $\mathbf{M} = \mathbf{r} \times \mathbf{F}$.

Another physical interpretation of the cross product can be made in terms of the vector projections. For example, the vector **C** in Figure 1.8a is given by

$$\mathbf{C} = |\mathbf{A}||\mathbf{B}| \sin \alpha \, \mathbf{a}_c$$

$$= \mathbf{A} \times \mathbf{B}_\perp = |\mathbf{A}||\mathbf{B}_\perp| \, \mathbf{a}_c$$

where $\mathbf{B}_\perp$ is the vector component of **B** perpendicular to **A**. This observation simply indicates that the cross product of two vectors involves the multiplication of one vector (e.g., **A**) by the component of the other perpendicular to it. Based on this observation, it is useful to note that the cross product of two vectors that are in the same direction (i.e., parallel vectors) is zero. This may be seen by either noting that the angle α between two parallel vectors is zero and hence $\sin \alpha = 0$, or by recognizing that for parallel vectors the component of one vector perpendicular to the other is zero. The usefulness of such observations will be clarified in later discussions.

From the right-hand rule of Figure 1.8b, it is rather straightforward to see that

$$\mathbf{B} \times \mathbf{A} = -\mathbf{C} = -\mathbf{A} \times \mathbf{B}$$

which means that the ordering of the vectors in the cross product is an important consideration because the cross product does not obey a commutative law.

1.3 COORDINATE SYSTEMS

The vectors and the vector relations given in the previous sections are not defined with respect to any particular coordinate system. Hence, all the previously indicated definitions of the dot product, cross product, and so forth are presented in graphical and general terms.

Having a certain reference system (known as the coordinate system), however, is important to describe *uniquely* the position of a point in space, and the magnitude and direction of a vector. Although several coordinate systems are available, we will restrict our discussion to the three simplest ones—namely, the so-called Cartesian, cylindrical, and spherical coordinate systems. Expressions for transforming a vector representation from one coordinate system to another will be derived and the previously defined vector algebraic relations will be given in these three coordinate systems.

To start with, each of the three coordinate systems is specified in terms of three independent variables. In the Cartesian coordinate system these independent variables are (x, y, z), whereas for the cylindrical and spherical coordinate systems these independent variables are (ρ, ϕ, z) and (r, θ, ϕ), respectively. In each coordinate system, we also set up three mutually orthogonal reference surfaces by letting each of the independent variables be equal to a constant. For example, in the Cartesian coordinate system, the three reference surfaces (planes in this case) are obtained by letting x be equal to a constant value, say x_1, y be equal to a constant value y_1, and z equal to z_1. As a result, these mutually orthogonal planes will intersect at a point denoted by (x_1, y_1, z_1) as shown in Figure 1.10a. The point of intersection of the three reference planes for which $x = 0$, $y = 0$, and $z = 0$ defines the origin of the coordinate system as shown in Figure 1.10b. After establishing the three reference surfaces in each

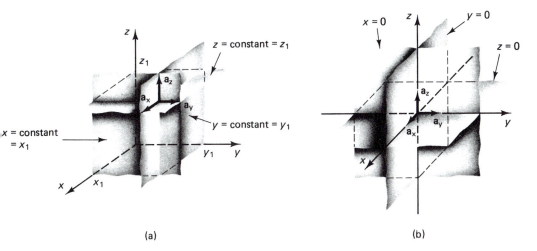

(a)   (b)

Figure 1.10   The Cartesian coordinate system. (a) The point (x_1, y_1, z_1) is generated at the intersection of $x = x_1$ plane with the $y = y_1$ and $z = z_1$ planes. (b) The origin is the point of intersection of $x = 0$, $y = 0$, and $z = 0$ planes. The base vectors $\mathbf{a}_x$, $\mathbf{a}_y$, and $\mathbf{a}_z$ are mutually orthogonal, and each is perpendicular to a reference plane.

coordinate system, we define three mutually orthogonal unit vectors, called the *base vectors*. The directions of these base vectors are chosen such that each base vector is perpendicular to a reference surface and oriented in the direction of increasing the independent variable. For example, the base vector $\mathbf{a}_x$ shown in Figure 1.10b is oriented perpendicular to the $x = $ constant plane and is in the direction of increasing x. Similarly the base vectors $\mathbf{a}_y$ and $\mathbf{a}_z$ are oriented perpendicular to the $y = $ constant and $z = $ constant planes, respectively. Any vector is represented in a coordinate system in terms of its components along the base vectors of that system. For example, in the Cartesian coordinate system, a vector $\mathbf{A}$ should be represented in terms of its components A_x, A_y, A_z along the unit (base) vectors $\mathbf{a}_x$, $\mathbf{a}_y$, and $\mathbf{a}_z$. These, as well as other characteristics of the three coordinate systems, will be described in the following sections.

1.3.1 Cartesian Coordinate System

As indicated earlier, the three independent variables in the Cartesian coordinate system are (x, y, z), and the three base vectors are $\mathbf{a}_x, \mathbf{a}_y$, and $\mathbf{a}_z$. The location of a point in this coordinate system is obtained by locating the point of intersection of the three reference planes. For example, the point (x_1, y_1, z_1) is the point of intersection of the three reference planes $x = x_1$, $y = y_1$, and $z = z_1$. The base vectors are mutually orthogonal, and each points in the direction of increase of an independent variable.

To obtain expressions for elements of length, surface, and volume in the Cartesian coordinate system, let us start from an arbitrarily located point P_1 of coordinates (x, y, z) and move to another closely placed point P_2 of coordinates $(x + dx, y +$

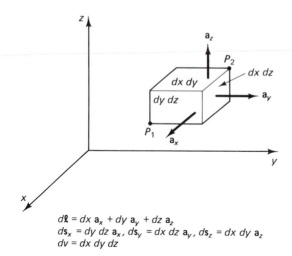

$d\boldsymbol{\ell} = dx\ \mathbf{a}_x + dy\ \mathbf{a}_y + dz\ \mathbf{a}_z$
$d\mathbf{s}_x = dy\ dz\ \mathbf{a}_x,\ d\mathbf{s}_y = dx\ dz\ \mathbf{a}_y,\ d\mathbf{s}_z = dx\ dy\ \mathbf{a}_z$
$dv = dx\ dy\ dz$

Figure 1.11 The elements of length, surface, and volume in the Cartesian coordinate system.

$dy, z + dz$) as shown in Figure 1.11. Thus, in moving from P_1 to P_2, we basically changed the values of the independent variables from x to $x + dx$, y to $y + dy$, and from z to $z + dz$. The element of volume, dv, generated from these incremental changes in the independent variables is given, as shown in Figure 1.11, by

$$dv = dx\ dy\ dz$$

The *vector* element of length, $d\boldsymbol{\ell}$, between P_1 and P_2, conversely, should be expressed, like any other vector, in terms of its components along the three mutually orthogonal base vectors. From Figure 1.11, it can be shown that $d\boldsymbol{\ell}$ has a component, dx, along the $\mathbf{a}_x$ base vector, dy along the $\mathbf{a}_y$, and dz along the $\mathbf{a}_z$ unit vector. Therefore, $d\boldsymbol{\ell}$ may be expressed as

$$d\boldsymbol{\ell} = dx\ \mathbf{a}_x + dy\ \mathbf{a}_y + dz\ \mathbf{a}_z$$

Regarding the elements of area, it is important to emphasize that each element of area should be accompanied by a unit vector specifying its orientation in the coordinate system. For example, it is not sufficient to indicate an element of area $d\mathbf{s}_x$ equal to $dy\ dz$ because it leaves the orientation or the direction of this element of area unspecified. As a result, we can specify three elements of areas in the Cartesian coordinate system as

$$d\mathbf{s}_x = dy\ dz\ \mathbf{a}_x$$

$$d\mathbf{s}_y = dx\ dz\ \mathbf{a}_y$$

$$d\mathbf{s}_z = dx\ dy\ \mathbf{a}_z$$

where each element of area is specified by a unit vector perpendicular to it. Actually the subscripts are not necessary to include in this case but are here just to emphasize that $d\mathbf{s}_x$ (subscript x) is an element of area in the $\mathbf{a}_x$ direction and so on.

It should be noted that the three coordinate axes x, y, and z are oriented with respect to each other according to the right-hand rule as shown in Figure 1.12 and that

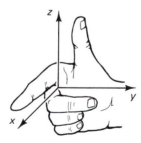

Figure 1.12   The coordinate axes in the Cartesian coordinate system are mutually orthogonal and the rotation from two of them toward the third axis follows the right-hand rule.

the directions of the base vectors in the Cartesian coordinate system are always the same at all points. In other words, the base vectors $\mathbf{a}_x$, $\mathbf{a}_y$, and $\mathbf{a}_z$ do not change their directions at various points in the coordinate system, a subject that we will fully explore when we describe the other coordinate systems.

1.3.2 Cylindrical Coordinate System

In this coordinate system the three independent variables are ρ, ϕ, and z. The three reference surfaces are a cylindrical surface generated by letting ρ = constant = ρ_1, and two plane surfaces obtained from ϕ = constant = ϕ_1 and z = constant = z_1. These three reference planes intersect at the coordinate point (ρ_1, ϕ_1, z_1). The origin of the coordinate system is the point of the intersection of the three reference planes for which the values of the independent variables are all equal to zero. Figure 1.13 shows the reference surfaces in the cylindrical coordinate system.

The three base vectors $\mathbf{a}_\rho$, $\mathbf{a}_\phi$, and $\mathbf{a}_z$ are also shown in Figure 1.13 where it is clear that these vectors are oriented perpendicular to the reference surfaces—that is, $\mathbf{a}_\rho$ is perpendicular to the ρ = constant cylindrical surface, $\mathbf{a}_\phi$ is perpendicular to the plane

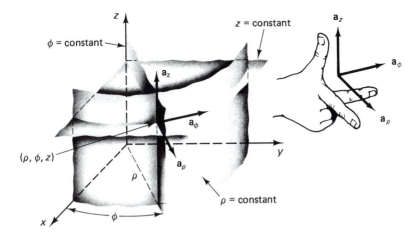

Figure 1.13   The cylindrical coordinate system. The three reference planes intersect at the point (ρ, ϕ, z), and the three base vectors are $\mathbf{a}_\rho$ normal to the cylindrical surface, ρ = constant, $\mathbf{a}_\phi$ normal to the ϕ = constant plane, and $\mathbf{a}_z$ is normal to the z = constant plane.

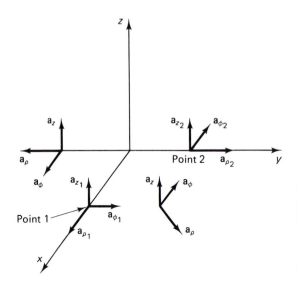

Figure 1.14   The base vectors in the cylindrical coordinate system change directions at the various points. The base vectors at point 1 are $\mathbf{a}_{\rho_1}$, $\mathbf{a}_{\phi_1}$, and $\mathbf{a}_{z_1}$, whereas $\mathbf{a}_{\rho_2}$, $\mathbf{a}_{\phi_2}$, and $\mathbf{a}_{z_2}$ are the base vectors at point 2.

ϕ = constant, and $\mathbf{a}_z$ is perpendicular to the z = constant plane—and that all the base vectors point in the direction of the increase in the independent variables. Figure 1.14 shows the directions of the base vectors at various points—that is, various values of ρ, ϕ, and z, in the cylindrical coordinate system. From Figure 1.14, it may be seen that, unlike the base vectors in the Cartesian coordinate system, the base vectors in the cylindrical coordinate system do not maintain their same directions at the various points. For example, we note that the base vectors at a point 1 along the x axis, that is, $\phi = 0$, are related to those at a point 2 along the y axis, that is, $\phi = \pi/2$, by $\mathbf{a}_{\rho_1} = -\mathbf{a}_{\phi_2}$, $\mathbf{a}_{\phi_1} = \mathbf{a}_{\rho_2}$, and $\mathbf{a}_{z_1} = \mathbf{a}_{z_2}$ where $\mathbf{a}_{\rho_1}$, $\mathbf{a}_{\phi_1}$, and $\mathbf{a}_{z_1}$ are the base vectors at point 1, whereas $\mathbf{a}_{\rho_2}$, $\mathbf{a}_{\phi_2}$, and $\mathbf{a}_{z_2}$ are the base vectors at point 2. Therefore, it is very important before we perform any vector operation in this coordinate system, such as addition of two vectors, that we make sure that the vectors are expressed with respect to the *same* base vectors at a specific point. This particular point will be further clarified in the section on the vector representation in the various coordinate systems.

For now, let us focus our attention on generating elements of length, surface, and volume in the cylindrical coordinate system. To generate an element of volume in the cylindrical coordinate system, we make incremental changes in the independent variables from ρ, ϕ, and z to $\rho + d\rho$, $\phi + d\phi$, and $z + dz$. This results in generating an element of volume dv as shown in Figure 1.15. Before we can calculate the volume of dv, however, it should be noted that the incremental changes in the independent variables $d\rho$ and dz are actually changes in elements of length, whereas the incremental change $d\phi$ is just a change in angle and not in length. To transform the incremental change in ϕ to a change in element of length, $d\phi$ must be multiplied by ρ and the corresponding change in the linear dimension will be $d\ell_\phi = \rho d\phi \, \mathbf{a}_\phi$ as shown in Figure 1.15. In other words, we multiplied $d\phi$ by ρ, which is called the metric coefficient to transform the change in the angle $d\phi$ to change in the linear dimension $d\ell_\phi$. With this

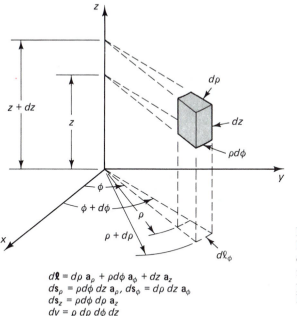

$$d\boldsymbol{\ell} = d\rho\, \mathbf{a}_\rho + \rho d\phi\, \mathbf{a}_\phi + dz\, \mathbf{a}_z$$
$$d\mathbf{s}_\rho = \rho d\phi\, dz\, \mathbf{a}_\rho,\ d\mathbf{s}_\phi = d\rho\, dz\, \mathbf{a}_\phi$$
$$d\mathbf{s}_z = \rho d\phi\, d\rho\, \mathbf{a}_z$$
$$dv = \rho\, d\rho\, d\phi\, dz$$

Figure 1.15   The elements of length, volume, and surface in the cylindrical coordinate system. $d\phi$ is not a length and should be multiplied by its metric coefficient ρ to have the dimension of length. $\therefore d\ell_\phi = \rho d\phi\, \mathbf{a}_\phi$.

in mind, it is easy to show that the element of volume dv, which resulted from incremental changes in the independent variables, is given by

$$dv = d\rho(\rho d\phi)dz = \rho\, d\rho\, d\phi\, dz$$

The resultant element of length $d\ell$ from P_1 to P_2 is given by

$$d\ell = d\rho\, \mathbf{a}_\rho + \rho d\phi\, \mathbf{a}_\phi + dz\, \mathbf{a}_z$$

whereas the resultant elements of area that are associated with unit vectors perpendicular to each of the areas to emphasize their orientations are given by

$$d\mathbf{s}_\rho = \rho d\phi\, dz\, \mathbf{a}_\rho$$

$$d\mathbf{s}_\phi = d\rho\, dz\, \mathbf{a}_\phi$$

$$d\mathbf{s}_z = \rho d\rho\, d\phi\, \mathbf{a}_z$$

1.3.3 Spherical Coordinate System

In this coordinate system the three independent variables are (r, θ, ϕ) as shown in Figure 1.16a. The three reference surfaces are: spherical surface obtained by letting the independent variable r = constant, conical surface obtained for θ = constant value, and a plane surface obtained for ϕ = constant value. These three reference surfaces intersect at the coordinate point $P(r, \theta, \phi)$ as shown in Figure 1.16b. The three base vectors $\mathbf{a}_r$, $\mathbf{a}_\theta$, and $\mathbf{a}_\phi$ are perpendicular to the spherical, conical, and the plane reference surfaces, respectively. These three base vectors are clearly mutually orthogonal, and

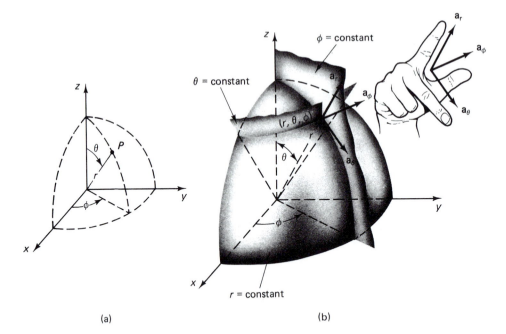

Figure 1.16   The spherical coordinate system. (a) The three independent variables (r, θ, ϕ) at point P. (b) The three reference surfaces are the spherical surface r = constant, the conical surface θ = constant, and the plane ϕ = constant. The three base vectors $\mathbf{a}_r$, $\mathbf{a}_\theta$, and $\mathbf{a}_\phi$ are mutually orthogonal and follow the right-hand rule.

they point in the directions of the increase of the independent variables. The orientation of the base vectors is in accordance to the right-hand rule as also shown in Figure 1.16b. The differential elements of volume, surface, and length are routinely generated by incrementally changing the independent variables from r, θ, and ϕ to $r + dr$, $\theta + d\theta$, and $\phi + d\phi$ as shown in Figure 1.17. Expressions for the differential elements are obtained by noting that the incremental changes in the independent variables $d\theta$ and $d\phi$ are not actual changes in elements of length, but instead are just changes in angles. To transform the change $d\theta$ into a change in a differential element of length, $d\theta$ must be multiplied by the metric coefficient which is, in this case, r. In other words, the incremental element of length $d\ell_\theta$ which is associated with the change of the angle θ by $d\theta$ is $d\ell_\theta = r\,d\theta$, whereas the element $d\ell_\phi$ associated with the change of the angle ϕ by $d\phi$ is given by $d\ell_\phi = r \sin\theta\,d\phi$. From Figure 1.17, it is clear that the metric coefficient $r \sin\theta$ is basically the projection of r in the x-y plane where the incremental change in the angle ϕ occurs. $d\ell_\phi$ is therefore obtained from the relation

$$d\ell_\phi = \text{projection of } r \text{ in the } x\text{-}y \text{ plane} \times d\phi$$

$$= r \sin\theta\,d\phi$$

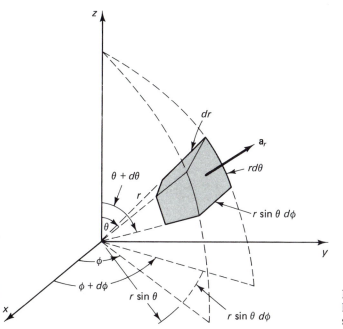

Figure 1.17 The elements of length, surface, and volume in the spherical coordinate system.

Based on the preceding discussion and from Figure 1.17, it is fairly straightforward to show that the incremental element of volume dv is given by

$$dv = dr(rd\theta)(r \sin \theta \, d\phi)$$

$$= r^2 \sin \theta \, dr \, d\theta \, d\phi$$

The element of length $d\ell$ from P_1 to P_2 is

$$d\ell = dr \, \mathbf{a}_r + d\ell_\theta \mathbf{a}_\theta + d\ell_\phi \mathbf{a}_\phi$$

$$= dr \, \mathbf{a}_r + rd\theta \, \mathbf{a}_\theta + r \sin \theta \, d\phi \, \mathbf{a}_\phi$$

The various elements of area are given by

$$d\mathbf{s}_r = r^2 \sin \theta \, d\theta \, d\phi \, \mathbf{a}_r$$

$$d\mathbf{s}_\theta = r \sin \theta \, dr \, d\phi \, \mathbf{a}_\theta$$

$$d\mathbf{s}_\phi = r \, dr \, d\theta \, \mathbf{a}_\phi$$

Clearly each element of area is associated with a unit vector perpendicular to it. In Figure 1.17, the unit vector $\mathbf{a}_r$ of the element of area $d\mathbf{s}_r$ is indicated.

A summary of the base vectors in the Cartesian, cylindrical, and the spherical coordinate systems is given in Figure 1.18. It should be noted that the base vectors in the spherical coordinate system are similar to those in the cylindrical coordinates insofar as they change their directions at various points in the coordinate system.

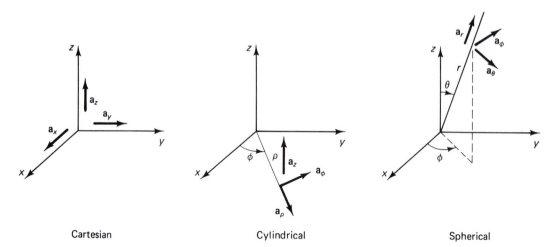

Cartesian   Cylindrical   Spherical

Figure 1.18 The base vectors of the three most commonly used coordinate systems.

1.4 VECTOR REPRESENTATION IN THE VARIOUS COORDINATE SYSTEMS

A vector quantity is completely specified in any coordinate system if the origin of the vector and its components (projections) in the directions of the three base vectors are known. For example, components of a vector **A** are designated by A_x, A_y, A_z in the Cartesian coordinate system, by A_ρ, A_ϕ, A_z in the cylindrical coordinate system, and by A_r, A_θ, A_ϕ in the spherical coordinate system. The vector **A** may then be represented in terms of its components as:

$$\mathbf{A} = A_x\,\mathbf{a}_x + A_y\,\mathbf{a}_y + A_z\,\mathbf{a}_z \text{ (Cartesian system)}$$

$$\mathbf{A} = A_\rho\,\mathbf{a}_\rho + A_\phi\,\mathbf{a}_\phi + A_z\,\mathbf{a}_z \text{ (Cylindrical system)}$$

$$\mathbf{A} = A_r\,\mathbf{a}_r + A_\theta\,\mathbf{a}_\theta + A_\phi\,\mathbf{a}_\phi \text{ (Spherical system)}$$

Let us now consider two vectors **A** and **B** that have origins at the same point in any one of these coordinate systems. It is important to note that the unit vectors are directed in the same directions at all points only in the Cartesian coordinate system. We illustrated in the previous sections that in the cylindrical and the spherical coordinate systems the unit vectors generally have different directions at different points. Therefore, in all the vector operations that we will describe in this section, it will be assumed that either the vectors are *originating from the same point* in the coordinate system and are thus expressed in terms of the same base vectors, or that the vectors are originating at different points and their components are all expressed in terms of a single set of the base vectors at either one of the two origins of the two vectors. What is important here is that the two vectors are expressed in terms of their components along the *same base vectors*. Let us now consider two vectors, **A** and **B**, expressed in terms of the same base vector, $\mathbf{u}_1, \mathbf{u}_2,$ and $\mathbf{u}_3$.

$$A = A_1 \mathbf{u}_1 + A_2 \mathbf{u}_2 + A_3 \mathbf{u}_3$$

$$B = B_1 \mathbf{u}_1 + B_2 \mathbf{u}_2 + B_3 \mathbf{u}_3$$

where $\mathbf{u}_1$, $\mathbf{u}_2$, and $\mathbf{u}_3$ stand for any set of three unit vectors $(\mathbf{a}_x, \mathbf{a}_y, \mathbf{a}_z)$, $(\mathbf{a}_\rho, \mathbf{a}_\phi, \mathbf{a}_z)$, or $(\mathbf{a}_r, \mathbf{a}_\theta, \mathbf{a}_\phi)$. The vector's addition or subtraction is given by

$$A \pm B = (A_1 \pm B_1)\mathbf{u}_1 + (A_2 \pm B_2)\mathbf{u}_2 + (A_3 \pm B_3)\mathbf{u}_3$$

Also, because the three base vectors are mutually orthogonal, therefore

$$\mathbf{u}_1 \cdot \mathbf{u}_2 = \mathbf{u}_1 \cdot \mathbf{u}_3 = \mathbf{u}_2 \cdot \mathbf{u}_3 = 0$$

and

$$\mathbf{u}_1 \cdot \mathbf{u}_1 = \mathbf{u}_2 \cdot \mathbf{u}_2 = \mathbf{u}_3 \cdot \mathbf{u}_3 = 1$$

The unity value in the dot product is indicated because the magnitudes of these base vectors are unity by definition. The dot product of two vectors *with origins at the same points* is, therefore,

$$A \cdot B = (A_1 \mathbf{u}_1 + A_2 \mathbf{u}_2 + A_3 \mathbf{u}_3) \cdot (B_1 \mathbf{u}_1 + B_2 \mathbf{u}_2 + B_3 \mathbf{u}_3)$$

$$= A_1 B_1 + A_2 B_2 + A_3 B_3$$

Furthermore, because the unit vectors are mutually orthogonal, we have the following relations for the cross products

$$\mathbf{u}_1 \times \mathbf{u}_2 = \mathbf{u}_3, \qquad \mathbf{u}_2 \times \mathbf{u}_3 = \mathbf{u}_1, \qquad \mathbf{u}_3 \times \mathbf{u}_1 = \mathbf{u}_2$$

and

$$\mathbf{u}_1 \times \mathbf{u}_1 = \mathbf{u}_2 \times \mathbf{u}_2 = \mathbf{u}_3 \times \mathbf{u}_3 = 0$$

The cross product of two **A** and **B** vectors may then be expressed in the form

$$A \times B = (A_1 \mathbf{u}_1 + A_2 \mathbf{u}_2 + A_3 \mathbf{u}_3) \times (B_1 \mathbf{u}_1 + B_2 \mathbf{u}_2 + B_3 \mathbf{u}_3)$$

$$= \mathbf{u}_1(A_2 B_3 - A_3 B_2) + \mathbf{u}_2(A_3 B_1 - A_1 B_3) + \mathbf{u}_3(A_1 B_2 - A_2 B_1)$$

which can be written in the form of a determinant:

$$A \times B = \begin{vmatrix} \mathbf{u}_1 & \mathbf{u}_2 & \mathbf{u}_3 \\ A_1 & A_2 & A_3 \\ B_1 & B_2 & B_3 \end{vmatrix}$$

which is an easier form to remember.

EXAMPLE 1.1

Which of the following sets of independent variables (coordinates) define a point in a coordinate system?

1. $x = 2, y = -4, z = 0$.
2. $\rho = -4, \phi = 0°, z = -1$.
3. $r = 3, \theta = -90°, \phi = 0°$.

Solution

Only the point in (a), because ρ in (b) and θ in (c) have to be positive, that is, $\rho \geqq 0$ and $0 \leqq \theta \leqq \pi$, which they are not.

EXAMPLE 1.2

Find a unit vector normal to the plane containing the following two vectors:

$$\mathbf{OA} = 4\,\mathbf{a}_x + 10\,\mathbf{a}_y$$

$$\mathbf{OB} = 4\,\mathbf{a}_x + 5\,\mathbf{a}_z$$

Solution

The cross product of two vectors **OA** and **OB** is a vector quantity whose magnitude is equal to the product of the magnitudes of **OA** and **OB** and the sine of the angle between them, and whose direction is perpendicular to the plane containing the two vectors. Hence,

$$\mathbf{OA} \times \mathbf{OB} = \begin{vmatrix} \mathbf{a}_x & \mathbf{a}_y & \mathbf{a}_z \\ 4 & 10 & 0 \\ 4 & 0 & 5 \end{vmatrix} = 50\,\mathbf{a}_x - 20\,\mathbf{a}_y - 40\,\mathbf{a}_z$$

The required unit vector is obtained by dividing $\mathbf{OA} \times \mathbf{OB}$ by its magnitude; hence,

$$\mathbf{a}_n = \frac{50\mathbf{a}_x - 20\mathbf{a}_y - 40\mathbf{a}_z}{|50\mathbf{a}_x - 20\mathbf{a}_y - 40\mathbf{a}_z|} = \frac{5\mathbf{a}_x - 2\mathbf{a}_y - 4\mathbf{a}_z}{\sqrt{25 + 4 + 16}}$$

$$= \frac{1}{3\sqrt{5}}(5\mathbf{a}_x - 2\mathbf{a}_y - 4\mathbf{a}_z)$$

EXAMPLE 1.3

Show that vectors $\mathbf{A} = \mathbf{a}_x + 4\mathbf{a}_y + 3\mathbf{a}_z$ and $\mathbf{B} = 2\mathbf{a}_x + \mathbf{a}_y - 2\mathbf{a}_z$ are perpendicular to each other.

Solution

The dot product consists of multiplying the magnitude of one vector by the projection of the second along the direction of the first. The dot product of two perpendicular vectors is therefore zero. For the two vectors given in this example,

$$\mathbf{A} \cdot \mathbf{B} = 2 + 4 - 6 = 0$$

so that **A** and **B** are perpendicular.

EXAMPLE 1.4

The two vectors **A** and **B** are given by

$$A = a_\rho + \pi \, a_\phi + 3 \, a_z$$

$$B = \alpha \, a_\rho + \beta \, a_\phi - 6 \, a_z$$

Determine α and β such that the two vectors are parallel.

Solution

For these two vectors to be parallel the cross product of **A** and **B** should be zero, that is,

$$A \times B = 0$$

$$= \begin{vmatrix} a_\rho & a_\phi & a_z \\ 1 & \pi & 3 \\ \alpha & \beta & -6 \end{vmatrix}$$

$$0 = a_\rho(-6\pi - 3\beta) + a_\phi(3\alpha + 6) + a_z(\beta - \pi\alpha)$$

For the vector that resulted from the cross product to be zero, each one of its components should be independently zero. Hence,

$$-6\pi - 3\beta = 0, \qquad \therefore \beta = -2\pi$$

and

$$3\alpha + 6 = 0, \qquad \therefore \alpha = -2$$

These two values of α and β clearly satisfy the remaining relation $\beta - \pi\alpha = 0$. The vector **B** is therefore given by

$$B = -2 \, a_\rho - 2\pi \, a_\phi - 6 \, a_z$$

———◆◆———

1.5 VECTOR COORDINATE TRANSFORMATION

The vector coordinate transformation is basically a process in which we change a vector representation from one coordinate system to another. This procedure is similar to scalar coordinate transformation with the additional necessity of transforming the individual components of the vector from being along the base vectors of the first coordinate system to components along the base vectors of the other coordinate system. Therefore, the transformation of a vector representation from one coordinate system to another involves a two-step process which includes the following:

a. *Changing the independent variables* (e.g., expressing x, y, and z of the rectangular coordinate system in terms of ρ, ϕ, and z of the cylindrical coordinate system or r, θ, ϕ of the spherical coordinate system).

b. *Changing the components of the vector* from those along the unit vectors of one coordinate system to those along the unit vectors of the other (e.g., changing the components from those along $\mathbf{a}_x$, $\mathbf{a}_y$, and $\mathbf{a}_z$ in the Cartesian coordinate system to components along $\mathbf{a}_\rho$, $\mathbf{a}_\phi$, and $\mathbf{a}_z$ in the cylindrical coordinate system).

In the following sections we shall describe specific transformation of the independent variables and the vector components from one coordinate system to another.

1.5.1 Cartesian-to-Cylindrical Transformation

The relation between the independent variables of these two coordinate systems is shown in Figure 1.19a. From Figure 1.19a it may be seen that

$$x = \rho \cos \phi \qquad \qquad \rho = \sqrt{x^2 + y^2}$$

$$y = \rho \sin \phi \qquad \qquad \phi = \tan^{-1}(y/x)$$

$$z = z \text{ (the same in both coordinates)}$$

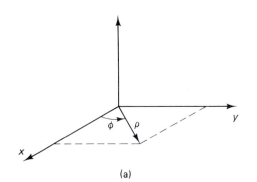

(a)

Figure 1.19a Relation between the independent variables in the Cartesian and cylindrical coordinate systems.

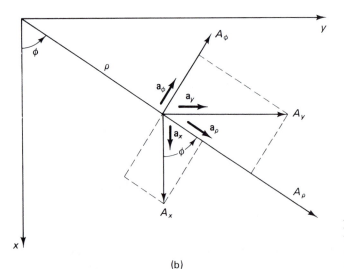

(b)

Figure 1.19b The relation between the vector components in the rectangular and cylindrical coordinate systems.

To illustrate the process of changing the components of a vector from being along the base vectors of one coordinate system (e.g., the Cartesian) to the other components along the base vectors of the other coordinate system (e.g., the cylindrical), let us solve the following example.

EXAMPLE 1.5

Transform the vector $\mathbf{A}$ given in the Cartesian coordinate system by

$$\mathbf{A} = A_x\,\mathbf{a}_x + A_y\,\mathbf{a}_y + A_z\,\mathbf{a}_z$$

to the form

$$\mathbf{A} = A_\rho\,\mathbf{a}_\rho + A_\phi\,\mathbf{a}_\phi + A_z\,\mathbf{a}_z$$

in the cylindrical coordinate system.

Solution

In transforming the vector $\mathbf{A}$ from the Cartesian coordinate system to the cylindrical one, it is required to obtain the components A_ρ, A_ϕ, and A_z of the vector $\mathbf{A}$ along the base vectors $\mathbf{a}_\rho$, $\mathbf{a}_\phi$, and $\mathbf{a}_z$ in the cylindrical coordinate systems. From the definition of the dot product, the component of $\mathbf{A}$ along $\mathbf{a}_\rho$ is given by

$$A_\rho = \mathbf{A}\cdot\mathbf{a}_\rho = (A_x\,\mathbf{a}_x + A_y\,\mathbf{a}_y + A_z\,\mathbf{a}_z) \cdot \mathbf{a}_\rho$$

$$= A_x\,\mathbf{a}_x \cdot \mathbf{a}_\rho + A_y\,\mathbf{a}_y\cdot\mathbf{a}_\rho + A_z\,\mathbf{a}_z \cdot \mathbf{a}_\rho$$

$\mathbf{a}_x\cdot\mathbf{a}_\rho$ from Figure 1.19b is equal to $\cos\phi$ because the magnitudes of both $\mathbf{a}_x$ and $\mathbf{a}_\rho$ are both equal to unity and the angle between them is ϕ. Similarly, $\mathbf{a}_y\cdot\mathbf{a}_\rho = \cos(\pi/2 - \phi) = \sin\phi$, and $\mathbf{a}_z\cdot\mathbf{a}_\rho = 0$. A_ρ is therefore given by

$$A_\rho = A_x \cos\phi + A_y \sin\phi$$

which is the same result previously obtained using the projections of the vector components. Similarly, the A_ϕ component may be obtained by

$$A_\phi = \mathbf{A}\cdot\mathbf{a}_\phi = A_x\,\mathbf{a}_x\cdot\mathbf{a}_\phi + A_y\,\mathbf{a}_y\cdot\mathbf{a}_\phi + A_z\,\mathbf{a}_z\cdot\mathbf{a}_\phi$$

$$= -A_x \sin\phi + A_y \cos\phi$$

The negative sign of the A_x component is included because the component $A_x \sin\phi$ is not along the positive $\mathbf{a}_\phi$ direction but instead along the negative $\mathbf{a}_\phi$ direction. Alternatively, the negative sign may be considered as a result of the fact that the angle between $\mathbf{a}_x$ and $\mathbf{a}_\phi$ is $(\pi/2 + \phi)$. The dot product $\mathbf{a}_x\cdot\mathbf{a}_\phi$ requires calculation of the cosine of the angle between them and $\cos(\pi/2 + \phi) = -\sin\phi$.

The A_z component of the vector will, of course, remain unchanged between the Cartesian and cylindrical coordinate systems.

1.5.2 Cartesian-to-Spherical Transformation

The relations between the independent variables can be obtained from Figure 1.20. It should be noted that $r_\parallel$ in Figure 1.20 is simply the projection of r in the x-y plane and

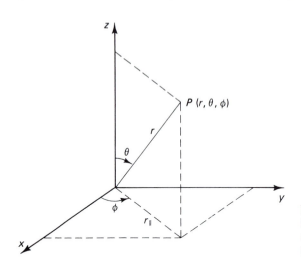

Figure 1.20 Relation between the independent variables in the spherical and Cartesian coordinate systems.

hence is given by $r_\parallel = r \sin\theta$. Once again, to illustrate expressing the vector components from one coordinate system to another, we will solve the following example.

EXAMPLE 1.6

Derive the vector components transformation from the Cartesian to the spherical coordinate systems and vice versa.

$$x = r_\parallel \cos\phi \qquad r = \sqrt{x^2 + y^2 + z^2}$$
$$ = r \sin\theta \cos\phi \quad \phi = \tan^{-1}(y/x)$$
$$y = r_\parallel \sin\phi \qquad \theta = \tan^{-1}\sqrt{(x^2 + y^2)}/z$$
$$ = r \sin\theta \sin\phi$$
$$z = r \cos\theta$$

Solution

The problem can be alternatively stated by considering the vector **A**, which is given in the Cartesian coordinate system $\mathbf{A} = A_x \mathbf{a}_x + A_y \mathbf{a}_y + A_z \mathbf{a}_z$, and it is required to find the vector components A_r, A_θ, and A_ϕ along the $\mathbf{a}_r$, $\mathbf{a}_\theta$, and $\mathbf{a}_\phi$ unit vectors in the spherical coordinate system. The relationship between the vector components is illustrated in Figure 1.21.

From Figure 1.21, it may be seen that the projections of the components A_x, A_y, and A_z along the direction $\mathbf{a}_r$ are given, respectively, by $\cos\phi \sin\theta$, $\sin\phi \sin\theta$, and $\cos\theta$. Therefore, the radial component A_r of the vector **A** is given by

$$A_r = A_x \cos\phi \sin\theta + A_y \sin\phi \sin\theta + A_z \cos\theta \qquad (1.1)$$

Following a similar procedure, we next find the projections of A_x, A_y, and A_z in the directions of $\mathbf{a}_\theta$ and $\mathbf{a}_\phi$. These are given, respectively, by $(\cos\phi \cos\theta, \sin\phi \cos\theta, -\sin\theta)$ along the $\mathbf{a}_\theta$ direction, and $(-\sin\phi, \cos\phi)$ along the $\mathbf{a}_\phi$ direction. Hence,

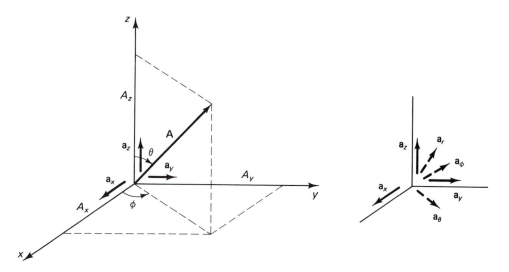

Figure 1.21   Transformation of the vector components from the Cartesian to the spherical coordinate system.

$$A_\theta = A_x \cos\phi \cos\theta + A_y \sin\phi \cos\theta - A_z \sin\theta \qquad (1.2)$$

$$A_\phi = -A_x \sin\phi + A_y \cos\phi \qquad (1.3)$$

and the vector **A** expressed in the spherical coordinates system is given by

$$\mathbf{A} = A_r \mathbf{a}_r + A_\theta \mathbf{a}_\theta + A_\phi \mathbf{a}_\phi$$

where the components A_r, A_θ, and A_ϕ are given in equations 1.1 to 1.3.

   To find the inverse transformation, we simply start with the vector **A** given in the spherical coordinate system by

$$\mathbf{A} = A_r \mathbf{a}_r + A_\theta \mathbf{a}_\theta + A_\phi \mathbf{a}_\phi$$

and then find the components of A_r, A_θ, and A_ϕ along the unit vectors $\mathbf{a}_x$, $\mathbf{a}_y$, and $\mathbf{a}_z$ of the Cartesian coordinate system. Alternatively, we can just solve the set of equations 1.1 to 1.3 simultaneously for A_x, A_y, and A_z. The result in both cases is

$$A_x = A_r \sin\theta \cos\phi + A_\theta \cos\theta \cos\phi - A_\phi \sin\phi \qquad (1.4)$$

$$A_y = A_r \sin\theta \sin\phi + A_\theta \cos\theta \sin\phi + A_\phi \cos\phi \qquad (1.5)$$

and

$$A_z = A_r \cos\theta - A_\theta \sin\theta \qquad (1.6)$$

Clearly, the vector **A** in the Cartesian coordinate system is given in terms of its components A_x, A_y, and A_z given in equations 1.4 to 1.6.

Alternative Procedure.   In the previous sections we described a process for making the vector coordinate transformation by dealing with each of the vector components in the "new" coordinate system and deriving expressions for the contributions of the vector components in the "old" coordinate system along the direction of the vector component of interest. In other words, the transformation basically involves finding the projections of the already available vector components along the various base vectors of the desired new coordinate system. In the following, we present an alternative procedure for finding the vector components along the desired base vectors. This can simply be achieved by taking the dot product of the vector by a unit vector along the desired direction. For example, in transforming the vector $\mathbf{A}$ given by

$$\mathbf{A} = A_x\, \mathbf{a}_x + A_y\, \mathbf{a}_y + A_z\, \mathbf{a}_z$$

to the spherical coordinate system, let us obtain A_r, A_θ, and A_ϕ from known values A_x, A_y, and A_z by performing the following dot products:

$$A_r = \mathbf{A}\cdot\mathbf{a}_r = A_x\, \mathbf{a}_x\cdot\mathbf{a}_r + A_y\, \mathbf{a}_y\cdot\mathbf{a}_r + A_z\, \mathbf{a}_z\cdot\mathbf{a}_r$$

From Figure 1.21, $\mathbf{a}_x\cdot\mathbf{a}_r = \cos\phi\,\sin\theta$, $\mathbf{a}_y\cdot\mathbf{a}_r = \sin\phi\,\sin\theta$, and $\mathbf{a}_z\cdot\mathbf{a}_r = \cos\theta$. Hence, A_r is given by

$$A_r = A_x\,\cos\phi\,\sin\theta + A_y\,\sin\phi\,\sin\theta + A_z\,\cos\theta$$

which is the same result we obtained in the previous section. Similarly, it can be shown that

$$A_\theta = \mathbf{A}\cdot\mathbf{a}_\theta = A_x\, \mathbf{a}_x\cdot\mathbf{a}_\theta + A_y\, \mathbf{a}_y\cdot\mathbf{a}_\theta + A_z\, \mathbf{a}_z\cdot\mathbf{a}_\theta$$

Once again from Figure 1.21, it is quite clear that $\mathbf{a}_x\cdot\mathbf{a}_\theta = \cos\phi\,\cos\theta$, $\mathbf{a}_y\cdot\mathbf{a}_\theta = \sin\phi\,\cos\theta$, and $\mathbf{a}_z\cdot\mathbf{a}_\theta = -\sin\theta$. Hence,

$$A_\theta = A_x\,\cos\phi\,\cos\theta + A_y\,\sin\phi\,\cos\theta - A_z\,\sin\theta$$

EXAMPLE 1.7

Express the vector

$$\mathbf{A} = z\,\cos\phi\,\mathbf{a}_\rho + \rho^2\,\sin\phi\,\mathbf{a}_\phi + 16\rho\,\mathbf{a}_z$$

in the Cartesian coordinates.

Solution

We first change the independent variables from ρ, ϕ, and z in the cylindrical coordinate system to x, y, and z in the Cartesian coordinate system. These changes are previously indicated as

$$\rho = \sqrt{x^2 + y^2}, \qquad \cos\phi = \frac{x}{\sqrt{x^2 + y^2}}, \qquad \sin\phi = \frac{y}{\sqrt{x^2 + y^2}}$$

Next we use the vector component transformation between the two coordinate systems. From the relations given in example 1.5, we obtain

$$A_x = A_\rho\,\cos\phi - A_\phi\,\sin\phi$$

$$= z\,\cos^2\phi - \rho^2\,\sin^2\phi = \frac{zx^2}{x^2 + y^2} - y^2$$

$$A_y = A_\rho \sin\phi + A_\phi \cos\phi$$

$$= z \sin\phi \cos\phi + \rho^2 \sin\phi \cos\phi = \frac{zxy}{x^2 + y^2} + xy$$

$$A_z = 16\rho = 16\sqrt{x^2 + y^2}$$

and, in vector notation, the vector **A** is given by:

$$\mathbf{A} = \left(\frac{zx^2}{x^2 + y^2} - y^2\right)\mathbf{a}_x + \left(\frac{xyz}{x^2 + y^2} + xy\right)\mathbf{a}_y + (16\sqrt{x^2 + y^2})\,\mathbf{a}_z$$

EXAMPLE 1.8

A vector **B** lies in the x-y plane, and is given by

$$\mathbf{B} = x\,\mathbf{a}_x + y\,\mathbf{a}_y$$

1. Obtain an expression for **B** in cylindrical coordinates.
2. Determine the magnitude and direction of **B** at the point $x = 3, y = 4$.

Solution

1. Using the coordinate transformation

$$x = \rho\cos\phi, \qquad y = \rho\sin\phi$$

and the vector transformation given in example 1.5, we obtain

$$B_\rho = B_x \cos\phi + B_y \sin\phi = x\cos\phi + y\sin\phi$$

$$= \rho\cos^2\phi + \rho\sin^2\phi = \rho$$

$$B_\phi = -B_x \sin\phi + B_y \cos\phi$$

$$= -\rho\sin\phi\cos\phi + \rho\sin\phi\cos\phi = 0$$

$$B_z = 0$$

and, in vector notation,

$$\mathbf{B} = \rho\,\mathbf{a}_\rho$$

2. At the point $x = 3, y = 4$, the radial distance is

$$\rho = x^2 + y^2 = 5$$

Hence,

$$\mathbf{B} = 5\,\mathbf{a}_\rho$$

EXAMPLE 1.9

Express the vector $\mathbf{A} = \dfrac{x^2 z}{y} \mathbf{a}_x$ in the spherical coordinate system.

Solution

Because the vector $\mathbf{A}$ has only an A_x component, its component A_r along the base vectors $\mathbf{a}_r$ in the spherical coordinate system is given by

$$A_r = A_x \cos\phi \sin\theta$$

$$A_r = \frac{x^2 z}{y} \cos\phi \sin\theta = \frac{(r\sin\theta\cos\phi)^2 \, r\cos\theta}{r\sin\theta\sin\phi} \cos\phi\sin\theta$$

$$= r^2 \frac{\sin^2\theta \cos\theta \cos^3\phi}{\sin\phi}$$

Similarly, the A_θ and A_ϕ components are given by

$$A_\theta = A_x \cos\phi \cos\theta = \frac{x^2 z}{y} \cos\phi \cos\theta$$

$$= \frac{(r^2 \sin^2\theta \cos^2\phi)(r\cos\theta) \cos\phi \cos\theta}{r\sin\theta\sin\phi}$$

$$= r^2 \frac{\sin\theta \cos^2\theta \cos^3\phi}{\sin\phi}$$

$$A_\phi = -A_x \sin\phi = -\frac{(r^2 \sin^2\theta \cos^2\phi)(r\cos\theta)}{r\sin\theta\sin\phi} \sin\phi$$

$$= -r^2 \sin\theta \cos\theta \cos^2\phi$$

It is rather surprising to see that a simple vector such as $\mathbf{A}$ that has only one A_x component in the Cartesian coordinate system actually has three components of complicated expressions in the spherical coordinate system

$$\mathbf{A} = A_r \mathbf{a}_r + A_\theta \mathbf{a}_\theta + A_\phi \mathbf{a}_\phi$$

This problem emphasizes the importance of choosing the right coordinate system that best fits the representation of a given vector.

1.6 ELECTRIC AND MAGNETIC FIELDS

Basic to our study of electromagnetics is an understanding of the concept of electric and magnetic fields. Before studying electromagnetic fields, however, we must first define what is meant by a field. A field is associated with a region in space, and we say that a field exists in the region if there is a physical phenomenon associated with points in that region. In other words, we can talk of the field of any physical quantity as being a description of how the quantity varies from one point to another in the region of the field. For example, we are familiar with the earth's gravitational field; we do not "see" the field, but we know of its existence in the sense that objects of given mass are acted on by the gravitational force of the earth.

1.6.1 Coulomb's Law and Electric Field Intensity

We are all familiar with Newton's law of universal gravitation, which states that every object of mass m in the universe attracts every other object m' with a force that is directly proportional to the product of the masses and inversely proportional to the square of the distance R between them—that is,

$$\mathbf{F} = G\frac{mm'}{R^2}\mathbf{a}$$

where G is the gravitational constant and $\mathbf{a}$ is a unit vector along the straight line joining the two masses. The equation above simply means that there is a gravitational force of attraction between bodies of given masses and that this force is along the line joining the two masses. In a similar manner, a force field known as the *electric field* is associated with bodies that are *charged*.

In the experiments conducted by Coulomb, he showed that for two charged bodies that are very small in size compared with their separation—so that they may be considered as *point charges*—the following hold:

1. The magnitude of the force is proportional to the product of the magnitudes of the charges.
2. The magnitude of the force is inversely proportional to the square of the distance between the charges.
3. The direction of the force is along the line joining the charges.
4. The magnitude of the force depends on the medium.
5. Like charges repel; unlike charges attract.

Hence, if we consider two point charges Q_1 and Q_2 separated by a distance R, the force is then given by:

$$\boxed{\mathbf{F} = k\frac{Q_1 Q_2}{R^2}\mathbf{a}_{12}}$$

where k is a proportionality constant and $\mathbf{a}_{12}$ is a unit vector along the line joining the two charges as indicated by the third observation in the experiment by Coulomb. If the international system of units (SI system) is used, then Q is measured in coulombs (C), R in meters (m), and the force should be in newtons (N) (see Appendix B). In this case, the constant of proportionality k will be

$$k = \frac{1}{4\pi\epsilon_o}$$

where ϵ_o is called the permittivity of air (vacuum) and has a value measured in farads per meter (F/m),

$$\epsilon_0 = 8.854 \times 10^{-12} \approx \frac{1}{36\pi} \times 10^{-9} \text{ F/m}$$

The direction of the force in the above equation should actually be defined in terms of two forces $\mathbf{F}_1$ and $\mathbf{F}_2$ experienced by Q_1 and Q_2, respectively. These two forces with their appropriate directions are given by

$$\mathbf{F}_1 = \frac{Q_1 Q_2}{4\pi\epsilon_o R^2} \mathbf{a}_{21}$$

$$\mathbf{F}_2 = \frac{Q_1 Q_2}{4\pi\epsilon_o R^2} \mathbf{a}_{12}$$

where $\mathbf{a}_{21}$ and $\mathbf{a}_{12}$ are unit vectors along the line joining Q_1 and Q_2 as shown in Figure 1.22.

Electric Field Intensity.   From Coulomb's law, if we let one of the two charges, say Q_2, be a small test charge q, we have

$$\mathbf{F}_2 = \frac{Q_1 q}{4\pi\epsilon_o R^2} \mathbf{a}_{12}$$

The electric field intensity $\mathbf{E}_2$ at the location of the test charge owing to the point charge Q_1 is defined as

$$\mathbf{E}_2 = \frac{\mathbf{F}_2}{q} = \frac{Q_1}{4\pi\epsilon_o R^2} \mathbf{a}_{12}$$

In general, the electric field intensity $\mathbf{E}$ *is defined as the vector force on a unit positive test charge.*

$$\boxed{\mathbf{E} = \frac{Q}{4\pi\epsilon_o R^2} \mathbf{a}_R}$$

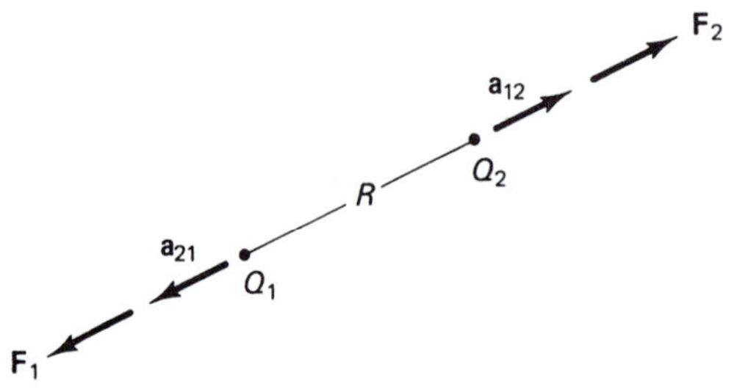

Figure 1.22 The electric force between two point charges Q_1 and Q_2.

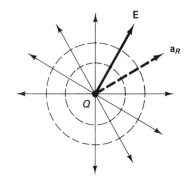

Figure 1.23 Direction lines and constant-magnitude surfaces of electric field owing to a point charge.

where $\mathbf{a}_R$ is a unit vector along the line joining the point charge Q and the test point wherever it is (the test point in this case is the point at which the value of the electric field intensity $\mathbf{E}$ is desired). The electric field intensity owing to a positive point charge is thus directed everywhere radially away from the point charge, and its constant magnitude surfaces are spherical surfaces centered at the point charge as shown in Figure 1.23.

If we have N point charges $Q_1, Q_2, \ldots, Q_N$ as shown in Figure 1.24, the force experienced by a test charger q placed at a point P is the vector sum of the forces experienced by the test charge owing to the individual charges, that is,

$$\boxed{\mathbf{E} = \sum_{i=1}^{N} \frac{Q_i}{4\pi\epsilon_o R_i^2} \mathbf{a}_{R_i}}$$

and

$$\mathbf{F} = q\mathbf{E}$$

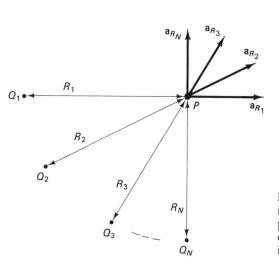

Figure 1.24 The total electric field intensity at point P owing to N point charges equals the vector sum of the electric field intensities owing to all of the charges.

EXAMPLE 1.10

A point charge $Q = 10^{-9}$ C is located at $(-0.5, -1, 2)$ in air.

1. What is the magnitude of the electric field intensity at a distance of 1 m from the charge?
2. Find the electric field **E** at the point $(0.9, 1.2, -2.4)$.

Solution

1. The electric field intensity is given by

$$\mathbf{E} = \frac{Q}{4\pi\epsilon_o R^2}\, \mathbf{a}_R$$

which is in the radial direction, and its magnitude $|\mathbf{E}|$ at $R = 1$ m is given by

$$|\mathbf{E}| = \frac{10^{-9}}{4\pi\dfrac{1}{36\pi} \times 10^{-9}} = 9 \text{ N/C}$$

2. A diagram illustrating the locations of the charge and the test point is shown in Figure 1.25. **E** at $(0.9, 1.2, -2.4)$ is

$$\mathbf{E} = \frac{Q}{4\pi\epsilon_o (QT)^2}\, \mathbf{a}_{QT}$$

$$\mathbf{QT} = \mathbf{OT} - \mathbf{OQ}$$

$$= (0.9 - (-0.5))\,\mathbf{a}_x + (1.2 - (-1))\,\mathbf{a}_y + (-2.4 - (2))\,\mathbf{a}_z$$

$$= 1.4\,\mathbf{a}_x + 2.2\,\mathbf{a}_y - 4.4\,\mathbf{a}_z$$

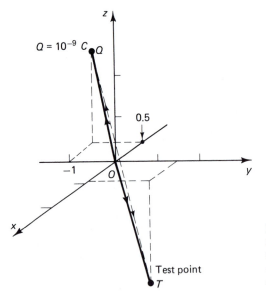

Figure 1.25 A diagram illustrating the location of a point charge Q and the coordinates of the point T at which the electric field is required.

$$\mathbf{a}_{QT} = \frac{1.4\,\mathbf{a}_x + 2.2\,\mathbf{a}_y - 4.4\,\mathbf{a}_z}{\sqrt{(1.4)^2 + (2.2)^2 + (4.4)^2}}$$

$$= 0.274\,\mathbf{a}_x + 0.43\,\mathbf{a}_y - 0.86\,\mathbf{a}_z$$

$$|\mathbf{QT}| = QT = \sqrt{26.16}$$

$$\mathbf{E} = \frac{10^{-9}}{4\pi\dfrac{1}{36\pi} \times 10^{-9} \times 26.16}\,\mathbf{a}_{QT}$$

$$= 0.094\,\mathbf{a}_x + 0.148\,\mathbf{a}_y - 0.296\,\mathbf{a}_z \text{ N/C}$$

——————◆◆◆——————

1.6.2 Flux Representation of Vector Field

As indicated in earlier sections, a vector quantity is completely specified in terms of its magnitude and direction. Therefore, the variation of a vector field in space can be graphically illustrated by drawing different vectors at various points in the field region as shown in Figure 1.26a. The magnitudes and the directions of these vectors represent the different values of the field (magnitude and direction) at the various points in space. Although the graphical representation in Figure 1.26a is possible and correct, it is a rather poor illustration and might get confusing for fields with rapid spatial variation. A widely adopted graphical representation of vector fields is in terms of their flux lines. In this procedure, a vector field is represented by arrows of the same length but of different separation between them. The direction of these arrows (flux lines) is in the direction of the vector field (or tangential to it). The magnitude of the field in this case, however, is not described in terms of the length of the arrow but instead in terms of the distance between the flux lines. The closer together the flux lines are, the larger the magnitude of the field and a further separation between these flux lines simply indicates a decrease in the magnitude of the field. Flux representations of uniform (of the same magnitude) and nonuniform fields are shown in Figures 1.26b and 1.26c, respectively.

It should be emphasized that the reason for our desire to develop such graphical representation is simply to help us visualize the quantitative properties of an existing field. For example, if we reexamine our previous representation of the electric field shown in Figure 1.26d which is due to a point charge Q, we can clearly see that this field is radially directed away from the point charge (as expected from Coulomb's law) and that the magnitude of this field is decreasing (as judged from the increase in the separation distance between the lines) with the increase in the distance away from the charge. This is also true according to Coulomb's law. From Figure 1.26d, it is also clear that the magnitude of the electric field is constant (equal distance between flux lines) at a fixed distance from the point charge.

For a more accurate description of the flux representation of electric fields, let us define a vector quantity $\mathbf{D}$ known as the *electric flux density*. $\mathbf{D}$ has the same direction as $\mathbf{E}$, the electric field intensity, and its magnitude is $\mathbf{D} = \epsilon_o \mathbf{E}$. From Coulomb's law, $\epsilon_o \mathbf{E}$ has the dimension of charge/area. Based on Gauss's law, which we will describe in later sections, the number of the flux lines emanating from a charge $+ Q$ is equal

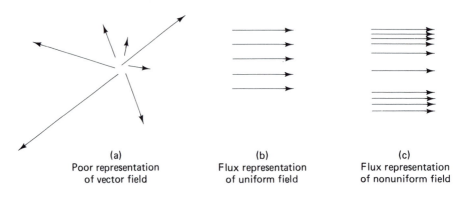

(a)	(b)	(c)
Poor representation	Flux representation	Flux representation
of vector field	of uniform field	of nonuniform field

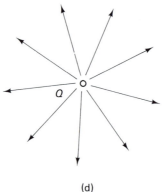

(d)
Map of flux lines
around the electric charge Q

Figure 1.26 Various graphical representations of fields.

to the value of the charge in the SI system of units. Hence, if ϵ is the total number of flux lines

$$\epsilon_{(\text{lines})} = Q(\text{C}) \text{ in the SI system of units}$$

The vector **D** is therefore equal to

$$\mathbf{D} = \frac{\text{charge } Q}{\text{area}} = \frac{\epsilon}{\text{area}} \equiv \text{electric flux density}$$

Hence, **D** is an important parameter in our graphical representation of the field simply because it indicates the number of the flux lines per unit area. This flux representation of a vector field will be further used in future discussions.

EXAMPLE 1.11

Use the flux representation to illustrate graphically the following vector fields:

1. $\mathbf{A} = K \mathbf{a}_x$
2. $\mathbf{B} = Kx \mathbf{a}_y$
3. $\mathbf{C} = Kx \mathbf{a}_x$
4. $\mathbf{D} = \dfrac{K}{\rho} \mathbf{a}_\rho$
5. $\mathbf{F} = K \mathbf{a}_\rho$

Solution

Before we start graphically representing the given vector fields, let us review the basic rules of the flux representation of a vector field.

1. The direction of the flux lines is in the direction of the vector field or tangential to it.
2. The distance between the flux lines is inversely proportional to the magnitude of the vector field. In other words, the larger the magnitude of the vector field, the smaller the distance between the flux lines will be. With these basic rules in mind, let us now make the desired flux representations.

   (a) $\mathbf{A} = K \mathbf{a}_x$ is an x-directed vector with uniform (equal) magnitude everywhere in the Cartesian coordinate system simply because it is independent of the x, y, z variables. A flux representation of the vector $\mathbf{A}$ is given in Figure 1.27a.

   (b) Vector $\mathbf{B}$ is in the y direction, and more important is that the magnitude of the vector increases with the increase of x. Vector $\mathbf{B}$, therefore, is not uniform and its magnitude increases with the increase in x. The flux lines representing this vector are hence drawn closer as the magnitude of the vector increases with the increase in x. Furthermore, the vector will be directed in the negative y direction for negative values of x. A flux representation of $\mathbf{B}$ is given in Figure 1.27b.

   (c) In this case, the vector $\mathbf{C}$ is also directed in the $\mathbf{a}_x$ direction for positive values of x and in the $-\mathbf{a}_x$ direction for negative values of x. Furthermore, the magnitude of the vector increases with the increase in x. This increase in the magnitude of vector $\mathbf{C}$ is represented graphically in Figure 1.27c by decreasing the distance between the flux lines, or in other words, by increasing the number of flux lines with the increase in x.

   (d) Vector field $\mathbf{D}$ is best graphically illustrated in the cylindrical coordinate system. It is an $\mathbf{a}_\rho$ directed vector, and its magnitude decreases with the increase in ρ. Figure 1.27d illustrates the flux representation of such a vector where it is clear that just by drawing the $\mathbf{a}_\rho$ directed flux lines, the distance between these lines increases with the increase in ρ, thus demonstrating the decrease in the magnitude of the vector $\mathbf{D}$ with the increase in ρ.

   (e) The flux representation of the vector $\mathbf{F}$ is also made in the cylindrical coordinate system because $\mathbf{F}$ is simply in the $\mathbf{a}_\rho$ direction. To illustrate the uniform magnitude of the vector $\mathbf{F}$, however, the distance between the flux lines should be maintained constant. This is achieved graphically by drawing more and more flux

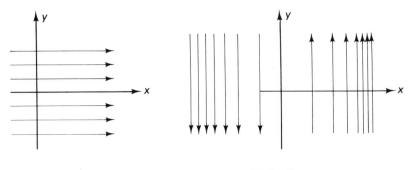

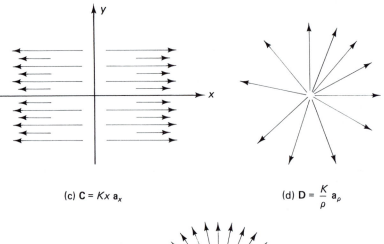

(a) $\mathbf{A} = K\,\mathbf{a}_x$

(b) $\mathbf{B} = Kx\,\mathbf{a}_y$

(c) $\mathbf{C} = Kx\,\mathbf{a}_x$

(d) $\mathbf{D} = \dfrac{K}{\rho}\,\mathbf{a}_\rho$

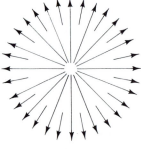

(e) $\mathbf{E} = K\,\mathbf{a}_\rho$

Figure 1.27 Flux representation of various vectors.

lines with the increase in ρ, as shown in Figure 1.27e, so as to maintain the density of the flux lines (i.e., number of flux lines per unit area) almost constant throughout Figure 1.27e.

1.6.3 Magnetic Field

The concept of a field should be familiar by now. Fields really possess no physical basis, because the physical measurements must always be in terms of the forces that result from these fields. As an example of these fields, we discussed in detail the electric field **E** and the forces associated with static electric charges. Another type of force is the *magnetic force* that may be produced by the steady magnetic field of a permanent magnet, an electric field changing with time, or a direct current. We might all be familiar with the magnetic field produced by a permanent magnet that can be recognized through its force of attraction on iron file placed in the neighborhood of the magnet. This phenomenon has been recognized and reported throughout history. It was only in 1820, however, that Oersted discovered that a magnet placed near a current-carrying wire will align itself perpendicular to the wire. This simply means that the steady electric currents exert forces on permanent magnets similar to those exerted by permanent magnets on each other. Ampere then showed that electric currents also exert forces on each other, and that a magnet can be replaced by an equivalent current with the same result. Biot and Savart quantified Ampere's observations, and in the following section we will discuss their findings. Before going to the next section, however, it is worth mentioning that the magnetic field produced by time-varying electric fields is just a mathematical discovery made by Maxwell through his attempt to unify the laws of electromagnetism available at that time. The hypothesis introduced by Maxwell postulating that time-varying electric fields produce magnetic fields will be discussed in detail later in this chapter. In this section we will focus our discussion on the production of magnetic fields by current-carrying conductors. The fundamental law in this study is Biot-Savart's law, which quantifies the magnetic flux density produced by a differential current element.

Biot-Savart's Law.   The Biot-Savart law quantifies the magnetic flux density $d\mathbf{B}$ produced by a differential current element $\mathbf{I}d\ell$. The experimental law was introduced to describe the force on a small magnet owing to the magnetic flux produced from a long conductor carrying current **I**. If each of the poles of a small magnet has a strength m, the force **F** caused by the flux **B** is given by

$$\mathbf{F} = m\mathbf{B}$$

This force law is clearly analogous to Coulomb's law for electrostatic field. In this case, the electric force is equal to the charge Q multiplied by the electric field intensity **E**. Hence, $\mathbf{F} = Q\mathbf{E}$.

To quantify the experimental observations by Biot and Savart, the force $d\mathbf{F}$ owing to the magnetic flux $d\mathbf{B}$ produced by a differential current element $\mathbf{I}d\ell$, as shown in Figure 1.28, is found to have the following characteristics.

1. It is proportional to the product of the current, the magnitude of the differential length, and the sine of the angle between the current element and the line connecting the current element to the observation point P.

2. It is inversely proportional to the square of the distance from the current element to the point P.

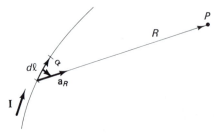

Figure 1.28 The magnetic field intensity at a point P owing to a current element $I d\ell$. $\mathbf{a}_R$ is a unit vector between the element and the observation point P.

3. The direction of the force is normal to the plane containing the differential current element and a unit vector from the current element to the observation point P. It also follows the right-hand rule from $I d\ell$ to the line from the filament to P. Hence,

$$|d\mathbf{F}| = |m d\mathbf{B}| = \frac{m\mu_o I \, d\ell \, \sin\alpha}{4\pi} \frac{}{r^2}$$

Because the direction of the force $d\mathbf{F}$ is perpendicular to $I d\ell$ and $\mathbf{a}_R$, a compact expression of the force may take the form

$$d\mathbf{F} = m d\mathbf{B} = m\mu_o \frac{I d\ell \times \mathbf{a}_R}{4\pi R^2}$$

where $\mu_o/4\pi$ is the constant of proportionality.

The following examples will illustrate the use of Biot-Savart's law in calculating magnetic fields from current carrying conductors.

EXAMPLE 1.12

Let us use Biot-Savart's law to find the magnetic flux density produced by a single turn loop carrying a current $\mathbf{I}$. We will limit the calculation to the magnetic field along the axis of the loop.

Solution

The magnetic field resulting from the current element 1 ($I d\ell_1$), which is located at an angle ϕ in Figure 1.29 is given according to Biot-Savart's law by

$$\mathbf{dB}_1 = \frac{\mu_o I d\ell \times \mathbf{a}_R}{4\pi R^2} = \frac{\mu_o I d\ell \mathbf{a}_\phi \times \mathbf{a}_R}{4\pi (a^2 + z^2)}$$

At the element 2, which is symmetrically located with respect to element 1, that is, located at the angle $\phi + \pi$ in Figure 1.29, the magnetic flux density is given by

$$d\mathbf{B}_2 = \frac{\mu_o I d\ell \, \mathbf{a}_\phi \times \mathbf{a}_{R_2}}{4\pi (a^2 + z^2)}$$

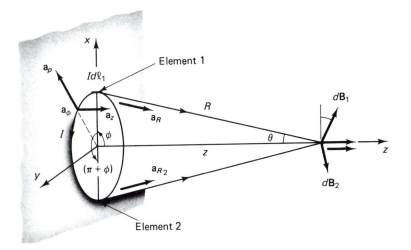

Figure 1.29   Magnetic flux density resulting from a current loop.

From Figure 1.29, it may be seen that the components of $d\mathbf{B}_1$ and $d\mathbf{B}_2$ perpendicular to the z axis cancel and the other components along the z axis, that is, $|d\mathbf{B}_1| \sin\theta$ and $|d\mathbf{B}_2| \sin\theta$, will add, hence,

$$dB_z = \frac{\mu_o\, Ia\, d\phi\, \sin\theta}{4\pi\,(a^2 + z^2)} = \frac{\mu_o\, Ia\, d\phi}{4\pi\,(a^2 + z^2)}\, \frac{a}{(a^2 + z^2)^{1/2}}$$

$$= \frac{\mu_o\, Ia^2\, d\phi}{4\pi\,(a^2 + z^2)^{3/2}}$$

The total magnetic flux density B_z is obtained by integrating dB_z with respect to ϕ from 0 to 2π. Because dB_z is independent of ϕ, we simply multiply dB_z by 2π, hence,

$$B_z = \frac{\mu_o\, Ia^2}{4\pi\,(a^2 + z^2)^{3/2}}\, 2\pi = \frac{\mu_o\, Ia^2}{2(a^2 + z^2)^{3/2}}$$

or

$$\mathbf{B} = \frac{\mu_o\, Ia^2}{2(a^2 + z^2)^{3/2}}\, \mathbf{a}_z$$

This result indicates that the direction of the current flow and the direction of the resulting magnetic field are according to the right-hand rule. When the fingers of the right hand are folded in the direction of the current flow, the thumb will point to the direction of the magnetic flux density $\mathbf{B}$.

EXAMPLE 1.13

An infinitely long conducting wire carrying a constant current $\mathbf{I}$ and is oriented along the z axis as shown in Figure 1.30. Determine the magnetic flux density at P.

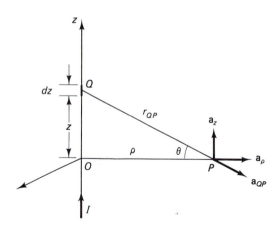

Figure 1.30 The magnetic flux density **B** resulting from an infinitely long conducting wire. The wire is oriented along the positive z axis.

Solution

Because the wire is infinitely long and due to symmetry around the wire, the resulting magnetic field should be independent of z and ϕ. Hence, without loss of generality, we will place P on the $z = 0$ plane. Let us consider an incremental current element $I\,dz$ located at Q, which is a distance z from the origin O. The unit vector in the direction joining the incremental current element to the field P is

$$\mathbf{a}_{QP} = \mathbf{a}_\rho \cos\theta - \mathbf{a}_z \sin\theta$$

$$= \mathbf{a}_\rho \frac{\rho}{r_{QP}} - \mathbf{a}_z \frac{z}{r_{QP}}$$

where $r_{QP} = \sqrt{z^2 + \rho^2}$. The magnetic field resulting from this current element is given according to Biot-Savart's law by

$$d\mathbf{B} = \frac{\mu_o I\,dz\ \mathbf{a}_z \times \mathbf{a}_{QP}}{4\pi\, r_{QP}^2}$$

Substituting $\mathbf{a}_{QP}$ and noting that $\mathbf{a}_z \times \mathbf{a}_\rho = \mathbf{a}_\phi$, and $\mathbf{a}_z \times \mathbf{a}_z = 0$, we obtain

$$d\mathbf{B} = \frac{\mu_o I \rho\, dz}{4\pi\, r_{QP}^3} \mathbf{a}_\phi$$

The total magnetic field from the current line is obtained by integrating the contributions from all elements along the line, hence,

$$B_\phi = \frac{\mu_o I \rho}{4\pi} \int_{-\infty}^{\infty} \frac{dz}{(z^2 + \rho^2)^{3/2}}$$

$$= \frac{\mu_o I \rho}{4\pi} \left. \frac{z}{\rho^2 (z^2 + \rho^2)^{1/2}} \right|_{z = -\infty}^{\infty}$$

$$B_\phi = \frac{\mu_o I}{2\pi\rho} \ \text{Wb/m}^2$$

or

$$\mathbf{B} = \frac{\mu_o I}{2\pi\rho} \mathbf{a}_\phi$$

Once again, we note that the direction of the magnetic flux lines and the direction of the current producing it obey the right-hand rule. With the fingers of the right hand folded in the direction of the flux lines, the thumb indicates the direction of the current flow.

1.6.4 Lorentz Force Equation

Let us now determine the forces exerted by the magnetic field on charges. We learned in previous sections that the electric field causes forces on charges that may be either stationary or in motion. This is because any charged particle (whether in motion or not) is capable of producing an electric field that interacts with the already existing electric field, resulting in exerting a force on the charged particle. We do not expect an electric field to exert forces on uncharged particles (e.g., particles of given masses) simply because such particles do not produce electric field, and hence there will be no interaction. Similarly, the magnetic field is capable of exerting a force only on moving charges. This result appears logical because we are considering magnetic fields produced by moving charges (currents) and therefore may exert forces on moving charges. The magnetic field cannot be produced from stationary charges and, hence, cannot exert any force on stationary charges.

The force exerted on a charged particle in motion in a magnetic field of flux density **B** is found experimentally to be the following:

1. Proportional to the charge Q, its velocity **v**, the flux density **B**, and to the sine of the angle between the vectors **v** and **B**.
2. The direction of the force is perpendicular to both **v** and **B**, and is given by a unit vector in the direction **v** × **B**. The force is, therefore, given by

$$\boxed{\mathbf{F} = Q\mathbf{v} \times \mathbf{B}}$$

The force on a moving charge as a result of combined electric and magnetic fields is obtained easily by the superposition of the separate electric and magnetic forces. Hence,

$$\boxed{\mathbf{F} = Q(\mathbf{E} + \mathbf{v} \times \mathbf{B})}$$

This equation is known as *Lorentz force* equation, and its solution is required in determining the motion of a charged particle in combined electric and magnetic fields.

1.6.5 Differences in Effect of Electric and Magnetic Fields on Charged Particles

The force exerted by the magnetic field is always perpendicular to the direction in which the particle is moving. This force, therefore, does not change the magnitude of the particle's velocity because the work dW done on the particle or the energy delivered to it by the magnetic field is always zero.

$$dW = \mathbf{F} \cdot d\boldsymbol{\ell} = q\mathbf{v} \times \mathbf{B} \cdot \mathbf{v}dt = 0$$

TABLE 1.1   COMPARISON BETWEEN THE ELECTRIC AND MAGNETIC FIELDS

Electric field	Magnetic field
1. Can be produced by charged particles moving or stationary.	Can be produced by direct current that can be attributed to only moving charges.
2. The direction of the force exerted is along the line joining the two charges and is, therefore, independent of the direction of motion of the charged particle.	The force is always perpendicular to the direction of the velocity of the particle.
3. Electric field force causes energy transfer between the field and the charged particle.	The work done on the charged particle is always equal to zero. This is because the magnetic force is always perpendicular to the velocity and hence cannot change the magnitude of the particle velocity.

The magnetic field may, however, deflect the trajectory of the particle's motion but not change the total energy or the total velocity.

The electric field, conversely, exerts a force on the particle that is independent of the direction in which the particle is moving. A velocity component along the direction of the electric field can be generated. The electric field, therefore, causes an energy transfer between the field and the particle. Some fundamental differences between the electric and magnetic fields are summarized in Table 1.1.

To enhance our understanding of the electric and magnetic fields and the nature of their interaction with charged particles further, let us solve the following additional examples.

EXAMPLE 1.14

Consider a particle of mass m and charge q moving in a magnetic field that is oriented in the z direction. The magnetic flux density is given by $\mathbf{B} = B_o \, \mathbf{a}_z$. If the particle has an initial velocity $\mathbf{v} = v \, \mathbf{a}_x$ (i.e., at $t = 0$), describe the motion of the particle under the influence of the magnetic field.

Solution

From Newton's law and Lorentz force

$$m\mathbf{a} = q \, \mathbf{v} \times \mathbf{B} \tag{1.7}$$

where $\mathbf{a}$ is the particle's acceleration. Expressing equation 1.7 in terms of its components, we obtain

$$m\left(\frac{dv_x}{dt}\mathbf{a}_x + \frac{dv_y}{dt}\mathbf{a}_y + \frac{dv_z}{dt}\mathbf{a}_z\right) = q\begin{vmatrix} \mathbf{a}_x & \mathbf{a}_y & \mathbf{a}_z \\ v_x & v_y & v_z \\ 0 & 0 & B_o \end{vmatrix} \tag{1.8}$$

$$= q(v_y B_o \, \mathbf{a}_x - v_x B_o \, \mathbf{a}_y + 0 \, \mathbf{a}_z)$$

Note that although the initial velocity of the particle is in the x direction, we considered all the velocity components in the $\mathbf{v} \times \mathbf{B}$ expression because the velocity of the particle under the influence of the magnetic field is unknown and it is likely that the magnetic field would deflect the particle's trajectory thus generating other velocity components.

Now equating the various components of equation 1.8, we obtain

$$m\frac{dv_x}{dt} = q\,v_y\,B_o \tag{1.9a}$$

$$m\frac{dv_y}{dt} = -q\,v_x\,B_o \tag{1.9b}$$

$$m\frac{dv_z}{dt} = 0 \tag{1.9c}$$

From equation 1.9c it is clear that by integrating with respect to time, we will obtain v_z = constant. Hence, if the particle has an initial velocity in the direction of the magnetic field (z direction), this component of velocity will continue to be constant and unchanged under the influence of the magnetic field. If, conversely, no component of the velocity is initially in the z direction, that is, along the magnetic field, this component will continue to be zero even after the interaction of the charged particle with the magnetic field.

With this in mind regarding the component of the velocity v_z, let us solve equations 1.9a and b for the other two components of the velocity v_x and v_y. Differentiating equation 1.9a once more with respect to t and substituting equation 1.9b for dv_y/dt, we obtain

$$m\frac{d^2 v_x}{dt^2} = -\frac{q^2 B_o^2}{m}v_x$$

or

$$\frac{d^2 v_x}{dt^2} + \frac{q^2 B_o^2}{m^2}v_x = 0 \tag{1.10}$$

A solution of equation 1.10 is in the form

$$v_x = A_1 \cos \omega_o t + A_2 \sin \omega_o t \tag{1.11}$$

where $\omega_o = qB_o/m$ and A_1 and A_2 are two unknown constants to be determined from the initial conditions of the velocity. Substituting v_x in equation 1.9b, we obtain v_y in the form

$$v_y = -\frac{qB_o}{m}\left(+ A_1 \frac{\sin \omega_o t}{\omega_o} - A_2 \frac{\cos \omega_o t}{\omega_o} \right) \tag{1.12}$$

$$= -A_1 \sin \omega_o t + A_2 \cos \omega_o t$$

To determine A_1 and A_2, let us use the initial conditions of the velocity. At $t = 0$, $\mathbf{v} = v\,\mathbf{a}_x$, and $v_y = 0$, substituting these initial conditions in equations 1.11 and 1.12, we obtain $A_2 = 0$ and $A_1 = v$. The expressions for v_x and v_y are therefore given by

$$v_x = v \cos \omega_o t \qquad \text{and} \qquad v_y = -v \sin \omega_o t$$

The particle's total velocity in the magnetic field is, therefore,

$$\mathbf{v} = v \cos \omega_o t\,\mathbf{a}_x - v \sin \omega_o t\,\mathbf{a}_y \tag{1.13}$$

If we plot the variation of the particle's velocity as a function of time, we can easily see that the particle is rotating in the clockwise direction around the magnetic field as shown in Figure 1.31.

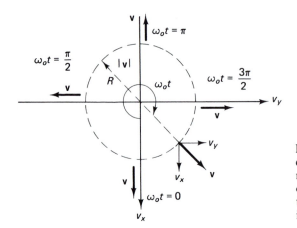

Figure 1.31 Motion of positively charged particle in a constant magnetic field. If the magnetic field is oriented along the positive z axis, the particle moves in a circular path in the clockwise direction.

TABLE 1.2 VELOCITY COMPONENTS AND DIRECTION AS A FUNCTION OF TIME

$\omega_o t$	$|\mathbf{v}|$	Direction
0	v	$\mathbf{a}_x$
$\dfrac{\pi}{2}$	v	$-\mathbf{a}_y$
π	v	$-\mathbf{a}_x$
$\dfrac{3\pi}{2}$	v	$\mathbf{a}_y$
2π	v	$\mathbf{a}_x$
ϕ	v	Has both $\mathbf{a}_x$ and $\mathbf{a}_y$ components

From Table 1.2 and by noting that

$$|\mathbf{v}| = v\sqrt{\cos^2 \omega_o t + \sin^2 \omega_o t}$$

$$= v$$

it is clear that the magnitude of the particle's velocity is always constant and is equal to the initial velocity. Its components, however, vary as the particle presses around the magnetic field vector in a circular trajectory. The angular velocity ω_o is called the *cyclotron frequency*. The radius of the circle in which the particle travels around the magnetic field is

$$R = \frac{v}{\omega_o}$$

This example simply emphasizes the statement made in the previous section that the magnetic field may deflect the particle's trajectory but not change its velocity—that is, causes no energy transfer from or to the particle.

EXAMPLE 1.15

A charge q of mass m is injected into a field region containing perpendicular electric and magnetic fields. When the charge velocity at any point along the motion path is $\mathbf{v} = v_x\, \mathbf{a}_x$, the observed acceleration is $\mathbf{a} = a_x\, \mathbf{a}_x + a_y\, \mathbf{a}_y$.

Find an $\mathbf{E}$ and $\mathbf{B}$ combination that would generate this acceleration $\mathbf{a}$.

Solution

When the velocity has only one component in the x direction, the acceleration was found to have two components.

$$\mathbf{a} = a_x\, \mathbf{a}_x + a_y\, \mathbf{a}_y$$

In the presence of both $\mathbf{E}$ and $\mathbf{B}$ fields, the force is given by

$$\mathbf{F} = m(a_x\, \mathbf{a}_x + a_y\, \mathbf{a}_y) = q(\mathbf{E} + \mathbf{v} \times \mathbf{B})$$

Because $\mathbf{v}$ has only one component in the x direction, then the magnetic field force cannot be responsible for the x component of the force. The electric field force is therefore the cause of the x component of the acceleration. Hence,

$$\mathbf{E} = \frac{m\, a_x}{q}\, \mathbf{a}_x$$

and

$$\mathbf{B} = -\frac{m\, a_y}{q\, v_x}\, \mathbf{a}_z$$

To explain further the reason for $\mathbf{B}$ to have only an $\mathbf{a}_z$ component, let us assume that $\mathbf{B}$ has B_y and B_z components. We should note that $\mathbf{B}$ has no B_x component because $\mathbf{E}$ (has only x component) and $\mathbf{B}$ are perpendicular to each other. Now if we assume that $\mathbf{B}$ has B_y and B_z components, from $\mathbf{v} \times \mathbf{B}$ determinant, there should be an $\mathbf{a}_z$ component of force or consequently an $\mathbf{a}_z$ component of acceleration.

$$\begin{vmatrix} \mathbf{a}_x & \mathbf{a}_y & \mathbf{a}_z \\ v_x & 0 & 0 \\ 0 & B_y & B_z \end{vmatrix} = -v_x B_z\, \mathbf{a}_y + v_x B_y\, \mathbf{a}_z$$

Because the $\mathbf{a}_z$ component of the acceleration is zero, B_y has therefore to be zero.

EXAMPLE 1.16

Two small balls of masses m have a charge Q each, and are suspended at a common point by thin filaments, each of length ℓ. Assuming that the charges are to be located approximately at the centers of the balls, find the angle α between the filaments. (Assume α to be small.) *Note*: Such a system can be used as a primitive device for measuring charges and potentials and is called an electroscope.

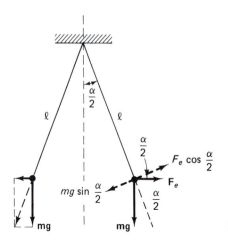

Figure 1.32   The electroscope.

Solution

Because the two balls are charged with similar charges, the repulsion force will cause them to separate away from each other. The two balls will reach the equilibrium position when the force perpendicular to each string becomes zero as shown in Figure 1.32. This is simply because this force is responsible for swinging the balls.

From Figure 1.32, it may be seen that the equilibrium position will occur when

$$mg\ \sin \alpha/2 = F_e\ \cos \alpha/2$$

The electric force between the two charged balls F_e is given by

$$F_e = \frac{Q^2}{4\pi\,\epsilon_o\,(2\ell\ \sin \alpha/2)^2}$$

Hence, the equilibrium equation reduces to

$$\frac{Q^2}{16\pi\,\epsilon_o\,\ell^2\,mg} = \frac{\sin^3 \alpha/2}{\cos \alpha/2}$$

For small α, $\cos \alpha/2 = 1$ and $\sin \alpha/2 = \alpha/2$. Thus,

$$\alpha^3 = \frac{Q^2}{2\pi\,\epsilon_o\,\ell^2\,mg}$$

In the previous sections we familiarized ourselves with the simple rules of vector algebra and the basic concepts of fields. In the following section we will continue our efforts to pave the way for the introduction of Maxwell's equations. Specifically, we will introduce the vector integration as a prerequisite to the discussion of Maxwell's equations in integral form.

1.7 VECTOR INTEGRATION

Besides the vector representation of the electromagnetic field quantities and the ability to transform such representation from one coordinate system to another, it is important that we develop a thorough understanding of basic vector integral and differential operations. Vector differential operations will be discussed in chapter 2 just before the introduction of Maxwell's equations in differential form. In preparation of the introduction of Maxwell's equation in integral form, we introduce vector integral operations next.

1.7.1 Line Integrals

The scalar line integral $\int_a^b A(\ell)\,d\ell$ (where ℓ is the length of the contour and a and b are the two end points along the path of integration) is defined as the limit of the sum $\sum_{i=1}^N A(\ell_i)\,\Delta\ell_i$ as $\Delta\ell_i \to 0$. $A(\ell_i)$ is the value of $A(\ell)$ evaluated at the point ℓ_i within the segment $\Delta\ell_i$. This simply means that in evaluating $\int_c A(\ell)\,d\ell$, we divide the contour of integration c into N segments, as shown in Figure 1.33, evaluate the scalar quantity $A(\ell_i)$ at the center of each element, multiply $A(\ell_i)$ by the length of the element $\Delta\ell_i$, and add the contributions from all the segments. The sum of these contributions will equal exactly the line integral of the scalar quantity in the limit when the lengths of these elements $\Delta\ell_i$ approach zero. Hence,

$$\int_c A(\ell)\,d\ell = \lim_{\substack{\Delta\ell_i \to 0 \\ N \to \infty}} \sum_{i=1}^N A(\ell_i)\,\Delta\ell_i$$

A simple example of this line integral is the evaluation of $\int_c d\ell$ where the contour c is given by the curve shown in Figure 1.34. The element of length $d\ell$ in this case is given by $\rho\,d\phi$ where $\rho = 1$ along the given contour c. Therefore,

$$\int_c d\ell = \int_0^{\pi/2} \rho|_{\rho=1}\,d\phi = \int_0^{\pi/2} d\phi = \frac{\pi}{2}$$

If we follow the physical reasoning behind the evaluation of the line integral of a scalar quantity as described earlier, it can be shown that the line integral of the form $\int_c d\ell$ is simply the length of the contour c. Hence, if c is given by the curve shown in Figure 1.34, then

$$\int_c d\ell = \frac{\text{Circumference of circle}}{4}$$

$$= \frac{2\pi(1)}{4} = \frac{\pi}{2}$$

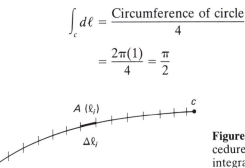

Figure 1.33 An approximate procedure for calculating a scalar line integral.

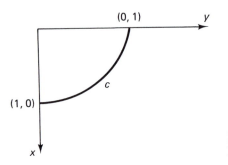

(0, 1)

y

c

(1, 0)

x

Figure 1.34 The contour c of the line integral $\int_c d\ell$.

which is the same result obtained by carrying out the integration. Another example illustrating the evaluation of a scalar line integral is given next.

EXAMPLE 1.17

Evaluate the line integral $\int_c (\cos \phi / \rho) \, dy$, where c is the straight line from $(a, 0)$, to (a, a).

Solution

The integrand is given in the cylindrical coordinates, whereas the integration contour and limits can easily be identified in the Cartesian coordinates, as shown in Figure 1.35. Hence, we transform the integrand to rectangular coordinates, and note that $x = a$ along the path c. Thus

$$\int_c \frac{\cos \phi}{\rho} \, dy = \int_0^a \frac{x}{\rho^2} \, dy = \int_0^a \frac{x}{x^2 + y^2} \, dy$$

$$= \int_0^a \frac{a}{a^2 + y^2} \, dy = \left[\tan^{-1} \frac{y}{a} \right]_0^a = \frac{\pi}{4}$$

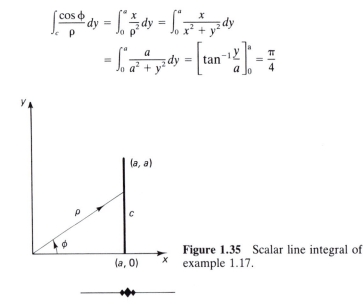

y

(a, a)

ρ

c

φ

(a, 0)   x

Figure 1.35 Scalar line integral of example 1.17.

Next is an illustration of the difference between the scalar line integral of the form $\int_c d\ell$ and a vector line integral of the form $\int_c d\boldsymbol{\ell}$ where $d\boldsymbol{\ell}$ is a differential vector element of length. If we follow the same procedure described earlier to evaluate the vector line

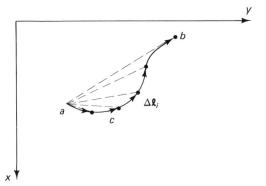

Figure 1.36 Approximate procedure for evaluating $\int_c d\ell$.

integral $\int_c d\ell$, we simply divide the contour c into small vector segments and the value of the line integral is the *vector* sum of the contributions from all the segments as shown in Figure 1.36. In other words,

$$\int_c d\ell = \lim_{|\Delta\ell_i| \to 0} \sum_{i=1}^{N} \Delta\ell_i$$

It is clear from Figure 1.36 that the value of the line integral is simply the vector **ab**.

EXAMPLE 1.18

Evaluate the line integral $\int_c d\ell$ where c is given by the contour *abd* as shown in Figure 1.37.

Solution

Based on the preceding discussion, the vector line integral $\int_c d\ell$ is simply given by the vector **ad**.

The vector

$$\mathbf{ad} = \mathbf{od} - \mathbf{oa}$$

$$= \mathbf{a}_x + 4\,\mathbf{a}_y - (\mathbf{a}_x + \mathbf{a}_y)$$

$$= 3\,\mathbf{a}_y$$

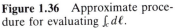

Figure 1.37 The contour c of the vector line integral $\int_c d\ell$ is given by the curve *abd*.

Hence,

$$\int_c d\boldsymbol{\ell} = 3\,\mathbf{a}_y$$

Besides the fact that the result is a vector rather than a scalar quantity, the value of the vector line integral is far from being equal to the length of the contour abd.

Of particular interest to our study in electromagnetics is the line integral that involves the *tangential component of some vector* $\mathbf{A}$ along the integration contour. These integrals are of the form:

$$\int_c \mathbf{A}\,(x,y,z)\cdot\mathbf{t}\,(x,y,z)\,d\ell$$

where $\mathbf{t}$ is a unit vector tangent to the contour c.

To illustrate the importance of a line integral of this form, let us consider an electric field $\mathbf{E}$ shown in terms of its flux representation in Figure 1.38. We wish to calculate the work done in moving a charge q along the contour c in the electric field $\mathbf{E}$.

First, we consider the element of length $\Delta\ell_i$ along the contour c. A unit vector tangential to this element is $\mathbf{t}$, as shown in Figure 1.38. In calculating the work done in moving the charge q along $\Delta\ell_i$, we need to obtain first the component of $\mathbf{E}$ (which is defined as the force per unit positive charge) along the element $\Delta\ell_i$. This component is given in Figure 1.38 by $E_i \cos \alpha_i$ where E_i is the magnitude of the electric field at the $\Delta\ell_i$ location and α_i is the angle between the direction of the electric field and $\mathbf{t}$. The work done ΔW_i in moving the positive charge q against the electric field lines along $\Delta\ell_i$ is then

$$\Delta W_i = q \underbrace{E_i \cos \alpha_i}_{\substack{\text{component}\\\text{of force}\\\text{along the}\\\text{the contour}}} \Delta\ell_i$$

The total work done W in moving the charge q is then given by

$$W = q \sum_{i=1}^{N} E_i \cos \alpha_i \, \Delta\ell_i$$

$$= q \sum_{i=1}^{N} \mathbf{E}_i \cdot \mathbf{t}\, \Delta\ell_i$$

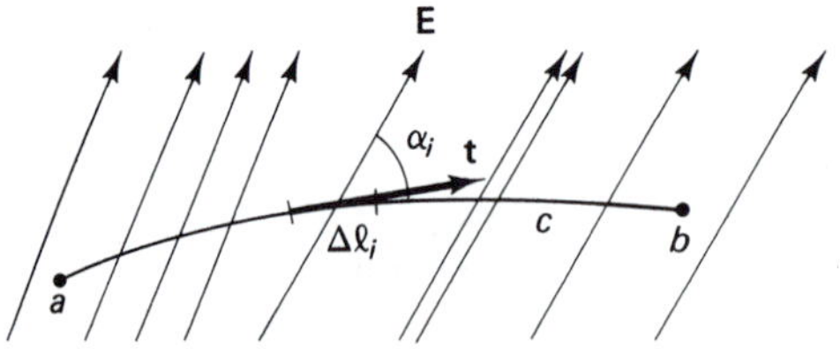

Figure 1.38 A charge q is moving along the contour c in the electric field.

where the dot product shorthand notation was used to substitute $E_i \cos \alpha_i$, which is simply the projection of $\mathbf{E}_i$ along the unit vector $\mathbf{t}$. In the limit as $\Delta \ell_i \to 0$, the work done can be expressed in the form

$$W = \operatorname*{Lim}_{\substack{\Delta \ell_i \to 0 \\ N \to \infty}} q \sum_{i=1}^{N} \mathbf{E}_i \cdot \mathbf{t} \Delta \ell_i$$

$$= q \int_c \mathbf{E} \cdot \mathbf{t} d\ell$$

which is the same form as the line integral $\int_c \mathbf{A} \cdot d\ell$ where $\mathbf{A}$ is any vector in this case. Although the preceding example illustrates the importance of line integrals of this form, it should be emphasized that such a line integral has the physical meaning of work done along the contour c only if $\mathbf{A}$ represents a force vector. Otherwise a line integral of this form may or may not have a physical meaning depending on the nature of $\mathbf{A}$. The only property that should always be kept in mind is that what we are actually integrating is the component of $\mathbf{A}$ tangential to the contour c.

Let us now focus our attention on evaluating an integral of the form $\int_c \mathbf{A} \cdot \mathbf{t} d\ell$. By noting that $\mathbf{t} d\ell$ is simply a differential element of length $d\ell$ along the contour c, it is clear that

$$\int_c \mathbf{A} \cdot \mathbf{t} d\ell = \int_c \mathbf{A} \cdot d\ell$$

where $d\ell$ is given in the various coordinate systems by

$$d\ell = dx\, \mathbf{a}_x + dy\, \mathbf{a}_y + dz\, \mathbf{a}_z \qquad \text{(Cartesian)}$$

$$= d\rho\, \mathbf{a}_\rho + \rho d\phi\, \mathbf{a}_\phi + dz\, \mathbf{a}_z \qquad \text{(Cylindrical)}$$

$$= dr\, \mathbf{a}_r + rd\theta\, \mathbf{a}_\theta + r \sin \theta\, d\phi\, \mathbf{a}_\phi \qquad \text{(Spherical)}$$

We discussed these various expressions of $d\ell$ when we were first introduced to the three coordinate systems.

These expressions of $d\ell$ significantly reduce the efforts in evaluating integrals of the form $\int_c \mathbf{A} \cdot d\ell$. To illustrate this, let us consider the vector $\mathbf{A}$ in the Cartesian coordinate system.

$$\mathbf{A} = A_x \mathbf{a}_x + A_y \mathbf{a}_y + A_z \mathbf{a}_z.$$

The integral $\int_c \mathbf{A} \cdot d\ell$ is then given by

$$\int_c \mathbf{A} \cdot d\ell = \int (A_x\, dx + A_y\, dy + A_z\, dz)$$

$$= \int_{x_1}^{x_2} A_x\, dx + \int_{y_1}^{y_2} A_y\, dy + \int_{z_1}^{z_2} A_z\, dz$$

which means that we have changed the original line integral into three much simpler scalar integrations.

In cylindrical coordinates, the results would be

$$\int_c \mathbf{A}\cdot d\ell = \int_{\rho_1}^{\rho_2} A_\rho \, d\rho + \int_{\phi_1}^{\phi_2} A_\phi \, \rho \, d\phi + \int_{z_1}^{z_2} A_z \, dz$$

And in spherical coordinates

$$\int_c \mathbf{A}\cdot d\ell = \int_{r_1}^{r_2} A_r \, dr + \int_{\theta_1}^{\theta_2} A_\theta \, r \, d\theta + \int_{\phi_1}^{\phi_2} A_\phi \, r \sin\theta \, d\phi$$

Let us now illustrate this procedure by solving the following examples.

EXAMPLE 1.19

Find the line integral of the vector $\mathbf{A} = y \, \mathbf{a}_x - x \, \mathbf{a}_y$ around the closed path in the x-y plane that follows the parabola $y = x^2$ from the point $x = -1, y = 1$ to the point $x = 2, y = 4$ and returns along the straight line $y = x + 2$. The integration path $c = c_1 + c_2$ is shown in Figure 1.39.

Solution

The integration along the closed contour c may be broken up into two paths, c_1 and c_2, which follow the parabola and the straight line, respectively. Then

$$\oint_c \mathbf{A}\cdot d\ell = \int_{c_1} \mathbf{A}\cdot d\ell + \int_{c_2} \mathbf{A}\cdot d\ell$$

$$\int_{c_1} \mathbf{A}\cdot d\ell = \int_{-1}^{2} A_x \, dx + \int_{1}^{4} A_y \, dy$$

$$= \int_{-1}^{2} y \, dx - \int_{1}^{4} x \, dy$$

We should be careful in substituting the integrands y and x according to the equation of the parabolic path $y = x^2$. In particular, in substituting x in the second integral $x = \pm\sqrt{y}$ and the appropriate plus or minus signs should be used depending on the region of

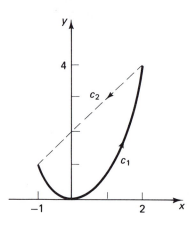

Figure 1.39 The contour of integration in example 1.19.

integration. We will use $x = -\sqrt{y}$ in the region of integration from $x = -1$ to 0, and $x = \sqrt{y}$ in the region of integration from $x = 0$ to $x = 2$. Hence,

$$\int_{c_1} \mathbf{A} \cdot d\boldsymbol{\ell} = \int_{-1}^{2} x^2 \, dx - \int_{1}^{0} -\sqrt{y} \, dy - \int_{0}^{4} \sqrt{y} \, dy = -3$$

or alternatively we can substitute $dy = 2x\,dx$, and in this case, we integrate with respect to x without having to substitute x in terms of y in the integrand.

In this case,

$$\int_{c_1} \mathbf{A} \cdot d\boldsymbol{\ell} = \int_{-1}^{2} x^2 \, dx - \int_{-1}^{2} x(2x\,dx) = -3$$

The integration along the contour c_2, conversely, is given by

$$\int_{c_2} \mathbf{A} \cdot d\boldsymbol{\ell} = \int_{2}^{-1} y\,dx - \int_{4}^{1} x\,dy$$

where x is related to y in this case by the equation of the straight line

$$y = x + 2$$

Substituting y in the first integrand and x in the second, we obtain

$$\int_{c_2} \mathbf{A} \cdot d\boldsymbol{\ell} = \int_{2}^{-1} (x + 2)\,dx - \int_{4}^{1} (y - 2)\,dy$$

$$= -6$$

so that

$$\oint_{c} \mathbf{A} \cdot d\boldsymbol{\ell} = -3 - 6 = -9$$

---◆◆◆---

EXAMPLE 1.20

Find the work done on moving a particle once around a circle c in the x-y plane, if the circle has center at the origin and radius 3, and if the force is given by

$$\mathbf{F} = (2x - y + z)\,\mathbf{a}_x + (x + y - z^2)\,\mathbf{a}_y$$
$$+ (3x - 2y + 4z)\,\mathbf{a}_z$$

Solution

In the plane $z = 0$,

$$\mathbf{F} = (2x - y)\,\mathbf{a}_x + (x + y)\,\mathbf{a}_y + (3x - 2y)\,\mathbf{a}_z$$

The work done is then given by

$$\int_{c} \mathbf{F} \cdot d\boldsymbol{\ell} = \int_{c} [(2x - y)\,\mathbf{a}_x + (x + y)\,\mathbf{a}_y + (3x - 2y)\,\mathbf{a}_z] \cdot [dx\,\mathbf{a}_x + dy\,\mathbf{a}_y]$$

$$= \int_{c} (2x - y)\,dx + \int_{c} (x + y)\,dy$$

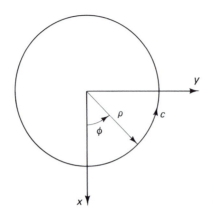

Figure 1.40   The contour of integration c of example 1.20.

From Figure 1.40, it can be seen that $x = 3 \cos \phi$, $y = 3 \sin \phi$, where ϕ varies from 0 to 2π. Furthermore, differentiating the x and y equations with respect to the new variable ϕ, we obtain

$$dx = -3 \sin \phi \, d\phi \qquad \text{and} \qquad dy = 3 \cos \phi \, d\phi$$

The line integral is then

$$\int_0^{2\pi} [2(3 \cos \phi) - 3 \sin \phi][-3 \sin \phi \, d\phi] + [3 \cos \phi + 3 \sin \phi][3 \cos \phi \, d\phi]$$

$$= \int_0^{2\pi} (9 - 9 \sin \phi \cos \phi) \, d\phi$$

$$= 18\pi$$

Alternative Solution.   Because of the symmetry of the contour of integration, we will use the cylindrical coordinates. In cylindrical coordinates, $d\ell$ is given by

$$d\ell = d\rho \, \mathbf{a}_\rho + \rho d\phi \, \mathbf{a}_\phi + dz \, \mathbf{a}_z$$

but along the contour c, $\rho = \text{constant} = 3$, and $z = \text{constant} = 0$. Hence,

$$d\ell = \rho d\phi \, \mathbf{a}_\phi$$

The work done is therefore

$$\int_c \mathbf{F} \cdot d\ell = \int (2x - y) \, \mathbf{a}_x + (x + y) \, \mathbf{a}_y + (3x - 2y) \, \mathbf{a}_z \cdot \rho d\phi \, \mathbf{a}_\phi$$

$$= \int (2x - y) \rho d\phi \, \mathbf{a}_x \cdot \mathbf{a}_\phi + \int (x + y) \rho d\phi \, \mathbf{a}_y \cdot \mathbf{a}_\phi$$

$$+ \int (3x - 2y) \rho d\phi \, \mathbf{a}_z \cdot \mathbf{a}_\phi$$

From the vector coordinate transformation described in the previous sections, it was noted that

$$\mathbf{a}_x \cdot \mathbf{a}_\phi = -\sin\phi$$

$$\mathbf{a}_y \cdot \mathbf{a}_\phi = \cos\phi$$

$$\mathbf{a}_z \cdot \mathbf{a}_\phi = 0$$

Also the following relations between the independent variables were noted:

$$x = \rho\cos\phi \qquad y = \rho\sin\phi$$

substituting all these relations in our integral and noting that along the contour of interest $\rho = 3$, we obtain

$$\int_c \mathbf{F}\cdot d\boldsymbol{\ell} = \int_{\phi=0}^{2\pi} (6\cos\phi - 3\sin\phi)\,3(-\sin\phi)\,d\phi$$

$$+ \int_0^{2\pi} (3\cos\phi + 3\sin\phi)\,3(\cos\phi)\,d\phi$$

$$= \int_0^{2\pi} -9\cos\phi\sin\phi\,d\phi + \int_0^{2\pi} 9\,d\phi$$

$$= 18\pi$$

which is the same value we obtained in the previous calculations.

1.7.2 Surface Integral

Across an infinitesimal area Δs in a large surface s, the flux of a vector field (e.g., magnetic field) may be assumed uniform. The flux distribution over the entire surface areas s may or may not be uniform. If the infinitesimal surface is oriented normal to the flux lines, as shown in Figure 1.41a, then the total flux crossing this area may be calculated by simply multiplying the *surface area* by the *flux density* (i.e., $|\mathbf{F}|\Delta s$). If, conversely, the infinitesimal surface is oriented parallel to the field flux lines, there will be no flux crossing the area Δs, as shown in Figure 1.41b. In general, the surface may be oriented at an angle α with respect to the flux lines, as shown in Figure 1.41c. In this case, the amount of flux crossing the surface is then determined by multiplying the normal component of the flux $\mathbf{F}$ by the surface area Δs.

The total flux crossing the area Δs in Figure 1.41c is given by

$$(|\mathbf{F}|\cos\alpha)\,\Delta s = |\mathbf{F}|\,\Delta s\,\cos\alpha$$

$$= \mathbf{F}\cdot\mathbf{n}\Delta s$$

where $\mathbf{n}$ is a unit vector normal to the area Δs.

An arbitrary surface area can always be divided into many infinitesimal areas, and the total field flux crossing the total area is then the sum of the field flux crossing all of these small areas, that is,

$$\text{The total flux} = \sum_{i=1}^{N} F_i\cos\alpha_i\,\Delta s_i$$

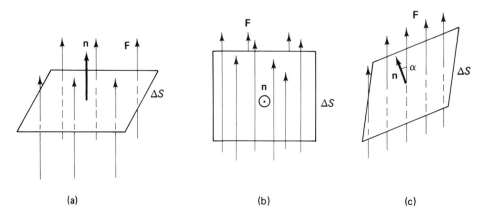

(a)   (b)   (c)

Figure 1.41   Calculation of the total vector flux crossing an element of area Δs. (a) The flux of **F** is perpendicular to the area (i.e., parallel to unit normal **n**) and the total flux crossing the area is $|\mathbf{F}|\,\Delta s$. (b) The flux is parallel to the area (i.e., perpendicular to unit normal **n**) and the total flux crossing the area is zero. (c) The flux is making an angle α with the unit normal to the area **n**. The total flux crossing the area in this case is $|\mathbf{F}|\,\cos\alpha\Delta s$.

In the limit, when the number of areas goes to ∞ and the value of each area approaches zero, the summation becomes an integral, that is,

$$\text{The total flux across the area } s = \int_s \mathbf{F}\cdot d\mathbf{s}$$

It should be noted that the surface integral is a double integral because ds is the product of two differential lengths.

EXAMPLE 1.21

If the magnetic flux density **B** is given by

$$\mathbf{B} = (x + 2)\,\mathbf{a}_x + (1 - 3y)\,\mathbf{a}_y + 2z\,\mathbf{a}_z$$

evaluate the total magnetic flux out of the box bounded by

$$x = 0, 1$$
$$y = 0, 1$$

and

$$z = 0, 1$$

Solution

The closed surface is shown in Figure 1.42. The total magnetic flux out of the box is given by

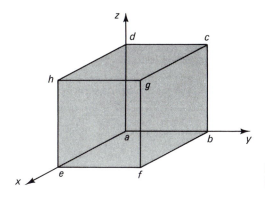

Figure 1.42 The closed surface of integration in example 1.21.

$$\oint_s \mathbf{B} \cdot d\mathbf{s}$$

where s is the closed surface bounding the box.

From Figure 1.42, it can be seen that

$$\oint_s \mathbf{B} \cdot d\mathbf{s} = \int_{abcd} \mathbf{B} \cdot d\mathbf{s} + \int_{efgh} \mathbf{B} \cdot d\mathbf{s} + \int_{aehd} \mathbf{B} \cdot d\mathbf{s}$$

$$+ \int_{bfgc} \mathbf{B} \cdot d\mathbf{s} + \int_{aefb} \mathbf{B} \cdot d\mathbf{s} + \int_{dhgc} \mathbf{B} \cdot d\mathbf{s}$$

Although each of these surface integrals may be evaluated individually, we should keep in mind that what we are trying to calculate is simply the net flux flowing out of the box. Therefore, the unit vectors normal to the integration surfaces should all point out of the box. For example,

$$\int_{abcd} \mathbf{B} \cdot d\mathbf{s} = \int_{z=0}^{1} \int_{y=0}^{1} [(x_0 + 2)\mathbf{a}_x + (1 - 3y)\mathbf{a}_y + 2z\,\mathbf{a}_z]$$

$$\cdot -dydz\,\mathbf{a}_x = -\int_{z=0}^{1} \int_{y=0}^{1} 2dydz = -2$$

where the element of area $d\mathbf{s}$ on the surface $abcd$ was taken as $-dydz\,\mathbf{a}_x$ to emphasize the fact that the unit vector is in a direction out of the box. For the surface $efgh$,

$$\int_{efgh} \mathbf{B} \cdot d\mathbf{s} = \int_{z=0}^{1} \int_{y=0}^{1} [(x + 2)\mathbf{a}_x + (1 - 3y)\mathbf{a}_y + 2z\,\mathbf{a}_z]$$

$$\cdot dydz\,\mathbf{a}_x = \int_{0}^{1} \int_{0}^{1} 3\,dydz = 3$$

For the surface $aehd$,

$$y = 0, \qquad \mathbf{B} = (x + 2)\mathbf{a}_x + 1\,\mathbf{a}_y + 2z\,\mathbf{a}_z$$

$$d\mathbf{s} = (-dx\,dz)\,\mathbf{a}_y$$

$$\int_{aehd} \mathbf{B} \cdot d\mathbf{s} = \int_{x=0}^{1} \int_{z=0}^{1} (-1)\,dxdz = -1$$

Similarly, the surface integrals over *bfgc*, *aefb*, and *dhgc* are given by $-2, 0$, and 2, respectively.

The overall surface integral that gives the total magnetic flux out of the box is the sum of all the six surface integrals and given by

$$\int_s \mathbf{B} \cdot d\mathbf{s} = -2 + 3 - 1 - 2 + 0 + 2 = 0$$

———— ◆◆ ————

EXAMPLE 1.22

Given a vector $\mathbf{A} = 10\,\mathbf{a}_\rho + 3\,\rho\mathbf{a}_\phi - 2z\rho\,\mathbf{a}_z$, find the total flux of this vector emanating from a cylindrical volume enclosed by a cylinder of radius 2, height 4, whose axis is the z axis and whose base lies in the $z = 1$ plane.

Solution

To calculate the total flux emanating from the cylindrical volume, we need to calculate the integration of $\mathbf{A}$ over the surface enclosing the cylindrical volume. From Figure 1.43, it can be seen that the contributions to the closed surface integral are from the top, bottom, as well as the curved surface of the cylinder.

At the top of the cylinder, the vector $\mathbf{A}$ is given by

$$\mathbf{A} = 10\,\mathbf{a}_\rho + 3\rho\,\mathbf{a}_\phi - 10\rho\,\mathbf{a}_z$$

and the element of area $d\mathbf{s}$ is

$$d\mathbf{s} = \rho\,d\rho\,d\phi\,\mathbf{a}_z$$

$$\int_{\text{top}} \mathbf{A} \cdot d\mathbf{s} = \int_0^2 \int_0^{2\pi} (-10\rho)\,\rho\,d\rho\,d\phi = -10 \int_0^{2\pi} \int_0^2 \rho^2\,d\rho\,d\phi$$

$$= -10(2\pi)\frac{\rho^3}{3}\bigg|_0^2 = -\frac{160\pi}{3}$$

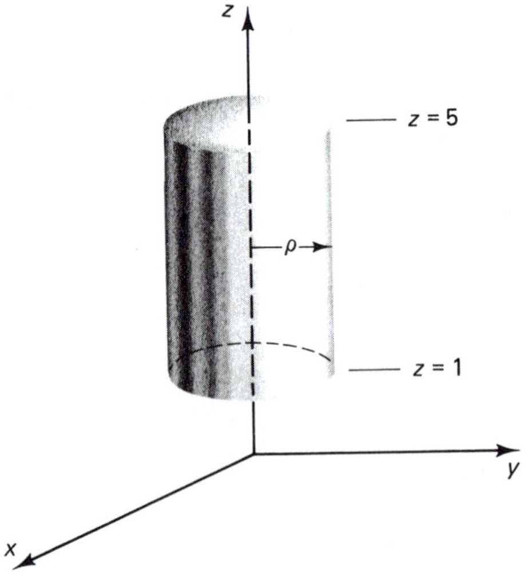

z = 5

ρ→

z = 1

Figure 1.43   The closed surface of integration in example 1.22.

At the bottom of the cylinder, $z = 1$ and, hence,

$$\mathbf{A} = 10\,\mathbf{a}_\rho + 3\rho\,\mathbf{a}_\phi - 2\rho\,\mathbf{a}_z$$

and the element of area ds is

$$d\mathbf{s} = \rho\,d\rho\,d\phi\,(\mathbf{a}_z)$$

Hence,

$$\int_{bottom} \mathbf{A}\cdot d\mathbf{s} = -\int_0^{2\pi}\int_0^2 (-2\rho)\,\rho\,d\rho\,d\phi$$

$$= 2(2\pi)\frac{8}{3} = \frac{32\pi}{3}$$

At the curved surface of the cylinder

$$\mathbf{A} = 10\,\mathbf{a}_\rho + 6\,\mathbf{a}_\phi - 4z\,\mathbf{a}_z$$

and ds is

$$d\mathbf{s} = 2d\phi\,dz\,\mathbf{a}_\rho$$

Hence,

$$\int_{curved\ surface} \mathbf{A}\cdot d\mathbf{s} = \int_0^{2\pi}\int_1^5 20\,d\phi\,dz = 20(2\pi)(4) = 160\pi$$

The total surface integral is the sum of the preceding three contributions and is equal to $352\pi/3$, which is the desired answer.

EXAMPLE 1.23

Show that $\oint_s \cos\theta\,\mathbf{a}_r\cdot d\mathbf{s} = 0$ if s is any sphere centered about the origin.

Solution

The element of area in the spherical coordinate system is given, as shown in Figure 1.44, by

$$r^2 \sin\theta\,d\theta d\phi\,\mathbf{a}_r$$

For a sphere of radius a, the surface integral is given by

$$\oint_s \cos\theta\,\mathbf{a}_r\cdot a^2 \sin\theta\,d\theta d\phi\,\mathbf{a}_r = \int_0^\pi\int_0^{2\pi} a^2 \cos\theta \sin\theta\,d\theta d\phi$$

$$= (2\pi)a^2 \int_0^\pi \cos\theta \sin\theta\,d\theta$$

$$= a^2\,\pi\left[-\frac{\cos 2\theta}{2}\right]_0^\pi = 0,\ \text{because } \cos 2\pi = \cos 0 = 1$$

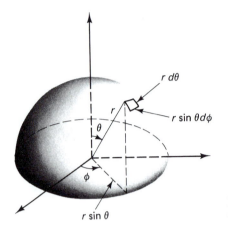

Figure 1.44   The spherical surface of integration of example 1.23.

1.8 MAXWELL'S EQUATIONS IN INTEGRAL FORM

Afer briefly reviewing some of the basic vector operations as well as the line and surface integrals of vectors, we are now prepared for the introduction of Maxwell's equations. In this chapter we will limit our discussion to Maxwell's equations in integral form simply because they are easier to understand physically. In the next chapter and after a brief discussion of the vector differential operations, we will introduce Maxwell's equations in differential form, which provide more powerful relationships in solving engineering electromagnetic problems.

This withstanding, let us start our discussion by indicating what these Maxwell's equations are. They are mathematical relations between the electric and magnetic fields, and their current and charge sources. We know that electric charges are the source of the electric field, whereas electric currents produce magnetic fields. There-fore, Maxwell's equations are the mathematical relations between the electric and magnetic fields, and their associated charge and current distributions. Four experimen-tally obtained laws that constitute Maxwell's equations are the following:

1. Gauss's law for electric field.
2. Gauss's law for magnetic field.
3. Faraday's law.
4. Ampere's circuital law.

These laws are all named after scientists who contributed to the discovery of these laws either by conducting experiments or by quantifying experimental observations made by others. All of these laws (with the exception of a single term in Ampere's law) were present before Maxwell. Maxwell, however, tried to unify these laws of electricity and magnetism, and he found that it was necessary to introduce an additional term in Ampere's law. This term turned out to be a history-making term simply because through that term it was possible to predict the phenomenon of wave propagation as will be described later. In honor of Maxwell's significant contribution they named this group of experimental laws after him. With this brief introduction, let us now closely examine each of these laws.

1.8.1 Gauss's Law for Electric Field

This law simply quantifies the electric field (static or time varying) in terms of the charge distribution associated with it. Michael Faraday conducted an experiment where he clamped two hemispheres around a charged smaller sphere and momentarily connected them to ground. He then showed that the total charge induced on the two hemispheres is equal in magnitude to the original charge on the inner sphere. As a result of this experiment it was suggested that some kind of electric flux is surrounding the charged objects. It was further indicated that the number of these electric flux lines is proportional to the value of the charge producing them. On these bases it was possible to explain the experimental results whereby equal total charges were experimentally measured on the larger and smaller spheres when placed concentrically. Gauss's law quantifies this observation. It states that the total electric flux emanating from a closed surface equals the electric charge enclosed by that surface. Mathematically, this means

$$\oint_s \epsilon_o \mathbf{E} \cdot d\mathbf{s} = Q$$

where Q is the total discrete charge enclosed by the surface s. If instead of a discrete charge, we have charge distribution, say a volume charge distribution ρ_v, Gauss's law in this case may be written in the form

$$\oint_s \epsilon_o \mathbf{E} \cdot d\mathbf{s} = \int_v \rho_v \, dv$$

where the volume v is enclosed by the surface s. The requirement that the volume v being enclosed by the surface s is further illustrated in Figure 1.45 where it is clear that the volume v_1 is enclosed by the surface s_1, and the total charge in the overall volume v should be included if the surface of integration is changed from s_1 to s.

   Also the statement that the charge enclosed by a surface is equal to the total electric flux emanating from that surface should be clear by noting that

$$\oint_s \epsilon_o \mathbf{E} \cdot d\mathbf{s}$$

is indeed the total electric flux of $\epsilon_o \mathbf{E}$ crossing the surface s. As indicated earlier, the dot product in the surface integral term emphasizes the fact that we are actually integrating the component of the electric flux density $\epsilon_o \mathbf{E}$ in the direction of a unit normal to the surface.

   Gauss' law provides a valuable tool to evaluate the electric field as a result of a given charge distribution. The solution procedure basically involves constructing a suitable Gauss' surface

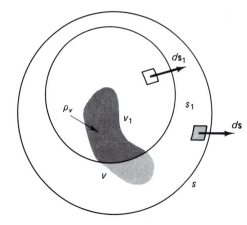

Figure 1.45 Consideration of the surface and volume integrals in Gauss's law. The total charge enclosed by the surface should be included in the volume integral. In this figure, the volume v_1 is enclosed by s_1, and the total volume v is enclosed by s.

surrounding the given charge distribution and calculating the total electric flux emanating from this surface. By equating the calculated emanating flux to the total charge enclosed by the Gauss' surface, we obtain an expression for the electric field. This electric field is evaluated at the location of the Gauss surface.

EXAMPLE 1.24

Determine the electric field inside and outside a spherical charge distribution of constant charge density ρ_v and radius r_o.

Solution

We first calculate the electric field outside the spherical charge distribution of radius r_o, that is, for $r > r_o$. From symmetry considerations, it can be seen that the electric field will be in the radial direction. If we add the electric fields of two symmetrically located point charges within the spherical charge distribution, we will find that the resulting electric field is in the radial direction. We then construct a Gauss surface that uses this symmetry property of the charge distribution and the resulting radial electric field. We choose a spherical Gauss surface of radius r and then calculate the total electric flux (radial) emanating from that surface as

$$\oint_s \epsilon_o E_r \, \mathbf{a}_r \cdot d\mathbf{s}$$

The element of area $d\mathbf{s}$ on the spherical surface is given by $d\mathbf{s} = r^2 \sin\theta \, d\theta \, d\phi \, \mathbf{a}_r$. The total flux emanating is therefore given by

$$\int_{\phi=0}^{2\pi}\int_{\theta=0}^{\pi} \epsilon_o E_r \, \mathbf{a}_r \cdot r^2 \sin\theta \, d\theta \, d\phi \, \mathbf{a}_r = \epsilon_o r^2 E_r (2\pi)[-\cos\theta]_0^\pi$$

$$= 4\pi \epsilon_o r^2 E_r$$

From Figure 1.46a, we may calculate that the total electric charge enclosed by the Gauss's surface s. This is given by

$$\int_v \rho_v \, dv = \rho_v \frac{4}{3}\pi r_o^3$$

where we simply multiplied the charge distribution by the spherical volume, because ρ_v is uniform throughout the volume v.

Hence, according to Gauss's law, we have

$$4\pi \epsilon_o r^2 E_r = \rho_v \frac{4}{3}\pi r_o^3$$

or

$$E_r = \frac{\rho_v r_o^3}{3\epsilon_o r^2} \ (\text{V/m}) \qquad r > r_o$$

The resulting electric field is in the radial direction and given by

$$\mathbf{E} = E_r \mathbf{a}_r = \frac{\rho_v r_o^3}{3\epsilon_o r^2} \mathbf{a}_r \ (\text{V/m}) \qquad r > r_o$$

It should be noted that if instead of the distributed spherical charge density, a discrete charge Q was enclosed by the surface s, Gauss's law provides that

$$4\pi\epsilon_o r^2 E_r \ (\text{total flux emanating from } s) = Q$$

Hence,

$$E_r = \frac{Q}{4\pi\epsilon_o r^2}$$

which is the statement of Coulomb's law.

We next consider the electric field inside the spherical volume of charge, that is, $r < r_o$. In this case, Gauss surface, which was also chosen to be a sphere to utilize the spherical symmetry of the given charge distribution, is drawn inside the spherical charge and at a distance $r < r_o$. It is desired to calculate the electric field at $r < r_o$, as shown in

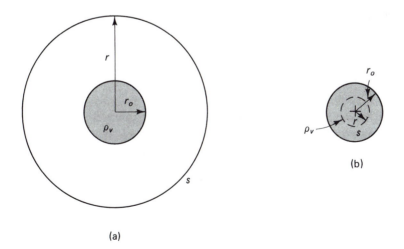

(a)

Figure 1.46   (a) A spherical Gauss's surface of radius $r > r_o$ surrounding a spherical charge distribution of constant charge density ρ_v and radius r_o. (b) Gauss's surface is a sphere of radius $r < r_o$ inside the spherical charge distribution.

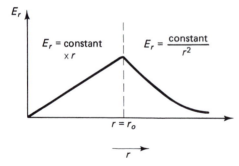

Figure 1.47 The variation of the radial electric field with r inside and outside the spherical charge distribution.

Figure 1.46b. Once again, because of the spherical symmetry, the electric field will be in the radial direction, that is, $\mathbf{E} = E_r \mathbf{a}_r$, and the total electric flux emanating from the closed surface will be given by $4\pi\epsilon_o r^2 E_r$, as previously illustrated. The total charge enclosed by the surface s in this case will be $\rho_v \, 4/3\pi r^3$. It should be noted that in this case we integrated over only the portion of the charge density enclosed by s of radius $r < r_o$. From Gauss's law, we therefore have

$$4\pi\epsilon_o r^2 E_r = \rho_v \frac{4}{3}\pi r^3$$

or

$$\mathbf{E} = E_r \mathbf{a}_r = \frac{\rho_v r}{3\epsilon_o} \mathbf{a}_r \text{ (V/m)} \qquad r < r_o$$

The electric field inside the spherical charge increases linearly with the radial distance r, whereas it is *inversely* proportional to r^2 outside the spherical charge. The variation of the radial field inside $r < r_o$ and outside $r > r_o$ the spherical charge density is shown in Figure 1.47. Other examples on the use of Gauss's law to calculate the electric field resulting from a given distribution of charge will be described later in this chapter.

1.8.2 Gauss's Law for Magnetic Field

We next describe a law that constitutes a constraint on the property of the magnetic flux lines. Gauss's law for the magnetic field states that the total magnetic flux lines emanating from a closed surface should be equal to zero, hence,

$$\oint_s \mathbf{B} \cdot d\mathbf{s} = 0$$

Once again, the dot product in the surface integral emphasizes the fact that we are concerned with the component of the magnetic flux in the direction of a unit outward normal to the surface s. In other words, the surface integral accounts for the total magnetic flux emanating from the closed surface s. As indicated earlier, this law mainly provides a constraint on the physical characteristic of the magnetic flux lines. This means that in solving an electromagnetic field problem, the obtained solution for the

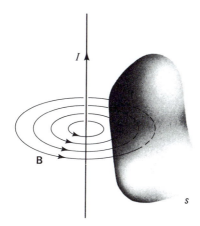

Figure 1.48   Magnetic flux lines associated with a constant current-carrying straight conductor. The directions of the flux lines **B** and the current **I** are according to the right-hand rule. The surface *s* is an arbitrarily located surface that intersects the flux lines.

magnetic field should satisfy this property. For example, let us consider the magnetic flux lines produced by the straight current-carrying conductor of example 1.13. The flux lines produced by a straight conductor carrying a constant current *I*, are shown once again for convenience in Figure 1.48.

If we consider a surface *s* arbitrarily placed outside the conductor, it can be seen from Figure 1.48 that the number of flux lines entering the surface *s* is equal to the number of flux lines outflowing from the surface. This can be shown to be true for any other configuration of the magnetic flux lines produced by various geometries of current-carrying conductors. In other words, what Gauss's law for the magnetic field really emphasizes is the fact that magnetic flux lines are closed lines. They are unlike the electric flux lines that originate on positive charges and terminate on negative charges. Magnetic flux lines are closed lines, and the number entering a closed surface *s* is equal to the number emanating from the surface. Comparison between this Gauss's law and Gauss's law for the electric field also emphasizes the fact that magnetic charges do not exist. This is because Gauss's law for the magnetic field can be obtained from the law in the electric field case by replacing the electric flux by the magnetic flux and also equating the enclosed charge to zero. In other words, if magnetic charges do exist, we should have been able to enclose them by a surface *s*, the matter that contradicts Gauss's law for the magnetic field that inherently assumes that the total enclosed magnetic charge to be zero.

Instead of solving other specific examples to illustrate further the validity of this constraint on the property of the magnetic flux lines (i.e., magnetic flux lines are closed) we will make it a point to emphasize this property in many of the other examples that we are going to solve later in this chapter.

1.8.3 Faraday's Law

After Oersted discovered in 1820 that current-carrying conductors produce or are associated with magnetic fields, Faraday tried experimentally to prove that the reverse phenomenon is also true, that is, the magnetic fields are capable of producing electric currents. His experimental arrangement is shown in Figure 1.49 where he used a

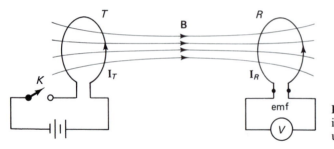

Figure 1.49   A schematic illustrating the experimental arrangement used by Faraday.

transmitting circular loop T connected to a battery through the switch K to generate the magnetic flux **B**. He then placed a receiving conducting loop R perpendicular to the magnetic flux lines and measured the induced currents using an ammeter A.

Alternatively, Faraday could have measured the induced voltage between the terminals of the receiving loop if a voltmeter was connected between these terminals. The experimental observations obtained from Faraday's experiment are shown in Figure 1.50 where it can be seen that a current would circulate in the receiving loop only on closing and opening the switch K. During the period when the switch was in the on position, there was no current circulating in the receiving loop even with the presence of the current in the transmitting loop and the associated magnetic flux density **B**. The conclusion reached by Faraday, therefore, was that only time-varying magnetic fields (i.e., during the build-up and decay of the magnetic field in the process of opening and closing the switch) produce currents in the receiving loop.

Actually a closer look at these experimental observations shows that time-varying magnetic field produces an electric field that circulates around the time-varying magnetic field. This induced electric field is present whether we have the receiving loop or not. We can, however, recognize the presence of the induced electric field by placing the receiving loop in the time-varying magnetic field. When we place a conducting loop perpendicular to the time-varying magnetic field, there will be a force applied by the induced electric field on the free electrons in the conducting loop. The force by the electric field will cause the free charges to circulate, thus generating electric current in the receiving loop. In summary, therefore, the experimental observations in Faraday's

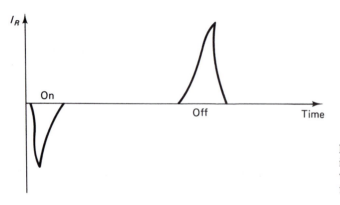

Figure 1.50   The induced current in the receiving loop owing to the variation with time of the magnetic flux **B** crossing this loop.

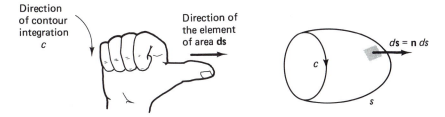

Figure 1.51   The right-hand rule that relates the direction of the contour of integration to the direction of the element of area ds.

experiments can be explained in terms of the electric field actually generated by the time-varying magnetic field. Quantitatively, these observations are mathematically expressed as follows:

$$emf = \oint_c \mathbf{E} \cdot d\boldsymbol{\ell} = -\frac{d}{dt}\int_s \mathbf{B} \cdot d\mathbf{s}$$

The first two terms of the left-hand side of this equation simply indicate that the induced *electromotive force* (*emf*) between the two terminals of the receiving loop is equal to the work done by the induced electric field in moving the free charges along the contour c of the receiving loop. The first and last terms of this equation emphasize the general characteristic of Faraday's law, which states that time-varying magnetic fields generate electric field and the fact that the detection of this field by placing the receiving loop nearby is nothing but a means to realize its presence.

Our final comment regarding Faraday's law is actually related to the physical significance of the negative sign in front of the rate of change of the magnetic flux term. Before we can examine the importance of this negative sign, however, we should emphasize that the direction of the contour c and the unit vector associated with the element of area ds should be related by the right-hand rule as shown in Figure 1.51. The area s can, of course, be any area as long as it is enclosed by the contour c. With this in mind, let us solve the following example to illustrate the significance of the negative sign in Faraday's law.

EXAMPLE 1.25

The rectangular current loop shown in Figure 1.52, is placed perpendicular to time-varying magnetic flux lines of density $\mathbf{B} = B_o \cos \omega t \, \mathbf{a}_z$. Determine the induced *emf* in the conducting loop, and examine its variation with time. Also compare the time variations of the induced *emf* with that of the magnetic flux. Comment on the direction of the induced *emf*.

Solution

Let us assume that the path of integration c is in the counterclockwise direction as shown in Figure 1.52. The corresponding unit vector associated with the element of area will then conventionally be out of the paper (right-hand rule), which is in the $\mathbf{a}_z$ direction as shown

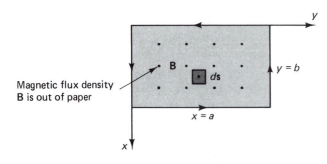

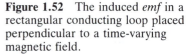

Magnetic flux density
B is out of paper

Figure 1.52 The induced *emf* in a
rectangular conducting loop placed
perpendicular to a time-varying
magnetic field.

in Figure 1.52. The total magnetic flux ψ_m enclosed by the loop and directed out of the
paper is given by

$$\psi_m = \int_s \mathbf{B} \cdot d\mathbf{s} = \int_{x=0}^{a} \int_{y=0}^{b} B_o \cos \omega t \, \mathbf{a}_z \cdot dy \, dx \, \mathbf{a}_z$$

$$\psi_m = B_o \cos \omega t \int_{x=0}^{a} \int_{y=0}^{b} dy \, dx = ab \, B_o \cos \omega t$$

The induced *emf* in the counterclockwise sense is then given according to Faraday's law
by

$$emf = \oint_c \mathbf{E} \cdot d\boldsymbol{\ell} = -\frac{d\psi_m}{dt} = ab \, B_o \, \omega \sin \omega t$$

The time variation of the magnetic flux enclosed by the loop ψ_m and the induced *emf* around
the loop are shcwn in Figure 1.53.

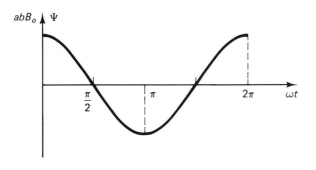

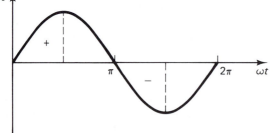

Figure 1.53 Illustration of the
variation of the total magnetic flux
and the induced *emf* as a function
of time.

It can be seen from Figure 1.53 that when the enclosed magnetic flux is decreasing with time during the first half-cycle of the cosine curve, the induced *emf* is positive, which means that the induced *emf* is indeed in the counterclockwise direction, which we originally assumed as shown in Figure 1.52. This *emf* produces a counterclockwise current that subsequently gives rise to a magnetic field in the direction out of the paper inside the loop. This induced magnetic field is therefore in the same direction as the originally present magnetic flux density **B** and hence acts to increase the magnetic flux enclosed by the loop. When the magnetic flux enclosed by the loop is increasing with time, conversely (the second half-cycle in the cosine curve), the induced *emf* is negative, which means that it is in the opposite direction to the counterclockwise one shown in Figure 1.52. This *emf* produces a clockwise current around the loop. This polarity of current gives rise to a magnetic field directed into the paper that is opposite to the direction of the original magnetic flux density **B**. Such a magnetic field, therefore, acts to decrease the magnetic flux enclosed by the loop. This observation is known as *Lenz's law*, which states that *the induced emf is in such a direction so as to oppose the change in the magnetic flux producing it*. If the magnetic flux is increasing with time, the induced *emf* will be in such a direction so as to produce a magnetic flux in the opposite direction, and when the magnetic flux is decreasing with time, the induced *emf* will be in such a direction so as to produce magnetic flux in the same direction as the originally present one. The minus sign on the right side of Faraday's law ensures that Lenz's law is always satisfied subject that we keep the right-hand rule relating the directions of the contour *c* and the element of area *d***s** in mind.

Lenz's law is not new to us, and many of us have actually encountered its effect in the electric circuit course when we talked about inductance. As we may recall, a physical property of the inductance is that it opposes any sudden change in the electric current passing through the inductor. In other words, the increase and decrease of the current passing through an inductor should be gradual. The reason for this is precisely described by Lenz's law. If the current in the inductor is increasing with time there will be an induced *emf* in the inductor to counter this increase, that is, slows it down, thus resulting in a gradual instead of an abrupt increase. If the current in the inductor is decreasing, conversely, the induced *emf* will be in such a direction so as to oppose this decrease, and it basically tries to help the current to stay by slowing down its decrease.

In summary, therefore, the physical property of the inductance that is related to its opposition to any sudden change in the electric current passing through it can be explained in terms of Lenz's law.

EXAMPLE 1.26

A rectangular loop of conducting wire consists of three fixed sides, and the fourth side is a conducting bar moving with a velocity $\mathbf{v} = v_o\,\mathbf{a}_y$ in the *y* direction. The loop is placed in a plane perpendicular to a uniform magnetic field of the density $\mathbf{B} = B_o\,\mathbf{a}_z$ as shown in Figure 1.54. Determine the induced *emf* and show that its direction is consistent with Lenz's law.

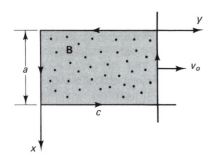

Figure 1.54 A rectangular loop with one of its sides moving with a velocity **v** in the y direction is placed perpendicular to a magnetic field of flux density **B**.

Solution

Although the magnetic flux density **B** is not varying with time in this example, there will still be an induced *emf* in the loop because the total magnetic flux crossing the area of the loop is increasing with time as the moving bar continues to move in the y direction. In other words, the increase in the magnetic flux enclosed by the area is, in this case, due to the increase of the enclosed area with time as the bar continues to move. If the initial location of the moving bar at $t = 0$ is y_o, the total magnetic flux ψ_m enclosed by the loop is given by

$$\psi_m = B_o \, \mathbf{a}_z \cdot [a(y_o + v_o t)] \, \mathbf{a}_z$$

$$= B_o a (y_o + v_o t)$$

It should be noted that with the decision we made to take the area in the $\mathbf{a}_z$ direction we should consequently take the direction of the induced *emf* in the counterclockwise direction as shown in the Figure 1.54 so as to satisfy the right-hand rule.

According to Faraday's law, the induced *emf* is then

$$emf = -\frac{d}{dt}\int_S \mathbf{B} \cdot d\mathbf{s} = -\frac{d\psi_m}{dt}$$

$$= -B_o a v_o$$

The negative sign in the value of the induced *emf* simply indicates that the direction of the induced *emf* is in a direction opposite to that indicated in Figure 1.54. This negative sign is consistent with Lenz's law because the induced *emf*, if it were in the direction shown in the Figure 1.54, will produce a magnetic field (according to the right-hand rule) in the same direction as the originally present one and, hence, enhances the total magnetic field. This clearly contradicts Lenz's law because the total magnetic flux enclosed by the loop, that is, crossing the area of the loop, is increasing with time as the area increases with time. Because the total magnetic flux is increasing, the induced *emf* should be (according to Lenz's law) in such a direction so as to oppose such an increase. This will clearly be the case if the induced *emf* is in the clockwise direction that is opposite to the direction indicated in Figure 1.54. The negative sign in the induced *emf* equation is, therefore, consistent with Lenz's law.

1.8.4 Ampere's Circuital Law

This is the fourth and final law in Maxwell's equations. Ampere's law states that the line integral of the magnetic flux density along a closed contour c is equal to the total current crossing the area s enclosed by this contour. This total current may be in the form of a conventional type current $\int_s \mathbf{J} \cdot d\mathbf{s}$, which is due to the flow of electric charges, or a new type of current introduced mathematically by Maxwell and is related to the time rate of change of the total electric flux crossing the area s. Mathematically, this law is expressed in the form

$$\oint_c \frac{\mathbf{B}}{\mu_o} \cdot d\ell = \text{total current crossing the area } s \text{ that is enclosed by the contour } c$$

$$= \underbrace{\int_s \mathbf{J} \cdot d\mathbf{s}}_{\substack{\text{current} \\ \text{resulting} \\ \text{from flow} \\ \text{of charges}}} + \underbrace{\frac{d}{dt}\int_s \epsilon_o \mathbf{E} \cdot d\mathbf{s}}_{\substack{\text{new current "introduced by Maxwell" resulting from the time rate of} \\ \text{change of the electric flux crossing the area } s \text{ enclosed by countour } c}}$$

The relation between the magnetic flux $\mathbf{B}$ and the electric current density $\mathbf{J}$ as well as the electric flux density $\epsilon_o \mathbf{E}$ is shown in Figure 1.55.

It should be noted that the directions of the contour c and the element of area $d\mathbf{s}$ should follow the right-hand rule. Also, both surface integrals must be evaluated on the same surface.

Ampere's law in its original form actually did not include the current term related to the rate of change of the electric flux that is called the "displacement current" term for reasons that will be clear shortly. Maxwell in his effort to unify the laws of electromagnetism found it necessary to include this term so that these laws will be consistent with other existing physical laws such as the law of conservation of charge. Before we get involved in examining the nature of this displacement current term, let us just point out the following facts about Ampere's law:

1. The quantity $\mathbf{J}$, the current flux density, is a vector current density due to actual flow of charges, and it has the dimension of (A/m^2). For example, if we have electric charges of density ρ which is flowing (moving) with an average velocity $\mathbf{v}$ in a region, then the current density $\mathbf{J}$ is defined at any point as

$$\mathbf{J} = \rho\mathbf{v}\left(\frac{\text{C}}{\text{m}^3} \cdot \frac{\text{m}}{\text{sec}}\right) \text{ or } \left(\frac{\text{C/sec}}{\text{m}^2}\right) \text{ or } \left(\frac{\text{A}}{\text{m}^2}\right)$$

Thus the quantity $\int_s \mathbf{J} \cdot d\mathbf{s}$, being the surface integral of $\mathbf{J}$ over s, has the meaning of current because of the flow of charges crossing the surface s.

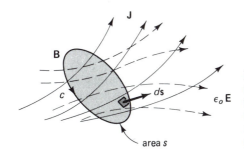

Figure 1.55 Ampere's law relates the line integral of the magnetic field along the contour c to the total current and electric flux densities crossing the area s enclosed by contour c.

2. From Coulomb's law, it is clear that $\mathbf{E}$ has the units of {charge/[(permittivity)(distance)^2]} and the displacement flux density $\epsilon_o \mathbf{E}$ thus has the units of charge per unit area (C/m^2). Hence the integration

$$\int_s \epsilon_o \mathbf{E} \cdot d\mathbf{s}$$

gives the total displacement flux and has the units of charge. Conversely, the quantity

$$\frac{d}{dt} \int_s \epsilon_o \mathbf{E} \cdot d\mathbf{s}$$

will have the units of d/dt (charge) or current and hence is known as displacement current. This current does not represent the flow of charges and hence physically it is not a current. Mathematically, however, it is equivalent to a current crossing the surface s. More discussion regarding the nature of this current will be presented later on in this section.

3. We learned from Faraday's law that the quantity

$$\oint_c \mathbf{E} \cdot d\boldsymbol{\ell}$$

has the physical meaning of work done in moving a unit positive test charge around the closed path c. In Ampere's circuital law, the quantity

$$\oint_c \frac{\mathbf{B}}{\mu_o} \cdot d\boldsymbol{\ell}$$

does not have a similar meaning. This is because, as indicated by Lorentz force, the magnetic force on a moving charge is directed perpendicular to both the direction of motion of the charge and the magnetic field. Hence, such a force (see example 1.14) does not change the energy of the moving charge. In other words, the force resulting from the magnetic field is not along $\mathbf{B}$ and, hence, $\oint_c \mathbf{B} \cdot d\boldsymbol{\ell}$ does not represent integrating a component of force tangential to the contour c.

1.9 DISPLACEMENT CURRENT

From Ampere's law it is clear that the displacement current (which results from the rate of change of the electric flux) can produce a time-varying magnetic field (but not a steady one) just as effectively as the conventional magnetic field production using flow of charges (i.e., conventional currents). We indicated earlier that the introduction of the displacement current term allowed Maxwell to unify the separate laws of electricity and magnetism into an electromagnetic theory.

The need for displacement current becomes apparent when we consider the physical interpretation of the current flow through a capacitor circuit. If we assume that the area of the capacitor plates is large compared with the separation distance so that the electric field is confined between the plates, Ampere's law, when applied in the part of the circuit shown in Figure 1.56a, gives

$$\oint_{c(A)} \frac{\mathbf{B}}{\mu_o} \cdot d\ell = I$$

where the area A is pierced by the wire that carries the conduction current I. In this case, there is no contribution from the displacement current term. For the case of Figure 1.56b, conversely, area A' is not pierced by a flow of conduction current and

$$\oint_{c(A')} \frac{\mathbf{B}}{\mu_o} \cdot d\ell = 0$$

If this is true, Ampere's law cannot be considered general because the result depends on the specific choice of the area A encircled by the same contour c. This is true unless a displacement current is assumed to exist between the capacitor's plates. In this case

$$\oint_{c(A')} \frac{\mathbf{B}}{\mu_o} \cdot d\ell = \iint_{A'} \frac{\partial(\epsilon_o \mathbf{E})}{\partial t} \cdot d\mathbf{s}$$

The argument attributed to Maxwell is as follows: The two areas, A and A', are bounded by the same path that is being used to calculate the magnetic field $\mathbf{B}$. Hence, the contour integrals should be equal, that is,

$$\oint_{c(A)} = \oint_{c(A')}$$

Therefore, current in the wire must equal the total displacement current in the capacitor, or

$$I = \iint_{A'} \frac{\partial(\epsilon_o \mathbf{E})}{\partial t} \cdot d\mathbf{s}$$

This is one of those rare cases in which purely mathematical reasoning has preceded and guided the way for experimentation.

The preceding arguments merely justify the importance of including the displacement current term in Ampere's law. Without it we would not be able to apply the law

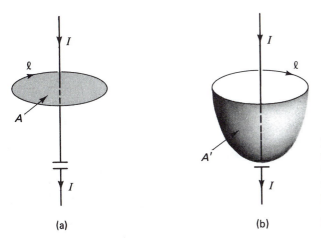

(a)   (b)

Figure 1.56 Two surfaces of integration for the same contour integral around c. (a) Surface A intersects the wire. (b) Surface A' passes between the capacitor plates and thus does not intercept current I.

successfully in the simple capacitor circuit. Such arguments, however, do not answer the question of why Maxwell chose to introduce his term in the specific form of time rate of change of the total electric flux crossing the area s. To answer this question we need to examine closely the mechanism that made the flow of the current possible in the capacitor circuit described earlier. As we know, the current actually did not cross the plates of the capacitor and the mechanism that completed the circuit is related to the ability of the capacitor to store the electric charge on its plates. If we break the circuit somewhere else along the wire connecting the capacitor to the source, there will obviously be no current, and this just emphasizes the fact that the ability of the capacitor to store charges actually played a key role in completing the circuit. Let us consider an alternating current flowing in a capacitor circuit. During the positive half of the cycle, positive charge will be accumulated on one plate of the capacitor, and negative charge will be accumulated on the other. In the second half of the cycle, the current will reverse its direction, and this simply means that positive charge will start accumulating on the originally negatively charged plate, and negative charge will accumulate on the originally positively charged plate. In other words, the continuous flow of current in the circuit is not related to an actual crossing of the charge between the plates of the capacitor but instead to the "displacement" of the accumulated charges between the two plates. Although this explanation still does not answer the question of why Maxwell proposed his term in its specific form, it does bring us a step closer regarding the nature of the displacement current term. We now know that this current in our simple capacitor circuit is related to the "displacement" of the charges between the plates of the capacitor. Therefore, to show why this current term was given its specific mathematical expression, let us consider an infinitely large plane surface, representing one of the plates of the capacitor, which is charged with a surface charge density $+\rho_s$ as shown in Figure 1.57. We are going to use Gauss's law to calculate the electric field as a result of this charge distribution. First let us examine the symmetry of the resulting electric field. If we consider the electric fields $\mathbf{E}_1$ and $\mathbf{E}_2$ of the two symmetrically located points 1 and 2 on a circular ring in the plane, it can be seen that on adding these electric

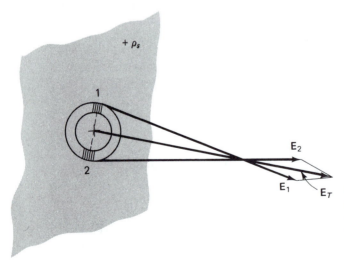

Figure 1.57 An infinitely large plane charged with a surface charge density ρ_s. This plane represents one of the plates of the parallel plate capacitor.

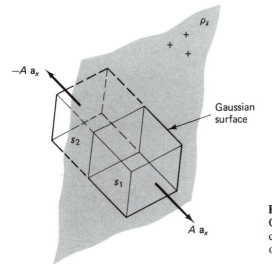

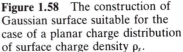

Figure 1.58 The construction of Gaussian surface suitable for the case of a planar charge distribution of surface charge density ρ_s.

fields the total field **E** will be in a direction perpendicular to the charged plane. Hence, the electric field in front of the plane is $\mathbf{E}_{\text{front}} = E_x\,\mathbf{a}_x$, and the electric field behind the charged plane will still be directed away from the plane, that is, $\mathbf{E}_{\text{back}} = -\mathbf{E}_x\,\mathbf{a}_x$. With the identification of this symmetry property of the electric field, we next construct a suitable Gaussian surface that will be used to apply Gauss's law. A suitable surface in this case may be in the form of a rectangular parallelepiped extending equally on both sides of the planar charge as shown in Figure 1.58.

By applying Gauss's law, it is clear that the contribution to the total electric flux emanating from a rectangular parallelepiped will be only from the surfaces s_1 and s_2, hence,

$$\int_{s_1} \epsilon_o E_x\,\mathbf{a}_x \cdot A\,\mathbf{a}_x + \int_{s_2} \epsilon_o(-E_x\,\mathbf{a}_x) \cdot -A\,\mathbf{a}_x = \text{total charge enclosed by the rectangular}$$
$$\text{parallelepiped}$$

$$2\epsilon_o E_x A = \rho_s A$$

$$\therefore E_x = \frac{\rho_s}{2\epsilon_o}$$

and in vector form

$$\mathbf{E} = E_x\,\mathbf{a}_x = \frac{\rho_s}{2\epsilon_o}\,\mathbf{a}_x \qquad x > 0$$

$$= -\frac{\rho_s}{2\epsilon_o}\,\mathbf{a}_x \qquad x < 0$$

Now, if we consider bringing two of these charged planes, one with positive charge density $+\rho_s$, the other with a negative charge density $-\rho_s$, at a distance d from each other, it is clear from Figure 1.59 that the electric field outside the planes will cancel

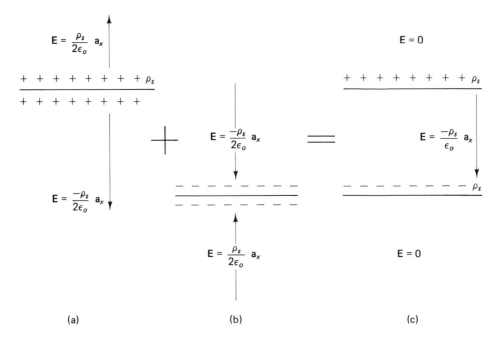

Figure 1.59 (a) The electric field in front of uniformly charged planes with positive charge density ρ_s. (b) The electric field if the plane is charged with negative charge density $-\rho_s$. (c) The superposition of (a) + (b) results in the electric field inside and outside a parallel plate capacitor.

out, thus resulting in a zero electric field, whereas the electric field between the planes will add up, thus resulting in a total electric field $\mathbf{E} = \rho_s/\epsilon_o \, \mathbf{a}_x$ as shown in Figure 1.59.

If we go back now and try to relate the charge accumulation on the plates of the parallel plate capacitor of the circuit in Figure 1.56 to the electric field between these plates we find that the circulating current I = the rate of the charge accumulation on the plates = dQ/dt. The total charge on one of the plates is equal to the charge density multiplied by the area, hence,

$$Q = (\rho_s)(\text{area})$$

$$\therefore I = \frac{d}{dt}(\rho_s \, \text{area})$$

The current density is then

$$\frac{I}{\text{area}} = J = \frac{d\rho_s}{dt}$$

On substituting ρ_s in terms of the electric field between the plates (see Figure 1.59c), we find

$$\mathbf{J} = \frac{d(\epsilon_o \, \mathbf{E})}{dt}$$

If we integrate over an arbitrary area

$$\int_s \mathbf{J} \cdot ds = \frac{d}{dt} \int_s \epsilon_o \mathbf{E} \cdot ds$$

which is the displacement current term introduced by Maxwell. It should be emphasized, however, that Maxwell, in introducing his term, did not follow an argument similar to ours. His specific procedure to derive the displacement current term will be described after introducing Maxwell's equations in differential form in chapter 2. The preceding derivation, however, does emphasize the nature of the displacement current term and arrives at a term of the same form as that introduced by Maxwell. Other examples illustrating the application of Maxwell's equation will be presented in the following sections.

.10 GENERAL CHARACTERISTICS OF MAXWELL'S EQUATIONS

Before we solve more examples illustrating the various applications of Maxwell's equations, let us review some general characteristics of these equations. To start with, a summary of the four equations is given by

1. Gauss's law for electric field.

$$\oint_s \epsilon_o \mathbf{E} \cdot ds = \int_v \rho_v \, dv \qquad (1.14)$$

2. Gauss's law for magnetic field.

$$\oint_s \mathbf{B} \cdot ds = 0 \qquad (1.15)$$

3. Faraday's law.

$$\oint_c \mathbf{E} \cdot d\ell = -\frac{d}{dt} \int_s \mathbf{B} \cdot ds \qquad (1.16)$$

4. Ampere's law.

$$\oint_c \frac{\mathbf{B}}{\mu_o} \cdot d\ell = \int_s \mathbf{J} \cdot ds + \frac{d}{dt} \int_s \epsilon_o \mathbf{E} \cdot ds \qquad (1.17)$$

We will examine specific properties of these equations for static and time-varying fields.

1.10.1 Static Fields

In this case all the derivatives with respect to time are zero, that is, $d/dt = 0$, and Maxwell's equations reduce to

$$\oint_{s} \epsilon_{o} \mathbf{E} \cdot d\mathbf{s} = \int_{v} \rho_{v} \, dv \dots \tag{1.18}$$

$$\oint_{s} \mathbf{B} \cdot d\mathbf{s} = 0 \dots \tag{1.19}$$

$$\oint_{c} \mathbf{E} \cdot d\ell = 0 \dots \tag{1.20}$$

$$\oint_{c} \frac{\mathbf{B}}{\mu_{o}} \cdot d\ell = \int_{s} \mathbf{J} \cdot d\mathbf{s} \dots \tag{1.21}$$

It is clear that equations 1.18 and 1.20 deal exclusively with the static electric field, whereas equations 1.19 and 1.21 are related only to the static magnetic field. Hence, the static electric and magnetic fields can be independently generated and quantified. Equation 1.18 indicates that the electric field is generated by and quantified in terms of the charge distribution ρ_{v}, whereas equation 1.21 indicates that current distributions $\mathbf{J}$ generate magnetic fields. The general conclusion from this discussion, therefore, is that the static electric and magnetic fields are uncoupled, which means that they are independent, do not generate each other, and can be treated separately.

1.10.2 Time-varying Fields

To examine an important property of the time-varying fields, let us focus our attention on Faraday's and Ampere's laws given in equations 1.16 and 1.17. From Faraday's law it may be seen that the time-varying magnetic field generates an electric field, and from Ampere's law and in particular the displacement current term, we may see that the time-varying electric field generates magnetic fields. In other words, the time-varying electric and magnetic fields generate each other. Therefore, once these fields are generated by a source, say an antenna, they have the ability subsequently to generate each other and hence propagate away from the source. This specific property is extremely important in explaining the phenomenon of wave propagation, which will be examined further in chapter 2. The displacement current term introduced by Maxwell predicted that time-varying electric fields would generate magnetic fields, and this is why it is considered a history-making term; it helped predict the phenomenon of wave propagation.

It should also be noted that for time-varying fields, the four Maxwell's equations shouid be solved simultaneously to determine the electric and magnetic fields in terms of their sources.

Let us now solve more examples to illustrate the applications of Maxwell's equations.

EXAMPLE 1.27

Use Gauss's law to determine the electric field intensity of an infinitely long straight line of charge, of linear charge density ρ_{ℓ}.

Solution

Let us assume that the line of charge is placed along the z axis. Based on symmetry consideration the total electric field intensity $\mathbf{E}_T$ resulting from symmetrically located elements 1 and 2 is perpendicular to the line of charge as shown in Figure 1.60a. Because the line of charge is infinitely long, it is clear that the resultant electric field must be in the radial direction perpendicular to the charge line. With this information from the symmetry consideration it is now convenient to construct a Gaussian surface as shown in Figure 1.60b.

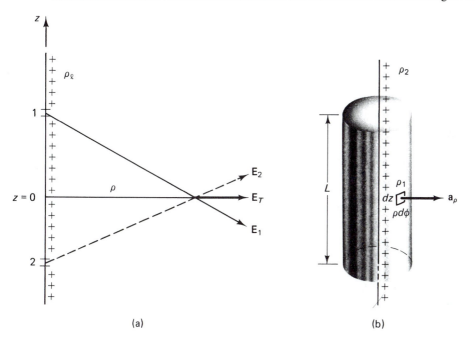

| | |
| (a) | (b) |

Figure 1.60   (a) The electric field intensity of a straight wire of charge. (b) Gaussian surface is a finite cylinder of length ℓ and radius ρ.

To use the cylindrical symmetry, we will choose a Gaussian surface in the form of a finite cylinder of length ℓ and radius ρ. Because the electric field is only in the radial direction, there will be no contribution from the top and bottom surfaces of the finite cylinder when we perform the integration over the closed surface as required by Gauss's law. Hence,

$$\oint_s \epsilon_o \mathbf{E} \cdot d\mathbf{s} = \int_{z=0}^{\ell} \int_{\phi=0}^{2\pi} \epsilon_o E_\rho \mathbf{a}_\rho \cdot \rho\, d\phi\, dz\, \mathbf{a}_\rho$$

$$= \epsilon_o E_\rho \rho\, 2\pi\, \ell$$

the total charge enclosed by the Gaussian surface is

$$\int_0^\ell \rho_\ell\, d\ell = \rho_\ell\, \ell$$

and according to Gauss's law,

$$\epsilon_o E_\rho \rho\, 2\pi = \rho_\ell$$

or

$$\mathbf{E} = E_\rho \mathbf{a}_\rho = \frac{\rho_\ell}{2\pi \epsilon_o \rho} \mathbf{a}_\rho$$

This example together with example 1.24 emphasizes the importance of using the symmetry present in a specific problem to construct a suitable Gaussian surface.

———————◆◆———————

We would like to consider next some examples illustrating the application of Ampere's law. One may wish to specify time-varying current density **J** and ask for the resulting magnetic field **B**. The situation in this case would be very complicated because the time-varying current density **J** generates time-varying magnetic field **B**, which consequently generates time-varying electric field as required by Faraday's law. This generated time-varying electric field would certainly impact the value of the initially calculated magnetic field **B** through the displacement current term in Ampere's law. Therefore, we conclude that if we have time-varying electric currents, Ampere's and Faraday's laws should be solved simultaneously subject to the constraints provided by the two Gauss's laws. In other words all four Maxwell's equations should be solved simultaneously when we deal with time-varying fields. The simultaneous solution of all Maxwell's equations will be simplified and hence possible in an introductory text in electromagnetics after the introduction of their point or differential forms at the end of chapter 2. Therefore, to help us solve some examples illustrating the application of Ampere's law, we resort to the special case of static electric currents. In this case, static electric and magnetic fields are uncoupled, and we may use only Ampere's law to determine the magnetic fields that result from a given static current distribution. We will soon see that there are many interesting applications that can be dealt with under this assumption. The comprehensive case of the simultaneous solution of all Maxwell's relations for time-varying fields will be postponed to the end of chapter 2.

EXAMPLE 1.28

Use Ampere's law for the special case of direct currents to determine the following:

1. The magnetic flux density inside and outside a straight conductor of radius a and carrying a static total current I.
2. The magnetic field inside and outside the core of an N turn closely wound toroid of a circular cross section and carrying a static current I.
3. The magnetic field inside and outside an infinitely long, closely wound solenoid having N turns per unit length and carrying a static current I.

Solution

Ampere's law for static fields is given by

$$\oint_c \frac{\mathbf{B}}{\mu_o} \cdot d\boldsymbol{\ell} = \int_s \mathbf{J} \cdot d\mathbf{s}$$

This equation relates the magnetic flux density **B** to the current distribution generating it. Therefore, the solution procedure in all the given configurations basically involves identifying first the direction of the magnetic field based on symmetry considerations. We then construct a suitable contour c at which the value of the magnetic field is required and carry out the line integral of the magnetic field along the specified contour c, that is, $\oint_c \mathbf{B}/\mu_o \cdot d\boldsymbol{\ell}$. By equating the result of the line integral with the total current crossing, the area s enclosed (encircled) by the contour c, we obtain a relationship between the magnetic flux density **B** and the given current distribution. This relationship is then solved to determine **B**. With this summary of the solution procedure, let us now determine the magnetic field in each of the given current configurations.

1. *Magnetic flux density inside and outside of an infinitely long circular conductor of radius a and carrying a total current I.* From example 1.13 where we used Biot-Savart's law to calculate the magnetic field of an infinitely long conducting wire, it was shown that the magnetic field is simply in the ϕ direction around the wire, that is, $\mathbf{B} = B_\phi \mathbf{a}_\phi$. It was further indicated that the direction of the magnetic field and the source current should follow the right-hand rule. To calculate the magnetic flux inside and outside the wire of radius a shown in Figure 1.61, we construct a suitable contour at the location where the flux is to be calculated. Based on this symmetry consideration, we construct the contours c_1 and c_2 to calculate the magnetic fields inside and outside the conductor.

   First inside the conductor

$$\oint_{c_1} \frac{\mathbf{B}}{\mu_o} \cdot \rho \, d\phi \, \mathbf{a}_\phi = \text{total current crossing the area } s_1$$
$$\text{enclosed by the contour } c_1$$

$$= I_1 = \frac{I}{\pi a^2}(\pi \rho^2)$$

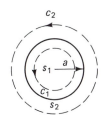

(b)

(a)

Figure 1.61 (a) Calculation of the magnetic field inside and outside an infinitely long conductor. (b) The cross section of the conductor, and the contours of integration c_1 and c_2 used to calculate the magnetic field inside and outside the conductor. s_1 and s_2 are the areas encircled by c_1 and c_2, respectively.

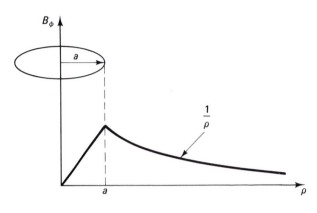

Figure 1.62   The cross section of the current-carrying conductor and the magnetic field variation with ρ inside and outside the conductor.

$$\therefore \int_0^{2\pi} \frac{B_\phi}{\mu_o} \mathbf{a}_\phi \cdot \rho \, d\phi \, \mathbf{a}_\phi = I_1$$

$$\therefore \frac{B_\phi}{\mu_o} \rho (2\pi) = \frac{I\rho^2}{a^2}$$

$$\mathbf{B} \text{ inside} = B_\phi \mathbf{a}_\phi = \frac{\mu_o I}{2\pi a^2} \rho \, \mathbf{a}_\phi$$

We then calculate **B** outside by integrating over the contour c_2

$$\oint_{c_2} \frac{\mathbf{B}}{\mu_o} \cdot d\boldsymbol{\ell} = \text{total current crossing the contour } c_2 = I$$

$$\mathbf{B} = B_\phi \mathbf{a}_\phi = \frac{\mu_o I}{2\pi \rho} \mathbf{a}_\phi$$

Examination of the spatial variation of the magnetic field inside and outside the conductor simply shows that the magnetic field inside the conductor varies linearly with ρ (the radial distance) and inversely with ρ, that is, $1/\rho$ outside the conductor. This variation of the magnetic field with ρ is shown in Figure 1.62.

2. *Magnetic flux density inside and outside a toroidal coil.* The geometry of the closely wound toroid is shown in Figure 1.63. Based on the circular symmetry of the toroidal coil, it can be seen that if we curl the fingers of our right hand, according to the right-hand rule, in the direction of the current flow, the thumb would be in the ϕ direction and, hence, the magnetic flux density would have only one component in the ϕ direction, as shown in Figure 1.63. To determine magnetic flux density, we use the symmetry and construct circular contours c_1, c_2, and c_3. The contours c_1 and c_3 will be used to calculate B_ϕ inside $\rho < a$ and outside $\rho > b$ the toroidal core, respectively, whereas the contour c_2 will be used to calculate the magnetic flux density within the toroidal core. First for $\rho < a$, we use the contour c_1 to apply Ampere's law.

$$\oint_{c_1} \frac{\mathbf{B}}{\mu_o} \cdot d\boldsymbol{\ell} = \text{total current crossing the area enclosed by } c_1$$

$$= 0$$

$$\therefore \mathbf{B} = 0 \qquad \rho < a$$

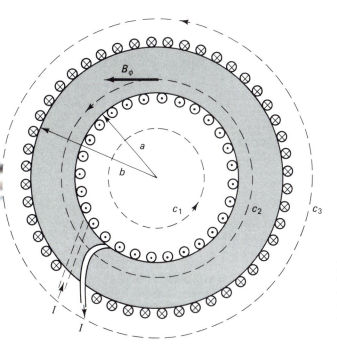

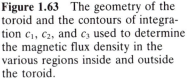

Figure 1.63 The geometry of the toroid and the contours of integration c_1, c_2, and c_3 used to determine the magnetic flux density in the various regions inside and outside the toroid.

For $\rho > b$, that is, outside the core, we use contour c_3, hence,

$$\oint_{c_3} \frac{\mathbf{B}}{\mu_o} \cdot d\ell = \text{total current crossing the area enclosed by } c_3$$

$$= \underbrace{-NI}_{\substack{\text{entering the} \\ \text{area (into} \\ \text{the paper)}}} + \underbrace{NI = 0}_{\substack{\text{leaving the} \\ \text{area (out of} \\ \text{the paper)}}}$$

$$\therefore \mathbf{B} = 0 \qquad \rho > b$$

The magnetic flux density within the core, that is, $a < \rho < b$, is calculated by using the contour c_2

$$\oint_{c_2} \frac{\mathbf{B}}{\mu_o} \cdot d\ell = NI$$

$$\int_{\phi=0}^{2\pi} \frac{B_\phi}{\mu_o} \mathbf{a}_\phi \cdot \rho \, d\phi \, \mathbf{a}_\phi = NI$$

$$\therefore B_\phi = \frac{\mu_o NI}{2\pi\rho}$$

where we have assumed that the toroid has an air core with permeability $\mu = \mu_o$.

3. *Magnetic flux density inside and outside an infinitely long solenoid.* If we assume that the solenoid is infinitely long and closely wound, it is clear (using the right hand rule) that the magnetic flux will be along the z axis of the coil. You can also verify this by examining the magnetic flux we calculated along the axis of a single turn loop in

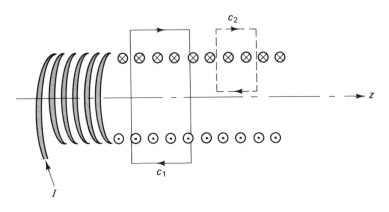

Figure 1.64 The geometry of a closely wound solenoid and the contours of integration used to determine the magnetic flux density.

example 1.15. With the identification of the direction of **B** based on the symmetry consideration, we construct two contours c_1 to calculate **B** outside the coil and c_2 to calculate **B** inside the coil, as shown in Figure 1.64. To determine **B** outside the coil, we use the contour c_1 and apply Ampere's law to find

$$\oint_{c_1} \frac{\mathbf{B}}{\mu_o} \cdot d\boldsymbol{\ell} = \text{total current crossing the area}$$
$$\text{enclosed by the contour } c_1$$

$$= 0$$

$$\therefore \mathbf{B} \text{ outside the coil} = 0$$

To determine **B** inside the coil, we use the contour c_2 shown also in Figure 1.64. From Ampere's law, we have

$$\oint_{c_2} \frac{\mathbf{B}}{\mu_o} \cdot d\boldsymbol{\ell} = \int_0^d \frac{B_z}{\mu_o} \mathbf{a}_z \bigg|_{\mathbf{B} \text{ outside the coil}} \cdot dz \, \mathbf{a}_z + \int_d^0 \frac{B_z(\mathbf{a}_z)}{\mu_o} \bigg|_{\mathbf{B} \text{ inside the coil}} \cdot dz \, \mathbf{a}_z$$

$$= 0 - \frac{B_z}{\mu_o d}$$

$$= \text{total current crossing the area encircled by } c_2$$

$$= NdI$$

where N is the number of turns per unit length and Nd is the total number of turns in a distance d,

$$\therefore B_z = -\mu_o NI$$

$$\mathbf{B} = B_z(-\mathbf{a}_z) = \mu_o NI(-\mathbf{a}_z)$$

The negative sign simply means that the magnetic flux density is in the $-\mathbf{a}_z$ direction, which certainly agrees with the right-hand rule relating the directions of the current flow and the magnetic flux density.

SUMMARY

Vector analysis including addition, multiplication, line, and surface integrations are essential to the study of electromagnetic fields. The electric and magnetic fields are vectors, and study of their characteristics and interactions requires knowledge of vector algebra. Therefore, this chapter started with a brief review of the various vector operations including the line and surface integration of vector quantities. We emphasized the physical meaning of a line integral of the form

$$\int_c \mathbf{A} \cdot d\boldsymbol{\ell}$$

It means the integration of the tangential component of $\mathbf{A}$ along the contour c. This line integral may have the physical meaning of work done if $\mathbf{A}$ represents a vector force.

The surface integral $\int_s \mathbf{A} \cdot d\mathbf{s}$, conversely, represents the calculation of the net outflow of the flux of the vector $\mathbf{A}$ from the surface s. The dot product in $\mathbf{A} \cdot d\mathbf{s}$ represents the fact that the component of $\mathbf{A}$ in the direction of $d\mathbf{s}$ (which is normal to the surface s) is being considered in the integration process.

To help us carry out the various vector operations we introduced three commonly used coordinate systems including the Cartesian, cylindrical, and the spherical coordinate systems. Each of these coordinate systems is identified in terms of three independent variables (e.g., ρ, ϕ, z) in the cylindrical coordinate system; three reference surfaces (each for a constant value of the independent variables); and three base vectors, each perpendicular to a reference surface and in the direction of increase of the independent variable. For example, $\mathbf{a}_\rho$ is a unit vector perpendicular to the $\rho = $ constant surface and is in the direction of increase ρ. We also noticed that except for the Cartesian coordinate system, the base vectors may change their directions at various points in space. This fact should be considered before performing any vector operations on vectors expressed in terms of base vectors at different points. Vectors may be represented in these coordinate systems in terms of their components along the various base vectors. The various vector operations may be performed by carrying out the desired operation on the vector components. Expressions for the transformation of a vector representation between the various coordinate systems are also given. It is important to emphasize that in addition to changing the independent variables, the vector components must also be changed from those along the base vectors of a given coordinate system to new components along the desired base vectors.

Also in this chapter we introduced the vector electric and magnetic fields. For the electric field $\mathbf{E}$ we introduced Coulomb's law to quantify $\mathbf{E}$ from a set of discrete charges. For the magnetic field $\mathbf{B}$, conversely, we initially used Biot-Savart's law and later on Ampere's law to quantify $\mathbf{B}$ that results from a current-carrying conductor. We then examined Lorentz's force law on a charged particle moving in electric and magnetic fields, and emphasized some differences between the electric and magnetic forces on charged particles. It is shown that magnetic forces are possible only on moving charges and that these forces are always perpendicular to the trajectory and hence do not cause any energy transfer to or from the moving charge.

With the acquired background on the basic vector operations, and the brief introduction of the electric and magnetic fields associated with static charges and dc currents, respectively, we introduced Maxwell's equations. These equations are general mathematical relations between the electric and magnetic fields, and their charge and current sources.

These equations are given by

GAUSS'S LAW FOR ELECTRIC FIELD

$$\oint_s \epsilon_o \mathbf{E} \cdot d\mathbf{s} = \int_v \rho_v \, dv = Q$$

GAUSS'S LAW FOR MAGNETIC FIELD

$$\oint_s \mathbf{B} \cdot d\mathbf{s} = 0$$

FARADAY'S LAW

$$\oint_c \mathbf{E} \cdot d\ell = -\frac{d}{dt} \int_s \mathbf{B} \cdot d\mathbf{s}$$

AMPERE'S LAW

$$\oint_c \frac{\mathbf{B}}{\mu_o} \cdot d\ell = \int_s \mathbf{J} \cdot d\mathbf{s} + \frac{d}{dt} \int_s \epsilon_o \mathbf{E} \cdot d\mathbf{s}$$

The following observations may be made based on careful examination of Maxwell's equations:

1. For time-varying fields we may recognize from Faraday's law that time-varying magnetic flux density **B** represents yet another source for the electric field **E**. This, of course, is in addition to the charge distribution source considered in Gauss's law. In other words both the charge ρ_v and time-varying magnetic flux are sources of the electric field **E**.

2. Ampere's law shows that both electric currents $\int_s \mathbf{J} \cdot d\mathbf{s}$ and time-varying electric field are sources of the magnetic flux density **B**. As discussed earlier, the fact that electric current is a source of magnetic flux was recognized first by Oersted. It was then quantified by Biot-Savart's law and later on in Ampere's law in its original form. Maxwell mathematically suggested the second term in Ampere's law and suggested time-varying electric field as a source of magnetic field.

3. The four Maxwell's equations demonstrate the coupling between the electric and magnetic fields (in Faraday's and Ampere's laws) and hence these four equations must be solved simultaneously. In their present integral form such a simultaneous solution may be difficult and hence will be deferred until we develop their differential forms in the next chapter.

4. To help familiarize ourselves further with Maxwell's equations, we considered the special case of static fields. Under this assumption, the electric- and magnetic-field quantities are separable. Therefore the electric field may be quantified in terms of the charge distribution using Gauss's law for electric fields, and the magnetic field may be quantified in terms of the current source using Ampere's law.

5. To calculate a static electric field from a given charge distribution we may use Coulomb's law or Gauss's law. To use Gauss's law we construct a suitable Gaussian surface that encloses the portion (or all) the charge source of interest and also takes advantage of the symmetry of the charge distribution. For example, for

spherical charge distribution we use spherical Gaussian surface, and for wire or cylindrical charge distribution we use cylindrical Gaussian surface. We place these surfaces at the location where we want to calculate the electric field. We then apply Gauss's law at the surface and specifically equate the electric flux ϵ_o **E** out flowing from the closed surface $\oint_s \epsilon_o \mathbf{E} \cdot d\mathbf{s}$ to the total electric charge in the volume enclosed by the Gaussian surface s (i.e., $\int_v \rho_v \, dv$). This gives us the desired relation for calculating the electric field from a given charge distribution.

6. In calculating the magnetic flux **B** from a given current distribution **J**, we use Ampere's law. For static fields, the displacement current term is zero, and **B** is directly related to **J**. The calculation procedure involves establishing a suitable amperian contour c at the location where **B** needs to be calculated. The shape of the contour should take advantage of the symmetry of the given geometry of the current distribution. Then we integrate **B** around the closed contour and equate the result to the total current crossing the area s encircled by the contour, that is, $\int_s \mathbf{J} \cdot d\mathbf{s}$.

7. We also used Faraday's law to introduce Lenz's law. It is shown that the negative sign in the term $-d/dt \int_s \mathbf{B} \cdot d\mathbf{s}$ was included to emphasize the fact that the $emf = \oint_c \mathbf{E} \cdot d\ell$ is induced in such a direction so as to oppose the change in the magnetic flux producing it. If the rate of change in the magnetic flux crossing the area s, that is, $d/dt \int_s \mathbf{B} \cdot d\mathbf{s}$ is positive, the emf will be negative; if $d/dt \int_s \mathbf{B} \cdot d\mathbf{s}$ is negative, the emf will be positive. In the former case the emf will produce a countermagnetic flux (i.e., in opposite direction to the original one), whereas in the second case the induced emf will produce magnetic flux that is in the same direction as the original one. In both cases the induced emf and the magnetic flux associated with it oppose the change (increase or decrease) in the original magnetic flux crossing the area s. Lenz's law is known to us from the circuit theory when the physical property of an inductor was introduced. Inductors oppose sudden change in currents. This is accomplished by a counterinduced emf governed by Lenz's law.

The other important topic we discussed in this chapter is the displacement current term in Ampere's law. This is a virtual current, and in addition to the conventional current density **J** it represents a source of the magnetic flux **B**. Maxwell added this term to Ampere's law on mathematical bases as we will discuss in more detail in chapter 2. We, however, emphasized the importance of this term in, for example, closing an electric circuit containing a capacitor. We showed that the accumulation of the charge on the plates of the capacitor and the associated electric field between the plates are responsible for closing the circuit and for effectively allowing the current to flow. No charges have actually crossed the plates, and this is why this current is known as virtual current.

Unfortunately, it is rather involved to solve simultaneously the four Maxwell's equations in their present integral forms. As we will see in chapter 2, such a simultaneous solution is much easier to obtain after the development of the differential forms of these equations.

PROBLEMS

1. In Figure P1-1, let $\mathbf{OP} = a\,\mathbf{a}_x + b\,\mathbf{a}_y + c\,\mathbf{a}_z$. The angles α, β, and γ are between $\mathbf{OP}$ and $\mathbf{a}_x$, $\mathbf{a}_y$, and $\mathbf{a}_z$, respectively. Cos α, cos β, and cos γ are the *direction cosines* of $\mathbf{OP}$. Let $p = |\mathbf{OP}|$.

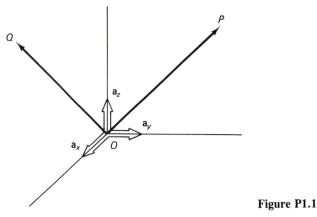

Figure P1.1

(a) Find a, b, and c in terms of p, and the direction cosines of $\mathbf{OP}$.
(b) Show that $\cos^2 \alpha + \cos^2 \beta + \cos^2 \gamma = 1$.
(c) Let $\mathbf{a}_p$ be a unit vector along $\mathbf{OP}$. Find the components of $\mathbf{a}_p$.
(d) Let $\mathbf{OQ}$ be another vector that makes an angle θ with $\mathbf{OP}$. If the direction cosines of $\mathbf{OQ}$ are $\cos \alpha_1$, $\cos \beta_1$, and $\cos \gamma_1$, using the dot product show that

$$\cos \theta = \cos \alpha \, \cos \alpha_1 + \cos \beta \, \cos \beta_1 + \cos \gamma \, \cos \gamma_1$$

2. Given a vector $\mathbf{A} = \mathbf{a}_x + 2\mathbf{a}_y - 3\mathbf{a}_z$ and a vector $\mathbf{B}(x, y, z)$ that is directed from $(2, 1, 3)$ to $(0, -2, 2)$, determine
(a) An expression for $\mathbf{B}$.
(b) The magnitude of the projection of $\mathbf{B}$ on $\mathbf{A}$.
(c) The smaller angle between $\mathbf{A}$ and $\mathbf{B}$.
(d) A unit vector perpendicular to the plane containing $\mathbf{A}$ and $\mathbf{B}$.

3. Given a vector $\mathbf{A}$

$$\mathbf{A} = x\mathbf{a}_x + y^2 \mathbf{a}_y + 3z\mathbf{a}_z$$

Find a unit vector along $\mathbf{A}$ at $x = 1$, $y = 2$, and $z = 4$.

4. Given

$$\mathbf{A} = 3\mathbf{a}_x + x\mathbf{a}_y + y\mathbf{a}_z$$

$$\mathbf{B} = x^2 \mathbf{a}_x + 4\mathbf{a}_y$$

(a) Determine $\mathbf{A} \cdot \mathbf{B}$ at the point $x = 2, y = 3$.
(b) Find the angle between $\mathbf{A}$ and $\mathbf{B}$.
(c) Find the projection of $\mathbf{B}$ along the direction of $\mathbf{A}$.

5. Given the vectors

$$\mathbf{A} = 5\mathbf{a}_x + 2\mathbf{a}_y + 3\mathbf{a}_z$$

$$\mathbf{B} = B_x \mathbf{a}_x + 2\mathbf{a}_y + B_z \mathbf{a}_z$$

$$\mathbf{C} = 3\mathbf{a}_x + C_y \mathbf{a}_y + \mathbf{a}_z$$

Find B_x, B_z, and C_y so that **A**, **B**, and **C** are mutually orthogonal (i.e., perpendicular to each other).

6. Given two vectors, $\mathbf{B} = 2\mathbf{a}_x + \mathbf{a}_y + 3\mathbf{a}_z$ and $\mathbf{C} = 2\mathbf{a}_y + 6\mathbf{a}_z$
   (a) Find a unit vector perpendicular to the plane containing **B** and **C**. Draw a schematic diagram illustrating your answer.
   (b) Find the area of the parallelogram constructed from **B** and **C**.

7. Two vectors **A** and **B** have the same origin and are given by

$$\mathbf{A} = \mathbf{a}_\rho + \mathbf{a}_\phi + 3\mathbf{a}_z$$

$$\mathbf{B} = \alpha\mathbf{a}_\rho + \beta\mathbf{a}_\phi - 6\mathbf{a}_z$$

   (a) Determine α and β such that the two vectors are parallel.
   (b) Express **A** in the Cartesian coordinate system.

8. Given

$$\mathbf{A} = \cos\phi\,\mathbf{a}_\rho + \sin\phi\,\mathbf{a}_{\phi'} + \rho\mathbf{a}_z$$

$$\mathbf{B} = \rho\mathbf{a}_\rho + \phi\mathbf{a}_\phi + 2\mathbf{a}_z$$

   with ϕ expressed in radians. Determine $\mathbf{A} \cdot \mathbf{B}$ at the point $x = 2, y = 3$. Assume that the origins of the cylindrical and Cartesian coordinates are the same, and the x axis coincides with $\phi = 0$. Express the vector **A** in Cartesian coordinates.

9. Given that $\mathbf{A} = 3\mathbf{a}_r - 7\mathbf{a}_\theta + 2\mathbf{a}_\phi$ has its origin at the point $(1, \pi/2, 0°)$ in the spherical coordinate system and $\mathbf{B} = -2\mathbf{a}_r - 4\mathbf{a}_\theta + 2\mathbf{a}_\phi$ has its origin at the point $(3, \pi/2, \pi/2)$. Determine $\mathbf{A} - \mathbf{B}$.

10. The vector **A** shown in Figure P1.10 is a unit vector in the x-y plane and makes an angle 45° with the x axis. **B** is a vector in the z-y plane of length equal to 2 and makes an angle 30° with the y axis.

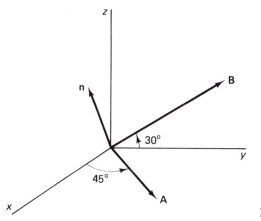

Figure P1.10

   (a) Obtain expressions for the vectors **A** and **B** in the Cartesian coordinate system.
   (b) Find the angle between **A** and **B**.
   (c) Find the unit vector **n** that is perpendicular to both **A** and **B** and such that **A**, **B**, and **n** follow the right-hand rule.
   (d) Determine the angle between **n** and the z axis.

11. Let the two vectors **A** and **B** be given by

$$\mathbf{A} = \mathbf{a}_x + b\mathbf{a}_y + c\mathbf{a}_z$$

$$\mathbf{B} = -\mathbf{a}_x + 3\mathbf{a}_y - 8\mathbf{a}_z$$

(a) Find the values of the constants b and c that will make **A** and **B** *parallel* to each other.
(b) Find the *relation* between b and c that makes **A** and **B** *perpendicular* to each other.

12. (a) Express the rectangular vector

$$\mathbf{A} = x^2 y\, \mathbf{a}_x + y^2 z\, \mathbf{a}_y + x^2 z\, \mathbf{a}_z$$

in the cylindrical coordinate system.
(b) Find the component of **A** along the $\mathbf{a}_r$ direction in the spherical coordinate system.

13. The vector **A** is given in the spherical coordinates by

$$\mathbf{A} = 2\,\mathbf{a}_r + \mathbf{a}_\theta - 3\,\mathbf{a}_\phi$$

with respect to the origin $(3, \pi/2, \pi/2)$. The vector **B** is given by

$$\mathbf{B} = -\mathbf{a}_r + 3\,\mathbf{a}_\theta + 2\,\mathbf{a}_\phi$$

with respect to the origin $(6, 0, \pi/2)$. Determine $\mathbf{A} \times \mathbf{B}$ expressed with respect to the origin of **B**.

14. Two point charges, $Q_1 = 1nC$ and $Q_2 = 3nC$, are located at the Cartesian points $(1, 1, 0)$ and $(0, 2, 1)$, respectively. Find **E** at $(3, 5, 5)$. Coordinates are given in meters and one $nC = 10^{-9}$ C.

15. What is the force of attraction between the electron and the nucleus of the hydrogen atom, which are spaced at approximately 10^{-10} m? The hydrogen atom has an electron of charge $e = -1.6 \times 10^{-19}$ C, and the nucleus has a charge equal but opposite in sign to that of the electron.

16. Two point charges $Q_A = 0.02 \times 10^{-9}$ C and $Q_B = 0.01 \times 10^{-9}$ C are located at $(0, 1, 0)$ and $(0, 4, 0)$, respectively. Calculate the total vector electric field at a point $P(0, 2.5, 2)$. All coordinates are given in centimeters.

17. Two point charges Q_1 and Q_2 are located as follows:

$$Q_1 = 10^{-9} \text{ C} \qquad \text{located at } (3, 1, 1)$$

$$Q_2 = 2 \times 10^{-9} \text{ C} \qquad \text{located at } (0, 1, 0)$$

Using vector addition, determine the electric field at the point $(6, 3, 2)$. All distances are given in centimeters.

18. A charge $+Q$ is placed at $(-a, 0, 0)$ and a charge $-2Q$ is placed at $(a, 0, 0)$. Is there a point in space where $\mathbf{E} = 0$?

19. Two conducting balls of equal radii are charged with charges Q_1 and Q_2. These two charges are placed at a distance d from each other. The balls are brought into contact, then placed back in their original positions. Determine the force between the two balls in both cases.

20. Two charges are arranged in the x-y plane as follows: $Q_1 = 10^{-9}$ C at $(0, 0)$, and $Q_2 = 4 \times 10^{-9}$ C at $(3, 0)$. Determine the electric field **E** at the points $(1, 0)$ and $(1, 2)$ by determining the **E** field resulting from each charge and adding the results vectorially. Dimensions are given in centimeters.

21. Two conducting spheres of negligible diameter have masses of 0.2 g each. Two nonconducting threads, each 1 m long and of negligible mass, are used to suspend the two spheres from a common support. After placing an equal charge on the spheres, it is found that they separate with an angle of $45°$ between the threads.

(a) If the gravitational force is 980×10^{-5} N/g, find the charge on each sphere.

(b) Find the angle between threads if the charge on each sphere is 0.5 μC.

22. Four charges of 1 μC each (1 μC = 10^{-6} C) are located in free space in a plane at $(\pm 1, \pm 1)$. Find **E** at $(3, 0)$. Coordinates are in centimeters.

23. A charged particle (charge q, mass m) is moving under the influence of a magnetic field

$$\mathbf{B} = B_o \, \mathbf{a}_z.$$

If the velocity vector of the particle is

$$\mathbf{v} = v_x \, \mathbf{a}_x + v_y \, \mathbf{a}_y + v_z \, \mathbf{a}_z$$

Show that the acceleration of this particle lies in the x-y plane.

24. A magnetic field $\mathbf{B} = B_o(\mathbf{a}_x + 2\mathbf{a}_y - 4\mathbf{a}_z)$ exists at a point. What should be the electric field at that point if the force experienced by a test charge moving with a velocity $\mathbf{v} = v_o(3\mathbf{a}_x - \mathbf{a}_y + 2\mathbf{a}_z)$ is to be zero?

25. An electron of velocity **v** travels into a region of constant electric field **E** in Figure P1.25. In what *direction* must a *magnetic field* be applied to cause a zero net force on the electron?

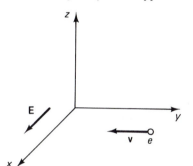

Figure P1.25

26. Consider the contour c shown in Figure P1.26 and the vector field

$$\mathbf{F} = 2\rho(z^2 + 1) \cos \phi \, \mathbf{a}_\rho - \rho(z^2 + 1) \sin \phi \, \mathbf{a}_\phi + 2\rho^2 z \, \cos \phi \, \mathbf{a}_z.$$

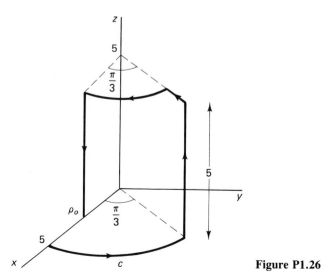

Figure P1.26

(a) Evaluate $\int_c \mathbf{F} \cdot d\ell$.

(b) Evaluate $\int_{c_1} \mathbf{F} \cdot d\ell$ where c_1 is the straight line joining $(\rho = \rho_o, \phi = 0, z = 0)$ to $(\rho = 5, \phi = 0, z = 0)$.

(c) Are the results of a and b consistent with the field $\mathbf{F}$ being conservative? (A field is conservative when its line integral along any closed contour is zero.)

27. Given an electric field $\mathbf{E} = (5xy - 6x^2)\mathbf{a}_x + (2y - 4x)\mathbf{a}_y$, find the work required to move a charge $q = 1 \times 10^{-6}\,\mathrm{C}$ along the curve c in the x-y plane given by $y = x^3$ from the point $(1, 1)$ to $(2, 8)$.

28. Find the vector line integral $\mathbf{A} = \int_\ell d\ell$, where ℓ is the path from P_1 to P_2 as shown in Figure P1.28.

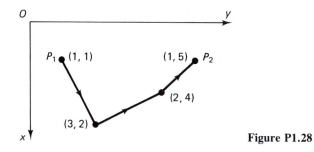

Figure P1.28

29. Evaluate the line integral

$$\int_c (\sin\phi\,\mathbf{a}_\rho + \rho\,\cos\phi\,\mathbf{a}_\phi + \tan\phi\,\mathbf{a}_z) \cdot d\ell$$

for the contour shown in Figure P1.29.

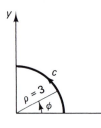

Figure P1.29   Contour of integration for problem 29.

30. If the electric field vector $\mathbf{E}$ is given in the spherical coordinate system by

$$\mathbf{E} = 3r^2 \cos\phi \sin\theta\,\mathbf{a}_r + r^2 \cos\theta \cos\phi\,\mathbf{a}_\theta - r^2 \sin\phi\,\mathbf{a}_\phi$$

evaluate the work done in moving a unit positive charge along the contour c shown in Figure P1.30.

31. Determine the net flux of the vector field $\mathbf{F}(r, \theta, \phi) = r \sin\theta\,\mathbf{a}_r + \mathbf{a}_\theta + \mathbf{a}_\phi$ emanating from a closed surface defined by $r = 1, 0 \leq \theta \leq \pi/2, 0 \leq \phi \leq 2\pi$. Hint: The closed surface consists of the hemisphere s_1 and the base plane s_2 shown in Figure P1.31.

32. The equation for a field is $\mathbf{B} = x\mathbf{a}_x + y\mathbf{a}_y + z\mathbf{a}_z$. Evaluate $\int_s \mathbf{B} \cdot d\mathbf{s}$ over a circular area s of radius 2 and that is centered on the z axis and is parallel to the x-y plane at $z = 4$ as shown in Figure P1.32.

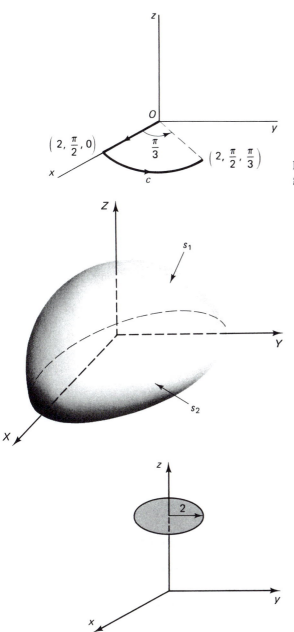

Figure P1.30 The contour of integration.

Figure P1.31 Surface of integration.

Figure P1.32 Circular area of integration.

33. If the magnetic flux density vector is given by

$$\mathbf{B} = zy\,\mathbf{a}_x + x\,\mathbf{a}_y + z^2 x\,\mathbf{a}_z$$

find the total magnetic flux emanating (passing through) the following surfaces.
   (a) The rectangular area shown in Figure P1.33a.
   (b) The cylindrical surface shown in Figure P1.33b.

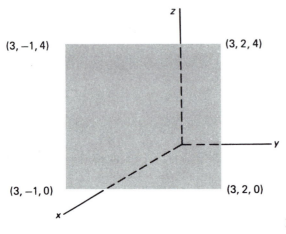

(a)

Figure P1.33a   The rectangular area through which the magnetic flux is passing.

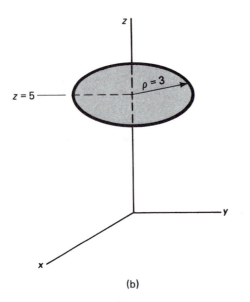

(b)

Figure P1.33b   The cylindrical area through which the magnetic flux is passing.

34. Consider two concentric cylindrical surfaces, shown in Figure P1.34, one having a radius a and charge density ρ_s, and the other having a radius b and a charge density $-\rho_s$. Find the electric field $\mathbf{E}$ in the following regions:
    (a) $\rho < a$.
    (b) $a < \rho < b$.
    (c) $\rho > b$.
    Consider the cylinders to be infinitely long and use Gauss's law on per unit length basis.

35. A spherical capacitor shown in Figure P1.35 consists of two concentric spherical surfaces, one having a radius a and charge density ρ_s, and the other having a radius b and a charge density $-\rho_s$. Find the electric field $\mathbf{E}$ in the following regions:

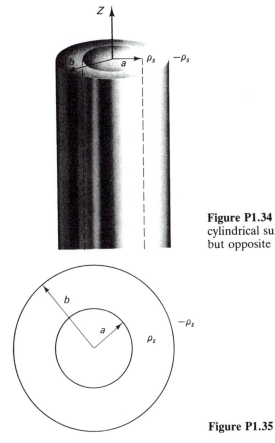

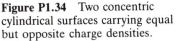

Figure P1.34   Two concentric cylindrical surfaces carrying equal but opposite charge densities.

Figure P1.35

(a) $r < a$.
(b) $a < r < b$.
(c) $r > b$.

36. A spherical capacitor consists of two concentric spherical shells of radii a and b as shown in Figure P1.36. The space between the two spheres is filled with air. If the surface charge density on the outer surface of the inside sphere is ρ_s C/m^2, and the outer sphere is grounded (i.e., the *total* charge on the outer sphere is equal *in magnitude* to the *total* charge on the inner sphere)

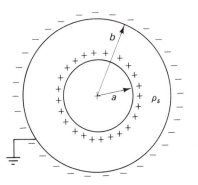

Figure P1.36   The geometry of a spherical capacitor with a surface charge density ρ_s on the inner sphere.

(a) Determine the electric field *in the region between the concentric spheres* (i.e., $a < r < b$) and in the region *outside the outer sphere* (i.e., $r > b$).

(b) If the capacitor is connected to a circuit so that the charge density ρ_s is given as a function of time by

$$\rho_s(t) = 2 \times 10^{-9} \cos(10^5 t) \text{ C/m}^2$$

determine the displacement current density in the spherical capacitor.

37. In Figure P1.37 a spherical cloud of charge in free space is characterized by a volume charge density

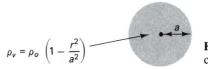

$$\rho_v = \rho_o \left(1 - \frac{r^2}{a^2}\right)$$

Figure P1.37 Spherical cloud of charge.

$$\rho_v = \begin{cases} \rho_o\left(1 - \frac{r^2}{a^2}\right) & r < a \\ 0 & r > a \end{cases}$$

where a is the surface radius and ρ_o is a constant. Solve for the electric field intensity for all values of r, that is, for $r < a$ and $r > a$.

38. A cylindrical beam of electrons consists of a uniform volume charge density moving at a constant velocity $v_o = 10^7$ m/sec. The total current carried by the beam that is of radius $a = 1$ mm is $I_o = 10^{-2}$ A. Use Gauss's law to calculate the electric field intensity inside and outside the electron beam.

39. For the electromagnet shown in Figure P1.39, the magnetic field is given by

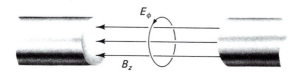

Figure P1.39 Magnetic flux between the poles of an electromagnet.

$$B_z = B_o\left[1 - \left(\frac{\rho}{0.15}\right)^2\right] \sin \omega t$$

where ρ is the radial distance from the symmetry axis.

(a) Use Faraday's law in integral form to find the electric field component E_ϕ.

(b) Sketch the variation of B_z and E_ϕ as a function of time. Show that the induced *emf* satisfies Lenz's law.

40. A loop conductor shown in Figure P1.40 lies in the $z = 0$ plane, has an area of 0.1 m^2 and a resistance of 5 Ω. Given $\mathbf{B} = 0.2 \sin 10^3 t\, \mathbf{a}_z$ (T), determine the current.

41. In Figure P1.41, an area of 0.65 m^2 in the $z = 0$ plane is enclosed by a filamentary conductor. Use Faraday's law to find the induced electromotive force ($emf = \oint_c \mathbf{E} \cdot d\ell$) if the magnetic field enclosed by the contour is given by

$$\mathbf{B} = \frac{0.05}{\sqrt{2}} \cos 10^3 t(\mathbf{a}_y + \mathbf{a}_z)$$

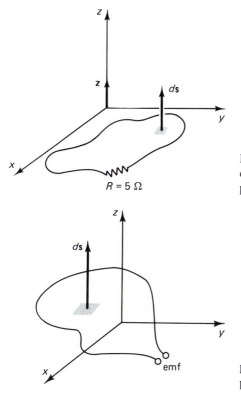

Figure P1.40   A conducting loop connected to a 5 Ω resistor and placed in $x - y$ plane.

Figure P1.41   A conducting loop placed at $z = 0$ plane.

42. The conducting loop shown in Figure P1.42 is placed perpendicular to a magnetic field of flux density **B** given by

$$\mathbf{B} = 5 \cos(10^4 t - 2.1x)\, \mathbf{a}_z \text{ Wb/m}^2$$

The loop is terminated on both sides by two resistors $R_1 = 15\ \Omega$ and $R_2 = 7\ \Omega$.

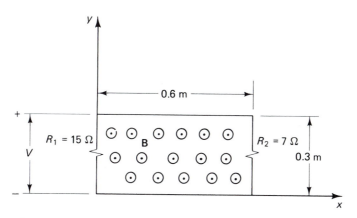

Figure P1.42   A conducting loop placed perpendicular to a magnetic flux density **B**.

If the induced *emf* is monitored by measuring the voltage V across R_1, determine the value of V.

43. In cylindrical coordinates in Figure P1.43 the current density is given by

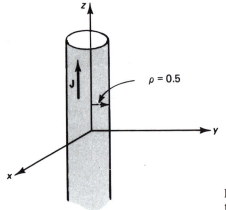

Figure P1.43 Cylindrical conductor carrying a current density **J**.

$$\mathbf{J} = 4.5e^{-2\rho}\,\mathbf{a}_z \text{ A/m}^2 \qquad \begin{array}{l} 0 < \rho \le 0.5 \\ 0 < \phi \le 2\pi \end{array}$$

and $\mathbf{J} = 0$ elsewhere. Use Ampere's law to find the magnetic field **B** both for $\rho < 0.5$ and $\rho \ge 0.5$.

44. In Figure P1.44, an infinitely long cylindrical conductor has thickness d and carries a current of density **J** given by

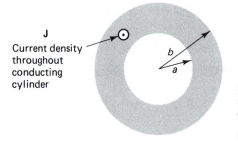

J
Current density throughout conducting cylinder

Figure P1.44 A cross section of an infinitely long cylindrical conductor. The current density **J** is given by $\mathbf{J} = 2\rho\,\mathbf{a}_z \text{ A/m}^2$.

$$\mathbf{J} = 2\rho\,\mathbf{a}_z \text{ A/m}^2$$

The cylindrical conductor has an inner and outer radii equal to a and b, respectively. Determine the magnetic flux density **B** in the following regions:

(a) $\rho < a$.

(b) $a < \rho < b$.

(c) $\rho > b$.

45. A long round wire of radius a and carrying a time-varying current $\mathbf{I} = I_o \cos \omega t\,\mathbf{a}_z$, is shown in Figure P1.45.

(a) Write down an expression for the magnetic field outside the wire $(\rho > a)$.

(b) If we located a rectangular conducting loop of sides a and b at distance d from the center of the wire as shown in Figure P-1.45, find

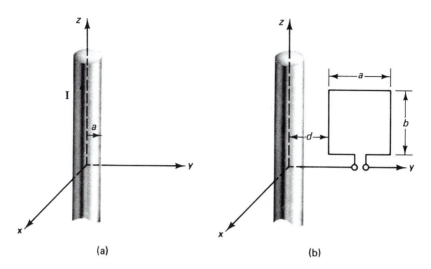

Figure P1.45 (a) Cylindrical conductor of radius a and carrying current **I**. (b) Rectangular loop placed at a distance d from the axis of the conductor.

(i) The total magnetic flux crossing the loop.
(ii) The induced *emf* at the terminals of the loop.

46. A long straight wire carrying a current $I \cos \omega t$ is along the z axis. Find the induced *emf* around a rectangular loop in the z-y plane as shown in Figure P1.46. Show that it satisfies Lenz's law.

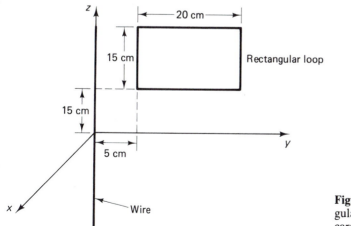

Figure P1.46 A conducting rectangular loop placed near a current carrying wire.

47. (a) In a region of space in the neighborhood of an electromagnetic plastic heat sealer, the magnetic field of the source is unknown and is assumed to be arbitrarily oriented. Suggest a practical procedure to measure this arbitrarily oriented magnetic field.

(b) The magnetic field intensity **B** of a short electric current source may be approximately given in the cylindrical coordinates by

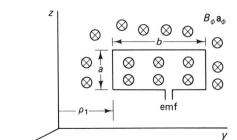

Figure P1.47 A rectangular loop placed in the *y-z* plane to measure the magnetic field.

$$\mathbf{B} = \left(K_1 \frac{1}{\rho^2} - \frac{K^2}{\rho} \right) \sin \omega t \ \mathbf{a}_\phi \, \text{Wb/m}^2$$

To measure this magnetic field, the rectangular conducting loop shown in Figure P1.47 is placed in the *y-z* plane.

(i) Calculate the induced *emf* at the terminals of the conducting loop.

(ii) Show that the variation of the induced *emf* and the magnetic flux satisfy Lenz's law.

48. (a) Explain what we mean by a field. Clarify your answer by describing the meaning of the electromagnetic field.

(b) Explain the differences between the action (force) of the electric and magnetic fields on charged particles.

(c) An electron ($e = -1.6 \times 10^{-19}$ C) is injected with an *initial* (at $t = 0$) velocity $\mathbf{v} = v_o \, \mathbf{a}_y$ into a region occupied by both electric $\mathbf{E}$ and magnetic $\mathbf{B}$ fields. Describe the motion of the electron if $\mathbf{E} = E_o \, \mathbf{a}_z$ and $\mathbf{B} = B_o \, \mathbf{a}_x$.

CHAPTER 2

MAXWELL'S EQUATIONS IN DIFFERENTIAL FORM

*.1 INTRODUCTION

In chapter 1 we introduced Maxwell's equations in integral form. These forms are easier to understand physically and were therefore used in our first acquaintance with them to emphasize the importance of the various terms that relate the electric and magnetic fields to their sources. The problem with the integral representation, however, is its restricted application to simple geometries in which the integrations may be carried out easily by taking advantage of various symmetries. You probably noticed in all the examples we solved in the previous chapter that symmetrical geometries (planes, cylinders, spheres, etc.) were considered to simplify the integrations. Undoubtedly, the derivation of alternative relations between the electric and magnetic fields and their sources that are satisfied at every point in space (i.e., point instead of integral relations) would have an advantage in solving a wider class of electromagnetic field problems. This will be clarified in this chapter and after we introduce Maxwell's equations in a point form. Fortunately, everything that we learned from the integral forms and the physical interpretation of the various terms will still be helpful in our understanding of the new point or differential forms. This is because the point relations will be derived by taking the limiting cases of the integral forms, where the surface, volume, and line integrals are reduced to those over infinitesimally small elements. The resulting point relations include various differential operators, and this is why these relations are often called Maxwell's equations in differential form.

In this chapter, we will derive differential expressions of Maxwell's equations. Before deriving these point relations, however, we will review some vector differential operations including the gradient, divergence, and curl of vectors. Several examples will be given to illustrate these operations. The divergence and Stokes's theorems will then be defined and employed to derive the differential forms of Maxwell's equations. As a simple example of solving these equations, we will discuss the propagation properties of a uniform plane wave in free space.

2.2 VECTOR DIFFERENTIATION

Consider a vector $\mathbf{A}(u)$ that is a function of the independent variable u. When the independent variable changes from u to $u + \Delta u$, the magnitude and direction of the vector will generally change. The vector $\mathbf{A}(u)$ and the resulting new vector $\mathbf{A}(u + \Delta u)$ are shown in Figure 2.1a.

The incremental change in the vector $\Delta \mathbf{A}$ may be obtained by constructing the conventional triangle arrangement of Figure 2.1b, where it may be seen that $\Delta \mathbf{A}$ is given by

$$\Delta \mathbf{A} = \mathbf{A}(u + \Delta u) - \mathbf{A}(u)$$

The differential change of the vector is known as the vector differentiation. It is defined as

$$\frac{d\mathbf{A}}{du} = \lim_{\Delta u \to 0} \frac{\Delta \mathbf{A}}{\Delta u} = \lim_{\Delta u \to 0} \frac{\mathbf{A}(u + \Delta u) - \mathbf{A}(u)}{\Delta u}$$

Figure 2.1 Incremental change in the vector $\mathbf{A}(u)$ by $\Delta \mathbf{A}$ as a result of changing the independent variable from u to $u + \Delta u$.

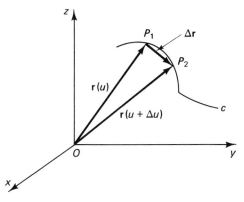

Figure 2.2   Position vector $\mathbf{r}(u)$ and its incremental variation $\Delta\mathbf{r}$ as a result of changing the independent variable u.

If the vector is, for example, a position vector $\mathbf{r}(u)$ measured from the origin of a coordinate system to point P_1, as shown in Figure 2.2, then

$$\mathbf{r}(u) = x(u)\,\mathbf{a}_x + y(u)\,\mathbf{a}_y + z(u)\,\mathbf{a}_z$$

The differential variation of $\mathbf{r}(u)$ with respect to u is given by

$$\frac{d\mathbf{r}}{du} = \lim_{\Delta u \to 0} \frac{\mathbf{r}(u + \Delta u) - \mathbf{r}(u)}{\Delta u}$$

$$= \lim_{\Delta u \to 0} \frac{(x(u + \Delta u)\mathbf{a}_x + y(u + \Delta u)\mathbf{a}_y + z(u + \Delta u)\mathbf{a}_z) - (x(u)\mathbf{a}_x + y(u)\mathbf{a}_y + z(u)\mathbf{a}_z)}{\Delta u}$$

$$= \lim_{\Delta u \to 0} \left[\frac{(x(u + \Delta u) - x(u))\mathbf{a}_x}{\Delta u} + \frac{(y(u + \Delta u) - y(u))\mathbf{a}_y}{\Delta u} + \frac{(z(u + \Delta u) - z(u))\mathbf{a}_z}{\Delta u} \right]$$

$$= \frac{dx}{du}\mathbf{a}_x + \frac{dy}{du}\mathbf{a}_y + \frac{dz}{du}\mathbf{a}_z$$

It can be seen from Figure 2.2 that $\lim_{\Delta u \to 0}(\Delta\mathbf{r}/\Delta u)$ represents a vector that is tangential to the contour c. Hence, $d\mathbf{r}/du$ is a vector in the direction of a tangent to the curve c.

EXAMPLE 2.1

Find a unit vector tangent to the curve given in terms of the parameter t by

$$x = 4t - 1,\ y = t^2 + 1,\ \text{and}\ z = t^2 - 6t\ \text{at}\ t = 2$$

Solution

A vector that is tangential to the given curve is given by

$$\frac{d\mathbf{r}}{dt} = \frac{d}{dt}\left[(4t - 1)\,\mathbf{a}_x + (t^2 + 1)\,\mathbf{a}_y + (t^2 - 6t)\,\mathbf{a}_z \right]$$

$$= 4\,\mathbf{a}_z + 2\,t\mathbf{a}_y + (2t - 6)\,\mathbf{a}_z$$

The required unit vector **n** is then

$$\mathbf{n} = \frac{4\,\mathbf{a}_x + 2\,ta_y + (2t - 6)\,\mathbf{a}_z}{\sqrt{(4)^2 + (2t)^2 + (2t - 6)^2}}$$

At $t = 2$, **n** is given by

$$\mathbf{n} = \frac{2}{3}\mathbf{a}_x + \frac{2}{3}\mathbf{a}_y - \frac{1}{3}\mathbf{a}_z$$

━━━━◆◆◆━━━━

EXAMPLE 2.2

Prove the following vector differential relation

$$\frac{d}{dt}(\mathbf{A} \cdot \mathbf{B}) = \mathbf{A} \cdot \frac{d\mathbf{B}}{dt} + \frac{d\mathbf{A}}{dt} \cdot \mathbf{B}$$

Solution

Let us express **A** and **B** in terms of their components along the three orthogonal unit vectors $\mathbf{a}_1$, $\mathbf{a}_2$, and $\mathbf{a}_3$

$$\mathbf{A} = A_1\,\mathbf{a}_1 + A_2\,\mathbf{a}_2 + A_3\,\mathbf{a}_3$$

$$\mathbf{B} = B_1\,\mathbf{a}_1 + B_2\,\mathbf{a}_2 + B_3\,\mathbf{a}_3$$

$$\mathbf{A} \cdot \mathbf{B} = A_1 B_1 + A_2 B_2 + A_3 B_3$$

$$\frac{d(\mathbf{A} \cdot \mathbf{B})}{dt} = \frac{d}{dt}(A_1 B_1 + A_2 B_2 + A_3 B_3)$$

$$= A_1\frac{dB_1}{dt} + B_1\frac{dA_1}{dt} + A_2\frac{dB_2}{dt} + B_2\frac{dA_2}{dt} + A_3\frac{dB_3}{dt} + B_3\frac{dA_3}{dt}$$

$$= \left(A_1\frac{dB_1}{dt} + A_2\frac{dB_2}{dt} + A_3\frac{dB_3}{dt} + B_1\frac{dA_1}{dt} + B_2\frac{dA_2}{dt} + B_3\frac{dA_3}{dt} \right)$$

The sum on the right-hand side may be expressed as two dot products, thus

$$\frac{d(\mathbf{A} \cdot \mathbf{B})}{dt} = \mathbf{A} \cdot \frac{d\mathbf{B}}{dt} + \mathbf{B} \cdot \frac{d\mathbf{A}}{dt}$$

━━━━◆◆◆━━━━

If a vector **A** depends on more than one independent variable, say on three variables (u_1, u_2, u_3), then the partial derivative of **A** with respect to any one of these variables (e.g., u_1) is given by

$$\frac{\partial \mathbf{A}}{\partial u_1} = \operatorname*{Lim}_{\Delta u_1 \to 0} \frac{\mathbf{A}(u_1 + \Delta u_1, u_2, u_3) - \mathbf{A}(u_1, u_2, u_3)}{\Delta u_1}$$

It should be noted that the change in the vector **A** resulted from an incremental change in variable u_1, whereas the remaining variables u_2 and u_3 were kept constant.

Rules for partial differentiation of vectors are similar to those in calculus for scalar functions. For example, if **A** and **B** are two vectors that are functions of the three independent variables u_1, u_2, and u_3, then

$$\frac{\partial}{\partial u_1}(\mathbf{A} \cdot \mathbf{B}) = \mathbf{A} \cdot \frac{\partial \mathbf{B}}{\partial u_1} + \frac{\partial \mathbf{A}}{\partial u_1} \cdot \mathbf{B}$$

and the higher order differentiations are also obtained using similar rules, thus,

$$\frac{\partial^2}{\partial u_1\, \partial u_2}(\mathbf{A} \cdot \mathbf{B}) = \frac{\partial}{\partial u_1}\left\{\frac{\partial}{\partial u_2}(\mathbf{A} \cdot \mathbf{B})\right\} = \frac{\partial}{\partial u_1}\left\{\mathbf{A} \cdot \frac{\partial \mathbf{B}}{\partial u_2} + \frac{\partial \mathbf{A}}{\partial u_2} \cdot \mathbf{B}\right\}$$

$$= \mathbf{A} \cdot \frac{\partial^2 \mathbf{B}}{\partial u_1\, \partial u_2} + \frac{\partial \mathbf{A}}{\partial u_1} \cdot \frac{\partial \mathbf{B}}{\partial u_2} + \frac{\partial \mathbf{A}}{\partial u_2} \cdot \frac{\partial \mathbf{B}}{\partial u_1} + \frac{\partial^2 \mathbf{A}}{\partial u_1\, \partial u_2} \cdot \mathbf{B}$$

The student should review the various differentiation rules as they will be handy in performing some important analyses in this chapter.

2.3 GRADIENT OF SCALAR FUNCTION

Besides the vector differentiation that we discussed in the previous section, there are other vector differential relations that we will use in obtaining the point (differential) forms of Maxwell's equations. The first of these vector operations is the gradient of a scalar function. To derive an expression for the gradient and also to investigate its various properties and its physical meaning, let us consider a scalar field such as the temperature of an extended region of space. The temperature at arbitrary point $P_1(x, y, z)$ is $T_1(x, y, z)$ and its value at a nearby point $P_2(x + \Delta x, y + \Delta y, z + \Delta z)$ is $T_2(x + \Delta x, y + \Delta y, z + \Delta z)$. We are interested in relating the temperature T_2 to T_1 assuming that the temperature T together with its derivatives are continuous functions of the coordinates (x, y, z).

Because the location of P_2 is obtained by changing the coordinates of P_1 by incremental amounts Δx, Δy, and Δz, and because we assumed no discontinuities in the temperature distribution, T_2 may be related to T_1 using the Taylor series expansion. Hence,

$$T_2 = T_1 + \left.\frac{\partial T}{\partial x}\right|_{P_1} \Delta x + \left.\frac{\partial T}{\partial y}\right|_{P_1} \Delta y + \left.\frac{\partial T}{\partial z}\right|_{P_1} \Delta z + \frac{1}{2}\left.\frac{\partial^2 T}{\partial x^2}\right|_{P_1} \Delta x^2$$

$$+ \text{ similar higher-order differential terms}$$

The higher-order differential terms will all be neglected because they are multiplied by terms such as Δx^2, Δy^2, $\Delta x \Delta y$, which are negligibly small because of the small incremental changes in the coordinates Δx, Δy, Δz, and as we are interested in the limiting case $\Delta x \to 0$, $\Delta y \to 0$, and $\Delta z \to 0$.

The difference in temperature $\Delta T = T_2 - T_1$ is therefore given by

$$\Delta T = \left.\frac{\partial T}{\partial x}\right|_{P_1} \Delta x + \left.\frac{\partial T}{\partial y}\right|_{P_1} \Delta y + \left.\frac{\partial T}{\partial z}\right|_{P_1} \Delta z$$

In the limit when the incremental changes in the coordinates reduce to zero, the differential change in temperature dT is given by

$$\underset{\substack{\Delta x \to 0 \\ \Delta y \to 0 \\ \Delta z \to 0}}{\text{Lim}} \Delta T = dT = \frac{\partial T}{\partial x} dx + \frac{\partial T}{\partial y} dy + \frac{\partial T}{\partial z} dz$$

$$= \left(\frac{\partial T}{\partial x} \mathbf{a}_x + \frac{\partial T}{\partial y} \mathbf{a}_y + \frac{\partial T}{\partial z} \mathbf{a}_z \right) \cdot (dx\, \mathbf{a}_x + dy\, \mathbf{a}_y + dz\, \mathbf{a}_z)$$

It is known that the differential change in the position vector $d\mathbf{r}$ is $d\mathbf{r} = dx\, \mathbf{a}_x + dy\, \mathbf{a}_y + dz\, \mathbf{a}_z$. Also calling the first term in the dot product the gradient of the temperature or simply grad T, the differential change in temperature is then

$$dT = \text{grad } T \cdot d\mathbf{r} = \nabla T \cdot d\mathbf{r} \tag{2.1}$$

where the vector differential operator ∇ (del) is given by

$$\nabla = \frac{\partial}{\partial x} \mathbf{a}_x + \frac{\partial}{\partial y} \mathbf{a}_y + \frac{\partial}{\partial z} \mathbf{a}_z \tag{2.2}$$

The substitution of the vector differential expression by the del operator is merely a shorthand notation introduced in the Cartesian coordinate system for convenience.

The expression for the gradient in other coordinate systems may be obtained by following similar analysis in the general orthogonal curvilinear coordinate system.

In this coordinate system, the three independent variables are u_1, u_2, and u_3. In general, these independent variables are not all elements of length such as x, y, and z in the Cartesian coordinates, ρ and z in the cylindrical, and r in the spherical coordinate systems. They may be angles such as θ and ϕ in the spherical coordinates. Hence, to make sure that we will have elements of length when we change the u_1, u_2, and u_3 independent variables, we multiply each by a "metric coefficient," h_1, h_2, and h_3. If the independent variable represents an element of length, the metric coefficient will be one. For cases where the independent variables do not represent elements of lengths, the metric coefficients will be conversion factors to convert the change in the independent variable to an element of length. In general, therefore, $h_1 u_1$, $h_2 u_2$, and $h_3 u_3$ represent three elements of length in the curvilinear coordinate system.

With this in mind, let us go back to our temperature scalar function T. T is a function of u_1, u_2, and u_3. We use the Taylor series to relate T_2 at a point P_2 to T_1 at a nearby point P_1, hence

$$T_2 = T_1 + \frac{\partial T}{\partial u_1} \Delta u_1 + \frac{\partial T}{\partial u_2} \Delta u_2 + \frac{\partial T}{\partial u_3} \Delta u_3 + \text{higher-order terms that will be neglected.}$$

$$\Delta T = T_2 - T_1 = \frac{\partial T}{\partial u_1} \Delta u_1 + \frac{\partial T}{\partial u_2} \Delta u_2 + \frac{\partial T}{\partial u_3} \Delta u_3$$

In the limit, as Δu_1, Δu_2, and Δu_3 go to zero, we obtain

$$\underset{\substack{\Delta u_1 \to 0 \\ \Delta u_2 \to 0 \\ \Delta u_3 \to 0}}{\text{Lim}} \Delta T = dT = \frac{\partial T}{\partial u_1} du_1 + \frac{\partial T}{\partial u_2} du_2 + \frac{\partial T}{\partial u_3} du_3$$

$$= \frac{1}{h_1} \frac{\partial T}{\partial u_1} (h_1 \, du_1) + \frac{1}{h_2} \frac{\partial T}{\partial u_2} (h_2 \, du_2) + \frac{1}{h_3} \frac{\partial T}{\partial u_3} (h_3 \, du_3)$$

$$= \left(\frac{1}{h_1} \frac{\partial T}{\partial u_1} \mathbf{a}_1 + \frac{1}{h_2} \frac{\partial T}{\partial u_2} \mathbf{a}_2 + \frac{1}{h_3} \frac{\partial T}{\partial u_3} \mathbf{a}_3 \right) \cdot (h_1 \, du_1 \, \mathbf{a}_1 + h_2 \, du_2 \, \mathbf{a}_2 + h_3 \, du_3 \, \mathbf{a}_3)$$

$$= \operatorname{grad} T \cdot d\mathbf{r}$$

where the differential distance $d\mathbf{r}$ is given by

$$d\mathbf{r} = d\ell_1 \mathbf{a}_1 + d\ell_2 \mathbf{a}_2 + d\ell_3 \mathbf{a}_3 = h_1 \, du_1 \, \mathbf{a}_1 + h_2 \, du_2 \, \mathbf{a}_2 + h_3 \, du_3 \, \mathbf{a}_3$$

The gradient of T in the curvilinear coordinate system is given by

$$\operatorname{grad} T = \frac{1}{h_1} \frac{\partial T}{\partial u_1} \mathbf{a}_1 + \frac{1}{h_2} \frac{\partial T}{\partial u_2} \mathbf{a}_2 + \frac{1}{h_3} \frac{\partial T}{\partial u_3} \mathbf{a}_3$$

where h_1, h_2, h_3 are the metric coefficients, u_1, u_2, u_3 are the independent variables, and $\mathbf{a}_1, \mathbf{a}_2, \mathbf{a}_3$ are the base vectors of the curvilinear coordinate system. Although the del operator ∇ was formerly introduced as a shorthand notation in the Cartesian coordinate system, it became customary to use grad T and ∇T interchangeably in all coordinate systems.

Considering the specific values of (u_1, u_2, u_3), (h_1, h_2, h_3), and $(\mathbf{a}_1, \mathbf{a}_2, \mathbf{a}_3)$ given in Appendix A.2, we obtain explicit values of the gradient in the various coordinate systems. Specifically, in the cylindrical coordinates we have

$$\nabla T = \frac{\partial T}{\partial \rho} \mathbf{a}_\rho + \frac{1}{\rho} \frac{\partial T}{\partial \phi} \mathbf{a}_\phi + \frac{\partial T}{\partial z} \mathbf{a}_z$$

and in the spherical coordinates

$$\nabla T = \frac{\partial T}{\partial r} \mathbf{a}_r + \frac{1}{r} \frac{\partial T}{\partial \theta} \mathbf{a}_\theta + \frac{1}{r \sin \theta} \frac{\partial T}{\partial \phi} \mathbf{a}_\phi$$

An easy way to remember these expressions is to note that the denominator of each term has the form of one of the components of $d\ell$ in that coordinate system, except that partial differentials replace ordinary differentials. For example, $\rho \, d\phi$ becomes $\rho \, \partial \phi$ in the cylindrical coordinate system and $r \sin \theta \, d\phi$ becomes $r \sin \theta \, \partial \phi$ in the spherical coordinate system. The gradient of a scalar field such as ∇T has some important properties that we are going to discuss next.

EXAMPLE 2.3

Show that if the points P_1 and P_2 are both on a surface of constant temperature, then ∇T is a vector perpendicular to that surface.

Solution

Going back to the earlier analysis in this section, and noting that the difference in temperature dT between P_1 and P_2 in a surface of constant temperature is zero, then

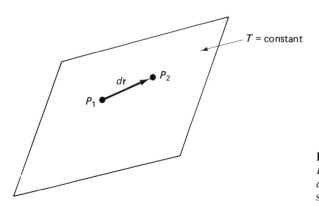

Figure 2.3 The two points P_1 and P_2 are on the T = constant surface. $d\mathbf{r}$ is therefore tangential to the surface.

$$0 = \frac{\partial T}{\partial x}dx + \frac{\partial T}{\partial y}dy + \frac{\partial T}{\partial z}dz$$

$$= \nabla T \cdot d\mathbf{r}$$

(2.3)

The vector $d\mathbf{r}$ is a differential position vector between P_1 and P_2. It therefore lies on the T = constant surface as shown in Figure 2.3. The zero dot product in equation 2.3 means that the vectors ∇T and $d\mathbf{r}$ are perpendicular to each other. The gradient of a scalar quantity is, therefore, a vector perpendicular to $d\mathbf{r}$ and consequently perpendicular to the surface T = constant. This is a general characteristic of the gradient and is often used to find a unit normal to a surface.

EXAMPLE 2.4

Find the angle between the surfaces $x^2 + y^2 + z^2 = 9$ and $z = x^2 + y^2 - 3$ at the point $(2, -1, 2)$.

Solution

The angle between the surfaces at a point is equal to the angle between the normals to the surfaces at that point.

A normal to the surface $s_1 = x^2 + y^2 + z^2 - 9$ at a point $(2, -1, 2)$ is given by

$$\nabla s_1 = \nabla(x^2 + y^2 + z^2 - 19) = 2x\,\mathbf{a}_x + 2y\,\mathbf{a}_y + 2z\,\mathbf{a}_z$$

$$= 4\,\mathbf{a}_x - 2\,\mathbf{a}_y + 4\,\mathbf{a}_z$$

A normal to $s_2 = x^2 + y^2 - z - 3$ at $(2, -1, 2)$ is

$$\nabla s_2 = 2x\,\mathbf{a}_x + 2y\,\mathbf{a}_y - \mathbf{a}_z = 4\,\mathbf{a}_x - 2\,\mathbf{a}_y - \mathbf{a}_z$$

$$(\nabla s_1) \cdot (\nabla s_2) = |\nabla s_1|\,|\nabla s_2|\cos\alpha,$$

where α is the angle between the normals to the two surfaces

$$(\nabla s_1) \cdot (\nabla s_2) = 16 + 4 - 4 = 16$$

$$|\nabla s_1| = \sqrt{4^2 + (-2)^2 + 4^2} = 6$$

$$|\nabla s_2| = \sqrt{4^2 + (-2)^2 + (-1)^2} = \sqrt{21}$$

$$\therefore \cos \alpha = \frac{16}{6\sqrt{21}} = 0.5819 \text{ or } \alpha \approx 54°$$

————◆◆————

EXAMPLE 2.5

If $P_1(x, y, z)$ and $P_2(x + \Delta x, y + \Delta y, z + \Delta z)$ are two nearby points on two surfaces of different temperatures T_1 and T_2 as shown in Figure 2.4 and the temperature on the first surface $T_1 = T(x, y, z)$, whereas the temperature T_2 on the other surface is $T_2 = (x + \Delta x, y + \Delta y, z + \Delta z)$.

1. Interpret physically the quantity $\Delta T/\Delta \ell$ where $\Delta \ell$ is the distance between P_1 and P_2.
2. Evaluate $\Delta T/\Delta \ell$ in the limiting case as $\Delta \ell \to 0$, and show that $dT/d\ell = \nabla T \cdot \mathbf{n}$ where $\mathbf{n}$ is a unit vector in the direction between P_1 and P_2.

Solution

1. Because ΔT is the change in temperature between the points P_1 and P_2, and $\Delta \ell$ is the distance between these points, then $\Delta T/\Delta \ell$ represents the average rate of change in temperature per unit length in the direction from P_1 to P_2.
2. From the earlier analysis in this section, the change in temperature ΔT is given by

$$\Delta T = \frac{\partial T}{\partial x} \Delta x + \frac{\partial T}{\partial y} \Delta y + \frac{\partial T}{\partial z} \Delta z$$

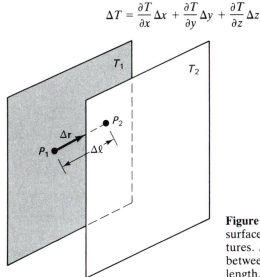

Figure 2.4   Points P_1 and P_2 are on surfaces of two different temperatures. $\Delta \mathbf{r}$ is an incremental vector between P_1 and P_2 and $\Delta \ell$ is its length.

The limiting value of the average change in temperature per unit length ($\Delta T/\Delta\ell$) as $\Delta\ell \to 0$ is known as the differential rate of change in temperature ($dT/d\ell$) which is given by

$$\lim_{\Delta\ell \to 0} \frac{\Delta T}{\Delta\ell} = \frac{dT}{d\ell} = \frac{\partial T}{\partial x}\frac{dx}{d\ell} + \frac{\partial T}{\partial y}\frac{dy}{d\ell} + \frac{\partial T}{\partial z}\frac{dz}{d\ell}$$

$$= \left(\frac{\partial T}{\partial x}\mathbf{a}_x + \frac{\partial T}{\partial y}\mathbf{a}_y + \frac{\partial T}{\partial z}\mathbf{a}_z\right) \cdot \left(\frac{dx}{d\ell}\mathbf{a}_x + \frac{dy}{d\ell}\mathbf{a}_y + \frac{dz}{d\ell}\mathbf{a}_z\right)$$

$$= \nabla T \cdot \frac{d\mathbf{r}}{d\ell}$$

But $d\mathbf{r}/d\ell$ is simply a unit vector in the direction of $d\mathbf{r}$ from P_1 to P_2. The rate of change of temperature with distance is, hence, the component of the gradient of T (∇T) in the direction of $\mathbf{n}$, which is a unit vector from P_1 to P_2.

EXAMPLE 2.6

Show that the maximum rate of change of T is in the direction of ∇T and that this maximum rate of change has the magnitude of ∇T.

Solution

From the previous example, we showed that the rate of temperature change $dT/d\ell = \nabla T \cdot \mathbf{n}$, which represents the projection of ∇T in the direction of $\mathbf{n}$. The dot product $\nabla T \cdot \mathbf{n}$ may take the form

$$\nabla T \cdot \mathbf{n} = |\nabla T||1| \cos\alpha$$

where α is the angle between the vector ∇T and the unit vector $\mathbf{n}$. The maximum rate of temperature change, therefore, occurs when the angle α is zero ($\cos\alpha = 1$) and this corresponds to the case when ∇T and $\mathbf{n}$ are in the same direction. Because ∇T is a vector normal to the surface T = constant (example 2.3), however, the maximum rate of change occurs along the direction of the normal to the surface T = constant. In this case when $\alpha = 0$, $|dT/d\ell| = |\nabla T|$. This simply means that the maximum rate of temperature change is equal to the magnitude of ∇T.

All of the preceding properties that are described in terms of the temperature variation are applicable to any scalar field such as the electric potential Φ, as will be described in chapter 4.

The following summarizes the general properties of the gradient of a scalar field:

1. *The gradient of a scalar field is a vector perpendicular to the surface characterized by a constant value of the scalar field.*
2. *The vector gradient of a scalar function has a magnitude equal to the maximum rate of change of the scalar function. The gradient of a scalar field is a vector in the direction of the maximum rate of change of the scalar field with respect to distance.*

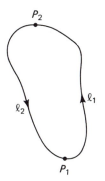

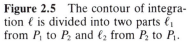

Figure 2.5 The contour of integration ℓ is divided into two parts ℓ_1 from P_1 to P_2 and ℓ_2 from P_2 to P_1.

Finally, it is important to note the *conservative property* of the gradient. The line integral of the gradient of any scalar function Φ over any closed path in space is zero, hence,

$$\oint_\ell \nabla\Phi \cdot d\mathbf{r} = 0$$

This property may be proved by noting that the differential change of the scalar function $d\Phi = \nabla\Phi \cdot d\mathbf{r}$ (equation 2.1). Hence,

$$\oint_\ell \nabla\Phi \cdot d\mathbf{r} = \oint_\ell d\Phi \qquad (2.4)$$

If the contour of integration is broken into two parts ℓ_1 and ℓ_2 as shown in Figure 2.5, equation 2.4 becomes

$$\oint_\ell d\Phi = \int_{\ell_1} d\Phi + \int_{\ell_2} d\Phi = \int_{P_1}^{P_2} d\Phi + \int_{P_2}^{P_1} d\Phi = (\Phi_{P_1} - \Phi_{P_2}) + (\Phi_{P_2} - \Phi_{P_1}) = 0$$

We can therefore conclude that any vector field that may be expressed as the gradient of a scalar function is a conservative field.

EXAMPLE 2.7

Consider a scalar, time-independent temperature field given in a region of space by

$$T = 3x + 2xyz - z^2 - 2$$

1. Determine a unit vector normal to the isotherms (constant temperature surfaces) at the point $(0, 1, 2)$.
2. Find the maximum rate of change of temperature at the same point.

Solution

1. The gradient of T (grad T) is a vector perpendicular to the isotherms. A unit vector in this direction is given by

$$
\mathbf{n} = \frac{\boldsymbol{\nabla} T}{|\boldsymbol{\nabla} T|} = \frac{\dfrac{\partial T}{\partial x}\,\mathbf{a}_x + \dfrac{\partial T}{\partial y}\,\mathbf{a}_y + \dfrac{\partial T}{\partial z}\,\mathbf{a}_z}{|\boldsymbol{\nabla} T|}
$$

$$
= \frac{(3 + 2yz)\,\mathbf{a}_x + 2xz\,\mathbf{a}_y + (2xy - 2z)\,\mathbf{a}_z}{|\boldsymbol{\nabla} T|}
$$

$$
\mathbf{n}\big|_{(0,1,2)} = \frac{7\,\mathbf{a}_x - 4\,\mathbf{a}_z}{\sqrt{65}} = \frac{7}{\sqrt{65}}\,\mathbf{a}_x - \frac{4}{\sqrt{65}}\,\mathbf{a}_z
$$

2. The maximum rate of temperature change equals the magnitude of $\boldsymbol{\nabla} T$. Maximum rate of temperate change at $(0, 1, 2)$ is therefore $\sqrt{65}$.

2.4 DIVERGENCE OF VECTOR FIELD

In chapter 1 we described the flux representation of a vector field. We also used the expression $\oint_s (\epsilon_o \mathbf{E}) \cdot ds$ to indicate the total electric flux of $(\epsilon_o \mathbf{E})$ emanating or crossing a closed surface s. The dot product in this expression emphasizes the fact that the integration is being carried out over the component of the vector field $(\epsilon_o \mathbf{E})$ normal to the closed surface s. In Gauss's law for the magnetic field, the zero surface integral $\oint_s \mathbf{B} \cdot ds = 0$ simply means that the total magnetic flux emanating from a closed surface s is equal to zero. Figure 2.6 illustrates some limiting cases in which $\oint_s \mathbf{F} \cdot ds$ is (1) zero, (2) positive, and (3) negative. For case 2, the number of flux lines of the vector $\mathbf{F}$ flowing out of the closed surface is larger than the number entering the surface and, therefore, it is assumed that a source is enclosed by the surface s. In case 3, the number of flux

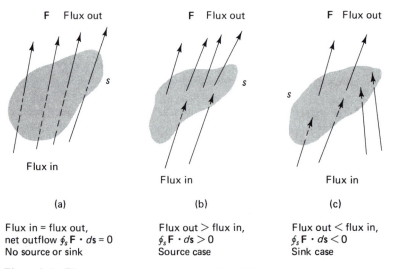

(a)	(b)	(c)
Flux in = flux out, net outflow $\oint_s \mathbf{F} \cdot d\mathbf{s} = 0$ No source or sink	Flux out > flux in, $\oint_s \mathbf{F} \cdot d\mathbf{s} > 0$ Source case	Flux out < flux in, $\oint_s \mathbf{F} \cdot d\mathbf{s} < 0$ Sink case

Figure 2.6 Flux representation of a vector $\mathbf{F}$ and illustration of the relation between source and sink cases to the closed surface integral $\oint_s \mathbf{F} \cdot d\mathbf{s}$.

lines entering the closed surface is larger than the number emanating and therefore a sink or a drain is assumed to be within the surface s.

In summary, the two Gauss's laws involve the evaluation of the net outflow of the electric and magnetic flux lines from closed surfaces. To help us evaluate this important vector property without having to carry out the integration over a specific closed surface, we define a point relation (i.e., a relation that is valid at a point) known as the divergence of a vector field. The divergence of a vector $\mathbf{F}$ is a measure of the outflow of flux from a small closed surface per unit volume as the volume shrinks to zero. A mathematical expression describing this operation is given by

$$\text{Divergence of } \mathbf{F} = \underset{\Delta v \to 0}{\text{Lim}} \frac{\oint_{\Delta s} \mathbf{F} \cdot d\mathbf{s}}{\Delta v} \tag{2.5}$$

The reason for carrying out the integration over a small closed surface Δs is to ensure that the integration would be independent of the specific shape of the surface in the limiting case as $\Delta v \to 0$. Dividing the closed surface integral by the volume Δv is introduced to make the resulting expression independent of the value of the volume Δv enclosed by the incremental closed surface Δs. As we take the limit $\Delta v \to 0$, the resulting expression for the divergence of $\mathbf{F}$ becomes a point relation describing the net outflow of the flux $\mathbf{F}$ at an infinitesimally small volume that diminishes to a point at its limit. In other words, the divergence of a vector field is a point relation that describes the net outflow of the vector flux from a closed surface. This relation, however, is independent of any specific shape of the surface or a specific value of the volume enclosed by the surface. Needless to say, the development of a mathematical expression for the divergence of a vector field will lead us to transforming Gauss's laws to their point or differential forms, which is one of our objectives in this chapter.

From equation 2.5, it can be seen that the derivation of an expression for the divergence of a vector field $\mathbf{F}$ involves the following steps: (1) identify an incremental closed surface Δs that encloses a small volume Δv, (2) calculate the total flux of the vector $\mathbf{F}$ that emanates from the closed surface Δs, (3) divide the obtained result by the element of volume Δv enclosed by Δs, and (4) take the limiting value of the result in step 3 as $\Delta v \to 0$. In the following discussion, we will carry out these outlined calculations in the Cartesian coordinate system.

Consider a vector field $\mathbf{F}$ given in terms of its components in the Cartesian coordinate system by $\mathbf{F} = F_x \mathbf{a}_x + F_y \mathbf{a}_y + F_z \mathbf{a}_z$. Let us calculate the net outflow of the flux of $\mathbf{F}$ from the incremental closed surface Δs enclosing the volume $\Delta v = \Delta x \, \Delta y \, \Delta z$, shown in Figure 2.7. The closed surface Δs consists of six surfaces and the closed surface integral must be expressed as the sum of these integrals, hence,

$$\oint_{\Delta s} \mathbf{F} \cdot d\mathbf{s} = \int_{\text{front}} \mathbf{F} \cdot d\mathbf{s} + \int_{\text{back}} \mathbf{F} \cdot d\mathbf{s} + \int_{\text{right}} \mathbf{F} \cdot d\mathbf{s} + \int_{\text{left}} \mathbf{F} \cdot d\mathbf{s} + \int_{\text{top}} \mathbf{F} \cdot d\mathbf{s} + \int_{\text{bottom}} \mathbf{F} \cdot d\mathbf{s}$$

Because of the dot product $\mathbf{F} \cdot d\mathbf{s}$, it is to be noted that only the component of $\mathbf{F}$ that is normal to the surface, that is, in the direction of $d\mathbf{s}$ of this particular surface, contributes to the surface integral. Every one of the six surfaces has a different unit normal pointing out of the closed surface, and thus different components of $\mathbf{F}$ would

$$\mathbf{F} = F_x \mathbf{a}_x + F_y \mathbf{a}_y + F_z \mathbf{a}_z$$

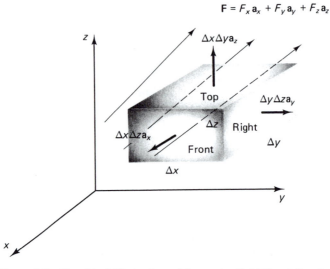

Figure 2.7   Graphical illustration of the vector field **F** and the element of area $\Delta\mathbf{s}$ enclosing the volume $\Delta v = \Delta x\, \Delta y\, \Delta z$.

contribute to each of the six surface integrals. Consider, for example, the contribution from the front and the back surfaces

$$\int_{\text{front}} \mathbf{F} \cdot d\mathbf{s} + \int_{\text{back}} \mathbf{F} \cdot d\mathbf{s} = \int_{\text{front}} \mathbf{F} \cdot \Delta y\, \Delta z\, (\mathbf{a}_x) + \int_{\text{back}} \mathbf{F} \cdot \Delta y\, \Delta z\, (-\mathbf{a}_x)$$

$$= F_{x_{\text{front}}} \Delta y\, \Delta z - F_{x_{\text{back}}} \Delta y\, \Delta z$$

where $F_{x_{\text{front}}}$ and $F_{x_{\text{back}}}$ are the x component of the vector field **F** evaluated at the front and back surfaces, respectively. Because we are considering an incremental element of volume, $F_{x_{\text{front}}}$ may be related to $F_{x_{\text{back}}}$ using Taylor expansion. With the field at the back surface being the reference, we obtain

$$F_{x_{\text{front}}} \approx F_{x_{\text{back}}} + \text{rate of change of } F_x \text{ with respect to } x \text{ multiplied by } \Delta x$$

$$+ \text{ higher-order terms that will be neglected, hence,}$$

$$F_{x_{\text{front}}} = F_{x_{\text{back}}} + \frac{\partial F_x}{\partial x} \Delta x$$

The contribution from the front and back surfaces is then

$$\int_{\text{front}} \mathbf{F} \cdot d\mathbf{s} + \int_{\text{back}} \mathbf{F} \cdot d\mathbf{s} \approx (F_{x_{\text{front}}} - F_{x_{\text{back}}}) \Delta y\, \Delta z$$

$$\approx \left(\frac{\partial F_x}{\partial x} \Delta x \right) \Delta y\, \Delta z$$

Following exactly the same procedure, we find that

$$\int_{\text{right}} \mathbf{F} \cdot d\mathbf{s} + \int_{\text{left}} \mathbf{F} \cdot d\mathbf{s} = \int_{\text{right}} \mathbf{F} \cdot \Delta x \, \Delta z \, \mathbf{a}_y + \int_{\text{left}} \mathbf{F} \cdot \Delta x \, \Delta y (-\mathbf{a}_y)$$

$$= (F_{y\,\text{right}} - F_{y\,\text{left}}) \, \Delta x \, \Delta z$$

$$\approx \frac{\partial F_y}{\partial y} \, \Delta y \, \Delta x \, \Delta z$$

and

$$\int_{\text{top}} \mathbf{F} \cdot d\mathbf{s} + \int_{\text{bottom}} \mathbf{F} \cdot d\mathbf{s} = \int_{\text{top}} \mathbf{F} \cdot \Delta x \, \Delta y \, \mathbf{a}_z + \int_{\text{bottom}} \mathbf{F} \cdot \Delta x \, \Delta y (-\mathbf{a}_z)$$

$$= (F_{z\,\text{top}} - F_{z\,\text{bottom}}) \, \Delta x \, \Delta y$$

$$\approx \frac{\partial F_z}{\partial z} \, \Delta z \, \Delta x \, \Delta y$$

These results are then combined to yield

$$\oint_{\Delta s} \mathbf{F} \cdot d\mathbf{s} \approx \left(\frac{\partial F_x}{\partial x} + \frac{\partial F_y}{\partial y} + \frac{\partial F_z}{\partial z} \right) \Delta x \, \Delta y \, \Delta z$$

$$\approx \left(\frac{\partial F_x}{\partial x} + \frac{\partial F_y}{\partial y} + \frac{\partial F_z}{\partial z} \right) \Delta v$$

If we now divide the closed surface integral by Δv, we obtain

$$\frac{\oint_{\Delta v} \mathbf{F} \cdot d\mathbf{s}}{\Delta v} \approx \frac{\partial F_x}{\partial x} + \frac{\partial F_y}{\partial y} + \frac{\partial F_z}{\partial z}$$

This approximate expression becomes better as Δv becomes smaller, and in the limit as Δv approaches zero, the expression becomes exact. Therefore

$$\lim_{\Delta v \to 0} \frac{\oint_{\Delta s} \mathbf{F} \cdot d\mathbf{s}}{\Delta v} = \frac{\partial F_x}{\partial x} + \frac{\partial F_y}{\partial y} + \frac{\partial F_z}{\partial z}$$

where the approximation has been replaced by an equals sign.

The divergence of the vector field $\mathbf{F}$ defined in equation 2.5 is then

$$\text{divergence of } \mathbf{F} = \frac{\partial F_x}{\partial x} + \frac{\partial F_y}{\partial y} + \frac{\partial F_z}{\partial z}$$

The divergence of $\mathbf{F}$ or, in short div $\mathbf{F}$, is symbolized in vector analysis by using the del operator in the form $\nabla \cdot \mathbf{F}$. ∇ is a vector differential operator given by $\nabla = \partial/\partial x \, \mathbf{a}_x + \partial/\partial y \, \mathbf{a}_y + \partial/\partial z \, \mathbf{a}_z$ in the Cartesian coordinate system. The dot product of ∇ with $\mathbf{F}$ provides the same result of div $\mathbf{F}$ as derived based on the calculation of the net outflow of the flux $\mathbf{F}$. Hence, the expression $\nabla \cdot \mathbf{F}$ is used as shorthand notation. The vector analysis symbolization $\nabla \cdot \mathbf{F}$ is, however, only a shorthand notation and the physical significance of such a mathematical operation is related to the value of the net flux of the vector $\mathbf{F}$ that outflows through a closed surface Δs per unit volume as $\Delta v \to 0$.

The dot product of the del operator given in equation 2.2 with the vector $\mathbf{F}$ in the Cartesian coordinate system gives

$$\nabla \cdot \mathbf{F} = \left(\frac{\partial}{\partial x} \mathbf{a}_x + \frac{\partial}{\partial y} \mathbf{a}_y + \frac{\partial}{\partial z} \mathbf{a}_z \right) \cdot (F_x \mathbf{a}_x + F_y \mathbf{a}_y + F_z \mathbf{a}_z)$$

$$= \frac{\partial F_x}{\partial x} + \frac{\partial F_y}{\partial y} + \frac{\partial F_z}{\partial z}$$

which is the expression for div $\mathbf{F}$ as obtained from the definition given in equation 2.5. The notations div $\mathbf{F}$ and $\nabla \cdot \mathbf{F}$ may, therefore, be used interchangeably.

Problem.   Starting with the element of volume in the generalized curvilinear coordinate system shown in Figure 2.8, where the independent variables are u_1, u_2, u_3 and the three orthogonal base vectors are $\mathbf{a}_1$, $\mathbf{a}_2$, $\mathbf{a}_3$, show that an expression for the div $\mathbf{F}$ is given by

$$\text{div } \mathbf{F} = \frac{1}{h_1 h_2 h_3} \left(\frac{\partial(F_1 h_2 h_3)}{\partial u_1} + \frac{\partial(F_2 h_1 h_3)}{\partial u_2} + \frac{\partial(F_3 h_1 h_2)}{\partial u_3} \right)$$

where h_1, h_2, and h_3 are the metric coefficients.
Also show that the expressions for div $\mathbf{F}$ in the cylindrical and spherical coordinates are given by

$$\text{div } \mathbf{F} = \frac{1}{\rho} \frac{\partial(F_\rho \rho)}{\partial \rho} + \frac{1}{\rho} \frac{\partial F_\phi}{\partial \phi} + \frac{\partial F_z}{\partial z} \qquad \text{(Cylindrical)}$$

$$\text{div } \mathbf{F} = \frac{1}{r^2} \frac{\partial(F_r r^2)}{\partial r} + \frac{1}{r \sin \theta} \frac{\partial(F_\theta \sin \theta)}{\partial \theta} + \frac{1}{r \sin \theta} \frac{\partial F_\phi}{\partial \phi} \qquad \text{(Spherical)}$$

It should be noted that the form of div $\mathbf{F}$ in the Cartesian coordinate system is the basis for introducing the notation $\nabla \cdot \mathbf{F}$ because it is possible to obtain div $\mathbf{F}$ from the dot product of ∇ and $\mathbf{F}$. Such a process, however, is not possible in other coordinate

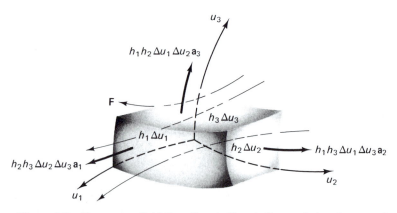

Figure 2.8   The vector field $\mathbf{F} = F_1 \mathbf{a}_1 + F_2 \mathbf{a}_2 + F_3 \mathbf{a}_3$ and the element of volume $\Delta v = h_1 h_2 h_3 \Delta u_1 \Delta u_2 \Delta u_3$ in the generalized curvilinear coordinate system.

systems. In other words, it is not possible to find an expression for ∇ in cylindrical or spherical coordinate systems such that when multiplied by $\mathbf{F}$ (dot product) will give the expression for div $\mathbf{F}$. The notation div $\mathbf{F}$ and $\nabla \cdot \mathbf{F}$ will, however, be used interchangeably just for convenience and regardless of the type of the coordinate system.

EXAMPLE 2.8

For each of the following vector fields, draw the flux representation and obtain an expression for the divergence. Use the flux representations to explain physically the results you obtained for the divergence of each vector field.

1. $\mathbf{A} = \mathbf{a}_y$
2. $\mathbf{B} = x\mathbf{a}_y$
3. $\mathbf{C} = x\mathbf{a}_x$
4. $\mathbf{D} = \dfrac{1}{\rho}\mathbf{a}_\rho$
5. $\mathbf{E} = \mathbf{a}_r$

Solution

The flux representation of each of these vectors are shown in Figure 2.9. The basic rules for such flux representation are given in example 1.11. As an example of such a procedure, we consider case 3 where it is clear that more flux lines are introduced as x increases because the magnitude of the vector field $\mathbf{C}$ is proportional to x. Also, the direction of $\mathbf{C}$ is in the positive x direction for positive values of x and in the negative x direction for negative values of x.

Let us now obtain expressions for the divergence of these vectors and try to explain the obtained results using the flux representations of Figure 2.9.

1. $\text{div } \mathbf{A} = \dfrac{\partial A_x}{\partial x} + \dfrac{\partial A_y}{\partial y} + \dfrac{\partial A_z}{\partial z}$

   $= 0$

Similarly,

2. $\text{div } \mathbf{B} = 0$
3. $\text{div } \mathbf{C} = 1$
4. $\text{div } \mathbf{D} = 0$
5. $\text{div } \mathbf{E} = \dfrac{1}{r^2}\dfrac{\partial}{\partial r}(r^2 E_r)$

   $= \dfrac{2}{r}$

The zero values of the divergences in cases 1, 2, and 4 may be easily explained by constructing small volumes, Δv, in each of the flux representations in Figure 2.9 and noting that the number of flux lines entering the element of volume is equal to the number emanating. In these cases, therefore, the total outflow of the flux lines from the element

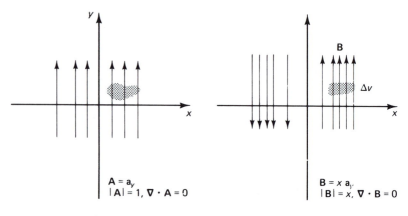

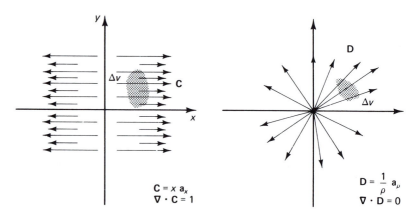

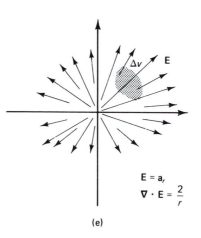

Figure 2.9 Flux representation of various vector fields. The direction of the flux lines is in the direction of the field, and the distance between the flux lines is proportional to the magnitude of the field.

of volume is equal to zero, that is, zero divergence and no source or sink is in the region of space where these vector fields are present. For the cases 3 and 5 where the divergences of the vector fields are not zero, it may be seen from Figures 2.9c and e that the number of flux lines outflowing from the element of volume is larger than the number inflowing. The net outflow of the flux lines is thus positive and this explains the reason for positive divergence in both cases.

The physical significance of the divergence of a vector field should by now be clear. It is a differential operator that provides the net outflow of the vector flux from a closed surface per unit volume as the volume enclosed by the surface shrinks to zero.

EXAMPLE 2.9

Apply Gauss's law for the electric field to a differential volume element and show that as the volume $\rightarrow 0$, Gauss's law becomes $\nabla \cdot \epsilon_o \mathbf{E} = \rho_v$ where ρ_v is the charge density per unit volume.

Solution

The integral form of Gauss's law for the electric field is given by

$$\oint_s \epsilon_o \mathbf{E} \cdot d\mathbf{s} = Q$$

where Q is the total charge enclosed by the surface s. If we now apply this law to a differential volume element of total surface Δs, which encloses a volume Δv, we obtain

$$\oint_{\Delta s} \epsilon_o \mathbf{E} \cdot d\mathbf{s} = Q$$

where Q now is the total charge enclosed by the element of volume Δv. Dividing by the element of volume, and taking the limit as $\Delta v \rightarrow 0$, we obtain

$$\underset{\Delta v \rightarrow 0}{\text{Lim}} \frac{\oint_{\Delta s} \epsilon_o \mathbf{E} \cdot d\mathbf{s}}{\Delta v} = \underset{\Delta v \rightarrow 0}{\text{Lim}} \frac{Q}{\Delta v} \qquad (2.6)$$

The limiting value of the left-hand side of equation 2.6 is simply $\nabla \cdot \epsilon_o \mathbf{E}$, whereas the limiting value of $Q/\Delta v$ as $\Delta v \rightarrow 0$ is simply the volume charge density ρ_v. Equation 2.6 then reduces to

$$\nabla \cdot \epsilon_o \mathbf{E} = \rho_v \qquad (2.7)$$

which is Gauss's law for the electric field in its differential form.

The point to be emphasized here is that although the differential form in equation 2.7 involves the divergence differential vector operator, it still maintains its same physical meaning as applied to a differential volume element as the volume $\rightarrow 0$. In other words, equation 2.7 still states that the electric flux per unit volume leaving a vanishingly small volume is equal to the volume charge density at a point surrounded by the vanishingly small element of volume. It is this characteristic of relating the flux leaving an infinitesimally small element of volume to the charge density at a point that led to identifying equation 2.7 as the point form of Gauss's law. It is also known as the differential form of Gauss's law because it involves the divergence differential operation over the flux density vector $\epsilon_o \mathbf{E}$.

EXAMPLE 2.10

If the vector $\mathbf{E}$ given by

$$\mathbf{E} = 3\rho \, \mathbf{a}_\rho + 6 \, \mathbf{a}_z$$

represents a static electric field, determine the charge density associated with this field.

Solution

From the differential form of Gauss's law

$$\nabla \cdot \epsilon_o \mathbf{E} = \rho_v$$

$$\therefore \frac{1}{\rho} \frac{\partial}{\partial \rho}(\rho E_\rho) + \frac{1}{\rho} \frac{\partial E_\phi}{\partial \phi} + \frac{\partial E_z}{\partial z} = \frac{\rho_v}{\epsilon_o}$$

Substituting the given E_ρ and E_z components of the electric field vector, we obtain

$$\rho_v = 6\epsilon_o$$

———◆◆◆———

EXAMPLE 2.11

In example 1.24, we obtained expressions for the electric field inside and outside a spherical charge distribution of constant charge density ρ_v and radius r_o. It was shown that

$$\mathbf{E} = \frac{\rho_v r_o^3}{3\epsilon_o r^2} \mathbf{a}_r \qquad \text{for } r > r_o$$

and

$$\mathbf{E} = \frac{\rho_v r}{3\epsilon_o} \mathbf{a}_r \qquad \text{for } r < r_o$$

Use these expressions in the different: ! form of Gauss's law to determine the charge density inside $r < r_o$ and outside $r > r_o$, :he spherical charge distribution.

Solution

The differential form of Gauss's law is given by

$$\nabla \cdot \epsilon_o \mathbf{E} = \rho_v$$

Substituting $\mathbf{E}$ for $r > r_o$, we obtain

$$\frac{1}{r^2} \frac{\partial}{\partial r}(\epsilon_o r^2 E_r) = \frac{1}{r^2} \frac{\partial}{\partial r}\left(\epsilon_o r^2 \frac{\rho_v r_o^3}{3\epsilon_o r^2}\right) = 0$$

which means that the charge density outside the spherical charge distribution $r > r_o$ is zero. Substituting $\mathbf{E}$ for $r < r_o$, conversely,

$$\frac{1}{r^2}\frac{\partial}{\partial r}(\epsilon_o r^2 E_r) = \frac{1}{r^2}\frac{\partial}{\partial r}\left(\frac{\rho_v}{3}r^3\right)$$

$$= \rho_v$$

which means that within the sphere $r < r_o$, we have a constant charge density distribution ρ_v. It is evident, therefore, that the charge distribution equals zero for $r > r_o$ and ρ_v for $r < r_o$. These are, of course, the same assumptions we started with when we obtained the various expressions for the electric field in example 1.24.

2.5 THE DIVERGENCE THEOREM

The divergence theorem relates the surface integral of the normal component of a vector field over a closed surface to the volume integral of the divergence of the vector throughout the volume enclosed by the surface. It states that

$$\oint_s \mathbf{F} \cdot d\mathbf{s} = \int_v \mathbf{\nabla} \cdot \mathbf{F} \, dv$$

This theorem applies to any vector that satisfies the restrictions required to obtain the divergence of the vector. It simply requires that the vector **F** together with its first partial derivatives be continuous throughout the volume v.

For cases where **F** or div **F** are not continuous, any singularities must be excluded from the region of integration, as will be illustrated by some examples later in this section.

Let us now consider a simple physical proof of the divergence theorem. The volume v shown in Figure 2.10a is enclosed by a surface s. If we divide the volume into a large number N of volume elements, and consider the flux diverging from each of these cells, we find that such a flux is entering the surfaces of the adjacent cells, except for the cells that contain portions of the outer surface. In the latter case, part of the flux emanating from the surface cells contributes to the flux emanating from the closed surface s. To quantify the relation between the flux diverging from each element of volume and the total flux diverging from the closed surface s, we consider two adjacent elements of volume, the ith and $i + 1$, elements shown in Figure 2.10b. From the definition of the divergence we note that the total flux emanating from a sufficiently small ith element is given (see equation 2.5) by

$$\oint_{\Delta s_i} \mathbf{F} \cdot d\mathbf{s} = (\text{div } \mathbf{F})_i \, \Delta v_i$$

where Δs_i is the surface enclosing the ith element of volume Δv_i.

The total flux emanating from the overall closed surface is simply the sum of the contributions from the subsurfaces. It is therefore equal to

$$\sum_{i=1}^{N} \oint_{\Delta s_i} \mathbf{F} \cdot d\mathbf{s} \qquad (2.8)$$

To evaluate this sum, we note that the flux emanating from the ith element through the common surface s_o is actually entering the $i + 1$ element through the same surface

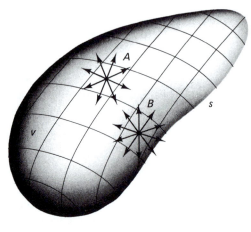

(a)

Figure 2.10a Geometry of a closed surface s enclosing the volume v. The volume was subdivided into a number of small elements of volumes. The flux diverging from cell A is totally entering the surfaces of the adjacent cells. Portions of the flux leaving the cell B that contains part of the external surface contributes to the total flux emanating from the closed surface s.

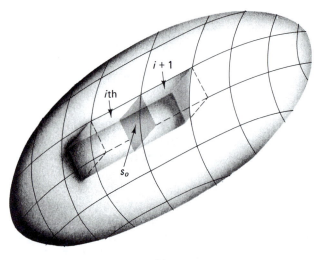

(b)

Figure 2.10b Two adjacent elements of volumes with a common surface s_o.

s_o. Applying the same principle to all the elements of volume, we find that the evaluation of the sum in equation 2.8 will contain contributions only from the outer surface s enclosing the volume v. Therefore,

$$\sum_{i=1}^{N} \oint_{\Delta s_i} \mathbf{F} \cdot d\mathbf{s} = \oint_s \mathbf{F} \cdot d\mathbf{s} \tag{2.9}$$

The sum of $(\operatorname{div}\mathbf{F})\,\Delta v_i$, conversely, over all the volume elements, as these volumes vanish to zero, simply results in integrating div $\mathbf{F}$ over the volume. Hence,

$$\operatorname*{Lim}_{\Delta v_i \to 0} \sum_{i=1}^{N} (\operatorname{div}\mathbf{F})\,\Delta v_i = \int_v \operatorname{div}\mathbf{F}\,dv \tag{2.10}$$

From equations 2.9 and 2.10, we obtain the following statement of the divergence theorem:

$$\oint_s \mathbf{F} \cdot d\mathbf{s} = \int_v \text{div} \, \mathbf{F} \, dv$$

This theorem is important because it relates the triple integration throughout a volume to a double integration over the surface enclosing the volume. It will also be used to derive the differential forms of Gauss's laws. Before we do this, however, let us emphasize some important features of the divergence theorem by solving the following example.

EXAMPLE 2.12

Illustrate the validity of the divergence theorem for the field $\mathbf{E} = k/\rho \, \mathbf{a}_\rho$ where k is a constant, by carrying out the integrations over a cylindrical volume of radius r_o and length ℓ as shown in Figure 2.11a and b.

Solution

The divergence theorem states that

$$\oint_s \mathbf{E} \cdot d\mathbf{s} = \int_v (\text{div} \, \mathbf{E}) \, dv$$

To illustrate its validity in the given geometry, let us start by calculating $\text{div} \, \mathbf{E}$.

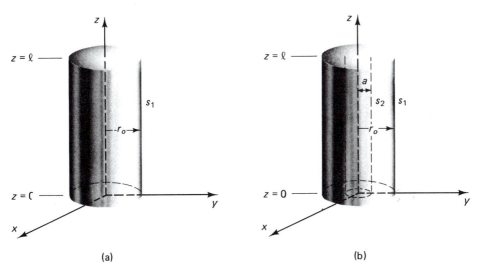

(a)   (b)

Figure 2.11a   The geometry of the cylindrical volume of example 2.12.
Figure 2.11b   The new geometry constructed to check the validity of the divergence theorem. It basically excludes the singularity at $\rho = 0$.

$$\mathbf{\nabla}\cdot\mathbf{E} = \frac{1}{\rho}\frac{\partial}{\partial\rho}[\rho E_\rho]$$

$$= \frac{1}{\rho}\frac{\partial}{\partial\rho}\left(\rho\frac{k}{\rho}\right)$$

$$= 0$$

The right-hand side of the divergence theorem will, therefore, be equal to zero,

$$\int_v \mathbf{\nabla}\cdot\mathbf{E}\,dv = 0 \tag{2.11}$$

The surface integral, conversely, is given by

$$\oint_s \mathbf{E}\cdot d\mathbf{s} = \int_{s_1} \frac{k}{\rho}\,\mathbf{a}_\rho\cdot\rho d\phi dz\,\mathbf{a}_\rho$$

where s_1 is the cylindrical (curved) surface, because the contributions from the top $z = \ell$ and the bottom $z = 0$ surfaces are zero because $\mathbf{E}$ is only in the $\mathbf{a}_\rho$ direction. Carrying out the surface integral we obtain

$$\oint_s \mathbf{E}\cdot d\mathbf{s} = \int_{s_1} kd\phi dz = 2\pi\ell k \tag{2.12}$$

Obviously equations 2.11 and 2.12 are not equal, and that is simply because $\mathbf{E}$ has a singularity (infinite value) at $\rho = 0$. In other words, the vector field $\mathbf{E}$ does not satisfy the conditions required for the validity of the divergence theorem throughout the volume v.

To overcome this singularity problem, we must exclude the discontinuity or the singularity at $\rho = 0$. This can be done by changing the region of integration to that bounded between the two cylindrical surfaces of radii a and r_o as shown in Figure 2.11b. In this case, the value of $\int_v \mathbf{\nabla}\cdot\mathbf{E}\,dv$ will still be zero, as given before, but the surface integral will include contributions from two cylindrical surfaces s_1 and s_2 at $\rho = r_o$ and $\rho = a$, respectively. Hence,

$$\oint_s \mathbf{E}\cdot d\mathbf{s} = \int_{s_1} \frac{k}{\rho}\,\mathbf{a}_\rho\cdot\rho d\phi dz\,\mathbf{a}_\rho + \int_{s_2} \frac{k}{\rho}\,\mathbf{a}_\rho\cdot\rho d\phi dz\,(-\mathbf{a}_\rho)$$

$$= \int_o^\ell\int_o^{2\pi} kd\phi dz - \int_o^\ell\int_o^{2\pi} kd\phi dz = 0$$

The divergence theorem is therefore valid if the discontinuity at $\rho = 0$ is excluded from the integration region.

After illustrating the required conditions for the divergence theorem to be valid, let us now use this theorem to obtain differential relations for two of Maxwell's equations.

2.6 DIFFERENTIAL EXPRESSIONS OF MAXWELL'S DIVERGENCE RELATIONS

The two Maxwell's divergence relations are Gauss's laws for the electric and magnetic fields. The integral forms of these relations are given by

$$\oint_s \epsilon_o \mathbf{E} \cdot d\mathbf{s} = \int_v \rho_v \, dv \tag{2.13}$$

and

$$\oint_s \mathbf{B} \cdot d\mathbf{s} = 0 \tag{2.14}$$

Our objective is to use the divergence theorem to obtain the differential or point relations of these equations. Let us apply the divergence theorem first to Gauss's law for the electric field

$$\oint_s \epsilon_o \mathbf{E} \cdot d\mathbf{s} = \int_v \mathbf{\nabla} \cdot (\epsilon_o \mathbf{E}) \, dv \tag{2.15}$$

where the volume v is enclosed by the surface s. Combining equations 2.13 and 2.15, we obtain

$$\int_v \mathbf{\nabla} \cdot (\epsilon_o \mathbf{E}) \, dv = \int_v \rho_v \, dv \tag{2.16}$$

If we carry out the integrations in equation 2.16 over an infinitesimally small volume, the integrands may be assumed constant within the small element of volume, hence,

$$\int_{\Delta v} \mathbf{\nabla} \cdot (\epsilon_o \mathbf{E}) \, dv = [\mathbf{\nabla} \cdot (\epsilon_o \mathbf{E})] \, \Delta v$$

$$= \rho_v \, \Delta v \tag{2.17}$$

Eliminating an equal element of volume from both sides of equation 2.17, we obtain

$$\boxed{\mathbf{\nabla} \cdot (\epsilon_o \mathbf{E}) = \rho_v} \tag{2.18}$$

which is the same point relation we obtained in example 2.9 based on the definition of the divergence of a vector field. Equation 2.18 relates the net outflow of the electric flux $\epsilon_o \mathbf{E}$ to the charge density at a point in space. We should recall that we obtained this relation in the limiting case when Gauss's surface was enclosing a vanishingly small element of volume, hence, the name point relation. It is also known as the differential form of Gauss's law for the electric field because it involves the divergence vector differential operator.

Following similar procedure in equation 2.14, it is easy to show that the differential form of Gauss's law for the magnetic field is given by

$$\boxed{\mathbf{\nabla} \cdot \mathbf{B} = 0} \tag{2.19}$$

Equation 2.19 shows that the magnetic flux has zero divergence, which implies that the magnetic flux lines are closed lines, and the net number of flux lines emanating from an element of volume is zero (i.e., number of flux lines entering = the number of lines leaving the element of volume). Comparing equations 2.18 and 2.19 shows that free magnetic charges are nonexistent simply because equation 2.19 with zero free magnetic charge density has to be satisfied at all points in space. The following examples will further illustrate the use of the differential forms of Gauss's laws of the electric and magnetic fields.

EXAMPLE 2.13

Show which of the following vector fields may represent a magnetic field:

1. $\mathbf{B} = \dfrac{1}{\rho}\mathbf{a}_\rho + \rho z\,\mathbf{a}_\phi + \cos\phi\,\mathbf{a}_z$

2. $\mathbf{B} = (x+2)\,\mathbf{a}_x + (1-3y)\,\mathbf{a}_y + 2z\,\mathbf{a}_z$

Solution

For a vector to represent a magnetic flux density it should satisfy Gauss's law for the magnetic field $\nabla \cdot \mathbf{B} = 0$. So we must carry out the divergence differential operation on each of the given vectors to see if it satisfies Gauss's law of equation 2.19.

1. In cylindrical coordinates

$$\nabla \cdot \mathbf{B} = \frac{1}{\rho}\frac{\partial}{\partial\rho}(\rho B_\rho) + \frac{1}{\rho}\frac{\partial B_\phi}{\partial\phi} + \frac{\partial B_z}{\partial z}$$

Substituting the B_ρ, B_ϕ, and B_z components of the given vector, we find $\nabla \cdot \mathbf{B} = 0$, which means the vector given in part 1 does satisfy Gauss's law and may therefore represent a magnetic field.

2. In Cartesian coordinates

$$\nabla \cdot \mathbf{B} = \frac{\partial B_x}{\partial x} + \frac{\partial B_y}{\partial y} + \frac{\partial B_z}{\partial z}$$

$$= 1 - 3 + 2 = 0$$

The vector $\mathbf{B}$ given in part 2 may also represent a magnetic flux density vector because it satisfies Gauss's law.

EXAMPLE 2.14

If the vector $\mathbf{E} = \cos\theta\,\mathbf{a}_r - \sin\theta\,\mathbf{a}_\theta + \cos\phi\,\mathbf{a}_\phi$ represents a static electric field, determine the charge density associated with this field in free space.

Solution

Because $\mathbf{E}$ represents static electric field, it satisfies Gauss's law:

$$\nabla \cdot \epsilon_o \mathbf{E} = \rho_v$$

In spherical coordinates,

$$\mathbf{V} \cdot \mathbf{E} = \frac{1}{r^2} \frac{\partial}{\partial r}(r^2 E_r) + \frac{1}{r \sin \theta} \frac{\partial}{\partial \theta}(E_\theta \sin \theta) + \frac{1}{r \sin \theta} \frac{\partial E_\phi}{\partial \phi} = \frac{\rho_v}{\epsilon_o}$$

Substituting $E_r = \cos \theta$, and $E_\theta = -\sin \theta$ and $E_\phi = \cos \phi$, we obtain

$$\mathbf{V} \cdot \mathbf{E} = \frac{2}{r} \cos \theta - \frac{2}{r} \cos \theta - \frac{\sin \phi}{r \sin \theta}$$

$$= \frac{\rho_v}{\epsilon_o}$$

$$\therefore \rho_v = \epsilon_o \left(-\frac{\sin \phi}{r \sin \theta} \right)$$

The value of the volume charge density at any point in space is obtained by substituting the (r, θ, ϕ) coordinates of that point.

EXAMPLE 2.15

A magnetic flux density **B** is described in the cylindrical coordinates in terms of its two components

$$B_z = B_o \rho^2 t \cos \frac{2\pi z}{L}$$

$$B_\phi = 0$$

and B_ρ is unknown. If we assume that there is no ϕ variation in the magnetic field, find the B_ρ component of this field.

Solution

We first try to find B_ρ component of the magnetic flux density using the fact that it should satisfy Gauss's law, $\mathbf{V} \cdot \mathbf{B} = 0$. Therefore, by letting $\partial B_\phi / \partial \phi = 0$, we obtain

$$\frac{1}{\rho} \frac{\partial}{\partial \rho}(\rho B_\rho) + \frac{\partial B_z}{\partial z} = 0$$

$$\frac{1}{\rho} \frac{\partial}{\partial \rho}(\rho B_\rho) = B_o \rho^2 t \frac{2\pi}{L} \sin \frac{2\pi z}{L}$$

$$\rho B_\rho = \frac{\rho^4}{4} B_o t \frac{2\pi}{L} \sin \frac{2\pi z}{L} + C$$

where C is the constant of integration,

$$B_\rho = \frac{\rho^3}{4} B_o t \frac{2\pi}{L} \sin \frac{2\pi z}{L} + \frac{C}{\rho}$$

but C must be zero so that B_ρ remains finite as $\rho \to 0$. Hence,

$$B_\rho = \frac{\rho^3}{4} B_o t \frac{2\pi}{L} \sin \frac{2\pi z}{L}$$

2.7 CURL OF VECTOR FIELD

Earlier in chapter 1 we discussed the physical meaning of a line integral of the form $\oint_c \mathbf{F} \cdot d\ell$. It was basically indicated that this form of a line integral integrates the *tangential* component of $\mathbf{F}$ along the closed contour c. If the vector $\mathbf{F}$ represents a force field, the line integral physically represents the work done by the force $\mathbf{F}$. If the line integral of the tangential component of $\mathbf{F}$ along the closed contour is zero, the field is called a conservative or irrotational field. If it is not zero, the field is described as having a circulation or rotation property. Ampere's and Faraday's laws involve line integrals of the form $\oint_c \mathbf{F} \cdot d\ell$ and it is helpful in obtaining the differential forms of these laws that we understand the relationship between the zero or nonzero value of the line integral and the circulation property of the field. To illustrate this, let us consider an example from the field of fluid dynamics in which the term "circulation" was actually first used. Figure 2.12 shows the water flow down in a river of width a and the velocity distribution of the water across the river $0 < y < a$ near the surface of the river. It can be seen that the water at the center of the river flows at maximum speed, whereas the flow velocity at the banks on both sides of the river is essentially zero. In an attempt to describe the nonuniformity of the water-velocity distribution across the river, we place a curl meter in the river as shown in Figure 2.13. This curl meter basically consists of a very small paddle wheel and a circular disk with a dot painted on it to show its rotation. If the curl meter is placed at the midpoint, $y = a/2$, near the surface of the river, it is clear from Figure 2.13a that the water velocities on both sides of the center line of the river are equal and, hence, the blades of the paddle wheel will be subjected to equal forces, thus resulting in a zero rotation of the curl meter as it travels down the stream. Conversely, if the curl meter is placed in the left side of the river (Figure 2.13b) the blades to the right would be subjected to a velocity larger than that at the location of the blades to the left. It is therefore expected that the curl meter would rotate in the counterclockwise direction as shown in Figure 2.13b. In other words, the direction of rotation would be pointing out of the paper that is exactly in the x direction as can be seen from Figure 2.12. Similarly, it can be shown that placing the curl meter in the

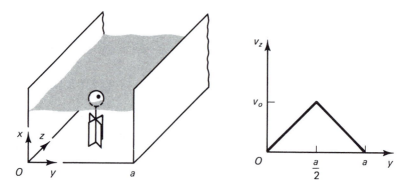

Figure 2.12   The velocity distribution in a tranverse section of the water flow near the surface of a river.

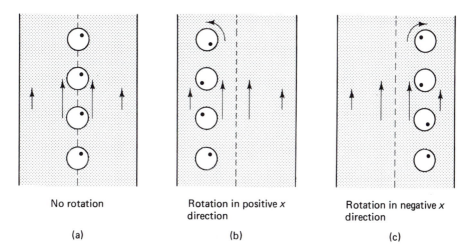

No rotation

(a)

Rotation in positive x
direction

(b)

Rotation in negative x
direction

(c)

Figure 2.13   Rotation of the curl meter indicating the nonuniformity of the flow velocity of the water near the surface of the river: (a) no rotation, (b) counterclockwise rotation, and (c) clockwise rotation.

right side of the river (Figure 2.13c) results in rotating the curl meter in the clockwise direction—that is, in the negative x direction as it moves down the stream in the river. In summary, the following facts should be observed from the curl meter example:

1. The curl meter will indicate rotation only if there is nonuniformity in the vector field. In other words, if the velocity distribution is uniform throughout the cross section of the river, the curl meter would not rotate at any location in the river. Some form of field nonuniformity is required for the rotation or circulation to occur.

2. The amount of rotation is proportional to the degree of nonuniformity of the vector field. The larger the difference between the velocities hitting the blades on both sides of the paddle wheel, the more the rotation would be.

3. The rotation of the curl meter cannot be described only in terms of the amount of rotation but should also be given a direction. The rotation is therefore a vector, and should be described in terms of its magnitude and direction.

The question now is: What is the relation between the rotation of the curl meter and the line integral of the form $\oint_c \mathbf{F} \cdot d\boldsymbol{\ell}$, which is part of Faraday's and Ampere's laws? In Figure 2.14, the vector velocity $\mathbf{v} = v_z \mathbf{a}_z$ is represented in terms of its flux. Integrating the velocity vector $\oint_{c_1} \mathbf{v} \cdot d\boldsymbol{\ell}$ along the closed contour c_1 will have only contributions from sides 1 and 2 of the contour where the velocity has a component along these sides. The line integral along c_1, however, will be zero because the contributions from these two sides are equal and in the opposite directions as shown in Figure 2.14. Having a zero value of the line integral $\oint_c \mathbf{v} \cdot d\boldsymbol{\ell}$ is therefore equivalent to having a zero rotation of the curl meter at the central location. The line integrals along contours c_2 and c_3, conversely, are not zero. The contributions are still due to sides 1 and 2, but these

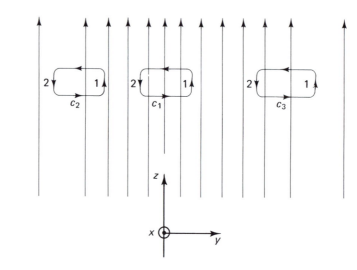

Figure 2.14   The flux representation of the vector velocity $\mathbf{v} = v_z \, \mathbf{a}_z$. The magnitude of the velocity is maximum at the center.

contributions are not equal because the magnitudes of the velocity at locations 1 and 2 are different. In particular, the velocity at location 1 in c_2 is larger than its magnitude at side 2, and similarly the magnitude of the velocity at side 2 in c_3 is larger than that at side 1 of the same contour. It may also be seen that because in the integration along c_2 the contribution from side 1 is larger than that of side 2, the net value of integration is such that a curl meter placed at this location will rotate in the positive x direction. Similarly a curl meter located at the center of c_3 will rotate in the negative x direction because the contribution from side 2 is larger than that of side 1.

The preceding discussion simply enforces the connection between the nonzero value of the line integral $\oint_c \mathbf{F} \cdot d\ell$ and the description that the vector $\mathbf{F}$ is rotational, or possessing circulation.

As we indicated earlier, the circulation of a vector field or the rotation of a curl meter is a vector, and should be described in terms of its direction in addition to its magnitude. Thus, for example, if we consider a longitudinal section of the water flow taken at the middle $y = a/2$ of the river, the curl meter would rotate in this case in the negative y direction because of the variation in the water velocity from maximum at the top to zero at the bottom as shown in Figure 2.15. The rotation of the curl meter in this case simply gives the y component of the circulation or the rotation property of the vector velocity field. Therefore, the *curl*, which is the mathematical description of the rotation or circulation of a vector, is a vector and any component of the curl is given in terms of a closed line integral of the vector about a small path in a plane normal to the direction of the desired component. In Cartesian coordinates, for example, the curl vector of a vector $\mathbf{F}$ is given by

$$\text{curl } \mathbf{F} = [\text{curl } \mathbf{F}]_x \, \mathbf{a}_x + [\text{curl } \mathbf{F}]_y \, \mathbf{a}_y + [\text{curl } \mathbf{F}]_z \, \mathbf{a}_z$$

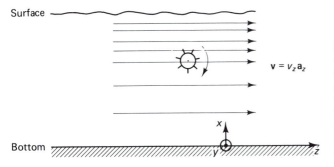

Figure 2.15 The velocity distribution of the water in a longitudinal section of a river. The orientation of the coordinate system was maintained the same as that of Figure 2.12. The rotation of the curl meter in this case gives the y component of the rotation of the velocity vector field.

where $[\text{curl}\,\mathbf{F}]_{x,y,z}$ are the x, y, and z components of the curl $\mathbf{F}$, respectively. Each of these components is calculated by integrating $\mathbf{F}$ along a closed contour surrounding an element of area whose direction is in the same direction in which the component of the curl is being calculated. For example, the x component of the curl is given by

$$[\text{curl}\,\mathbf{F}]_x = \lim_{\Delta y, \Delta z \to 0} \frac{\oint_{c_1} \mathbf{F} \cdot d\boldsymbol{\ell}}{\Delta y\, \Delta z}$$

The contour c_1 shown in Figure 2.16 surrounds the element of area $\Delta y\, \Delta z$, which has a unit vector in the $\mathbf{a}_x$ direction along which the curl component is being calculated. It should be emphasized that the mathematical derivation of the curl of a vector carries the same physical information regarding the vector field. It involves the line integral of the vector field that as indicated earlier emphasizes the presence of inhomogeneities or nonuniformities in the vector field. The importance of dividing the line integral by the enclosed area is to make the result independent of a specific value of the area enclosed. Taking the limit as the element of area goes to zero makes the obtained expression independent of the specific shape of the chosen contour as well as the value of the enclosed element of area.

    In summary, the curl of a vector is a point (i.e., in the limit at a point) form expression of the line integral of a vector field around a closed contour. Let us now

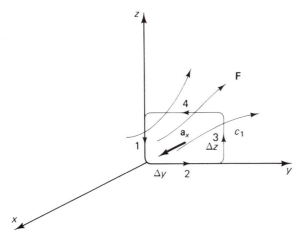

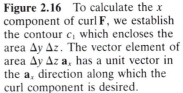

Figure 2.16 To calculate the x component of curl $\mathbf{F}$, we establish the contour c_1 which encloses the area $\Delta y\, \Delta z$. The vector element of area $\Delta y\, \Delta z\, \mathbf{a}_x$ has a unit vector in the $\mathbf{a}_x$ direction along which the curl component is desired.

develop a complete mathematical expression for the curl of a vector field in the Cartesian coordinate system. We assume the vector field $\mathbf{F}$ to be given at the origin of the Cartesian coordinate system by

$$\mathbf{F} = F_{x_o} \mathbf{a}_x + F_{y_o} \mathbf{a}_y + F_{z_o} \mathbf{a}_z \tag{2.20}$$

The x component of the curl is given by

$$[\operatorname{curl} \mathbf{F}]_x = \lim_{\Delta y, \Delta z \to 0} \frac{\oint_{c_1} \mathbf{F} \cdot d\boldsymbol{\ell}}{\Delta y \, \Delta z}$$

$$= \lim_{\Delta y, \Delta z \to 0} \frac{\int_1 \mathbf{F} \cdot -dz \, \mathbf{a}_z + \int_2 \mathbf{F} \cdot dy \, \mathbf{a}_y + \int_3 \mathbf{F} \cdot dz \, \mathbf{a}_z + \int_4 \mathbf{F} \cdot -dy \, \mathbf{a}_y}{\Delta y \, \Delta z}$$

The values of $\mathbf{F}$ along the sides 1 and 2 are given by the components in equation 2.20. Assuming the contour c to be infinitesimally small, the values of $\mathbf{F}$ along sides 3 and 4, conversely, may be obtained in terms of their values at the origin and the rate of their change with y and z using Taylor expansion. Thus,

$$F_z|_{\text{at } 3} = F_{z_o} + \frac{\partial F_z}{\partial y} \Delta y$$

$$F_y|_{\text{at } 4} = F_{y_o} + \frac{\partial F_y}{\partial z} \Delta z$$

The x component of the curl is then

$$[\operatorname{curl} \mathbf{F}]_x = \lim_{\Delta y, \Delta z \to 0} \frac{-\int_1 F_{z_o} \, dz + \int_2 F_{y_o} \, dy + \int_3 \left[F_{z_o} + \frac{\partial F_z}{\partial y} \Delta y \right] dz - \int_4 \left[F_{y_o} + \frac{\partial F_y}{\partial z} \Delta z \right] dy}{\Delta y \, \Delta z}$$

$$= \lim_{\Delta y, \Delta z \to 0} \frac{\int_3 \left(\frac{\partial F_z}{\partial y} \Delta y \right) dz - \int_4 \left(\frac{\partial F_y}{\partial z} \Delta z \right) dy}{\Delta y \, \Delta z}$$

If the value of the integral is evaluated approximately by multiplying the integrand by the length of the side of integration, we obtain

$$[\operatorname{curl} \mathbf{F}]_x = \lim_{\Delta y, \Delta z \to 0} \frac{\dfrac{\partial F_z}{\partial y} \Delta y \, \Delta z - \dfrac{\partial F_y}{\partial z} \Delta y \, \Delta z}{\Delta y \, \Delta z}$$

$$= \frac{\partial F_z}{\partial y} - \frac{\partial F_y}{\partial z}$$

Similarly, if we choose closed paths c_2 and c_3 in Figure 2.17, which are oriented perpendicular to the $\mathbf{a}_y$ and $\mathbf{a}_z$ unit vectors, and follow analogous procedures, we obtain the y and z components of the curl in the form

$$[\operatorname{curl} \mathbf{F}]_y = \frac{\partial F_x}{\partial z} - \frac{\partial F_z}{\partial x}$$

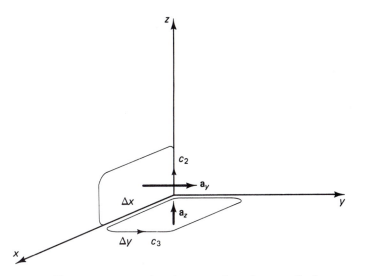

Figure 2.17   The contours c_2 and c_3 that are oriented perpendicular to $\mathbf{a}_y$ and $\mathbf{a}_z$ should be used to obtain the y and z components of [curl F], respectively.

$$[\operatorname{curl} \mathbf{F}]_z = \frac{\partial F_y}{\partial x} - \frac{\partial F_x}{\partial y}$$

An expression for the curl F is therefore

$$\operatorname{curl} \mathbf{F} = \left(\frac{\partial F_z}{\partial y} - \frac{\partial F_y}{\partial z}\right)\mathbf{a}_x + \left(\frac{\partial F_x}{\partial z} - \frac{\partial F_z}{\partial x}\right)\mathbf{a}_y + \left(\frac{\partial F_y}{\partial x} - \frac{\partial F_x}{\partial y}\right)\mathbf{a}_z$$

This result may be written in a determinant form which is easier to remember as

$$\operatorname{curl} \mathbf{F} = \begin{vmatrix} \mathbf{a}_x & \mathbf{a}_y & \mathbf{a}_z \\ \dfrac{\partial}{\partial x} & \dfrac{\partial}{\partial y} & \dfrac{\partial}{\partial z} \\ F_x & F_y & F_z \end{vmatrix}$$

The del ($\boldsymbol{\nabla}$) vector operator may once again be used in the Cartesian coordinates to express the curl in a form of a cross product, thus,

$$\operatorname{curl} \mathbf{F} = \boldsymbol{\nabla} \times \mathbf{F} = \begin{vmatrix} \mathbf{a}_x & \mathbf{a}_y & \mathbf{a}_z \\ \dfrac{\partial}{\partial x} & \dfrac{\partial}{\partial y} & \dfrac{\partial}{\partial z} \\ F_x & F_y & F_z \end{vmatrix}$$

Problem.   Let the vector **F** be given in terms of its components at the origin of the generalized curvilinear coordinate system by

$$\mathbf{F} = F_1\,\mathbf{a}_1 + F_2\,\mathbf{a}_2 + F_3\,\mathbf{a}_z$$

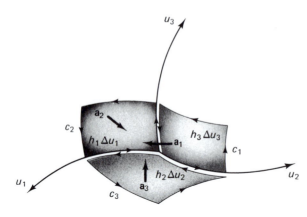

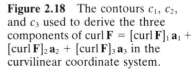

Figure 2.18 The contours c_1, c_2, and c_3 used to derive the three components of curl $\mathbf{F}$ = [curl $\mathbf{F}$]$_1$ $\mathbf{a}_1$ + [curl $\mathbf{F}$]$_2$ $\mathbf{a}_2$ + [curl $\mathbf{F}$]$_3$ $\mathbf{a}_3$ in the curvilinear coordinate system.

Kander 2.18

If we select the three contours c_1, c_2, c_3 shown in Figure 2.18, each surrounding an element of area that is perpendicular to the base vectors $\mathbf{a}_1, \mathbf{a}_2, \mathbf{a}_3$, and if we integrate $\mathbf{F}$ along each one of these contours to obtain the component of the curl in the direction of the unit vector perpendicular to the enclosed element of area, show that the expression of curl $\mathbf{F}$ in the generalized curvilinear coordinate system is given by

$$\text{curl } \mathbf{F} = \frac{1}{h_2 h_3}\left[\frac{\partial(F_3 h_3)}{\partial u_2} - \frac{\partial(F_2 h_2)}{\partial u_3}\right]\mathbf{a}_1$$

$$+ \frac{1}{h_1 h_3}\left[\frac{\partial(F_1 h_1)}{\partial u_3} - \frac{\partial(F_3 h_3)}{\partial u_1}\right]\mathbf{a}_2$$

$$+ \frac{1}{h_1 h_2}\left[\frac{\partial(F_2 h_2)}{\partial u_1} - \frac{\partial(F_1 h_1)}{\partial u_2}\right]\mathbf{a}_3$$

This expression may be easier to remember in the following determinant form:

$$\text{curl } \mathbf{F} = \begin{vmatrix} \dfrac{\mathbf{a}_1}{h_2 h_3} & \dfrac{\mathbf{a}_2}{h_1 h_3} & \dfrac{\mathbf{a}_3}{h_1 h_2} \\ \dfrac{\partial}{\partial u_1} & \dfrac{\partial}{\partial u_2} & \dfrac{\partial}{\partial u_3} \\ h_1 F_1 & h_2 F_2 & h_3 F_3 \end{vmatrix}$$

where h_1, h_2, and h_3 are the metric coefficients. Also you may use the Taylor expansion to find the value of $\mathbf{F}$ at the various sides of the contour in terms of its given value at the origin.

Problem.　Make the appropriate substitution in the expression for curl $\mathbf{F}$ in the generalized curvilinear coordinate system (see Appendix A-2) to show that curl $\mathbf{F}$ in the cylindrical and spherical coordinate systems is given by

$$\text{curl } \mathbf{F} = \left[\frac{1}{\rho}\frac{\partial F_z}{\partial \phi} - \frac{\partial F_\phi}{\partial z}\right]\mathbf{a}_\rho + \left[\frac{\partial F_\rho}{\partial z} - \frac{\partial F_z}{\partial \rho}\right]\mathbf{a}_\phi + \left[\frac{1}{\rho}\frac{\partial}{\partial \rho}(\rho F_\phi) - \frac{1}{\rho}\frac{\partial F_\rho}{\partial \phi}\right]\mathbf{a}_z$$

$$\operatorname{curl} \mathbf{F} = \frac{1}{r \sin \theta}\left[\frac{\partial}{\partial \theta}(F_\phi \sin \theta) - \frac{\partial F_\theta}{\partial \phi}\right]\mathbf{a}_r + \frac{1}{r}\left[\frac{1}{\sin \theta}\frac{\partial F_r}{\partial \phi} - \frac{\partial}{\partial r}(rF_\phi)\right]\mathbf{a}_\theta$$

$$+ \frac{1}{r}\left[\frac{\partial}{\partial r}(rF_\theta) - \frac{\partial F_r}{\partial \theta}\right]\mathbf{a}_\phi$$

To emphasize the physical properties of a curl of a vector field further, let us solve the following examples.

EXAMPLE 2.16

A vector $\mathbf{F}$ is called irrotational if $\operatorname{curl} \mathbf{F} = 0$. Find the constants a, b, c, so that

$$\mathbf{F} = (x + 2y + az)\mathbf{a}_x + (bx - 3y - z)\mathbf{a}_y + (4x + cy + 2z)\mathbf{a}_z$$

is irrotational.

Solution

$$\operatorname{curl} \mathbf{F} = \nabla \times \mathbf{F} = \begin{vmatrix} \mathbf{a}_x & \mathbf{a}_y & \mathbf{a}_z \\ \dfrac{\partial}{\partial x} & \dfrac{\partial}{\partial y} & \dfrac{\partial}{\partial z} \\ x + 2y + az & bx - 3y - z & 4x + cy + 2z \end{vmatrix}$$

$$= (c + 1)\mathbf{a}_x + (a - 4)\mathbf{a}_y + (b - 2)\mathbf{a}_z$$

For the vector $\nabla \times \mathbf{F}$ to be exactly equal to zero, each of its components should be zero. This may be achieved if $a = 4$, $b = 2$, and $c = -1$. The vector $\mathbf{F}$ is hence given by

$$\mathbf{F} = (x + 2y - 4z)\mathbf{a}_x + (2x - 3y - z)\mathbf{a}_y + (4x - y + 2z)\mathbf{a}_z$$

◆◆

EXAMPLE 2.17

Determine the curl of the vector $\mathbf{M}$ given by

$$\mathbf{M} = ky\,\mathbf{a}_x$$

and explain physically the nonzero value of $\nabla \times \mathbf{M}$.

Solution

Curl $\mathbf{M}$ in the Cartesian coordinate system is given by

$$\operatorname{curl} \mathbf{M} = \begin{vmatrix} \mathbf{a}_x & \mathbf{a}_y & \mathbf{a}_z \\ \dfrac{\partial}{\partial x} & \dfrac{\partial}{\partial y} & \dfrac{\partial}{\partial z} \\ ky & 0 & 0 \end{vmatrix} = -k\,\mathbf{a}_z$$

To explain physically the nonzero value and the direction of the curl, let us plot the flux representation of the vector $\mathbf{M}$. Figure 2.19 shows such a representation in the Cartesian coordinate system.

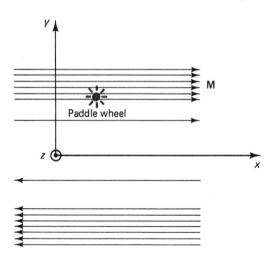

Figure 2.19 Flux representation of the vector **M** of example 2.17.

From Figure 2.19 it can be seen that a curl meter would rotate if placed in this field. Figure 2.19 also illustrates that the rotation would be along the negative z direction, thus confirming the nonzero magnitude and the direction of the result obtained for curl **M**.

———————◆◆———————

EXAMPLE 2.18

In chapter 1 we determined the magnetic flux density inside and outside a long round wire. It was shown that

$$\mathbf{B} = \frac{\mu_o I}{2\pi a^2} \rho \, \mathbf{a}_\phi \qquad \rho < a$$

$$= \frac{\mu_o I}{2\pi \rho} \mathbf{a}_\phi \qquad \rho > a$$

where a is the radius of the wire. Obtain expressions for the curl **B** inside and outside the conductor.

Solution

Inside the wire, curl **B** is given by

$$\nabla \times \mathbf{B} = \begin{vmatrix} \dfrac{\mathbf{a}_\rho}{\rho} & \mathbf{a}_\phi & \dfrac{\mathbf{a}_z}{\rho} \\[2mm] \dfrac{\partial}{\partial \rho} & \dfrac{\partial}{\partial \phi} & \dfrac{\partial}{\partial z} \\[2mm] 0 & \rho\left(\dfrac{\mu_o I \rho}{2\pi a^2}\right) & 0 \end{vmatrix} = \frac{\mu_o I}{\pi a^2} \mathbf{a}_z$$

Outside the wire, curl **B** is given by

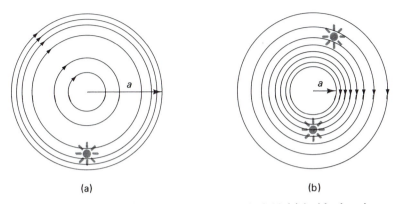

(a)   (b)

Figure 2.20   Flux representation of **B** in example 2.18 (a) inside the wire and (b) outside the wire.

$$\nabla \times \mathbf{B} = \begin{vmatrix} \dfrac{\mathbf{a}_\rho}{\rho} & \mathbf{a}_\phi & \dfrac{\mathbf{a}_z}{\rho} \\ \dfrac{\partial}{\partial \rho} & \dfrac{\partial}{\partial \phi} & \dfrac{\partial}{\partial z} \\ 0 & \rho\dfrac{\mu_o I}{2\pi\rho} & 0 \end{vmatrix} = 0$$

A simple flux representation of **B** inside the wire as shown in Figure 2.20a may be used to explain the nonzero value of the curl **B** inside the wire. The nonuniform distribution of the flux in Figure 2.20a justifies the nonzero value of the curl **B**. It is more difficult, however, to show the zero value of $\nabla \times \mathbf{B}$ from the flux representation of **B** outside the wire as shown in Figure 2.20b. The curl meter placed in this field of curved lines shows that a larger number of blades has a clockwise force exerted on them. This force is, in general, smaller than the counterclockwise force exerted on the smaller number of blades closer to the wire. It is therefore possible that if the curvature of the flux lines is just right, the net torque on the paddle wheel may be zero. Similar arguments were not necessary to make in case (a) because the larger number of blades was in the larger value of the field, and thus rotation is expected.

Another important point can be learned from this example. Although the divergences of **B** inside and outside the wire are zero, indicating closed flux lines in both cases, the curl values are different. We therefore conclude that the divergence and the curl of a vector field describe two different properties of this field. The divergence describes the net outflow of the vector flux at a point, whereas the curl describes the circulation property of the vector field that is related to its inhomogeneity.

2.8 STOKES'S THEOREM

Before we can use the curl differential operator to obtain the point "differential" forms of Ampere's and Faraday's laws, we need to devote some time to learn Stokes's theorem. This theorem provides a relation between the line integral over a closed

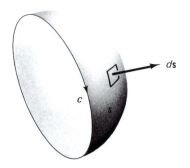

Figure 2.21 Stokes's theorem applied to an open surface s surrounded by the contour c.

contour and the surface integral over a surface enclosed by the contour. As we recall, the divergence theorem provides a relation between the surface and volume integrals. In Stokes's theorem, the surface integral of the component of the curl of a vector $\mathbf{F}$ in the direction of a unit vector perpendicular to an open surface is equal to the line integral of the vector over a closed contour surrounding (bounding) the surface. Thus

$$\int_s \boldsymbol{\nabla} \times \mathbf{F} \cdot d\mathbf{s} = \oint_c \mathbf{F} \cdot d\boldsymbol{\ell}$$

where c surrounds s as shown in Figure 2.21. The dot product, once again, emphasizes the fact that we are taking the component of $\boldsymbol{\nabla} \times \mathbf{F}$ in the direction of the element of area $d\mathbf{s}$. The sense in which we take c and the direction of the element of area $d\mathbf{s}$ should obey the right-hand rule. Next, we will present a physical proof of Stokes's theorem and discuss some of the limitations that should be observed when applying this theorem.

To prove this theorem, let us consider dividing the surface s into a large number N of surface elements as shown in Figure 2.22. With this division we will be able to use the definition of the curl in each of these surface elements. In a typical ith element we have

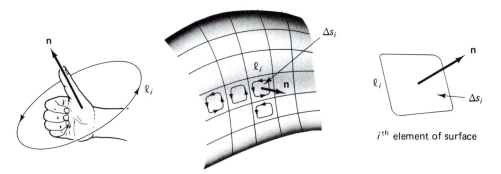

i^{th} element of surface

Figure 2.22 Geometry used in proving Stokes's theorem. The ith element of surface has an area Δs_i, bounded by the contour ℓ_1, and $\mathbf{n}$ is a unit vector normal to the surface.

$$\text{Lim}_{\Delta s_i \to 0} \frac{\oint_{\ell_i} \mathbf{F} \cdot d\boldsymbol{\ell}}{\Delta s_i} = [\text{curl} \, \mathbf{F}] \cdot \mathbf{n} \tag{2.21}$$

where $\mathbf{n}$ is a unit vector normal to the surface element Δs_i.

The preceding expression simply indicates that integrating $\mathbf{F}$ over the contour ℓ_i and dividing the result by the area Δs_i simply provides us with the component of the curl normal to the element of surface. Equation 2.21 may be rewritten in the form

$$\oint_{\ell_i} \mathbf{F} \cdot d\boldsymbol{\ell} = [\text{curl} \, \mathbf{F}] \cdot \Delta \mathbf{s}_i \tag{2.22}$$

subject to the fact that $\Delta \mathbf{s}_i$ is sufficiently small.

If we sum the left side of equation 2.22 over all closed contours ℓ_i, $i = 1$ to N, and by noting that the common edges of adjacent elements are considered twice and in opposite directions, it is clear that the contour integrations over all the elements will cancel everywhere except on the outer contour ℓ of the open surface s. Hence,

$$\sum_{i=1}^{N} \left[\oint_{\ell_i} \mathbf{F} \cdot d\boldsymbol{\ell} \right] = \oint_{\ell} \mathbf{F} \cdot d\boldsymbol{\ell} \tag{2.23}$$

Summing the right side of equation 2.22, conversely, results in the limit as $\Delta s_i \to 0$, or as the number of elements N approaches ∞, in integrating over the surface s.

$$\text{Lim}_{\Delta s_i \to 0} \sum_{i=1}^{N} \text{curl} \, \mathbf{F} \cdot \Delta \mathbf{s}_i = \int_{s} \text{curl} \, \mathbf{F} \cdot d\mathbf{s} \tag{2.24}$$

The limiting expressions in equations 2.23 and 2.24 provide a proof for Stokes's theorem,

$$\boxed{\oint_{\ell} \mathbf{F} \cdot d\boldsymbol{\ell} = \int_{s} \text{curl} \, \mathbf{F} \cdot d\mathbf{s}}$$

Before solving some examples illustrating the application of Stokes's theorem, it should be pointed out that because this theorem involves the curl of a vector field, its application should be limited to domains where $\mathbf{F}$ together with its first derivative are continuous. Infinite (singular) values of $\mathbf{F}$ or its first derivative should be excluded from the domain of integration by surrounding them with appropriate contours and excluding the bounded areas from the surface integration. With this in mind let us now solve some illustrative examples.

EXAMPLE 2.19

Given the ϕ directed electric field,

$$\mathbf{E} = k\rho^2 z \, \mathbf{a}_\phi, \quad k \text{ is a constant}$$

illustrate the validity of Stokes's theorem by evaluating the surface integral over the open surface shown in Figure 2.23 and the line integral about the closed contour bounding s.

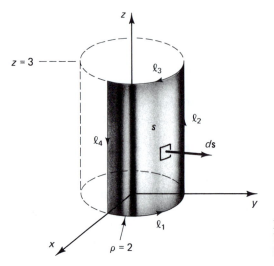

Figure 2.23   The surface s and its bounding contour that are used in Stokes's theorem.

Solution

To evaluate the surface integral, let us first evaluate $\nabla \times \mathbf{E}$.

$$\nabla \times \mathbf{E} = \begin{vmatrix} \dfrac{\mathbf{a}_\rho}{\rho} & \mathbf{a}_\phi & \dfrac{\mathbf{a}_z}{\rho} \\[6pt] \dfrac{\partial}{\partial \rho} & \dfrac{\partial}{\partial \phi} & \dfrac{\partial}{\partial z} \\[6pt] 0 & \rho k \rho^2 z & 0 \end{vmatrix}$$

$$= -\frac{\mathbf{a}_\rho}{\rho} k\rho^3 + \frac{\mathbf{a}_z}{\rho} 3\rho^2 kz = -\mathbf{a}_\rho k\rho^2 + \mathbf{a}_z 3k\rho z$$

The element of area $ds = \rho \, d\phi \, dz \, \mathbf{a}_\rho$. Therefore,

$$\int_s (\nabla \times \mathbf{E}) \cdot d\mathbf{s} = \int_0^3 \int_0^{\pi/2} (-k\rho^2 \mathbf{a}_\rho + 3 k\rho z \, \mathbf{a}_z) \cdot \rho \, d\phi \, dz \, \mathbf{a}_\rho$$

$$= \int_0^3 \int_0^{\pi/2} -k\rho^3 \, d\phi \, dz = -8k \int_0^3 dz \int_0^{\pi/2} d\phi$$

$$= -12\,\pi k$$

To evaluate the line integral, conversely, we consider the closed contour ℓ bounding s. It should be noted that the direction of the contour integration and the element of area should follow the right-hand rule.

$$\oint_\ell \mathbf{E} \cdot d\boldsymbol{\ell} = \int_{\ell_1} \mathbf{E} \cdot \rho \, d\phi \, \mathbf{a}_\phi + \int_{\ell_2} \mathbf{E} \cdot dz \, \mathbf{a}_z + \int_{\ell_3} \mathbf{E} \cdot \rho \, d\phi \, \mathbf{a}_\phi + \int_{\ell_4} \mathbf{E} \cdot dz \, \mathbf{a}_z$$

On ℓ_1, the independent variable $z = 0$, hence $\mathbf{E} = 0$ on ℓ_1, and the line integral contribution vanishes. Also $d\boldsymbol{\ell}$ on each of ℓ_2 and ℓ_4 is perpendicular to $\mathbf{E}$ and therefore $\mathbf{E} \cdot d\boldsymbol{\ell}$ will be zero and has no contribution to the overall line integral. Hence,

$$\int_\ell \mathbf{E} \cdot d\ell = \int_{\ell_3} \mathbf{E} \cdot \rho \, d\phi \, \mathbf{a}_\phi$$

$$= \int_{\pi/2}^{0} k\rho^2 z \, \rho \, d\phi$$

$$= 8k(3)\left(-\frac{\pi}{2}\right)$$

$$= -12\pi k$$

which is equal to the value obtained from evaluating $\int_s \nabla \times \mathbf{E} \cdot d\mathbf{s}$, thus proving the validity of Stokes's theorem.

2.9 AMPERE'S AND FARADAY'S LAWS IN POINT (DIFFERENTIAL) FORM

We are now in a position to derive the long-awaited for differential forms of Ampere's and Faraday's laws.

Ampere's law in its integral form in free space is given by

$$\oint_c \frac{\mathbf{B}}{\mu_o} \cdot d\ell = \int_s \mathbf{J} \cdot d\mathbf{s} + \frac{d}{dt} \int_s \epsilon_o \mathbf{E} \cdot d\mathbf{s} \tag{2.27}$$

To obtain an expression that is valid at a point, we start by integrating $\mathbf{B}$ over an infinitesimally small contour Δc, which bounds an element of area Δs. We further divide the result by Δs to help us use the definition of the curl,

$$\frac{\oint_{\Delta c} \frac{\mathbf{B}}{\mu_o} \cdot d\ell}{\Delta s} = \frac{\int_{\Delta s} \mathbf{J} \cdot d\mathbf{s}}{\Delta s} + \frac{\int_{\Delta s} \frac{\partial \epsilon_o \mathbf{E}}{\partial t} \cdot d\mathbf{s}}{\Delta s} \tag{2.28}$$

where the total differentiation d/dt in equation 2.27 was replaced by the partial differentiation $\partial/\partial t$ in equation 2.28 because $\mathbf{E}$ is, in general, a function of space (x, y, z) and time. After integrating $\mathbf{E}$ over the surface as in equation 2.27, however, the result is only a function of time, and we therefore use the total differentiation d/dt. The left side of equation 2.27 in the limit as $\Delta s \to 0$ is simply the component of curl $\mathbf{B}/\mu_o$ in a direction of a unit vector normal to Δs. Taking the element of area $d\mathbf{s}$ to be in the $\mathbf{a}_x$ direction (i.e., $d\mathbf{s} = ds \, \mathbf{a}_x$), we obtain

$$\left[\text{curl}\frac{\mathbf{B}}{\mu_o}\right]_x = J_x + \frac{\partial \epsilon_o E_x}{\partial t} \tag{2.29}$$

where J_x and E_x are the x components of $\mathbf{J}$ and $\mathbf{E}$, respectively. Following a similar procedure and choosing elements of areas in the y and z directions, we obtain

$$\left[\text{curl}\frac{\mathbf{B}}{\mu_o}\right]_y = J_y + \frac{\partial \epsilon_o E_y}{\partial t} \tag{2.30}$$

$$\left[\text{curl}\frac{\mathbf{B}}{\mu_o}\right]_z = J_z + \frac{\partial \epsilon_o E_z}{\partial t} \tag{2.31}$$

Combining the x, y, and z components of the curl from equation 2.29 to equation 2.31, we obtain

$$\left[\text{curl}\,\frac{\mathbf{B}}{\mu_o}\right]_x \mathbf{a}_x + \left[\text{curl}\,\frac{\mathbf{B}}{\mu_o}\right]_y \mathbf{a}_y + \left[\text{curl}\,\frac{\mathbf{B}}{\mu_o}\right]_z \mathbf{a}_z = (J_x\,\mathbf{a}_x + J_y\,\mathbf{a}_y + J_z\,\mathbf{a}_z) + \frac{\partial \epsilon_o E_x}{\partial t}\mathbf{a}_x$$
$$+ \frac{\partial \epsilon_o E_y}{\partial t}\mathbf{a}_y + \frac{\partial \epsilon_o E_z}{\partial t}\mathbf{a}_z$$

which may be expressed in compact form

$$\boxed{\text{curl}\,\frac{\mathbf{B}}{\mu_o} = \mathbf{J} + \frac{\partial \epsilon_o \mathbf{E}}{\partial t}} \tag{2.32}$$

Equation 2.32 is the point or differential form of Ampere's law. Equation 2.32 states that the curl of $\mathbf{B}/\mu_o$, which is related to its circulation at a point, is equal to the total current density at that point. The total current density is a combination of the current resulting from flow of charges $\mathbf{J}$ and the displacement current density $\partial \epsilon_o\,\mathbf{E}/\partial t$.

An alternative approach for obtaining the point form of Ampere's law is to use Stokes's theorem. Starting from equation 2.27 and using Stokes's theorem, we have

$$\oint_c \frac{\mathbf{B}}{\mu_o}\cdot d\boldsymbol{\ell} = \int_s \boldsymbol{\nabla}\times\frac{\mathbf{B}}{\mu_o}\cdot d\mathbf{s} = \int_s \mathbf{J}\cdot d\mathbf{s} + \int_s \frac{\partial \epsilon_o \mathbf{E}}{\partial t}\cdot d\mathbf{s}$$

By taking the element of area to be in the x direction (i.e., $d\mathbf{s} = ds\,\mathbf{a}_x$) and assuming that the area is sufficiently small so that the integration over the area may be substituted by multiplying the integrand by the area, we obtain

$$\left[\boldsymbol{\nabla}\times\frac{\mathbf{B}}{\mu_o}\right]_x \Delta s = J_x\,\Delta s + \frac{\partial \epsilon_o E_x}{\partial t}\Delta s \tag{2.33}$$

Repeating this procedure and taking the differential elements of area in the y and z directions, we obtain equations 2.30 and 2.31, which when combined with equation 2.33 provide the point form of Ampere's law in equation 2.32.

Faraday's law in integral form, conversely, is given by

$$\oint_c \mathbf{E}\cdot d\boldsymbol{\ell} = -\frac{d}{dt}\int_s \mathbf{B}\cdot d\mathbf{s}$$

using Stokes's theorem to change the closed line integral of $\mathbf{E}$ to a surface integral of $\boldsymbol{\nabla}\times\mathbf{E}$, we obtain

$$\oint_c \mathbf{E}\cdot d\boldsymbol{\ell} = \int_s \boldsymbol{\nabla}\times\mathbf{E}\cdot d\mathbf{s} = -\int_s \frac{\partial \mathbf{B}}{\partial t}\cdot d\mathbf{s}$$

If we carry out the integration over a sufficiently small area Δs so that the value of integration may be obtained by multiplying the integrand by Δs, we obtain

$$[\boldsymbol{\nabla}\times\mathbf{E}]_n\,\Delta s = -\frac{\partial B_n}{\partial t}\Delta s$$

where $[\mathbf{\nabla} \times \mathbf{E}]_n$ and B_n are the components of $\mathbf{\nabla} \times \mathbf{E}$ and $\mathbf{B}$, respectively, in the direction of a unit vector $\mathbf{n}$ normal to the element of area. Choosing $\Delta\mathbf{s}$ to be in the x, y, and z directions, and adding vectorially the resulting three components of $\mathbf{\nabla} \times \mathbf{E}$ and $\mathbf{B}$, we obtain

$$\mathbf{\nabla} \times \mathbf{E} = -\frac{\partial \mathbf{B}}{\partial t} \tag{2.34}$$

which is the point form of Faraday's law. It simply states that the circulation of the electric field at any point is equal to the time rate of decrease of the magnetic flux density at that point.

An important special case is when we consider static electric and magnetic fields. For static fields, the operator $\partial/\partial t$ should be set to zero, Ampere's and Faraday's laws then reduce to

$$\boxed{\begin{array}{l} \mathbf{\nabla} \times \dfrac{\mathbf{B}}{\mu_o} = \mathbf{J} \\[2mm] \mathbf{\nabla} \times \mathbf{E} = 0 \end{array}} \qquad \text{(Static electric and magnetic fields)}$$

These as well as other interesting properties of Maxwell's equations will be described in the following sections.

.10 SUMMARY OF MAXWELL'S EQUATIONS IN DIFFERENTIAL FORMS

Table 2.1 summarizes the four Maxwell's equations in differential form. This table, in addition to providing expressions for the different laws, emphasizes the basic definitions of the divergence and curl of a vector field. It is clear that in all cases the derivations of the divergence and the curl were introduced as limiting cases of carrying out the surface and line integrals over elements of areas and over the contours bounding them, respectively. Obtaining these limiting cases achieves the following:

1. The resulting equations are point relations and therefore do not require specifying surfaces, volumes, or contours to carry out the integration. These equations should therefore be satisfied at each point in space where the fields exist.

2. Because they are just limiting relations, the physical understanding we developed in chapter 1 of Maxwell's equations should still hold for the point relations. It is often expressed that it is difficult to understand what $\mathbf{\nabla} \times \mathbf{E}$ means. By recalling that $\mathbf{\nabla} \times \mathbf{E}$ is just a limiting (point) expression for $\oint_c \mathbf{E} \cdot d\boldsymbol{\ell}$ per unit area as the contour c shrinks to zero, it will be apparent that $\mathbf{\nabla} \times \mathbf{E}$ is still related to the circulation of $\mathbf{E}$ along a closed contour, and its magnitude describes the inhomogeneity of the $\mathbf{E}$ field as indicated in our explanation of the curl operator. Similarly, $\mathbf{\nabla} \cdot \mathbf{B}$ may be difficult to understand physically, particularly if we hurry and plug the given $\mathbf{B}$ field expression in the differential operator. From the basic definition of the divergence as given in Table

TABLE 2.1 SUMMARY OF MAXWELL'S EQUATIONS IN DIFFERENTIAL FORM

1. Gauss's Law for Electric Field

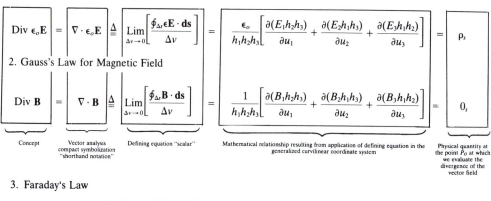

2. Gauss's Law for Magnetic Field

| Concept | Vector analysis compact symbolization "shorthand notation" | Defining equation "scalar" | Mathematical relationship resulting from application of defining equation in the generalized curvilinear coordinate system | Physical quantity at the point P_O at which we evaluate the divergence of the vector field |

3. Faraday's Law

4. Ampere's Law

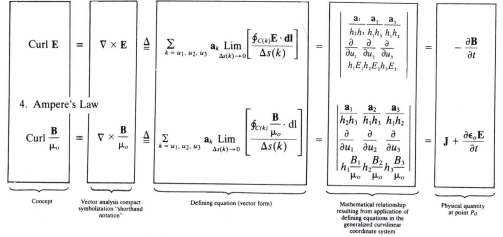

| Concept | Vector analysis compact symbolization "shorthand notation" | Defining equation (vector form) | Mathematical relationship resulting from application of defining equations in the generalized curvilinear coordinate system | Physical quantity at point P_O |

2.1, however, it is clear that $\nabla \cdot \mathbf{B}$ is just a limiting case of $\oint_{\Delta s} \mathbf{B} \cdot d\mathbf{s}$ per unit volume. $\nabla \cdot \mathbf{B}$ thus has the same physical meaning as $\oint_{\Delta s} \mathbf{B} \cdot d\mathbf{s}$, which indicates the total magnetic flux emanating from the closed surface Δs. Next, we shall illustrate the use of these point forms by solving some examples.

EXAMPLE 2.20

Describe which of the following vectors may represent a static electric field. If your answer is yes for any of the given vectors, determine the volume charge density associated with it.

1. $\mathbf{E} = y\,\mathbf{a}_x - x\,\mathbf{a}_y$
2. $\mathbf{E} = \cos\theta\,\mathbf{a}_r - \sin\theta\,\mathbf{a}_\theta$

Solution

For a static electric field $\partial/\partial t = 0$, Faraday's law is then given by

$$\nabla \times \mathbf{E} = 0$$

which is the equation that should be satisfied by any static electric field.

1. $\nabla \times \mathbf{E} = \begin{vmatrix} \mathbf{a}_x & \mathbf{a}_y & \mathbf{a}_z \\ \dfrac{\partial}{\partial x} & \dfrac{\partial}{\partial y} & \dfrac{\partial}{\partial z} \\ y & -x & 0 \end{vmatrix} = -2\,\mathbf{a}_z$

The vector in part 1, therefore, does not represent a static electric field.

2. $\nabla \times \mathbf{E} = \begin{vmatrix} \dfrac{\mathbf{a}_r}{r^2 \sin \theta} & \dfrac{\mathbf{a}_\theta}{r \sin \theta} & \dfrac{\mathbf{a}_\phi}{r} \\ \dfrac{\partial}{\partial r} & \dfrac{\partial}{\partial \theta} & \dfrac{\partial}{\partial \phi} \\ \cos \theta & -r \sin \theta & 0 \end{vmatrix}$

$$= 0$$

The vector in part 2 may represent a static electric field.

To determine the charge density associated with the electric field vector in part 2, we use Gauss's law for the electric field

$$\nabla \cdot \epsilon_o \mathbf{E} = \rho_v$$

In spherical coordinates we have,

$$\frac{1}{r^2}\frac{\partial}{\partial r}(r^2 E_r) + \frac{1}{r \sin \theta}\frac{\partial}{\partial \theta}(E_\theta \sin \theta) + \frac{1}{r \sin \theta}\frac{\partial E_\phi}{\partial \phi} = \frac{\rho_v}{\epsilon_o}$$

Substituting $E_r = \cos \theta$, $E_\theta = -\sin \theta$, $E_\phi = 0$, we obtain

$$\rho_v = 0$$

which simply means that the expression for the electric field (if it truly represents an electric field) is given in the region *outside* the charge distribution source producing it. Similar conclusions may be obtained for the true electric field expression in the region outside a spherical charge distribution, which is given in example 1.24.

In this case, E_r outside the spherical charge is given by

$$E_r = \frac{\rho_v r_o^3}{3\epsilon_o r^2} \qquad r > r_o$$

It may be shown that

$$\nabla \cdot \epsilon_o \mathbf{E} = 0$$

which means, as we already know, that the charge distribution outside the spherical charge, that is, $r > r_o$, is zero.

If we use the electric field expression inside the spherical charge, conversely,

$$E_r = \frac{\rho_v r}{3\epsilon_o} \qquad r < r_o$$

where r_o is the radius of the spherical charge distribution (see example 1.24), we obtain

$$\nabla \cdot \epsilon_o \mathbf{E} = \frac{\epsilon_o}{r^2} \frac{\partial}{\partial r} \left(\frac{\rho_v r^3}{3\epsilon_o} \right) = \rho_v$$

which is the same value of the uniform charge density we started with in example 1.24.

———————◆◆◆———————

EXAMPLE 2.21

Determine which of the following vectors may represent a static magnetic field. If so, calculate the current density associated with it.

1. $\mathbf{B} = x\,\mathbf{a}_x - y\,\mathbf{a}_y$
2. $\mathbf{B} = \rho\,\mathbf{a}_\phi$
3. $\mathbf{B} = r\cos\phi\,\mathbf{a}_r - 3r\sin\theta\sin\phi\,\mathbf{a}_\phi$

Solution

For a vector field to represent a magnetic flux density vector, it should satisfy Gauss's law, hence,

1. $$\nabla \cdot B = 0$$

$$\nabla \cdot \mathbf{B} = \frac{\partial B_x}{\partial x} + \frac{\partial B_y}{\partial y} + \frac{\partial B_z}{\partial z}$$

$$= 0$$

The **B** vector in part 1 may therefore represent a magnetic flux density vector. The current density associated with it is obtained using Ampere's law, which for static ($\partial/\partial t = 0$) fields reduces to

$$\nabla \times \frac{\mathbf{B}}{\mu_o} = \mathbf{J}$$

$$\therefore \begin{vmatrix} \mathbf{a}_x & \mathbf{a}_y & \mathbf{a}_z \\ \dfrac{\partial}{\partial x} & \dfrac{\partial}{\partial y} & \dfrac{\partial}{\partial z} \\ x & -y & 0 \end{vmatrix} = \mu_o \mathbf{J}$$

Carrying out the curl analysis, we obtain

$$\mathbf{J} = 0$$

which means that the **B** expression in part 1 is given in the region outside the current distribution causing it.

2.
$$\nabla \cdot \mathbf{B} = \frac{1}{\rho}\frac{\partial}{\partial\rho}(\rho B_\rho) + \frac{1}{\rho}\frac{\partial B_\phi}{\partial\phi} + \frac{\partial B_z}{\partial z}$$

$$= 0$$

The vector **B** in part 2 may also represent magnetic flux density vector. The current density **J** is obtained from

$$\nabla \times \mathbf{B} = \mu_o \mathbf{J}$$

$$= \left[\frac{1}{\rho}\left(\frac{\partial B_z}{\partial\phi} - \frac{\partial B_\phi}{\partial z}\right)\right]\mathbf{a}_\rho + \left[\frac{\partial B_\rho}{\partial z} - \frac{\partial B_z}{\partial\rho}\right]\mathbf{a}_\phi + \left[\frac{1}{\rho}\frac{\partial}{\partial\rho}(\rho B_\phi) - \frac{1}{\rho}\frac{\partial B_\rho}{\partial\phi}\right]\mathbf{a}_z$$

Substituting B_ρ, B_ϕ, and B_z of this vector, we obtain

$$\mathbf{J} = \frac{2}{\mu_o}\mathbf{a}_z$$

3.
$$\nabla \cdot \mathbf{B} = \frac{1}{r^2}\frac{\partial}{\partial r}(r^2 B_r) + \frac{1}{r\sin\theta}\frac{\partial}{\partial\theta}(B_\theta \sin\theta) + \frac{1}{r\sin\theta}\frac{\partial B_\phi}{\partial\phi}$$

$$= 3\cos\phi + \frac{1}{r\sin\theta}(-3r\sin\theta\cos\phi)$$

$$= 0$$

B may represent magnetic flux density vector.

$$\nabla \times \mathbf{B} = \frac{1}{r\sin\theta}\left[\frac{\partial}{\partial\theta}(B_\phi \sin\theta) - \frac{\partial B_\theta}{\partial\phi}\right]\mathbf{a}_r$$

$$+ \frac{1}{r}\left[\frac{1}{\sin\theta}\frac{\partial B_r}{\partial\phi} - \frac{\partial}{\partial r}(rB_\phi)\right]\mathbf{a}_\theta + \frac{1}{r}\left[\frac{\partial}{\partial r}(rB_\theta) - \frac{\partial B_r}{\partial\theta}\right]\mathbf{a}_\phi$$

$$= \mu_o \mathbf{J}$$

Substituting B_r, B_θ, and B_ϕ components, we obtain

$$\mathbf{J} = \frac{1}{\mu_o}\left[-6\cos\theta\sin\phi\,\mathbf{a}_r + \left(6\sin\theta\sin\phi - \frac{\sin\phi}{\sin\theta}\right)\mathbf{a}_\theta\right]$$

EXAMPLE 2.22

If the magnetic flux density in a region of free space (**J** = 0) is given by

$$\mathbf{B} = B_o z \cos\omega t\, \mathbf{a}_y$$

and if it is known that the time-varying electric field associated with it has only an x component:

1. Use Faraday's law to find $\mathbf{E} = E_x\,\mathbf{a}_x$.
2. Use the obtained value of **E** in Ampere's law to determine the magnetic flux density **B**.
3. Compare the obtained result in part 2 with the original expression of the magnetic field. Comment on your answer.

Solution

1. Let us first use Faraday's law to obtain the electric field associated with the given time-varying magnetic field

$$\nabla \times \mathbf{E} = -\frac{\partial \mathbf{B}}{\partial t}$$

Because it is given that $\mathbf{E}$ has only an x component, we obtain

$$\begin{vmatrix} \mathbf{a}_x & \mathbf{a}_y & \mathbf{a}_z \\ \dfrac{\partial}{\partial x} & \dfrac{\partial}{\partial y} & \dfrac{\partial}{\partial z} \\ E_x & 0 & 0 \end{vmatrix} = \omega B_o z \sin \omega t \, \mathbf{a}_y$$

Equating the y component on both sides

$$\left(\frac{\partial E_x}{\partial z} \right) \mathbf{a}_y = \omega B_o z \sin \omega t \, \mathbf{a}_y$$

$$\therefore E_x = \omega B_o \frac{z^2}{2} \sin \omega t + c$$

If we use the initial condition that $E_x = 0$ at $t = 0$, we will find that the constant of integration $c = 0$. The electric field is therefore given by

$$\mathbf{E} = \omega B_o \frac{z^2}{2} \sin \omega t \, \mathbf{a}_x$$

2. Ampere's law in a region of space that is free from current sources is given by

$$\nabla \times \frac{\mathbf{B}}{\mu_o} = \frac{\partial \epsilon_o \mathbf{E}}{\partial t}$$

Because $\mathbf{B}$ has only a y component, we obtain

$$\begin{vmatrix} \mathbf{a}_x & \mathbf{a}_y & \mathbf{a}_z \\ \dfrac{\partial}{\partial x} & \dfrac{\partial}{\partial y} & \dfrac{\partial}{\partial z} \\ \dfrac{B_x}{\mu_o} = 0 & \dfrac{B_y}{\mu_o} & \dfrac{B_z}{\mu_o} = 0 \end{vmatrix} = \epsilon_o \omega^2 B_o \frac{z^2}{2} \cos \omega t \, \mathbf{a}_x$$

Equating the x components on both sides of the equation,

$$-\frac{\partial (B_y/\mu_o)}{\partial z} = \omega^2 \epsilon_o B_o \frac{z^2}{2} \cos \omega t$$

$$\therefore B_y = -\epsilon_o \mu_o \omega^2 B_o \frac{z^3}{6} \cos \omega t$$

or

$$\mathbf{B} = -\epsilon_o \mu_o \omega^2 B_o \frac{z^3}{6} \cos \omega t \, \mathbf{a}_y$$

3. The obtained expression for **B** in part 2 is different from the originally given expression $\mathbf{B} = B_o z \cos \omega t \, \mathbf{a}_y$ of the time-varying magnetic flux density. This discrepancy means that the given expression for **B** is not a solution for Maxwell's equations. Any complete solution of a given electromagnetic fields problem should be obtained by solving all Maxwell's equations *simultaneously*. A solution cannot be obtained just by solving Faraday's or Ampere's laws separately. Because of the coupling between the electric and magnetic fields and the constraints on these fields (i.e., $\nabla \cdot \mathbf{B} = 0$, and $\nabla \cdot \epsilon_o \mathbf{E} = \rho_v$), these four equations should be solved simultaneously. Let us just emphasize this point by indicating the importance of considering the coupling between these fields in solving Faraday's and Ampere's laws. If we start with an arbitrary value of **B** and use Faraday's law to determine **E**, and then use the resulting **E** in Ampere's law to solve for **B**, we will generally observe that the resulting **B** is different from the value with which we originally started. If we use the value of **B** that resulted from Ampere's law to modify the originally assumed one and use the modified value of **B** in Faraday's law to solve for **E** and substitute the new value of **E** in Ampere's law to once again solve for **B**, the process should continue until we find new values of **E** and **B** that simultaneously satisfy both Faraday's and Ampere's laws. In practice, we, of course, do not follow the alternating procedure described earlier, but instead we try to solve Maxwell's equations *simultaneously*.

Problem

1. Show that the electric and magnetic fields given in free space (i.e., $\rho_v = 0 = \mathbf{J}$) by

$$\mathbf{E} = E_m \sin x \, \sin \omega t \, \mathbf{a}_y$$

$$\mathbf{B} = \frac{E_m}{\omega} \cos x \, \cos \omega t \, \mathbf{a}_z$$

satisfy Faraday's law and the two laws of Gauss, but do not satisfy Ampere's law.

2. Are these fields valid solutions of Maxwell's equations? Explain.

2.11 CONTINUITY EQUATION AND MAXWELL'S DISPLACEMENT CURRENT TERM

In this section, we will describe the argument that led Maxwell to introduce his history-making term, the *displacement current*. Before we can do this, however, we need to introduce a physical law that describes the conservation of charge. Electric charge, like mass, is conserved, which means that it cannot be destroyed or created. The law that mathematically states this conservative property of the electric charge is known as the continuity equation. To obtain an expression for this law, let us consider the current density flux emanating from a closed surface as shown in Figure 2.24. The outflow of the current flux from the closed surface results in a reduction of the total positive charge enclosed by the surface s. As a matter of fact, the total current flux

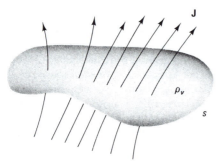

Figure 2.24 The current density flux emanating from the closed surface s equals the rate of decrease of the positive charge density enclosed by s.

emanating from the closed surface s equals precisely the rate of decrease of the positive charge enclosed by s. Hence,

$$\oint_s \mathbf{J} \cdot d\mathbf{s} = -\frac{d}{dt} \int_v \rho_v \, dv \qquad (2.35)$$

where v is the volume of the charge density within s. Equation 2.35 is the integral form of the continuity equation. To obtain the differential form of this law, we use the divergence theorem and consider the case in which the current density is emanating from an element of volume Δv.

$$\oint_{\Delta s} \mathbf{J} \cdot d\mathbf{s} = \int_{\Delta v} \mathbf{\nabla} \cdot \mathbf{J} \, dv = -\int_{\Delta v} \frac{\partial \rho_v}{\partial t} \, dv$$

In the limit as Δv approaches zero, the volume integrals may be evaluated by multiplying the integrand by Δv, hence,

$$\mathbf{\nabla} \cdot \mathbf{J} \, \Delta v = -\frac{\partial \rho_v}{\partial t} \Delta v$$

$$\mathbf{\nabla} \cdot \mathbf{J} = -\frac{\partial \rho_v}{\partial t} \qquad (2.36)$$

Equation 2.36 is a more convenient and easy-to-use mathematical statement of the continuity equation. It simply states that the divergence (net outflow from a differential element of volume per unit volume) of the current density owing to the outflow of charges from the element of volume is equal to the time rate of decrease of the charge density at the point to which the element of volume shrinks. This continuity equation is a physical law and should be satisfied at all times. It is this law that led Maxwell to introduce his displacement current term and to derive Ampere's law in its general form.

Ampere's law, when it was first formulated in 1820, basically related the magnetic flux density **B** to the electric current producing it

$$\mathbf{\nabla} \times \frac{\mathbf{B}}{\mu_o} = \mathbf{J} \qquad (2.37)$$

without accounting for any coupling between the electric and magnetic field. If we take the divergence of both sides of equation 2.37, we obtain

$$\mathbf{\nabla} \cdot \mathbf{\nabla} \times \frac{\mathbf{B}}{\mu_o} = \mathbf{\nabla} \cdot \mathbf{J} \tag{2.38}$$

According to vector identities, the divergence of the curl of any vector is always zero. Therefore, the left side of equation 2.38 is always zero. Equation 2.38 reduces to

$$\mathbf{\nabla} \cdot \mathbf{J} = 0$$

which violates the continuity equation in equation 2.36. Because equation 2.36 is a physical law and cannot be violated, Maxwell concluded that Ampere's law in equation 2.37 is not in its most general form. Anticipating a more general value of **B** does not help resolve the discrepancy because the divergence of the curl of any vector **B** will still be zero, and the continuity equation will still be violated. Maxwell used the only remaining alternative and anticipated the presence of some additional current term $\mathbf{J}_D$. He postulated the following general form of Ampere's law

$$\mathbf{\nabla} \times \frac{\mathbf{B}}{\mu_o} = \mathbf{J} + \mathbf{J}_D \tag{2.39}$$

Taking the divergence of both sides of equation 2.39, we obtain

$$\mathbf{\nabla} \cdot \mathbf{J} = -\mathbf{\nabla} \cdot \mathbf{J}_D$$

$$= -\frac{\partial \rho_v}{\partial t} \qquad \text{from the continuity equation}$$

$$\mathbf{\nabla} \cdot \mathbf{J}_D = \frac{\partial \rho_v}{\partial t}$$

To determine further the nature of the new current term $\mathbf{J}_D$, Maxwell used Gauss's law for the electric field, thus,

$$\mathbf{\nabla} \cdot \mathbf{J}_D = \frac{\partial \rho_v}{\partial t} = \frac{\partial}{\partial t}(\mathbf{\nabla} \cdot \epsilon_o \mathbf{E}) = \mathbf{\nabla} \cdot \frac{\partial \epsilon_o \mathbf{E}}{\partial t} \tag{2.40}$$

The equality in equation 2.40 leads to the determination of $\mathbf{J}_D$ in the form

$$\mathbf{J}_D = \frac{\partial \epsilon_o \mathbf{E}}{\partial t}$$

which is the displacement current term introduced by Maxwell. Ampere's law in its general form is now given by

$$\boxed{\mathbf{\nabla} \times \frac{\mathbf{B}}{\mu} = \mathbf{J} + \frac{\partial \epsilon_o \mathbf{E}}{\partial t}}$$

As indicated in chapter 1, the introduction of this term was significant because it simply introduced the time-varying electric field as a legitimate source of the magnetic field.

This term together with Faraday's law shows the ability of the time-varying electric and magnetic fields to generate each other; thus the introduction of the phenomenon of wave propagation. It is also this coupling between the electric and magnetic fields that made it necessary to solve all of Maxwell's equations simultaneously. The simplest possible solution of these equations is the case of plane wave propagation in free space that we are going to discuss next.

2.12 WAVE EQUATION IN SOURCE FREE REGION

Thus far we have introduced the general mathematical relations between the electromagnetic fields, and their current and charge sources. These relations are complete and may be used to solve any electromagnetic fields problem in the absence of a material medium. Their modification to include the *induced* charge and current distributions in a material region is the subject of chapter 3. For propagation in vacuum or air, however, these equations are complete and adequate for describing the propagation characteristics of the electric and magnetic fields.

In most wave propagation problems, we are interested in the propagation properties of the electric and magnetic fields away from their sources. In radiation problems the solution starts by specifying the current or the charge distributions on the source and the radiated fields are then calculated. In propagation problems, electromagnetic fields are studied under the assumption that the space of propagation is source free. Another way of looking at it is to assume that the sources are sufficiently far away from the propagation region of interest so that in examining the desired propagation properties we pay no attention to where or how these electromagnetic fields were generated. Such an argument justifies our careful examination of Maxwell's equations in a source-free region. With the charge ρ_v and current $\mathbf{J}$ distributions both set to zero, Maxwell's equations reduce to

$$\nabla \cdot \epsilon_o \mathbf{E} = 0 \tag{2.41a}$$

$$\nabla \cdot \mathbf{B} = 0 \tag{2.41b}$$

$$\nabla \times \mathbf{E} = -\frac{\partial \mathbf{B}}{\partial t} \tag{2.41c}$$

$$\nabla \times \frac{\mathbf{B}}{\mu_o} = \frac{\partial \epsilon_o \mathbf{E}}{\partial t} \tag{2.41d}$$

These are all first order differential equations in two unknown variables, the electric **E** and magnetic **B** fields. Because of the coupling between these fields (equations 2.41c and d), it is desirable to obtain separate equations for the electric and magnetic fields. To do this, we apply the curl operator to equation 2.41c to obtain

$$\nabla \times \nabla \times \mathbf{E} = -\frac{\partial}{\partial t} \nabla \times \mathbf{B}$$

Substituting $\nabla \times \mathbf{B}$ from equation 2.41d, we obtain

$$\nabla \times \nabla \times \mathbf{E} = -\frac{\partial}{\partial t}\left(\mu_o \epsilon_o \frac{\partial \mathbf{E}}{\partial t}\right) = -\mu_o \epsilon_o \frac{\partial^2 \mathbf{E}}{\partial t^2} \tag{2.42}$$

which is a second-order differential equation in the electric field only. This equation may be further simplified by noting the following vector identity

$$\nabla \times \nabla \times \mathbf{E} = \nabla(\nabla \cdot \mathbf{E}) - \nabla^2 \mathbf{E} \tag{2.43}$$

In mathematics courses we learned that the Laplacian operator on a scalar function Φ is given by

$$\nabla^2 \Phi = \frac{\partial^2 \Phi}{\partial x^2} + \frac{\partial^2 \Phi}{\partial y^2} + \frac{\partial^2 \Phi}{\partial z^2} \tag{2.44}$$

The Laplacian operator on a vector quantity such as $\mathbf{E}$ in the Cartesian coordinate system may be expressed in terms of three expressions, each is similar to equation 2.44 and is operating on a single component of $\mathbf{E} = E_x \mathbf{a}_x + E_y \mathbf{a}_y + E_z \mathbf{a}_z$. Hence, we obtain the following three scalar operations:

$$\nabla^2 E_x = \frac{\partial^2 E_x}{\partial x^2} + \frac{\partial^2 E_x}{\partial y^2} + \frac{\partial^2 E_x}{\partial z^2}$$

$$\nabla^2 E_y = \frac{\partial^2 E_y}{\partial x^2} + \frac{\partial^2 E_y}{\partial y^2} + \frac{\partial^2 E_y}{\partial z^2} \tag{2.45}$$

$$\nabla^2 E_z = \frac{\partial^2 E_z}{\partial x^2} + \frac{\partial^2 E_z}{\partial y^2} + \frac{\partial^2 E_z}{\partial z^2}$$

We also note that equation 2.43 may be further simplified by setting $\nabla \cdot \mathbf{E} = 0$, because we are dealing with charge-free regions. Equation 2.42 then reduces to

$$\nabla^2 \mathbf{E} - \mu_o \epsilon_o \frac{\partial^2 \mathbf{E}}{\partial t^2} = 0 \tag{2.46}$$

which is known as the *homogeneous vector wave equation* of the electric field in a *source-free region*. Similarly, we may obtain an equation for the magnetic flux density **B** in the form

$$\nabla^2 \mathbf{B} - \mu_o \epsilon_o \frac{\partial^2 \mathbf{B}}{\partial t^2} = 0 \tag{2.47}$$

Equation 2.47 is the *homogeneous vector wave equation* for the magnetic field in a source-free region. Each of the vector fields in equations 2.46 and 2.47 may be decomposed into three scalar components and wave equations with components from equation 2.45, are then obtained. Solutions to these equations provide propagation properties of waves as will be discussed next.

2.13 TIME HARMONIC FIELDS AND THEIR PHASOR REPRESENTATION

In many engineering applications, sinusoidal time functions are used because they are easy to generate. Solutions involving sinusoidal functions are also useful because arbitrary periodic time functions can be expanded into Fourier series of harmonic sinusoidal components. If sinusoidal source excitations are assumed, the current and charge distributions vary periodically with time as $\cos(\omega t + \theta)$ or $\sin(\omega t + \theta')$ where θ and θ' are arbitrary phase constants. Because $\cos(\omega t + \theta) = \sin(\omega t + \theta + \frac{\pi}{2})$, it is immaterial which function we use; however, once a decision is made to use a specific function—for example, the cosine—we have to stick with it throughout the solution. Unless otherwise indicated, we will use the cosine time function in our analysis.

Because of the linearity of Maxwell's equations, sinusoidal time variations of source functions of a given frequency produce steady-state sinusoidal variations of **E** and **B** of the same frequency. Therefore, in our analysis of time harmonic fields we will be dealing with instantaneous expressions of electric and magnetic fields in the form of cosine functions of the same frequency as that of the source. Maxwell's equations involve the differentiation of these fields with respect to time. The differentiation of cosine is sine, and hence it is likely in our analysis that we will be dealing with sine and cosine time functions in the same equation. Carrying these functions throughout the analysis is cumbersome, and combining them is tedious. An alternative formulation that avoids all these limitations may be achieved if the fields are represented in terms of their phasors or complex forms. To start with, the time variation is assumed to be in the form $e^{j\omega t}$ instead of the cosine function. This does not mean that there is an $e^{j\omega t}$ time function source, but such an assumed time form is more convenient for analysis. In the following, we will show that such an assumption will help us reduce the field functions of space and time to functions of space only, thus eliminating the problem of carrying sinusoidal time functions throughout the analysis.

Consider the current and charge sources $\mathbf{J}(\mathbf{r}, t)$ and $\rho(\mathbf{r}, t)$, which are, in general, functions of space $\mathbf{r}$ and time t. Assume that these sources have the complex time variation $e^{j\omega t}$, thus $\mathbf{J}(\mathbf{r}, t)$ may be replaced with $\hat{\mathbf{J}}(\mathbf{r})\, e^{j\omega t}$, and $\rho(\mathbf{r}, t)$ by $\hat{\rho}(\mathbf{r})\, e^{j\omega t}$. Because of the linearity of Maxwell's equations, the resulting electric and magnetic fields at steady state are given by $\hat{\mathbf{E}}(\mathbf{r})\, e^{j\omega t}$ and $\hat{\mathbf{B}}(\mathbf{r})\, e^{j\omega t}$. Substituting these source and field expressions in Maxwell's equations, we obtain

$$\nabla \cdot \epsilon_o(\hat{\mathbf{E}}(\mathbf{r}))e^{j\omega t} = \hat{\rho}\, e^{j\omega t}$$

$$\nabla \cdot (\hat{\mathbf{B}}(\mathbf{r})\, e^{j\omega t}) = 0$$

$$\nabla \times (\hat{\mathbf{E}}(\mathbf{r})\, e^{j\omega t}) = -j\omega(\hat{\mathbf{B}}(\mathbf{r})\, e^{j\omega t})$$

$$\nabla \times \left(\frac{\hat{\mathbf{B}}(\mathbf{r})\, e^{j\omega t}}{\mu_o} \right) = \hat{\mathbf{J}}(\mathbf{r})\, e^{j\omega t} + j\omega\epsilon_o(\hat{\mathbf{E}}(\mathbf{r})\, e^{j\omega t})$$

Eliminating the $e^{j\omega t}$ factor, we obtain the time harmonic Maxwell's equation in terms of the complex vector (phasor) fields and sources, thus,

$$\nabla \cdot (\epsilon_o\, \hat{\mathbf{E}}(\mathbf{r})) = \hat{\rho}$$

$$\nabla \cdot \hat{\mathbf{B}}(\mathbf{r}) = 0$$

$$\nabla \times \hat{\mathbf{E}}(\mathbf{r}) = -j\omega\hat{\mathbf{B}}(\mathbf{r}) \qquad (2.48)$$

$$\nabla \times \frac{\hat{\mathbf{B}}(\mathbf{r})}{\mu_o} = \hat{\mathbf{J}}(\mathbf{r}) + j\omega\epsilon_o\, \hat{\mathbf{E}}(\mathbf{r})$$

The important observation to be made here is regarding the absence of the time (t) variable; therefore, the time derivatives need no further consideration. The set of Maxwell's equations in equation 2.48 are much easier to solve because they are only functions of the space coordinates ($\mathbf{r}$). The obtained solutions from equation 2.48, however, are not complete because they lack the time information. The fields resulting from the solution of equation 2.48, for example, are not suitable for examining the propagation characteristics where the variation of the fields with the space coordinates and time is required. The obtained solutions need to be converted back to the real-time forms of the fields in which the time variable is restored. Similar to the procedure used in scalar voltage and current phasor analysis, the real-time forms may be obtained by multiplying the complex forms of the fields by $e^{j\omega t}$ and taking the real part of the result. Hence,

$$\mathbf{E}(\mathbf{r}, t) = Re(\hat{\mathbf{E}}(\mathbf{r}) e^{j\omega t})$$
$$\mathbf{B}(\mathbf{r}, t) = Re(\hat{\mathbf{B}}(\mathbf{r}) e^{j\omega t})$$

(2.49)

Taking the real part in equation 2.49 only emphasizes the fact that we are still sticking with our earlier decision to use cosine time functions exclusively. For example, if as a result of solving for the x component of the electric field in equation 2.48 we obtained a complex value of the form

$$\hat{E}_x(\mathbf{r}) = E_o e^{j\theta}$$

the real-time form of this component would be

$$E_x(\mathbf{r}, t) = Re(E_o e^{j\theta} e^{j\omega t})$$
$$= E_o \cos(\omega t + \theta)$$

which is the familiar sinusoidal time variation often used in engineering applications.

In summary, therefore, we will use the time harmonic Maxwell's equations in equation 2.48 to avoid carrying a cumbersome sine and cosine time function. These equations are easier to use in solving for the electric and magnetic fields because they involve only space ($\mathbf{r}$) variations. The resulting solutions, although easier to obtain, are not suitable for examining the propagation characteristics because the time variable is missing. To restore the time variable, we use equation 2.49. This procedure will be used extensively in our field analysis throughout this text.

Before concluding this section, let us derive expressions for the homogeneous vector wave equation for time harmonic fields. In a source-free region, we set $\hat{\rho}$ and $\hat{\mathbf{J}}$ equal to zero, and equations 2.48 reduce to

$$\boldsymbol{\nabla} \cdot \epsilon_o \hat{\mathbf{E}} = 0 \tag{2.50a}$$

$$\boldsymbol{\nabla} \cdot \hat{\mathbf{B}} = 0 \tag{2.50b}$$

$$\boldsymbol{\nabla} \times \hat{\mathbf{E}} = -j\omega \hat{\mathbf{B}} \tag{2.50c}$$

$$\boldsymbol{\nabla} \times \frac{\hat{\mathbf{B}}}{\mu_o} = j\omega \epsilon_o \hat{\mathbf{E}} \tag{2.50d}$$

The space argument **r** was eliminated for simplicity. Once again, taking the curl of both sides of (2.50c), using the vector identity equation 2.43, and substituting $\nabla \times \hat{\mathbf{B}}$ nd $\nabla \cdot \hat{\mathbf{E}}$ from equations 2.50d and a, respectively, we obtain

$$\boxed{\nabla^2 \hat{\mathbf{E}} + \omega^2 \mu_o \epsilon_o \hat{\mathbf{E}} = 0} \qquad (2.51a)$$

Following a similar procedure for the magnetic field, we obtain

$$\boxed{\nabla^2 \hat{\mathbf{B}} + \omega^2 \mu_o \epsilon_o \hat{\mathbf{B}} = 0} \qquad (2.51b)$$

Equations 2.51a and 2.51b are known as the homogeneous vector wave equations for the complex time harmonic fields in empty or free space. In the next section, we will discuss the solution of equation 2.51 for a plane wave propagating in empty space.

2.14 UNIFORM PLANE WAVE PROPAGATION IN FREE SPACE

Discussion of the propagation characteristics of a uniform plane wave in free or empty space provides the simplest solution of Maxwell's equations that considers the coupling between the electric and magnetic fields. The words used in the title of this section are quite important in emphasizing the various assumptions that will be used in this analysis. For example, we will be discussing waves that have wave fronts in the form of infinitely large plane surfaces. This property distinguishes the plane waves from other types of waves such as the cylindrical and the spherical waves. Cylindrical waves may be generated by infinitely long wire sources, and the spherical waves may be generated by three-dimensional sources such as short-wire and aperture antennas. Unlike the plane waves, the cylindrical waves have cylindrical wave fronts, and the spherical waves have wave fronts in the form of concentric spheres surrounding the source. The plane wave case is the simplest to solve for because, in this case, the familiar Cartesian coordinate system may be conveniently used. The other key word in the title is related to the uniform property of the wave. We will assume that the electric and magnetic fields associated with this wave are uniform throughout the infinite plane wave front surface. Finally, we will discuss the propagation characteristics of this wave in free space, which means the medium of propagation is free from any external charge and current distributions, and does not contain any material that may result in induced charge and current distributions.

To summarize and quantify these various assumptions, let us consider a plane wave propagation in the z direction. The plane wave fronts perpendicular to the z direction of propagation are shown in Figure 2.25. Because the x and y axes are in the plane of the wave fronts, the various assumptions described earlier may be summarized as follows:

1. For uniform plane waves the variation of the fields in the plane wave front—that is, with x and y—is equal to zero, hence,

$$\frac{\partial}{\partial x}(\hat{\mathbf{E}}, \hat{\mathbf{B}}) = \frac{\partial}{\partial y}(\hat{\mathbf{E}}, \hat{\mathbf{B}}) = 0$$

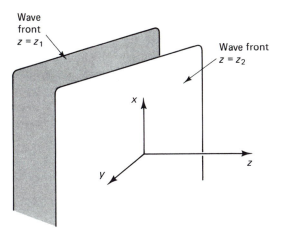

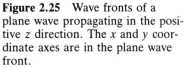

Figure 2.25   Wave fronts of a plane wave propagating in the positive z direction. The x and y coordinate axes are in the plane wave front.

2. For free (empty) space of propagation $\hat{\mathbf{J}} = \hat{\rho} = 0$. This assumption means that the sources of these fields are outside the region of propagation under consideration, and that we are not interested in how these fields were generated. It should be noted, however, that for generating waves of infinitely large plane wave fronts, an infinitely large source should be used.

Let us now examine the impact of these assumptions on Maxwell's equations. For free space, Maxwell's equations reduce to

$$\nabla \cdot \epsilon_o \hat{\mathbf{E}} = 0 \qquad\qquad \nabla \cdot \hat{\mathbf{B}} = 0$$
$$\nabla \times \hat{\mathbf{E}} = -j\omega\hat{\mathbf{B}} \qquad \nabla \times \frac{\hat{\mathbf{B}}}{\mu_o} = j\omega\epsilon_o\,\hat{\mathbf{E}} \tag{2.52}$$

For uniform plane wave, Faraday's equation may be simplified as

$$\begin{vmatrix} \mathbf{a}_x & \mathbf{a}_y & \mathbf{a}_z \\ \dfrac{\partial}{\partial x}=0 & \dfrac{\partial}{\partial y}=0 & \dfrac{\partial}{\partial z} \\ \hat{E}_x & \hat{E}_y & \hat{E}_z \end{vmatrix} = -j\omega(\hat{B}_x\,\mathbf{a}_x + \hat{B}_y\,\mathbf{a}_y + \hat{B}_z\,\mathbf{a}_z)$$

Equating the $\mathbf{a}_x$, $\mathbf{a}_y$, and $\mathbf{a}_z$ components on both sides of the equation, we obtain

$$-\frac{\partial \hat{E}_y}{\partial z} = -j\omega\hat{B}_x \tag{2.53a}$$

$$\frac{\partial \hat{E}_x}{\partial z} = -j\omega\hat{B}_y \tag{2.53b}$$

and

$$0 = -j\omega\hat{B}_z \tag{2.53c}$$

Similarly examining the curl equation of Ampere's law, we obtain

$$
\begin{vmatrix}
\mathbf{a}_x & \mathbf{a}_y & \mathbf{a}_z \\
0 & 0 & \dfrac{\partial}{\partial z} \\
\hat{B}_x & \hat{B}_y & \hat{B}_z
\end{vmatrix}
= j\omega\mu_o\,\epsilon_o(\hat{E}_x\,\mathbf{a}_x + \hat{E}_y\,\mathbf{a}_y + \hat{E}_z\,\mathbf{a}_z)
$$

The various components of this equation are

$$
-\frac{\partial \hat{B}_y}{\partial z} = j\omega\epsilon_o\,\mu_o\,\hat{E}_x
\tag{2.54a}
$$

$$
\frac{\partial \hat{B}_x}{\partial z} = j\omega\epsilon_o\,\mu_o\,\hat{E}_y
\tag{2.54b}
$$

$$
0 = j\omega\epsilon_o\,\mu_o\,\hat{E}_z
\tag{2.54c}
$$

By carefully examining the two sets of equations 2.53 and 2.54, we observe the following:

1. $\hat{B}_z = \hat{E}_z = 0$. Hence, for our uniform plane wave there are no electric or magnetic field components along the z direction of propagation. It should be noted that the uniform property of the wave, that is, setting $\frac{\partial}{\partial x} = \frac{\partial}{\partial y} = 0$ in the curl equations played an important role in arriving at this conclusion.

2. From equations 2.53a and 2.54b, it is clear that $\hat{B}_x$ and $\hat{E}_y$ are related to each other. In equation 2.53a, $\hat{B}_x$ acts like a source for generating $\hat{E}_y$, and $\hat{E}_y$ in equation 2.54b acts like a source for generating $\hat{B}_x$. There is also a similar relation between $\hat{B}_y$ and $\hat{E}_x$ as can be seen from equations 2.53b and 2.54a. To summarize this observation, we identify two independent (uncoupled) pairs of the electric and magnetic fields $(\hat{E}_x, \hat{B}_y)$ and $(\hat{E}_y, \hat{B}_x)$. Without loss of generality, we will consider the presence of only the $(\hat{E}_x, \hat{B}_y)$ pair and hence set $\hat{E}_y$ and $\hat{B}_x$ equal to zero.

It is desired, therefore, to solve for $\hat{E}_x$ and $\hat{B}_y$ using equations 2.53b and 2.54a. We may alternatively set $\hat{E}_x$, $\hat{B}_y$ equal to zero and examine the propagation characteristics of $\hat{E}_y$ and $\hat{B}_x$. This will be left for the student in one of the exercise problems. The propagation properties, in the general case when we have both pairs of the fields, may be described using the superposition. The polarization aspects of the resulting waves are discussed in section 2.15.

Differentiating equation 2.53b with respect to z and substituting $\partial \hat{B}_y/\partial z$ from equation 2.54a, we obtain the following second-order differential equation for the electric field component $\hat{E}_x$,

$$
\frac{\partial^2 \hat{E}_x}{\partial z^2} = -j\omega\frac{\partial \hat{B}_y}{\partial z} = -\omega^2\,\mu_o\,\epsilon_o\,\hat{E}_x
$$

or

$$
\frac{\partial^2 \hat{E}_x}{\partial z^2} + \omega^2\,\mu_o\,\epsilon_o\,\hat{E}_x = 0
\tag{2.55a}
$$

which is the scalar wave equation for the $\hat{E}_x$ component of the electric field. For uniform fields, $\hat{E}_x$ is a function only of z and hence the partial derivative may be replaced by the ordinary derivative.

$$\frac{d^2 \hat{E}_x}{dz^2} + \omega^2 \mu_o \epsilon_o \hat{E}_x = 0 \tag{2.55b}$$

The general solution of equation 2.55b may be expressed in the form

$$\hat{E}_x = \hat{C}_1 e^{-j\beta_o z} + \hat{C}_2 e^{j\beta_o z} \tag{2.56}$$

where $\hat{C}_1$ and $\hat{C}_2$ are complex constants and $\beta_o = \omega\sqrt{\mu_o \epsilon_o}$. A simple way to verify this solution is to substitute it back into equation 2.55b and see if it satisfies the differential equation. From equation 2.56, it will be shown shortly that as the wave propagates along the z direction, the amount of change in phase $e^{-j\beta_o z}$ depends on the value of β_o; this constant is therefore called the phase constant $\beta_o = \omega\sqrt{\mu_o \epsilon_o}$. We will also shortly show that the first term on the right side of equation 2.56 represents a wave traveling along the positive z direction, whereas the second term, $e^{j\beta_o z}$, represents a wave traveling along the negative z direction. For now, it is, however, appropriate to replace $\hat{C}_1$ by $\hat{E}_m^+$ and $\hat{C}_2$ by $\hat{E}_m^-$ to emphasize the anticipated property of propagation along the $+z$ (hence $\hat{E}_m^+$) and $-z(\hat{E}_m^-)$ of these waves. $\hat{E}_m^+$ and $\hat{E}_m^-$ are otherwise arbitrary (complex) amplitudes of these waves. Employing the amplitude symbols $\hat{E}_m^+$ and $\hat{E}_m^-$ in equation 2.56, we obtain

$$\hat{E}_x = \hat{E}_m^+ e^{-j\beta_o z} + \hat{E}_m^- e^{j\beta_o z} \tag{2.57}$$

Although it is generally more appropriate to maintain the amplitude constants $\hat{E}_m^+$ and $\hat{E}_m^-$ to be complex, let us, for the sake of simplifying the discussion, assume that these constants are real numbers E_m^+ and E_m^-, representing the amplitudes of the waves in V/m. Hence,

$$\hat{E}_x = E_m^+ e^{-j\beta_o z} + E_m^- e^{j\beta_o z} \tag{2.58}$$

We indicated in our discussion of the time harmonic fields that complex (phasor) expressions of these fields such as in equation 2.58 are not suitable for examining the propagation characteristics. Instead, the real-time forms that may be obtained from equation 2.49 should be used. The real-time form of E_x in equation 2.58 is, hence,

$$\begin{aligned}
E_x(z,t) &= Re(\hat{E}_x e^{j\omega t}) \\
&= Re[E_m^+ e^{j(\omega t - \beta_o z)} + E_m^- e^{j(\omega t + \beta_o z)}] \\
&= \underbrace{E_m^+ \cos(\omega t - \beta_o z)}_{\substack{\text{Wave traveling in the} \\ \text{positive } z \text{ direction}}} + \underbrace{E_m^- \cos(\omega t + \beta_o z)}_{\substack{\text{Wave traveling in the} \\ \text{negative } z \text{ direction}}}
\end{aligned} \tag{2.59}$$

To illustrate these positive z and negative z propagation directions, let us assume for a moment that E_x consists of only the first term of equation 2.59, and Figure 2.26 shows the variation of this term $E_x = E_m^+ \cos(\omega t - \beta_o z)$ with z for various specific values of t.

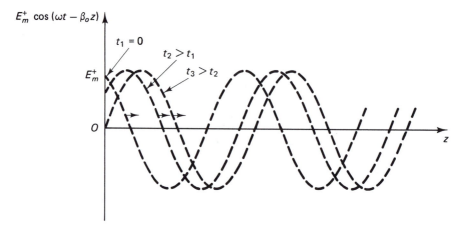

Figure 2.26   Plot of the variation of $E_m^+ \cos(\omega t - \beta_o z)$ as a function of z for different values of time t. It is clear that with the increase in time, the cosine wave of E_x continues to shift (travel) in the positive z direction.

At $t_1 = 0$ (first curve in Figure 2.26), for example,

$$E_x = E_m^+ \cos(-\beta_o z) = E_m^+ \cos \beta_o z$$

which is simply a cosine curve with its peak amplitude E_m^+ occurring at $z = 0$. At $t_2 > t_1$, E_x is given by

$$E_x = E_m^+ \cos(\omega t_2 - \beta_o z)$$

which is once again a cosine curve with its first peak value occurring at $\omega t_2 - \beta_o z = 0$, so that the cosine value would attain its maximum value of 1. This means that this peak occurs at $z = \omega t_2/\beta_o$, which is a positive value of z. It is clear that with the increase in time t from t_2 to t_3 ($t_3 > t_2$) the value of z increases and the peak value of the cosine curve continues to shift in the positive z direction, thus emphasizing the property of this wave as traveling along the positive z direction. We must therefore conclude that the first term in equation 2.59 represents a cosine wave of amplitude E_m^+ traveling along the positive z direction.

The second term to the right of equation 2.59 may be examined similarly as shown in Figure 2.27, where it is clear that the wave in this case is traveling in the negative z direction. For example, at $t = t_1 = 0$, the electric field wave is given by $E_m^- \cos(\beta_o z)$, which is a cosine wave with its peak value E_m^- occurring at the origin. At $t_2 > t_1$, the electric field will be $E_m^- \cos(\omega t_2 + \beta_o z)$ which attains its peak value when $\omega t_2 + \beta_o z = 0$, hence, at negative z value given by $z = -\omega t_2/\beta_o$. As is clear from Figure 2.27, this specific peak will continue to move in the negative z direction with the increase in time from t_2 to t_3, and so forth.

In summary, therefore, the general solution of the scalar wave equation in E_x given in equation 2.59 consists of two parts that describe two waves, one traveling in the positive z and the other in the negative z direction. It should be noted that equation

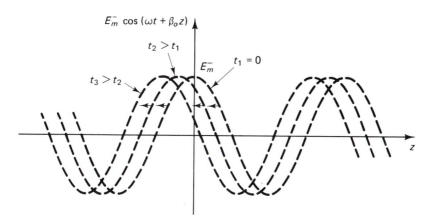

Figure 2.27 Plot of the variation of $E_m^-\cos(\omega t + \beta_o z)$ as a function of z for different values of t. With the time increase, the cosine wave representing the electric field continues to shift (travel) in the negative z direction.

2.59 represents the general solution of the wave equation, and any one component of the solution—for example, $E_x = E_m^+\cos(\omega t - \beta_o z)$ for the wave traveling in the positive z direction—may exist alone and without having to have the other wave traveling in the opposite direction.

Let us now turn our attention to the propagation properties of the magnetic field associated with this wave. The magnetic flux density may be obtained by substituting the complex expression for the electric field in equation 2.53b. Hence, for a wave traveling in the positive z direction, we have

$$\frac{\partial(E_m^+ e^{-j\beta_o z})}{\partial z} = -j\omega\hat{B}_y$$

$$(-j\beta_o)E_m^+ e^{-j\beta_o z} = -j\omega\hat{B}_y$$

or

$$\hat{B}_y = \frac{\beta_o}{\omega}E_m^+ e^{-j\beta_o z} = \frac{\omega\sqrt{\mu_o\epsilon_o}}{\omega}E_m^+ e^{-j\beta_o z}$$

$$= \frac{E_m^+}{c}e^{-j\beta_o z} = \frac{\hat{E}_x}{c} \tag{2.60}$$

where $c = 1/\sqrt{\epsilon_o\mu_o} \approx 3 \times 10^8$ m/s, which is the velocity of light in vacuum or air. Equation 2.60 simply states that for a wave traveling in free space along the positive z direction, the ratio between the electric field intensity $\hat{E}_x$ and the magnetic flux density $\hat{B}_y$ is real and equals c, the velocity of light in free space. The real ratio is important because it simply emphasizes that the electric and magnetic fields associated with a plane wave propagating in free space are *in phase*.

For a wave traveling along the negative z direction, the ratio between $\hat{E}_x$ and $\hat{B}_y$ is given using equation 2.53b, by

$$\frac{\partial(E_m^- e^{j\beta_o z})}{\partial z} = -j\omega \hat{B}_y$$

$$\therefore \hat{B}_y = -\frac{E_m^-}{c} e^{j\beta_o z} = -\frac{\hat{E}_x}{c} \tag{2.61}$$

The ratio between $\hat{E}_x$ and $\hat{B}_y$ is therefore still c, and the negative sign simply emphasizes the fact that either $\hat{E}_x$ has to reverse its direction from $(+\mathbf{a}_x)$ to $(-\mathbf{a}_x)$, or $\hat{B}_y$ should reverse its direction from $(+\mathbf{a}_y)$ to $(-\mathbf{a}_y)$. In other words, although an electric field in the $\mathbf{a}_x$ direction and magnetic field in the $\mathbf{a}_y$ direction are suitable for accompanying a wave traveling in the positive z direction, either $\hat{E}_x(-\mathbf{a}_x)$ and $\hat{B}_y \mathbf{a}_y$, or $\hat{E}_x(\mathbf{a}_x)$ and $\hat{B}_y(-\mathbf{a}_y)$ are suitable for accompanying a wave propagating in the negative z direction. In all cases, the ratio between the electric field and the magnetic flux density is equal to the speed of light in free space c.

A more commonly known ratio between the electric and magnetic fields is expressed in terms of the electric $\mathbf{E}$ and magnetic $\mathbf{H}$ field intensities. In MKS system of units, E is given in V/m, and the H units are in A/m. The ratio E/H is, therefore, V/A = ohms. The ratio of E/H has the units of ohms and is hence known as the intrinsic wave impedance η_o of free space. To obtain a value of this ratio we use equation 2.60 and replace $\hat{B}_y$ by $\mu_o \hat{H}_y$, therefore

$$\mu_o \hat{H}_y = \frac{\hat{E}_x}{c}$$

or

$$\frac{\hat{E}_x}{\hat{H}_y} = \mu_o c = \frac{\mu_o}{\sqrt{\mu_o \epsilon_o}} = \sqrt{\frac{\mu_o}{\epsilon_o}} = \eta_o \approx 120\pi = 377\Omega \tag{2.62}$$

It should be noted that the intrinsic wave impedance η_o is real, and once again this emphasizes that the electric and magnetic field intensities are in phase as shown in Figure 2.28. From equation 2.61 it can be shown that $\hat{E}_x/\hat{H}_y$ for a wave traveling in the negative z direction is equal to $-\eta_o = -120\pi$.

In summary, therefore, the electric and magnetic fields associated with a uniform plane wave propagating in free space have the following properties:

1. E $= \hat{E}_x \mathbf{a}_x$ and **H** $= \hat{H}_y \mathbf{a}_y$ are perpendicular to each other and also perpendicular to the direction of propagation. The direction of propagation is obtained by applying the right-hand rule from **E** to **H**. In other words, by curling our fingers from **E** to **H**, the thumb will point toward the direction of propagation, as shown in Figure 2.28.

2. The ratio between $\hat{E}_x$ and $\hat{H}_y$ is real and equals the intrinsic wave impedance η_o. These electric and magnetic fields are therefore in phase, which means that they reach their peak and zero values simultaneously, as also shown in Figure 2.28.

The following are additional important parameters that describe the characteristics of a propagating wave:

1. *Wavelength* λ. From the complex expressions of the electric or the magnetic fields (equations 2.57 and 2.60) it can be seen that as the wave propagates in the z

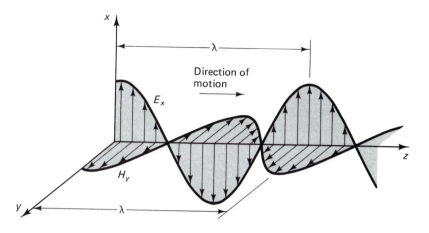

Figure 2.28 The electric and magnetic fields associated with a wave propagating along the positive z direction.

direction the phase term $e^{\pm j\beta_o z}$ changes by the amount $\beta_o z$. The distance z that the wave must travel so that the phase changes by 2π radians (one complete cycle) is of special interest and is called the wavelength λ. Hence,

$$\beta_o \lambda = 2\pi$$

or the wavelength

$$\lambda = \frac{2\pi}{\beta_o} = \frac{2\pi}{\omega\sqrt{\mu_o \epsilon_o}} = \frac{c}{f} \qquad \text{(meter)} \qquad (2.63)$$

for a wave propagating in free space. The wavelength λ is indicated in Figure 2.28.

   2. *Phase velocity* v_p. In an attempt to measure the velocity of propagation, an observer riding on a specific point in the wave, as shown in Figure 2.29, measures the time required for him to travel a distance z'. Because this observer is occupying a specific position in the wave, he experiences no phase change and moves along with the wave at a velocity known as the phase velocity v_p. To obtain an expression for v_p we note that occupying a specific position in the wave is mathematically equivalent to setting the argument of the cosine function in $E_m^+ \cos(\omega t - \beta_o z)$ equal to constant. Thus,

$$\omega t - \beta_o z = \text{constant}$$

and

$$\frac{dz}{dt} = v_p = \frac{\omega}{\beta_o} \text{ m/s.}$$

For a wave propagating in free space $\beta_o = \omega\sqrt{\mu_o \epsilon_o}$, hence,

$$v_p = \frac{1}{\sqrt{\mu_o \epsilon_o}} = c, \qquad \text{the speed of light} \qquad (2.64)$$

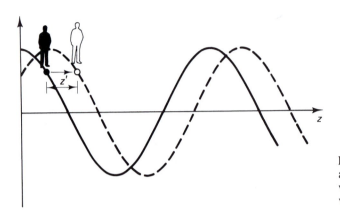

Figure 2.29 An observer riding along the wave moves with a velocity known as the phase velocity v_p.

   This concludes our discussion of the various properties of plane wave propagating in free space. These characteristics will be clarified by solving the following examples.

EXAMPLE 2.23

The time domain (instantaneous) expression for the magnetic field intensity of a uniform plane wave propagating in the positive z direction in free space is given by

$$\mathbf{H}(z,t) = 4 \times 10^{-6} \cos(2\pi \times 10^7 t - \beta_o z)\, \mathbf{a}_y \ \text{A/m}$$

1. Determine the phase constant β_o, and the wavelength λ.
2. Write a time domain expression for the electric field intensity $\mathbf{E}(z,t)$.

Solution

1. An important procedure for obtaining the wave parameters such as β_o, λ, and so forth is to compare the given expressions with those derived in our analysis. For example, from equation 2.62, we know that for a wave traveling along the positive z direction, the complex expression for the magnetic field intensity is given by

$$\hat{H}_y(z) = \frac{\hat{E}_x}{\eta_o} = \frac{\hat{E}_m^+}{\eta_o} e^{-j\beta_o z}$$

and the real-time form is, hence,

$$\mathbf{H}(z,t) = Re(\hat{H}_y(z)\, e^{j\omega t}) = Re\left(\frac{E_m^+}{\eta_o} e^{-j\beta_o z}\, e^{j\omega t} \right)$$

$$= \frac{E_m^+}{\eta_o} \cos(\omega t - \beta_o z)\, \mathbf{a}_y$$

Now comparing this expression with the given one for $\mathbf{H}$, it is clear that

$$\omega = 2\pi \times 10^7 \ \text{rad/s}$$

and

$$f = \frac{\omega}{2\pi} = 10^7 = 10 \ \text{MHz}$$

The wavelength $\lambda = c/f = 3 \times 10^8/10^7 = 30$ m. The phase constant β_o is obtained by noting from equation 2.64 that the phase velocity $v_p = c = \omega/\beta_o$.

$$\therefore \beta_o = \frac{\omega}{c} = \frac{2\pi \times 10^7}{3 \times 10^8} = 0.21 \text{ rad/m}$$

The phase constant may be alternatively obtained from equation 2.63,

$$\beta_o = \frac{2\pi}{\lambda} = \frac{2\pi}{\lambda} = 0.21 \text{ rad/m}$$

2. We know from the previous discussion that for a wave traveling along the positive z direction and with a magnetic field in the $\mathbf{a}_y$ direction, the electric field will be in the $\mathbf{a}_x$ direction. We also know that the amplitude of the electric field is related to that of the magnetic field by a constant ratio called intrinsic wave impedance η_o. Hence,

$$\mathbf{E}(z,t) = \eta_o \times 4 \times 10^{-6} \cos(2\pi \times 10^7 t - \beta_o z)\, \mathbf{a}_x \text{ V/m}$$

$$= 1.51 \times 10^{-3} \cos(2\pi \times 10^7 t - 0.21z)\, \mathbf{a}_x \text{ V/m}$$

There is, of course, no need to rederive these relations every time you are asked to use them. It is only important to recall the specific relations that may help you obtain the desired quantity or expression.

EXAMPLE 2.24

A uniform plane wave is traveling in the positive z direction in free space. The amplitude of the electric field E_x is 100 V/m, and the wavelength is 25 cm.

1. Determine the frequency of the wave.
2. Write complete-complex and time-domain expressions for the electric and magnetic field vectors.

Solution

1. For a plane wave propagating in free space, the wavelength is related to the frequency in equation 2.63 by

$$\lambda = \frac{c}{f} \qquad \text{or} \qquad f = \frac{c}{\lambda}$$

The frequency is then

$$f = \frac{3 \times 10^8}{0.25} = 12 \times 10^8 \text{ Hz}$$

$$= 1.2 \text{ GHz}$$

2. The complex expression for the electric field is

$$\hat{\mathbf{E}}(z) = 100 e^{-j\beta_o z}\, \mathbf{a}_x$$

To determine β_o, we use equation 2.63,

$$\beta_o = \frac{2\pi}{\lambda} = 8\pi$$

$$\therefore \hat{\mathbf{E}}(z) = 100e^{-j8\pi z}\,\mathbf{a}_x$$

The complex expression for the magnetic field intensity is

$$\hat{\mathbf{H}}(z) = \frac{100}{\eta_o}e^{-j8\pi z}\,\mathbf{a}_y$$

or

$$\hat{\mathbf{H}}(z) = \frac{100}{377}e^{-j8\pi z}\,\mathbf{a}_y$$

$$= 0.27e^{-j8\pi z}\,\mathbf{a}_y$$

The real-time forms of these fields are obtained from equation 2.59, hence,

$$\mathbf{E}(z,t) = 100\,\cos(24\pi \times 10^8\,t - 8\pi z)\,\mathbf{a}_x$$

and

$$\mathbf{H}(z,t) = 0.27\,\cos(24\pi \times 10^8\,t - 8\pi z)\,\mathbf{a}_y$$

EXAMPLE 2.25

A uniform plane wave with an electric field oriented in the x direction is propagating in free space along the positive z direction. If the frequency of the wave is $f = 10^8$ Hz and if the electric field has a maximum magnitude of 10 V/m at $t = 0$ and $z = 1/8$ m, obtain

1. The complex form expression for the electric field $\hat{E}_x$. Include numerical values for all the unknown constants.
2. Time-domain expression for the magnetic field intensity.

Solution

1. In our previous discussion we indicated that the amplitudes of the electric fields obtained from the solution of the scalar wave equation $\hat{E}_m^+$ and $\hat{E}_m^-$ are generally complex. We assumed them to be real to simplify the analysis and clarify the propagation characteristics of these waves, however. If we retained the complex amplitude $\hat{E}_m^+ = E_m^+ e^{j\phi^+}$ for the wave traveling along the positive z direction in equation 2.57, we obtain

$$\hat{E}_x(z) = E_m^+\,e^{j\phi^+}\,e^{-j\beta_o z}$$

where E_m^+ and ϕ^+ are the magnitude and phase of the complex amplitude $\hat{E}_m^+$. The real-time form for the electric field is

$$\mathbf{E}(z,t) = Re\big(E_m^+\,e^{j\phi^+}\,e^{-j\beta_o z}\,e^{j\omega t}\big)\,\mathbf{a}_x$$

$$= Re\big(E_m^+\,e^{j(\omega t - \beta_o z + \phi^+)}\big)\,\mathbf{a}_x$$

$$= E_m^+\,\cos(\omega t - \beta_o z + \phi^+)\,\mathbf{a}_x$$

This expression is certainly similar to the first term in equation 2.59 except for the phase term ϕ^+. The effect of this phase term is causing a shift in the initial location of the peak value of the electric field at $t = 0$. In equation 2.59, the maximum amplitude is at $t = 0$ and $z = 0$ (i.e., at the origin), because the cosine function attains its first peak value when the argument is zero. With the introduction of the phase term ϕ^+ the first peak value of the cosine function for $t = 0$ will be at $-\beta_o z + \phi^+ = 0$ or at a location $z = \phi^+/\beta_o$. In other words, the effect of this phase term is simply to shift the original cosine function at $t = 0$ by a positive z distance $= \phi^+/\beta_o$. The propagation characteristics from there on, that is, $t > 0$, are otherwise identical except with the additional phase shift ϕ^+ being carried throughout the analysis.

   In the present example, information was given to determine this phase constant ϕ^+. The electric field has its maximum value at $t = 0$, and $z = \frac{1}{8}$ m. Hence,

$$\omega t - \beta_o z + \phi^+ = 0 \text{ at } t = 0 \qquad \text{and} \qquad z = \frac{1}{8} \text{ m}$$

or

$$\phi^+ = \beta_o z = \beta_o \left(\frac{1}{8}\right)$$

To determine β_o, we note that $f = 10^8$ Hz, hence,

$$\lambda = \frac{c}{f} = \frac{3 \times 10^8}{10^8} = 3 \text{ m}$$

and

$$\beta_o = \frac{2\pi}{\lambda} = \frac{2\pi}{3} \text{ m}^{-1}$$

ϕ^+ is therefore

$$\phi^+ = \frac{2\pi}{3}\frac{1}{8} = \frac{\pi}{12} \text{ rad}$$

The complex expression for $\hat{E}_x$ is then

$$\hat{E}_x(z) = (10e^{j\pi/12})e^{-j2\pi/3\,z}$$

where $10e^{j\pi/12}$ is now the complex amplitude of the electric field.

2. The amplitude of the magnetic field is related to that of the electric field by $\eta_o = 120\pi$. Therefore,

$$\hat{H}_y(z) = \frac{10}{120\pi}e^{j\pi/12}e^{-j2\pi/3\,z}$$

The time-domain form of the magnetic field is then

$$\mathbf{H}(z,t) = 0.027 \cos\left(2\pi \times 10^8 t - \frac{2\pi}{3}z + \frac{\pi}{12}\right)\mathbf{a}_y \text{ A/m}$$

2.15 POLARIZATION OF PLANE WAVES

Earlier in Section 2.14, we indicated that there are two independent pairs of the electric and magnetic fields $(\hat{E}_x, \hat{H}_y)$ and $(\hat{E}_y, \hat{H}_x)$. We then continued our discussion based on the presence of the $\hat{E}_x, \hat{H}_y$ pair and often drew the parallel characteristics of the $\hat{E}_y, \hat{H}_x$ set. The question now is: What happens if we have both pairs? This brings up the polarization aspects of plane waves.

Let us consider the propagation characteristics of a plane wave in which the electric field has two components in the x and y directions. If the wave is propagating along the z axis, the electric field may be expressed as

$$\hat{E} = (\hat{A}\,\mathbf{a}_x + \hat{B}\,\mathbf{a}_y)e^{-j\beta z}$$

where the amplitudes $\hat{A}$ and $\hat{B}$ may be complex, that is,

$$\hat{A} = |\hat{A}|e^{ja}, \qquad \hat{B} = |\hat{B}|e^{jb}$$

To examine the polarization characteristics in this case, let us consider the following situations:

1. $\hat{A}$ and $\hat{B}$ have the same phase angle, that is, $a = b$. In this case, the x and y components of the electric field will be in phase, hence,

$$\hat{E} = (|\hat{A}|\,\mathbf{a}_x + |\hat{B}|\,\mathbf{a}_y)e^{-j(\beta z - a)} \tag{2.65}$$

and the real-time form of this field is given by

$$\mathbf{E} = (|\hat{A}|\,\mathbf{a}_x + |\hat{B}|\,\mathbf{a}_y)\cos(\omega t - \beta z + a) \tag{2.66}$$

From equations 2.65 or 2.66, it is clear that the electric field will always lie in the plane perpendicular to the z axis but be inclined at an angle θ, whose tangent is $|\hat{A}|/|\hat{B}|$ from the $x = 0$ plane, as shown in Figure 2.30. If $|\hat{B}| = 0$, the wave is clearly polarized in the x direction, whereas if $|\hat{A}| = 0$, the wave will be polarized in the y direction. These are the two special cases we discussed in the earlier sections. In all cases, however, and if we view the wave in the direction of propagation, we will find that the tip of the **E** vector follows a line, hence, the name *linear polarization*. A linearly polarized plane wave in which the **E** field oscillates along a line that makes a 45° angle with the x axis is shown in Figure 2.31. It should be emphasized that the wave linearly polarized even

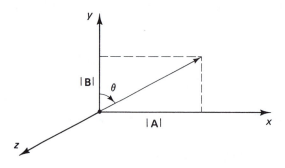

Figure 2.30 The x and y components of the electric field associated with a linearly polarized wave.

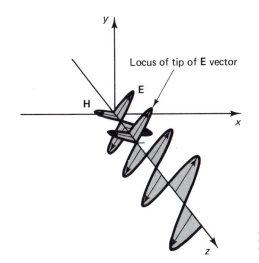

Locus of tip of **E** vector

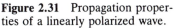

Figure 2.31 Propagation properties of a linearly polarized wave.

though the electric field contains two components in the x and y direction, as long as these components are in phase.

2. $\hat{A}$ and $\hat{B}$ have different phase angles. In this case, **E** will no longer remain in one plane, as will be illustrated in the following analysis:

$$\hat{\mathbf{E}} = \mathbf{a}_x |\hat{A}| e^{j(a - \beta z)} + \mathbf{a}_y |\hat{B}| e^{j(b - \beta z)}$$

where a and b are different phase angles. In the real-time form,

$$E_x = |\hat{A}| \cos(\omega t + a - \beta z)$$
$$E_y = |\hat{B}| \cos(\omega t + b - \beta z)$$

To examine the propagation characteristics of this electric field, let us consider the special case where $a = 0, b = \pi/2$. Then

$$E_x(z,t) = |\hat{A}| \cos(\omega t - \beta z)$$
$$E_y(z,t) = -|\hat{B}| \sin(\omega t - \beta z)$$
$$\mathbf{E}(z,t) = E_x(z,t) \mathbf{a}_x + E_y(z,t) \mathbf{a}_y$$

A plot of $\mathbf{E}(z,t)$ in the $z = 0$ plane is shown in Figure 2.32, where it is clear that the locus of the end point of the electric field vector $\mathbf{E}(0,t)$ will trace out an ellipse once each cycle, giving *elliptical polarization*.

3. For the special case in which $\hat{A}$ and $\hat{B}$ are equal in magnitude and differ in phase angle by $\pi/2$, the ellipse becomes a circle, and we have the case of *circular polarization*. Figure 2.33 shows the counterclockwise rotation of the electric field in a circularly polarized wave as it progresses along the z axis.

In summary, therefore, having two components of the electric field may change the polarization of the plane wave. If the two components of the electric field are in

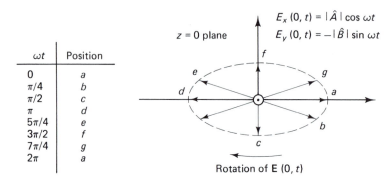

ωt	Position
0	a
$\pi/4$	b
$\pi/2$	c
π	d
$5\pi/4$	e
$3\pi/2$	f
$7\pi/4$	g
2π	a

Figure 2.32 Rotation of the electric field vector associated with an elliptically polarized plane wave.

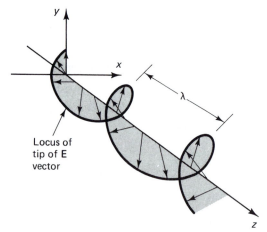

Figure 2.33 A circularly polarized wave in which the **E** field rotates as the wave advances.

phase, the plane wave will maintain its linear polarization status. Having a phase difference between the components of the electric field results in changing the polarization to elliptical polarization. If the magnitudes of the **E** field components are equal and the phases are different by $\pi/2$, the wave is said to be circularly polarized.

SUMMARY

In this chapter we derived the differential forms of Maxwell's equations from their integral forms. These differential forms relate the spatial variations of the electric and magnetic fields at a point to their time derivatives, and the charge and current densities at that point. To help us develop these forms we introduced some vector differential operations such as the divergence and the curl of a vector field. The basic definition of the divergence is

$$\text{div } \mathbf{F} = \lim_{\Delta v \to 0} \left[\frac{\oint_{\Delta s} \mathbf{F} \cdot d\mathbf{s}}{\Delta v} \right]$$

The divergence of a vector field at a point is thus a scalar quantity and is equal to the net outflow of the flux of the vector field emanating from a closed surface per unit volume, in the limit when the volume shrinks to zero. The basic definition of the curl, conversely, is

$$[\mathbf{\nabla} \times \mathbf{F}] \cdot \mathbf{n} = [\mathbf{\nabla} \times \mathbf{F}]_n = \underset{\Delta s \to 0}{\mathrm{Lim}} \left[\frac{\oint_{\Delta c} \mathbf{F} \cdot d\boldsymbol{\ell}}{\Delta s} \right]$$

where $[\mathbf{\nabla} \times \mathbf{F}]_n$ is the component of the curl in the direction of the unit vector $\mathbf{n}$, which is normal to the element of area Δs. The curl is the vector sum of three of such components. The magnitude of the curl is related to the circulation property of a vector field. This may be observed from the fact that each component of the curl is evaluated in terms of the circulation or the contour integration of the vector $\mathbf{F}$ around a closed contour Δc.

It should be emphasized that understanding the physical meaning of the divergence and the curl cannot be gained from the symbolic vector operations $\mathbf{\nabla} \cdot \mathbf{F}$ and $\mathbf{\nabla} \times \mathbf{F}$, respectively. Instead, their definitions in terms of the integrals of the vector should be kept in mind to help us understand the physical significance of these vector differential operators.

We also learned two theorems associated with the divergence and the curl. They are the divergence theorem and the Stokes's theorem, which are given, respectively, by

$$\oint_s \mathbf{A} \cdot d\mathbf{s} = \int_v (\mathbf{\nabla} \cdot \mathbf{A}) \, dv$$

and

$$\oint_c \mathbf{A} \cdot d\boldsymbol{\ell} = \int_s (\mathbf{\nabla} \times \mathbf{A}) \cdot d\mathbf{s}$$

The divergence theorem enables us to replace the surface integral of a vector over a closed surface by the volume integral of the divergence of the vector over the volume bounded by the closed surface and vice versa. Stokes's theorem, conversely, enables us to replace the line integral of a vector around a closed path by the surface integral of the curl of that vector over any surface bounded by that closed path. Using these vector differential operators and theorems, we obtained the following Maxwell's equations in differential forms.

Gauss's Law for Electric Field. The divergence of the electric flux density ($\epsilon_o \mathbf{E}$) at a point equal to the charge density at that point, that is,

$$\mathbf{\nabla} \cdot (\epsilon_o \mathbf{E}) = \rho_v$$

Gauss's Law for Magnetic Field. The divergence of the magnetic flux density at any point is equal to zero, that is,

$$\mathbf{\nabla} \cdot \mathbf{B} = 0$$

Faraday's Law.   The curl (or the circulation) of the electric field density at a point is equal to the time rate of decrease of the magnetic flux density at this point, that is,

$$\nabla \times \mathbf{E} = -\frac{\partial \mathbf{B}}{\partial t}$$

Ampere's Law.   The curl of the magnetic flux density at a point is equal to the total current density at this point. The total current density consists of two parts, one owing to actual flow of charges ($\mathbf{J}$), and the other is the displacement or virtual current density, which is the time derivative of the electric flux density, $\epsilon_o \mathbf{E}$, that is,

$$\nabla \times \frac{\mathbf{B}}{\mu_o} = \mathbf{J} + \frac{\partial(\epsilon_o \mathbf{E})}{\partial t}$$

We also examined the principles of uniform plane wave propagation in free space. Plane waves are the simplest possible simultaneous solution of the four Maxwell's equations. It, therefore, includes the coupling between the electric and magnetic fields in Maxwell's curl equations. We learned that uniform plane waves have their electric and magnetic fields perpendicular to each other and to the direction of propagation. The fields are uniform in the planes perpendicular to the direction of propagation (wave fronts). It is shown that $\cos(\omega t - \beta z)$ represents a wave motion in the positive z direction, whereas $\cos(\omega t + \beta z)$ represents a wave motion in the negative z direction.

The quantity $\beta = \omega\sqrt{\mu_o \epsilon_o}$ is called the phase constant and represents the amount of change of phase per unit distance along the direction of propagation. Waves propagating in air or vacuum only change in phase (no change in magnitude) as they propagate along the z direction.

The phase velocity is the velocity with which a constant phase (particular point on the wave) progresses along the direction of propagation. It is given by $v_p = \omega/\beta$. $v_p = c$ (the velocity of light) in vacuum or air.

The wavelength is the distance along the direction of propagation in which the phase of the wave changes by 2π radians, hence, $\lambda = 2\pi/\beta$.

The wavelength is related to the frequency f by $v_p = \lambda f = c$, where $c = 3 \times 10^8$ m/s.

The quantity $\eta_o = \sqrt{\mu_o/\epsilon_o}$ is the intrinsic wave impedance of free space. It is the ratio of the magnitudes of the $\mathbf{E}$ and $\mathbf{H}$ fields. $\eta_o = E_x/H_y = 120\pi$ Ω for a wave propagating in the positive z direction and -120π Ω for a wave propagating in the negative z direction. Negative sign means that E_x and H_y cannot be simultaneously associated with negative z traveling wave. Either $-E_x$ and H_y, or E_x and $-H_y$ are required.

We concluded this chapter with a brief discussion of the polarization of plane waves. Waves are linearly polarized when both components of the electric field E_x and E_y are in phase. Waves are circularly or elliptically polarized when there is a phase difference between the two field components E_x and E_y. Circular polarization is for the case when the magnitudes of E_x and E_y are equal, while elliptical polarization is for the case when the magnitudes of E_x and E_y are different.

PROBLEMS

1. Consider the function

$$\psi = E_o\left[1 - \left(\frac{a}{\rho}\right)^3\right] z \cos \phi$$

Evaluate $\nabla\psi$ in the cylindrical coordinate system.

2. Find $\nabla \times \mathbf{A}$ for the vector $\mathbf{A}$ given by

$$\mathbf{A} = -y\,\mathbf{a}_x + x\,\mathbf{a}_y$$

Evaluate $\oint_c \mathbf{A}\cdot d\boldsymbol{\ell}$ around the curve $x^2 + y^2 = 1$ (in the $z = 0$ plane). Also evaluate $\oint_s \nabla \times \mathbf{A}\cdot d\mathbf{s}$ over the surface s bounded by the contour $x^2 + y^2 = 1$. Use the obtained results to verify Stokes's theorem.

3. Evaluate both sides of the divergence theorem for the field vector $\mathbf{F} = 2\rho\,\mathbf{a}_\rho$. The volume of integration is bounded by $\rho = 3$, $0 \le \phi \le 2\pi$, and $0 \le z \le 2$.

4. If T and M are scalar fields, prove in the Cartesian coordinates that

$$\nabla(TM) = T\nabla M + M\nabla T$$

5. Calculate the curl of each of the following vectors:
  (a) $\mathbf{A} = \rho^2\,\mathbf{a}_\rho - z\,\mathbf{a}_z$
  (b) $\mathbf{B} = 3xz\,\mathbf{a}_x - y\,\mathbf{a}_y - x^2\,\mathbf{a}_z$
  (c) $\mathbf{C} = \dfrac{1}{\sqrt{r}}\,\mathbf{a}_r$
  (d) $\mathbf{D} = \rho\,\mathbf{a}_\rho + \rho^2\,\mathbf{a}_z$
  (e) $\mathbf{E} = xz\,\mathbf{a}_x + yz\,\mathbf{a}_y - y^2\,\mathbf{a}_z$
  (f) $\mathbf{F} = Kr^n\,\mathbf{a}_r$, where K is a constant

6. Verify Stokes's theorem for the vector $\mathbf{F}$ given by

$$\mathbf{F} = z^2\,\mathbf{a}_x - y^2\,\mathbf{a}_y$$

and the contour c being a square of side 1 lying in the x-z plane as shown in Figure P2.6.
  (a) Consider the area s to be the area of the square s_1, bounded by the contour c.
  (b) Consider the area s to be that of the five squares s_2, s_3, s_4, s_5, s_6, which are also bounded by the contour c.

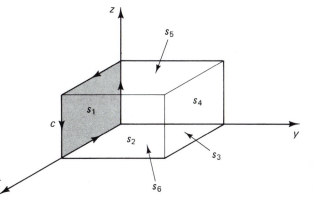

Figure P2.6 Surfaces and the contour c for integrations in Stokes's theorem.

7. Verify Stokes's theorem for the vector

$$\mathbf{F} = r\,\mathbf{a}_r$$

The contour c consists of the three-quarter circle arcs c_1, c_2, and c_3 as shown in Figure P2.7, and the area s is the octant of the sphere $r = 1$ bounded by the three arcs.

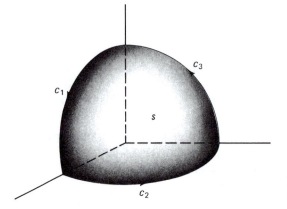

Figure P2.7   Surface s is the octant of the sphere $r = 1$.

8. Verify Stokes's theorem for the vector

$$\mathbf{F} = \rho\,\mathbf{a}_\phi - z\,\mathbf{a}_z$$

Choose the contour of integration c to be the circular path in the x-y plane as shown in Figure P2.8. Make the surface integral calculations for the following two surfaces:

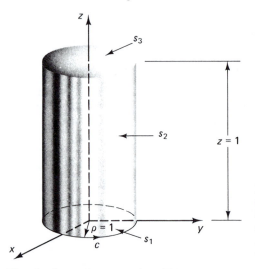

Figure P2.8 s_1 is the surface of a circle of unit radius in the x-y plane (bottom of the cylinder). s_2 and s_3 are the curved cylindrical surface and the top of the cylinder, respectively.

   (a) The circular surface s_1 enclosed by c.
   (b) The cylindrical surface $s_2 + s_3$, which is also enclosed by c.

9. Calculate the divergence of each of the following functions:
   (a) $\mathbf{A} = yz\,\mathbf{a}_x + xz\,\mathbf{a}_y + xy\,\mathbf{a}_z$
   (b) $\mathbf{B} = \rho\,\mathbf{a}_z$

(c) $\mathbf{C} = r\,\mathbf{a}_r$

(d) $\mathbf{D} = 2r^2\,\mathbf{a}_r$ (at $r = 3$)

(e) $\mathbf{E} = 3x\,\mathbf{a}_x + (y - 3)\,\mathbf{a}_y + (2 - z)\,\mathbf{a}_z$

10. Verify the divergence theorem for the vector

$$\mathbf{F} = \rho\,\mathbf{a}_\rho + z\,\mathbf{a}_z$$

Choose the surface of integration to be the quarter of a cylinder of radius a and height h as shown in Figure P2.10. The volume v is the volume enclosed by the surface s.

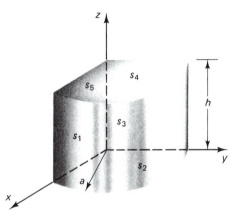

Figure P2.10   The closed surface of integration $s = s_1 + s_2 + s_3 + s_4$ and the volume of v is enclosed by s.

11. The line integral $\oint_c 2\rho^3(z + 1)\sin^2\phi\,d\phi$ is to be evaluated in the counterclockwise direction along the closed path $\rho = 2$, $z = 1$, and $0 \le \phi \le 2\pi$. Determine the result of this integration using Stokes's theorem and without actually evaluating the line integral. You may, however, want to check your result by carrying out the line integral and verifying Stokes's theorem.

12. Sketch each of the following vector fields and calculate their divergences. Use the sketches to explain the values of the obtained divergences.

(a) $\mathbf{A} = \rho\,\mathbf{a}_\phi$

(b) $\mathbf{B} = \phi\,\mathbf{a}_\rho$

(c) $\mathbf{C} = y\,\mathbf{a}_x$

(d) $\mathbf{D} = r\,\mathbf{a}_\phi$

(e) $\mathbf{E} = r\,\mathbf{a}_r$

(f) $\mathbf{F} = z\,\mathbf{a}_\rho$

13. In our discussion of the divergence and the curl of vector fields, we indicated that the vector flux representation of these vectors may help us physically understand the divergence and circulation properties of these vectors. For each of the following vectors, draw the flux representation and use your sketch to describe its divergence and circulation properties. Confirm your answers by actually calculating the divergence and the curl of these vectors.

(a) $\mathbf{A} = Kx\,\mathbf{a}_x$

(b) $\mathbf{B} = ky\,\mathbf{a}_x$

(c) $\mathbf{C} = K\rho\,\mathbf{a}_\phi$

(d) $\mathbf{D} = K\,\mathbf{a}_\rho$

14. (a) For the vector $\mathbf{A}$ given by

$$\mathbf{A} = K\cot\theta\,\mathbf{a}_\phi$$

in which K is constant, show that illustrating the validity of Stokes's theorem by using the contour c and the enclosed area s shown in Figure P2.14 is not possible.

(b) Explain the reason for violating Stokes's theorem under the condition specified in part a.

(c) Change the contour so as to overcome this difficulty and illustrate the validity of Stokes's theorem under the new conditions.

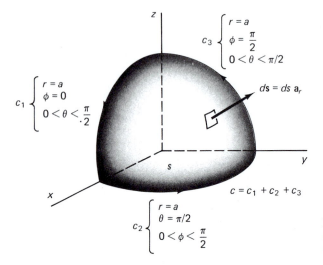

$$c_3 \begin{cases} r = a \\ \phi = \dfrac{\pi}{2} \\ 0 < \theta < \pi/2 \end{cases}$$

$$c_1 \begin{cases} r = a \\ \phi = 0 \\ 0 < \theta < \dfrac{\pi}{2} \end{cases}$$

$$ds = ds\ \mathbf{a}_r$$

$$c = c_1 + c_2 + c_3$$

$$c_2 \begin{cases} r = a \\ \theta = \pi/2 \\ 0 < \phi < \dfrac{\pi}{2} \end{cases}$$

Figure P2.14 Surface s and contour c for integration in Stokes's theorem.

15. Show that the vector

$$\mathbf{E} = \frac{\rho_v}{3} \frac{r}{\epsilon_o} \mathbf{a}_r$$

may represent a static electric field. Determine the charge density associated with it.

16. (a) Show that the following electric and magnetic fields are solutions to Maxwell's equations for time harmonics fields in free space $(\mathbf{J} = \rho_v = 0)$.

$$\hat{\mathbf{E}} = -2jE_o \sin \beta z\ \mathbf{a}_x \qquad \text{V/m}$$

$$\hat{\mathbf{B}} = 2\frac{E_o}{c} \cos \beta z\ \mathbf{a}_y \qquad \text{Wb/m}^2$$

where β is the phase constant $\beta = \omega \sqrt{\mu_o\,\epsilon_o}$, and $c = 1/\sqrt{\mu_o\,\epsilon_o}$ is the velocity of light.

(b) Obtain the real-time form for both these electric and magnetic fields.

17. Show that the vector

$$\mathbf{B} = \frac{1}{r^2} \sin \phi\ \cos^2 \theta\ \mathbf{a}_r$$

may represent a static magnetic flux density vector. Determine the current density associated with it.

18. An alternating voltage was connected between the plates of a parallel plate capacitor. The resulting electric field between the plates is given by

$$\mathbf{E} = 10 \cos \omega t\ \mathbf{a}_z \qquad \text{V/m}$$

If the medium between the plates is air (μ_o, ϵ_o), determine the total current crossing a square area of 0.1 m side length and placed perpendicular to the electric field in Figure P2.18.

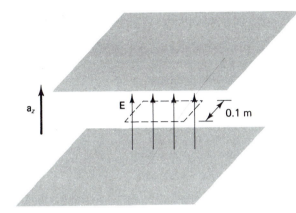

Figure P2.18 Parallel plate capacitor with an electric field **E** between the plates.

19. The vector **E** expressed in the cylindrical coordinate system by

$$\mathbf{E} = 3\rho^2 \mathbf{a}_\rho + \rho \cos\phi\, \mathbf{a}_\phi + \rho^3 \mathbf{a}_z$$

represents a static electric field. Calculate the volume charge density associated with this electric field at the point $(0.5, \pi/3, 0)$.

20. If the vector **B** represents a magnetic flux density in free space, find the component B_r of this vector

$$\mathbf{B} = B_r\, \mathbf{a}_r + \sin\theta\, \cos\phi\, \mathbf{a}_\theta + r\, \sin\phi\, \mathbf{a}_\phi$$

(specify an integration constant so that B_r remains finite as $r \to 0$).

21. Which of the following vectors can be a static electric field? Determine the charge density associated with it.

(a) $\mathbf{E} = ax^2 y^2 \mathbf{a}_x$, a is constant

(b) $\mathbf{E} = \dfrac{a}{\rho^2}[\mathbf{a}_\rho(1 + \cos\phi) + \mathbf{a}_\phi \sin\phi]$

22. Given the electric field $\mathbf{E} = E_o z\, \cos\omega t\, \mathbf{a}_x$ and that the current density **J** is zero in the region of interest, determine the magnetic flux density **B** using Ampere's law. Assume that **B** has only one component in the y direction, and that its value is zero at the coordinate origin.

23. Given a cylindrical electron beam of radius a, and if the electric field inside the beam is given by $\mathbf{E} = (\rho_o/4\epsilon_o a^2)\rho^3 \mathbf{a}_\rho$, where ρ_o is a constant, find the charge density in the beam. Also if the magnetic field inside the beam is given by $\mathbf{B} = (\mu_o J_o/3a)\rho^2 \mathbf{a}_\phi$ where J_o is a constant, determine the current density **J**.

24. (a) Starting from Ampere's law in its original form,

$$\nabla \times \frac{\mathbf{B}}{\mu_o} = \mathbf{J}$$

show why Maxwell found it necessary to introduce his displacement current term.

(b) Use Ampere's law in part a and the continuity equation 2.36 to obtain an expression for the displacement current term.

(c) Explain briefly why Maxwell's displacement current term was considered a history-making expression.

25. The electric field **E** induced owing to a time-varying magnetic flux density **B** is given by

$$\mathbf{E} = E_o z^2\, \cos\omega t\, \mathbf{a}_x$$

(a) Assuming that **B** has only $\mathbf{a}_y$ component, and that its initial value is zero, find **B** by using Faraday's law in differential form.

(b) Draw curves showing the variations of the electric and magnetic fields with time, and show that they satisfy Lenz's law.

26. (a) The vector **E** expressed in the cylindrical coordinate system

$$\mathbf{E} = 3\rho^2\,\mathbf{a}_\rho + \rho\,\cos\phi\,\mathbf{a}_\phi + \rho^3\,\mathbf{a}_z$$

represents a static electric field. Calculate the volume charge density associated with this electric field at the point $(0.5, \pi/3, 0)$ in free space.

(b) If the vector **B** represents a magnetic flux density in free space,

$$\mathbf{B} = B_r\,\mathbf{a}_r + \sin\theta\,\cos\phi\,\mathbf{a}_\theta + r\,\sin\phi\,\mathbf{a}_\phi$$

find the component B_r of this vector. (Specify an integration constant so that B_r remains finite as $r \to 0$.)

27. The **E** and **B** vectors are given by $\mathbf{E} = K_1 r\,\mathbf{a}_r$ (spherical coordinates) and $\mathbf{B} = K_2\rho\,\mathbf{a}_\phi$ (cylindrical coordinates).

(a) Show that these vectors may represent static electric and magnetic fields, respectively.

(b) Determine the charge density associated with the electric field and the current density associated with the magnetic field.

(c) Use the flux representation of these vectors to illustrate the zero or nonzero values of the various curl and divergence *vector operations* you used in parts a and b.

28. (a) Define and give expressions for wavelength, phase factor, phase velocity, and intrinsic wave impedance.

(b) The electric field intensity of a uniform plane wave is given by

$$\mathbf{E}(z,t) = 37.7\,\cos(6\pi \times 10^8 t + 2\pi z)\,\mathbf{a}_x$$

Find the following:

   (i) Frequency.

   (ii) Wavelength.

   (iii) Phase velocity.

   (iv) Direction of propagation.

   (v) Associated magnetic field intensity vector $\mathbf{H}(z,t)$.

29. An **H** field travels in the positive z direction in free space with a phase constant $\beta_o = 30$ rad/m and an amplitude of $1/3\pi$ A/m. If this field is in the $\mathbf{a}_y$ direction, and has its maximum value at $t = 0$ and $z = 0$, write a suitable expression for **E** and **H**. Determine the frequency and wavelength.

30. The electric field associated with a uniform plane wave propagating in free space is described by

$$\mathbf{E} = 50e^{j\pi/4}\,e^{-j\pi z/3}\,\mathbf{a}_x \qquad \text{V/m}$$

Find the following:

(a) Direction (polarization) of electric field.

(b) Direction in which wave travels.

(c) Wavelength.

(d) Frequency and period.

(e) Magnetic field vector.

(f) Real-time form of electric field.

31. A uniform plane wave in free space has the electric field,

$$\hat{E}e^{j\omega t} = 30\pi e^{j(10^8 t + \beta z)} \mathbf{a}_x \qquad \text{V/m}$$

and the magnetic field intensity,

$$\hat{H}e^{j\omega t} = H_m e^{j(10^8 t + \beta z)} \mathbf{a}_y \qquad \text{A/m}$$

(a) Determine the direction of propagation of this wave.
(b) Find H_m, β, and the wavelength λ.

32. Give a general real-time expression for $\mathbf{E}$ for a 200-MHz uniform plane wave propagating in the positive z direction. The electric field is oriented parallel to the x axis and reaches its positive maximum of 150 V/m at $z = 1$ and $t = 0$. The medium of propagation is free space. Give numerical values for all constants in your expression.

33. (a) In our discussion of plane wave propagation, we indicated that there are two uncoupled pairs of the electric and magnetic fields that may be solved independently. Following a procedure similar to that used in determining $\hat{E}_x$ and $\hat{B}_y$, obtain a solution for the other pair $\hat{E}_y$ and $\hat{B}_x$. Specifically show that

$$\hat{E}_y(z) = \hat{E}_m^+ e^{-j\beta_o z} + \hat{E}_m^- e^{j\beta_o z}$$

$$\hat{B}_x(z) = -\frac{\hat{E}_m^+}{c} e^{-j\beta_o z} + \frac{\hat{E}_m^-}{c} e^{j\beta_o z}$$

$$\hat{H}_x(z) = -\frac{\hat{E}_m^+}{\eta_o} e^{-j\beta_o z} + \frac{\hat{E}_m^-}{\eta_o} e^{j\beta_o z}$$

(b) Compare these results with those obtained earlier for the $\hat{E}_x$ and $\hat{B}_y$ fields.
(c) Also describe the significance of the negative sign in the magnetic fields expressions. Sketch a real-time wave plot similar to that of Figure 2.28 to illustrate your answer.

34. The electric field intensity of a uniform plane wave is given by

$$\mathbf{E} = 15 \cos\left(\pi \times 10^8 t + \frac{\pi}{3} z\right)\mathbf{a}_y \qquad \text{V/m}$$

Determine the following:
(a) Direction (polarization) of electric field.
(b) Direction of propagation.
(c) Frequency and wavelength.
(d) Magnetic field intensity vector $\mathbf{H}$. Specifically indicate the direction of $\mathbf{H}$.

35. The superposition of two uniform plane waves of equal magnitudes and propagating in opposite directions results in a composite wave having electric and magnetic fields given by

$$\mathbf{E}(z,t) = 2E_m \sin\beta_o z \, \sin\omega t \, \mathbf{a}_x$$

$$\mathbf{H}(z,t) = 2\frac{E_m}{\eta_o} \cos\beta_o z \, \cos\omega t \, \mathbf{a}_y$$

Show that these fields satisfy all Maxwell's equations and the scalar wave equations for time-harmonic electric and magnetic fields.

36. (a) Summarize in *four points* the important basic properties of a uniform plane wave propagating in free space ($\epsilon = \epsilon_o, \mu = \mu_o$).
(b) A uniform plane wave is propagating in the positive z direction in free space. If the wavelength of the wave is measured to be $\lambda = 3$ cm:

(i) Determine the frequency f and the phase constant β of this wave.

(ii) If the amplitude of the x polarized electric field associated with this wave is

$$\hat{E}_m = 200e^{j\pi/4} \qquad \text{V/m}$$

obtain a real-time expression for this electric field.

(iii) Obtain a *phasor* and a *real-time* expression for the magnetic field associated with this wave.

CHAPTER 3

MAXWELL'S EQUATIONS AND PLANE WAVE PROPAGATION IN MATERIALS

1 INTRODUCTION

In chapter 2, we introduced the differential forms of Maxwell's equations and used them to describe the plane wave propagation in free space. The space of propagation was assumed to be free from any charges, currents, or material media. In this chapter, we will introduce Maxwell's equations in conductive medium. We will learn that as a result of the interaction between the electric and magnetic fields and the material media, additional current and charge terms will be introduced and will therefore be considered in Maxwell's equations. To appreciate the need for these additional terms, let us briefly examine the classical atomic structure of materials. We learned in our elementary physics courses that all matter consists of atoms, a very large number of atoms, and that these atoms consist of a positively charged nucleus surrounded by a cloud of electrons. The electrons are constantly orbiting around the nucleus, whereas the nucleus is spinning around itself. In other words, according to the atomic structure of materials, all matter essentially consists of charged particles, and we therefore expect that the presence of the electric and magnetic fields in these materials will result in exerting forces on these charges. These forces may subsequently result in induced accumulation of charges and circulation of currents in the material. All these induced charges and currents should be included in Maxwell's equations, and this is why we indicated earlier that there will be modifications in these equations when they describe the fields in materials. In summary, we should therefore remember that materials contain charged particles that respond to applied electric and magnetic fields, and give rise to currents and charges that consequently modify the properties of the original fields.

We shall also learn that three basic phenomena result from the interaction of the charged particles in materials with the electric and magnetic fields. These are conduction, polarization, and magnetization. Although a given material may exhibit all three properties, it is classified as a conductor, a dielectric, or a magnetic material depending on the predominant phenomenon. These three kinds of materials will be introduced and the added effects, such as the generation of bound charge density, polarization, or conduction currents, will be included in Maxwell's equations. After including all the necessary modifications in Maxwell's equations, the integral form of these equations will be used to derive a set of boundary conditions at an interface separating two different material regions. These boundary conditions are basically mathematical relations that describe the transitional properties of the electric and magnetic fields from one material region to another. After learning about the various properties of materials, it will be easy to see that even if we have the same external sources, fields are different in different material regions. At the boundaries between any two different material regions, the fields have to change their properties from those in one region to those in the other. The laws of electromagnetic fields that provide the quantitative relations between these fields are called the boundary conditions, which will be described in this chapter.

Finally, the modified Maxwell's equations in their differential forms will be used to discuss uniform plane wave propagation in material media. Specific and very important differences between plane wave propagation in free space and in materials of various properties will then be indicated.

3.2 CHARACTERIZATION OF MATERIALS

We will start our discussion in this chapter by characterizing the various properties of materials. This characterization will be made based on the reaction of the material to the applied electric and magnetic fields. In general, materials can be divided into three types.

Conductors.   These conductors are characterized by the presence of many free conduction electrons. These free electrons are constantly in motion under thermal agitation, being released from their atom at one point and captured by another atom at a different point. Under the influence of an external electric field $\mathbf{E}$ these electrons experience a force, and the resulting flow of electrons is equivalent to an induced current known as the conduction current. In the following section, we will quantify the conduction current in terms of the applied electric field.

Dielectric materials.   These materials are basically insulators that are characterized by the presence of many bound, rather than free, charges. On the application of an external electric field, therefore, these charges will not be free to move but instead they will be only displaced from their original positions. As we know, applying an electric force that causes only the displacement of charges is equivalent to applying a mechanical force to stretch a spring. Both actions result in storing energies. Therefore, the dielectric materials are basically characterized by their ability to store electric energy. We will see in the following sections that as a result of applying an external electric field on a dielectric material, there will be induced charge and current distributions. These induced sources are known as the polarization charges and currents, and will be included in Maxwell's equations together with the external sources.

Magnetic materials.   Magnetic materials are generally characterized by their ability to store magnetic energy. To illustrate this effect, let us once again consider the model of an atom. We all know that the positively charged nucleus spins around itself, whereas the surrounding cloud of electrons is orbiting around the nucleus and the electrons are also spinning around themselves. The motion of a charge, such as the orbiting of electrons around the nucleus, is equivalent to the flow of an electric current in a loop. We will see in the following sections that the application of an external magnetic field tends to align these current loops in the direction of the magnetic field. In other words, there will be an additional magnetic energy stored in the material as a result of the work done in aligning the current loops in the direction of the magnetic field. We will also learn that as a result of the process of aligning the current loops (say equivalent to the orbiting electrons), there will be induced currents known as magnetization currents. These currents will, of course, result in an additional modification of Maxwell's equations. All these effects including the induced charges and the new induced current sources will be described in detail in the following sections.

3.3 CONDUCTORS AND CONDUCTION CURRENTS

On the application of an external electric field $\mathbf{E}$ to a conducting material, one may expect that the free electrons will accelerate under the influence of the electric field force. We will soon show that this is not the case simply because the electrons are not in free space. For example, if an electron of charge $(-e)$C and mass m (kg) is moving in an electric field $\mathbf{E}$ (V/m), the motion of this electron may be described according to Newton's law by

$$m\frac{d\mathbf{v}}{dt} = -e\,\mathbf{E}$$

Solving for the velocity **v** and assuming an initial condition of zero velocity at $t = 0$, we obtain

$$\mathbf{v} = -\frac{e\,\mathbf{E}}{m}t \qquad \text{m/s}$$

If we have n electrons per unit volume, then the charge density per unit volume is $n(-e)$. The current density resulting from the flow of the charge $(-ne)$ with a velocity **v** is given by

$$\mathbf{J} = (-ne)\mathbf{v} = \frac{ne^2\,\mathbf{E}}{m}t \qquad \text{A/m}^2$$

This current density quantity is obviously unrealistic simply because it indicates that the current will indefinitely increase with time t as long as the electric field is still applied. In other words, this current density quantity is not experimentally verifiable, and, hence, our assumed free-electron model is not correct. The reason for the invalidity of our assumed model is simply related to the fact that the conduction electrons, although called free electrons, are not actually moving in free space. They are, instead, under constant collision with the atomic lattice and, as indicated earlier, are being released from one atom and captured by another. As a result of their continuous collisions and because of the friction mechanisms in crystalline material, these conduction electrons do not accelerate under the influence of the electric field but instead they drift with an average velocity proportional to the magnitude of the applied electric field. The motion of the free electrons in the absence and under the influence of an external electric field is illustrated in Figure 3.1a and b, respectively. With this new picture describing the motion of the free electrons, we go back to Newton's law, which in this case states that the average change in the momentum of a free electron (or the average momentum transfer) equals the applied force. Hence,

$$\frac{m\mathbf{v}_a}{\tau_c} = -e\,\mathbf{E}$$

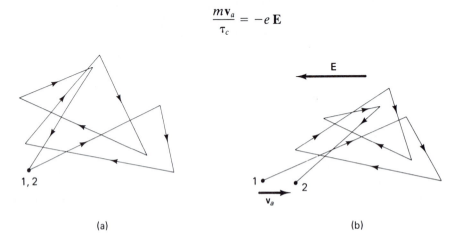

(a)   (b)

Figure 3.1   The random motion of free electrons. (a) In the absence of an electric field, there is no net average velocity drift. (b) In the presence of electric field where the electron effectively drifted from position 1 to 2 with an average velocity $\mathbf{v}_a$.

TABLE 3.1 CONDUCTIVITIES OF SEVERAL
CONDUCTING AND INSULATING MATERIALS

Material	Conductivity (S/m)
Silver	6.1×10^7
Copper	5.7×10^7
Gold	4.1×10^7
Aluminum	3.5×10^7
Tungsten	1.8×10^7
Sea water	4
Wet earth	10^{-3}
Silicon	3.9×10^{-4}
Distilled water	10^{-4}
Glass	10^{-12}
Mica	10^{-15}
Wax	10^{-17}

where τ_c is called the mean free time, which basically denotes the average time interval between collisions. v_a is the average drift velocity. This mean drift velocity is therefore given by

$$\mathbf{v}_a = -\frac{e\tau_c}{m}\mathbf{E} \qquad \text{m/s}$$

and the current density $\mathbf{J}$ associated with the flow of these electronic charges is hence

$$\mathbf{J} = (-ne)\,\mathbf{v}_a = \frac{ne^2\tau_c}{m}\mathbf{E} \qquad \text{A/m}^2 \qquad\qquad (3.1\text{a})$$

$$= \sigma\mathbf{E} \qquad \text{A/m}^2 \qquad\qquad (3.1\text{b})$$

The quantity σ is called the conductivity of the material. From equations 3.1a and b, it is clear that the conductivity is a physical characteristic of a material because it depends on the number of free electrons per unit volume of the material n and the mean free time τ_c. Table 3.1 provides values of σ for various commonly used materials. Additional conductivities of materials are given in Appendix D. Also the relation given in equation 3.1b between the current density $\mathbf{J}$, the conductivity of the material, and the applied electric field $\mathbf{E}$ is known as Ohm's law in a point form.

.4 DIELECTRIC MATERIALS AND THEIR POLARIZATION

In the previous section we learned that conductors are characterized by abundance of "conduction" or *free* electrons that give rise to conduction current under the influence of an applied electric field. In dielectric materials, however, *bound* electrons are predominant and their basic reaction to the application of an external electric field is therefore related to the displacement of the bound charges rather than to their drift. In other words, the common characteristic that all dielectric materials have is their ability to store electric energy because of the shifts in the relative positions of the internal positive and negative charges against the normal molecular and atomic forces. The mechanism of charge displacement is called polarization, and it may take different forms in various dielectric materials. There are, however, three basic mechanisms that may result in dielectric polarization.

Electronic polarization.   It results from the displacement of the bound electrons of an atom such that the center of the cloud of electrons is separated from the center of the nucleus as shown in Figure 3.2b. An electric dipole is therefore created, and the atom is said to be polarized. The *electric dipole* is the name given to two point charges $+Q$ and $-Q$ of equal magnitudes and opposite signs, separated by a small distance. If the vector length directed from $-Q$ to $+Q$ is **d**, as shown in Figure 3.2c, the dipole moment is defined as $Q\mathbf{d}$ and is given the symbol **p**. Thus $\mathbf{p} = Q\mathbf{d}$.

Orientational polarization.   In some dielectric materials known as polar substances, such as water, polarization may exist in the molecular structure even if there is no external electric field. In the absence of an external field, however, the polarizations of individual atoms are randomly oriented and hence the net polarization on a macroscopic scale is zero as shown in Figure 3.3a. The application of an external field results in torques acting on the *microscopic* dipoles (see Figure 3.3b) so as to orient them in the direction of the applied field as shown in Figure 3.3c.

Ionic polarization.   Certain materials such as sodium chloride (NaCl) consist of positive and negative ions that are electrically bound together. These ions are formed as a result of the transfer of electrons from one atom to another. On the application

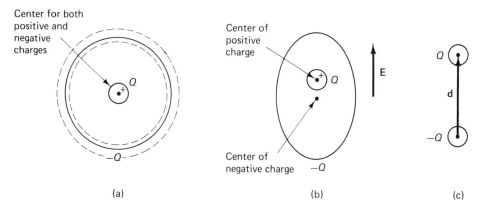

(a)   (b)   (c)

Figure 3.2 Microscopic view illustrating the formation of an electric dipole owing to the application of an electric field **E**. (a) In the absence of the electric field, the centeroids of the positive and negative charges are the same, and hence the dipole moment is equal to zero. (b) On the application of an electric field **E**, the centers of the positive and negative charges shift, thus resulting in the formation of an electric dipole. (c) The representation of an electric dipole moment configuration, $\mathbf{p} = Q\mathbf{d}$, where **d** is the vector distance from the negative charge to the positive charge.

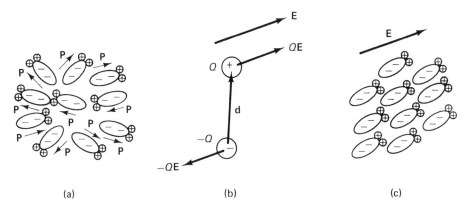

Figure 3.3 Orientational polarization in polar dielectric material. (a) Macroscopic view showing that the dipole moments are already existing in a polar dielectric. They, however, are arbitrarily oriented in the absence of an external electric field. (b) A microscopic view of the torque applied on each dipole moment in the presence of the external electric field **E**. (c) Macroscopic view of the torques in (b) that tend to orient the electric dipoles in the direction of the electric field, thus resulting in a total induced polarization.

of an external electric field, these ions separate and thus form electric dipoles. The resulting electric polarization is known as ionic polarization.

In our discussion of the dielectric materials we have so far quantified their polarization in the presence of an external electric field in terms of the electric dipole moment **p**. This dipole moment clearly describes the microscopic property of the material that is not only difficult but also inadequate for an overall macroscopic description of the dielectric material. Electric dipole moment varies from one atom to another, and the use of the dipole moment concept requires the knowledge of the spatial location of each atom or molecule in the material. It is, therefore, more adequate to characterize dielectric materials in terms of polarization which is a quantity that provides a macroscopic description of the electric dipole moment per unit volume. Thus, if n is the number of atoms or molecules per unit volume (per m^3) of the material, then the polarization **P** is given by

$$\mathbf{P} = \lim_{\Delta v \to 0} \frac{1}{\Delta v} \sum_{i=1}^{n\Delta v} \mathbf{p}_i = n\,\mathbf{p}_a = nQ\mathbf{d}_a = \rho_+ \mathbf{d}_a$$

where $n\Delta v$ is the number of dipoles in a volume Δv. $\mathbf{p}_a$ is the average dipole moment per molecule, and $\mathbf{d}_a$ is the average vector separation distance (displacement) between the center of the positive and negative charges. $\rho_+ = nQ$ and is the density of the positive charge (charge per m^3) generated in the polarized region. It should be emphasized that although the dipole moment **p** provides microscopic information about the polarization of the material, the polarization **P** quantifies the electric polarization of a material on an average or macroscopic basis. Thus, in the absence of an external electric field, the dipole moment in a polar dielectric is not zero, whereas the polarization is zero. The polarization concept is therefore more adequate (on an average basis) for describing the status of a bulk piece of a dielectric material and will be frequently used in the following sections to quantify the induced polarization charges and currents in a dielectric material.

3.4.1 Polarization Current

In this section, we will use the macroscopic concept called polarization to quantify the induced polarization current in a dielectric material. Let us assume a time-varying electric field **E** that is applied to a dielectric material.

$$\mathbf{E} = E_o \cos \omega t \, \mathbf{a}_x$$

This electric field may, for example, be due to a wave propagating in the dielectric medium. As a result of the presence of this electric field in the dielectric, there will be induced electric dipoles, as shown in Figure 3.4. These induced dipoles will also be oscillating with the time variation of the electric field, as shown in Figure 3.4. For example, at $\omega t = 0$, Figure 3.4 shows that there is a maximum polarization in the positive x direction, whereas at $\omega t = \pi/2$, the applied electric field equals to zero, and hence the induced polarization vanishes. The direction of the polarization also reverses with the reversal of the direction of the applied electric field. Across an infinitesimal element of area $\Delta y \, \Delta z$ that is perpendicular to the direction of the electric field, there will be positive charges crossing this area periodically with time. This flow of charge is clearly equivalent to an induced oscillating current called polarization current. To quantify this polarization current, let us assume a linear dielectric—that is, isotropic material—in which the polarization **P** is linearly proportional to the applied electric field, that is,

$$\mathbf{P} = \epsilon_o \chi_e \mathbf{E}$$

where χ_e is the constant of proportionality and is called the electric susceptibility of the material. χ_e simply describes the ability of the dielectric material to be polarized in the presence of an electric field. For these types of isotropic materials, the polarization is clearly in the direction of the applied electric field. For the time-varying electric field, the polarization is given by

$$\mathbf{P} = \epsilon_o \chi_e E_o \cos \omega t \, \mathbf{a}_x \tag{3.2}$$

Because the polarization is defined as the dipole moment per unit volume, the total dipole moment in the volume $d\Delta z \Delta y$, shown in Figure 3.4, is given by

$$\mathbf{P}\Delta v = \epsilon_o \chi_e \, d\Delta z \Delta y E_o \cos \omega t \, \mathbf{a}_x$$

Equivalently, we may think of this total dipole moment as resulting from two large time-varying charges $Q = \epsilon_o \chi_e \Delta z \, \Delta y \, E_o \cos \omega t$ separated by a distance d, as shown in Figure 3.5. The current associated with these time-varying charges is then $I = dQ/dt$. Hence, the induced current density is given by

$$\mathbf{J} = \frac{\mathbf{I}}{\Delta z \, \Delta y} = \frac{1}{\Delta z \, \Delta y} \frac{dQ}{dt} \mathbf{a}_x$$

$$= -\omega \epsilon_o \chi_e \sin \omega t \, \mathbf{a}_x$$

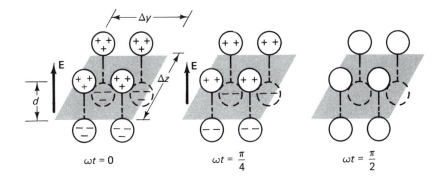

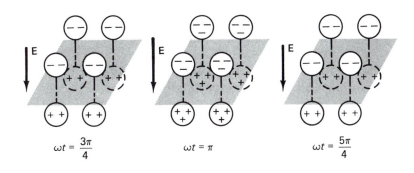

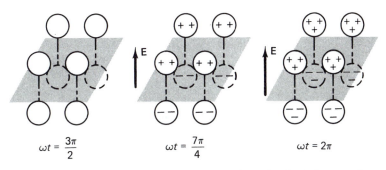

Figure 3.4   Induced electric dipoles in a dielectric under the influence of a time-varying electric field.

This induced current density term is identically equal to $\partial \mathbf{P}/\partial t$ where $\mathbf{P}$ is the polarization given by equation 3.2. Therefore, we conclude that the induced polarization current density $\mathbf{J}_p$ is equal to the rate of change of the polarization $\mathbf{P}$, that is,

$$\mathbf{J}_p = \frac{\partial \mathbf{P}}{\partial t}$$
$$= \frac{\partial (\epsilon_o \chi_e \mathbf{E})}{\partial t}$$

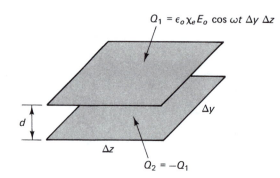

$$Q_1 = \epsilon_o \chi_e E_o \cos \omega t \, \Delta y \, \Delta z$$

Δy

d

Δz

$Q_2 = -Q_1$

Figure 3.5 Equivalent arrangement of the total dipole moment in the volume ($\Delta z \, \Delta y \, d$).

for linear dielectrics. Let us see now how this new polarization current will modify Ampere's law. Ampere's law in free (empty) space is given by

$$\nabla \times \frac{\mathbf{B}}{\mu_o} = \mathbf{J} + \frac{\partial \epsilon_o \mathbf{E}}{\partial t}$$

We should now add to the left-hand side of this equation the new current term called polarization current. Ampere's law is therefore given by

$$\nabla \times \frac{\mathbf{B}}{\mu_o} = \mathbf{J} + \frac{\partial \epsilon_o \mathbf{E}}{\partial t} + \frac{\partial \mathbf{P}}{\partial t}$$

$$= \mathbf{J} + \frac{\partial \epsilon_o \mathbf{E}}{\partial t} + \frac{\partial \epsilon_o \chi_e \mathbf{E}}{\partial t}$$

If we combine the displacement and polarization current terms, we obtain

$$\nabla \times \frac{\mathbf{B}}{\mu_o} = \mathbf{J} + \frac{\partial}{\partial t} [(\chi_e + 1)\epsilon_o \mathbf{E}]$$

The quantity $(\chi_e + 1)$ is referred to as the relative dielectric constant ϵ_r, hence,

$$\chi_e + 1 = \epsilon_r$$

ϵ_r is a physical property of the material, and it basically describes the susceptibility of the material to the storage of electric energy as a result of the induced polarization.

Some representative values of ϵ_r for several dielectric materials are given in Table 3.2. Additional dielectric constants are given in Appendix D.

TABLE 3.2 DIELECTRIC CONSTANT ϵ_r OF SEVERAL MATERIALS

Material	ϵ_r
Air	1.006
Glass	6.0
Lucite	3.2
Polystyrene	2.5
Dry soil	3.0
Teflon	2.1
Distilled water	81.0

Ampere's law now reduces to

$$
\nabla \times \frac{\mathbf{B}}{\mu_o} = \mathbf{J} + \frac{\partial(\epsilon_o \epsilon_r \mathbf{E})}{\partial t}
$$

$$
= \mathbf{J} + \frac{\partial(\epsilon \mathbf{E})}{\partial t}
$$

$$
= \mathbf{J} + \frac{\partial \mathbf{D}}{\partial t}
$$

where it can be seen that we effectively replaced ϵ_o of free space by $\epsilon = \epsilon_o \epsilon_r$ of the material. The quantity $\mathbf{D}$ is known as the electric flux density

$$
\mathbf{D} = \epsilon_o \epsilon_r \mathbf{E} \tag{3.3}
$$

EXAMPLE 3.1

If an electric field $\mathbf{E} = 0.1 \cos 2\pi \times 10^6 t\, \mathbf{a}_x$ V/m is applied to a dielectric material, determine the current density crossing a 1 m² area perpendicular to the x direction for the following types of dielectrics:

1. Polystyrene $\epsilon_r = 2.5$.
2. Distilled water $\epsilon_r = 81$.

Solution

The polarization is given by

$$
\mathbf{P} = 0.1 \chi_e \epsilon_o \cos 2\pi \times 10^6 t\, \mathbf{a}_x
$$

and the polarization current density is

$$
\mathbf{J}_p = \frac{\partial \mathbf{P}}{\partial t} = -0.1(2\pi \times 10^6)\chi_e\, \epsilon_o \sin 2\pi \times 10^6 t\, \mathbf{a}_x
$$

1. For polystyrene, $\chi_e = \epsilon_r - 1 = 1.5$, and

$$
\mathbf{J}_p = 0.1(2\pi \times 10^6)1.5\epsilon_o \sin 2\pi \times 10^6 t\, \mathbf{a}_x
$$

$$
= -0.3\pi \times 10^6 \epsilon_o \sin 2\pi \times 10^6 t\, \mathbf{a}_x \text{ A/m}^2
$$

2. For distilled water, $\chi_e = \epsilon_r - 1 = 80$, and

$$
\mathbf{J}_p = -16\pi \times 10^6 \epsilon_o \sin 2\pi \times 10^6 t\, \mathbf{a}_x \text{ A/m}^2
$$

3.4.2 Polarization Charge Density

As indicated earlier whenever an external electric field is applied to a dielectric material, there will be induced dipole moments, and the material is said to be polarized. As a result of this polarization, there may be induced polarization charge density inside

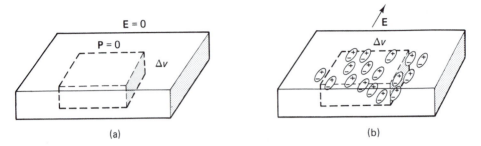

Figure 3.6 (a) In nonpolar material, the dipole moments and the polarization are equal to zero in the absence of external **E** field. (b) As a result of an applied electric field **E**, there will be induced dipole moments and net polarization per unit volume.

the material. Our objective in this section is to obtain an expression that quantifies the induced polarization charge density inside a slab of a dielectric material.

To start with, let us consider the slab of dielectric material shown in Figure 3.6a. If the material is nonpolar dielectric, there will be no dipole moments of any kind inside the material and, in the absence of the external electric field, the total polarization will be equal to zero, as shown in Figure 3.6a. On the application of the electric field, however, there will be induced dipole moments and the net polarization will be nonzero, as shown in Figure 3.6b.

To obtain an expression for the induced polarization charge density, let us consider the element of volume Δv inside the dielectric slab, as shown in Figure 3.7. From Figure 3.7, it is clear that the induced polarization within the element of volume Δv contributes zero additional charges inside the volume. This is because each induced electric dipole consists of spaced equal positive and negative charges and, hence, as long as these *dipoles are completely enclosed* by the element of volume, there will be no additional charges induced inside Δv as a result of the polarization. From Figure 3.7, it is also clear that there may be an increase or decrease of the total charge enclosed within the element of volume Δv because of the induced or oriented dipoles near the

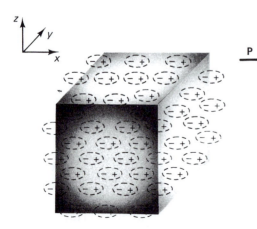

Figure 3.7 Total charge enclosed within a differential volume of dipoles has contribution only from the dipoles that are cut by the surfaces. All totally enclosed dipoles contribute a *zero* net charge enclosed (equal number of positive and negative charges) by the differential volume.

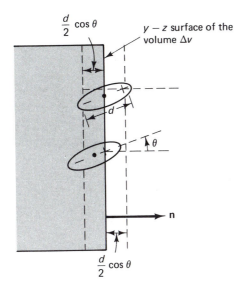

$\frac{d}{2} \cos \theta$

$y - z$ surface of the volume Δv

d

θ

n

$\frac{d}{2} \cos \theta$

Figure 3.8 A detailed examination of the induced dipole crossing the y-z surface of the element of volume Δv. Analysis of the situation shows that only those dipoles within a distance $\mathbf{d}/2 \cdot \mathbf{n}$ above or below the surface may contribute to the change in the total charge enclosed by the volume Δv.

surface of Δv. For example, the induced dipole moment along the x direction, shown in Figure 3.7, causes a negative charge to cross in the inward direction of the y-z surface of the element of volume Δv, if the induced dipole is just outside the surface. If this induced dipole is just inside the surface, conversely, a positive charge will cross this element of volume Δv in the outward direction. In both cases, it is clear that there may be net charge accumulation in the volume Δv as a result of the induced dipoles *near* the surface enclosing the element of volume Δv. To examine the situation further, let us consider one surface (e.g., the y-z surface) of the volume Δv, as shown in Figure 3.8. The outward directed unit vector **n** normal to that element of surface is in the $\mathbf{a}_x$ direction in this case.

If we assume that the average separation distance between the two charges constituting the electric dipole is d and that the induced dipoles make an angle θ with respect to the unit normal to the surface **n**, it is clear that only those dipoles with their centers within a distance $d/2 \cos \theta$ to the left or to the right of the surface will contribute to the charge crossing through this surface. In other words, the only contribution to the change in the charge enclosed by the volume element may result from those dipoles with their centers located a distance $(\mathbf{d}/2 \cdot \mathbf{n})$ outside or inside the surface enclosing the element of volume Δv.

With this background information, let us consider an incremental element of surface $\Delta \mathbf{s}_1$ in the direction shown in Figure 3.9a. Each of the induced dipoles that has its center within a distance $d/2 \cos \theta$ above the surface contributes a negative charge that crosses the element of area $\Delta \mathbf{s}_1$. If n is the number of dipoles per unit volume, then the number of dipoles with their centers within a distance $d/2 \cos \theta$ from the surface will be $n(d/2 \cos \theta)\Delta \mathbf{s}_1$. The number of negative charges that will flow into the volume partially enclosed by $\Delta \mathbf{s}_1 = \Delta s_1 \mathbf{a}_1$ ($\mathbf{a}_1$ is a unit vector in the direction of $\Delta \mathbf{s}_1$) is given by

$$n(-Q)\frac{d}{2} \cos \theta(-\Delta s_1)$$

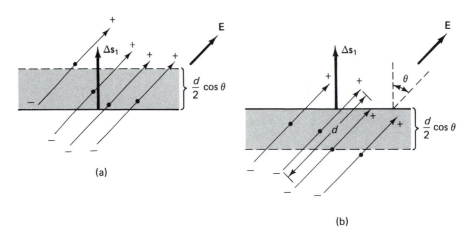

Figure 3.9 Quantification of the polarization charge crossing the area $\Delta\mathbf{s}_1$. (a) Negative charge crossing $\Delta\mathbf{s}_1$ as a result of the polarization. (b) Positive charge crossing $\Delta\mathbf{s}_1$.

Although the negative sign in front of Q is included simply because negative charges are crossing $\Delta\mathbf{s}_1$, the negative sign in front of Δs_1 is included because the direction of flow of the charges is into and not out of the surface. The change in the amount of charge partially enclosed by $\Delta\mathbf{s}_1$ is therefore

$$nQ\frac{d}{2}\cos\theta\,\Delta s_1$$

Similarly each of the induced dipoles that has its center within a distance $d/2\cos\theta$ below the surface $\Delta\mathbf{s}_1$ contributes a positive charge leaving (in the outward direction) the element of area as shown in Figure 3.9b. The total positive charge outflowing from the element of volume partially enclosed by $\Delta\mathbf{s}_1$ is therefore given by

$$n(+Q)\left[\frac{d}{2}\cos\theta(+\Delta s_1)\right] = nQ\frac{d}{2}\cos\theta\,\Delta s_1 \tag{3.4}$$

The two positive signs are included in this case to emphasize the fact that in this case we have positive charge crossing the element of area in the outward direction. The total increase in the *negative* charge in the element of volume Δv partially enclosed by $\Delta\mathbf{s}_1$ is hence

$$2\left(nQ\frac{d}{2}\cos\theta\,\Delta s_1\right) = nQd\cos\theta\,\Delta s_1$$

The factor of 2 is included because the inward crossing of negative charge and the outflow of positive charge are both equivalent to an increase of negative charges inside the element of volume Δv. The total increase in the negative charge density in the element of volume Δv enclosed by a surface Δs (i.e., $-\rho_p\Delta v$) is hence related to the polarization by

$$nQd \cos \theta \, \Delta s = \mathbf{P} \cdot \Delta \mathbf{s}$$

$$= -\rho_p \, \Delta v \tag{3.5}$$

where $\mathbf{P}$ is the polarization defined as the total dipole moment per unit volume ($nQ\mathbf{d}$) and ρ_p is the polarization positive charge density. The volume Δv is enclosed by the surface Δs. In a slab of dielectric material the induced polarization charge may be related to the polarization by subdividing the slab into small elements and summing the result of equation 3.5 over all the subvolumes. In the limit when the elements of volume and areas are reduced to infinitesimally small differential elements, the result of equation 3.5 reduces to

$$\oint_s \mathbf{P} \cdot d\mathbf{s} = -\int_v \rho_p \, dv \tag{3.6}$$

where the volume v is enclosed by the surface s. If we apply the divergence theorem to the left-hand side of equation 3.6, we obtain

$$\int_v \mathbf{\nabla} \cdot \mathbf{P} dv = -\int_v \rho_p \, dv$$

In the limiting case when the volume v reduces to an infinitesimal one, the point relation of equation 3.6 is obtained in the form

$$\mathbf{\nabla} \cdot \mathbf{P} = -\rho_p \tag{3.7}$$

Equation 3.7 simply indicates that the net outflow of the polarization flux density at a point (i.e., divergence of $\mathbf{P}$) is equal to the net polarization negative charge at this point.

.5 GAUSS'S LAW FOR ELECTRIC FIELD IN MATERIALS

With the identification of this new source of charge distribution—that is, the polarization charge—we should now modify Gauss's law for the electric field so as to include this new term. Gauss's law in this case will be given by

$$\mathbf{\nabla} \cdot \epsilon_o \mathbf{E} = \rho_v + \rho_p$$

where ρ_v is the *external* free charge distribution owing to an external source and ρ_p is the *induced* charge distribution resulting from the application of an electric field to the dielectric material. Substituting $\rho_p = -\mathbf{\nabla} \cdot \mathbf{P}$ (as given by equation 3.7), we obtain

$$\mathbf{\nabla} \cdot \epsilon_o \mathbf{E} = \rho_v - \mathbf{\nabla} \cdot \mathbf{P}$$

$$\mathbf{\nabla} \cdot (\epsilon_o \mathbf{E} + \mathbf{P}) = \rho_v$$

where ρ_v is once again the free charge density distribution produced by an external source.

For linear dielectrics—that is, isotropic materials—the polarization $\mathbf{P}$ and the electric field $\mathbf{E}$ are in the same direction. As indicated earlier, in this case the polarization is linearly proportional to the electric field, hence,

$$\mathbf{P} = \chi_e \, \epsilon_o \, \mathbf{E}$$

where χ_e is the constant of proportionality. Substituting $\mathbf{P}$ in Gauss's law, we obtain

$$\nabla \cdot (\epsilon_o \, \mathbf{E} + \chi_e \, \epsilon_o \, \mathbf{E}) = \rho_v$$

$$\nabla \cdot \epsilon_o \, \mathbf{E}(1 + \chi_e) = \rho_v$$

or

$$\nabla \cdot \mathbf{D} = \rho_v \tag{3.8}$$

where $\mathbf{D}$ as previously defined in equation 3.3 is given by

$$\mathbf{D} = \epsilon_o(1 + \chi_e)\mathbf{E} = \epsilon_o \epsilon_r \, \mathbf{E}$$

From equations 3.8 and 3.3, it is clear that the induced polarization charges and currents can be accounted for in Gauss's law for the electric field and in Ampere's law, respectively, simply by replacing $\epsilon_o \, \mathbf{E}$ by $\epsilon_o \, \epsilon_r \, \mathbf{E}$ or equivalently by $\mathbf{D}$. In other words, the effect of the polarization charge and current distributions is reflected in these equations through a change in the free space dielectric constant ϵ_o to the dielectric constant of the material $\epsilon = \epsilon_o \epsilon_r$ where ϵ_r is the relative dielectric constant of the material.

EXAMPLE 3.2

Compare the magnitudes of the conduction current $\mathbf{J}$ and the displacement current $\partial \mathbf{D}/\partial t$ for the materials sea water ($\sigma = .4, \epsilon = 81\epsilon_o$) and earth ($\sigma = 10^{-3}, \epsilon = 10\epsilon_o$) at frequencies 60 Hz, 1 kHz, and 1 GHz.

Solution

The ratio between the conduction and displacement currents are given by

$$\frac{\mathbf{J}}{\mathbf{J}_D} = \frac{\sigma \mathbf{E}}{\dfrac{\partial \epsilon \mathbf{E}}{\partial t}}$$

For $e^{j\omega t}$ time variation, the ratio between the complex forms of these currents is given by

$$\frac{\hat{\mathbf{J}}}{\hat{\mathbf{J}}_D} = \frac{\sigma \hat{\mathbf{E}}}{j\omega\epsilon \hat{\mathbf{E}}},$$

where $\epsilon = \epsilon_r \epsilon_o$

$$\left| \frac{\hat{\mathbf{J}}}{\hat{\mathbf{J}}_D} \right| = \frac{\sigma}{\omega\epsilon_o \epsilon_r}$$

The values of this ratio as a function of frequency are given in Table 3.3.

TABLE 3.3 RATIO BETWEEN MAGNITUDES
OF CONDUCTION AND DISPLACEMENT
CURRENTS AS A FUNCTION OF FREQUENCY

f	Sea water	Earth
60 Hz	1.48×10^7	3.0×10^4
1 kHz	8.9×10^5	1.8×10^3
1 GHz	0.89	0.0018

The values in Table 3.3 clearly demonstrate that although the conduction current is dominant at lower frequencies, the displacement current starts to dominate at the higher frequencies.

.6 MAGNETIC MATERIALS AND THEIR MAGNETIZATION

Similar to the case of dielectric materials, we will start our discussion of magnetic materials by examining the reaction of these materials to an externally applied magnetic field. We will learn that the prominent characteristic of these materials may be described in terms of its "magnetization," which is related to the alignment of the atomic magnetic dipole moments along the direction of the applied magnetic field.

To understand this effect, let us start by reexamining the atomic structure of materials. As we recall from previous discussions, materials are composed of many atoms, and each atom consists of a positively charged nucleus surrounded by a cloud of electrons. These orbiting electrons around the nucleus are *equivalent* to a current circulating along the electronic orbit. These currents, therefore, encircle a surface area $d\mathbf{s}$. The microscopic reaction of a magnetic material can be described in terms of a concept called *magnetic dipole moment* $\mathbf{m}$, which is a vector defined by the magnitude of the circulating current I multiplied by the differential element of area $d\mathbf{s}$ encircled by it, hence,

$$\boxed{\mathbf{m} = I\,d\mathbf{s}}$$

This magnetic dipole moment $\mathbf{m}$ is a useful concept and will play an important role in quantifying the reaction of magnetic materials to an applied magnetic field that is similar to the role played by the electric dipole concept in illustrating the polarization properties of dielectric materials. It may be worth mentioning that based on quantum mechanics considerations, there will also be other sources of magnetic dipole moments including those resulting from the electron and nucleus spins that may also be characterized by the same concept of the magnetic dipole moment. Unlike the electric polarization case in which there may or may not be electric dipoles in the absence of an external electric field, in all materials there are always magnetic dipoles because of the presence of orbiting electrons as well as spinning electrons and nuclei. Figure 3.10a illustrates the magnetic dipole moments in a slab of magnetic material. The total magnetic dipole in the element of volume is the vector sum of all the magnetic dipole

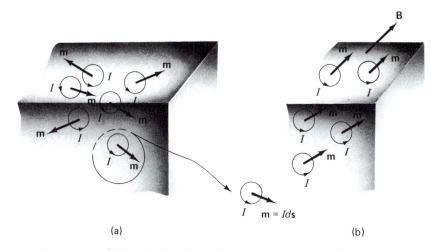

Figure 3.10   (a) Randomly oriented magnetic dipoles in a slab of magnetic material. (b) In the presence of an external magnetic field **B**, the magnetic dipoles will be oriented in the direction of the magnetic field, and the material is said to be magnetized.

moments in that volume. Hence, it is clear that in the absence of an external magnetic field, these magnetic dipole moments are randomly oriented, and the total magnetic dipole moment in that volume will be zero. When an external magnetic field is applied, as shown in Figure 3.10b, however, a torque will be exerted on these dipole moments as we will see in the following section. As a result of this torque, the magnetic dipole moments will tend to be oriented along the direction of the applied magnetic field. There will be, therefore, a net magnetic moment in the element of volume Δv. To describe quantitatively the total magnetic moment or the lack of it in an element of volume Δv, we introduce the magnetization concept. Magnetization **M** is defined as the total magnetic moment per unit volume, hence,

$$\mathbf{M} = \operatorname*{Lim}_{\Delta v \to \infty} \frac{1}{\Delta v} \sum_{i=1}^{n\Delta v} \mathbf{m}_i = n\mathbf{m}_a = nI\,d\mathbf{s}$$

where n is the number of dipoles per unit volume and $\mathbf{m}_a$ is the average magnetic dipole moment. It should be noted that, similar to the polarization concept, the magnetization describes the presence of total magnetic dipole moments on a macroscopic basis. This is why although the microscopic magnetic dipoles exist in the absence of an external magnetic field (see Figure 3.10a), the total magnetic moment (i.e., the magnetization **M**) is zero in this case because the dipole moments are randomly oriented. In the presence of an external magnetic field and because of the alignment of these microscopic magnetic dipole moments in the direction of the field, however, the magnetization is not zero, as shown in Figure 3.10b. We will use the magnetization concept to quantify further the effects of applying an external magnetic field to a slab of magnetic material. Before we can proceed further, however, we need to quantitatively describe the torque exerted on the magnetic dipole moments by an external magnetic field. The obtained expression will simply show that the torque exerted on the magnetic dipoles tends to orient them in the direction of the magnetic field. We will then use the magnetization of magnetic materials to derive an expression for the induced magnetization current in these materials.

3.6.1 Force and Torque on Current Loops

Consider a differential current loop placed in a magnetic field of flux density **B** as shown in Figure 3.11a. Because the differential current loop is essentially very small, we may consider a rectangular loop of the same area to simplify the analysis without any loss in the generality of the desired expression for the torque. The rectangular loop is oriented in the *x-y* plane as shown in Figure 3.11b. Further simplifications in the obtained expression are obtained by assuming the magnetic field to be constant along the sides of the loop.

The objective of the analysis in this section is to obtain an expression for the torque exerted by the magnetic field **B** on the differential rectangular loop carrying a current *I* as shown in Figure 3.11b. Before we can obtain this expression, however, we need to quantify the force because of the magnetic field on each side of the conducting loop that carries a current *I*. From Lorentz force, we know that the force exerted on a charge dQ moving with a velocity **v** in a magnetic field **B** is given by

$$d\mathbf{F} = dQ\,\mathbf{v} \times \mathbf{B}$$

The charge dQ may be due to a volume charge distribution ρ_v, a surface charge density ρ_s, or a linear charge density ρ_ℓ along the filimentary conductor $d\ell$. In the latter case, $dQ = \rho_\ell\,d\ell$.

Lorentz force in this case is given by

$$d\mathbf{F} = \rho_\ell\,d\ell\,\mathbf{v} \times \mathbf{B}$$

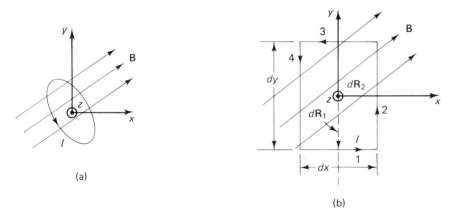

(a)

(b)

Figure 3.11 (a) A current loop oriented in the *x-y* plane and placed in a magnetic field of flux density **B**. (b) The simplified geometry used in the analysis where the loop is assumed rectangular and the magnetic field is assumed constant along the sides of the loop.

The linear charge density ρ_ℓ when moving with a velocity **v** is equivalent to a current **I**, hence,

$$d\mathbf{F} = I d\ell \times \mathbf{B} = I d\ell \times \mathbf{B} \tag{3.9}$$

Equation 3.9 provides an expression for the force exerted on a differential current element $I d\ell$ when placed in a magnetic field **B**.

Returning now to our current loop shown in Figure 3.11b. The vector force exerted on the current element labeled side 1 in the rectangular loop is given by

$$d\mathbf{F}_1 = I dx \, \mathbf{a}_x \times \mathbf{B} \tag{3.10}$$

where the magnetic field vector is assumed to be arbitrarily oriented and hence has three components in the Cartesian coordinate system, that is, $\mathbf{B} = B_x \, \mathbf{a}_x + B_y \, \mathbf{a}_y + B_z \, \mathbf{a}_z$.

Substituting **B** in equation 3.10 and performing the cross product, we obtain

$$d\mathbf{F}_1 = I dx (B_y \, \mathbf{a}_z - B_z \, \mathbf{a}_y)$$

If we consider the axis of rotation to be along the z axis, the torque arm from side 1 of the loop to the axis of rotation is given by

$$d\mathbf{R}_1 = -\frac{dy}{2} \mathbf{a}_y$$

The torque on side 1 is hence given by

$$d\mathbf{T}_1 = d\mathbf{R}_1 \times d\mathbf{F}_1 = -\frac{dy}{2} \mathbf{a}_y \times I dx (B_y \, \mathbf{a}_z - B_z \, \mathbf{a}_y)$$

$$= -\frac{1}{2} dx \, dy \, I \, B_y \, \mathbf{a}_x$$

It can be easily shown that the torque on side 3 equals that on side 1, hence,

$$d\mathbf{T}_3 = d\mathbf{T}_1$$

The contribution from both sides 1 and 3 of the rectangular loop to the total torque on the loop is then

$$d\mathbf{T}_1 + d\mathbf{T}_3 = -dx \, dy \, I \, B_y \, \mathbf{a}_x \tag{3.11}$$

For side 2, the force resulting from the magnetic field is given by

$$d\mathbf{F}_2 = I dy \, \mathbf{a}_y \times \mathbf{B}$$

$$= I dy (B_z \, \mathbf{a}_x - B_x \, \mathbf{a}_z)$$

The torque on side 2 is

$$d\mathbf{T}_2 = d\mathbf{R}_2 \times d\mathbf{F}_2 = \frac{dx}{2} \mathbf{a}_x \times d\mathbf{F}_2$$

$$= \frac{1}{2} dx \, dy \, I \, B_x \, \mathbf{a}_y$$

The torque on sides 2 and 4 is, hence,

$$d\mathbf{T}_2 + d\mathbf{T}_4 = I\,dx\,dy\,B_x\,\mathbf{a}_y \tag{3.12}$$

The total torque exerted on the loop $d\mathbf{T}$ is simply obtained by adding equations 3.11 and 3.12.

$$\begin{aligned} d\mathbf{T} &= I\,dx\,dy(B_x\,\mathbf{a}_y - B_y\,\mathbf{a}_x) \\ &= I\,dx\,dy\,\mathbf{a}_z \times \mathbf{B} \tag{3.13} \\ &= I\,d\mathbf{s} \times \mathbf{B} \end{aligned}$$

where $d\mathbf{s} = dx\,dy\,\mathbf{a}_z$ is the vector area of the differential current loop. From the result of equation 3.13, we can draw the following conclusions:

1. There is indeed a torque exerted on the differential current loop when placed in a magnetic field.
2. This torque will continue to exist until the element of area is aligned along the direction of the magnetic field **B**. In this case, $d\mathbf{s}$ will be along **B** and hence $d\mathbf{s} \times \mathbf{B} = 0$. The same conclusion can be restated by simply saying that the torque resulting from the magnetic field tends to orient the current loop in the direction of **B**.

Conclusions 1 and 2 provide the bases for the prominent characteristic of magnetic materials, "their magnetization" that results from the alignment of the magnetic dipoles in the direction of the magnetic field.

The expression for the torque in equation 3.13 may be put in a more familiar form by noting that $I\,d\mathbf{s}$ is simply the magnetic dipole moment as defined earlier in this section. The resulting torque may then be expressed as

$$\boxed{d\mathbf{T} = \mathbf{m} \times \mathbf{B}}$$

which more clearly indicates that the torque tends to align the magnetic dipole in the direction of the magnetic field.

3.6.2 Magnetization Current Density

Our overall objective from characterizing magnetic materials and in particular quantifying their reaction to an externally applied magnetic field is to identify and quantify any induced charges and current distributions that should be included in Maxwell's equations. From the previous discussion, it is clear that there are no induced magnetic charges, they simply do not exist, and that there may be induced currents as a result of the alignment of the magnetic dipoles in the direction of the magnetic field—that is, as a result of the magnetization. These induced currents are therefore called magnetization currents.

To quantify the magnetization current, let us consider a slab of magnetic material

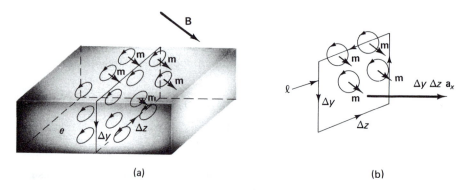

Figure 3.12 (a) Magnetic dipoles in a slab of magnetic material. (b) Procedure to calculate the component of the magnetization current in the $\mathbf{a}_x$ direction.

under the influence of an external magnetic field as shown in Figure 3.12a. To quantify the magnetization current, it is necessary not only to determine the magnitude of this induced current but also its direction because the current is a vector quantity. For example, in the slab of the magnetic material shown in Figure 3.12a, to quantify the induced magnetization current in the x direction it is necessary to construct a contour ℓ that encircles an area ds oriented in the $\mathbf{a}_x$ direction. The x-directed component of the magnetization current is then evaluated simply by determining the net current crossing the element of area $\Delta y \, \Delta z$ in the $\mathbf{a}_x$ direction as shown in Figure 3.12b.

From Figure 3.12b, it is clear that regardless of the direction of the magnetic dipoles, all the dipoles that are completely encircled by the contour ℓ will have no contribution to the component of the magnetization current in question. This is simply because all the magnetic dipoles that are completely enclosed by ℓ cross the element of area twice and hence result in a zero contribution to the total current crossing this element of area. Only those dipole elements that are on the edges of the element of area may, on their orientation in the direction of the magnetic field, contribute to the total current crossing the area. Let us now focus our attention on the magnetic dipoles along the edges of the differential path $d\ell$ shown in Figure 3.13. Because we are dealing with a differential path, all the magnetic moments will be assumed of the same magnitude and are aligned along the same direction that makes an angle θ with the differential path $d\ell$. If n is the number of dipole moments per unit volume, there will be $n \, ds \, \cos \theta \, d\ell$ or $n \, d\mathbf{s} \cdot d\ell$ magnetic dipoles in the small volume $(ds \cos \theta) d\ell$ around the differential path $d\ell$. In changing from a random orientation to the particular alignment along the direction of the magnetic field $\mathbf{B}$, the bound current crossing the surface enclosed by the path (to our left as we travel in the direction of $d\ell$) should increase by the value I for each of the $n \, d\mathbf{s} \cdot d\ell$ dipoles on the edge of $d\ell$. Thus, the total increase in the *current component in the direction of the element of area ds* enclosed by the contour $d\ell$ resulting from the magnetization of the material is given by

$$\mathbf{J}_m \cdot d\mathbf{s} = n \, I \, d\mathbf{s} \cdot d\ell = \mathbf{M} \cdot d\ell$$

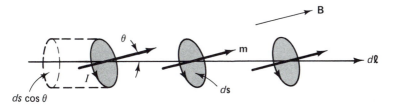

Figure 3.13   Magnetic dipoles along the differential path $d\ell$. The dipoles are all equal and make an angle θ with the direction of the differential path $d\ell$.

where **M** is the magnetization as described in the previous sections. Integrating over a closed contour c, we obtain the component of the magnetization current crossing the area s enclosed by c

$$\int_s \mathbf{J}_m \cdot d\mathbf{s} = \oint_c \mathbf{M} \cdot d\ell \tag{3.14}$$

Applying Stokes's theorem, we obtain

$$\int_s \mathbf{J}_m \cdot d\mathbf{s} = \int_s \boldsymbol{\nabla} \times \mathbf{M} \cdot d\mathbf{s} \tag{3.15}$$

If we consider the current crossing an element of area $\boldsymbol{\Delta}\mathbf{s}$, and on equating both sides of equation 3.15 on a component-by-component basis, we ultimately obtain

$$\mathbf{J}_m = \boldsymbol{\nabla} \times \mathbf{M} \tag{3.16}$$

Equation 3.16 is the desired expression for the bound magnetization current density in terms of the induced magnetization inside the magnetic material.

3.6.3 Characterization of Magnetic Materials

After our detailed discussion of the reaction of magnetic materials to an externally applied magnetic field and the quantification of the magnetization current, it is important to sit back, reflect on these interactions, and realize that not all magnetic materials react similarly. As a matter of fact, magnetic materials may be classified into six different categories including ferromagnetic, ferrimagnetic, antiferromagnetic, diamagnetic, paramagnetic, and superparamagnetic. These different categories are defined depending on the following:

1. The level of interaction of the material with the externally applied magnetic field. This may range from very strong to very small or negligible.

2. The residual effect of the external magnetic field on the material. Some materials return to their original state after the removal of the external magnetic field, whereas others, such as ferromagnetic materials, maintain changes. Permanent magnets are made mainly of ferromagnetic materials with composition percentages determined so as to increase remnant magnetic flux in the absence of an external magnetic field.

To explain these different levels and types of interactions clearly, it is important to recall that magnetic dipoles result from orbiting electrons and are, more important, due to electrons and nuclear spins. These various contributions may be individually large or small, and they may collectively be in the same direction, thus adding to a larger effect; they may be in opposite directions, and the net result would be a small or negligible magnetic moment. In general, the net magnetic moment is the vector sum of the electronic orbital moments and the spin moments.

For example, in ferromagnetic materials such as iron, the net atomic magnetic dipole moment is relatively large. Because of interatomic forces, these magnetic dipoles line up in parallel fashion in regions known as *domains*. The direction of the magnetic dipoles is, however, different in the different domains. The sizes and shapes of these domains may also be different depending on several factors including type, shape, size, and magnetic history of the material. These domains are separated by what is known as domain walls that consist of atoms whose atomic moments make small angles with neighboring atoms. For virgin ferromagnetic materials, the strong magnetic moment in the various domains are arbitrarily oriented so that the overall magnetic moment in the whole material sample is zero. On the application of an external magnetic field, the domains with magnetic dipoles in the direction of the applied magnetic field expand at the expense of the domains in different directions, thus resulting in a significant increase in the magnetic flux **B** inside the material as compared with the **B** outside the material. The removal of the external magnetic field, however, does not return the orientation of the magnetic dipoles in the different domains to the random distribution again, which is characterized with a zero value of average magnetic moment. Instead, residual magnetization is attained, and remnant average value of the magnetic dipole remains. The fact that a remnant magnetic dipole remains after the removal of the external field is known as hysteresis, derived from a Greek word which means "to lag."

As mentioned earlier, permanent magnets are mostly made of ferromagnetic materials with composition percentages chosen such as to increase the remnant magnetic flux. For example, a common type of permanent magnet material is Alnico 5, the composition of which is 24 percent cobalt, 14 percent nickel, 8 percent aluminum (paramagnetic, which will be described later), 3 percent copper, and 51 percent iron. Alnico 5 has a remnant magnetic flux density of 12,500 G.

Diamagnetic and paramagnetic materials, conversely, have an effectively zero net magnetic dipole moment in the absence of an external magnetic field. These materials, therefore, have a small or negligible interaction with the external magnetic field. In diamagnetic materials, the magnetic moments as a result of orbiting and spinning electrons as well as the nuclear spin cancel each other, and the net atomic magnetic moment is zero—hence, the negligible interaction with the external magnetic field. In paramagnetic materials, however, the atomic magnetic dipoles are not zero, but they are randomly oriented in the absence of an external magnetic field—hence, the zero average magnetic moment throughout a sample of the material. The presence of an external magnetic field helps align the atomic dipole moments—hence, an effective increase in **B** inside the material. Examples of diamagnetic materials include gold, silicon, and inert gases, whereas examples of paramagnetic materials include tungsten and potassium.

Ferrimagnetic and antiferromagnetic materials are, in a sense, in between the other two classes. In both cases, the interatomic forces cause the atomic moments to line up in antiparallel directions. In ferrimagnetic materials, such as nickel ferrite, the adjacent atomic moments are not equal, and a relatively large response (not as large as in ferromagnetic materials) is expected on the application of an external magnetic field. In antiferromagnetic materials, such as nickel oxide, conversely, the magnetic dipoles of adjacent atoms are almost equal, and the net magnetic moment is, hence, zero. The antiferromagnetic materials thus react only slightly to the presence of an external magnetic field. Ferrites are a subgroup of ferrimagnetic materials. Commercial ferrite materials are ceramic semiconductors. As a result, the electrical conductivity in these materials is five to fifteen orders of magnitude lower than the conductivity of metallic ferromagnets. The usefulness of ferrites in applications arises mainly from this fact. They can be formed for use in inductor cores without the need for laminations. In commercial fabrication, great use is made of mixed ferrites to enhance certain desirable properties and suppress undesirable ones. Disadvantages arise mainly from low permeability values ranging from 100 to 1000, which is smaller than the 4000 permeability value for pure iron and 100,000 for Mumetal.

The sixth and remaining category is that of a superparamagnetic material. A good example of this is the magnetic tape used in audio recorders. These materials are composed of ferromagnetic particles in a nonferromagnetic material. Ferromagnetic particles react strongly to the presence of an external magnetic field (as described earlier), but these reactions do not propagate throughout the material because of the nonmagnetic nature of the host material.

The preceding discussion simply summarizes various ways by which materials may interact with an externally applied magnetic field. The basic mechanisms of interaction are related to the orbiting electrons and the spin moments. Various interactions result from the relative strength of these various moment components and the overall average value of the magnetic moment in a material. As described in the previous section, the net interaction may be quantified in terms of magnetization current. The next section describes the impact of the magnetization current term on Ampere's law.

.7 AMPERE'S LAW AND MAGNETIZATION CURRENT

We learned that there will be an induced magnetization-bound current if an external magnetic field is applied to a magnetic material. This induced current should, in turn, modify the applied magnetic field. Ampere's law, which relates the magnetic field to the various currents producing it, should therefore include the induced magnetization current term. Hence,

$$\nabla \times \frac{\mathbf{B}}{\mu_o} = \mathbf{J} + \frac{\partial \mathbf{D}}{\partial t} + \mathbf{J}_m \qquad (3.17)$$

It should be noted that in equation 3.17 we expressed the displacement current as $\partial \mathbf{D}/\partial t$ to account for any dielectric polarization effects that might be present in the material region as described in the previous sections. $\mathbf{J}$ is the current density resulting from an external source, and $\mathbf{J}_m$ is the induced magnetization current. $\mathbf{J}_m$ can be replaced by $\nabla \times \mathbf{M}$ according to equation 3.16, and equation 3.17 and may then be expressed in the form

$$\nabla \times \frac{\mathbf{B}}{\mu_o} = \mathbf{J} + \frac{\partial \mathbf{D}}{\partial t} + \nabla \times \mathbf{M}$$

or

$$\nabla \times \left(\frac{\mathbf{B}}{\mu_o} - \mathbf{M} \right) = \mathbf{J} + \frac{\partial \mathbf{D}}{\partial t} \tag{3.18}$$

Let us now define

$$\frac{\mathbf{B}}{\mu_o} - \mathbf{M} = \mathbf{H}$$

where $\mathbf{H}$ is the magnetic field intensity. The magnetic flux density $\mathbf{B}$ is then given by

$$\mathbf{B} = \mu_o(\mathbf{H} + \mathbf{M}) \tag{3.19}$$

and Ampere's law in equation 3.18 reduces to

$$\boxed{\nabla \times \mathbf{H} = \mathbf{J} + \frac{\partial \mathbf{D}}{\partial t}} \tag{3.20}$$

Equation 3.20 is a general form of Ampere's law in material regions because it includes the induced magnetization and polarization currents. As we recall, $\mathbf{D} = \epsilon_o \epsilon_r \mathbf{E}$ includes the polarization current effect because ϵ_o of the dielectric constant of free space is replaced by $\epsilon_o \epsilon_r$ where ϵ_r is the relative dielectric constant of the material region. Also, $\mathbf{H} = \mathbf{B}/\mu_o - \mathbf{M}$ accounts for the induced magnetization current through the newly introduced magnetization term $\mathbf{M}$. A simplified expression for $\mathbf{H}$ may be obtained by eliminating the magnetization vector $\mathbf{M}$ from the equation. For a linear magnetic material—that is, isotropic materials—the magnetization $\mathbf{M}$ is along the direction of $\mathbf{H}$ and has a magnitude that is linearly proportional to $\mathbf{H}$. $\mathbf{M}$ is hence given by

$$\mathbf{M} = \chi_m \mathbf{H}$$

where χ_m is the constant of proportionality and is called the magnetic susceptibility of the material. Substituting $\mathbf{M}$ in equation 3.19 we obtain

$$\mathbf{B} = \mu_o(\mathbf{H} + \chi_m \mathbf{H}) = \mu_o(1 + \chi_m)\mathbf{H}$$
$$= \mu_o \mu_r \mathbf{H} \tag{3.21}$$

where μ_r, the relative permeability of the material, is equal to $1 + \chi_m$. If we define $\mu = \mu_o \mu_r$ as the permeability of the material, equation 3.21 reduces to

$$\mathbf{B} = \mu \mathbf{H}$$

TABLE 3.4 RELATIVE PERMEABILITY μ_r OF
VARIOUS MATERIALS

Material	μ_r
Silver	0.99998
Copper	0.999991
Nickel	600
Mild steel	2000
Iron	5000

which is a general relation between the magnetic flux density and the magnetic field
intensity that considers the magnetization of the material. Values for the relative
permeability of several materials are given in Table 3.4. Additional values of μ_r are
given in Appendix D.

EXAMPLE 3.3

The very long solenoid shown in Figure 3.14 contains two coaxial magnetic rods of radii
a and b, and permeabilities $\mu_1 = 2\mu_o$ and $\mu_2 = 3\mu_o$, respectively.
    If the solenoid has n turns every d meters along the axis and carries a steady current
I, determine the following quantities assuming that the windings are closely spaced:

1. Magnetic field intensity **H** in the three regions shown.
2. Magnetic flux density **B** in these three regions.
3. Magnetization in all the three regions.
4. Magnetization current in region 1.

Solution

For static fields, Ampere's law of equation 3.20 reduces to

$$\nabla \times \mathbf{H} = \mathbf{J}$$

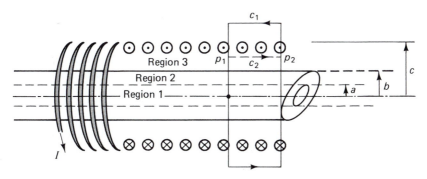

Figure 3.14   Long solenoid with a coaxial two-layer rod of magnetic materials.

To obtain an integral form for this expression, we simply integrate over an area ds and use Stokes's theorem. Hence,

$$\int_s \nabla \times \mathbf{H} \cdot d\mathbf{s} = \int_s \mathbf{J} \cdot d\mathbf{s} \tag{3.22}$$

Using Stokes's theorem we obtain

$$\oint_c \mathbf{H} \cdot d\ell = \int_s \mathbf{J} \cdot d\mathbf{s} \tag{3.23}$$

1. From the symmetry of the geometry and based on the other indicated assumptions such as an infinitely long and closely wound solenoid, it is clear that the magnetic flux will be along the axis (z direction) of the coil (see example 1.28). To determine the magnetic field intensity outside the coil, we construct the contour c_1 shown in Figure 3.14. Applying Ampere's law using the contour c_1 we obtain

$$\oint_{c_1} \mathbf{H} \cdot d\ell = 0 \tag{3.24}$$

In equation 3.24, the total enclosed current is identically zero simply because in each turn of the coil the current crosses the area s_1 enclosed by c_1 twice in the opposite direction. From equation 3.24, it is clear that the magnetic field intensity outside the coil is hence zero.

Next we determine the magnetic field intensity inside the coil. For this purpose, we construct a second Amperian contour c_2 that passes through the region in which the value of the magnetic field intensity is desired. Applying Ampere's law (equation 3.23) in this case and noting that $\mathbf{H}$ is zero outside the coil, we obtain

$$\oint_{c_2} \mathbf{H} \cdot d\ell = \int_{P_1}^{P_2} H_z \, \mathbf{a}_z \cdot dz \, \mathbf{a}_z = \int_o^d H_z \, dz = H_z \, d$$

$$= nI$$

$$\therefore \mathbf{H}_{\text{inside}} = H_z \, \mathbf{a}_z = \frac{nI}{d} \, \mathbf{a}_z$$

where n is the number of turns in a d (m) distance along the axis of the coil. Because nI is the total current enclosed by the contour c_2 regardless of whether P_1 and P_2 fall within any of the regions 1, 2, or 3, hence,

$$H_z = \frac{nI}{d}$$

in all the three regions enclosed by the coil. The magnetic field intensity $\mathbf{H}$ then depends on the parameters of the external source such as the number of turns n and the current in each turn I, but it does not reflect the effects induced by the magnetic materials. These effects, such as the magnetization, are reflected in the value of μ_r, which is included in the expression of the magnetic flux density $\mathbf{B}$.

2. The magnetic flux density in the ith region is given by $\mathbf{B}_i = \mu_i \mathbf{H}_i$ where μ_i and $\mathbf{H}_i$ are the permeability and the magnetic field intensity of the ith region.
   In region 1, $\mu_1 = 2\mu_o$

$$\therefore \mathbf{B}_1 = 2\frac{\mu_o \, nI}{d} \, \mathbf{a}_z$$

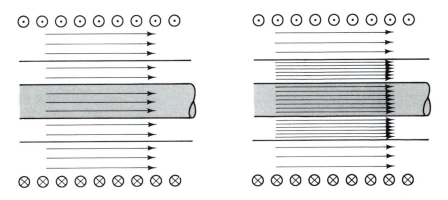

Figure 3.15   The magnetic field intensity **H** (uniform) and the magnetic flux density **B** (nonuniform) inside the core of the solenoidal coil.

in region 2, $\mu_2 = 3\mu_o$

$$\therefore \mathbf{B}_2 = 3\frac{\mu_o nI}{d}\mathbf{a}_z$$

and in region 3, $\mu_3 = \mu_o$, hence,

$$\mathbf{B}_3 = \frac{\mu_o nI}{d}\mathbf{a}_z$$

From the preceding equations, it can be seen that although the magnetic field intensity **H** is constant throughout the interior region of the coil, the magnetic flux density **B** varies depending on the permeability of the region. Regions with higher permeability such as regions 1 and 2 tend to concentrate the magnetic flux lines. Figure 3.15 illustrates the magnetic field intensity and the magnetic flux line distributions inside the coil.

3. The magnetization

$$\mathbf{M} = \chi_m\,\mathbf{H} = (\mu_r - 1)\mathbf{H}$$

$$\mathbf{M}_1 = (2-1)\mathbf{H}_1 = \mathbf{H}_1 = \frac{nI}{d}\mathbf{a}_z$$

$$\mathbf{M}_2 = (3-1)\mathbf{H}_2 = 2\frac{nI}{d}\mathbf{a}_z$$

and the magnetization of region 3 (air) is zero because $\mu_{r3} = 1$. This is also physically understandable because air does not react to or store magnetic energy.

4. The magnetization current density in region 1 is given by

$$\mathbf{J}_{m1} = \nabla \times \mathbf{M}_1$$

$$= \nabla \times \frac{nI}{d}\mathbf{a}_z = 0$$

The zero value of the magnetization current in all the regions can also be appreciated based on a simple flux representation of the magnetization vector. The magnetization vector is constant in each region, and its flux representation will therefore

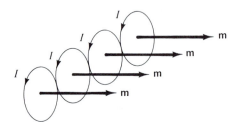

Figure 3.16 Zero magnetization current in a material results when the magnetization vector is uniform inside the material.

include flux lines that are uniformly spaced in each region. With the uniformly distributed flux lines, the placement of a curl meter in this region will not cause rotation because of the uniformity of the magnetization vector. Physically constant values of the magnetization at all points in the region of interest simply means that the magnetic dipole moments have the same value at all points. When aligned in the direction of the external magnetic field, the circulating currents associated with these magnetic dipole moments cancel each other (see Figure 3.16), thus resulting in a zero net-bound magnetization current within the material.

3.8 MAXWELL'S EQUATIONS IN MATERIAL REGIONS

In the previous sections we identified and quantified the various interactions of the electric and magnetic fields with materials. We specifically described the mechanisms of inducing additional currents such as the conduction current in conductors, the polarization current in dielectrics, and the magnetization current in magnetic materials. Also, the possibility of inducing charge distributions was discussed, and it is shown that polarization charges may result when a dielectric material is subjected to an externally applied electric field. All these induced sources will, of course, affect the applied electric and magnetic fields that originally produced them. Therefore, to develop unique mathematical relations between the fields and their sources, Maxwell's equations should be modified to include the induced sources in material regions. In this section, we will briefly indicate the various modifications in Maxwell's equations as a result of the inclusion of the induced charge and current sources. In preparation for using these modified Maxwell's equations to derive the boundary conditions at an interface between two different materials, the integral form of Maxwell's equations will also be developed.

3.8.1 Gauss's Law for Electric Field

This Gauss's law in the absence of material regions simply relates the electric field $\mathbf{E}$ or the electric flux density $\epsilon_o \mathbf{E}$ to the free charges obtained from an external source ρ_v. This Gauss's law is given by

$$\boldsymbol{\nabla} \cdot \epsilon_o \mathbf{E} = \rho_v \qquad \oint_s \epsilon_o \mathbf{E} \cdot d\mathbf{s} = \int_v \rho_v \, dv$$

The integral form simply indicates that the total electric flux density $\epsilon_o \mathbf{E}$ emanating from the closed surface s is equal to the total free charge enclosed by the surface. The point form, conversely, indicates that the net outflow of the electric flux density at a point (definition of the divergence) is equal to the charge density at that point. With the introduction of the polarization charge density ρ_p as a result of the polarization of the dielectric materials in an externally applied $\mathbf{E}$ field, Gauss's law should be modified so as to include this polarization charge. In point form, Gauss's law is given by

$$\nabla \cdot \epsilon_o \mathbf{E} = \rho_v + \rho_p = \rho_v - \nabla \cdot \mathbf{P} \tag{3.25}$$

where the polarization charge density ρ_p is substituted by $-\nabla \cdot \mathbf{P}$. For linear dielectric materials, the polarization is linearly proportional to the electric field, and equation 3.25 reduces to

$$\boxed{\nabla \cdot \mathbf{D} = \rho_v} \tag{3.26}$$

where $\mathbf{D} = \epsilon_o \epsilon_r \mathbf{E}$ is the electric flux density in a material that has a relative dielectric constant ϵ_r. To obtain the same relation in integral form, we integrate (equation 3.26) over a volume v and use the divergence theorem to obtain

$$\boxed{\oint_s \mathbf{D} \cdot d\mathbf{s} = \int_v \rho_v \, dv} \tag{3.27}$$

In both equations 3.26 and 3.27, it should be noted that the free charge distribution ρ_v is due to only the external source. The polarization-bound charges are included in the electric flux density $\mathbf{D}$ as previously described.

3.8.2 Gauss's Law for Magnetic Field

Through our study of the dielectric and magnetic properties of materials, we identified the presence of some induced sources such as the polarization charge and magnetization current, but we did not identify the presence of any induced magnetic charges. Magnetic charges, free or induced, simply do not exist physically in any material. Therefore, on modifying Gauss's law for the magnetic field, we should only include the effect of the induced magnetization currents. Under these conditions and because the vector $\mathbf{B} = \mu_o \mu_r \mathbf{H}$ includes the presence of the magnetization current, Gauss's law for the magnetic field in any material region is given by

$$\boxed{\nabla \cdot \mathbf{B} = 0 \qquad \oint_s \mathbf{B} \cdot d\mathbf{s} = 0}$$

The preceding equations indicate that the divergence of the magnetic field flux is still zero at any point in a material region. As in the case of free space and by comparing Gauss's laws for the electric and magnetic fields, this simply means that magnetic charges do not exist and that the magnetic flux lines are closed on themselves. The integral form of this Gauss's law implies that the net outflow of the magnetic flux lines from a closed surface s is zero, which once again leads to the same conclusion of closed magnetic flux lines.

3.8.3 Faraday's Law

Faraday's law simply relates the circulation of the electric charges under the influence of the induced electric field to its cause, which is the time-varying magnetic field $\nabla \times \mathbf{E} = -\partial \mathbf{B}/\partial t$. In the presence of material regions, we learned that $\mathbf{B}$ should be modified, and in most cases intensified, because of the alignment of the magnetic dipoles. Thus a modified value of $\mathbf{B} = \mu_o \mu_r \mathbf{H}$ should account for the magnetization of the magnetic materials, and we need not account for any extra terms in the right-hand side of Faraday's law. The left-hand side of Faraday's law, conversely, describes the circulation of the free electric charges as a result of the force exerted on them by the induced electric field. Because we did not discover any induced free magnetic charges in materials, we will not, therefore, require any modifications in the left-hand side of Faraday's law either. In other words, Faraday's law maintains its form in material regions. In integral form, this law may be written as

$$\oint_c \mathbf{E} \cdot d\ell = -\frac{d}{dt} \int_s \mathbf{B} \cdot ds$$

where the area s is enclosed by the contour c.

3.8.4 Ampere's Law

Three types of induced currents were identified as a result of the interaction of the electric and magnetic fields with materials. The conduction current $\mathbf{J}_c$ induced as a result of the drift of free electrons, polarization current $\mathbf{J}_p$ results if a time-varying electric field is applied to dielectric materials, and the magnetization current $\mathbf{J}_m$ that is induced when a magnetic material is magnetized by an externally applied magnetic field. Considering all these types of currents in addition to the current density $\mathbf{J}$ resulting from an external source, we may write Ampere's law in the form

$$\nabla \times \frac{\mathbf{B}}{\mu_o} = \mathbf{J} + \mathbf{J}_c + \mathbf{J}_m + \mathbf{J}_p + \frac{\partial \epsilon_o \mathbf{E}}{\partial t}$$

On substituting $\mathbf{J}_c = \sigma \mathbf{E}$,

$$\mathbf{J}_m = \nabla \times \mathbf{M} = \nabla \times \chi_m \mathbf{H} \qquad \text{(for linear magnetic material)}$$

and

$$\mathbf{J}_p = \frac{\partial \mathbf{P}}{\partial t} = \frac{\partial \epsilon_o \chi_e \mathbf{E}}{\partial t} \qquad \text{(for linear dielectric material)}$$

We obtain the general form of Ampere's law in material regions as

$$\nabla \times \mathbf{H} = \mathbf{J} + \sigma \mathbf{E} + \frac{\partial \epsilon \mathbf{E}}{\partial t} \qquad (3.28)$$

where

$$\mathbf{H} = \frac{\mathbf{B}}{\mu_o} - \mathbf{M} = \frac{\mathbf{B}}{\mu_o} - \chi_m \mathbf{H}$$

and

$$\epsilon = \epsilon_o \epsilon_r = \epsilon_o (1 + \chi_e)$$

χ_e and χ_m are the electric and magnetic susceptibilities, respectively. In integral form, Ampere's law may be given in the form

$$\oint_c \mathbf{H} \cdot d\ell = \int_s \mathbf{J} \cdot d\mathbf{s} + \int_s \sigma \mathbf{E} \cdot d\mathbf{s} + \frac{d}{dt} \int_s \epsilon \mathbf{E} \cdot d\mathbf{s} \qquad (3.29)$$

where, once again, s is any area enclosed by c. This integral form may be obtained by integrating equation 3.28 over any area s—for example, $\int_s \mathbf{J} \cdot d\mathbf{s}$ and using Stokes's theorem to change the integration over the curl term to an integration of $\mathbf{H}$ over the contour c enclosing the area s. Ampere's law given by equation 3.28 or 3.29 is a general form, and can be deduced to any special case of interest. For example, if there is no external current source $\mathbf{J} = 0$. Also if the electrical and magnetic fields are in nonconductive material (i.e., $\sigma = 0$) and, hence, $\mathbf{J}_c = 0$. Ampere's law under these conditions is given by

$$\nabla \times \mathbf{H} = \frac{\partial \epsilon \mathbf{E}}{\partial t}$$

Another important special case is obtained when we have static fields in nonconductive material region. In this case, $\partial \epsilon \mathbf{E}/\partial t = 0$ and $\mathbf{J}_c = 0$. Ampere's law is hence reduced to

$$\nabla \times \mathbf{H} = \mathbf{J}$$

where $\mathbf{J}$ is the current density of an external source.

In the following section, we will use these general forms of Maxwell's equations to obtain a set of mathematical relations that describe the transitional properties of the electric and magnetic fields at interfaces between two different material regions. This set of relations is known as the boundary conditions.

3.9 BOUNDARY CONDITIONS

Thus far in our study, we assumed that the fields are present in an unbounded space of specific electric and magnetic properties. After discussing the modifications in the electric and magnetic fields as a result of their presence in a material region, one may ask the question of how these fields adjust their properties at an interface between two different materials. It is the objective of this section to answer such a question. Specifically, we are going to obtain mathematical relations that describe the transitional properties of the electric and magnetic fields from one region to another. Furthermore, separate relations will be obtained for the tangential and normal components of both the electric and magnetic fields. In all cases, we will use Maxwell's equations in integral form to obtain these relations.

3.9.1 Boundary Conditions for Normal and Tangential Components of Electric Field

Normal components of electric field.   Let us consider first the boundary condition of the normal component of the electric field. Figure 3.17 shows an interface between two materials where region 1 has a permittivity ϵ_1 and region 2 has a permittivity ϵ_2. **n** is a unit vector normal to the interface and pointing from region 2 to region 1. The boundary conditions on the normal components of the electric field are found by applying Gauss's law to a small pillbox of area Δs and height δh placed at the interface between the two dielectric regions as shown in Figure 3.17. Gauss's law, hence, reduces to

$$\oint_{\Delta s} \mathbf{D} \cdot d\mathbf{s} = \int_v \rho_v \, dv$$

which requires that the total electric flux emanating from the "pillbox" be equal to the total charge within the box. Because we are interested in the boundary condition at the interface—that is, the special case as $\delta h \to 0$—the total electric flux emanating from the pillbox will have contributions only from the top and bottom. The contribution from the curved cylindrical surface will ultimately be zero as $\delta h \to 0$, which is the limiting case of interest. Applying Gauss's law to the top and bottom surfaces of the pillbox, we then obtain

$$D_{1n} \Delta s - D_{2n} \Delta s = \rho_v \Delta s \, \delta h \qquad (3.30)$$

It should be noted that the term $D_{1n} \Delta s$ is positive because it accounts for an *outflow* of the electric flux from the top surface, whereas a negative sign preceded the second term $D_{2n} \Delta s$ because it represents an electric flux *entering* the bottom of the box. The term $\rho_v \Delta s \, \delta h$ accounts for the total charge enclosed by the pillbox.

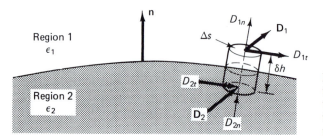

Figure 3.17   The boundary condition for the normal component of the electric field at the interface between regions 1 and 2, which have permittivities ϵ_1 and ϵ_2.

In the limit as the incremental height δh goes to zero, the enclosed charge term reduces to

$$\underset{\delta h \to 0}{\text{Lim}} \rho_v \, \Delta s \, \delta h = \rho_s \, \Delta s \qquad (3.31)$$

where ρ_s is a surface charge density on the interface (surface) between the two dielectrics. Substituting equation 3.31 in equation 3.30, we obtain

$$D_{1n} - D_{2n} = \rho_s \qquad (\text{C/m}^2) \qquad (3.32)$$

which means that the normal component of **D** is discontinuous at the interface between two materials by the amount of the free surface charge ρ_s that may be present at the interface. In a vector notation, D_n may be expressed in the form $\mathbf{n} \cdot \mathbf{D}$, which also means the component of **D** in the direction of the unit vector **n**. Equation 3.32 may then be expressed as

$$\boxed{\mathbf{n} \cdot (\mathbf{D}_1 - \mathbf{D}_2) = \rho_s \qquad \text{C/m}^2} \qquad (3.33)$$

which is the final form of the required boundary condition. Two special cases of (equation 3.33) will be considered next.

Boundary Condition at Interface between Two Perfect Dielectrics.   The free charge in any perfect dielectric is zero. Therefore, at the interface between two of such dielectrics, ρ_s should be zero, and the boundary condition in Eq. 3.33 reduces to

$$\mathbf{n} \cdot (\mathbf{D}_1 - \mathbf{D}_2) = 0 \qquad (3.34)$$

or

$$D_{n1} = D_{n2}$$

Equation 3.34 means that the normal component of the electric flux density should be continuous across the boundary between two perfect dielectrics.

Boundary Condition at Interface between Perfect Dielectric and Good Conductor. In general, conductors are characterized by the presence of free electrons. Therefore, it is expected that free charges would exist at the interface separating the perfect dielectric and the conductor. The boundary condition in this case may then be expressed in terms of the general equation 3.33. For this boundary condition, however, there are more interesting special cases that need to be considered further. This includes the following:

1. *Boundary condition for static fields.* For static fields, the electric field inside the conductor is zero. This is because any localized static charge distribution inside a conductor vanishes under the influence of its own electric forces. Therefore, at steady state (when the static condition of interest is reached), the static charges will redistribute themselves at the surface of the conductor (because of their inability to escape beyond that surface) in such a way that the electric field inside the conductor would be zero. It is this condition of zero field inside the conductor that would result in the final distribution of the charge on the surface of the conductor—hence, the steady-state condition.

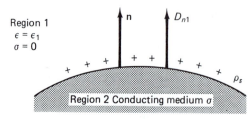

Figure 3.18 The boundary condition of static **D** at the surface between perfect dielectric and a conducting medium.

It may be worth emphasizing at this point that for static fields, the electric and magnetic fields are uncoupled (see general comments on Maxwell's equations in chapter 1); hence, a magnetic field may exist independently inside the conducting medium even if the electric field is zero inside that medium. Therefore, for static electric fields, the boundary condition at the interface between a conducting and perfect dielectric medium as shown in Figure 3.18 is given by

$$\mathbf{n} \cdot \mathbf{D}_1 = \rho_s$$

2. *Boundary condition for time-varying fields.* In this case, the electric and magnetic fields are coupled. As we will see in our discussion of plane wave propagation in conductive medium in the next section, the electric and magnetic fields may be present in such a medium. The boundary condition in this case is, therefore, described by the general form of equation 3.33. Of particular interest, however, is the case in which region 2 is a perfectly conducting one—that is, $\sigma_2 \rightarrow \infty$. In this case, the electric and magnetic fields will not penetrate the perfectly conducting medium, thus resulting in zero electric and magnetic fields inside the medium. Because the time-varying electric field is zero inside the perfectly conducting medium, the boundary condition of equation 3.33 reduces to

$$\mathbf{n} \cdot \mathbf{D}_1 = \rho_s$$

which is illustrated in Figure 3.19. It should be noted that although in the static case the electric field was zero inside the conducting medium even if it is only finitely conducting, for time-varying fields, the electric field was zero only for the case when region 2 was *perfectly* conducting.

Tangential component of electric field.   To derive the condition that should be satisfied by the tangential component of the electric field at the interface between two material regions, we use Faraday's law in integral form, which is given by

$$\oint_c \mathbf{E} \cdot d\ell = -\frac{d}{dt} \int_s \mathbf{B} \cdot d\mathbf{s} \tag{3.35}$$

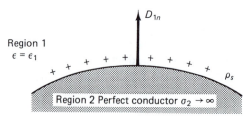

Figure 3.19 Boundary condition for time-varying **D** at the interface between a perfect dielectric and perfect conductor.

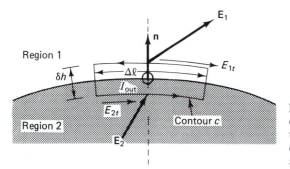

Figure 3.20 The contour c used to develop the boundary condition of the tangential components of the electric fields $\mathbf{E}_1$ and $\mathbf{E}_2$ in regions 1 and 2, respectively.

To apply this law, we construct a contour c, integrate the electric field around the contour, and equate the result with the negative time rate of change of the total magnetic flux crossing the area s enclosed by the contour c. For this purpose, we establish the rectangular contour c shown in Figure 3.20 between the two dielectric regions. Once again, because we are interested in relating the fields at the boundary, the side δh will be considered very small and our result will be obtained from the limiting case as $\delta h \to 0$. If we neglect the small contribution from δh to the line integral of equation 3.35, we obtain

$$E_{2t}\,\Delta\ell - E_{1t}\,\Delta\ell = -\frac{d}{dt}(\mathbf{B}\cdot\Delta\ell\,\delta h\,\mathbf{a}_{\text{out}}) \tag{3.36}$$

where $\mathbf{a}_{\text{out}}$ is a unit vector normal to the area $\Delta\ell\,\delta h$ and in our case pointing out of the plane of the paper. In equation 3.36, the first term $E_{2t}\,\Delta\ell$ is positive because E_{2t} is along the direction of integration, and the negative sign preceding $E_{1t}\,\Delta\ell$ is there because the contribution of $E_{1t}\,\Delta\ell$ to the line integral is opposite in direction to that of $E_{2t}\,\Delta\ell$. The term $\mathbf{B}\cdot\Delta\ell\,\delta h\,\mathbf{a}_{\text{out}}$ simply indicates the component of the magnetic flux crossing the enclosed area $\Delta\ell\,\delta h\,\mathbf{a}_{\text{out}}$.

As $\delta h \to 0$, the total flux crossing the enclosed area will be zero, and equation 3.36 reduces to

$$E_{2t} - E_{1t} = 0$$

or

$$E_{2t} = E_{1t}$$

In a vector term, this equation may be written as

$$\boxed{\mathbf{n}\times(\mathbf{E}_1 - \mathbf{E}_2) = 0} \tag{3.37}$$

where $\mathbf{n}$ is a unit vector normal to the interface. The cross product of $\mathbf{n}\times\mathbf{E}_1 = \mathbf{n}\times(E_{1n}\,\mathbf{n} + E_{1t}\,\mathbf{t}) = E_{1t}$. This is because $\mathbf{n}\times\mathbf{n} = 0$, and $\mathbf{n}\times\mathbf{t} = 1$, where $\mathbf{t}$ is a unit vector tangential to the interface and, hence, is normal to $\mathbf{n}$. Equation 3.37 is the boundary condition that should be satisfied by the tangential components of the electric fields at the boundary between two material regions.

The following are examples illustrating the use of the previously described boundary conditions.

EXAMPLE 3.4

The electric field intensity $\mathbf{E}_2$ in region 2 has a magnitude of 10 V/m and makes an angle $\theta_2 = 30°$ with the normal at the dielectric interface between regions 1 and 2, as shown in Figure 3.21. Calculate the magnitude of $\mathbf{E}_1$ and the angle θ_1 for the case when $\epsilon_1 = 1/2\epsilon_2$.

Solution

Because we have boundary conditions for the tangential and normal components of the electric field, we will start our solution by obtaining these components for $\mathbf{E}_1$ and $\mathbf{E}_2$.

	TANGENTIAL COMPONENTS	NORMAL COMPONENTS
Region 1	$E_1 \sin \theta_1$	$E_1 \cos \theta_1$
Region 2	$E_2 \sin \theta_2$	$E_2 \cos \theta_2$

where E_1 and E_2 are the magnitudes of $\mathbf{E}_1$ and $\mathbf{E}_2$, respectively. From equation 3.37, we know that the tangential components are continuous across the interface between regions 1 and 2, hence,

$$E_1 \sin \theta_1 = E_2 \sin \theta_2 = 10 \sin 30° \qquad (3.38)$$

Also because we have perfect dielectrics in both regions 1 and 2, the boundary condition for the normal component of the field is given according to equation 3.34 by

$$\epsilon_1 E_1 \cos \theta_1 = \epsilon_2 E_2 \cos \theta_2$$
$$= \epsilon_2 E_2 \cos 30° \qquad (3.39)$$

From equations 3.38 and 3.39, we obtain

$$\frac{\tan \theta_1}{\epsilon_1} = \frac{\tan \theta_2}{\epsilon_2}$$

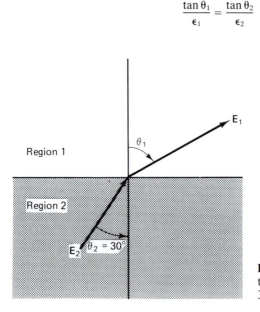

Region 1

Region 2

θ_1

E_1

$\mathbf{E}_2$ $\theta_2 = 30°$

Figure 3.21   Interface between the two dielectric regions of example 3.4.

or

$$\theta_1 = \tan^{-1}\left[\frac{\epsilon_1}{\epsilon_2}\tan\theta_2\right] = 16.1°$$

From equation 3.38, E_1 is given by

$$E_1 = \frac{10\sin 30°}{\sin 16.1°} = 18 \text{ V/m}$$

◆◆◆

EXAMPLE 3.5

Consider an interface between regions 1 and 2 as shown in Figure 3.22. Region 1 is air $\epsilon_1 = \epsilon_o$, and region 2 has a dielectric constant $\epsilon_2 = 2\epsilon_o$. Let us also assume that we have a surface charge density distribution $\rho_s = 0.2$ C/m² at the interface between these two regions. Determine the electric flux density $\mathbf{D}_2$ in region 2 if $\mathbf{D}_1$ is given by

$$\mathbf{D}_1 = 3\mathbf{a}_x + 4\mathbf{a}_y + 3\mathbf{a}_z \qquad (\text{C/m}^2)$$

Solution

Because of the presence of a surface charge density ρ_s, the boundary condition for the normal component of $\mathbf{D}$ is given in this case by

$$\mathbf{n}\cdot(\mathbf{D}_1 - \mathbf{D}_2) = \rho_s$$

the unit vector normal to the surface is given by

$$\mathbf{n} = \mathbf{a}_z,$$

hence,

$$\mathbf{a}_z\cdot(\mathbf{D}_1 - \mathbf{D}_2) = \rho_s$$

or

$$D_{1z} - D_{2z} = 0.2$$

$$D_{2z} = 3 - 0.2 = 2.8 \qquad \text{C/m}^2$$

D_{1x} and D_{1y} constitute the components of $\mathbf{D}_1$ tangential to the interface. To obtain D_{2x} and D_{2y}, we need to apply the boundary conditions for the tangential component of the electric field. From equation 3.37, we have

$$E_{1x} = E_{2x}$$

and

$$E_{1y} = E_{2y}$$

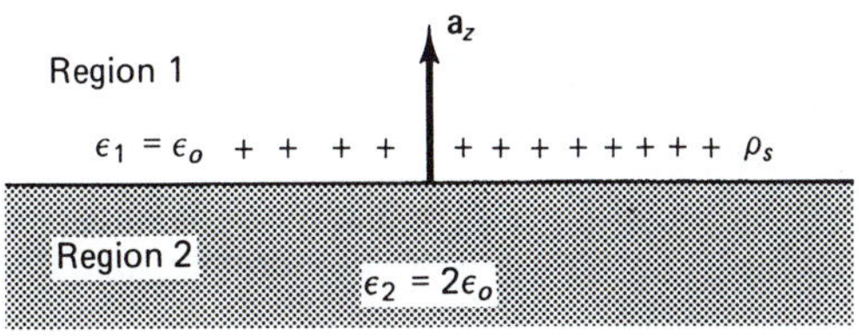

Figure 3.22 Surface charge density ρ_s at the interface between the two regions of example 3.5.

Hence,

$$\frac{D_{1x}}{\epsilon_1} = \frac{D_{2x}}{\epsilon_2}$$

and

$$\frac{D_{1y}}{\epsilon_1} = \frac{D_{2y}}{\epsilon_2}$$

or

$$D_{2x} = \frac{2\epsilon_o}{\epsilon_o} D_{1x} = 6$$

and

$$D_{2y} = \frac{2\epsilon_o}{\epsilon_o} D_{1y} = 8$$

The electric flux density in region 2 is therefore given by

$$\mathbf{D}_2 = 6\mathbf{a}_x + 8\mathbf{a}_y + 2.8\mathbf{a}_z \qquad (\text{C/m}^2)$$

3.9.2 Boundary Conditions for Normal and Tangential Components of Magnetic Field

Similar to the case of deriving the boundary condition for the normal component of the electric field, we will use Gauss's law to obtain the boundary condition for the normal component of the magnetic field. We will derive the boundary condition for the tangential component of the magnetic field, conversely, by using the integral form of Ampere's law.

Normal component of magnetic field.   Consider the interface between the two material regions 1 and 2 shown in Figure 3.23. The magnetic flux densities in regions 1 and 2 are $\mathbf{B}_1$ and $\mathbf{B}_2$, respectively. To obtain the boundary condition on the normal component of $\mathbf{B}$ we use Gauss's law for the magnetic field

$$\oint_s \mathbf{B} \cdot d\mathbf{s} = 0$$

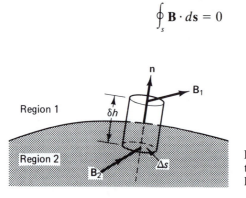

Figure 3.23 The boundary condition on the normal component of **B**.

This law requires that the total magnetic flux emanating from a closed surface s be equal to zero. For this purpose, we construct a small "pillbox" as shown in Figure 3.23 and calculate the total flux emanating from that enclosed surface. Once again, because we are interested in the boundary condition at the interface, the desired result will be obtained in the limiting case as $\delta h \to 0$. Therefore, in calculating the net magnetic flux outflowing from the pillbox, we will count only the contributions from the top and bottom surfaces. Hence,

$$\oint_s \mathbf{B} \cdot d\mathbf{s} = B_{1n} \Delta s - B_{2n} \Delta s = 0$$

or

$$B_{1n} = B_{2n} \tag{3.40a}$$

Equation 3.40a may be written in the form

$$\boxed{\mathbf{n} \cdot (\mathbf{B}_1 - \mathbf{B}_2) = 0} \tag{3.40b}$$

where $\mathbf{n}$ is a unit vector normal to the interface from region 2 to region 1. Equations 3.40a and b simply indicate that the normal component of the magnetic flux density is always continuous at the interface between two material regions.

Tangential component of magnetic field.   As indicated earlier, we will use Ampere's law to obtain this boundary condition on the tangential component of the magnetic field. Ampere's law is given by

$$\oint_c \mathbf{H} \cdot d\boldsymbol{\ell} = \int_s \mathbf{J} \cdot d\mathbf{s} + \frac{d}{dt} \int_s \mathbf{D} \cdot d\mathbf{s}$$

It states that the line integral of the magnetic field around the closed contour c equals the total current crossing the area s enclosed by c. This current may be due to an external source, displacement current as a result of time-varying electric field, or induced conduction or polarization currents. It should be noted that the magnetization current, conversely, is included in the $\mathbf{H}$ term.

To apply Ampere's law, we therefore construct a contour c as shown in Figure 3.24 and integrate $\mathbf{H}$ around this contour. The obtained result is then equated to the

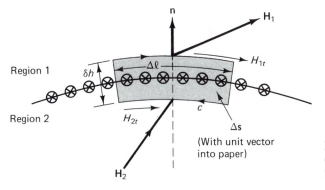

Figure 3.24   The contour c used to obtain the boundary condition on the tangential component of **H**.

total current crossing the area Δs enclosed by c. In carrying out the line integral we will, as we always do, consider the limiting case as $\delta h \to 0$ because we are interested in relating H_{1t} to H_{2t} at the interface. From Ampere's law we obtain

$$H_{1t}\,\Delta\ell - H_{2t}\,\Delta\ell = \mathbf{J}\cdot\Delta\mathbf{s} + \frac{d}{dt}(\mathbf{D}\cdot\Delta\mathbf{s}) \qquad (3.41)$$

The dot product in the current terms is maintained in equation 3.41 to emphasize the fact that we are considering only the component of these currents crossing the area $\Delta\mathbf{s}$—that is, into the plane of the paper. Equation 3.41 may then be rewritten in the form

$$H_{1t}\,\Delta\ell - H_{2t}\,\Delta\ell = J_{\text{in}}\,\Delta\ell\,\delta h + \frac{d}{dt}D_{\text{in}}\,\Delta\ell\,\delta h \qquad (3.42)$$

where J_{in} and D_{in} represent the components of the $\mathbf{J}$ and $\mathbf{D}$ vectors crossing (i.e., normal and into) the area Δs in the direction into the plane of the paper. In the limiting case as $\delta h \to 0$, the electric flux $D_{\text{in}}\,\Delta\ell\,\delta h$ crossing the area Δs will be zero. Also the total current $J_{\text{in}}\,\Delta\ell\,\delta h$ crossing this area will be zero except for cases in which we have free surface current density at the interface. In this case,

$$\underset{\delta h \to 0}{\text{Lim}}\ J_{\text{in}}\,\Delta\ell\,\delta h = J_{s(\text{in})}\Delta\ell$$

where $J_{s(\text{in})}$ is the component of the surface current density normal to the area Δs and tangential to the interface between the two media. Equation 3.42 then reduces to

$$H_{t1} - H_{t2} = J_{s(\text{in})} \qquad (3.43)$$

which states that the tangential component of $\mathbf{H}$ is discontinuous at the interface by the amount of the surface current density that may be present. In a vector form, equation 3.43 may be expressed as

$$\boxed{\mathbf{n} \times (\mathbf{H}_1 - \mathbf{H}_2) = \mathbf{J}_s} \qquad (3.44)$$

where $\mathbf{n}$ is a unit vector normal to the interface as shown in Figure 3.24, and $\mathbf{J}_s$ is the surface current density. Equation 3.44 is a convenient form for expressing this boundary condition for the following two reasons:

1. If we express $\mathbf{H}_1$ in terms of its normal H_{1n} and tangential H_{1t} components to the interface, it is clear that $\mathbf{n} \times \mathbf{H}_1$ will only involve H_{1t} because the cross product of $\mathbf{n}$ with the component of $\mathbf{H}_1$ normal to the interface—that is, $\mathbf{n} \times H_{1n}\,\mathbf{n} = 0$. Therefore, equation 3.44 is identical to equation 3.43 insofar as dealing with the tangential component of $\mathbf{H}$. The real advantage in using equation 3.44 will be clear from the next point.

2. In equation 3.43, we emphasized that $J_{s(\text{in})}$ is the component of the surface current crossing the element of area Δs into the plane of the paper. Equation 3.44 automatically takes such consideration into account because the cross product $\mathbf{n} \times \mathbf{H}$ provides a vector normal to $\mathbf{n}$ and $\mathbf{H}$, and in our case will be normal to the plane of the paper. Hence, in using equation 3.44, we do not have to memorize that $\mathbf{J}_s$ is actually perpendicular to $\mathbf{n}$ and $\mathbf{H}$, because the direction of $\mathbf{J}_s$ comes out of the cross product. The direction of the current $\mathbf{J}_s$ is hence determined by the right-hand rule from $\mathbf{n}$ to $\mathbf{H}$.

Before we conclude this section, let us consider an important special case when we have time-varying fields and when region 2 is a *perfectly conducting* one. As we indicated earlier, time-varying electric and magnetic fields are coupled and cannot penetrate a perfectly conducting medium of $\sigma = \infty$. Hence, the boundary condition of equation 44 reduces to

$$\mathbf{n} \times \mathbf{H}_1 = \mathbf{J}_s \qquad \text{(A/m)}$$

This equation simply indicates that the magnitude of the tangential component of the magnetic field intensity is equal to the current density on the surface of a perfectly conducting region. The direction of the surface current is determined by the right-hand rule from $\mathbf{n}$ to $\mathbf{H}$.

EXAMPLE 3.6

The magnetic field intensity $\mathbf{H}_2$ at the interface shown in Figure 3.25 between medium 1 of $\mu_1 = \mu_o$ and medium 2 of $\mu_2 = 3.1\mu_o$ is given by

$$\mathbf{H}_2 = 2\mathbf{a}_x + 5\mathbf{a}_y + 5\mathbf{a}_z$$

Determine the magnetic flux density $\mathbf{B}_1$ in region 1.

Solution

From the boundary condition of the normal component of the magnetic field we have

$$\mathbf{n} \cdot (\mathbf{B}_1 - \mathbf{B}_2) = 0$$

$$\mathbf{a}_z \cdot (\mathbf{B}_1 - \mu_2 \mathbf{H}_2) = 0$$

$$\therefore B_{1z} = \mu_2 H_{1z} = 3.1\mu_o(5) = 15.5\mu_o$$

For the boundary condition of the tangential component of the magnetic field, we have

$$\mathbf{n} \times (\mathbf{H}_1 - \mathbf{H}_2) = 0$$

because the surface current density at the interface is equal to zero. This simply means that, in this case, the tangential component of the magnetic field intensity is continuous across the boundary. Hence,

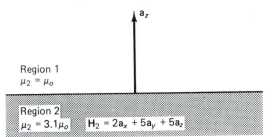

Region 1
$\mu_2 = \mu_o$

Region 2
$\mu_2 = 3.1\mu_o$ $H_2 = 2a_x + 5a_y + 5a_z$

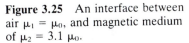

Figure 3.25   An interface between air $\mu_1 = \mu_0$, and magnetic medium of $\mu_2 = 3.1 \ \mu_0$.

$$H_{1x} = H_{2x} \qquad \therefore H_{1x} = 2$$

$$H_{1y} = H_{2y} \qquad \therefore H_{1y} = 5$$

$\mathbf{B}_1$ is therefore given by

$$\mathbf{B}_1 = \mu_o H_{1x} \mathbf{a}_x + \mu_o H_{1y} \mathbf{a}_y + B_{1z} \mathbf{a}_z$$

$$= 2\mu_o \mathbf{a}_x + 5\mu_o \mathbf{a}_y + 15.5\mu_o \mathbf{a}_z$$

───────◆◆◆───────

EXAMPLE 3.7

A cylinder of radius 7 cm is made of magnetic material for which $\mu_r = 5$. The region outside the cylinder $\rho > 7$ is air. If the magnetic field intensity $\mathbf{H}$ inside the cylinder is given at the point $(7, \pi/2, 0)$ by

$$\mathbf{H}_{in} = 2\mathbf{a}_x - \mathbf{a}_y - 3\mathbf{a}_z$$

and if we assume a surface current density

$$\mathbf{J}_s = 0.3\mathbf{a}_z,$$

determine the magnetic field intensity just outside the cylinder $\mathbf{H}_{out}$ at the same surface point $(7, \pi/2, 0)$.

Solution

The geometry of the problem is illustrated in Figure 3.26 where it is clear that without transforming the magnetic field vector into the cylindrical coordinate system, a unit vector normal to the interface at the point of interest P is $\mathbf{a}_y$, whereas $\mathbf{a}_x$ and $\mathbf{a}_z$ are both tangential to the cylindrical surface at that point. The boundary condition for the normal component of the magnetic field requires that

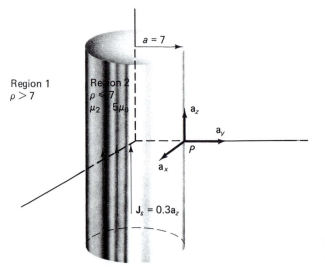

Figure 3.26 The geometry of the magnetic cylinder of example 3.7.

$$\mathbf{n} \cdot (\mathbf{B}_1 - \mathbf{B}_2) = 0$$

$$\mathbf{a}_y \cdot (\mathbf{B}_1 - \mathbf{B}_2) = 0$$

$$B_{1y} = B_{2y}$$

$$= \mu_2 H_{2y} = 5\mu_o(-1) = -5\mu_o$$

$$\therefore H_{1y} = \frac{B_{1y}}{\mu_o} = -5$$

The boundary condition for the tangential component of the magnetic field is, conversely, more complicated because of the presence of the surface current density. From equation 3.44, we obtain

$$\mathbf{a}_y \times [(H_{1x}\,\mathbf{a}_x + H_{1y}\,\mathbf{a}_y + H_{1z}\,\mathbf{a}_z) - (2\mathbf{a}_x - \mathbf{a}_y - 3\mathbf{a}_z)] = 0.3\mathbf{a}_z$$

$$\therefore \mathbf{a}_y \times [(H_{1x} - 2)\mathbf{a}_x + (H_{1y} + 1)\mathbf{a}_y + (H_{1z} + 3)\mathbf{a}_z] = 0.3\mathbf{a}_z$$

Carrying out the cross product we obtain

$$-(H_{1x} - 2)\mathbf{a}_z + (H_{1z} + 3)\mathbf{a}_x = 0.3\mathbf{a}_z$$

Equating each of the $\mathbf{a}_y$ and $\mathbf{a}_x$ components we obtain

$$-(H_{1x} - 2) = 0.3 \qquad \therefore H_{1x} = 1.7$$

and

$$H_{1z} + 3 = 0 \qquad \therefore H_{1z} = -3$$

The magnetic field intensity vector $\mathbf{H}_1$ that is just outside the cylinder is therefore given by

$$\mathbf{H}_1 = \mathbf{H}_{out} = 1.7\mathbf{a}_x - 5\mathbf{a}_y - 3\mathbf{a}_z$$

————————◆◆◆————————

EXAMPLE 3.8

A sphere of magnetic material of $\mu = 600\mu_o$ and radius $a = 0.1$ m is subjected to an external magnetic field. The induced magnetic flux density at the point $(0.1, \pi/2, \pi/2)$ (see Figure 3.27) just inside the surface of the sphere is given by

$$\mathbf{B}_{in} = 7\mathbf{a}_x + 2\mathbf{a}_y - 3\mathbf{a}_z$$

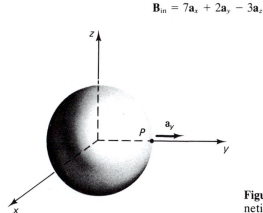

Figure 3.27 The sphere of magnetic material of example 3.8.

There is also an induced surface current density given by

$$\mathbf{J}_s = 0.5\mathbf{a}_x + 0.1\mathbf{a}_z$$

Determine the magnetic flux density at the same point $(0.1, \pi/2, \pi/2)$ in air $(\mu = \mu_o)$ just outside the sphere.

Solution

At the specific point of interest $P(0.1, \pi/2, \pi/2)$, the unit normal to the spherical surface is $\mathbf{a}_y$, whereas the tangential unit vectors are $\mathbf{a}_x$ and $\mathbf{a}_z$. Because of the simplicity of the geometry in this case, there is no need to transform the vector magnetic flux density $\mathbf{B}_{in}$ to the spherical coordinates so as to be able to identify the tangential and normal components of $\mathbf{B}$ at the point of interest. It should be noted, however, that in general it may be necessary to transform $\mathbf{B}$ to the spherical coordinate system using the formulas developed in chapter 1.

We now apply the boundary conditions for the normal component of $\mathbf{B}$, $\mathbf{n} \cdot (\mathbf{B}_{out} - \mathbf{B}_{in}) = 0$, where $\mathbf{B}_{out}$ is the magnetic flux density just outside the sphere. Hence,

$$\mathbf{a}_y \cdot (\mathbf{B}_{out} - \mathbf{B}_{in}) = 0$$

or

$$B_{out\,y} = B_{in\,y} = 2$$

In other words, the y component of the magnetic flux density outside the sphere $B_{out\,y}$ is equal to 2. To obtain the rest of the components of $\mathbf{B}_{out}$, we apply the boundary condition for the tangential components of the magnetic field intensity $\mathbf{H}$

$$\mathbf{n} \times (\mathbf{H}_{out} - \mathbf{H}_{in}) = \mathbf{J}_s$$

$$\mathbf{a}_y \times \left[\left(\frac{B_{out\,x}}{\mu_o} \mathbf{a}_x + \frac{B_{out\,y}}{\mu_o} \mathbf{a}_y + \frac{B_{out\,z}}{\mu_o} \mathbf{a}_z \right) - \left(\frac{7}{600\mu_o} \mathbf{a}_x + \frac{2}{600\mu_o} \mathbf{a}_y - \frac{3}{600\mu_o} \mathbf{a}_z \right) \right]$$
$$= 0.5\mathbf{a}_x + 0.1\,\mathbf{a}_z$$

$$\therefore \mathbf{a}_y \times \left[\left(\frac{B_{out\,x}}{\mu_o} - \frac{7}{600\mu_o} \right) \mathbf{a}_x + \left(\frac{B_{out\,y}}{\mu_o} - \frac{2}{600\mu_o} \right) \mathbf{a}_y + \left(\frac{B_{out\,z}}{\mu_o} + \frac{3}{600\mu_o} \right) \mathbf{a}_z \right]$$
$$= 0.5\mathbf{a}_x + 0.1\mathbf{a}_z$$

It should be noted that the cross product on the left-hand side should be performed before we equate the various components of the vector quantities on both sides of the equation. Hence,

$$\left(\frac{B_{out\,x}}{\mu_o} - \frac{7}{600\mu_o} \right)(-\mathbf{a}_z) + \left(\frac{B_{out\,z}}{\mu_o} + \frac{3}{600\mu_o} \right) \mathbf{a}_x = 0.5\mathbf{a}_x + 0.1\mathbf{a}_z$$

Now equating the z components, we obtain

$$\therefore B_{out\,x} = \frac{7}{600} - 0.1\mu_o$$

and by equating the x component we obtain

$$B_{out\,z} = \frac{-3}{600} + 0.5\mu_o$$

The magnetic flux density vector just outside the sphere is hence given by

$$\mathbf{B}_{out} = \left(\frac{7}{600} - 0.1\mu_o\right)\mathbf{a}_x + 2\mathbf{a}_y + (-0.005 + 0.5\mu_o)\,\mathbf{a}_z$$

---◆◆◆---

3.9.3 Other Boundary Conditions

In modifying Maxwell's equations so as to account for the charge and current distributions induced as a result of the interaction between the electric and magnetic fields and the materials, we identified, among other points, two important new induced sources. These are the polarization charge and the magnetization current densities. The polarization charge density ρ_p is given by

$$\boxed{\nabla \cdot \mathbf{P} = -\rho_p} \tag{3.45}$$

whereas the magnetization current density may be calculated from

$$\boxed{\nabla \times \mathbf{M} = \mathbf{J}_m} \tag{3.46}$$

where $\mathbf{P}$ and $\mathbf{M}$ are the polarization and the magnetization vectors, respectively.

   In this section, we will derive expressions describing the boundary conditions that should be satisfied by the polarization and magnetization vectors at interfaces between different material regions.

   Boundary Condition for the Polarization $\mathbf{P}$.   Let us consider the integral form of the expression relating the polarization to the induced polarization charge

$$\oint_s \mathbf{P} \cdot d\mathbf{s} = -\int_v \rho_p \, dv \tag{3.47}$$

This expression may be obtained from its differential form simply by integrating both sides of the differential form in equation 3.45 over a volume v and using the divergence theorem to convert the volume integral of $\nabla \cdot \mathbf{P}$ to the surface integral of $\mathbf{P}$ over the surface s enclosing v as given by equation 3.47.

   Equation 3.47 requires that the total polarization vector flux emanating from a closed surface s be equal to the total polarization negative charge in the volume v enclosed by s. To obtain the desired boundary condition, we therefore construct a small "pillbox" between the two media of interest as shown in Figure 3.28 and calculate the total polarization vector flux emanating from it. Because we are interested in relating $\mathbf{P}_1$ to $\mathbf{P}_2$ at the interface between regions 1 and 2, our result will be obtained in the limiting case as $\delta h \to 0$. In calculating the total flux of $\mathbf{P}$ emanating from the "pillbox," we will neglect the contribution from the curved surface of the box. Equation 3.47 then reduces to

$$P_{1n}\,\Delta s - P_{2n}\,\Delta s = -\rho_p\,\Delta s\,\delta h \tag{3.48}$$

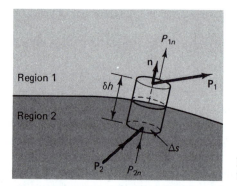

Figure 3.28 Boundary condition for the polarization vector **P**.

In the limit as $\delta h \to 0$, the total polarization charge term $\rho_p \, \Delta s \, \delta h$ goes to zero except if we have an induced surface polarization charge density ρ_{ps} defined as

$$\underset{\delta h \to 0}{\text{Lim}} \, \rho_p \, \Delta s \, \delta h = \rho_{ps} \, \Delta s$$

Equation 3.48 then reduces to

$$P_{1n} - P_{2n} = -\rho_{ps}$$

or in vector notation

$$\boxed{\mathbf{n} \cdot (\mathbf{P}_1 - \mathbf{P}_2) = -\rho_{ps}} \tag{3.49}$$

It should be noted that ρ_{ps} is due to surface bound charge density at the interface between the two media. Examples illustrating the calculation of ρ_{ps} will follow the next section on the boundary condition for the magnetization **M**.

***Boundary Condition for Magnetization* M.**   We will use equation 3.14 that is the integral form relating the induced magnetization current density $\mathbf{J}_m$ to the magnetization **M** to obtain this boundary condition. This equation is given by

$$\oint_c \mathbf{M} \cdot d\boldsymbol{\ell} = \int_s \mathbf{J}_m \cdot d\mathbf{s}$$

and simply requires integrating the magnetization vector over a closed contour c and equating the result to the total magnetization current crossing the area s enclosed by the contour c. To evaluate both sides of this equation at the interface, we construct the contour c shown in Figure 3.29 between the two media. Integrating **M** over the contour c and keeping in mind that we are interested in the limiting case as $\delta h \to 0$, we obtain

$$M_{1t} \, \Delta \ell - M_{2t} \, \Delta \ell = \mathbf{J}_m \cdot \delta h \, \Delta \ell \, \mathbf{a}_{\text{in}} \tag{3.50}$$

The area $\delta h \, \Delta \ell \, \mathbf{a}_{\text{in}}$ is enclosed by the contour c and the unit vector $\mathbf{a}_{\text{in}}$ is directed into the plane of the paper so that the directions of the line integral and the element of area become in accordance with the right-hand rule. Equation 3.50 may be written in the form

$$M_{1t} \, \Delta \ell - M_{2t} \, \Delta \ell = J_{mn} \, \Delta \ell \, \delta h \tag{3.51}$$

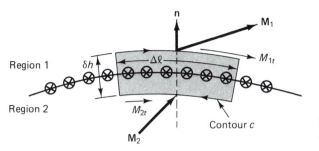

Figure 3.29 Boundary condition on the magnetization vector **M**.

where J_{mn} is the component of the magnetization current crossing the area $\Delta\ell\,\delta h\,\mathbf{a}_{in}$. In the limit as $\delta h \to 0$, the total current term in equation 3.51 reduces to zero, except for the case in which we have surface magnetization current density at the interface. In this case,

$$\lim_{\delta h \to 0} J_{mn}\,\Delta\ell\,\delta h = J_{msn}\,\Delta\ell$$

where J_{msn} is the component of the magnetization surface current density $\mathbf{J}_{ms}$ crossing the differential element of area $\Delta\ell\,\delta h$ in the limit as $\delta h \to 0$. Equation 3.51 then reduces to

$$M_{1t} - M_{2t} = J_{msn} \tag{3.52}$$

Similar to the case of the boundary condition for the tangential component of the magnetic field, equation 3.52 may be presented in the following more convenient vector form

$$\mathbf{n} \times (\mathbf{M}_1 - \mathbf{M}_2) = \mathbf{J}_{ms} \tag{3.53}$$

Once again, it should be noted that equation 3.53 emphasizes the fact that we are dealing with the tangential component of **M** and that the direction of $\mathbf{J}_{ms}$ is obtained according to the right-hand rule from **n** to **M**. The following examples illustrate the application of these as well as the other boundary conditions.

EXAMPLE 3.9

A metallic sphere of radius a is charged with a total charge Q. The sphere is also uniformly coated with a layer of dielectric material of dielectric constant $\epsilon_1 = \epsilon_o\,\epsilon_r$ as shown in Figure 3.30.

1. Obtain expressions for the electric flux density **D**, the electric field intensity **E**, and the polarization **P** in regions 1 and 2.
2. Find the polarization surface charge density ρ_{ps} at the interfaces $r = a$ and $r = b$.

Solution

To obtain the electric flux density everywhere, we use Gauss's law

$$\oint_s \mathbf{D} \cdot d\mathbf{s} = \int_v \rho_v\,dv$$

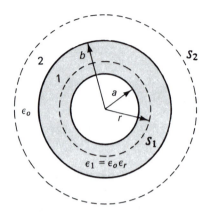

Figure 3.30   Illustration of the geometry of the metallic sphere of example 3.9.

For this purpose we establish the Gaussian surface s_1 in region 1 to determine $\mathbf{D}_1$ and the surface s_2 in region 2 to determine $\mathbf{D}_2$. Both s_1 and s_2 are shown in Figure 3.30. Hence,

$$\oint_{s_1} \mathbf{D}_1 \cdot ds = \int_{\theta=0}^{\pi} \int_{\phi=0}^{2\pi} D_{1r}\,\mathbf{a}_r \cdot \mathbf{r}^2 \sin\theta\, d\theta\, d\phi\, \mathbf{a}_r = Q$$

$$D_{1r} = \frac{Q}{4\pi r^2} \qquad \text{or} \qquad \mathbf{D}_1 = \frac{Q}{4\pi r^2}\mathbf{a}_r$$

We have, of course, used the symmetry of the problem and concluded that $\mathbf{D}_1$ has only an $\mathbf{a}_r$ component. Similarly, on using the Gaussian surface s_2 we obtain

$$\mathbf{D}_2 = \frac{Q}{4\pi r^2}\mathbf{a}_r$$

From the preceding discussion, it is clear that $\mathbf{D}_1$ and $\mathbf{D}_2$ have the same expression that is independent of the properties of the medium. The electric fields in both regions, conversely, are given by

$$\mathbf{E}_1 = \frac{Q}{4\pi\epsilon_1 r^2}\mathbf{a}_r, \qquad a < r < b,$$

$$\mathbf{E}_2 = \frac{Q}{4\pi\epsilon_o r^2}\mathbf{a}_r \qquad r > b$$

which shows that, unlike the electric flux densities, the electric field intensities depend on the properties of the medium.

The polarization for a linear medium is given by

$$\mathbf{P} = \epsilon_o \chi_e \mathbf{E}$$

Hence

$$\mathbf{P}_1 = \epsilon_o(\epsilon_r - 1)\mathbf{E}_1 = \epsilon_o(\epsilon_r - 1)\frac{Q}{4\pi\epsilon_1 r^2}\mathbf{a}_r$$

and $\mathbf{P}_2 = 0$ simply because the susceptibility χ_e in region 2 (air) is zero.

To determine the polarization surface charge at $r = a$, we use the polarization boundary condition

$$\mathbf{n} \cdot (\mathbf{P}_1 - \mathbf{P}_2) = -\rho_{ps}$$

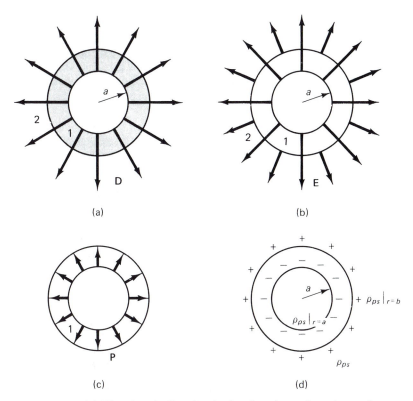

Figure 3.31 (a) The electric flux density in all regions where it may be seen that **D** is continuous. (b) The electric field intensity **E** in all regions where it may be seen that **E** is larger in region 2 and hence was presented by closer flux lines. (c) The polarization **P** that only exists in region 1 where we have dielectric material and (d) the polarization surface charge on both surfaces of region 1.

Hence,

$$\mathbf{a}_r \cdot \left(\frac{\epsilon_o(\epsilon_r - 1)Q}{4\pi\epsilon_1 a^2} \mathbf{a}_r - 0 \right) = -\rho_{ps}$$

and

$$\rho_{ps} \bigg|_{\text{at } r = a} = -\frac{\epsilon_o(\epsilon_r - 1)Q}{4\pi\epsilon_1 a^2}$$

Similarly at $r = b$

$$\mathbf{a}_r \cdot \left(0 - \frac{\epsilon_o(\epsilon_r - 1)Q}{4\pi\epsilon_1 b^2} \mathbf{a}_r \right) = -\rho_{ps}$$

and the polarization surface charge at $r = b$ is given by

$$\rho_{ps} \bigg|_{r = b} = \frac{\epsilon_o(\epsilon_r - 1)Q}{4\pi\epsilon_1 b^2}$$

Figure 3.31 illustrates the electric flux density, the electric field intensity, the polarization, and the surface polarization charge in all regions.

EXAMPLE 3.10

In example 3.3, we obtained expressions for the magnetic field intensity **H**, the magnetic flux density **B**, and the magnetization **M** in the core of a long solenoid. Determine the magnetization surface current densities at the interfaces between the various regions.

Solution

From example 3.3, it is shown that the magnetization vectors in the three regions are given by

$$\mathbf{M}_1 = \frac{nI}{d}\,\mathbf{a}_z \qquad r < a$$

$$\mathbf{M}_2 = \frac{2nI}{d}\,\mathbf{a}_z \qquad a < r < b$$

and

$$\mathbf{M}_3 = 0 \qquad b < r < c$$

To determine the magnetization surface current, we use the boundary condition

$$\mathbf{n} \times (\mathbf{M}_1 - \mathbf{M}_2) = \mathbf{J}_{ms}$$

At $\rho = a$

$$\mathbf{a}_\rho \times (\mathbf{M}_2 - \mathbf{M}_1) = \mathbf{J}_{ms}$$

Hence,

$$\mathbf{a}_\rho \times \left(\frac{2nI}{d}\,\mathbf{a}_z - \frac{nI}{d}\,\mathbf{a}_z\right) = \mathbf{J}_{ms}$$

$$-\frac{nI}{d}\,\mathbf{a}_\phi = \mathbf{J}_{ms},$$

At $\rho = b$

$$\mathbf{a}_\rho \times (\mathbf{M}_3 - \mathbf{M}_2) = \mathbf{J}_{ms}$$

$$\mathbf{a}_\rho \times \left(0 - \frac{2nI}{d}\,\mathbf{a}_z\right) = \mathbf{J}_{ms}$$

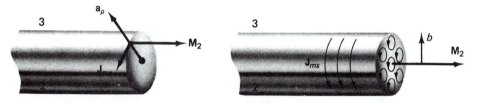

Figure 3.32   Illustration of the magnetization $\mathbf{M}_2$ in region 2 and the magnetization surface current $\mathbf{J}_{ms}$ at the surface $\rho = b$.

Hence,

$$\mathbf{J}_{ms} = \frac{2nI}{d} \mathbf{a}_\phi$$

It is clear from the surface magnetization current equations that although the magnetization vector is in the $\mathbf{a}_z$ direction, the magnetization surface currents are in the $\mathbf{a}_\phi$ direction as shown in Figure 3.32.

3.10 SUMMARY OF BOUNDARY CONDITION FOR ELECTRIC AND MAGNETIC FIELDS

Although boundary conditions are described clearly in the previous sections, a brief summary of these conditions will be provided for a quick and clear reference:

General case.   Normal components of **D** are discontinuous by surface charge layer ρ_s, where ρ_s is a layer of *free charge* (bound charge, such as polarization, does not count for it has already been included in **D**), hence

$$\boxed{\mathbf{n} \cdot (\mathbf{D}_1 - \mathbf{D}_2) = \rho_s}$$
$$(3.54)$$

Normal components of **B** are continuous because no magnetic charge exists, hence

$$\boxed{\mathbf{n} \cdot (\mathbf{B}_1 - \mathbf{B}_2) = 0}$$
$$(3.55)$$

Tangential components of **E** are always continuous at the interface between two media, hence

$$\boxed{\mathbf{n} \times (\mathbf{E}_1 - \mathbf{E}_2) = 0}$$
$$(3.56)$$

Tangential components of **H** are discontinuous by surface current density $\mathbf{J}_s$, hence

$$\boxed{\mathbf{n} \times (\mathbf{H}_1 - \mathbf{H}_2) = \mathbf{J}_s}$$
$$(3.57)$$

As shown in Figure 3.33, the unit normal **n** points from medium 2 into medium 1. The relationships $\mathbf{D} = \epsilon \mathbf{E}$ and $\mathbf{B} = \mu \mathbf{H}$ can be used to express other variations of the preceding boundary conditions.

Dielectrics.   Because the dielectric materials are dominated by "bound" rather than "free" charges, the major action of an electric field in a dielectric is related to forcing the positive and negative charges of all molecules to separate slightly and form dipoles throughout the interior of the material. It is therefore important to realize that the free charge density and the surface current density $\mathbf{J}_s$ are zero for a dielectric. Hence, we have

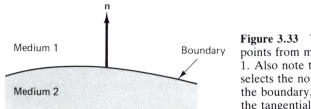

Figure 3.33 The unit normal **n** points from medium 2 into medium 1. Also note that the operation **n** · selects the normal component to the boundary, whereas **n**× selects the tangential component.

$$\mathbf{n} \cdot (\mathbf{D}_1 - \mathbf{D}_2) = 0 \tag{3.58}$$

$$\mathbf{n} \cdot (\mathbf{B}_1 - \mathbf{B}_2) = 0 \tag{3.59}$$

$$\mathbf{n} \times (\mathbf{E}_1 - \mathbf{E}_2) = 0 \tag{3.60}$$

$$\mathbf{n} \times (\mathbf{H}_1 - \mathbf{H}_2) = 0 \tag{3.61}$$

which simply states that the normal components of **D** and **B**, as well as the tangential components of **E** and **H**, are continuous. Other components, however, such as $E_n = D_n/\epsilon$ and $B_t = \mu H_t$ may be discontinuous. For example, at an interface between two dielectric media, although the normal component of **D** is continuous, the normal components of **E** are discontinuous and given by

$$E_{n_1} = \frac{\epsilon_2}{\epsilon_1} E_{n_2}$$

where ϵ_1 and ϵ_2 are the dielectric constants of the two media.

Good conductors.   In practical problems, we often treat good conductors as though they were perfect conductors. This is a good approximation because metallic conductors, such as copper, do have high conductivities $\sigma = 6 \times 10^7$ S/m or mho/m. Only superconductors, however, have infinite conductivity and are truly perfect conductors. When discussing boundary conditions involving good conductors, it is best to separate these conditions into static (time-independent) and time-dependent cases. For static fields we have

$$\mathbf{n} \cdot \mathbf{D}_1 = \rho_s \tag{3.62}$$

$$\mathbf{n} \cdot (\mathbf{B}_1 - \mathbf{B}_2) = 0 \tag{3.63}$$

$$\mathbf{n} \times \mathbf{E}_1 = 0 \tag{3.64}$$

and

$$\mathbf{n} \times (\mathbf{H}_1 - \mathbf{H}_2) = 0 \tag{3.65}$$

where subscript 2 denotes the conducting medium.

There are a few important characteristics that can be obtained from equations 3.62 to 3.65. These include the following:

1. Electrostatic field inside a good conducting medium is zero. Free charge can exist on the surface of a conductor, which makes the normal component of **D** discontinuous being zero inside the conductor and nonzero outside. The tangential component of **E** just inside the conductor must be zero even if the surface is charged. Hence, according to equation 3.64, the tangential component of the electric field just outside a good conductor is also zero.

To help us understand the physical property of a conducting material that indicates that the static electric field is zero everywhere inside the conductor, let us suppose that there suddenly appears a number of electrons in the interior of a conductor. The electric field set up by these electrons is not counteracted by any positive charges, and the electrons therefore begin to accelerate from each other. This continues until the electrons reach the surface of the conductor, and there the outward progress of the electrons will stop, for the material surrounding the conductor is an insulator. From the preceding discussion, it is clear that there is no charge density within a conductor, and it also follows that no electrostatic field can exist inside a conductor. Electrons will move in response to any remaining field until they have arranged themselves to produce a zero field everywhere inside the conductor.

2. The electric and magnetic fields in the static case are independent. A static magnetic field can thus exist inside a metallic body, even though an **E** field cannot. The normal component of **B** and the tangential components of **H** are thus continuous across the interface.

For time-varying fields, the boundary conditions for good (perfect) conductors are

$$\mathbf{n} \cdot \mathbf{D} = \rho_s \tag{3.66}$$

$$\mathbf{n} \cdot \mathbf{B} = 0 \tag{3.67}$$

$$\mathbf{n} \times \mathbf{E} = 0 \tag{3.68}$$

$$\mathbf{n} \times \mathbf{H} = \mathbf{J}_s \tag{3.69}$$

where the subscripts have been deleted because in this case the only nonvanishing fields are those outside the conducting body. It will be shown later in this chapter that the *depth of penetration* of the fields inside a perfect conductor is zero. Thus, all fields are excluded from the interior of a good conductor, and current flow is confined in a thin layer at the surface.

The following are two more examples on the properties of the fields in material regions and the boundary conditions.

EXAMPLE 3.11

The dielectric in a parallel plate capacitor consists of two slabs, of permitivities ϵ_1 and ϵ_2, as shown in Figure 3.34. The capacitor is connected to a potential difference V. Determine (1) the electric field intensities and the electric flux density vectors in both slabs, (2) the surface charge density of free charges on the electrodes, and (3) the surface charge density of polarization charges on the surfaces of the slabs.

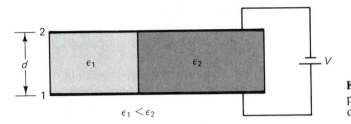

Figure 3.34 The geometry of the parallel plate capacitor with two dielectric slabs of example 3.11.

Solution

1. The electric field intensity vector **E** is clearly normal to the plates at all points. The work done in moving a unit positive charge from plate 2 to 1 is equal to the potential difference V, hence

$$\int_2^1 \mathbf{E} \cdot d\ell = Ed = V$$

therefore

$$E = E_1 = E_2 = \frac{V}{d}$$

The boundary condition for the tangential components of the electric field intensity at the boundary surface of the dielectric—that is, $E_1 = E_2$—is clearly satisfied. To determine the free surface charge density, let us consider the flux density vectors in both dielectrics.

$$D_1 = \epsilon_1 E_1 \text{ in region 1,} \qquad \text{and} \qquad D_2 = \epsilon_2 E_2 \text{ in region 2}$$

therefore

$$D_1 = \epsilon_1 \frac{V}{d} \qquad \text{and} \qquad D_2 = \epsilon_2 \frac{V}{d}$$

2. At the surface of the conductor, the free surface charge density ρ_s is given in terms of the normal component D_n of **D** by

$$D_n = \rho_s$$

Therefore, the surface charge density of free charges on the part of the electrodes in contact with dielectric 1 is

$$\rho_{s_1} = D_{1n} = \epsilon_1 \frac{V}{d}$$

On the parts of the electrodes that are in contact with dielectric 2, the surface charge density is

$$\rho_{s_2} = \mathbf{n} \cdot \mathbf{D}_2 = \epsilon_2 \frac{V}{d}$$

as shown in Figure 3.35.

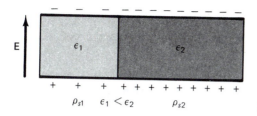

Figure 3.35   The free charge density distribution on the plates of the parallel plate capacitor.

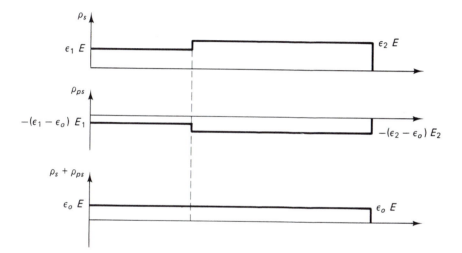

Figure 3.36   Free surface charge density distribution ρ_s and the polarization charge density distribution ρ_{ps} on the interface between the conducting planes and the dielectric materials. The total charge density on the surface of the conductor is constant and equal to $\epsilon_o E$.

3. The surface charge density of polarization charges on the surfaces of slab 1 is

$$-\rho_{ps_1} = p_{1n} = (\epsilon_1 - \epsilon_o)E_1 = (\epsilon_1 - \epsilon_o)\frac{V}{d}$$

and on the surface of slab 2

$$-\rho_{ps_2} = p_{2n} = (\epsilon_2 - \epsilon_o)E_2 = (\epsilon_2 - \epsilon_o)\frac{V}{d}$$

so that the total surface charge density (the sum of free and polarization charges) is constant over the surfaces of the electrodes as shown in Figure 3.36.

EXAMPLE 3.12

To measure the ac current flowing in a single conductor wire, circular magnetic loops made of high μ material are often used. Figure 3.37 illustrates the use of one of these devices.

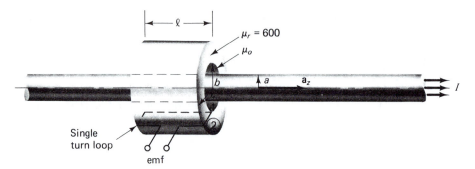

Figure 3.37 A magnetic loop placed coaxially with a long conductor carrying a current I.

1. Use Ampere's law to determine the magnetic field intensity in regions 1: $a < \rho < b$, and 2: $b < \rho < c$.
2. Determine the magnetic flux density **B**, magnetization **M**, and the magnetization current density $\mathbf{J}_m$ in region 2: $b < \rho < c$.
3. Determine the magnetization surface current $\mathbf{J}_{ms}$ at the surface $\rho = b$.
4. If the current flowing in the long conductor is given by $\mathbf{I} = I_o \cos \omega t \, \mathbf{a}_z$, use Faraday's law to determine the induced *emf* across the terminals of the single turn loop shown in Figure 3.37.

Solution

1. We first apply Ampere's law to determine the magnetic field intensity **H** in all regions outside the current carrying conductor. Because we have time-varying current, the resulting magnetic field will also be time varying, and we expect coupling between the electric and magnetic fields. In this case, all four Maxwell's equations should be solved simultaneously to determine the values of **H** and **E**. If we assume that the time variation ω is small enough, however, the coupling between the **E** and **H** field may be assumed negligible, and the displacement current term in Ampere's law may hence be neglected. Under the assumption of slow time-varying fields, Ampere's law is given by

$$\oint_c \mathbf{H} \cdot d\ell = I \qquad \left(\text{neglect the displacement current term } \frac{d}{dt}\int_s \mathbf{D} \cdot d\mathbf{s} \right)$$

By establishing a suitable contour c around the current carrying conductor, integrating **H** around this contour and equating the result to the total current I crossing the area surrounded by the contour c we obtain

$$H_\phi(2\pi\rho) = I$$

or

$$\mathbf{H} = \frac{I}{2\pi\rho} \mathbf{a}_\phi$$

The magnetic field intensity depends only on the parameters of the external source, hence,

$$\mathbf{H}_1 = \frac{I}{2\pi\rho} \mathbf{a}_\phi, \qquad \mathbf{H}_2 = \frac{I}{2\pi\rho} \mathbf{a}_\phi$$

2. The magnetic flux density is obtained simply by multiplying the magnetic field intensity **H** by the appropriate permeability of the medium in region 2,

$$\mathbf{B}_2 = 600\mu_o \frac{I}{2\pi\rho} \mathbf{a}_\phi \qquad b \leq \rho \leq c$$

$$\mathbf{M}_2 = \chi_m \mathbf{H}_2 = (\mu_r - 1)\mathbf{H}_2 = 599 \frac{I}{2\pi\rho} \mathbf{a}_\phi$$

The magnetization current is given by

$$\mathbf{J}_m = \nabla \times \mathbf{M} = \begin{vmatrix} \dfrac{\mathbf{a}_\rho}{\rho} & \mathbf{a}_\phi & \dfrac{\mathbf{a}_z}{\rho} \\ \dfrac{\partial}{\partial\rho} & \dfrac{\partial}{\partial\phi} & \dfrac{\partial}{\partial z} \\ 0 & \rho M_\phi & 0 \end{vmatrix} = 0$$

3. To obtain an expression for the magnetization surface current density, we apply the boundary condition on the tangential component of **M**. Hence,

$$\mathbf{n} \times (\mathbf{M}_2 - \mathbf{M}_1) = \mathbf{J}_{sm}$$

$$\mathbf{a}_\rho \times \left(\frac{599I}{2\pi\rho} \mathbf{a}_\phi - 0 \right) = \mathbf{J}_{sm} \qquad (3.70)$$

where $\mathbf{M}_1$, the magnetization of air, is zero. Carrying out the cross product in equation 3.70, we obtain

$$\mathbf{J}_{sm} = \frac{599I}{2\pi\rho} \mathbf{a}_z$$

4. To determine the induced *emf*, we calculate first the total magnetic flux crossing the area enclosed by the loop and then use Faraday's law to find the *emf*. The magnetic flux density crossing the area of the loop is

$$\mathbf{B} = 600\mu_o \frac{I}{2\pi\rho} \mathbf{a}_\phi = \frac{300}{\pi} \mu_o \frac{I_o}{\rho} \cos \omega t \, \mathbf{a}_\phi$$

The total magnetic flux crossing the area of the loop is hence,

$$\psi_m = \iint_s \mathbf{B} \cdot d\mathbf{s} = \int_{\rho=b}^c \int_{z=0}^\ell \mathbf{B} \cdot d\rho \, dz \, \mathbf{a}_\phi$$

$$= \int_{\rho=b}^c \int_{z=0}^\ell \frac{300\mu_o}{\pi} \frac{I_o}{\rho} \cos \omega t \, \mathbf{a}_\phi \cdot d\rho \, dz \, \mathbf{a}_\phi$$

$$= \frac{300\mu_o}{\pi} I_o \ell \cos \omega t \, \ell n \frac{c}{b}$$

According to Faraday's law, the *emf* is given by

$$emf = -\frac{d\psi_m}{dt} = \frac{300\omega}{\pi} \mu_o I_o \, \ell n \frac{c}{b} \sin \omega t$$

As indicated earlier, this *emf* is taken as a measure of the magnitude of the current I_o flowing in the long conductor.

———————◆◆———————

3.11 UNIFORM PLANE WAVE PROPAGATION IN CONDUCTIVE MEDIUM

In the introduction to this chapter, we indicated that after modifying Maxwell's equations to account for the properties of the medium, the developed new equations will then be used to describe the propagation characteristics of plane waves in conductive medium. Several assumptions were made when we first described in chapter 2 the properties of the uniform plane wave propagation in free space. This includes an infinite plane wavefront perpendicular to the positive z direction of propagation, constant (uniform) values of **E** and **H** in the plane of the wavefront, that is,

$$\frac{\partial(\mathbf{E} \text{ or } \mathbf{H})}{\partial x} = 0 = \frac{\partial(\mathbf{E} \text{ or } \mathbf{H})}{\partial y}$$

and the fact that the propagation medium is free from any charge or current distributions ($\mathbf{J} = \rho_v = 0$). In studying the properties of plane wave propagating in conductive medium, several other additional assumptions are needed to simplify the analysis. All these new assumptions are related to the properties of the medium and include the following:

1. Medium of propagation is homogeneous. This simply means that the properties of the medium do not vary from one location to another.

2. The medium is linear. This means that **E** and **D** in this medium are related by a constant number describing the polarization properties of the medium (i.e., $\mathbf{D} = \epsilon\mathbf{E}$). Also, **H** and **B** are related by a constant number describing the magnetic properties of the medium (i.e., $\mathbf{B} = \mu\mathbf{H}$). It should be noted that such an assumption of linear medium was inherently included in deriving the modifications in Maxwell's equations. Specifically, in obtaining equation 3.8, we used the fact that for linear dielectrics, $\mathbf{P} = \epsilon_o \chi_e \mathbf{E}$, which means the polarization is linearly proportional to the electric field. We also simplified the relation between **B** and **H** (equation 3.20) by assuming that for linear magnetic material, the magnetization **M** is linearly proportional to the magnetic field intensity **H**. In summary, the linearity property of the propagation medium is inherently used in arriving at Maxwell's equations for material media.

3. We will also assume that the propagation medium is isotropic, which means that the various properties of the medium μ, ϵ, and σ do not change selectively in one direction from that in another.

With these assumptions in mind, let us assume that the properties of the propagation medium are described by the following constants ϵ, μ, and σ.

To help us develop the various propagation properties of a plane wave in conductive medium, we shall compare Ampere's and Faraday's equations in free space and in our linear, homogeneous, and isotropic medium. This comparison is given in Table 3.5.

It may be seen that these two sets of equations are similar and may be interchanged by making the following substitutions:

TABLE 3.5 COMPARISON BETWEEN MAXWELL'S
CURL EQUATIONS IN FREE SPACE AND IN
CONDUCTIVE MEDIUM

Free space	Conductive medium
$\nabla \times \mathbf{E} = -j\omega\mu_o\,\mathbf{H}$ $\nabla \times \mathbf{H} = j\omega\epsilon_o\,\mathbf{E}$	$\nabla \times \mathbf{E} = -j\omega\mu\mathbf{H}$ $\nabla \times \mathbf{H} = \sigma\mathbf{E} + j\omega\epsilon\mathbf{E}$ $\qquad = j\omega\left(\epsilon - j\dfrac{\sigma}{\omega}\right)\mathbf{E}$

$$\mu_o \leftrightarrow \mu(\mu_o\,\mu_r) \tag{3.71}$$

$$\epsilon_o \leftrightarrow \left(\epsilon - j\frac{\sigma}{\omega}\right) \tag{3.72}$$

Therefore, the new propagation characteristics of a plane wave in a conductive medium may be obtained from those we previously developed for waves propagating in free space in chapter 2 simply by making the substitutions in equations 3.71 and 3.72.

It should be noted that replacing μ_o by $\mu_o\,\mu_r$ is rather simple because it just involves multiplying μ_o by a real number μ_r. The second substitution, conversely, is rather involved because ϵ_o is real while $\left(\epsilon - j\dfrac{\sigma}{\omega}\right)$ is a complex number. This latter substitution will result in new and interesting properties of the plane wave propagation in conductive medium.

From chapter 2, the complex form of the electric field associated with a plane wave in free space is given by

$$\hat{E}_x(z) = \hat{E}_m^+ e^{-j\beta_o z} + \hat{E}_m^- e^{j\beta_o z} \tag{3.73}$$

where the propagation constant β_o is a real number and is given by

$$j\beta_o = j\omega\sqrt{\mu_o\,\epsilon_o} \tag{3.74}$$

In conductive medium, a similar expression for the electric field is given by

$$\hat{E}_x(z) = \hat{E}_m^+ e^{-\hat{\gamma}z} + \hat{E}_m^- e^{\hat{\gamma}z} \tag{3.75}$$

where $\hat{\gamma}$ is now a complex number obtained by making the substitutions in equations 3.71 and 3.72. Hence,

$$\hat{\gamma} = j\omega\sqrt{\mu_o\,\mu_r\left(\epsilon - j\frac{\sigma}{\omega}\right)}$$

$$= \alpha + j\beta$$

where α and β are the real and imaginary parts, respectively, of the complex *propagation constant* $\hat{\gamma}$.

It is left for the student to prove that α and β are given by and

$$\alpha = \frac{\omega\sqrt{\mu\epsilon}}{\sqrt{2}}\left[\sqrt{1 + \left(\frac{\sigma}{\omega\epsilon}\right)^2} - 1\right]^{1/2} \quad \text{Np/m} \tag{3.76a}$$

$$\beta = \frac{\omega\sqrt{\mu\epsilon}}{\sqrt{2}}\left[\sqrt{1 + \left(\frac{\sigma}{\omega\epsilon}\right)^2} + 1\right]^{1/2} \quad \text{rad/m} \tag{3.76b}$$

The electric field in equation 3.75 may then be expressed in the form

$$\hat{E}_x(z) = \hat{E}_m^+ e^{-\alpha z} e^{-j\beta z} + \hat{E}_m^- e^{\alpha z} e^{j\beta z} \tag{3.77}$$

As we know by now, the propagation properties cannot be studied using the complex forms of the fields, but instead the real-time forms should be used. Hence,

$$E_x(z,t) = Re(\hat{E}_x(z)e^{j\omega t})$$
$$= E_m^+ e^{-\alpha z} \cos(\omega t - \beta z + \phi^+) + E_m^- e^{\alpha z} \cos(\omega t + \beta z + \phi^-) \tag{3.78}$$

where the complex amplitudes $\hat{E}_m^+$ and $\hat{E}_m^-$ were replaced in terms of their magnitudes and phases, respectively, that is, $\hat{E}_m^+ = E_m^+ e^{j\phi^+}$ and $\hat{E}_m^- = E_m^- e^{j\phi^-}$.

Plotting the first term in equation 3.78 as a function of z for various values of t as shown in Figure 3.38 simply shows that this term represents a cosine wave moving in the positive z direction with the increase in time.

This is clearly similar to the case described in chapter 2 of a plane wave propagating in free space. The only difference that we may notice from Figure 3.38, however, is that the amplitude of the wave decreases as it propagates along the positive z direction because of the attenuation factor $e^{-\alpha z}$. Such attenuation (decrease in magnitude) naturally resulted from replacing the propagation constant $j\beta_o$ in free space by the complex propagation constant $\hat{\gamma} = \alpha + j\beta$ in the conductive medium. Hence, the presence of amplitude attenuation for waves propagating in conductive medium repre-

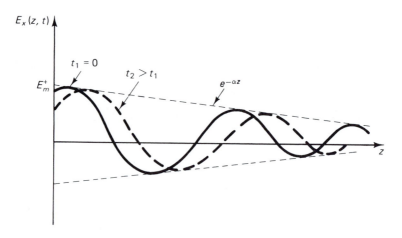

Figure 3.38   The electric field associated with a plane wave propagating along the positive z direction.

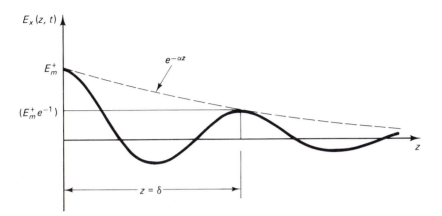

Figure 3.39 Depth of penetration of a plane wave in conductive medium.

sents one of the fundamental differences between wave propagation in free space and in conductive medium. We define the distance z at which the magnitude of the electric field decreases to e^{-1} of its value E_m^+ at $z = 0$ as the "depth of penetration δ." Hence, δ in Figure 3.39 is given in meters by

$$\delta = \frac{1}{\alpha} \text{ m}$$

Let us examine the following two special values of δ of interest:

1. Medium of propagation is a perfect conductor.
For this case, $\sigma \to \infty$ and α from equation 3.76a becomes an infinitely large number. δ is hence zero, and this indicates that plane waves cannot penetrate a perfectly conducting medium.

2. Medium of propagation is nonconductive, that is, $\sigma = 0$.
From equation 3.76a, it is clear that α is also zero when $\sigma = 0$. The depth of penetration δ is therefore infinitely large, which simply indicates that the wave does not attenuate while propagating in nonconductive medium even if the medium of propagation is not free space, that is, $\mu \neq \mu_o$ and $\epsilon \neq \epsilon_o$.

The wave impedance defined in chapter 2 as the ratio between the electric and magnetic fields is obtained by making the substitutions in equations 3.71 and 3.72 in the wave impedance expression given in chapter 2.

$$\frac{\hat{E}_x}{\hat{H}_y} = \hat{\eta} = \sqrt{\frac{\mu}{\left(\epsilon - j\frac{\sigma}{\omega}\right)}} = \frac{\sqrt{\frac{\mu}{\epsilon}}}{\left[1 + \left(\frac{\sigma}{\omega\epsilon}\right)^2\right]^{1/4}} e^{j\frac{1}{2}\tan^{-1}\left(\frac{\sigma}{\omega\epsilon}\right)} \tag{3.79}$$

Unlike the real number ratio of $\hat{E}_x/\hat{H}_y$ in free space, the wave impedance $\hat{\eta}$ in conductive medium is a complex number given by equation 3.79. This simply means that the electric and magnetic fields are not in phase and instead the phase angle between them is given by

$$\theta = \frac{1}{2}\tan^{-1}\left(\frac{\sigma}{\omega\epsilon}\right)$$

The phase velocity defined in chapter 2 as the velocity of propagation of a specific point in the wave (i.e., constant phase) is given by

$$v_p = \frac{\omega}{\beta}$$

Because at the same angular frequency ω, β in conductive medium given by Eq. 3.76b is larger than $\beta_o = \omega\sqrt{\mu_o\epsilon_o}$ in free space, hence,

$$v_p < \frac{\omega}{\beta_o},$$

which is equal to the velocity of light c in free space. In other words, the wave travels in conductive medium at a speed less than the velocity of light c, which is the propagation velocity in free space. Also, because $\beta > \beta_o$, the wavelength λ in the conductive medium $\lambda = 2\pi/\beta$ is shorter than the wavelength λ_o in free space at the same frequency. Table 3.6 summarizes the similarities and differences between plane waves propagating in free space and in conductive medium.

TABLE 3.6 SIMILARITIES AND DIFFERENCES BETWEEN THE PROPAGATION OF UNIFORM PLANE WAVES IN FREE SPACE AND CONDUCTIVE MEDIUM

Similarities
1. In both cases, the electric and magnetic fields are uniform in the plane perpendicular to the direction of propagation.
2. The electric and magnetic fields are perpendicular to each other, and to the direction of propagation. In other words, no component of either the electric or the magnetic field is in the direction of propagation.

Differences	
Free Space	Conductive Medium
1. **E** and **H** vectors are in phase, that is, the intrinsic wave impedance η_o is a real number (see Figure 3.40a).	**E** and **H** vectors are not in phase, that is, the intrinsic wave impedance $\hat{\eta}$ is a complex number (see Figure 3.40b).
2. The phase velocity is equal to the velocity of light, $v_p = c$.	The phase velocity is less than the velocity of light $v_p < c$.
3. For a plane wave of a given frequency, the wavelength λ_o is longer than the wavelength in the material medium, $\lambda_o = c/f$.	The wavelength in a conductive medium is shorter than the corresponding wavelength of a plane wave of the same frequency propagating in free space, $\lambda = 2\pi/\beta < \lambda_o$.
4. Does not attentuate in magnitude as it propagates.	It exponentially attenuates, with the skin depth given by $\delta = 1/\alpha$.

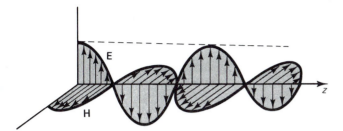

(a) **E** and **H** in phase

(b) **E** and **H** out of phase

Figure 3.40   The electric and magnetic fields associated with plane wave propagation (a) in free space and (b) in conductive medium.

EXAMPLE 3.13

A 5-GHz uniform plane wave is propagating in a dielectric material that is characterized by $\epsilon_r = 2.53$, $\mu_r = 1$. If the electric field is given by

$$\mathbf{E}(z,t) = 10 \cos(\omega t - \beta z)\, \mathbf{a}_x$$

determine

1. The phase constant β, the wavelength λ, and the phase velocity v_p.
2. The amplitude of the magnetic field intensity.
3. Write a time-domain expression for the magnetic field intensity.

Solution

1. In the given expression of the electric field, the absence of the exponential decay term $e^{-\alpha z}$ simply indicates that $\alpha = 0$. From equation 3.76a, it is clear that α would be zero when $\sigma = 0$, regardless of the values of ϵ and μ. In this example, the conductivity σ is hence zero, and β is given by

$$\beta = \omega\sqrt{\mu\epsilon} = 2\pi f \sqrt{\mu_o \epsilon_o \epsilon_r} = 166.57 \text{ rad/m}$$

$$\lambda = \frac{2\pi}{\beta} = 3.77 \text{ cm}$$

It should be noted that because the medium is not free space with ϵ_o and μ_o, λ would be different from $\lambda_o = c/f$. The velocity of propagation v_p is given by

$$v_p = \frac{\omega}{\beta} = \frac{1}{\sqrt{\mu_o \, \epsilon_o \, \epsilon_r}} = 1.89 \times 10^8 \text{ m/s}$$

2. When $\sigma = 0$, the wave impedance $\hat{\eta}$ will be real and is given by

$$\eta = \sqrt{\frac{\mu_o}{\epsilon_o \, \epsilon_r}} = 237 \; \Omega$$

The magnitude of the magnetic field is hence

$$|H_y| = \frac{|E_x|}{\eta} = 0.042 \text{ A/m}$$

3. The time-domain expression of the magnetic field is

$$\mathbf{H} = 0.042 \cos(10\pi \times 10^9 t - 166.57z) \, \mathbf{a}_y \text{ A/m}$$

EXAMPLE 3.14

The electric field of a uniform plane wave propagating in a dielectric nonconducting medium having $\mu = \mu_o$ is given by

$$\mathbf{E} = 10 \cos(6\pi \times 10^7 t - 0.4\pi z) \, \mathbf{a}_x \text{ V/m}$$

Find the following:

1. Frequency.
2. Wavelength.
3. Phase velocity.
4. Permittivity of medium.
5. Associated magnetic field vector **H**.

Solution

1. Comparing the given electric field expression with the general form given in equation 3.78, we find that

$$\alpha = 0 \text{ (for a nonconductive medium } \sigma = 0)$$

$$\omega = 6\pi \times 10^7$$

$$\therefore f = 30 \text{ MHz}$$

2. Similar comparison between the given expression and equation 3.78, we obtain

$$\beta = 0.4\pi$$

$$\therefore \lambda = \frac{2\pi}{\beta} = 5 \text{ m}$$

3. $v_p = \dfrac{\omega}{\beta} = \dfrac{6\pi \times 10^7}{0.4\pi} = 1.5 \times 10^8$ m/s

4. Having an attenuation constant α equal to zero in the electric field expression does not mean that the wave is propagating in free space. It only means that the medium is lossless and has a zero conductivity σ. With $\sigma = 0$, the expression for the phase constant β reduces to

$$\beta = \omega\sqrt{\mu_o \epsilon_o \epsilon_r} = 0.4\pi$$

Substituting $\beta = 0.4\pi$ and $\omega = 6\pi \times 10^7$, we obtain

$$\epsilon_r = 4$$

5. The wave impedance for this medium is given by

$$\eta = \sqrt{\frac{\mu_o}{\epsilon_o \epsilon_r}} = \frac{120\pi}{\sqrt{4}} = 60\pi$$

The magnitude of the magnetic field $|\hat{H}_y|$ is hence,

$$|\hat{H}_y| = \frac{|\hat{E}_x|}{\eta} = \frac{10}{60\pi} = 0.053 \text{ A/m}$$

The time-domain expression of the vector magnetic field is

$$\mathbf{H} = 0.053 \cos(6\pi \times 10^7 t - 0.4\pi z)\, \mathbf{a}_y \text{ A/m}$$

EXAMPLE 3.15

1. Define the skin depth (δ) of a transverse electromagnetic (TEM) plane wave propagating in a conductive medium.
2. A 1-MHz plane wave is propagating in a conductive medium of $\epsilon_r = 8$, $\sigma = 4.8 \times 10^{-2}$ S/m, and $\mu = \mu_o$.
   (a) Find the ratio between the magnitudes of the conduction to the displacement currents.
   (b) With the aid of the result obtained in part a, calculate the skin depth (δ).
   (c) If the maximum magnitude of the sinusoidal variation of the x-directed electric field is 100 V/m at $t = 0$ and $z = 0.3\pi$, give an expression for E_x in the *real-time* form. Substitute numerical values for the propagation coefficients.

Solution

1. The skin depth δ of a plane wave propagating in a conductive medium is defined as the distance z the wave travels before its amplitude decays to e^{-1} of its value at a reference surface at $z = 0$.
2. (a) From example 3.2, the magnitude of the ratio of the conduction to displacement currents is given by

$$\left|\frac{\text{conduction current}}{\text{displacement current}}\right| = \frac{\sigma}{\omega\epsilon} = 108$$

This ratio is clearly much greater than 1, and the propagation phenomenon in this conductive medium is dominated by the conduction current.

(b) Since $\dfrac{\sigma}{\omega\epsilon} \gg 1$, the expressions for the attenuation α and phase constant β may be simplified from equations 3.76a and b to

$$\alpha = \frac{\omega\sqrt{\mu\epsilon}}{\sqrt{2}}\left[\sqrt{1 + \left(\frac{\sigma}{\omega\epsilon}\right)^2} - 1\right]^{1/2} = \frac{\omega\sqrt{\mu\epsilon}}{\sqrt{2}}\left(\frac{\sigma}{\omega\epsilon}\right)^{1/2} = \sqrt{\frac{\omega\sigma\mu}{2}} = \sqrt{\pi f \sigma\mu}$$

Following a similar procedure to calculate an approximate value for β, we obtain

$$\alpha = \beta = \sqrt{\pi f \sigma\mu} = 0.435$$

The skin depth is then

$$\delta = \frac{1}{\alpha} = 2.3 \text{ m}$$

(c) The general time-domain expression of the positive z propagating electric field is given by

$$\mathbf{E}(z,t) = E_m^+ e^{-\alpha z} \cos(\omega t - \beta z + \phi^+)\mathbf{a}_x \text{ V/m}$$

To determine the phase constant ϕ^+, we use the given initial conditions. At $t = 0$ and $z = 0.3\pi$, $\mathbf{E}$ will have its maximum value when the angle of the cosine function is zero. Hence,

$$-\beta(0.3\pi) + \phi^+ = 0$$

or

$$\phi^+ = 0.3\pi\beta = 0.41 \text{ rad}$$

The maximum amplitude of 100 V/m occurs at $z = 0.3\pi$, hence

$$100 = E_m^+ e^{-0.41}$$

$$\therefore E_m^+ = 151 \text{ V/m}$$

The final expression for the electric field is, therefore,

$$\mathbf{E}(z,t) = 151 e^{-0.435z} \cos(2\pi \times 10^6 t - 0.435z + 0.41)\mathbf{a}_x$$

––––––◆◆◆––––––

EXAMPLE 3.16

A microwave engineer designed a $\lambda/2$ dipole antenna to operate in free space at 600 MHz.

1. Calculate the length of this antenna.
2. If he still wants to use a $\lambda/2$ dipole antenna at the same frequency for underwater communication ($\sigma = 4$ S/m, $\epsilon = 81\epsilon_o$), calculate the necessary change in the dimensions of the antenna.

Solution

At 600 MHz, the wavelength in free space is

$$\lambda = \frac{c}{f} = \frac{3 \times 10^8}{600 \times 10^6} = 0.5 \text{ m}$$

1. The length of the antenna in free space ℓ_{air} is

$$\ell_{air} = \frac{\lambda}{2} = 0.25 \text{ m}$$

2. For underwater communication, the ratio of conduction to displacement current at 600 MHz is

$$\frac{\sigma}{\omega\epsilon} = 1.48$$

The sea water in this case is a moderate conductor and no approximation should be used in calculating β.

$$\beta = \frac{\omega\sqrt{\mu\epsilon}}{\sqrt{2}}\left[\sqrt{1 + \left(\frac{\sigma}{\omega\epsilon}\right)^2} + 1\right]^{1/2} = 133.02 \text{ rad/m}$$

$$\lambda_{seawater} = \frac{2\pi}{\beta} = 0.0472 \text{ m}$$

The required antenna length is, hence,

$$\ell_{water} = \frac{\lambda_{seawater}}{2} = 0.0236 \text{ m}$$

The ratio of the two lengths is

$$\frac{\ell_{water}}{\ell_{air}} = \frac{0.0236}{0.25} = 0.094$$

which means that ℓ_{water} is less than 10 percent of ℓ_{air}.

EXAMPLE 3.17

An antenna engineer is designing a communication system for the navy. The system consists of a transmitter located far above the surface of the sea and a receiver located deep in the sea water. He also chose the operating frequency to be 10 kHz, (see Figure 3.41).

1. If he designed both the transmitting and receiving antennas to be 0.05 λ, where λ is the wavelength in the respective media, determine the lengths in meters of both antennas.

2. Assuming a plane wave propagation inside the sea water, determine the skin depth.

3. If the amplitude of the electric field transmitted in the sea water is E_o (just beneath the water surface), how far will the wave penetrate before reaching 5 percent of its surface value?

Solution

1. Following a procedure similar to that in example 3.16, it is rather straightforward to show that

$$\ell_1 \text{ (in air)} = 1.5 \text{ km}$$

$$\ell_2 \text{ (in sea water)} = 0.785 \text{ m}$$

Transmitter

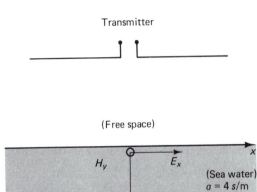

(Free space)

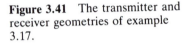

Figure 3.41 The transmitter and receiver geometries of example 3.17.

Receiver

2. The ratio of the conduction to displacement current in sea water at 10 kHz is

$$\frac{\sigma}{\omega\epsilon} = 0.89 \times 10^5 \gg 1$$

Using $\sigma/\omega\epsilon \gg 1$ approximation in the expressions for α and β, we obtain

$$\alpha = \beta = \sqrt{\pi f \mu \sigma} = 0.4$$

The skin depth $\delta = 1/\alpha = 2.5$ m.

3. The magnitude of the electric field in the sea water decreases as the wave penetrates the conductive medium

$$|\mathbf{E}(z)| = E_o e^{-\alpha z}$$

where E_o is the reference magnitude at $z = 0$.

$$\frac{|\mathbf{E}(z)|}{E_o} = e^{-\alpha z} = 0.05$$

$\therefore$ the required distance $z = \dfrac{\ln 0.05}{-\alpha} = 7.49$ m $\approx 3\delta$

In other words, the depth required for the magnitude of the wave to decrease to 5 percent of its reference value is approximately equal to three skin depths.

3.12 ELECTROMAGNETIC POWER AND POYNTING THEOREM

In examining the propagation characteristics of plane waves thus far, we discussed the electric and magnetic fields associated with these waves and other characteristics such as the wave impedance and the propagation phase and attenuation constants. It is known that electromagnetic energy is also associated with the propagation of these waves, and quantifying this energy in terms of the electric and magnetic fields is the objective of this section. In 1884, an English physicist, John H. Poynting, was the first to realize that the vector $\mathbf{E} \times \mathbf{H}$ would play an important role in quantifying the electromagnetic energy. This is because such a vector has the dimension of $\mathbf{E}$ multiplied by $\mathbf{H}$ (V/m · A/m) which is power density W/m². Also, based on our experience with the plane wave propagation, the vector $\mathbf{E} \times \mathbf{H}$ results in a vector along the direction of propagation that is the direction of the energy flow. This is why the power density vector, $\mathbf{E} \times \mathbf{H}$, which is also along the direction of the electromagnetic energy flow, is known as the Poynting vector. In the next section, we will use the Poynting vector to obtain an electromagnetic power balance equation known as the *Poynting theorem*.

3.12.1 Poynting Theorem

Poynting theorem is just an expression of the electromagnetic power balance that includes the relationship between the generated, transmitted, stored, and dissipated electromagnetic powers. To obtain a mathematical expression of this theorem, let us start with a vector identity that involves the divergence of the Poynting vector

$$\nabla \cdot (\mathbf{E} \times \mathbf{H}) = \mathbf{H} \cdot \nabla \times \mathbf{E} - \mathbf{E} \cdot \nabla \times \mathbf{H}$$

Substituting $\nabla \times \mathbf{E}$ and $\nabla \times \mathbf{H}$ from Maxwell's equations and if we assume that the medium of propagation does not contain any external charge and current distributions, we obtain

$$\nabla \cdot (\mathbf{E} \times \mathbf{H}) = \mathbf{H} \cdot \left(-\frac{\partial \mathbf{B}}{\partial t} \right) - \mathbf{E} \cdot \left(\mathbf{J}_c + \frac{\partial \mathbf{D}}{\partial t} \right) \tag{3.80}$$

where $\mathbf{J}_c$ is the conduction current. To explain the physical meaning of each of the terms in equation 3.80, we need to rearrange them in more familiar forms. For example, from the vector calculus briefly reviewed in chapter 2, we know that

$$\frac{\partial}{\partial t} \left(\frac{\mathbf{H} \cdot \mathbf{B}}{2} \right) = \frac{1}{2} \mathbf{H} \cdot \frac{\partial \mathbf{B}}{\partial t} + \frac{1}{2} \frac{\partial \mathbf{H}}{\partial t} \cdot \mathbf{B}$$

Substituting $\mathbf{B} = \mu\mathbf{H}$ and assuming the magnetic property of the medium to be independent of time, we obtain

$$\frac{\partial}{\partial t} \left(\frac{\mathbf{H} \cdot \mathbf{B}}{2} \right) = \frac{1}{2} \mathbf{H} \cdot \mu \frac{\partial \mathbf{H}}{\partial t} + \frac{1}{2} \frac{\partial \mathbf{H}}{\partial t} \cdot \mu \mathbf{H}$$

$$= \mathbf{H} \cdot \frac{\partial \mu \mathbf{H}}{\partial t} = \mathbf{H} \cdot \frac{\partial \mathbf{B}}{\partial t} \tag{3.81}$$

Similarly, if ϵ is assumed to be independent of time, we obtain

$$\frac{\partial}{\partial t} \left(\frac{\mathbf{E} \cdot \mathbf{D}}{2} \right) = \mathbf{E} \cdot \frac{\partial \mathbf{D}}{\partial t} \tag{3.82}$$

Substituting equations 3.81 and 3.82 in equation 3.80,

$$\nabla \cdot (\mathbf{E} \times \mathbf{H}) = -\frac{\partial}{\partial t} \left(\frac{\mathbf{H} \cdot \mathbf{B}}{2} \right) - \frac{\partial}{\partial t} \left(\frac{\mathbf{E} \cdot \mathbf{D}}{2} \right) - \mathbf{E} \cdot \mathbf{J}_c \tag{3.83}$$

Integrating equation 3.83 over a closed volume in which we would like to examine the power balance equation, that is, the relation between the power input to and output from this volume, we obtain

$$-\int_v \mathbf{\nabla} \cdot (\mathbf{E} \times \mathbf{H}) dv = \int_v \frac{\partial}{\partial t}\left(\frac{\mathbf{H} \cdot \mathbf{B}}{2}\right) dv + \int_v \frac{\partial}{\partial t}\left(\frac{\mathbf{E} \cdot \mathbf{D}}{2}\right) dv + \int_v \mathbf{E} \cdot \mathbf{J}_c \, dv$$

Using the divergence theorem,

$$-\oint_s (\mathbf{E} \times \mathbf{H}) \cdot d\mathbf{s} = \frac{\partial}{\partial t}\left[\int_v \left(\frac{\mathbf{H} \cdot \mathbf{B}}{2}\right) dv + \int_v \left(\frac{\mathbf{E} \cdot \mathbf{D}}{2}\right) dv\right] + \int_v \mathbf{E} \cdot \mathbf{J}_c \, dv \qquad (3.84)$$

If the element of area $d\mathbf{s}$ in the left-hand-side term of equation 3.84 is pointing in the outward direction, this term may be physically interpreted as the power density $(\mathbf{E} \times \mathbf{H})$ entering the surface s enclosing the volume of interest v. We recall that taking the dot product of the $(\mathbf{E} \times \mathbf{H})$ and $d\mathbf{s}$ vectors basically emphasizes that we are considering the component of the power density vector $(\mathbf{E} \times \mathbf{H})$ crossing (i.e., entering) the surface s. Also, the two terms

$$\int_v \left(\frac{\mathbf{E} \cdot \mathbf{D}}{2}\right) dv \qquad \text{and} \qquad \int_v \left(\frac{\mathbf{H} \cdot \mathbf{B}}{2}\right) dv$$

may be interpreted in terms of the electric and magnetic energies, respectively, in the volume. For example, based on electrostatic considerations, it may be shown that

$$\int_v \left(\frac{\mathbf{E} \cdot \mathbf{D}}{2}\right) dv$$

is indeed related to the amount of work needed to be done to elevate the electric energy in a region of space, for example, by assembling a system of static charges. Similarly, the presence or the increase of the magnetic field in a region of space results in an increase in the magnetic energy in the space. The quantity

$$\int_v \left(\frac{\mathbf{H} \cdot \mathbf{B}}{2}\right) dv$$

is equal to the energy required to establish a magnetic field intensity $\mathbf{H}$ and to sustain it against the conductor losses as well as to other forms of decay. Therefore, the two terms on the right-hand side of equation 3.84

$$\frac{\partial}{\partial t}\left[\int_v \left(\frac{\mathbf{E} \cdot \mathbf{D}}{2}\right) dv + \int_v \left(\frac{\mathbf{H} \cdot \mathbf{B}}{2}\right) dv\right]$$

describe the rate of increase of the electric and magnetic energies stored within the volume v.

Finally, we are going to show that

$$\int_v \mathbf{E} \cdot \mathbf{J}_c \, dv$$

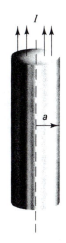

Figure 3.42   The power dissipated in a cylindrical conductor equal to the ohmic losses $= I^2 R$.

describes the power dissipated (ohmic losses) within the volume v. Consider a cylindrical conductor of conductivity σ, cross sectional area A, and carrying a current I as shown in Figure 3.42. The current density in the conductor is

$$\mathbf{J} = \frac{I}{A}\,\mathbf{a}_z$$

The electric field in the conductor is hence

$$\mathbf{E} = \frac{\mathbf{J}}{\sigma} = \frac{I}{A\sigma}\,\mathbf{a}_z$$

With this information, let us evaluate

$$\int_v \mathbf{E}\cdot\mathbf{J}\,dv$$

in a cylindrical volume of length ℓ of the conductor

$$\int_{z=0}^{\ell}\int_{\phi=0}^{2\pi}\int_{\rho=0}^{a}\mathbf{E}\cdot\mathbf{J}\,dv = \int_0^{\ell}\int_0^{2\pi}\int_0^{a}\frac{I}{\sigma A}\left(\mathbf{a}_z\cdot\frac{I}{A}\,\mathbf{a}_z\right)\rho\,d\phi\,d\rho\,dz$$

$$= \frac{I^2}{\sigma A^2}\,\pi a^2\ell = \frac{I^2\ell}{\sigma A} = I^2 R$$

where $R = \ell/\sigma A$ is the resistance of a cylindrical conductor of conductivity σ, area A, and length ℓ. Hence,

$$\int_v \mathbf{E}\cdot\mathbf{J}\,dv = I^2 R$$

and the term $\int_v \mathbf{E}\cdot\mathbf{J}\,dv$, therefore, represents the total Ohmic losses in the volume v. The following is a summary of the physical interpretation of the various terms in the power balance equation 3.84, which is known as the Poynting theorem:

$$-\oint_s \mathbf{E} \times \mathbf{H} \cdot ds = \frac{\partial}{\partial t}\int_v \left[\int \frac{\mathbf{E} \cdot \mathbf{D}}{2} dv + \int_v \frac{\mathbf{H} \cdot \mathbf{B}}{2} dv\right] + \int_v \mathbf{E} \cdot \mathbf{J}\, dv$$

The power density entering (because of negative sign) the closed surface s; ds is an element of area pointing to the outward of the surface	The rate of increase of the electric and magnetic energies stored in the volume v enclosed by the surface s	Power dissipated "ohmic losses" in the volume v	(3.85)

It should be emphasized that equation 3.85 is the Poynting theorem that describes the power balance equation for the receiver case in which the electromagnetic fields are generated outside the closed surface s. In the transmitter case, the generator of the electromagnetic energy will be inside the volume, and the generator current should be included in Ampere's law in the form

$$\nabla \times \mathbf{H} = \mathbf{J}_g + \mathbf{J}_c + \frac{\partial \mathbf{D}}{\partial t}$$

where $\mathbf{J}_c$ is the familiar conduction current term and $\mathbf{J}_g$ is the generator current density. Introducing the source term in the Poynting theorem, we obtain

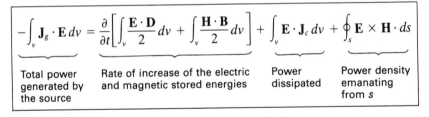

$$-\int_v \mathbf{J}_g \cdot \mathbf{E}\, dv = \frac{\partial}{\partial t}\left[\int_v \frac{\mathbf{E} \cdot \mathbf{D}}{2} dv + \int_v \frac{\mathbf{H} \cdot \mathbf{B}}{2} dv\right] + \int_v \mathbf{E} \cdot \mathbf{J}_c\, dv + \oint_s \mathbf{E} \times \mathbf{H} \cdot ds$$

Total power generated by the source	Rate of increase of the electric and magnetic stored energies	Power dissipated	Power density emanating from s

It should be noted that the term

$$\int_v \mathbf{E} \cdot \mathbf{J}_c\, dv$$

is related to power dissipated in a conductive medium, whereas

$$-\int_v \mathbf{E} \cdot \mathbf{J}_g\, dv$$

is characterized as the power generated by the source. The negative sign in the latter term is due to the fact that $\mathbf{E}$ and $\mathbf{J}_g$ in the generator are in the opposite direction. This may be illustrated by examining the current flow resulting from the chemical reactions within an electric battery and the direction of the electric field that results from the polarization of the positive and negative charges on the battery's terminals.

EXAMPLE 3.18

If a direct current is flowing in the cylindrical conductor of Figure 3.42, show that both sides of the Poynting theorem can be applied to calculate the ohmic losses in the conductor.

Solution

At steady state, because we are considering static fields, the rate of increase ($\partial/\partial t$) of the electric and magnetic energies within the conductor is zero. Hence, the statement of Poynting theorem in equation 3.84 reduces to

$$-\oint_s (\mathbf{E} \times \mathbf{H}) \cdot d\mathbf{s} = \int_v \mathbf{E} \cdot \mathbf{J}_c \, dv$$

We already showed that

$$\int_v \mathbf{E} \cdot \mathbf{J}_c \, dv = I^2 R,$$

the ohmic losses in the conductor. It remains to show that

$$-\oint_s (\mathbf{E} \times \mathbf{H}) \cdot d\mathbf{s}$$

is also equal to the ohmic losses. From Ampere's law, the magnetic field surrounding the current-carrying conductor is

$$\mathbf{H} = \frac{I}{2\pi\rho} \mathbf{a}_\phi$$

At the surface of the conductor $\rho = a$, $\mathbf{H} = (I/2\pi a)\, \mathbf{a}_\phi$. The Poynting vector

$$\mathbf{E} \times \mathbf{H} = \frac{I}{\sigma A} \mathbf{a}_z \times \frac{I}{2\pi a} \mathbf{a}_\phi = \frac{I^2}{2\pi a\sigma A}(-\mathbf{a}_\rho)$$

The surface integral,

$$-\oint_s (\mathbf{E} \times \mathbf{H}) \cdot d\mathbf{s} = \underset{\uparrow}{\int (\mathbf{E} \times \mathbf{H}) \cdot d\mathbf{s}} + \underset{\uparrow}{\int (\mathbf{E} \times \mathbf{H}) \cdot d\mathbf{s}}$$
$$\text{Cylindrical surface} \qquad \text{End caps}$$

Because the Poynting vector is in the $\mathbf{a}_\rho$ direction, the contribution from the end caps will be zero. Hence,

$$-\oint_s \mathbf{E} \times \mathbf{H} \cdot d\mathbf{s} = \int_0^\ell \int_0^{2\pi} \frac{I^2}{2\pi a\sigma A} \mathbf{a}_\rho \cdot a \, d\phi \, dz \, \mathbf{a}_\rho = \frac{I^2}{\sigma A} \ell = I^2 R$$

which is the same result obtained by integrating the ohmic losses term of the Poynting theorem.

———————◆◆◆———————

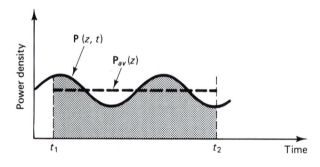

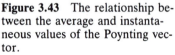

Figure 3.43 The relationship between the average and instantaneous values of the Poynting vector.

3.12.2 Time-average Poynting Vector

The Poynting vector $\mathbf{E} \times \mathbf{H}$ describes the instantaneous power density associated with the propagation of electromagnetic waves. This quantity, although valuable in discussing the power balance equation "Poynting theorem" at any instant of time, is of limited value from the practical measurement viewpoint. If one tries to measure the power density associated with an electromagnetic wave, he or she will be certainly limited to an averaged value over a reasonable period that depends on the instrumentation response time. In other words, from the practical viewpoint, it is more appropriate to discuss time average rather than the instantaneous Poynting vector.

To obtain the time-average Poynting vector, it is necessary to integrate the instantaneous values $\mathbf{P}(z,t)$ over a period, for example, from t_1 to t_2 and equate the result to an average value $\mathbf{P}_{av}(z)$ multiplied by the period as shown in Figure 3.43.

$$\mathbf{P}_{av}(z) = \frac{1}{t_2 - t_1} \int_{t_1}^{t_2} \mathbf{P}(z,t) \, dt \qquad (3.86)$$

For periodic functions, such as a sinusoidal excitation, it is necessary only to integrate over a time interval equal to one period, hence,

$$\mathbf{P}_{av}(z) = \frac{1}{T} \int_{0}^{T} \mathbf{P}(z,t) dt$$

where T is the period of the periodic source.

EXAMPLE 3.19

Obtain an expression for the time-average power density associated with plane waves propagating along the positive z direction in the following:

1. Free space.
2. Conductive medium.

Solution

1. For a plane wave propagating in free space, the electric and magnetic fields are given by

$$\mathbf{E} = E_m^+ \cos(\omega t - \beta_o z)\, \mathbf{a}_x$$

$$\mathbf{H} = \frac{E_m^+}{\eta_o} \cos(\omega t - \beta_o z)\, \mathbf{a}_y$$

The instantaneous Poynting vector $\mathbf{P}(z,t)$ is given by

$$\mathbf{P}(z,t) = \mathbf{E} \times \mathbf{H} = \frac{E_m^{+2}}{\eta_o} \cos^2(\omega t - \beta_o z)\, \mathbf{a}_z$$

and the time average Poynting vector is, hence,

$$\mathbf{P}_{av}(z) = \frac{1}{T} \int_0^T \mathbf{P}(z,t)\,dt$$

$$= \frac{E_m^{+2}}{\eta_o} \left[\frac{1}{T} \int_0^T \cos^2(\omega t - \beta_o z)\,dt \right] \mathbf{a}_z$$

$$= \frac{E_m^{+2}}{\eta_o} \left[\frac{1}{T} \int_0^T \frac{1}{2}\{1 + \cos 2(\omega t - \beta_o z)\}\,dt \right] \mathbf{a}_z$$

$$= \frac{E_m^{+2}}{2\eta_o}\, \mathbf{a}_z \tag{3.87}$$

It should be noted that

$$\int_0^T \cos 2(\omega t - \beta_o z)\,dt = 0$$

because it involves integrating a sinusoidal function of double the original frequency over a complete period. The relation between the instantaneous Poynting vector and the time-average value for a wave propagating in free space is shown in Figure 3.44.

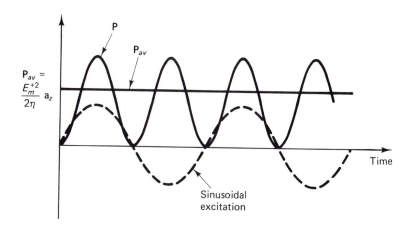

Figure 3.44 The relationship between the sinusoidal excitation, the instantaneous Poynting vector $\mathbf{P}(z,t)$ at double the frequency and the time-average value $\mathbf{P}_{av}(z)$.

2. The electric and magnetic fields of a plane wave propagating in conductive medium are

$$\mathbf{E} = E_m^+ e^{-\alpha z} \cos(\omega t - \beta z)\, \mathbf{a}_x$$

$$\mathbf{H} = \frac{E_m^+}{|\hat{\eta}|} e^{-\alpha z} \cos(\omega t - \beta z - \theta)\, \mathbf{a}_y$$

$$\mathbf{P}(z,t) = \frac{E_m^{+2}}{|\hat{\eta}|} e^{-2\alpha z} \cos(\omega t - \beta z) \cos(\omega t - \beta z - \theta)\, \mathbf{a}_z$$

$$= \frac{1}{2} \frac{E_m^{+2}}{|\hat{\eta}|} e^{-2\alpha z}[\cos\theta + \cos(2\omega t - 2\beta z - \theta)]\, \mathbf{a}_z$$

The time-average Poynting vector is then

$$\mathbf{P}_{av}(z) = \frac{1}{T}\int_0^T \mathbf{P}(z,t)dt = \frac{1}{2}\frac{E_m^{+2}}{|\hat{\eta}|} e^{-2\alpha z} \cos\theta\, \mathbf{a}_z \qquad (3.88)$$

The relation between the instantaneous Poynting vector $\mathbf{P}(z,t)$, and the electric and magnetic fields in this case is shown in Figure 3.45.

The main differences between expressions in equations 3.87 and 3.88 for the average Poynting vectors are the following:

1. The time-average power density associated with the free-space propagation is a constant value simply because of the absence of the attenuation in the medium of propagation. In a conductive medium, the power density attenuates with the factor $e^{-2\alpha z}$ as the wave propagates along the positive z direction.

2. The attenuation factor $e^{-2\alpha z}$ is not the only source of power loss in the conductive medium. The $\cos\theta$ term that resulted from having an angle θ between the electric

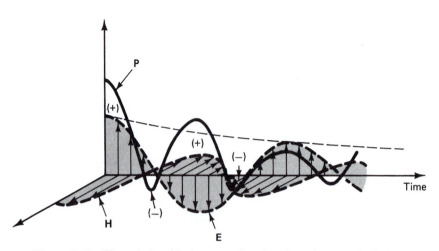

Figure 3.45 The relationship between the electric and magnetic fields, and the instantaneous Poynting vector for a plane wave propagating in a conductive medium.

and magnetic fields in conductive medium (i.e., complex value of the wave impedance $\hat{\eta}$) contributes further reduction in the time-average value of the power density in this case. Figure 3.45 shows that such a reduction in the time-average value results from the oscillation of the instantaneous Poynting vector between positive and negative values. These positive and negative values of $\mathbf{P}(z,t)$ are obtained because the electric and magnetic fields are not in phase at all times.

3.12.3 Complex Poynting Vector for Time Harmonic Fields

In the previous section, we obtained the time-average Poynting vector for periodic excitation by integrating the instantaneous expression over a complete period. In this case, time-domain or instantaneous expressions of the electric and magnetic fields were used to obtain the expression for the instantaneous Poynting vector. In this section, we will derive a simple procedure for directly calculating the time-average Poynting vector. The advantages of the present procedure include eliminating the need for performing integrations and the direct use of the complex expressions for the electric and magnetic fields that are readily available from the solution of time harmonic Maxwell's equations.

For time harmonic fields, the time-domain forms of the electric and magnetic fields, $\mathbf{E}(z,t)$, $\mathbf{H}(z,t)$, are given in terms of their complex (phasor) values, $\hat{E}(z)$, $\hat{H}(z)$, by

$$\mathbf{E}(z,t) = Re[\hat{E}(z)e^{j\omega t}]\,\mathbf{a}_x$$

$$\mathbf{H}(z,t) = Re[\hat{H}(z)e^{j\omega t}]\,\mathbf{a}_y$$

Expressing the complex fields by their real and imaginary parts, that is,

$$\hat{E}(z) = E_r + jE_i$$

$$\hat{H}(z) = H_r + jH_i$$

we obtain

$$\mathbf{E}(z,t) = Re[(E_r + jE_i)e^{j\omega t}]\,\mathbf{a}_x = (E_r \cos \omega t - E_i \sin \omega t)\,\mathbf{a}_x$$

$$\mathbf{H}(z,t) = Re[(H_r + jH_i)e^{j\omega t}]\,\mathbf{a}_y = (H_r \cos \omega t - H_i \sin \omega t)\,\mathbf{a}_y$$

The instantaneous Poynting vector is, hence,

$$\mathbf{P}(z,t) = \mathbf{E} \times \mathbf{H} = [E_r H_r \cos^2 \omega t + E_i H_i \sin^2 \omega t$$
$$- E_r H_i \sin \omega t \cos \omega t - E_i H_r \sin \omega t \cos \omega t]\,\mathbf{a}_z$$

The time-average Poynting vector is

$$\mathbf{P}_{av}(z) = \frac{1}{T}\int_0^T \mathbf{P}(z,t)dt = \left[\frac{1}{2}E_r H_r + \frac{1}{2}E_i H_i\right]\mathbf{a}_z \qquad (3.89)$$

An alternative procedure for obtaining equation 3.89 is to use directly the complex expressions of the electric and magnetic fields in the form

$$\mathbf{P}_{av}(z) = \frac{1}{2}Re(\hat{\mathbf{E}} \times \hat{\mathbf{H}}^*)$$

$$= \frac{1}{2}Re[(E_r + jE_i)\,\mathbf{a}_x \times (H_r - jH_i)\,\mathbf{a}_y] \qquad (3.90)$$

$$= \frac{1}{2}(E_r H_r + E_i H_i)\,\mathbf{a}_z$$

which is the same form as equation 3.89. It should be noted that the complex conjugate operator * resulted in the negative sign in the imaginary part of the magnetic field expression.

EXAMPLE 3.20

Use complex (phasor) forms of the electric and magnetic fields to obtain expressions for the time-average Poynting vectors of plane waves propagating along the positive z direction in the following:

1. Free space.
2. Conductive medium.

Solution

1. For a plane wave propagating along the positive z direction in free space, the complex forms of the electric and magnetic fields are

$$\hat{\mathbf{E}}(z) = E_m^+ e^{-j\beta_o z}\,\mathbf{a}_x$$

$$\hat{\mathbf{H}}(z) = \frac{E_m^+}{\eta_o}e^{-j\beta_o z}\,\mathbf{a}_y$$

The time-average Poynting vector from equation 3.90 is

$$\mathbf{P}_{av}(z) = \frac{1}{2}Re(\hat{\mathbf{E}} \times \hat{\mathbf{H}}^*) = \frac{1}{2}Re\left(E_m^+ e^{-j\beta_o z}\,\mathbf{a}_x \times \frac{E_m^+}{\eta_o}e^{j\beta_o z}\,\mathbf{a}_y\right)$$

$$= \frac{1}{2}Re\left(\frac{E_m^{+2}}{\eta_o}\,\mathbf{a}_z\right) = \frac{E_m^{+2}}{2\eta_o}\,\mathbf{a}_z \qquad (3.91)$$

2. For a plane wave propagating in conductive medium, the complex expressions of the fields are

$$\hat{\mathbf{E}}(z) = E_m^+ e^{-\alpha z}\,e^{-j\beta z}\,\mathbf{a}_x$$

$$\hat{\mathbf{H}}(z) = \frac{E_m^+}{|\hat{\eta}|}e^{-\alpha z}\,e^{-j\beta z}\,e^{-j\theta}\,\mathbf{a}_y$$

where the complex wave impedance was expressed as $\hat{\eta} = |\hat{\eta}|e^{j\theta}$.

$$\mathbf{P}_{av}(z) = \frac{1}{2} Re \left[(E_m^+ e^{-\alpha z} e^{-j\beta z}) \mathbf{a}_x \times \frac{E_m^+}{|\hat{\eta}|} e^{-\alpha z} e^{j\beta z} e^{j\theta} \mathbf{a}_y \right]$$

$$= \frac{1}{2} \frac{E_m^{+2}}{|\hat{\eta}|} e^{-2\alpha z} Re(e^{j\theta}) \mathbf{a}_z$$

$$= \frac{1}{2} \frac{E_m^{+2}}{|\hat{\eta}|} e^{-2\alpha z} \cos \theta \, \mathbf{a}_z \qquad (3.92)$$

Both the average power expressions in equations 3.91 and 3.92 are the same as those obtained previously in equations 3.87 and 3.88 by integrating the instantaneous Poynting vectors.

———◆◆◆———

EXAMPLE 3.21

The free space safety standard of the electromagnetic radiation, (i.e., the level of radiation that should not be exceeded to avoid any hazardous effects) is 10 mW/cm². Express this safety standard in terms of the electric and magnetic fields of a plane wave in free space.

Solution

In SI system of units, the power density safety standard is

$$P_{av} = 10 \text{ mW/cm}^2 = \frac{10 \times 10^{-3} \text{ W}}{10^{-4} \text{ m}^2} = 100 \text{ W/m}^2$$

The average power density of a plane wave in free space is

$$P_{av} = \frac{E_m^{+2}}{2\eta_o} = \frac{E_m^{+2}}{2(120\pi)} = 100$$

$$E_m^+ = 274.5 \text{ V/m}$$

The mean square value of the electric field for sinusoidal excitation is

$$E_{rms}^+ = \frac{E_m^+}{\sqrt{2}} = 194.1 \text{ V/m}$$

The corresponding magnetic field is

$$H_{rms}^+ = \frac{E_{rms}^+}{\eta_o} = 0.52 \text{ A/m}$$

———◆◆◆———

EXAMPLE 3.22

A uniform plane wave is propagating in a lossless (nonconducting) dielectric medium along the positive z direction. The time-domain form of the x-directed electric field is given by

$$\mathbf{E}(z,t) = 377 \cos\left(\omega t - \frac{4\pi}{3} z + \frac{\pi}{6} \right) \mathbf{a}_x \text{ V/m}$$

The measured time-average power density associated with this wave was 377 W/m².

1. Determine the properties of the propagation medium assuming $\mu = \mu_o$.
2. Determine the frequency of the wave.
3. Write down a time-domain expression of the magnetic field.

Solution

1. In a lossless medium the attenuation constant $\alpha = 0$, and the wave impedance is a real number and its angle θ is zero. Hence,

$$\eta = \sqrt{\frac{\mu_o}{\epsilon_o \, \epsilon_r}} \qquad (\sigma = 0)$$

The expression for the time-average power is, hence,

$$P_{av} = \frac{1}{2} \frac{E_m^{+2}}{\eta} = 377 \text{ W/m}^2$$

$$= \frac{1}{2} \frac{(377)^2}{\eta}$$

For $P_{av} = 377$ W/m², η is, hence,

$$\eta = \frac{377}{2} = \sqrt{\frac{\mu_o}{\epsilon_o \, \epsilon_r}} \approx \frac{377}{\sqrt{\epsilon_r}}$$

and

$$\epsilon_r = 4$$

2. From the given expression of the electric field,

$$\beta = \frac{4}{3}\pi = \omega \sqrt{\mu_o \, \epsilon_o \, \epsilon_r} = \frac{\omega \sqrt{\epsilon_r}}{3 \times 10^8}$$

Substituting $\beta = \dfrac{4\pi}{3}$, we obtain

$$\omega = \frac{4\pi/3(3 \times 10^8)}{2}$$

$$= 2\pi \times 10^8 \text{ rad/sec}$$

The frequency

$$f = 10^8 \text{ Hz}$$

$$= 100 \text{ MHz}$$

3.
$$\mathbf{H}(z,t) = \frac{377}{\eta} \cos\left(2\pi \times 10^8 t - \frac{4\pi}{3}z + \frac{\pi}{6}\right) \mathbf{a}_y$$

$$= 2 \cos\left(2\pi \times 10^8 t - \frac{4\pi}{3}z + \frac{\pi}{6}\right) \mathbf{a}_y.$$

———————◆◆◆———————

SUMMARY

In this chapter we modified Maxwell's equations so as to include additional induced sources that result from materials interactions with electromagnetic fields. These general Maxwell's equations were then solved to describe the propagation properties of uniform plane waves in conductive media, and a set of boundary conditions was derived to describe the transitional relationship between fields at an interface between two different media. The chapter was concluded with the description of the electromagnetic power balance equation known as the Poynting theorem, and the derivation of an expression for the time-average power density associated with time harmonic electromagnetic fields. The following is a summary of some of the major topics discussed in this chapter.

Materials Interactions with Electromagnetic Fields and Summary of Induced Sources

Materials interactions with electromagnetic fields are characterized in terms of conduction, polarization, and magnetization.

Conduction   Related to the "drift" (not acceleration) of the free conduction electrons under the influence of externally applied electric field. The conduction current density $\mathbf{J} = \sigma\mathbf{E}$, results from this interaction where σ is the conductivity, and $\mathbf{E}$ is the external applied field.

Polarization   Electric dipoles $\mathbf{p} = q\,\mathbf{d}$ are either induced or parallelly oriented when an electric field is applied to a dielectric or insulating material. As a result of this interaction, two new sources may be induced. The polarization current density $\mathbf{J} = \partial\mathbf{P}/\partial t$, where the polarization (electric dipole per unit volume) $\mathbf{P} = \epsilon_o \chi_e \mathbf{E}$ and $\mathbf{E}$ is a time-varying field. The induced polarization charge density ρ_p is given by $\rho_p = -\nabla \cdot \mathbf{P}$.

Magnetization   Magnetic dipoles $\mathbf{m} = I\,d\mathbf{s}$ are oriented in the direction of an externally applied magnetic field intensity $\mathbf{H}$. Magnetization current $\mathbf{J}_m = \nabla \times \mathbf{M}$, where $\mathbf{M} = \chi_m \mathbf{H}$ is the magnetization (magnetic dipole per unit volume) that results from this interaction.

Maxwell's Equations in Materials

On adding all the induced currents and charge new sources to Maxwell's equations in the vacuum described in earlier chapters, we obtain

	INTEGRAL FORM	DIFFERENTIAL FORM
Gauss's law of electric field	$\oint_s \mathbf{D} \cdot d\mathbf{s} = \int_v \rho_v \, dv$	$\nabla \cdot \mathbf{D} = \rho_v$
Gauss's law of magnetic field	$\oint_s \mathbf{B} \cdot d\mathbf{s} = 0$	$\nabla \cdot \mathbf{B} = 0$

	INTEGRAL FORM	DIFFERENTIAL FORM

Faraday's law
$$\oint_c \mathbf{E} \cdot d\ell = -\frac{d}{dt} \int_s \mathbf{B} \cdot d\mathbf{s} \qquad \nabla \times \mathbf{E} = -\frac{\partial \mathbf{B}}{\partial t}$$

Ampere's law
$$\oint_c \mathbf{H} \cdot d\ell = \int_s \mathbf{J} \cdot d\mathbf{s} \qquad \nabla \times \mathbf{H} = \mathbf{J} + \frac{\partial \mathbf{D}}{\partial t} + \sigma \mathbf{E}$$

$$+ \int_s \sigma \mathbf{E} \cdot d\mathbf{s} + \frac{d}{dt} \int_s \mathbf{D} \cdot d\mathbf{s}$$

where $\mathbf{D} = \epsilon_o \epsilon_r \mathbf{E}$ and $\mathbf{B} = \mu_o \mu_r \mathbf{H}$

It should be emphasized that ρ_v is the free charge density and the polarization charge density ρ_p is included in the $\mathbf{D}$ term specifically by including $\epsilon_r = 1 + \chi_e$.

Boundary Conditions

Electric and magnetic fields interact differently with different materials. It is, therefore, important that we derive mathematical relations that describe the transitional properties of these fields across an interface between two media.

ELECTRIC FIELD BOUNDARY CONDITIONS	MAGNETIC FIELD BOUNDARY CONDITIONS
$\mathbf{n} \cdot (\mathbf{D}_1 - \mathbf{D}_2) = \rho_v$	$\mathbf{n} \cdot (\mathbf{B}_1 - \mathbf{B}_2) = 0$
$\mathbf{n} \times (\mathbf{E}_1 - \mathbf{E}_2) = 0$	$\mathbf{n} \times (\mathbf{H}_1 - \mathbf{H}_2) = \mathbf{J}_s$

where $\mathbf{n}$ is a unit vector normal to the interface and directed from region 2 to 1. For both the electric and magnetic fields, separate boundary conditions are available for the tangential and normal components. In addition, we derived the following two auxiliary boundary conditions for the polarization surface charge, ρ_{ps}, and magnetization surface currents, $\mathbf{J}_{ms}$

$$\mathbf{n} \cdot (\mathbf{P}_1 - \mathbf{P}_2) = -\rho_{ps}$$

$$\mathbf{n} \times (\mathbf{M}_1 - \mathbf{M}_2) = \mathbf{J}_{ms}$$

Uniform Plane Wave Propagation in Conductive Media

Maxwell's equations that include the induced sources were then solved to describe the characteristics of plane wave propagation in conductive media. Table 3.6 summarizes similarities and differences between the plane wave propagation in a vacuum and in a conductive medium. Some of the important differences include the following:

1. The wave attenuation in conductive media. The depth penetration (skin depth) δ is defined as the distance in which the amplitude of the wave ($\mathbf{E}$ or $\mathbf{H}$ field) drops to e^{-1} or 36.8 percent of its initial value.

2. Changes in the propagation parameters, including reduction in the phase velocity $v_p < c$, decrease in the wave length, $\lambda < \lambda_o$, and the out-of-phase ratio of the electric and magnetic fields. λ_o and c are the wavelength and phase velocity in vacuum and approximately in air. The intrinsic impedance $\hat{\eta}$ is complex—hence, the out-of-phase relationship between the electric and magnetic fields. The propagation parameters in conductive media are given by

$$\hat{\gamma} = \alpha + j\beta, \text{ where}$$

$$\begin{matrix} \beta \\ \alpha \end{matrix} = \frac{\omega\sqrt{\mu\epsilon}}{\sqrt{2}}\left[\sqrt{1 + \left(\frac{\sigma}{\omega\epsilon}\right)^2} \pm 1\right]^{1/2}$$

$$\hat{\eta} = \sqrt{\frac{\mu}{\left(\epsilon - j\frac{\sigma}{\omega}\right)}} = \frac{\sqrt{\frac{\mu}{\epsilon}}}{\left[1 + \left(\frac{\sigma}{\omega\epsilon}\right)^2\right]^{1/4}}e^{j\frac{1}{2}\tan^{-1}\frac{\sigma}{\omega\epsilon}}$$

$$\delta = \frac{1}{\alpha}$$

Electromagnetic Power

In addition to the Poynting theorem for receiver (equation 3.85) and transmitter cases, an important expression for calculating the time-average power density associated with time harmonic electromagnetic fields was derived. This is given by

$$\mathbf{P}_{av} = \frac{1}{2}Re(\hat{\mathbf{E}} \times \hat{\mathbf{H}}^*)$$

where $\hat{\mathbf{E}}$ and $\hat{\mathbf{H}}$ are the phasor expressions of the fields. For plane waves propagating in the positive z direction, the power density is given by

$$\mathbf{P}_{av} = \frac{1}{2}\frac{|\hat{E}_m|^2}{\eta_o}\mathbf{a}_z \qquad \text{(Vacuum or air)}$$

$$\mathbf{P}_{av} = \frac{1}{2}\frac{|\hat{E}_m|^2}{|\hat{\eta}|}e^{-2\alpha z}\cos\theta\mathbf{a}_z \qquad \text{(Conductive medium)}$$

It should be emphasized that in addition to attenuation, the out-of-phase relationship (θ) between the electric and magnetic fields results in further reduction in the time-average power density associated with a plane wave propagating in conductive medium.

PROBLEMS

1. The electric field $\mathbf{E} = 3z^2 y \cos(10^8 t) \mathbf{a}_x$ is applied to a dielectric material (Lucit) of $\epsilon_r = 2.56$. Determine the following:
   (a) The polarization **P**.

(b) The induced polarization charge density ρ_p. Explain physically the reason for the zero value of ρ_p.

(c) The polarization current density $\mathbf{J}_p$.

2. A coaxial power cable has a core (conductor) of radius a. The region between the inner and outer conductors is filled with two concentric layers of dielectrics $\epsilon_1 = 1.5\epsilon_o$ and $\epsilon_2 = 4.5\epsilon_o$ as shown in Figure P3.2. If the outer conductor is grounded while the inner conductor is raised to a voltage that produces a linear charge density distribution ρ_ℓ, determine the following:

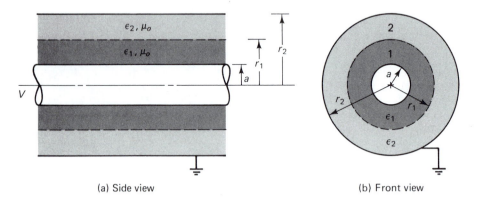

(a) Side view   (b) Front view

Figure P3.2   A power coaxial cable.

(a) The electric flux density, the electric field intensity, and the polarization in the two regions inside and the air outside the cable.

(b) The polarization surface charge at $\rho = a$ and at $\rho = r_1$.

(c) The polarization charge density in region 2.

3. An N turn toroid of rectangular cross section is shown in Figure P3.3. The core consists of three regions: region 1, an iron core of relative permeability $\mu_{r_1} = 3000$; region 2, an in-

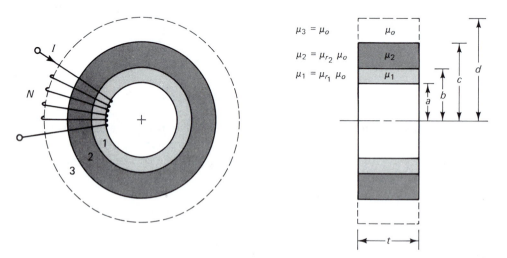

Figure P3.3   An N turn toroid with a core that consists of three regions.

homogeneous material of relative permeability $\mu_{r_2} = 1 + 2/\rho$, where ρ is in meters; region 3 is air.

(a) Find the magnetic field intensity **H**, the magnetic flux density **B**, and the magnetization **M** in the three regions.

(b) Find the magnetization current density $\mathbf{J}_m$ in region 2.

(c) Use the boundary conditions to find the magnetization surface current $\mathbf{J}_{ms}$ on the surface between regions 1 and 2, and between regions 2 and 3.

4. A perfect conductor medium occupying the region $x \geq 5$ has a surface charge density

$$\rho_s = \frac{\rho_o}{\sqrt{y^2 + z^2}}$$

Write expressions for **E** and **D** in air just outside the conductor.

5. Consider the two solid concentric cylinders shown in Figure P3.5. The current density in the inside cylinder is given by

$$\mathbf{J}_1 = 0.5\mathbf{a}_z \text{ A/m}^2 \qquad (\rho \leq a)$$

whereas the current density in the outside cylinder is given by

$$\mathbf{J}_2 = \frac{0.5\rho}{a}(-\mathbf{a}_z) \text{ A/m}^2 \qquad a < \rho \leq b$$

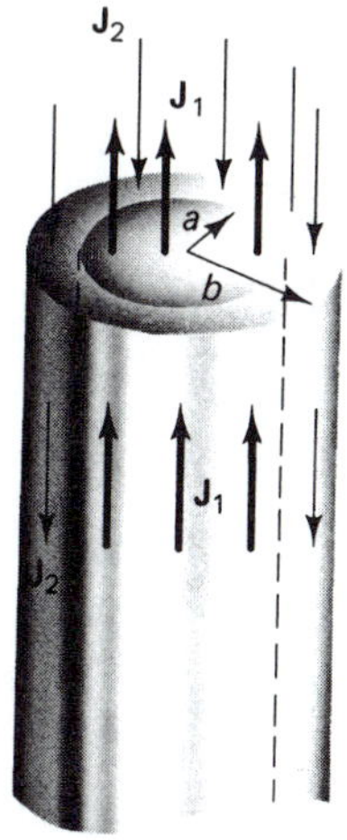

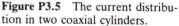

Figure P3.5 The current distribution in two coaxial cylinders.

The permeabilities of the materials of the inside and outside cylinders are given by $\mu_1 = \mu_o \mu_{r1}$ and $\mu_2 = \mu_o \mu_{r2}$, respectively. Determine the following:

(a) The magnetic field intensity **H** in the regions, $\rho \leq a$ and $a < \rho \leq b$.

(b) The magnetic flux density **B** in the inside and outside cylinders.

(c) The magnetization **M** and the magnetization current $\mathbf{J}_m$ in the *outside* cylinder, $a < \rho \leq b$.

(d) Also show that the boundary condition for the magnetic field intensity **H** at the cylindrical surface, $\rho = a$, is satisfied.

6. The cylindrical surface $\rho = a$ is charged with a surface charge density $\rho_s = 0.2 \times 10^{-6} \text{ C/m}^2$. The electrical properties of the materials inside $\rho \leq a$ and outside $\rho > a$ the cylindrical surface are given by

INSIDE THE CYLINDER ($\rho \le a$)   OUTSIDE THE CYLINDER ($\rho > a$)

$\epsilon_1 = 6.1\epsilon_o$ $\epsilon_2 = \epsilon_o$

$\mu_1 = \mu_o$ $\mu_2 = \mu_o$

The electric flux $\mathbf{D}_1$ in the region inside the dielectric cylinder is given by

$$\mathbf{D}_1 = 0.5\mathbf{a}_\rho - 3\mathbf{a}_\phi + 2\mathbf{a}_z$$

(a) Determine the polarization vector $\mathbf{P}$ and the polarization charge density ρ_p inside the dielectric cylinder.

(b) Use the boundary conditions to determine the electric flux density outside the dielectric cylinder.

7. In Figure P3.7, a cylindrical conductor of radius a carries a direct current that exponentially varies throughout the cross section of the conductor. If the current density is given by

$$\mathbf{J} = Ke^{-\left(1-\frac{\rho}{a}\right)}\mathbf{a}_z \ \text{A/m}^2$$

and if the conductor is made of magnetic material of $\mu = \mu_o\,\mu_1$, determine

Figure P3.7   Cylindrical conductor of radius a and carrying nonuniform current density $\mathbf{J}$.

(a) The magnetic field intensity $\mathbf{H}$ inside and outside the conductor.

(b) The magnetic flux density $\mathbf{B}$ and the magnetization $\mathbf{M}$ inside and outside the conductor.

(c) The magnetization current $\mathbf{J}_m$ within the conductor and the magnetization surface current density $\mathbf{J}_{ms}$ at the surface of the cylindrical conductor $\rho = a$.

In your integrations, make use of the following relations:

$$\int xe^{ax}\,dx = \frac{e^{ax}}{a^2}(ax - 1)$$

$$\int e^{-x}\,dx = -e^{-x}$$

8. (a) In characterizing materials according to their reactions to externally applied electric and magnetic fields, we in general identified three different types of materials.

(i) Indicate these three types of materials and explain (in a few words) their basic characteristics.

(ii) Identify the induced charge and current sources as a result of the interaction of external electric and magnetic fields with these materials.

(iii) Explain the impact of these new induced sources on Maxwell's equations.

(b) A spherical conductor of radius a is charged with a total positive charge Q. If the conductor is coated with two different dielectric materials of radii r_1 and r_2 as shown in Figure P3.8, determine

(i) The electric flux density **D**, the electric field intensity **E**, the polarization **P**, and the polarization charge density ρ_p in regions 1, 2, and 3.

(ii) The polarization surface charge density ρ_{ps} at the interface between regions 1 and 2 (i.e., at $r = r_1$).

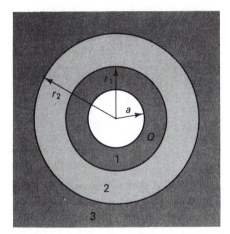

Region 1: $\epsilon_1 = \epsilon_o\, \epsilon_{r_1}, \mu = \mu_o$

Region 2: $\epsilon_2 = \epsilon_o\, \epsilon_{r_2}, \mu = \mu_o$

Region 3: $\epsilon_3 = \epsilon_o, \mu = \mu_o$

Figure P3.8   A conducting sphere of radius a is coated with two different dielectrics of radii r_1 and r_2.

9. The interface between regions 1 and 2 is charged with a surface charge density $\rho_s = 0.2\ \text{C/m}^2$. Region 1 ($z > 0$) is air, whereas region 2 ($z < 0$) is a material with $\epsilon_2 = 2\epsilon_o$ and $\mu_2 = 3.1\mu_o$. If the electric flux density in region 1 is given by $\mathbf{D}_1 = 3\mathbf{a}_x + 4\sqrt{y}\,\mathbf{a}_y + 3\mathbf{a}_z$, and the magnetic field intensity in region 2 is

$$\mathbf{H}_2 = 4\mathbf{a}_x + 3y^2\,\mathbf{a}_y + 5\mathbf{a}_z,$$

determine the electric flux density ($\mathbf{D}_2$) and the magnetic flux density $\mathbf{B}_1$ at the interface between regions 1 and 2 in Figure P3.9.

10. The electric and magnetic fields inside a rectangular wave guide made of approximately perfectly conducting material are given by

$$\hat{\mathbf{E}} = -j\omega\mu\frac{a}{\pi}H_o\,\sin\frac{\pi x}{a}\,\mathbf{a}_y$$

$$\hat{\mathbf{H}} = j\beta\frac{a}{\pi}H_o\,\sin\frac{\pi x}{a}\,\mathbf{a}_x + H_o\,\cos\frac{\pi x}{a}\,\mathbf{a}_z$$

where ω, μ, H_o, and β are all constants, and a is the larger dimension of the rectangular cross section as shown in Figure P3.10. Use the boundary conditions to determine (1)

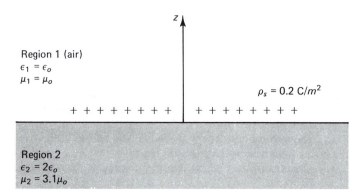

Figure P3.9 An interface between two media of different dielectric and magnetic properties.
Iskander P3.9

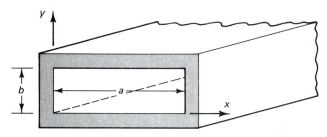

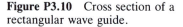

Figure P3.10 Cross section of a rectangular wave guide.

the *surface charge densities* and (2) the *surface current densities* on the *four* walls $(x = 0, x = a, y = 0, y = b)$ of the wave guide. The medium inside the wave guide is air ϵ_o, μ_o.

11. A uniform plane wave is propagating in a material medium that is characterized by the following parameters:

$$\epsilon_r = 6.3 \qquad \mu_r = 1.98$$

If the frequency of propagation is 1 GHz and the electric field is given by $\mathbf{E}(z,t) = 100 \cos(\omega t - \beta z)\mathbf{a}_x$, determine
(a) β, λ, and the phase velocity v_p.
(b) The characteristic impedance of the medium.
(c) A time-domain expression for the magnetic field associated with the wave.

12. Determine the frequency range for which the conduction current exceeds the displacement current by a factor of at least 100 in sea water ($\sigma = 4$ S/m, $\mu_r = 1$, and $\epsilon_r = 81$).

13. For an applied magnetic field $\hat{\mathbf{B}} = 10^{-6} \cos 2\pi z \, \mathbf{a}_y$ Wb/m^2, find the magnetization current crossing an area of 1 m^2 normal to the x direction in a magnetic material of $\chi_m = 10^{-3}$.

14. A 30-MHz plane wave is propagating along the positive z axis in a conductive medium of $\epsilon_r = 4$, $\mu_r = 1$, $\sigma = 0.8$ S/m.
(a) If the x-directed electric field reaches its maximum magnitude of 200 V/m at $t = 0$ and $z = 0$, give expressions for E_x in the *phasor* (complex) and *real-time* forms.
(b) Determine the magnetic field intensity $\mathbf{H}$ associated with this plane wave.
(c) Draw a sketch illustrating the electric and magnetic fields associated with this positive z traveling wave in conductive medium. Emphasize the directions of the electric and magnetic fields, and their magnitude and phase relationships.

(d) If we assume that the conductivity $\sigma \rightarrow 0$, determine the magnetic field intensity in this case. By drawing a sketch similar to that of part c, illustrate the differences between the plane wave propagation in conductive and nonconductive media.

15. (a) State the similarities and differences between the plane wave propagation in free space and in a conductive medium. Draw sketches illustrating some basic differences between the propagation characteristics in both cases. Also compare the time-average power density transmitted by the waves in both cases.

(b) A plane wave is incident normal to the surface of sea water having the following constants: $\mu_r = 1$, $\epsilon_r = 79$, and $\sigma = 3$ S/m. The electric field is parallel to the surface, and its magnitude is 10 V/m just *inside* the surface of the water. At what depth would it be possible for a submarine to receive a signal if the sub's receiver requires a field intensity of 10μ V/m. Make your calculations at the following two frequencies:

(i) 20 kHz. (Can the displacement current be neglected?)

(ii) 20 GHz. (Can the conduction current be neglected?)

(c) In part b, determine the time-average power density at the location of the submarine.

16. An alternating voltage was connected between the plates of a parallel plate capacitor. The resulting electric field between the plates is given by

$$\mathbf{E} = 10 \cos \omega t \, \mathbf{a}_z$$

(a) If the medium between the plates is air (ϵ_o), determine the total current crossing a square area of 0.1 m side length and placed perpendicular to the electric field.

(b) If we substitute sea water for the air in the region between the parallel plates (for sea water $\epsilon = 80\epsilon_o$ and $\sigma = 4$ S/m), determine the ratio between the conduction and displacement currents in the sea water between the parallel plates at 100 MHz. Also calculate the total current crossing the square area of Figure P3.16 in this case.

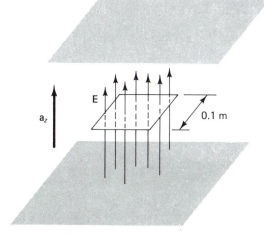

Figure P3.16 The displacement and conduction currents in the regions between parallel plates.

17. A high-voltage wire of radius a is insulated with an insulation coating of radius b and dielectric constant $\epsilon = \epsilon_o \epsilon_r$. The insulated high-voltage wire is suspended at the center of a grounded pipe of radius c. The geometry of the suspended cable is shown in Figure P3.17. When a high-voltage V is applied between the center wire and the ground pipe, a charge ρ_s (per unit area) was added to the center conductor.

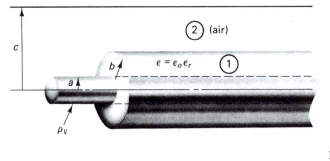

Figure P3.17   The geometry of the insulated suspended voltage wire.

(a) Determine the electric flux density **D** inside the insulation (region 1) and in the air (region 2).
(b) Determine the electric field intensity **E** and the polarization **P** in regions 1 and 2.
(c) Determine the free charge density ρ_s at the surface of the ground pipe of radius c.
(d) Determine the induced surface polarization charge at the interface between regions 1 and 2, that is, at $\rho = b$.
(e) Plot the electric field **E** as a function of ρ for $a < \rho < c$.
(f) Repeat part e for the case in which we replace region 1 by air and place the dielectric material $\epsilon = \epsilon_o \, \epsilon_r$ in region 2.
(g) As a result of the plots in parts e and f, which case is better from the insulation viewpoint?

18. A perfectly conducting cylindrical pipe (wave guide) of rectangular cross sections is shown in Figure P3.18. If the electric and magnetic fields inside the pipe are given in terms of their complex expressions by

$$\hat{\mathbf{E}} = E_o \sin \frac{\pi x}{a} \mathbf{a}_y$$

$$\hat{\mathbf{H}} = -\frac{E_o}{Z} \sin \frac{\pi x}{a} \mathbf{a}_x - \frac{jE_o \lambda}{2a\eta} \cos \frac{\pi x}{a} \mathbf{a}_z$$

where Z, λ, and η are all constants. Determine the following:

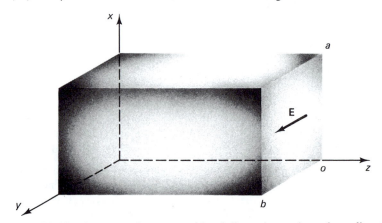

Figure P3.18   A rectangular wave guide of dimension a along the x direction and b along the y direction.

(a) The time-average power density (Poynting vector) associated with the transmission of these fields down the wave guide.

(b) Calculate the time-average *total* power transmitted along the wave guide.

19. A uniform plane wave is traveling in the x direction in a lossless medium with the 50 V/m electric field in the z direction. If the wavelength is 25 cm, and the velocity of propagation is 2×10^8 m/s, determine the following:

(a) The frequency of the wave and the relative permittivity of the medium if the medium is characterized by free space permeability.

(b) Write the phasor and complete the time-domain expressions for the electric and magnetic field vectors. Assume that these fields have maximum amplitude at $t = 0$ and $x = 0$.

20. A nonuniform time-varying electric field given in the cylindrical coordinates by

$$\mathbf{E} = \left(3\rho^2 \cot \phi \, \mathbf{a}_\rho + \frac{\cos \phi}{\rho} \mathbf{a}_\phi \right) \sin 3 \times 10^8 t \text{ V/m}$$

is applied to the following homogeneous, isotropic dielectric materials:

$$\text{Teflon } \mu = \mu_o, \, \epsilon = 2.1\epsilon_o, \quad \text{and} \quad \sigma = 0$$

$$\text{Glass } \mu = \mu_o, \, \epsilon = 6.3\epsilon_o, \quad \text{and} \quad \sigma = 0$$

$$\text{Sea water } \mu = \mu_o, \, \epsilon = 81\epsilon_o, \quad \text{and} \quad \sigma = 4 \text{ S/m}$$

Determine the following:

(a) The polarization vector, the polarization current density, and the polarization charge density for the preceding materials.

(b) The ratio of the conduction to the displacement currents in the sea water.

21. A magnetic field given in the spherical coordinates by

$$\mathbf{H} = \left(r^2 \sin \theta \, \mathbf{a}_r + \frac{1}{r} \cos \theta \cos \phi \, \mathbf{a}_\theta + r^2 \sin \theta \, \mathbf{a}_\phi \right)$$

is applied to the following magnetic materials:

$$\text{Aluminum } \mu_r = 1.000021$$

$$\text{Cobalt } \mu_r = 250$$

$$\text{Nickel } \mu_r = 600$$

$$\text{Iron } \mu_r = 2 \times 10^5$$

Determine the induced magnetization vector $\mathbf{M}$ and the magnetization current density for all the preceding materials.

22. The magnetization curve of commercial iron is shown in Figure P3.22. The permeability at any point along the curve is given by the slope dB/dH.

(a) Draw a curve showing the variation of relative permeability with the magnetic field intensity H.

(b) Determine the magnetization $\mathbf{M}$ at $B = 1.0$ Wb/m^2 and at $B = 1.6$ Wb/m^2.

23. Three coaxial cylinders separated by two different dielectric are charged as follows:

The inner cylinder of radius a has a positive linear charge density $\rho_{\ell 1}$ C/m.

The outer cylinder of radius c has a negative linear charge density $-\rho_{\ell 2}$ C/m.

The middle cylinder is connected to ground as shown in Figure P3.23.

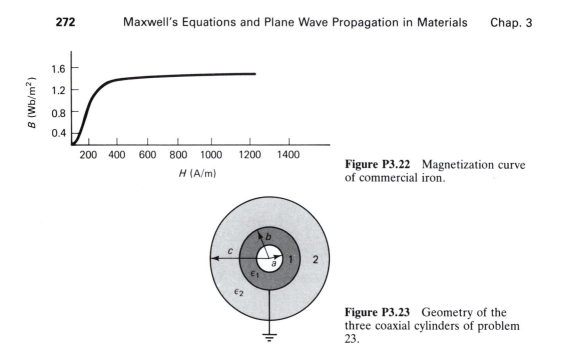

Figure P3.22 Magnetization curve of commercial iron.

Figure P3.23 Geometry of the three coaxial cylinders of problem 23.

(a) Determine and draw sketches showing the variation of the electric flux density and the electric field intensity in regions 1 and 2 between the cylinders, and region 3 outside them.

(b) Determine the induced surface charge on the middle conductor.

24. (a) A plane wave at 24 MHz, traveling through a lossy material, has a phase shift of 1 (rad/m), and its amplitude is reduced 50 percent for every meter traveled. Find α, β, λ, the phase velocity v_p, and the skin depth.

(b) A uniform plane wave is propagating in lossless (nonconductive) dielectric medium along the positive z direction. The phasor form of the x directed electric field is given by

$$\hat{\mathbf{E}}(z) = (40\pi\, e^{j\frac{4\pi}{6}}) e^{-j\frac{4\pi}{3}z}\, \mathbf{a}_x$$

The measured time-average power density associated with this wave is 377 W/m^2.

(i) If $\mu = \mu_o$ in the medium of propagation, determine the relative dielectric constant of this medium.

(ii) Determine the frequency of the wave.

(iii) Write a time-domain expression of the vector magnetic field intensity $\mathbf{H}(z,t)$ associated with this wave.

CHAPTER 4

STATIC ELECTRIC AND MAGNETIC FIELDS

.1 INTRODUCTION

In the previous chapters, we focused on dynamic (time-varying) electromagnetic fields. In this case, the electric and magnetic fields are coupled, and the four Maxwell's equations should be solved simultaneously. As an example of such solutions, we described the propagation characteristics of plane waves in air and in conductive media.

In this chapter, we consider the special case of static electric and magnetic fields. It will be shown that, in this case, the electric and magnetic fields are uncoupled; hence, solutions for electric fields in terms of their charge distribution sources will be separate from solutions of magnetic fields in terms of their direct current sources. Furthermore, after a brief discussion of interesting concepts such as the electric potential, capacitance, magnetic vector potential and inductance, we will describe solution procedures of electrostatic and magnetostatic problems. Besides describing some simple analytical solutions, numerical methods such as the finite difference and method of moments will also be discussed. This chapter basically includes interesting analyses of electrostatic and magnetostatic fields.

4.2 MAXWELL'S EQUATIONS FOR STATIC FIELDS

In the previous chapters, we introduced Maxwell's equations in their integral and differential forms, and developed their more general formulation in materials. The following is a summary of these general forms in integral and differential forms:

INTEGRAL FORMS   DIFFERENTIAL FORMS

$$\oint_s \epsilon_o \epsilon_r \mathbf{E} \cdot d\mathbf{s} = \int_v \rho_s \, dv \qquad\qquad \nabla \cdot \epsilon_o \epsilon_r \mathbf{E} = \rho_s \qquad (4.1)$$

$$\oint_s \mathbf{B} \cdot d\mathbf{s} = 0 \qquad\qquad\qquad \nabla \cdot \mathbf{B} = 0 \qquad (4.2)$$

$$\oint_c \mathbf{E} \cdot d\ell = -\frac{d}{dt}\int_s \mathbf{B} \cdot d\mathbf{s} \qquad \nabla \times \mathbf{E} = -\frac{\partial \mathbf{B}}{\partial t} \qquad (4.3)$$

$$\oint_c \mathbf{H} \cdot d\ell = \int_s \mathbf{J} \cdot d\mathbf{s} + \frac{d}{dt}\int_s \epsilon_o \epsilon_r \mathbf{E} \cdot d\mathbf{s} \qquad \nabla \times \mathbf{H} = \mathbf{J} + \frac{\partial \epsilon_o \epsilon_r \mathbf{E}}{\partial t} \qquad (4.4)$$

These Maxwell's equations, together with the boundary conditions described in chapter 3, provide complete and unique solutions for the electric and magnetic fields.

For static fields, conversely, $\partial/\partial t$ in the preceding equations will be set to zero. This results in the following Maxwell's equations for the electric and magnetic fields.

$$\text{Electric field} \begin{cases} \oint_s \epsilon_o \epsilon_r \mathbf{E} \cdot d\mathbf{s} = \int_v \rho_s \, dv \qquad \nabla \cdot \epsilon_o \epsilon_r \mathbf{E} = \rho_s \qquad (4.5) \\[2ex] \oint_c \mathbf{E} \cdot d\ell = 0 \qquad\qquad\qquad \nabla \times \mathbf{E} = 0 \qquad (4.6) \end{cases}$$

$$\text{Magnetic field} \begin{cases} \oint_s \mathbf{B} \cdot d\mathbf{s} = 0 \qquad\qquad\qquad \nabla \cdot \mathbf{B} = 0 \qquad (4.7) \\[2ex] \oint_c \mathbf{H} \cdot d\ell = \int_s \mathbf{J} \cdot d\mathbf{s} \qquad\qquad \nabla \times \mathbf{H} = \mathbf{J} \qquad (4.8) \end{cases}$$

Equations 4.5 and 4.6 describe the electric field in terms of its source charge density distribution ρ_s, whereas equations 4.7 and 4.8 describe the magnetic field in terms of its source current distribution $\mathbf{J}$. Equations 4.5 to 4.8, unlike equations 4.1 to 4.4, do not include any coupling between the electric and magnetic fields. Equations 4.5 and 4.6 together with the electric field boundary conditions,

$$\mathbf{n} \cdot (\epsilon_o \epsilon_{r1} \mathbf{E}_1 - \epsilon_o \epsilon_{r2} \mathbf{E}_2) = \rho_s \qquad (4.9)$$

$$\mathbf{n} \times (\mathbf{E}_1 - \mathbf{E}_2) = 0 \qquad (4.10)$$

where $\mathbf{n}$ is a unit vector perpendicular to the boundary surface, should provide complete and unique solution for the electric field. Equations 4.7 and 4.8, conversely, together with the magnetic field boundary conditions,

$$\mathbf{n} \cdot (\mathbf{B}_1 - \mathbf{B}_2) = 0 \qquad (4.11)$$

$$\mathbf{n} \times (\mathbf{H}_1 - \mathbf{H}_2) = \mathbf{J}_s \qquad (4.12)$$

should provide complete and unique solutions for the magnetic field quantities. The following is a discussion of the electric and magnetic fields solutions under static ($\partial/\partial t = 0$) considerations.

.3 ELECTROSTATIC FIELDS

Based on the preceding discussion, solutions for electrostatic fields may be obtained by solving equations 4.5 and 4.6 subject to the boundary conditions of equations 4.9 and 4.10. From equation 4.6, however, further simplifications are possible, and solutions based on such procedures are the subject of this section.

Consider the integral form of equation 4.6, hence,

$$\oint_c \mathbf{E} \cdot d\boldsymbol{\ell} = 0$$

Because the integration of the tangential component of $\mathbf{E}$ along a *closed* contour c is equal to zero, we may conclude, based on Figure 4.1, that

$$\oint_c \mathbf{E} \cdot d\boldsymbol{\ell} = \int_a^b \mathbf{E} \cdot d\boldsymbol{\ell} + \int_b^a \mathbf{E} \cdot d\boldsymbol{\ell} = 0$$

or

$$\int_a^b \mathbf{E} \cdot d\boldsymbol{\ell} \bigg|_{\text{along } c_1} = -\int_b^a \mathbf{E} \cdot d\boldsymbol{\ell} \bigg|_{\text{along } c_2} = \int_a^b \mathbf{E} \cdot d\boldsymbol{\ell} \bigg|_{\text{along } -c_2} \qquad (4.13)$$

From equation 4.13, it is clear that the integration of static field from a to b is independent of the specific shape of the contour that we follow in the integration. The

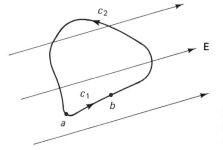

Figure 4.1   Closed contour $c = c_1 + c_2$ for integrating the static electric field $\mathbf{E}$.

length and shape of c_1, is certainly different from that of c_2, yet equation 4.13 indicates that $\int_a^b \mathbf{E} \cdot d\boldsymbol{\ell} = $ constant. The vector field that satisfies equation 4.6 or, equivalently, the condition that $\int_a^b \mathbf{E} \cdot d\boldsymbol{\ell} = $ constant, that is independent of the specific shape of the contour, is known as conservative field. The question now is: How may such an observation help us solve electrostatic field problems? To answer this question, we consider the electrostatic field $\mathbf{E}$ shown in Figure 4.2.

It is desired that we solve (describe) the electric field $\mathbf{E}$ everywhere in space in terms of its static charge distribution source ρ_v. Because the source is static, the resulting electric field at any point in Figure 4.2 satisfies equation 4.6 and, consequently,

$$\int_a^O \mathbf{E} \cdot d\boldsymbol{\ell} = \text{constant} \tag{4.14}$$

Because $\mathbf{E} \cdot d\boldsymbol{\ell}$ is a scalar quantity and represents the work done by electric forces in moving a unit positive charge from a to O, it seems easier to use the work (scalar) concept rather than the vector $\mathbf{E}$ to describe the static electric field resulting from the charge distribution ρ_v. It should be emphasized that the use of scalar work or energy concept is possible due to the fact that the work described in equation 4.14 is constant and independent of the specific shape of the contour from a to O. In other words, under the condition of equation 4.14 and if we choose the point O to be a reference, it is possible to describe the field at any point in terms of a scalar quantity called the electric potential Φ. We define Φ_a, the scalar potential at a, as the work done by electric forces in moving a unit positive charge from a to the reference point O. Hence,

$$\Phi_a = \int_a^O \mathbf{E} \cdot d\boldsymbol{\ell} = -\int_O^a \mathbf{E} \cdot d\boldsymbol{\ell} \tag{4.15}$$

Similarly, the electric potential at another point b is defined as

$$\Phi_b = \int_b^O \mathbf{E} \cdot d\boldsymbol{\ell} \tag{4.16}$$

From equations 4.15 and 4.16, it is clear that the electric potential is defined in terms of a reference point O. Subtracting equation 4.16 from equation 4.15, we obtain

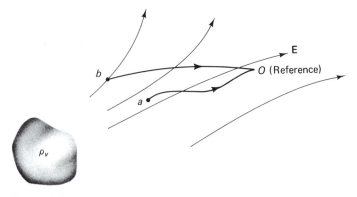

Figure 4.2   Electrostatic field $\mathbf{E}$ and the concept of electric potential.

$$\Phi_a - \Phi_b = \int_a^b \mathbf{E} \cdot d\boldsymbol{\ell} \qquad (4.17)$$

which is independent of the reference potential. Once again, describing the field resulting from static charges in terms of the electric potential is certainly attractive because it makes it possible to describe vector fields in terms of scalar quantity, the scalar electric potential. Such a description is possible because of the independency of the work done values in equations 4.15 and 4.16 on the specific shape of the contour from the point of interest to reference. In general, therefore, the scalar electric potential at any point A in the electrostatic field is defined as

$$\Phi_A = \int_A^O \mathbf{E} \cdot d\boldsymbol{\ell}$$

and is known as the electrostatic potential at A. Thus, by having A be an arbitrarily located point in the electric field region, we may uniquely describe each point in terms of its potential Φ_A with respect to an overall reference point O.

EXAMPLE 4.1

To illustrate the usefulness of the potential concept, let us calculate the potential resulting from a single point charge.

Solution

From Coulomb's law, the electric field resulting from a point charge is given by

$$\mathbf{E} = \frac{Q}{4\pi\epsilon_o r^2} \mathbf{a}_r$$

The electric potential at any point—for example, a—in Figure 4.3 with respect to a reference point is, hence,

$$\Phi_a = \int_a^O \mathbf{E} \cdot d\boldsymbol{\ell}$$

Expressing $d\boldsymbol{\ell}$ in the spherical coordinate system and maintaining only $dr\,\mathbf{a}_r$ because of the dot product, we obtain

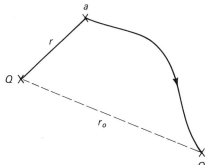

Figure 4.3 Electric potential resulting from a point charge.

$$\Phi_a = \int_a^O \frac{Q}{4\pi\epsilon_o r^2} \, \mathbf{a}_r \cdot dr \, \mathbf{a}_r = \left[-\frac{Q}{4\pi\epsilon_o r} \right]_a^O = \frac{Q}{4\pi\epsilon_o r} - \frac{Q}{4\pi\epsilon_o r_o}$$

The location of the reference point is completely arbitrary, so we can take it at infinity. At r_o equals to infinity, the reference electrostatic potential is zero. Hence, the potential Φ_a at a due to a point charge Q is given by

$$\Phi_a = \frac{Q}{4\pi\epsilon_o a}$$

Because a is an arbitrarily chosen point a distance r from the point charge, we may conclude that the electric potential at a distance r from the point charge is given by

$$\boxed{\Phi = \frac{Q}{4\pi\epsilon_o r}} \tag{4.18}$$

It is, of course, understood that scalar potential is taken with respect to a reference point of zero potential at infinity.

———◆◆———

Based on the preceding discussion, we can determine the electrostatic potential resulting from an arbitrary distribution of charges. The potential owing to a set of point charges is given by

$$\boxed{\Phi = \sum_{k=1}^N \frac{Q_k}{4\pi\epsilon_o r_k}} \tag{4.19}$$

where N is the number of charges, and r_k is the distance from the charge Q_k to the observation point at which Φ is being evaluated.

If, instead of N discrete point charges, we have a volume ρ_v, surface ρ_s, or linear ρ_ℓ charge distribution, the electrostatic potential Φ may be obtained as a limiting case of equation 4.19. Hence,

$$\Phi = \int_v \frac{\rho_v \, dv}{4\pi\epsilon_o r}$$

or

$$= \int_s \frac{\rho_s \, ds}{4\pi\epsilon_o r}$$

or

$$= \int_\ell \frac{\rho_\ell \, d\ell}{4\pi\epsilon_o r}$$

r in this case is not the distance from the origin, but instead is the distance from the charge point in a given distribution to the observation point at which the potential is being evaluated. Hence, r in this case will be variable as we integrate over the domain of the charge distribution v, s, or ℓ.

EXAMPLE 4.2

A circular disk of radius a is uniformly charged with a charge density ρ_s. Determine the electric potential along the axis of the circular disk.

Solution

The geometry of the circular disk of charge density ρ_s and the observation point P at which the potential is to be calculated is shown in Figure 4.4. If we consider an element of area $\rho \, d\rho \, d\phi$ with charge density ρ_s, the electric potential $d\Phi$ is given by

$$d\Phi = \frac{\rho_s \, \rho \, d\rho \, d\phi}{4\pi\epsilon_o R}$$

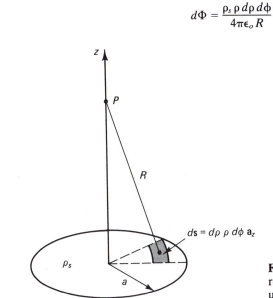

Figure 4.4   The electric potential resulting from a circular disk of uniform charge distribution.

The total electric potential at P is hence

$$\Phi_P = \int_{\phi=0}^{2\pi} \int_{\rho=0}^{a} \frac{\rho_s \, \rho \, d\rho \, d\phi}{4\pi\epsilon_o \sqrt{z^2 + \rho^2}}$$

$$= \frac{\rho_s}{2\epsilon_o} \left(\sqrt{z^2 + a^2} - z \right)$$

◆◆◆

4.4 EVALUATION OF ELECTRIC FIELD E FROM ELECTROSTATIC POTENTIAL Φ

From the differential relation that describes the conservative property of the static electric field, equation 4.6 states that

$$\nabla \times \mathbf{E} = 0$$

Any vector field with zero curl could always be expressed as a gradient of a scalar field. Hence, the vector static electric field **E** may be defined as

$$\mathbf{E} = -\nabla \Phi \tag{4.20}$$

where Φ is the scalar potential, and the reason for the negative sign will be explained shortly. From our previous discussion of the force electric field, we indicated that the quantity

$$\int_{P_1}^{P_2} \mathbf{E} \cdot d\ell$$

is equal to the work done in moving a unit positive charge from P_1 to P_2. Hence,

$$\int_{P_1}^{P_2} \mathbf{E} \cdot d\ell = \Delta W \tag{4.21}$$

Substituting equation 4.20 into equation 4.21, we obtain

$$\Delta W = \int_{P_1}^{P_2} \mathbf{E} \cdot d\ell = -\int_{P_1}^{P_2} \nabla\Phi \cdot d\ell = -\int_{P_1}^{P_2} \frac{d\Phi}{d\ell} d\ell = \Phi_{P_1} - \Phi_{P_2}$$

which is the same difference in potentials between two points, as described in equation 4.17. Equation 4.20 suggests a useful procedure for calculating the vector electric field from a given charge density distribution. First, the scalar electric potential may be calculated from the charge distribution, as described in the previous section, then the vector electric field may be obtained from Φ using equation 4.20. Before we solve examples illustrating this procedure, we still need to explain the reason for the negative sign in equation 4.20. From equation 4.21, it is clear that positive work should be done in moving a positive charge against the electric field lines. For example, if the electric field lines point from P_2 to P_1, there would be positive work that need to be done to move a unit positive charge from P_1 to P_2 against the electric field lines. Because positive work is being done in this case, it is expected that Φ_{P_2} be larger than Φ_{P_1} by the amount of the work done. Hence,

$$\Phi_{P_2} - \Phi_{P_1} \text{ (positive)} = -\int_{P_1}^{P_2} \mathbf{E} \cdot d\ell \tag{4.22}$$

The negative sign in equation 4.22 emphasizes that positive work needs to be done in moving against the direction of the electric field. In other words, the negative sign in equation 4.22 and also in equation 4.20 clarifies the fact that, in moving against the electric field, the potential increases. This also agrees with our understanding of what happens in an electric battery. The electric field lines are directed from the positive to the negative plates, whereas the electric potential increases from the negative to positive—that is, in the opposite direction. The following examples illustrate the solution procedure for obtaining the vector electric field from, the easier to calculate, scalar electric potential Φ.

EXAMPLE 4.3

For the circular disk problem discussed in example 4.2, determine the electric field along the axis directly from the given uniform charge distribution. Compare the obtained results with that calculated using $\mathbf{E} = -\nabla\Phi$, where Φ is the electric potential along the axis, as given in example 4.2.

Solution

From Figure 4.5, the electric field at point P along the axis resulting from the total charge $\rho_s \, \rho \, d\rho \, d\phi$ at element 1 is given (from Coulomb's law) by

$$d\mathbf{E}_1 = \frac{\rho_s \, \rho \, d\rho \, d\phi}{4\pi\epsilon_o(z^2 + \rho^2)} \mathbf{a}_1$$

where $\mathbf{a}_1$ is a unit vector in the direction of the electric field $d\mathbf{E}_1$.
    For the symmetrically located element 2, the electric field is given by

$$d\mathbf{E}_2 = \frac{\rho_s \, \rho \, d\rho \, d\phi}{4\pi\epsilon_o(z^2 + \rho^2)} \mathbf{a}_2$$

From the symmetry around the z axis, the components of $d\mathbf{E}_2$ along the z axis will add, whereas the components perpendicular to it will cancel. The total electric field is then

$$dE_z = dE_1 \cos\alpha + dE_2 \cos\alpha$$

Substituting

$$\cos\alpha = \frac{z}{\sqrt{z^2 + \rho^2}}$$

we obtain

$$dE_z = \frac{\rho_s \, \rho \, d\rho \, d\phi(2z)}{4\pi\epsilon_o(z^2 + \rho^2)^{3/2}}$$

The total electric field at a point along the axis is then

$$\mathbf{E} = \int_{\phi=0}^{\pi} \int_{\rho=0}^{a} \frac{2\rho_s \, z \, \rho \, d\rho \, d\phi}{4\pi\epsilon_o(z^2 + \rho^2)^{3/2}} \mathbf{a}_z$$

The integration over ϕ is carried from $\phi = 0$ to $\phi = \pi$ (not 2π) because in the expression for dE_z, we added the contributions from two symmetrically located elements. Carrying out the integration, we obtain

$$\mathbf{E} = \frac{\rho_s}{2\epsilon_o}\left[1 - \frac{z}{(z^2 + a^2)^{1/2}}\right]\mathbf{a}_z \tag{4.23}$$

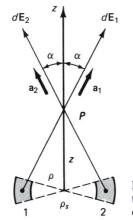

Figure 4.5 Calculation of the electric field along the axis of a circular disk.

From example 4.2, conversely, we calculated the electric potential Φ as

$$\Phi = \frac{\rho_s}{2\epsilon_o}\left(\sqrt{z^2 + a^2} - z\right)$$

The electric field according to equation 4.20 is then

$$\mathbf{E} = -\boldsymbol{\nabla}\Phi = -\frac{\rho_s}{2\epsilon_o}\frac{\partial}{\partial z}\left(\sqrt{z^2 + a^2} - z\right)\mathbf{a}_z = -\frac{\rho_s}{2\epsilon_o}\left(\frac{2z}{2\sqrt{z^2 + a^2}} - 1\right)\mathbf{a}_z$$

$$= \frac{\rho_s}{2\epsilon_o}\left[1 - \frac{z}{\sqrt{z^2 + a^2}}\right]\mathbf{a}_z \qquad (4.24)$$

Equations 4.23 and 4.24 clearly provide the same answers for the electric field.

◆◆

EXAMPLE 4.4

Consider the electric dipole shown in Figure 4.6. Two equal, but opposite, charges $+q$ and $-q$ are separated by a distance d. Calculate the electric field $\mathbf{E}$ at a point P located at a far distance $R \gg d$ from the dipole.

Solution

The geometry of the electric dipole is shown in Figure 4.6. The electric potential at P is given by

$$\Phi_p = \frac{q}{4\pi\epsilon_o R_1} - \frac{q}{4\pi\epsilon_o R_2} = \frac{q}{4\pi\epsilon_o}\left(\frac{R_2 - R_1}{R_1 R_2}\right)$$

For an observation point located at a large distance from the dipole,

$$\frac{1}{R_1} \approx \frac{1}{R_2} \approx \frac{1}{r}$$

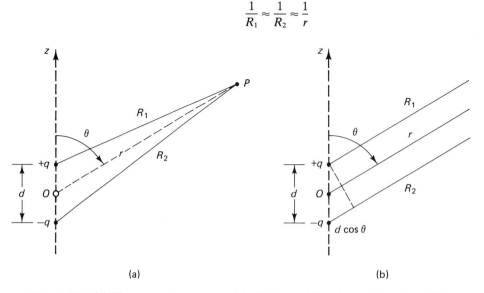

(a)   (b)

Figure 4.6 (a) The geometry of an electric dipole and the observation point. (b) Far distance approximation $1/R_1 \approx 1/R_2 \approx 1/r$, and $R_2 - R_1 = d\cos\theta$.

In this case, Figure 4.6b shows that $R_2 - R_1 \approx d \cos \theta$ because R_1, R_2, and r may be assumed parallel.

Substituting these far field approximations in Φ_p, we obtain

$$\Phi_p = \frac{q}{4\pi\epsilon_o} \frac{d \cos \theta}{r^2}$$

To calculate the electric field at P, we use the gradient relationship in spherical coordinates. Hence,

$$E_p = -\nabla\Phi_p = -\left(\frac{\partial\Phi_p}{\partial r}\mathbf{a}_r + \frac{1}{r}\frac{\partial\Phi_p}{\partial\theta}\mathbf{a}_\theta + \frac{1}{r \sin\theta}\frac{\partial\Phi_p}{\partial\phi}\mathbf{a}_\phi\right)$$

$$= \frac{qd}{2\pi\epsilon_o}\frac{\cos\theta}{r^3}\mathbf{a}_r + \frac{qd}{4\pi\epsilon_o}\frac{\sin\theta}{r^3}\mathbf{a}_\theta$$

$$= \frac{qd}{4\pi\epsilon_o r^3}(2\cos\theta\,\mathbf{a}_r + \sin\theta\,\mathbf{a}_\theta)$$

The preceding expression for the vector electric field at a distance point was obtained in a rather straightforward and simple manner. Any student who may attempt to calculate E_p directly from the vector sum of the electric fields from each of the two charges will discover quickly that the process is too long and involved, the matter that brings an immediate appreciation for the scalar potential solution procedure.

EXAMPLE 4.5

For the line charge of charge density ρ_ℓ and length L, obtain an expression for the electric field at a point along its axis.

Solution

The geometry of the line charge is shown in Figure 4.7. We use the scalar potential solution procedure. The potential at P along the axis of the line is given by

$$\Phi_p = \int_{-L/2}^{L/2} \frac{\rho_\ell \, dx'}{4\pi\epsilon_o(x - x')}$$

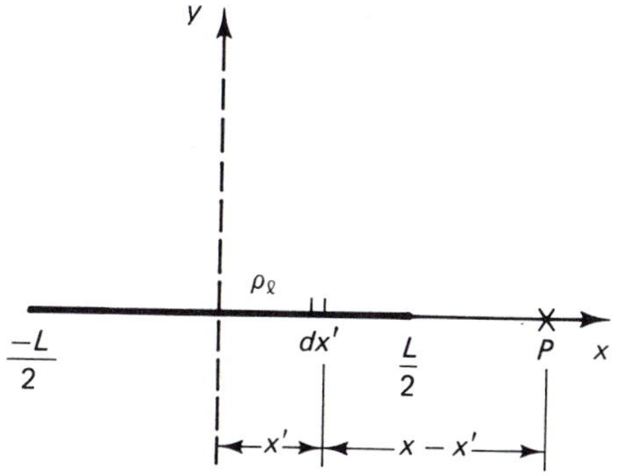

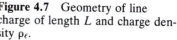

Figure 4.7 Geometry of line charge of length L and charge density ρ_ℓ.

The integration is being carried out over x' (the source point), which extends from $-L/2$ to $L/2$. Hence,

$$\Phi_p = -\frac{\rho_\ell}{4\pi\epsilon_o} \ell n \, (x - x') \Big|_{-L/2}^{L/2} = \frac{\rho_\ell}{4\pi\epsilon_o} \ell n \left[\frac{x + \dfrac{L}{2}}{x - \dfrac{L}{2}} \right] \qquad \left(x > \frac{L}{2} \right)$$

The electric field $\mathbf{E}_p$ is then

$$\mathbf{E}_p = -\nabla\Phi_p = -\frac{\partial\Phi_p}{\partial x}\mathbf{a}_x = \frac{\rho_\ell L}{4\pi\epsilon_o[x^2 - (L/2)^2]}\mathbf{a}_z \qquad \left(x > \frac{L}{2} \right)$$

———◆◆———

Once again, in this case, the vector electric field was obtained in a straightforward manner from the scalar electric potential. Calculation of the vector electric field directly from the given charge distribution would have been cumbersome. Furthermore, unlike the infinitely long line charge case (example 1.27), if one would attempt to apply Gauss's law, he or she will quickly discover that it is not possible to construct a suitable Gaussian surface that can be integrated over easily. From examples 4.4 and 4.5, we therefore conclude that the electric potential solution procedure for the vector electric field provides a valuable tool that complements others such as those based on calculating the electric field directly from the given charge distribution or the application of Gauss's law.

4.5 CAPACITANCE

Capacitance is a property of a geometric configuration usually of two conducting objects surrounded by a homogeneous dielectric. It is a measure of the amount of charge a particular configuration of two conductors is able to retain per unit voltage applied between them. In other words, the capacitance describes the ability of a given configuration of two conductors to store electrostatic energy. Let us, for example, consider the two conductors geometry shown in Figure 4.8. The two conductors are surrounded by a homogeneous dielectric of dielectric constant $\epsilon = \epsilon_o \epsilon_r$. The total charge on each conductor is Q. Conductor 1 carries a total charge $+Q$, whereas the charge on conductor 2 is $-Q$. As a result of this charge arrangement, there will be electric flux emanating from the positive charge and terminating at the negative one. Also, there will be work that must be done to carry a unit-positive charge from the negatively charged conductor to the positively charged one. This amount of work defines the potential difference V between the two conductors. The capacitance of this two-conductor system is then defined as the ratio of the amount of positive charge to the resulting potential difference between the conductors.

$$C = \frac{Q}{V}$$

The capacitance, hence, is independent of the specific amount of charge on the conductors and the specific value of the potential difference between them. It depends

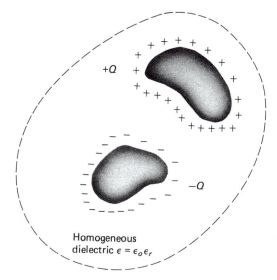

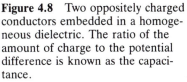

Figure 4.8   Two oppositely charged conductors embedded in a homogeneous dielectric. The ratio of the amount of charge to the potential difference is known as the capacitance.

only on the ratio between these quantities. The capacitance, however, depends on the geometrical arrangement of the two-conductor systems, their dimensions, and the type of dielectric material surrounding them. The capacitance is measured in farads, which is defined as one coulomb per volt.

For a given geometry of the two-conductor system, there are generally two procedures for calculating C. The first starts by assuming a charge Q and $-Q$ on the two conductors, and then Gauss's law or some other means is used to calculate the resulting electric field from these charges. The potential difference between the two conductors is then found using $V = -\int_c \mathbf{E} \cdot d\ell$, and the capacitance is calculated as $C = Q/V$. The other solution procedure starts in a reverse order by assuming the potential difference between the conductors, and then the total charge on the conductors is calculated. This latter procedure requires the solution of Laplace's equation, which will be described in later sections of this chapter. Therefore, in the following examples, we will limit the discussion to simple geometries in which the first procedure may be applied.

EXAMPLE 4.6

Determine the capacitance of a spherical capacitor that consists of two concentric spheres of radii a and b, as shown in Figure 4.9. The space between the two spherical conductors is filled with a dielectric of permittivity ϵ.

Solution

Following the first procedure suggested earlier, we assume two equal, opposite charges on the two conductors. If the charge on the inner conductor is Q, the resulting electric field in the space between the two conductors may be determined using Gauss's law. Because

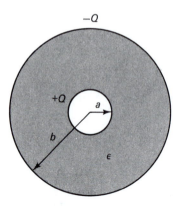

Figure 4.9 The geometry of a spherical capacitor.

of the spherical symmetry, we construct a spherical Gaussian surface s of radius $a < r < b$ and apply Gauss's law, hence,

$$\oint_s \epsilon \mathbf{E} \cdot d\mathbf{s} = \text{total charge enclosed}$$

$$= Q$$

Therefore,

$$\mathbf{E} = \frac{Q}{4\pi\epsilon r^2} \mathbf{a}_r$$

The potential difference between the conductors is then

$$V = -\int_b^a \mathbf{E} \cdot d\boldsymbol{\ell}$$

It is important to note the negative sign and the limits of integration from b to a. This is determined by the fact that positive work must be done to move a unit-positive charge from b to a against the electric field lines. If we take the integration path along the radial direction $d\boldsymbol{\ell} = dr\,\mathbf{a}_r$, the potential difference V is then given by

$$V = -\int_b^a \frac{Q}{4\pi\epsilon r^2} \mathbf{a}_r \cdot dr\,\mathbf{a}_r = \frac{Q}{4\pi\epsilon}\left[\frac{1}{a} - \frac{1}{b}\right]$$

The capacitance C of the spherical capacitor is then

$$C = \frac{Q}{V} = \frac{4\pi\epsilon}{\left[\dfrac{1}{a} - \dfrac{1}{b}\right]}$$

◆◆

EXAMPLE 4.7

Consider the parallel plate capacitor with two dielectric layers in the space between the conductors, as shown in Figure 4.10. Calculate its capacitance and show that the result is equivalent to the series connection of two capacitances, each with a homogeneous dielectric.

Figure 4.10 Parallel plate capacitor with two dielectric layers.

Solution

We will solve this example under the assumption that we have infinitely large parallel plates. This means that we will neglect the fringing capacitance at the ends of the plates. Accounting for such effects has to await the introduction of the numerical solution of Laplace's equation, which will be described in following sections. If we assume a charge density ρ_s C/m² on the lower plate and an equal but opposite charge on the upper one, then establishing Gaussian surface similar to the one used in the section on displacement current in chapter 1, we obtain the following expressions for the electric fields in the two dielectrics:

$$\mathbf{E}_2 = \frac{\rho_s}{\epsilon_2}\mathbf{a}_z \qquad \text{and} \qquad \mathbf{E}_1 = \frac{\rho_s}{\epsilon_1}\mathbf{a}_z$$

The electric flux densities in the two dielectrics $\mathbf{D}_2 = \epsilon_2 \mathbf{E}_2 = \rho_s$ and $\mathbf{D}_1 = \epsilon_1 \mathbf{E}_1 = \rho_s$ is clearly continuous across the dielectric interface as required by the boundary conditions. Furthermore, because $\mathbf{D}$ is normal to the conductor, its value is equal to the charge density ρ_s, which is also consistent with the boundary conditions described in chapter 3.

The potential difference V between the parallel plates is then calculated as

$$V = -\int_{d_2}^{O} \mathbf{E}_2 \cdot d\boldsymbol{\ell} - \int_{d_1+d_2}^{d_2} \mathbf{E}_1 \cdot d\boldsymbol{\ell} = \frac{\rho_s d_2}{\epsilon_2} + \frac{\rho_s d_1}{\epsilon_1} = \rho_s\left(\frac{d_2}{\epsilon_2} + \frac{d_1}{\epsilon_1}\right)$$

For a section of the parallel plate capacitor of area A, the total charge $Q = \rho_s A$. The capacitance is, hence,

$$C = \frac{Q}{V} = \frac{A}{\left(\dfrac{d_2}{\epsilon_2} + \dfrac{d_1}{\epsilon_1}\right)} = \frac{\epsilon_1 A}{\left(d_1 + \dfrac{\epsilon_1}{\epsilon_2}d_2\right)}$$

If we consider two series capacitances, each with homogeneous dielectrics, as shown in Figure 4.11, the total potential difference

$$V = V_1 + V_2$$

$$= \frac{Q}{C_1} + \frac{Q}{C_2}$$

where Q is assumed the same because of the series connection. Hence, the total capacitance $C = Q/V$ is given by

$$\frac{V}{Q} = \frac{1}{C} = \frac{1}{C_1} + \frac{1}{C_2}$$

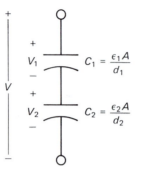

Figure 4.11 Series connection of two capacitors.

or

$$C = \frac{C_1 C_2}{C_1 + C_2} = \frac{\left(\dfrac{\epsilon_1 A}{d_1}\right)\left(\dfrac{\epsilon_2 A}{d_2}\right)}{\dfrac{\epsilon_1 A}{d_1} + \dfrac{\epsilon_2 A}{d_2}} = \frac{\epsilon_1 A}{\left(\dfrac{\epsilon_1}{\epsilon_2} d_2 + d_1\right)}$$

which is the same result obtained from the routine calculation of the two-dielectric system of Figure 4.10.

EXAMPLE 4.8

Consider the coaxial cable of Figure 4.12. The space between the two cylindrical conductors of radii a and c is filled with two homogeneous dielectrics of ϵ_1 and ϵ_2. The interface between the two dielectrics is of radius b. Calculate the capacitance per unit length of this coaxial cable.

Solution

The solution procedure for calculating the capacitance in this case closely follows those used in the previous examples. It starts by assuming equal and opposite charge distributions on the conductors, and we then use Gauss's law to calculate the electric field. If ρ_s is the

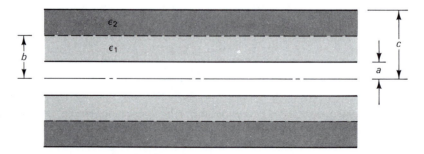

Figure 4.12 Coaxial cylindrical capacitor with two dielectric materials.

assumed charge density (per unit area) on the center conductor, application of Gauss's law yields

$$\int_{\phi=0}^{2\pi}\int_{z=0}^{\ell}\epsilon_1\mathbf{E}_1\cdot d\mathbf{s}=\int_{\phi=0}^{2\pi}\int_{z=0}^{\ell}\rho_s\,ds$$

Due to the cylindrical symmetry, the electric field is in the radial direction, hence,

$$\int_0^{2\pi}\int_0^{\ell}\epsilon_1 E_{\rho_1}\mathbf{a}_\rho\cdot\rho\,d\phi\,dz\,\mathbf{a}_\rho=\int_0^{2\pi}\int_0^{\ell}\rho_s\,a\,d\phi\,dz$$

Hence,

$$\mathbf{E}_1=E_{\rho_1}\mathbf{a}_\rho=\frac{\rho_s\,a}{\epsilon_1\rho}\mathbf{a}_\rho$$

Similarly,

$$\mathbf{E}_2=\frac{\rho_s\,a}{\epsilon_2\rho}\mathbf{a}_\rho$$

The potential difference between the two conductors,

$$V=-\int_b^a\mathbf{E}_1\cdot d\ell-\int_c^b\mathbf{E}_2\cdot d\ell$$

Taking the path of integration $d\ell$ in the radial direction $d\ell=d\rho\,\mathbf{a}_\rho$, we obtain

$$V=\frac{\rho_s\,a}{\epsilon_1}\ell n\frac{b}{a}+\frac{\rho_s\,a}{\epsilon_2}\ell n\frac{c}{b}$$

The charge per unit length along the inner conductor is $\rho_\ell=2\pi a\rho_s$, where ρ_s is the charge per unit area. The capacitance per unit length is then

$$C=\frac{2\pi a\rho_s}{V}=\frac{2\pi}{\dfrac{1}{\epsilon_1}\ell n\dfrac{b}{a}+\dfrac{1}{\epsilon_2}\ell n\dfrac{c}{b}}$$

$$=\frac{2\pi\epsilon_1}{\ell n\dfrac{b}{a}+\dfrac{\epsilon_1}{\epsilon_2}\ell n\dfrac{c}{b}}$$

It can be shown that this multidielectric capacitance arrangement is equivalent to two cylindrical capacitors connected in series.

EXAMPLE 4.9

Determine the capacitance per unit length for the two-conductor transmission line shown in Figure 4.13. The two wires are of radius a, and the separation distance is d.

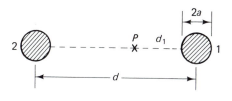

Figure 4.13 Geometry of two parallel wires' transmission line.

Solution

We assume equal and opposite charges ρ_ℓ per unit length on the two conductors. If conductor 1 on the right-hand side is assumed to have $+\rho_\ell$, application of Gauss's law yields the following expression for the electric field:

$$\int_s \epsilon \mathbf{E}_1 \cdot d\mathbf{s} = \int_c \rho_\ell \, d\ell$$

Hence,

$$\int_{\phi=0}^{2\pi} \int_{z=0}^{\ell} \epsilon E_{1\rho} \mathbf{a}_\rho \cdot \rho \, d\phi \, dz \, \mathbf{a}_\rho = \rho_\ell \ell$$

$$E_{1\rho} = \frac{\rho_\ell}{2\pi\epsilon\rho}$$

The electric field at a point P, which is a distance d_1 from conductor 1 and along the line joining the two conductors, is given by

$$\mathbf{E}_1 = \frac{\rho_\ell}{2\pi\epsilon d_1} \mathbf{a}_\rho$$

The electric field at the same point resulting from the negative charge on conductor 2 may be similarly calculated and is given by

$$\mathbf{E}_2 = \frac{\rho_\ell}{2\pi\epsilon(d - d_1)} (-\mathbf{a}_\rho)$$

The total electric field along the line joining the two conductors is given by

$$\mathbf{E} = \left(\frac{\rho_\ell}{2\pi\epsilon d_1} - \frac{\rho_\ell}{2\pi\epsilon(d - d_1)} \right) \mathbf{a}_\rho$$

To determine the potential difference between the two wires, it is sufficient to determine the potential difference between any two points on the two wires. The simplest case is obtained by considering the line between the two conductors. The potential difference is hence

$$V = -\int_{d-a}^{a} \mathbf{E} \cdot d\ell$$

$d\ell$ in this case is taken along the $\mathbf{a}_\rho$. Therefore,

$$V = -\int_{d-a}^{a} \left[\frac{\rho_\ell}{2\pi\epsilon\rho} - \frac{\rho_\ell}{2\pi\epsilon(d - \rho)} \right] \mathbf{a}_\rho \cdot d\rho \, \mathbf{a}_\rho$$

The arbitrarily located electric field point P at a distance d_1 from conductor 1 was replaced by a variable point at distance ρ from the conductor. The potential difference is then

$$V = \frac{\rho_\ell}{2\pi\epsilon} \left[\ell n \frac{d - a}{a} - \ell n \frac{a}{d - a} \right] = \frac{\rho_\ell}{\pi\epsilon} \ell n \frac{d - a}{a}$$

The capacitance per unit length is then

$$C = \frac{\rho_\ell}{V} = \frac{\pi\epsilon}{\ell n \dfrac{d - a}{a}}$$

If the separation distance d is much larger than the radius of the wire a, the value of the capacitance reduces to

$$C = \frac{\pi \epsilon}{\ell n \dfrac{d}{a}}$$

4.6 ELECTROSTATIC ENERGY DENSITY

Thus far, we discussed electrostatic fields from the electric field and electric potential viewpoint. From a given charge distribution, we learned how to determine the resulting electric field, either directly or through the calculation of the auxiliary scalar potential function. In introducing electrostatic potentials, we defined them in terms of the work that must be done to move a unit-positive charge from one point to another. Clearly doing work on electrostatic charges will result in an increase in their potential energy. In this section, we will quantify the potential energy present in a system of charges by determining the amount of work that must be done to assemble such a system of charge distribution. To simplify the discussion, we will deal first with a discrete system of charges and will then generalize the result to a system of charge distribution.

To start with, let us assume that we have a space that is completely empty and does not contain any charges. Bringing a single point charge Q_1 from infinity, where we have zero potential to a specific location in our empty space, does not require any work to be done. This is because there are no other forces in our empty space that oppose bringing in such a charge. We may also add that this is subject to neglecting the amount of energy that was spent to assemble such a charge in the first place. The situation does not continue like this, however, because on bringing a second charge Q_2 in the space where Q_1 is contained, there will be an opposition force because of the electric field of Q_1. The amount of work W_2 required to bring Q_2 a distance d from Q_1 is given by

$$W_2 = Q_2 \Phi_2^1$$

where Φ_2^1 is the electric potential at the location of Q_2 owing to Q_1. The work done W_2 is equal to Q_2 multiplied by ϕ_2^1, because the electric potential is defined as the work done on a unit-positive charge.

As we continue to assemble the system of charges, more work will be required. For example, bringing a third charge Q_3 will now be opposed by the fields resulting from both charges Q_1 and Q_2. Hence, the required work W_3 will be given by

$$W_3 = Q_3 \Phi_3^1 + Q_3 \Phi_3^2$$

The total work that needs to be done to assemble n point charge is

$$W = Q_2 \Phi_2^1 + Q_3(\Phi_3^1 + \Phi_3^2) + Q_4(\Phi_4^1 + \Phi_4^2 + \Phi_4^3)$$
$$+ \cdots \cdot Q_n(\Phi_n^1 + \Phi_n^2 + \cdots \cdot + \Phi_n^{n-1}) \tag{4.25}$$

To help us put the preceding expression in a more compact form, we specify expressions for the potentials Φ_2^1, Φ_3^1, Φ_3^2, and so on. From equation 4.18, the electrostatic potential resulting from a point charge,

$$\Phi_2^1 = \frac{Q_1}{4\pi\epsilon_o R_{12}}$$

where R_{12} is the distance between the charges Q_1 and Q_2. Therefore, the work $W_2 = Q_2 \Phi_2^1$ may be written as

$$W_2 = Q_2 \Phi_2^1 = Q_2 \frac{Q_1}{4\pi\epsilon_o R_{12}} = Q_1 \frac{Q_2}{4\pi\epsilon_o R_{12}} = Q_1 \Phi_1^2$$

where Φ_1^2 represents the potential at the location of Q_1 owing to charge Q_2. Carrying out similar substitutions, we may rewrite equation 4.25 in the form

$$W = Q_1 \Phi_1^2 + Q_1 \Phi_1^3 + Q_1 \Phi_1^4 + \cdots + Q_1 \Phi_1^n$$
$$+ Q_2(\Phi_2^3 + \Phi_2^4 \cdots + \Phi_2^n)$$
$$+ Q_3(\Phi_3^4 + \cdots \Phi_3^n)$$
$$+ \cdots \cdots \tag{4.26}$$

Adding equation 4.25 to equation 4.26, we obtain

$$2W = Q_1(\Phi_1^2 + \Phi_1^3 + \Phi_1^4 + \cdots \Phi_1^n)$$
$$+ Q_2(\Phi_2^1 + \Phi_2^3 + \cdots + \Phi_2^n)$$
$$+ \cdots \cdots$$
$$+ Q_n(\Phi_n^1 + \Phi_n^2 + \cdots + \Phi_n^{n-1}) \tag{4.27}$$

Each term between parentheses represents the total electric potential at the location of the charge in front of it. For example,

$$(\Phi_2^1 + \Phi_2^3 + \Phi_2^4 + \cdots \Phi_2^n)$$

is equal to the total potential at Q_2. We may hence rewrite equation 4.27 in the form

$$2W = Q_1 \Phi_1 + Q_2 \Phi_2 + \cdots + Q_i \Phi_i + \cdots + Q_n \Phi_n \tag{4.28}$$

where Φ_i is the total electric potential at the ith charge owing to all other charges ($i = 1$ to n, except the ith charge itself). The work done in assembling the system of n charges is therefore

$$W = \frac{1}{2} \sum_{i=1}^{n} Q_i \Phi_i \tag{4.29}$$

If, instead of n discrete charges, we have a continuous charge distribution, W will be given by

$$W = \frac{1}{2} \int_v \rho_v \Phi \, dv \tag{4.30}$$

W is the work done and also represents the potential energy (i.e., the energy stored) in a system of charge distribution ρ_v in a volume v. Clearly, if instead of volume charge distribution ρ_v, we have surface or line charge distribution, equation 4.30 reduces to

$$W = \frac{1}{2} \int_c \rho_\ell \Phi \, d\ell$$

and

$$W = \frac{1}{2} \int_s \rho_s \Phi \, ds$$

It is often desirable to obtain an expression equivalent to equation 4.30, but instead expressed in terms of the $\mathbf{E}$ and $\mathbf{D}$ fields. To do so, we first replace ρ_v by $\nabla \cdot \mathbf{D}$ (according to Gauss's law) and then use the following vector identity,

$$\nabla \cdot (\Phi \mathbf{D}) = \Phi(\nabla \cdot \mathbf{D}) + \mathbf{D} \cdot (\nabla \Phi) \tag{4.31}$$

Making these substitutions, equation 4.30 becomes

$$W = \frac{1}{2} \int_v (\nabla \cdot \mathbf{D}) \Phi \, dv = \frac{1}{2} \int_v [\nabla \cdot (\Phi \mathbf{D}) - \mathbf{D} \cdot \nabla \Phi] \, dv \tag{4.32}$$

Using the divergence theorem, the first term in the right-hand side of equation 4.32 reduces to

$$\int_v \nabla \cdot (\Phi \mathbf{D}) \, dv = \oint_s \Phi \mathbf{D} \cdot ds \tag{4.33}$$

According to equation 4.33, the volume v contains all the charges in space. Hence, the surface s should enclose all these charges and might as well be taken at infinity. Because of the manner in which Φ varies as $1/r$ and $\mathbf{D}$ as $1/r^2$, and the element of area ds varies as r^2, it turns out that

$$\oint_s \Phi \mathbf{D} \cdot ds = 0$$

over a spherical surface at infinity. Equation 4.32 then reduces to

$$W = -\frac{1}{2} \int_v \mathbf{D} \cdot \nabla \Phi \, dv = \frac{1}{2} \int_v \mathbf{D} \cdot \mathbf{E} \, dv \tag{4.34}$$

where the relation $\mathbf{E} = -\nabla \Phi$ was used in equation 4.34. For linear dielectrics, $\mathbf{D} = \epsilon \mathbf{E}$, where ϵ is scalar, and the electrostatic energy density becomes $\frac{1}{2} \epsilon E^2$. The total electrostatic energy is given by

$$W = \frac{1}{2} \int_v \epsilon E^2 \, dv \tag{4.35}$$

EXAMPLE 4.10

In example 4.6, we calculated the electric field between the conductors of a spherical capacitor to be

$$\mathbf{E} = \frac{Q}{4\pi\epsilon r^2}\, \mathbf{a}_r$$

Determine the electrostatic energy stored.

Solution

$$W = \frac{1}{2}\int_v \mathbf{D}\cdot\mathbf{E}\,dv = \frac{1}{2}\int_{\phi=0}^{2\pi}\int_{\theta=0}^{\pi}\int_{r=a}^{b} \frac{Q^2}{(4\pi)^2\epsilon r^4}r^2\sin\theta\,dr\,d\theta\,d\phi$$

$$= \frac{Q^2}{32\pi^2\,\epsilon}\,2\pi\,[-\cos\theta]_0^{\pi}\left[-\frac{1}{r}\right]_a^b$$

$$= \frac{Q^2}{8\pi\epsilon}\left[\frac{1}{a}-\frac{1}{b}\right]$$

The same result may be obtained using equation 4.30. First, let us calculate the potential difference Φ,

$$\Phi = -\int_b^a \mathbf{E}\cdot d\ell = -\int_b^a \frac{Q}{4\pi\epsilon r^2}\,\mathbf{a}_r\cdot dr\,\mathbf{a}_r$$

$$= \frac{Q}{4\pi\epsilon}\left[\frac{1}{a}-\frac{1}{b}\right]$$

The electrostatic energy stored is then

$$W = \frac{1}{2}\int_v \Phi\rho_v\,dv = \frac{1}{2}\int_{\phi=0}^{2\pi}\int_{\theta=0}^{\pi}\int_{r=a}^{b} \frac{Q}{4\pi\epsilon}\left[\frac{1}{a}-\frac{1}{b}\right]\rho_v\,dv \qquad (4.36)$$

The charge density is zero in the region between the conductors, and the electric potential is zero at the outer conductor. The integration in equation 4.36 should therefore be carried out over the surface of the inner conductor, and the result should be equal to the total charge on this conductor multiplied by the potential. Hence,

$$W = \frac{Q^2}{8\pi\epsilon}\left[\frac{1}{a}-\frac{1}{b}\right]$$

which is the same answer obtained using equation 4.34.

EXAMPLE 4.11

In example 4.8, we considered a coaxial cable with two dielectric layers between the inner and outer conductors. Determine the electrostatic energy stored per unit length of this coaxial cable.

Solution

From example 4.8, the electric fields in the two dielectric regions $a < \rho < b$ and $b < \rho < c$ are given by

$$\mathbf{E}_1 = \frac{\rho_s\, a}{\epsilon_1\, \rho}\, \mathbf{a}_\rho \qquad a < \rho < b$$

$$\mathbf{E}_2 = \frac{\rho_s\, a}{\epsilon_2\, \rho}\, \mathbf{a}_\rho \qquad b < \rho < c$$

The electrostatic energies per unit length in both regions are given by

$$W_1 = \frac{1}{2} \int_{\rho\, =\, a}^{b} \int_{\phi\, =\, 0}^{2\pi} \int_{z\, =\, 0}^{1} \epsilon_1 \left(\frac{\rho_s\, a}{\epsilon_1\, \rho} \right)^2 \rho\, d\rho\, d\phi\, dz \qquad a < \rho < b$$

$$= \frac{1}{2} \frac{\rho_s^2\, a^2}{\epsilon_1}\, (2\pi)\, \ell n\, \frac{b}{a}$$

$$= \frac{\pi a^2\, \rho_s^2}{\epsilon_1}\, \ell n\, \frac{b}{a}$$

Similarly,

$$W_2 = \frac{\pi a^2\, \rho_s^2}{\epsilon_2}\, \ell n\, \frac{c}{b}$$

Total energy stored,

$$W = \pi a^2\, \rho_s^2 \left[\frac{1}{\epsilon_1}\, \ell n\, \frac{b}{a} + \frac{1}{\epsilon_2}\, \ell n\, \frac{c}{b} \right]$$

Once again, the same result could have been obtained from equation 4.30. The total potential difference is

$$\Phi = -\int_c^b \frac{\rho_s\, a}{\epsilon_2\, \rho}\, \mathbf{a}_\rho \cdot d\rho\, \mathbf{a}_\rho - \int_b^a \frac{\rho_s\, a}{\epsilon_1\, \rho}\, \mathbf{a}_\rho \cdot d\rho\, \mathbf{a}_\rho$$

$$= \frac{\rho_s\, a}{\epsilon_2}\, \ell n\, \frac{c}{b} + \frac{\rho_s\, a}{\epsilon_1}\, \ell n\, \frac{b}{a}$$

and

$$W = \frac{1}{2} \int_v \Phi \rho_v\, dv$$

Recalling that ρ_v, the free charge density is zero in the region between the two conductors and that $\Phi = 0$ on the outer conductor, hence,

$$W = \frac{1}{2} \int_{\phi\, =\, 0}^{2\pi} \int_{z\, =\, 0}^{1} \rho_s\, a \left[\frac{1}{\epsilon_2}\, \ell n\, \frac{c}{b} + \frac{1}{\epsilon_1}\, \ell n\, \frac{b}{a} \right] \rho_s\, a\, d\phi\, dz$$

$$= \pi \rho_s^2\, a^2 \left[\frac{1}{\epsilon_2}\, \ell n\, \frac{c}{b} + \frac{1}{\epsilon_1}\, \ell n\, \frac{b}{a} \right]$$

which are the same results obtained in terms of the electric field.

4.7 LAPLACE'S AND POISSON'S EQUATIONS

One of the frustrating experiences in teaching an introductory electromagnetic course is the inability to use the fundamental knowledge and the developed skills to solve practical engineering problems. For example, throughout our study thus far, we were able to solve electrostatic field problems under strict symmetry considerations. We calculated capacitances for parallel plate, coaxial, and spherical capacitors, all of which possess symmetries that highly simplify the analysis. In many engineering problems, it is not possible to use such symmetry considerations, and it is very valuable and highly motivating to see that our developed knowledge is suitable for solving practical engineering problems. In this section we focus on development of analytical techniques for solving Laplace's and Poisson's equations. Unlike solutions of electrostatic problems that start from knowledge of the charge distribution, solutions of Laplace's and Poisson's equations do not require prior knowledge of the distribution of electrostatic charges. Instead, the value of the electrostatic potential is required at conducting boundaries. In a sense, the solution procedure based on Laplace's and Poisson's equations is complementary to those based on knowledge of the charge distribution, as described in the previous sections. Furthermore, Laplace's and Poisson's equations lend themselves to numerical solutions that make their application general and suitable to many engineering problems.

To begin with, let us develop Poisson's and Laplace's equations. Substituting equation 4.20 in Gauss's law $\nabla \cdot \epsilon \mathbf{E} = \rho_v$, we obtain

$$\nabla \cdot \epsilon(-\nabla \Phi) = \rho_v$$

For a homogeneous medium, ϵ is constant and, hence, may be taken out of the divergence operation. Hence,

$$\nabla^2 \Phi = -\frac{\rho_v}{\epsilon} \tag{4.37}$$

Equation 4.37 is known as Poisson's equation. ∇^2, the Laplacian operator, is given in the Cartesian, cylindrical, and spherical coordinates by

$$\nabla^2 = \frac{\partial^2}{\partial x^2} + \frac{\partial^2}{\partial y^2} + \frac{\partial^2}{\partial z^2} \qquad \text{(Cartesian)}$$

$$\nabla^2 = \frac{1}{\rho} \frac{\partial}{\partial \rho}\left(\rho \frac{\partial}{\partial \rho}\right) + \frac{1}{\rho^2} \frac{\partial^2}{\partial \phi^2} + \frac{\partial^2}{\partial z^2} \qquad \text{(Cylindrical)}$$

$$\nabla^2 = \frac{1}{r^2} \frac{\partial}{\partial r}\left(r^2 \frac{\partial}{\partial r}\right) + \frac{1}{r^2 \sin \theta} \frac{\partial}{\partial \theta}\left(\sin \theta \frac{\partial}{\partial \theta}\right) + \frac{1}{r^2 \sin^2 \theta} \frac{\partial^2}{\partial \phi^2} \qquad \text{(Spherical)}$$

At points in space where there is no charge distribution $\rho_v = 0$, equation 4.37 reduces to

$$\nabla^2 \Phi = 0 \tag{4.38}$$

which is known as Laplace's equation. In the following, we will present some examples illustrating the solution of Laplace's equation in the various coordinate systems.

EXAMPLE 4.12

In the spherical capacitor shown in Figure 4.14, the inner conductor is maintained at a potential V, whereas the outer conductor is grounded ($\Phi = 0$). Use Laplace's equation (equation 4.38) and the given boundary conditions to solve for the potential and electric field in the space between the conductors.

Solution

Because of the spherical symmetry, the electric potential is independent of ϕ and θ, and the Laplacian in spherical coordinates simplifies to

$$\nabla^2 \Phi = \frac{1}{r^2} \frac{d}{dr} \left(r^2 \frac{d\Phi}{dr} \right) = 0$$

The general solution of this equation is of the form

$$\Phi = \frac{A}{r} + B$$

where A and B are two arbitrary constants to be determined from the boundary conditions. At the surface of the inner conductor, $r = a$, $\Phi = V$.

$$V = \frac{A}{a} + B$$

The second boundary condition requires that $\Phi = 0$ at $r = b$, hence,

$$0 = \frac{A}{b} + B$$

Solving for A and B, we obtain

$$V = A \left(\frac{1}{a} - \frac{1}{b} \right)$$

or

$$A = \frac{V}{\left(\frac{1}{a} - \frac{1}{b} \right)}$$

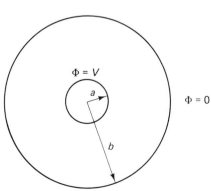

Figure 4.14   Geometry of spherical capacitor.

and the constant B is given by

$$B = -\frac{A}{b} = -\frac{V}{\left(\dfrac{b}{a} - 1\right)}$$

The potential Φ is hence,

$$\Phi = \frac{V}{\left(\dfrac{1}{a} - \dfrac{1}{b}\right)}\left[\frac{1}{r} - \frac{1}{b}\right]$$

The electric field $\mathbf{E}$ may then be obtained from $\mathbf{E} = -\boldsymbol{\nabla}\Phi$.

$$\mathbf{E} = -\left[\frac{\partial\Phi}{\partial r}\mathbf{a}_r + \frac{1}{r}\frac{\partial\Phi}{\partial\theta}\mathbf{a}_\theta + \frac{1}{r\sin\theta}\frac{\partial\Phi}{\partial\phi}\mathbf{a}_\phi\right]$$

Because Φ varies only with r, $\mathbf{E}$ is given by

$$\mathbf{E} = -\frac{\partial\Phi}{\partial r}\mathbf{a}_r = \frac{V}{\left(\dfrac{1}{a} - \dfrac{1}{b}\right)}\frac{1}{r^2}\mathbf{a}_r$$

which has the $1/r^2$ variation expected from previous solutions based on knowledge of the charge distribution. The total charge on the inner conductor may be calculated from Gauss's law and knowledge of $\mathbf{E}$ in the space between the conductors.

$$\oint_s \epsilon\mathbf{E}\cdot d\mathbf{s} = Q$$

Integrating over the surface of the inner sphere, we obtain

$$Q = \frac{\epsilon 4\pi V}{\left(\dfrac{1}{a} - \dfrac{1}{b}\right)}$$

The capacitance is then

$$C = \frac{Q}{V} = \frac{4\pi\epsilon}{\left(\dfrac{1}{a} - \dfrac{1}{b}\right)}$$

which is the same result obtained in example 4.6.

———◆◆◆———

EXAMPLE 4.13

Consider the parallel plate capacitor with two dielectric layers discussed in example 4.7. Solve for the potential, the electric field, and the capacitance, using Laplace's equation and the boundary condition,

$$\Phi = V \quad \text{at} \quad z = 0 \quad \text{and} \quad \Phi = 0 \quad \text{at} \quad z = d = d_1 + d_2$$

Solution

Because it is assumed that the parallel plates are of infinite extent, the variation of the potential Φ with x and y should be zero. Laplace's equation then reduces to

$$\frac{d^2 \Phi}{dz^2} = 0$$

Solution for Φ is then given by

$$\Phi_1(z) = A z + B \qquad \text{in region 1 with } \epsilon = \epsilon_1$$

and

$$\Phi_2(z) = D z + G \qquad \text{in region 2 with } \epsilon = \epsilon_2$$

The four unknown coefficients A, B, D, and G are to be determined from the boundary conditions.

$$\Phi_1(z = 0) = V, \qquad \Phi_2(z = d) = 0$$
$$\Phi_1(z = d_1) = \Phi_2(z = d_1)$$

and

$$D_{1z}(z = d_1) = D_{2z}(z = d_1)$$

The last boundary condition enforces the continuity of the normal component of the electric flux density $\mathbf{D}$ at the interface between two perfect dielectric media (i.e., no free-surface charge).

Substituting these boundary conditions in $\Phi_1(z)$ and $\Phi_2(z)$, we obtain the following expression for the unknown constants

$$A = \frac{-V}{d_1 + d_2 \dfrac{\epsilon_1}{\epsilon_2}}$$

$$B = V$$

$$D = A \frac{\epsilon_1}{\epsilon_2}$$

$$G = -(d_1 + d_2)D$$

The potentials $\Phi_1(z)$ and $\Phi_2(z)$ are then given by

$$\Phi_1(z) = \frac{-Vz}{d_1 + \dfrac{\epsilon_1}{\epsilon_2}d_2} + V$$

and

$$\Phi_2(z) = \frac{-\dfrac{\epsilon_1}{\epsilon_2}V[z - (d_1 + d_2)]}{d_1 + d_2 \dfrac{\epsilon_1}{\epsilon_2}}$$

The electric fields in both regions are given by

$$\mathbf{E}_1 = -\nabla\Phi_1(z) = \frac{V}{d_1 + \dfrac{\epsilon_1}{\epsilon_2}d_2}\,\mathbf{a}_z$$

$$\mathbf{E}_2 = -\nabla\Phi_2(z) = \frac{\dfrac{\epsilon_1}{\epsilon_2}V}{d_1 + d_2\dfrac{\epsilon_1}{\epsilon_2}}\,\mathbf{a}_z$$

The positive charge density on the bottom plate is obtained from the boundary condition as

$$\epsilon_1 E_{1z} = \rho_s$$

and the capacitance per unit area

$$C = \frac{\rho_s}{V} = \frac{\epsilon_1}{d_1 + \dfrac{\epsilon_1}{\epsilon_2}d_2}$$

which is the same result obtained from example 4.7.

———————◆◆———————

4.8 NUMERICAL SOLUTION OF POISSON'S AND LAPLACE'S EQUATIONS—FINITE DIFFERENCE METHOD

The preceding examples illustrated the solution of Laplace's or Poisson's equations in electrostatic problems of simple geometries. In all cases, the symmetry played an important role in simplifying the solution. There are other solution techniques for boundary value problems formulated in terms of Laplace's or Poisson's equations. These include the method of separation of variables, which most students are probably familiar with through their elementary mathematics and physics classes. Instead of spending more time discussing this and other related simple solution techniques, we will now focus our attention on developing a powerful and truly general solution technique known as the finite difference method. With the proliferation of computers on university campuses, and the need of preparing working engineers to be able to solve problems of more and more complex geometries, our time will be better invested in developing these powerful computational methods than on further discussing analytical solutions of limited applications. In this section, the finite difference solution procedure will be discussed and it will be further shown that many quantities of engineering interest, such as the capacitance, may be subsequently obtained from the solution of the potential Φ.

4.8.1 Finite Difference Representation of Laplace's Equation

This derivation is available in other texts* and will be included here for completeness. Let us consider the potential function $\Phi(x)$, which varies with only one independent

*Peter Silvester, *Modern Electromagnetic Fields* (Englewood Cliffs, N.J.: Prentice Hall, 1968).

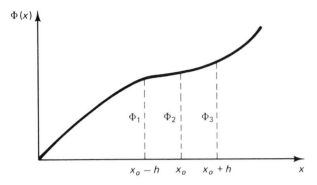

Figure 4.15 Geometry used in derivation of the difference equations.

variable x, as shown in Figure 4.15. The first-order derivative of Φ, that is, $d\Phi/dx$, may be expressed in terms of its discrete values Φ_1, Φ_2, Φ_3 at the neighboring points $x_o - h$, x_o, and $x_o + h$, respectively. x_o is the central point and $\pm h$ is the difference between the chosen values of x. According to Figure 4.15, the first-order derivative may be given by

$$\left.\frac{d\Phi}{dx}\right|_{x_o} \approx \frac{\Phi_3 - \Phi_2}{h} \qquad \text{(Forward difference method)}$$

$$\left.\frac{d\Phi}{dx}\right|_{x_o} \approx \frac{\Phi_2 - \Phi_1}{h} \qquad \text{(Backward difference method)}$$

$$\left.\frac{d\Phi}{dx}\right|_{x_o} \approx \frac{\Phi_3 - \Phi_1}{2h} \qquad \text{(Central difference method)}$$

The central difference method may be interpreted as the average between the forward and backward difference equations. In the preceding difference equations, the approximate sign $\approx$ was used instead of the equals sign to emphasize the approximate nature of these representations. To estimate the errors involved in these representations, we use Taylor series expansion of the potential Φ, hence,

$$\Phi(x_o + h) = \Phi(x_o) + h\left.\frac{d\Phi}{dx}\right|_{x_o} + \frac{h^2}{2}\left.\frac{d^2\Phi}{dx^2}\right|_{x_o} + \frac{h^3}{3 \times 2}\frac{d^3\Phi}{dx^3} + \text{higher-order terms} \quad (4.39)$$

Therefore, neglecting the third order derivative and the higher-order terms we obtain

$$\left.\frac{d\Phi}{dx}\right|_{x_o} = \frac{\Phi(x_o + h) - \Phi(x_o)}{h} - \frac{h}{2}\left.\frac{d^2\Phi}{dx^2}\right|_{x_o} \qquad (4.40)$$

It is clear that equation 4.40 agrees with the forward difference method with the exception of an error term,

$$\frac{h}{2}\left.\frac{d^2\Phi}{dx^2}\right|_{x_o}$$

plus higher-order terms. The leading error term is of the order of h, which means that the smaller the value of h, the better the forward difference representation would be.

Similarly,

$$\Phi(x_o - h) = \Phi(x_o) - h\left.\frac{d\Phi}{dx}\right|_{x_o} + \frac{h^2}{2}\left.\frac{d^2\Phi}{dx^2}\right|_{x_o} - \frac{h^3}{6}\left.\frac{d^3\Phi}{dx^3}\right|_{x_o} + \cdots \tag{4.41}$$

and

$$\left.\frac{d\Phi}{dx}\right|_{x_o} = \frac{\Phi(x_o) - \Phi(x_o - h)}{h} + \frac{h}{2}\left.\frac{d^2\Phi}{dx^2}\right|_{x_o} \tag{4.42}$$

Once again, equation 4.42 agrees with the backward difference method, with the exception of a leading error term that is of the order of h. We observe that the error in both the forward and backward finite difference representations are of the order of h.

Subtracting equation 4.41 from equation 4.39, we obtain

$$\Phi\left(x_o + h\right) - \Phi\left(x_o - h\right) \approx 2h\left.\frac{d\Phi}{dx}\right|_{x_o} + \frac{h^3}{3}\left.\frac{d^3\Phi}{dx^3}\right|_{x_o}$$

Hence,

$$\left.\frac{d\Phi}{dx}\right|_{x_o} = \frac{\Phi\left(x_o - h\right) - \Phi\left(x_o + h\right)}{2h} - \frac{h^2}{6}\left.\frac{d^3\Phi}{dx^3}\right|_{x_o} \tag{4.43}$$

Equation 4.43 is similar to the central difference equation, with the exception of the leading error term,

$$-\frac{h^2}{6}\left.\frac{d^3\Phi}{dx^3}\right|_{x_o}$$

which is of the order of h^2. For small values of h, which are often used in engineering problems (i.e., 0.1, 0.01, 0.2, etc.), the error in using the central difference equation is smaller than those of the forward or backward difference equations.

It is, therefore, more advantageous to use the central difference method. From Figure 4.15 and using the central difference method, an expression for the second-order derivative $d^2\Phi/dx^2$ may be obtained as

$$\left.\frac{d^2\Phi}{dx^2}\right|_{x_o} \approx \frac{\left.\frac{d\Phi}{dx}\right|_{x_o + h/2} - \left.\frac{d\Phi}{dx}\right|_{x_o - h/2}}{h}$$

$$\approx \frac{\dfrac{\Phi_3 - \Phi_2}{h} - \dfrac{\Phi_2 - \Phi_1}{h}}{h}$$

$$\approx \frac{\Phi_3 + \Phi_1 - 2\Phi_2}{h^2} \tag{4.44}$$

If the potential Φ is a function of two independent variables x and y, the second-order derivatives,

$$\frac{\partial^2\Phi}{\partial x^2} \qquad \text{and} \qquad \frac{\partial^2\Phi}{\partial y^2}$$

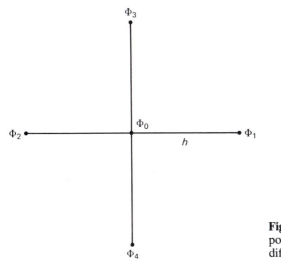

Figure 4.16 Geometry of the five-point star used in two-dimensional difference equations.

may be obtained using Figure 4.16 as

$$\frac{\partial^2 \Phi}{\partial x^2} = \frac{\Phi_1 + \Phi_2 - 2\Phi_o}{h^2}$$

$$\frac{\partial^2 \Phi}{\partial y^2} = \frac{\Phi_3 + \Phi_4 - 2\Phi_o}{h^2}$$

Laplace's equation in two dimensions may then be expressed as

$$\nabla^2 \Phi = \frac{\partial^2 \Phi}{\partial x^2} + \frac{\partial^2 \Phi}{\partial y^2} = \frac{\Phi_1 + \Phi_2 + \Phi_3 + \Phi_4 - 4\Phi_o}{h^2} \qquad (4.45)$$

Equation 4.45 is known as the five-point equal arm difference equation.

The finite difference solution procedure of Poisson's or Laplace's equations may then be summarized as follows:

1. Divide the domain of interest (in which the potential is to be determined) into suitably fine grid. Instead of a solution for $\Phi(x, y)$, which provides its continuous variation for a given charge distribution $\rho(x, y)$, the finite difference solution will provide discrete values of Φ at the "nodes" of the established grid.
2. Apply the difference equation 4.45 at each node of the grid to obtain, for example, N equations in the N unknown node potentials.
3. Solve the resulting system of equations, either iteratively or using one of the direct methods.

4.8.2 Difference Equation at Interface between Two Dielectric Media

In many engineering applications, interfaces between two different dielectric media are encountered. For this, we will derive a special case difference equation that should be satisfied at nodes on the interface between two dielectrics. Figure 4.17 illustrates the

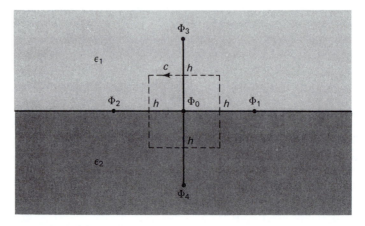

Figure 4.17 Geometry of grid nodes at the interface between medium 1 of ϵ_1 and medium 2 of ϵ_2.

geometry of an interface, and the difference equation in this case may be obtained from Gauss's law for the electric field,

$$\oint_s \epsilon \mathbf{E} \cdot d\mathbf{s} = q = 0 \tag{4.46}$$

$q = 0$ in equation 4.46 because there is no "free" charge enclosed by the surface s. Substituting $\mathbf{E} = -\nabla\Phi$, we obtain

$$\oint_s \epsilon \nabla\Phi \cdot d\mathbf{s} = \oint_c \epsilon \nabla\Phi \cdot d\mathbf{c} = \oint_c \epsilon \frac{\partial \Phi}{\partial n}\, dc \tag{4.47}$$

The surface integration on the left-hand side of equation 4.47 was replaced by a contour integration, because in Figure 4.17 we are dealing with a two-dimensional case, and the solution of Φ is independent of the axial independent variable z. $\partial\Phi/\partial n$ denotes the normal derivative of Φ on the contour c.

Carrying out detailed integration of $\partial\Phi/\partial n$ along c, we obtain

$$\oint_c \frac{\partial \phi}{\partial n}(\epsilon\, dc) = \frac{\Phi_1 - \Phi_o}{h}\left(\epsilon_2 \frac{h}{2} + \epsilon_1 \frac{h}{2}\right) + \frac{\Phi_3 - \Phi_o}{h}(\epsilon_1 h)$$

$$+ \frac{(\Phi_2 - \Phi_o)}{h}\left(\epsilon_1 \frac{h}{2} + \epsilon_2 \frac{h}{2}\right) + \frac{\Phi_4 - \Phi_o}{h}(\epsilon_2 h) \tag{4.48}$$

Rearranging the terms in equation 4.48, we obtain

$$2\epsilon_1 \Phi_3 + 2\epsilon_2 \Phi_4 + (\epsilon_1 + \epsilon_2)\Phi_1 + (\epsilon_1 + \epsilon_2)\Phi_2 - 4(\epsilon_1 + \epsilon_2)\Phi_o = 0 \tag{4.49}$$

The difference equation (equation 4.49) may be used at interfaces between two dielectrics. Its use in engineering problems will be illustrated by the following examples.

EXAMPLE 4.14

Consider the rectangular region shown in Figure 4.18a. The electric potential is specified on the conducting boundaries. Use the finite difference representation to solve for the potential distribution within this region.

Solution

The electric potential everywhere in the rectangular region should satisfy Laplace's equation. In this case of simple geometry, the analytical solution (e.g., using separation of variables) of Laplace's equation is possible. We will, however, use this simple example to develop a numerical solution procedure that may be used to solve much more complicated geometries. Using a numerical solution means we will define Φ in the rectangular region of interest by calculating its values at discrete points, the nodes of a mesh. The step-by-step solution procedure includes the following:

1. Layout a coarse square mesh and identify the nodes at which the electric potential is to be calculated. The geometry of a 2×4 mesh is shown in Figure 4.18b. The value of h (mesh size) in this case is $h = 5$ cm.

2. Replace Laplace's equation by its finite difference representation.

$$\frac{1}{h^2}(\Phi_{i+1,j} + \Phi_{i-1,j} + \Phi_{i,j+1} + \Phi_{i,j-1} - 4\Phi_{i,j}) = 0$$

$\Phi_{i,j}$ are the discrete values of the potential at points (nodes) within the domain of interest.

3. Apply the difference equation in step 2 at each node. At node 1,

$$\frac{1}{(0.05)^2}(0 + 0 + 0 + \Phi_2 - 4\Phi_1) = 0$$

or

$$4\Phi_1 = \Phi_2 \qquad (4.50)$$

At node 2,

$$\frac{1}{(0.05)^2}(\Phi_1 + \Phi_3 + 0 + 0 - 4\Phi_2) = 0$$

or

$$\Phi_1 - 4\Phi_2 + \Phi_3 = 0 \qquad (4.51)$$

At node 3,

$$\frac{1}{(0.05)^2}(\Phi_2 + 100 - 4\Phi_3) = 0$$

or

$$\Phi_2 - 4\Phi_3 = -100 \qquad (4.52)$$

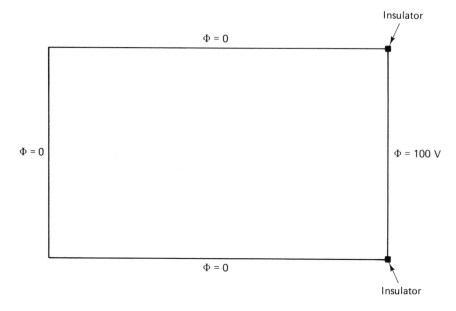

(a)

Figure 4.18a   Rectangular geometry and boundary condition for the electric potential problem of example 4.14.

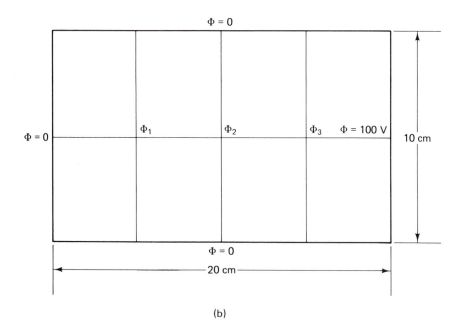

(b)

Figure 4.18b   Geometry of 2×4 finite difference mesh.

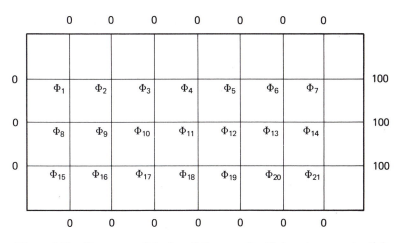

Figure 4.19   Geometry of the $h = 2.5$-cm mesh with twenty-one potential nodes.

4. Equations 4.50 to 4.52 are three equations in the three unknowns, Φ_1, Φ_2, and Φ_3. These three equations may be solved using one of the methods described in calculus courses. The results are

$$\Phi_1 = 1.79, \qquad \Phi_2 = 7.14, \qquad \Phi_3 = 26.79$$

5. With the coarse mesh we used, we do not expect to get accurate final results. Redoing the problem with a smaller value of h should improve the accuracy of the solution. Figure 4.19 shows the mesh geometry for $h = 2.5$ cm, which is half the mesh size used in the previous calculations. In this case, however, we have twenty-one unknown values of the potential at the various nodes.

6. Once again, applying the difference equation of step 2 at the various nodes results in the following 21×21 matrix:

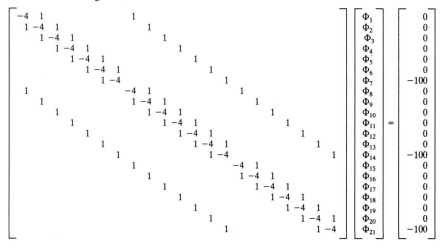

TABLE 4.1 COMPARISON BETWEEN FINITE DIFFERENCE AND ANALYTICAL
RESULTS

	Potential values	Percentage error	Potential values	Percentage error	Analytical solution
	(h = 5 cm)		(h = 2.5 cm)		
Φ_9	1.786	63	1.289	17.8	1.094
Φ_{11}	7.143	30	6.019	9.7	5.489
Φ_{13}	26.786	2.7	26.289	0.75	26.094

The solution for the electric potential at the various nodes is given by

$$\Phi_1 = 0.353 \qquad \Phi_8 = 0.499 \qquad \Phi_{15} = 0.353$$

$$\Phi_2 = 0.913 \qquad \Phi_9 = 1.289 \qquad \Phi_{16} = 0.913$$

$$\Phi_3 = 2.010 \qquad \Phi_{10} = 2.832 \qquad \Phi_{17} = 2.010$$

$$\Phi_4 = 4.296 \qquad \Phi_{11} = 6.019 \qquad \Phi_{18} = 4.296$$

$$\Phi_5 = 9.153 \qquad \Phi_{12} = 12.654 \qquad \Phi_{19} = 9.153$$

$$\Phi_6 = 19.663 \qquad \Phi_{13} = 26.289 \qquad \Phi_{20} = 19.663$$

$$\Phi_7 = 43.210 \qquad \Phi_{14} = 53.177 \qquad \Phi_{21} = 43.210$$

Table 4.1 compares the results for h = 5 cm, and h = 2.5 cm. From this comparison, it is clear that the h = 2.5 cm results agree better with the analytical solution available for this simple geometry. As expected, the accuracy of the finite difference results improves with the reduction in the mesh size h. Clearly, any further reduction in h results in a larger-size matrix; hence, a compromise should be made between the desired accuracy and the computational time and effort required.

EXAMPLE 4.15

In the 6 × 8 m² rectangular region shown in Figure 4.20a, the electric potential is zero on the boundaries. The charge distribution, however, is uniform and given by $\rho_v = 2\epsilon_o$. Solve Poisson's equation to determine the potential distribution in the rectangular region.

Solution

To determine the potential distribution in the rectangular region, we use Poisson's equation.

$$\nabla^2 \Phi = -\frac{\rho_v}{\epsilon_o} = -2$$

with zero potential $\Phi = 0$ on the boundaries. By establishing the rectangular grid shown in Figure 4.20b, we realize that we have six nodes and, hence, six unknown potentials for which to solve. Replacing $\nabla^2 \Phi$ by its finite difference representation, we obtain

$$\frac{1}{h^2}(\Phi_{i+1,j} + \Phi_{i-1,j} + \phi_{i,j+1} + \phi_{i,j-1} - 4\phi_{i,j}) + 2 = 0 \qquad (4.53)$$

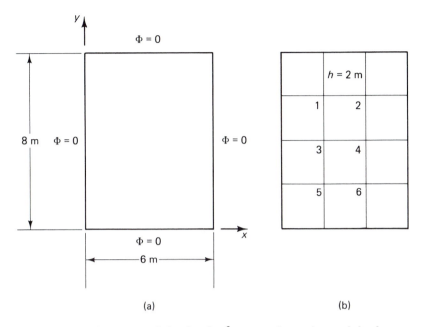

Figure 4.20   Geometry of the 6×8 m^2 rectangular region and the $h = 2$ m mesh.

It should be noted that although the mesh size was not explicitly used in solving Laplace's equation in the previous example, h is included as a part of the matrix formation in solving Poisson's equation. In SI system of units, h should be in meters. By applying the preceding difference equation at the various nodes in Figure 4.20b, we obtain the following matrix equation:

$$
\begin{bmatrix}
-4 & 1 & 1 & 0 & 0 & 0 \\
1 & -4 & 0 & 1 & 0 & 0 \\
1 & 0 & -4 & 1 & 1 & 0 \\
0 & 1 & 1 & -4 & 0 & 1 \\
0 & 0 & 1 & 0 & -4 & 1 \\
0 & 0 & 0 & 1 & 1 & -4
\end{bmatrix}
\begin{bmatrix}
\Phi_1 \\ \Phi_2 \\ \Phi_3 \\ \Phi_4 \\ \Phi_5 \\ \Phi_6
\end{bmatrix}
=
\begin{bmatrix}
-8 \\ -8 \\ -8 \\ -8 \\ -8 \\ -8
\end{bmatrix}
$$

Instead of solving the resulting six equations, we may note some symmetry considerations in Figure 4.20b. It is clear that

$$\Phi_1 = \Phi_2 = \Phi_5 = \Phi_6 \qquad \text{and that} \qquad \Phi_3 = \Phi_4$$

Taking these symmetry considerations into account, the number of equations reduces to two, and we obtain the following solution:

$$\Phi_1 = 4.56, \qquad \Phi_3 = 5.72$$

To improve the accuracy of the potential distribution, finer mesh such as the one shown in Figure 4.21 is required. Because of the large number of nodes in this case, symmetry should be used, and a solution for only one-quarter of the rectangular geometry is desired. The application of the difference equation at nodes 1, 2, 4, and 5 should proceed

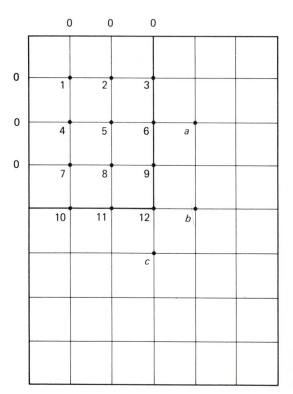

Figure 4.21 The finer mesh solution and symmetry consideration of example 4.15.

routinely, whereas special care should be exercised at the boundary nodes 3, 6, 9, 10, and 11, and also at the corner node 12.

For example, applying the difference equation at node 6 yields

$$\frac{1}{h^2}(\Phi_a + \Phi_3 + \Phi_9 + \Phi_5 - 4\Phi_6) + 2 = 0$$

or

$$\frac{1}{h^2}(\Phi_3 + \Phi_9 + 2\Phi_5 - 4\Phi_6) + 2 = 0 \qquad (4.54)$$

In equation 4.54, symmetry was used to complete the five-point star difference equation. Specifically the potential at node a to the right of 6 was taken equal to Φ_5. Similarly at the corner node 12, we obtain

$$\frac{1}{h^2}(\Phi_{11} + \Phi_9 + \Phi_b + \Phi_c - 4\Phi_{12}) + 2 = 0$$

Because of symmetry, $\Phi_{11} = \Phi_b$ and $\Phi_9 = \Phi_c$, hence,

$$\frac{1}{h^2}(2\Phi_{11} + 2\Phi_9 - 4\Phi_{12}) + 2 = 0$$

The matrix equation for the twelve nodes shown in Figure 4.21 is then

$$\begin{bmatrix} -4 & 1 & & 1 & & & & & & & & \\ 1 & -4 & 1 & & 1 & & & & & & & \\ & 2 & -4 & & & 1 & & & & & & \\ 1 & & & -4 & 1 & & 1 & & & & & \\ & 1 & & 1 & -4 & 1 & & 1 & & & & \\ & & 1 & & 2 & -4 & & & 1 & & & \\ & & & 1 & & & -4 & 1 & & 1 & & \\ & & & & 1 & & 1 & -4 & 1 & & 1 & \\ & & & & & 1 & & 2 & -4 & & & 1 \\ & & & & & & 2 & & & -4 & 1 & \\ & & & & & & & 2 & & 1 & -4 & 1 \\ & & & & & & & & 2 & & 2 & -4 \end{bmatrix} \begin{bmatrix} \Phi_1 \\ \Phi_2 \\ \Phi_3 \\ \Phi_4 \\ \Phi_5 \\ \Phi_6 \\ \Phi_7 \\ \Phi_8 \\ \Phi_9 \\ \Phi_{10} \\ \Phi_{11} \\ \Phi_{12} \end{bmatrix} = \begin{bmatrix} -2 \\ -2 \\ -2 \\ -2 \\ -2 \\ -2 \\ -2 \\ -2 \\ -2 \\ -2 \\ -2 \\ -2 \end{bmatrix} \qquad (4.55)$$

The "2" coefficient in the coefficient matrix (to the left) of equation (4.55) appears whenever symmetry consideration is used at boundary and corner nodes. It should be noted that the 12 × 12 coefficient matrix in equation 4.55 is the same for both Laplace's and Poisson's equations. The constant vector on the right-hand side of equation 4.55, however, depends on the charge distribution within and the potential at the boundaries of the region of interest. Furthermore, if instead of a uniform charge distribution we have a given charge distribution $\rho_v(x, y)$, the constants vector on the right-hand side of equation 4.55 should reflect the value of $\rho_v(x, y)$ calculated at each node. Solution of equation 4.55 gives

$$\Phi_1 = 2.04, \qquad \Phi_2 = 3.05, \qquad \Phi_3 = 3.35,$$
$$\Phi_4 = 3.12, \qquad \Phi_5 = 4.79, \qquad \Phi_6 = 5.32,$$
$$\Phi_7 = 3.66, \qquad \Phi_8 = 5.69, \qquad \Phi_9 = 6.34,$$
$$\Phi_{10} = 3.82, \qquad \Phi_{11} = 5.96, \qquad \Phi_{12} = 6.65.$$

4.8.3 Capacitance Calculation Using Finite Difference

The question that is often asked by students is how to relate the calculated values of the node potential to quantities of engineering interest, such as the capacitance of a system of conductors of complex geometry. Calculation of the capacitance from the obtained potential distribution may be done through Gauss's law, as follows:

$$\oint_s \epsilon \mathbf{E} \cdot d\mathbf{s} = q = -\oint_s \epsilon \nabla \Phi \cdot d\mathbf{s} \qquad (4.56)$$

In equation 4.56, the electric field $\mathbf{E}$ was substituted by the electric potential $\mathbf{E} = -\nabla \Phi$. In a two-dimensional problem such as the conductor arrangement shown in Figure 4.22, the potential is independent of the axial coordinate, and the closed surface s may be replaced by the closed contour c. Hence,

$$-\oint_s \epsilon \nabla \Phi \cdot d\mathbf{c} = -\oint_c \epsilon \frac{\partial \Phi}{\partial n} dc = q \qquad (4.57)$$

where q in this case is the charge per unit length in coulombs per meter. $\nabla \Phi \cdot d\mathbf{c}$ in equation 4.57 is replaced by the normal derivative $\partial \Phi / \partial n$ integrated over the closed

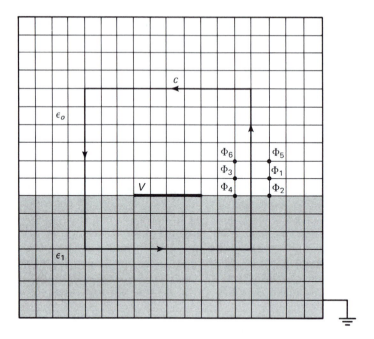

Figure 4.22 Calculation of capacitance in partially filled strip capacitor with dielectric of ϵ_1, using finite difference.

contour c. Evaluating equation 4.57 by using the discrete node values of Φ in Figure 4.22, we obtain

$$\epsilon_o\left(\frac{\Phi_1 - \Phi_3}{2h}\right)h + \epsilon_o\left(\frac{\Phi_5 - \Phi_6}{2h}\right)h + \epsilon_o\left(\frac{\Phi_2 - \Phi_4}{2h}\right)\frac{h}{2} + \epsilon_1\left(\frac{\Phi_2 - \Phi_4}{2h}\right)\frac{h}{2} + \cdots$$

(Contributions from other nodes along the closed contour c) $= -q$ (4.58)

and the capacitance per unit length is

$$C = \frac{q}{V}$$

where V is an initially assumed potential difference between the center conductor and the grounded outer one. This potential is used to solve for Φ, using the finite difference representation of Laplace's equation.

Exercise

For the strip capacitor shown in Figure 4.23, write a computer program to calculate the capacitance per unit length. Input to your program are the dimensions of the strip capacitor, the dielectric constant of the insulation between the two center conductors, and the mesh size h. You will assume that the two center conductors are at the same potential $V = 10$ V (split feed) and the outer conductor is grounded $V = 0$. In choosing

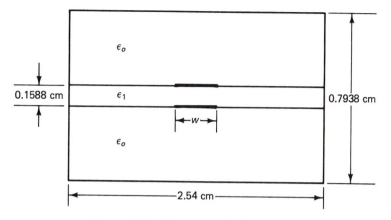

Figure 4.23 Geometry of strip capacitor.

the Gaussian contour to calculate the total charge, it is numerically preferred to take it along the finite difference mesh, midway between the center and outer conductors. After you write your program, you may compare your results with those given in Table 4.2, which illustrates the variation of C with the width of the center strip W and the dielectric constant ϵ_1.

TABLE 4.2 CAPACITANCE PER UNIT LENGTH pF/m FOR VARIOUS VALUES OF W AND ϵ_1

Strip width W (cm)	Teflon $\epsilon_r = 2.05$	Rexolite $\epsilon_r = 2.65$	Air
0.558	58.97	61.83	53.86
0.7145	67.62	70.51	62.34
0.8733	76.77	79.72	71.29

4.9 NUMERICAL SOLUTION OF ELECTROSTATIC PROBLEMS—METHOD OF MOMENTS

In section 4.3, we introduced the concept of electric potential and obtained integral equations that relate the potential to a given charge distribution. For example, if $\rho_s(\mathbf{r}')$ is a surface charge distribution, it is shown that $\Phi(\mathbf{r})$ is given by

$$\Phi(\mathbf{r}) = \frac{1}{4\pi\epsilon_o} \int_s \frac{\rho_s \, ds'}{|\mathbf{r} - \mathbf{r}'|} \tag{4.59}$$

where $|\mathbf{r} - \mathbf{r}'|$ is the distance from the charge distribution $\mathbf{r}'$ to the point $\mathbf{r}$ at which the potential Φ is to be evaluated. In all the examples we solved to illustrate the application of equation 4.59, we assumed the charge distribution in simple geometries and evaluated equation 4.59 to calculate Φ at specific locations. In many engineering problems, including the determination of capacitance of a system of conductors of complex geometry, the charge distribution is not known and instead the potentials of the conductors are known. We wish to solve for the charge distribution in equation 4.59 from given potentials on conductors. To emphasize the importance of developing this numerical solution procedure, we reexamine the problem of determining the capacitance of a parallel plate capacitor discussed in example 4.7. It is shown that under the assumption of having infinitely large parallel plates, that is, the dimensions of the plates are much larger than the separation distance, we found that the capacitance is given by

$$C = \frac{\epsilon_o A}{d}$$

where the total separation between the plates $d = d_1 + d_2$ and the medium is assumed air $\epsilon_o = \epsilon_1 = \epsilon_2$. As the separation d increases, the parallel plate approximation becomes less and less accurate because of the fringing field effects. Figure 4.24 shows the variation in the capacitance C as a function of the separation distance d. It is shown that as d reaches the side length w of the square parallel plate capacitor, C becomes more than three times larger than C_o, which is based on the infinite parallel plate assumption. This is certainly a significant difference, and more care should therefore be taken in calculating C. The problem is if such a simplifying assumption is not used, it is difficult to relate the potential to the charge distribution in a straightforward manner. Instead, numerical solutions should be used. We will illustrate the method of moments solution procedure by solving the following examples.

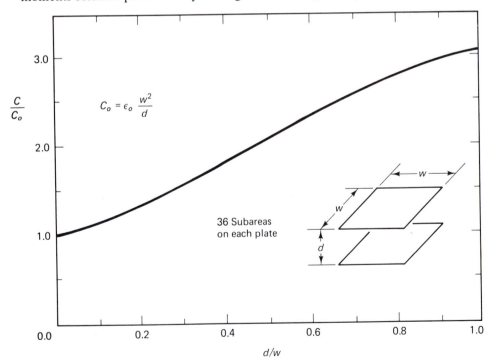

Figure 4.24 Variation of parallel plate capacitance with the increase of the separation distance d.

EXAMPLE 4.16

Consider a cylindrical conductor of radius a and length L. If the conductor is kept at a constant potential of 1 V, determine the charge distribution ρ_s on its surface.

Solution

The geometry of the cylindrical conductor is shown in Figure 4.25. The potential at any point $\mathbf{r}$ in space is related to the charge distribution $\rho_s(\mathbf{r}')$ by

$$\Phi(\mathbf{r}) = \frac{1}{4\pi\epsilon_o} \int_s \frac{\rho_s(\mathbf{r}')ds'}{|\mathbf{r} - \mathbf{r}'|} \qquad (4.60)$$

To help us solve this problem numerically, we divide the conductor into N small sections. These sections are assumed to be sufficiently small so that the charge distribution is constant in each. Equation 4.60 then reduces to

$$\Phi(\mathbf{r}) = \frac{1}{4\pi\epsilon_o} \sum_{i=1}^{N} \frac{\rho_{si}\, \Delta s_i}{|\mathbf{r} - \mathbf{r}_i'|} \qquad (4.61)$$

Figure 4.26 shows the sectioned conductor and total charge ($\rho_{si}\, \Delta s_i$) on each section. Equation 4.61 is an equivalent representation of the potential Φ at an observation point $\mathbf{r}$ owing to N point charges ($\rho_{si}\, \Delta s_i$) located at the center points ($\mathbf{r}_i'$) of the N sections. The surface area of each section is $\Delta s_i = 2\pi a \Delta\ell_i$, where $\Delta\ell_i = L/N$ is the length of each section. In equation 4.61, the N values of the charge density $\rho_{si}, i = 1, \ldots, N$ are unknown. In other words, equation 4.61 is one equation in the N unknown values of ρ_{si}. To determine these unknowns, we need N equations, which may be obtained by enforcing the validity of equation 4.61 at N points at which the potential Φ is known. Well, we know that $\Phi = 1$ V on the conductor. This means that $\Phi = 1$ V at the N centers of the sections. Therefore,

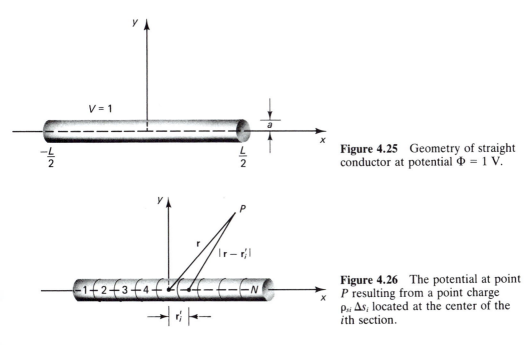

Figure 4.25 Geometry of straight conductor at potential $\Phi = 1$ V.

Figure 4.26 The potential at point P resulting from a point charge $\rho_{si}\, \Delta s_i$ located at the center of the ith section.

we may obtain the N desired equations by evaluating $\Phi = 1$ V at the N centers of the sections shown in Figure 4.26. Hence,

$$1 = \frac{1}{4\pi\epsilon_o}\left[\frac{\rho_{s1}\,\Delta s_1}{|x_1 - x_1|} + \frac{\rho_{s2}\,\Delta s_2}{|x_1 - x_2|} + \frac{\rho_{s3}\,\Delta s_3}{|x_1 - x_3|} + \cdots + \frac{\rho_{sN}\,\Delta s_N}{|x_1 - x_N|}\right]$$

$$1 = \frac{1}{4\pi\epsilon_o}\left[\frac{\rho_{s1}\,\Delta s_1}{|x_2 - x_1|} + \frac{\rho_{s2}\,\Delta s_2}{|x_2 - x_2|} + \frac{\nabla_{s3}\,\Delta s_3}{|x_2 - x_3|} + \cdots + \frac{\rho_{sN}\,\Delta s_N}{|x_2 - x_N|}\right]$$

$$\vdots$$

$$1 = \frac{1}{4\pi\epsilon_o}\left[\frac{\rho_{s1}\,\Delta s_1}{|x_N - x_1|} + \frac{\rho_{s2}\,\Delta s_2}{|x_N - x_2|} + \frac{\rho_{s3}\,\Delta s_3}{|x_N - x_3|} + \cdots + \frac{\rho_{sN}\,\Delta s_N}{|x_N - x_N|}\right] \quad (4.62)$$

In the set of N equations in equation 4.62, the first equation is obtained by enforcing the condition that $\Phi = 1$ V at the center x_1 (i.e., $\mathbf{r} = x_1$) of the first section. The second equation in equation 4.62 is obtained by enforcing $\Phi = 1$ V at the center x_2 of the second section. The charge $(\rho_{s1}\,\Delta s_1)$ is assumed to be located at the center x_1 of the first section, the charge $\rho_{s2}\,\Delta s_2$ is assumed located at the center x_2 of the second section, and so on. Equation 4.62 provides the desired N equations in the N unknown values of the charges $\rho_{s1}, \rho_{s2}, \ldots, \rho_{sN}$, assumed to be located at the center of the N sections.

Careful examination of equation 4.62, however, reveals that the self terms,

$$\frac{\rho_{s1}\,\Delta s_1}{|x_1 - x_1|}, \frac{\rho_{s2}\,\Delta s_2}{|x_2 - x_2|}, \frac{\rho_{s3}\,\Delta s_3}{|x_3 - x_3|}, \cdots \cdots \frac{\rho_{sN}\,\Delta s_N}{|x_N - x_N|}$$

are singular. These terms result from calculating the potential Φ at the center of each section because of its own charge—that is, the source point $\mathbf{r}'$ and observation point $\mathbf{r}$ are the same. Solution for the set of equations in equation 4.62 is hence not possible, and some additional effort is required to remedy the singular behavior of the self terms. To do this, we examine closely the situation in the ith section shown in Figure 4.27. In the diagonal elements in equation 4.62, we were trying to evaluate the potential at the center of each section resulting from its own charge, which is also assumed to be located at the center of the section. To overcome the singularity problem, therefore, we do not assume that the charge is located at the center but instead distributed at the surface of the conductor. This is actually the situation to begin with, and what resulted in the singularity is our desire to use the point charges approximation in equation 4.61. From Figure 4.27, we may evaluate the potential at the center x_i of the ith cell as

$$\Phi = 1 = \frac{1}{4\pi\epsilon_o} \int_{-\Delta\ell_i/2}^{\Delta\ell_i/2} \int_0^{2\pi} \frac{\rho_{si}\,a\,d\phi\,dx'}{\sqrt{a^2 + x'^2}} \quad (4.63)$$

In equation 4.63, the distance from the charge point on the surface to the observation point at the center is taken $|\mathbf{r} - \mathbf{r}'| = \sqrt{a^2 + x'^2}$. Integrating equation 4.63, we obtain

$$1 = \frac{1}{4\pi\epsilon_o}\rho_{si}(2\pi a)\,\ell n\left[x' + \sqrt{a^2 + x'^2}\right]\Bigg|_{-\Delta\ell_i/2}^{\Delta\ell_i/2}$$

$$= \frac{\rho_{si}\,a}{2\epsilon_o}\,\ell n\,\frac{\dfrac{\Delta\ell_i}{2} + \sqrt{a^2 + \left(\dfrac{\Delta\ell_i}{2}\right)^2}}{-\dfrac{\Delta\ell_i}{2} + \sqrt{a^2 + \left(\dfrac{\Delta\ell_i}{2}\right)^2}}$$

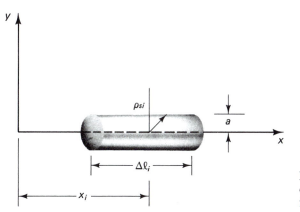

Figure 4.27 The potential at the center of the ith section resulting from its own surface charge.

For a conductor of small radius as compared to the length $\Delta\ell_i$ of each section, we obtain

$$1 = \frac{\rho_{si} a}{2\epsilon_o} \ell n \frac{\Delta\ell_i}{-\frac{\Delta\ell_i}{2} + \frac{\Delta\ell_i}{2}\left[1 + \frac{1}{2}\frac{a^2}{\left(\frac{\Delta\ell_i}{2}\right)^2}\right]}$$

$$= \frac{\rho_{si} a}{\epsilon_o} \ell n \frac{\Delta\ell_i}{a} \tag{4.64}$$

Equation 4.64 may be used for the diagonal (self) terms in equation 4.62, thus resulting in the following matrix equation:

$$
\begin{bmatrix}
4\pi\epsilon_o \\
4\pi\epsilon_o \\
4\pi\epsilon_o \\
\vdots \\
4\pi\epsilon_o
\end{bmatrix}
=
\begin{bmatrix}
4\pi a\,\ell n\dfrac{\Delta\ell_1}{a} & \dfrac{2\pi a\,\Delta\ell_2}{|x_1 - x_2|} & \dfrac{2\pi a\,\Delta\ell_3}{|x_1 - x_3|} & \cdots & \dfrac{2\pi a\,\Delta\ell_N}{|x_1 - x_N|} \\
\dfrac{2\pi a\,\Delta\ell_1}{|x_2 - x_1|} & 4\pi a\,\ell n\dfrac{\Delta\ell_2}{a} & \dfrac{2\pi a\,\Delta\ell_3}{|x_2 - x_3|} & \cdots & \dfrac{2\pi\Delta\ell_N}{|x_2 - x_N|} \\
\dfrac{2\pi a\,\Delta\ell_1}{|x_3 - x_1|} & \dfrac{2\pi a\,\Delta\ell_2}{|x_3 - x_2|} & 4\pi a\,\ell n\dfrac{\Delta\ell_3}{a} & \cdots & \dfrac{2\pi\Delta\ell_N}{|x_3 - x_N|} \\
\dfrac{2\pi a\,\Delta\ell_1}{|x_N - x_1|} & \dfrac{2\pi a\,\Delta\ell_2}{|x_N - x_2|} & \dfrac{2\pi\Delta\ell_3}{|x_N - x_3|} & \cdots & 4\pi a\,\ell n\dfrac{\Delta\ell_N}{a}
\end{bmatrix}
\begin{bmatrix}
\rho_{s1} \\
\rho_{s2} \\
\rho_{s3} \\
\rho_{sN}
\end{bmatrix}
$$

$$(4.65)$$

Equation 4.65, which can be further simplified if we choose all the sections of the same length $\Delta\ell_1 = \Delta\ell_2 = \cdots\cdots = \Delta\ell_N$, may now be solved by a simple matrix inversion subroutine. The result is the charge distribution $\rho_{s1}, \rho_{s2}, \ldots, \rho_{sN}$ on the surface of the conductor.

Exercise

Write a computer program that solves for the charge distribution on the surface of a conductor of radius a and length L. In addition to the geometry of the conductor, the input to the program should include the number of sections and the dielectric properties

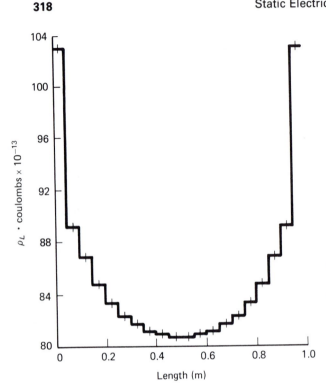

$\rho_L \cdot \text{coulombs} \times 10^{-13}$

Length (m)

Figure 4.28   Charge distribution on a cylindrical conductor of radius $a = 1$ mm and length $L = 1$ m. The conductor is at a potential of 1 V. (See L. L. Tsai and C. E. Smith, Moment Method in Electromagnetics for Undergraduates, IEEE Trans. on Education, vol. E-21, pp. 14–22, 1979)

ϵ of the surrounding medium. Check your results with those shown in Figure 4.28 for a conductor of radius $a = 1$ mm and of length $L = 1$ m. The conductor is at a potential of 1 V, and twenty sections were used to obtain the results shown in Figure 4.28.

One may wonder if it was really worth going through all of this to determine the charge distribution on the surface of a cylindrical conductor. The answer is that although it seems too much effort for this simple geometry, the method of moment solution procedure is general and certainly capable of handling much more complicated conductor geometries at no additional effort than what we spent in the last example. The large variation of the charge distribution on the conductor as shown in Figure 4.28 also justifies the spent effort. Many of the electrostatic problems at the end of this chapter cannot be handled using rigorous solution procedures, and use of numerical ones such as the method of moment is appropriate. The following example will illustrate the application of the method of moment to multiconductor problems.

EXAMPLE 4.17

Consider the parallel plate capacitor shown in Figure 4.29a. We wish to use the method of moments to determine the capacitance as a function of the separation distance between the parallel plates.

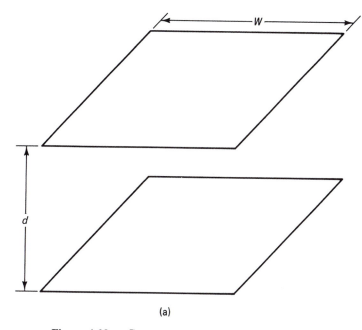

Figure 4.29a   Geometry of parallel plate capacitor.

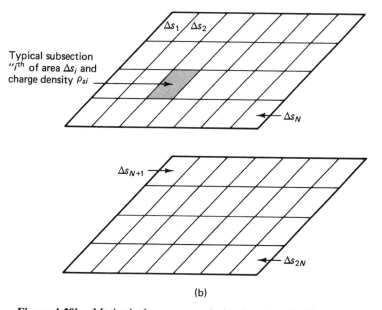

Figure 4.29b   Method of moments solution involves dividing the parallel plates in $2N$ small sections each of area Δs.

Solution

The solution procedure follows steps similar to those used in the previous example with the additional step of calculating the total charge by integrating the surface charge density, hence,

$$q = \int_s \rho_s \, ds$$

where s is the surface of one of the conductors. Determination of the capacitance then follows,

$$C = \frac{q}{V}$$

where V is the potential difference between the two capacitor plates. Let us, therefore, proceed with our step-by-step solution procedure for determining the charge distribution on the parallel plates as a result of a potential difference V applied between them. To simplify the solution by reducing the number of unknowns, we introduce symmetry (without any loss of generality) by making the top plate at the potential of 1 V and the lower plate at -1 V. The potential at any point in space $(\mathbf{r})$ is given in terms of the charge distribution on the two plates by

$$\Phi(\mathbf{r}) = \frac{1}{4\pi\epsilon_o} \int_s \frac{\rho_s \, ds'}{|\mathbf{r} - \mathbf{r}'|} \tag{4.66}$$

where $s = s_1 + s_2$, the total surface of the two plates.

1. Dividing the two plates into $2N$ subsections each of area $\Delta s_1 = \Delta s_2 = \cdots = \Delta s_{2N} = \Delta s$, Eq. 4.66 reduces to

$$\Phi(\mathbf{r}) = \frac{1}{4\pi\epsilon_o} \sum_{i=1}^{2N} \int_{\Delta s_i} \frac{\rho_{si} \, ds_i'}{|\mathbf{r} - \mathbf{r}_i'|} \tag{4.67}$$

In equation 4.67, ρ_{si} is the charge distribution within the ith subsection, and it is integrated over Δs_i, the area of this ith subsection.

2. At this point, we make the decision of what type of charge distribution we would like to choose in each subsection. If the areas of the subsections $\Delta s_i, i = 1, \ldots, 2N$ are chosen sufficiently small, we may consider ρ_{si} to be constant within each subsection. Hence,

$$\Phi(\mathbf{r}) = \frac{1}{4\pi\epsilon_o} \sum_{i=1}^{2N} \rho_{si} \int_{\Delta s_i} \frac{ds_i'}{|\mathbf{r} - \mathbf{r}'|} \tag{4.68}$$

3. Even at this stage, carrying out the integration in equation 4.68 over each of the subsections is not a simple task. Interested students should consult the book by Adams.* If we, however, further simplify the situation by assuming that the charge $\rho_{si} \, \Delta s_i$ is localized at the center of each subregion, we obtain

$$\Phi(\mathbf{r}) = \frac{1}{4\pi\epsilon_o} \sum_{i=1}^{2N} \frac{\rho_{si} \, \Delta s_i}{|\mathbf{r} - \mathbf{r}_i'|} \tag{4.69}$$

---

* A. T. Adams, *Electromagnetics for Engineers* (New York: The Ronald Press Company, 1971).

Once again, equation 4.69 indicates that the potential at $\mathbf{r}$ resulting from $2N$ point charges $(\rho_{si} \Delta s_i)$ each located at $\mathbf{r}'_i$ is simply the algebraic sum of the individual potentials $\Phi_i(\mathbf{r})$.

$$\Phi_i(\mathbf{r}) = \frac{1}{4\pi\epsilon_o} \frac{(\rho_{si} \Delta s_i)}{|\mathbf{r} - \mathbf{r}'_i|}$$

4. Equation 4.69 contains $2N$ unknowns $\rho_{si} \Delta s_i, i = 1, 2, \ldots, 2N$. To determine these unknowns, we need to know the potential at $2N$ locations. We choose these locations to be the centers of the $2N$ subsections. The potential is assumed to be 1 V at the upper conductor and -1 V at the lower one. Hence,

$$1 = \frac{1}{4\pi\epsilon_o} \sum_{i=1}^{2N} \frac{(\rho_{si} \Delta s_i)}{|\mathbf{r}_1 - \mathbf{r}'_i|}$$

$$1 = \frac{1}{4\pi\epsilon_o} \sum_{i=1}^{2N} \frac{(\rho_{si} \Delta s_i)}{|\mathbf{r}_2 - \mathbf{r}'_i|}$$

$$\vdots \qquad \vdots$$

$$-1 = \frac{1}{4\pi\epsilon_o} \sum_{i=1}^{2N} \frac{(\rho_{si} \Delta s_i)}{|\mathbf{r}_{N+1} - \mathbf{r}'_i|}$$

$$\vdots \qquad \vdots$$

$$-1 = \frac{1}{4\pi\epsilon_o} \sum_{i=1}^{2N} \frac{(\rho_{si} \Delta s_i)}{|\mathbf{r}_{2N} - \mathbf{r}'_i|} \tag{4.70}$$

The first equation in equation 4.70 is obtained by enforcing the condition of 1 V potential at the center of the first subsection, whereas the last is obtained by substituting $\Phi = -1\,V$ at the center of the last subsection $2N$. Equation 4.70 contains $2N$ equations in the $2N$ unknown charges $(\rho_{si} \Delta s_i), i = 1, \ldots, 2N$. We may write equation 4.70 as

$$\sum_{i=1}^{2N} (\rho_{si} \Delta s_i) \Phi_{ij} = \begin{matrix} 1 \\ -1 \end{matrix} \begin{matrix} 1 \leq i \leq N \\ N + 1 \leq i \leq 2N \end{matrix} \tag{4.71}$$

where

$$\Phi_{ij} = \frac{1}{4\pi\epsilon_o |\mathbf{r}_j - \mathbf{r}'_i|}$$

$(\rho_{si} \Delta s_i)\Phi_{ij}$ is the potential at the jth point resulting from the ith charge.

5. As expected, the system of equation 4.70 is singular along the diagonal terms. This is due to the point charge approximation and the fact that the diagonal elements involve evaluating the potential at the center of each subregion because of its own charge, which is also assumed to be located at the center. To overcome this problem, we approximate each subregion by a circular disk of the same area and having a uniform charge distribution. The potential at the center of the disk, owing to its own uniform charge distribution of density ρ_s, is given by

$$\Phi = \int_0^{2\pi} \int_0^a \frac{\rho_s \, \rho d\rho \, d\phi}{4\pi\epsilon_o \, \rho} = \frac{\rho_s}{4\pi\epsilon_o} 2\pi a = \frac{\rho_s}{2\epsilon_o} a$$

$$= \frac{\rho_s}{2\epsilon_o \sqrt{\pi}} \sqrt{A} = \frac{0.282}{\epsilon_o} \rho_s \sqrt{\Delta s} \tag{4.72}$$

where $A = \Delta s$ is the area of the circular disk ($A = \pi a^2$) representing the area of the subsection Δs. Therefore, in equation 4.71, expressions for Φ_{ij} will be used for $i \neq j$, whereas equation 4.72 will be used to calculate the diagonal elements when $i = j$. The matrix equation (equation 4.71) may therefore be written as

$$
\begin{bmatrix}
\dfrac{0.282\sqrt{\Delta s_1}}{\epsilon_o} & \dfrac{\Delta s_2}{4\pi\epsilon_o|\mathbf{r}_1 - \mathbf{r}_2'|} & \dfrac{\Delta s_3}{4\pi\epsilon_o|\mathbf{r}_1 - \mathbf{r}_3'|} & \cdots \\[2ex]
\dfrac{\Delta s_1}{4\pi\epsilon_o|\mathbf{r}_2 - \mathbf{r}_1'|} & \dfrac{0.282\sqrt{\Delta s_2}}{\epsilon_o} & \dfrac{\Delta s_3}{4\pi\epsilon_o|\mathbf{r}_2 - \mathbf{r}_3'|} & \cdots \\[2ex]
\vdots & \vdots & \vdots & \ddots \\[2ex]
\dfrac{\Delta s_1}{4\pi\epsilon_o|\mathbf{r}_{2N} - \mathbf{r}_1'|} & \dfrac{\Delta s_2}{4\pi\epsilon_o|\mathbf{r}_{2N} - \mathbf{r}_2'|} & \dfrac{0.282\sqrt{\Delta s_{2N}}}{\epsilon_o} &
\end{bmatrix}
\begin{bmatrix}
\rho_{s1} \\ \rho_{s2} \\ \rho_{s3} \\ \vdots \\ \rho_{sN} \\ \rho_{sN+1} \\ \vdots \\ \rho_{s2N}
\end{bmatrix}
=
\begin{bmatrix}
1 \\ 1 \\ \cdot \\ \cdot \\ 1 \\ -1 \\ -1 \\ -1 \\ -1 \\ -1
\end{bmatrix}
$$

$$(4.73)$$

The areas $\Delta s_1, \Delta s_2, \ldots, \Delta s_{2N}$ may, of course, be assumed equal if a regular grid is used to subdivide the capacitor geometry. Furthermore, if we use symmetry, we note that $\rho_{s_1} = -\rho_{s_{N+1}}, \rho_{s_2} = -\rho_{s_{N+2}}, \ldots, \rho_{s_N} = -\rho_{s_{2N}}$, and we may solve for only the first N terms in equation 4.73.

6. Once the ρ_s's are determined, we find the total charge, say, on the top plate, by adding up ρ_s's in the form

$$
q = \sum_{i=1}^{N} \rho_{si} \, \Delta s_i
$$

and the capacitance is then calculated from

$$
C = \frac{q}{V} = \frac{1}{2} \sum_{i=1}^{N} \rho_{si} \, \Delta s_i
$$

where V, the potential difference between the plates, is chosen to be 2 V in the initial setting of the problem.

Exercise

Write a computer program to calculate the capacitance of a parallel plate capacitor, such as that shown in Figure 4.29. You may just program equation 4.73 and include a matrix inversion subroutine to solve the $2N$ system of equations, but you should also allow for some flexibility to solve for unsymmetrical capacitors. By writing this program, you will gain valuable experience in calculating the charge distribution on and the capacitance of a system of conductors of general geometry.

10 MAGNETOSTATIC FIELDS AND MAGNETIC VECTOR POTENTIAL

An important result of dealing with static fields is the separation of the electric and magnetic field quantities. This simply means it is possible to determine electrostatic quantities such as charge distribution, electric fields, and potentials, and, conversely, the magnetostatic quantities such as current distributions, magnetic flux, and stored magnetic energy without having to deal with the coupling between the electric and magnetic quantities. Furthermore, the laws of electrostatic fields and, in particular, the fact that the line integral of the electric field along closed contour is zero—that is, $\oint_c \mathbf{E} \cdot d\ell = 0$—allowed us to introduce a scalar quantity Φ, "the scalar electric potential" in terms of which all other electric quantities may be obtained. The introduction of this "scalar" function Φ was certainly most rewarding because, through it, it was possible to solve for "vector" field quantities. As a matter of fact, most of the electrostatic problems we discussed in the previous sections were solved in terms of the scalar electric potential, either directly by expressing the electrostatic potential in terms of charge distributions or by solving boundary value problems involving Laplace's or Poisson's equations.

Turning our attention now to magnetostatic fields, one should wonder at first if it would be possible to introduce another scalar function that will be as useful in solving magnetostatic problems as the scalar electric potential Φ is in electrostatic fields. Examination of the magnetostatic laws reveals that the contour integration of the magnetic field intensity is equal to the enclosed current, hence,

$$\oint_c \mathbf{H} \cdot d\ell = \int_s \mathbf{J} \cdot d\mathbf{s} \qquad (4.74)$$

This is unlike the electrostatic case in which $\oint_c \mathbf{E} \cdot d\ell = 0$. Hence, we conclude that it is not possible to introduce a scalar function that will play an important role in solving magnetostatic fields as that played by the electrostatic potential Φ. This is certainly unfortunate because Φ significantly simplified solutions to electrostatic problems. To point out what is possible, however, we examine Gauss's law for the magnetic fields. We have

$$\oint_s \mathbf{B} \cdot d\mathbf{s} = 0 \qquad \text{or equivalently} \qquad \nabla \cdot \mathbf{B} = 0$$

Because the magnetic flux density vector $\mathbf{B}$ is solenoidal—that is, $\nabla \cdot \mathbf{B} = 0$—we may express $\mathbf{B}$ as a curl of any other vector $\mathbf{A}$. Hence,

$$\mathbf{B} = \nabla \times \mathbf{A} \qquad (4.75)$$

Equation 4.75 is possible because the vector identity

$$\nabla \cdot \nabla \times \mathbf{A} = 0$$

is always true. We define $\mathbf{A}$ as the magnetic vector potential. Substituting equation 4.75 into Ampere's law in point or differential form (equation 4.8), we obtain the following differential equation for $\mathbf{A}$,

$$\nabla \times \mathbf{H} = \nabla \times \frac{\mathbf{B}}{\mu_o} = \nabla \times \nabla \times \frac{\mathbf{A}}{\mu_o} = \mathbf{J}$$

or

$$\nabla \times \nabla \times \mathbf{A} = \mu_o \mathbf{J} \qquad (4.76)$$

Equation 4.76 relates **A** to the source current distribution **J**. Once a solution for **A** is obtained from equation 4.76, we may directly determine other magnetic field quantities, such as **B** from equation 4.75. In a sense, **A** plays a role similar to that of Φ. **A** is directly related to the current distribution **J**, and once a solution for **A** is obtained, it is straightforward to determine **B** and **H**. Before we illustrate such a solution procedure, however, it is important to learn more about the characteristics of the vector potential **A**.

From Biot-Savart's law, we learned that the magnetic flux density is given in terms of the current I in an element of length $d\ell$ by

$$d\mathbf{B} = \frac{\mu_o}{4\pi} \frac{I \, d\ell \times \mathbf{a}_R}{R^2} \qquad (4.77)$$

Instead of the linear current I (in amperes), we may have a current density **J** (in amperes/meter squared) in a differential volume dv. In this case, equation 4.77 reduces to

$$d\mathbf{B} = \frac{\mu_o}{4\pi} \frac{\mathbf{J} \, dv \times \mathbf{a}_R}{R^2} \qquad (4.78)$$

The magnetic flux resulting from the current in the total volume v is, hence,

$$\mathbf{B} = \frac{\mu_o}{4\pi} \int_v \frac{\mathbf{J} \, dv \times \mathbf{a}_R}{R^2} \qquad (4.79)$$

R is the distance between the source point $\mathbf{r}'$ and the observation point $\mathbf{r}$.

$$R = |\mathbf{r} - \mathbf{r}'| = [(x - x')^2 + (y - y')^2 + (z - z')^2]^{1/2}$$

$\mathbf{a}_R$ is a unit vector from the source to the observation point. To obtain the desired relation between **A** and **J**, we observe the following:

$$\nabla\left(\frac{1}{R}\right) = -\nabla'\left(\frac{1}{R}\right) = -\frac{\mathbf{a}_R}{R^2}$$

Hence, the integrand in equation 4.79 may be written as

$$\frac{\mathbf{J} \, dv \times \mathbf{a}_R}{R^2} = -\mathbf{J} \, dv \times \nabla\left(\frac{1}{R}\right) = \nabla\left(\frac{1}{R}\right) \times \mathbf{J} \, dv \qquad (4.80)$$

From the vector identity,

$$\nabla \times (f\mathbf{F}) = \nabla f \times \mathbf{F} + f \nabla \times \mathbf{F}$$

we may express equation 4.80 as

$$\frac{\mathbf{J} \, dv \times \mathbf{a}_R}{R^2} = \nabla \times \left(\frac{1}{R} \mathbf{J} \, dv\right) - \frac{1}{R} \nabla \times \mathbf{J} \, dv \qquad (4.81)$$

The curl operator ∇ is a differential operator over the coordinates of the observation point $\mathbf{r}(x, y, z)$. $\mathbf{J}(\mathbf{r}')$, conversely, is a function of the source coordinates $\mathbf{r}'(x', y', z')$. Hence, $\nabla \times \mathbf{J}(\mathbf{r}') = 0$. This reduces Eq. 4.81 to

$$\frac{\mathbf{J}\,dv \times \mathbf{a}_R}{R^2} = \nabla \times \left(\frac{1}{R}\mathbf{J}\,dv\right) \tag{4.82}$$

Substituting equation 4.82 into equation 4.79, we obtain

$$\mathbf{B} = \frac{\mu_o}{4\pi} \int_v \nabla \times \left[\frac{1}{R}\mathbf{J}(\mathbf{r}')\,dv'\right]$$

$$= \frac{\mu_o}{4\pi} \nabla \times \int_v \frac{1}{R}\mathbf{J}(\mathbf{r}')\,dv' \tag{4.83}$$

In equation 4.83, it was possible to take the curl operator outside the integral because ∇ operates on the unprimed coordinates of the observation point, whereas the integral is over the source coordinates. $\mathbf{B}$ in equation 4.83 can also be expressed as $\mathbf{B} = \nabla \times \mathbf{A}$, hence,

$$\mathbf{A}(r) = \frac{\mu_o}{4\pi} \int_v \frac{\mathbf{J}(\mathbf{r}')}{R}\,dv' \tag{4.84}$$

Equation 4.84 is an important relation that directly relates the magnetic vector potential $\mathbf{A}$ to the source current distribution $\mathbf{J}$. It should be noted that besides its vector nature, equation 4.84 is similar to the one that relates the scalar electric potential Φ to the source charge distribution,

$$\Phi(\mathbf{r}) = \frac{1}{4\pi\epsilon} \int \frac{\rho(\mathbf{r}')}{R}\,dv' \tag{4.85}$$

Equations 4.84 and 4.85 bring us to the following important observation regarding the solution of Maxwell's equations. Instead of directly solving for the electric and magnetic fields, we introduce two auxiliary potential functions the scalar electric potential Φ and the magnetic vector potential $\mathbf{A}$. These potential functions are directly and more simply related to the source charge and current distributions. Hence, on solving equations 4.84 and 4.85, other electric and magnetic field components may be obtained through straightforward differential operations. This observation is also true for time-varying fields, except that the field quantities will be expressed in terms of both potential functions because of the coupling between the fields. The following examples will illustrate the use of the magnetic vector potential $\mathbf{A}$ in solving magnetostatic problems.

EXAMPLE 4.18

Consider the differential current element $d\ell$ that carries a total current $\mathbf{I}$ and is oriented along the z axis. Determine the magnetic flux $\mathbf{B}(\mathbf{r})$ at the observation point P, as shown in Figure 4.30.

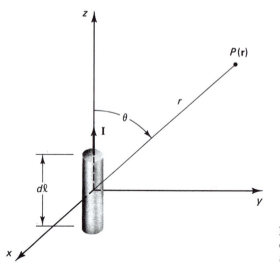

Figure 4.30 A differential current element $\mathbf{I}\,d\ell$ oriented along the z axis.

Solution

We will try to relate the current to the magnetic vector potential $\mathbf{A}$, and will then determine $\mathbf{B}$ from $\mathbf{A}$. From equation 4.84, $\mathbf{A}$ is related to the current distribution by

$$\mathbf{A}(\mathbf{r}) = \frac{\mu_o}{4\pi} \int_{-d\ell/2}^{d\ell/2} \frac{I\,dz}{r}\, \mathbf{a}_z \tag{4.86}$$

In equation 4.86, the current and, hence, $\mathbf{A}$ are oriented along the z direction, and the distance $R = r$ because the current element is small and located at the origin of the coordinate system. If the current I is assumed uniform along the z axis, equation 4.86 reduces to

$$\mathbf{A}(\mathbf{r}) = \frac{\mu_o}{4\pi} \frac{I\,d\ell}{r}\, \mathbf{a}_z \tag{4.87}$$

The magnetic flux is subsequently obtained as

$$\mathbf{B} = \nabla \times \mathbf{A} = \begin{bmatrix} \dfrac{\mathbf{a}_r}{r^2 \sin\theta} & \dfrac{\mathbf{a}_\theta}{r \sin\theta} & \dfrac{\mathbf{a}_\phi}{r} \\[2mm] \dfrac{\partial}{\partial r} & \dfrac{\partial}{\partial \theta} & \dfrac{\partial}{\partial \phi} \\[2mm] A_r & rA_\theta & r \sin\theta\, A_\phi \end{bmatrix} \tag{4.88}$$

The components of $\mathbf{A}$ in the spherical coordinate system are given by

$$A_r = A_z \cos\theta, \qquad A_\theta = -A_z \sin\theta, \qquad A_\phi = 0$$

Substituting these components into equation 4.88, we obtain

$$\mathbf{B} = \frac{\mu_o}{4\pi r^2} I\,d\ell\, \sin\theta\, \mathbf{a}_\phi \tag{4.89}$$

For this simple case of a differential current element, equation 4.89 may be directly obtained by applying Biot-Savart's law, hence,

$$dB = \frac{\mu_o}{4\pi} \frac{I\,d\ell \times \mathbf{a}_R}{r^2}$$

The total magnetic flux is then

$$\mathbf{B} = \frac{\mu_o}{4\pi} \int_{-d\ell/2}^{d\ell/2} \frac{I\,d\ell \times \mathbf{a}_R}{r^2}$$

Substituting $d\ell = dz\,\mathbf{a}_z$ and $\mathbf{a}_R = \mathbf{a}_r$ from Figure 4.30, we obtain

$$\mathbf{B} = \frac{\mu_o}{4\pi} \frac{I\,d\ell}{r^2}\,\mathbf{a}_z \times \mathbf{a}_r$$

$$= \frac{\mu_o}{4\pi} \frac{I\,d\ell}{r^2}\,[(\mathbf{a}_r\,\cos\theta - \mathbf{a}_\theta\,\sin\theta) \times \mathbf{a}_r]$$

$$= \frac{\mu_o}{4\pi} \frac{I\,d\ell}{r^2}\,\sin\theta\,\mathbf{a}_\phi$$

which is the same answer in equation 4.89.

EXAMPLE 4.19

Let us extend the calculations made in the previous example to the case in which we have a current-carrying conductor of length L. The conductor carries a current I and extends along the z axis from $-L/2$ to $L/2$.

Solution

From Figure 4.31, the magnetic vector potential at an observation point $P(\mathbf{r})$ is given by

$$A_z(\rho, z) = \frac{\mu_o}{4\pi} \int_{-L/2}^{L/2} \frac{I\,dz'}{\sqrt{\rho^2 + (z - z')^2}} \tag{4.90}$$

In equation 4.90, cylindrical coordinates were used to take advantage of the cylindrical symmetry in the given geometry. Carrying out the integration in equation 4.90, we obtain

$$A_z(\rho, z) = \frac{\mu_o I}{4\pi}\,\ell n\left[(z - z') + \sqrt{\rho^2 + (z - z')^2}\right]_{-L/2}^{L/2}$$

$$= \frac{\mu_o I}{4\pi}\,\ell n \frac{\left(z - \dfrac{L}{2}\right) + \sqrt{\rho^2 + \left(z - \dfrac{L}{2}\right)^2}}{\left(z + \dfrac{L}{2}\right) + \sqrt{\rho^2 + \left(z + \dfrac{L}{2}\right)^2}}$$

The magnetic flux density is then

$$\mathbf{B} = \nabla \times \mathbf{A} = -\frac{\partial A_z}{\partial \rho}\,\mathbf{a}_\phi \tag{4.91}$$

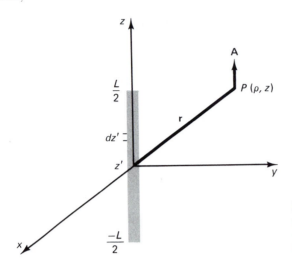

Figure 4.31 The magnetic vector potential at $P(\rho, z)$ resulting from a current-carrying conductor of length L.

The simplified expression in equation 4.91 was possible because **A** has only an A_z component, which is independent of ϕ. For an observation point $P(\rho, 0)$ in the bisecting plane $z = 0$, the result of equation 4.91 becomes

$$\mathbf{B} = \frac{\mu_o I}{2\pi\rho} \frac{\dfrac{L}{2}}{\sqrt{\rho^2 + \left(\dfrac{L}{2}\right)^2}} \mathbf{a}_\phi \qquad (4.92)$$

A similar expression could have been obtained from Biot-Savart's law, as illustrated in the previous example. In the special case of an infinitely long wire—that is, $L/2 \gg \rho$—equation 4.92 reduces to

$$\mathbf{B} = \frac{\mu_o I}{2\pi\rho} \mathbf{a}_\phi \qquad (4.93)$$

The result in equation 4.93 is for the special case of an infinitely long current-carrying conductor. This result was obtained in chapter 1 using Ampere's law and using the cylindrical symmetry of the geometry. Hence, we may conclude that to solve magnetostatic problems—that is, determine magnetic field quantities **B** and **H** from a given current distribution—we may use one of the following procedures:

1. Use Biot-Savart's law,

$$d\mathbf{B} = \frac{\mu I d\boldsymbol{\ell} \times \mathbf{a}_R}{4\pi R^2}$$

and integrate over the geometry of the current distribution of interest, that is,

$$\mathbf{B} = \frac{\mu}{4\pi} \int_c \frac{I d\boldsymbol{\ell} \times \mathbf{a}_R}{R^2} \qquad \text{or} \qquad \mathbf{B} = \frac{\mu}{4\pi} \int_v \frac{\mathbf{J} dv \times \mathbf{a}_R}{R^2}$$

2. Use Ampere's law,

$$\oint_c \mathbf{H} \cdot d\boldsymbol{\ell} = \int_s \mathbf{J} \cdot d\mathbf{s}$$

where c is the Amperian contour established at the location where it is desired to calculate $\mathbf{H}$.

3. Relate the magnetic vector potential $\mathbf{A}$ to the given current distribution by

$$\mathbf{A} = \frac{\mu}{4\pi} \int_v \frac{\mathbf{J}\,dv'}{R}$$

and then determine the magnetic field quantities from $\mathbf{A}$ using

$$\mathbf{B} = \nabla \times \mathbf{A} \qquad \text{and} \qquad \mathbf{H} = \frac{\mathbf{B}}{\mu}$$

The following example will further illustrate the solution procedure using the third option described earlier. This example also illustrates that from an observation point located at a large distance from a small circular loop carrying a current I, the current loop of area $d\mathbf{s}$ behaves as a magnetic dipole of moment $\mathbf{m} = I\,d\mathbf{s}$.

EXAMPLE 4.20

Consider a small circular loop of radius a and carrying a current I. Determine the magnetic flux density at an observation point P located at a large distance r from the loop.

Solution

The magnetic vector potential $\mathbf{A}$ resulting from the current I in the circular loop is given by

$$\mathbf{A} = \frac{\mu_o}{4\pi} \oint_c \frac{I\,d\boldsymbol{\ell}}{R} \qquad\qquad (4.94)$$

where $R = |\mathbf{r} - \mathbf{r}'|$ is the distance from the current element $I\,d\boldsymbol{\ell}$ to the observation point P. c is the closed contour of the loop. For a current loop placed in the x-y plane, $d\boldsymbol{\ell} = a\,d\phi\,\mathbf{a}_\phi$, and equation 4.94 reduces to

$$A_\phi = \frac{\mu_o I a}{4\pi} \int_0^{2\pi} \frac{d\phi}{R}$$

Because of the symmetry of the geometry under consideration, we may choose the observation point P in the y-z plane (see Figure 4.32) without loss of generality. Furthermore, if we consider two current elements symmetrically located with respect to the y axis, their corresponding magnetic vector potentials, $d\mathbf{A}_1$ and $d\mathbf{A}_2$ at P, are shown in Figure 4.32 $d\mathbf{A}_1$ and $d\mathbf{A}_2$ are in the same directions as $I\,d\boldsymbol{\ell}_1$ and $I\,d\boldsymbol{\ell}_2$, as illustrated in example 4.18. Expressing $d\mathbf{A}_1$ and $d\mathbf{A}_2$ in terms of their components along the y and x axes, we note that the components in the y direction cancel, whereas the components in the x direction add. The x components of the magnetic vector potentials are given by

$$dA_x = -\frac{\mu_o I a}{4\pi} \frac{d\phi}{R} \sin\phi$$

The x components of the magnetic vector potentials resulting from the two symmetrically located current elements are, hence,

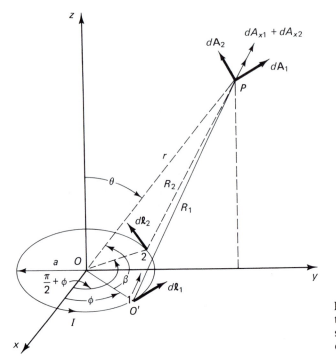

Figure 4.32 Magnetic vector potential **A** at a distance point from a small circular loop of radius a and carrying current I.

$$dA_{x1} + dA_{x2} = -\frac{\mu_o Ia \ \sin\phi}{4\pi} \ d\phi\left[\frac{1}{R_1} + \frac{1}{R_2}\right]$$

$$= -\frac{\mu_o Ia \ \sin\phi}{2\pi R} \ d\phi \qquad (4.95)$$

Regarding the distance R, we note from Figure 4.32 and specifically from the triangle POO' that

$$R_{1,2}^2 = r^2 + a^2 - 2ra \ \cos\beta$$

$r \cos\beta$ is simply the projection of r along the OO' line. One way of obtaining such a projection is to project r along the y axis ($r \sin\theta$) and then project the latter along the OO' lines ($r \sin\theta$) $\sin\phi$. Hence,

$$R = \sqrt{r^2 + a^2 - 2ra \ \sin\theta \ \sin\phi}$$

For a distant observation point $r \gg a$, we obtain

$$R \approx r\left[1 - 2\frac{a}{r} \ \sin\theta \ \sin\phi\right]^{1/2} \qquad (4.96a)$$

$$\approx r\left[1 - \frac{a}{r} \ \sin\theta \ \sin\phi\right] \qquad (4.96b)$$

where equation 4.96b is obtained by using the binomial approximation of equation 4.96a. Similarly, we approximate $1/R$ as

$$\frac{1}{R} = \frac{1}{r\left[1 - \dfrac{a}{r}\sin\theta\sin\phi\right]} = \frac{1}{r}\left[1 - \frac{a}{r}\sin\theta\sin\theta\right]^{-1}$$

$$\approx \frac{1}{r}\left[1 + \frac{a}{r}\sin\theta\sin\theta\right] \approx \frac{1}{r} + \frac{a}{r^2}\sin\theta\sin\phi$$

Substituting the $1/R$ approximation into equation 4.95 and noting that $d\mathbf{A}$ is generally in the ϕ direction (which is the same as the $-x$ direction for an observation point on the y-z plane), we obtain the following expression for $\mathbf{A}$:

$$\mathbf{A} = \frac{\mu_o I a}{4\pi}\int_0^{2\pi}\left[\frac{1}{r} + \frac{a}{r^2}\sin\theta\sin\phi\right]\sin\phi\, d\phi\, \mathbf{a}_\phi \qquad (4.97)$$

Integrating the first term $\sin\phi/r$ from 0 to 2π is zero, and the final result for $\mathbf{A}$ from equation 4.97 is

$$\mathbf{A} = \frac{\mu_o I a^2}{4r^2}\sin\theta\, \mathbf{a}_\phi \qquad (4.98)$$

The magnetic flux $\mathbf{B}$ is then

$$\mathbf{B} = \nabla \times \mathbf{A} = \frac{\mu_o I a^2}{4r^3}(2\cos\theta\, \mathbf{a}_r + \sin\theta\, \mathbf{a}_\theta) \qquad (4.99a)$$

Before concluding this section, it is of interest to point out the similarity between equation 4.99 and the electric field result in example 4.4 of this chapter.

To emphasize the analogy between these quantities, we rearrange equation 4.99a in a more appropriate form and rewrite the electric field result owing to an electric dipole (see example 4.4), hence,

$$\mathbf{B} = \frac{\mu_o I(\pi a^2)}{4\pi r^3}(2\cos\theta\, \mathbf{a}_r + \sin\theta\, \mathbf{a}_\theta)$$

$$= \frac{\mu_o I(ds)}{4\pi r^3}(2\cos\theta\, \mathbf{a}_r + \sin\theta\, \mathbf{a}_\theta) \qquad (4.99b)$$

where $ds = \pi a^2$ is the area of the current loop.

$$\mathbf{E} = \frac{qd}{4\pi\epsilon_o r^3}(2\cos\theta\, \mathbf{a}_r + \sin\theta\, \mathbf{a}_\theta) \qquad (4.100)$$

From equation 4.99b and 4.100, we observe the analogies between the electric and magnetic field quantities given in Table 4.3.

It is clear at this point that the solution procedure using the magnetic vector potential does not have a clear advantage over others, including the use of Ampere's law or Biot-Savart's law. It certainly does not possess the many advantages of its counterpart, the scalar electric potential. It is not scalar and, hence, as difficult to solve for as the field quantities themselves. It is also not associated with a physical quantity similar to the work or energy property associated with the scalar electric potential. The

TABLE 4.3　ANALOGIES BETWEEN ELECTRIC AND
MAGNETIC FIELD QUANTITIES

Electric quantities	Magnetic quantities
$1/\epsilon_o$ |Electric dipole| $= qd$ Distant (far) electric field **E**	μ_o |Magnetic dipole| $= I\,ds$ Distant (far) magnetic flux **B**

lack of physical interpretation and the vector nature of the magnetic vector potential clearly limit its usefulness as a solution tool. In certain cases, however, the simple relationship between the current source and the magnetic vector potential is very desirable, and solution procedures using **A** become attractive. An example of these situations in which the use of **A** is useful is the solution of the time-varying radiation fields from linear wire antennas. The calculation of the radiation characteristics of wire antennas based on knowledge of their current distributions is described in detail in chapter 9.

4.11 MAGNETIC CIRCUITS

In chapter 3, we introduced magnetic materials and described in detail their reaction to externally applied magnetic fields. These materials are used in many practical applications including electromagnets, transformers, electric machines, and so on. In our discussion of solution procedures for magnetostatic problems, thus far, we have not described a procedure that is suitable for solving this kind of complicated yet practical problems. In this section, we will introduce an approximate method that may be used to solve these kinds of problems. As suggested by the title of this section, this approximate method is based on an analogy between some devices made of magnetic materials and electric circuits. To introduce the solution procedure, identify the analogous quantities, and clearly explain the various assumptions made in this approximate solution procedure, we start by solving the following simple example.

EXAMPLE 4.21

Consider the ferromagnetic toroid of inner and outer radii a and b, respectively, and of thickness d, as shown in Figure 4.33. An N turn coil carrying current I is used to generate the magnetic flux within the toroid. Determine the magnetic field intensity **H** and the magnetic flux ψ_m within the toroid.

Solution

Based on the symmetry and the direction of the current, we may use the right-hand rule to show that the magnetic flux will be mainly circulating in the ϕ direction, as shown in Figure 4.33. For this simple geometry, we may apply Ampere's law, hence,

$$\oint_c \mathbf{H} \cdot d\ell = NI$$

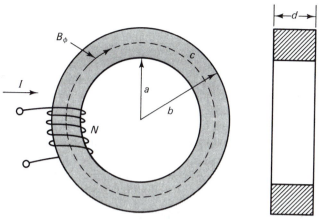

Figure 4.33 Ferromagnetic toroid of square cross section.

where c is the contour shown in Figure 4.33. Considering a contour of general radius ρ, $a < \rho < b$, we obtain

$$\mathbf{H} = \frac{NI}{2\pi\rho}\mathbf{a}_\phi \tag{4.101}$$

We may now calculate the total flux within the toroid simply by integrating $\mathbf{B} = \mu\mathbf{H}$ over the cross section. Hence,

$$\psi_m = \int_s \mathbf{B} \cdot d\mathbf{s} = \frac{\mu NI}{2\pi} \int_{z=0}^{d} \int_a^b \frac{1}{\rho}\mathbf{a}_\phi \cdot dz\, d\rho\, \mathbf{a}_\phi$$

$$= \frac{\mu NI}{2\pi} d \ln \frac{b}{a} \tag{4.102}$$

As can be seen for this simple geometry and in the absence of air gaps, closed form solution was possible. To help us develop an approximate but more widely applicable solution procedure, we will solve this example in an alternative way and examine the errors in the obtained result.

If we substitute $\mathbf{H}$ in terms of the total flux and assume that the flux is uniformly distributed through the toroid cross section, that is, $\mathbf{B} = \psi_m/s\, \mathbf{a}_\phi$, we obtain

$$\mathbf{H} = \frac{\mathbf{B}}{\mu} = \frac{\psi_m}{\mu s}\mathbf{a}_\phi$$

where s is the cross-sectional area of the toroid $s = (b - a)d$. Ampere's law then becomes

$$\oint_c \frac{\psi_m}{\mu s}\mathbf{a}_\phi \cdot d\ell\, \mathbf{a}_\phi = NI$$

In our case, the cross-sectional area s is not a function of the contour c; therefore, we may carry out the integration by multiplying by an average length ℓ, hence,

$$\psi_m \frac{\ell}{\mu s} = NI$$

or

$$\psi_m = \frac{NI}{\left(\dfrac{\ell}{\mu s}\right)} \tag{4.103}$$

Multiplying $\psi_m/\mu s$ by an average contour length ℓ is equivalent to assuming that the magnetic flux is concentrated in a mean path of length ℓ. The difference between equations 4.102 and 4.103 is only the assumption that the magnetic flux is uniformly distributed throughout the cross section, instead of the $1/\rho$ dependence given in equation 4.101. It is important to quantify the error that may result from this approximation. To do this, we assume $N = 200$ turns, $I = 1$ A, $d = 3$ cm, and that the radii a and b are equal to 15 and 18 cm, respectively. The ferromagnetic material has $\mu = 5000\mu_o$. From equation 4.102,

$$\psi_m = \frac{5000(4\pi \times 10^{-7})200 \times 3 \times 10^{-2}}{2\pi} \, \ell n \, \frac{0.18}{0.15} \text{ Wb}$$

$$= 0.1094 \times 10^{-2} \text{ Wb}$$

Using the approximate solution in equation 4.103 and substituting $\ell = 2\pi\left(\dfrac{a+b}{2}\right)$, we obtain

$$\psi_m = \frac{200}{\dfrac{2\pi(16.5)10^{-2}}{5000 \times 4\pi \times 10^{-7} \times 9 \times 10^{-4}}}$$

$$= 0.1091 \times 10^{-2} \text{ Wb}$$

The approximate result in equation 4.103 is in error of about 0.27 percent compared with equation 4.102.

This example clearly illustrates an acceptable accuracy of the approximate solution procedure in equation 4.103. We have not, however, indicated the advantages of using equation 4.103. To do this, we compare the magnetic and electric circuits shown in Figure 4.34. We are all familiar with the electromotive force (emf) V that generates

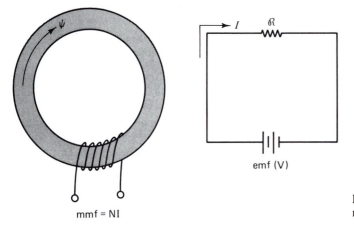

Figure 4.34 Comparison between magnetic and electric circuits.

the circulating current I in the electric circuit. According to Ohm's law, the emf is related to the current by

$$I = \frac{V}{R}$$

where R is the total resistance in the electric circuits. By analogy, equation 4.103 suggests that the flux circulating in the magnetic circuit ψ_m is related to the magneto-motive force (mmf) $\oint \mathbf{H} \cdot d\ell = NI$ by

$$\psi_m = \frac{\text{mmf} = NI}{\mathcal{R}} \tag{4.104}$$

where $\mathcal{R}$ is known as the reluctance of the magnetic circuit,

$$\mathcal{R} = \frac{\ell}{\mu s} \tag{4.105}$$

It should be emphasized that $\oint_c \mathbf{H} \cdot d\ell$ was defined as magnetomotive force by analogy to the electromotive force $= \oint_c \mathbf{E} \cdot d\ell$. $\oint_c \mathbf{H} \cdot d\ell$, however, is not a force and is only a source for the magnetic flux circulating in the magnetic circuit. Furthermore, it is known that the electrical resistance of a cylindrical conductor of area s, length ℓ, and conductivity σ is given by

$$R = \frac{\ell}{\sigma s} \tag{4.106}$$

Comparison between equations 4.105 and 4.106 shows that the conductivity σ in an electric circuit is analogous to the permeability μ in a magnetic circuit. Table 4.4 summarizes the analogous quantities in electric and magnetic circuits.

    After pointing out the analogy between the electric and magnetic circuits, it is possible now to discuss the true reason for using the approximations that led to equation 4.103. If we neglect the flux leakage and assume that the total magnetic flux is confined within the ferromagnetic core, it is possible to solve magnetic circuits in an analogous manner to the familiar procedure for solving for electric circuits. We merely assume that the magnetic flux is circulating in a mean path of length that can be calculated based on the geometry of the magnetic circuit, calculate the reluctance, and obtain a value

TABLE 4.4 ANALOGOUS QUANTITIES IN ELECTRIC AND MAGNETIC CIRCUITS

Electric	Magnetic
Electromotive force $\equiv \oint_c \mathbf{E} \cdot d\ell$ $= V$	Magnetomotive force $\equiv \oint_c \mathbf{H} \cdot d\ell$ $= NI$
Current I	Flux ψ_m
Resistance $R = \dfrac{\ell}{\sigma s}$	Reluctance $\mathcal{R} = \dfrac{\ell}{\mu s}$
Conductivity σ	Permeability μ

for the circulating flux from equation 4.104. This solution procedure has great advantage in cases of more general magnetic circuits that may include air gaps as well as series of sections of various magnetic materials. As long as the assumptions of having all the flux circulating within the core of the magnetic circuit (i.e., neglect leakage flux), and of neglecting the fringing effects in case of air gaps are valid, the accuracy of the solution will be highly acceptable. The following examples illustrate the use of the solution procedure in more general magnetic circuits.

EXAMPLE 4.22

Consider the magnetic circuit shown in Figure 4.35. It consists of a toroidal core of ferromagnetic material and an air gap of length ℓ_g. Assuming that the mmf is due to a coil of N turns carrying a current I, determine the magnetic flux ψ_m, and the magnetic flux density in the ferromagnetic core and in the air gap.

Solution

If we directly apply Ampere's law,

$$\oint_c \mathbf{H} \cdot d\ell = NI$$

we will quickly find out that, in this case, we have two **H** fields to deal with: $\mathbf{H}_c$ in the ferromagnetic core and $\mathbf{H}_g$ in the air gap. Hence,

$$\int_{\text{core contour}} \mathbf{H}_c \cdot d\ell + \int_{\text{gap}} \mathbf{H}_g \cdot d\ell = NI \qquad (4.107)$$

To solve this equation with two unknowns, $\mathbf{H}_c$ and $\mathbf{H}_g$, we need to make some assumptions. In addition to neglecting the "leakage flux," which means that we assume that the total flux is confined within the ferromagnetic core, we will also neglect the "fringing effect." The latter means that we will assume that the air gap is sufficiently narrow so that the total

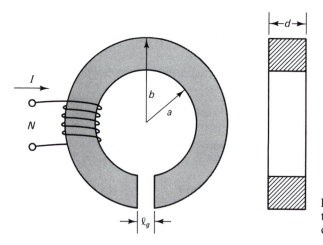

Figure 4.35 A magnetic circuit that consists of a ferromagnetic core with an air gap.

flux in the ferromagnetic core will continue to flow across the air gap without any loss resulting from fringing effects. Hence,

$$\psi_c = \psi_g \tag{4.108}$$

Because the gap and the core have the same cross-sectional area s, we have

$$B_c = \mu H_c = \frac{\psi_c}{s} \tag{4.109a}$$

and

$$B_g = \mu_o H_g = \frac{\psi_g}{s} \tag{4.109b}$$

From equations 4.108 and 4.109, we obtain

$$\mathbf{H}_g = \frac{\mu}{\mu_o} \mathbf{H}_c \tag{4.110}$$

Substituting equation 4.110 into equation 4.107 and assuming that the total flux is uniform over the cross section s, we obtain

$$H_c \ell_c + \frac{\mu}{\mu_o} H_c \ell_g = NI$$

Hence,

$$\mathbf{H}_c = \frac{NI}{\left(\ell_c + \dfrac{\mu}{\mu_o} \ell_g \right)} \mathbf{a}_\phi \tag{4.111}$$

In equation 4.111, the contour integration is approximated by assuming that the flux is circulating in a mean path $\ell_c = [2\pi(a + b)/2 - \ell_g]$ at the center of the ferromagnetic core.

The magnetic flux density **B** and magnetic flux ψ_m follow directly from equation 4.111,

$$\mathbf{B}_c = \mu \mathbf{H}_c = \frac{\mu\, NI}{\left(\ell_c + \dfrac{\mu}{\mu_o} \ell_g \right)}$$

$$\mathbf{B}_g = \mu_o \mathbf{H}_g = \mathbf{B}_c$$

and

$$\psi_c = \psi_g = \frac{s\,\mu\, NI}{\left(\ell_c + \dfrac{\mu}{\mu_o} \ell_g \right)} \tag{4.112}$$

where s, the cross-sectional area, is given by $s = d(b - a)$, where d is the thickness of the toroid.

Alternatively, we could have directly used the analogy between the electric and magnetic circuits. In this case, the magnetic circuit in Figure 4.35 is equivalent to the series resistive circuit shown in Figure 4.36. The reluctances $\mathcal{R}_c$ and $\mathcal{R}_g$ are those of the ferromagnetic core and the air gap, respectively. These quantities, according to the analogy described in the previous example, are given by

$$\mathcal{R}_c = \frac{\ell_c}{\mu s}, \qquad \mathcal{R}_g = \frac{\ell_g}{\mu_o s}$$

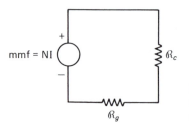

Figure 4.36 An approximate equivalent circuit of the toroid with an air gap.

The total flux ψ_m is, hence,

$$\psi_m = \frac{\text{mmf}(NI)}{\mathcal{R}_c + \mathcal{R}_g} \tag{4.113}$$

Once again neglecting the fringing effects, this flux will be assumed the same in the ferromagnetic core and in the air gap. Substituting the appropriate values of the reluctances $\mathcal{R}_c$ and $\mathcal{R}_g$, we see that equations 4.113 and 4.112 are identical. We, therefore, conclude that to solve for the flux circulating in magnetic circuits, we may directly use the equivalent circuit, and the analogy between the electric and magnetic circuits.

Before we solve more complicated examples of magnetic circuits, let us briefly indicate the various assumptions under which this solution procedure is possible.

1. *Flux leakage*: In magnetic circuits, we assume that the total flux is confined within the ferromagnetic material and that the leakage to the surrounding air is negligible. In electric circuits, we also assume that the current is confined within the conductors. There is a difference, however, between the two cases. The air has practically zero conductivity compared with that of metals used as electric wires. The conductivity of copper, for example, is 5.7×10^7, and the ratio of the conductivity of copper to that of a typical insular surrounding the wire (e.g., rubber) is over 10^{20}. This clearly justifies the accuracy of assuming the confinement of the conduction current in the copper wire rather than the surrounding insulator. In magnetic circuits, conversely, the permeability of ferromagnetic material is typically 5000 to 8000 times that of air $\mu_o = 4\pi \times 10^{-7}$ A/m. Hence, in magnetic circuits, we are typically dealing with $1000 < \mu/\mu_o < 10{,}000$, which is much smaller than the 10^{20} conductivity ratio. Hence, although μ/μ_o is sufficiently large to justify the use of the analogy between the electric and magnetic circuits, neglecting the flux leakage is not justifiable to the same accuracy as that of neglecting the conduction in insulators surrounding wires in electric circuits.

2. *Fringing effects*: In magnetic circuits containing air gaps, we neglected the fringing effects and assumed that the total flux density is the same in both the ferromagnetic core and in the air gap. The accuracy of this approximation may be acceptable for narrow air gaps (i.e., ℓ_g is much smaller than the linear cross-sectional dimension of the core). Clearly, for practical values of ℓ_g, the fringing effects cannot be completely neglected and, instead, it is customary to consider the total flux in the air gap crossing an area slightly larger than that of the core. The cross-sectional area of the air gap is obtained by adding the gap length to each of the linear dimensions of the ferromagnetic

core. The following example will illustrate the application of the electric circuit analogies to solve more complicated magnetic circuits, as well as a procedure for accounting
for the fringing effects.

EXAMPLE 4.23

Consider the magnetic circuit shown in Figure 4.37a. The magnetic flux is generated by
a mmf = NI (ampere turns). The ferromagnetic core has μ = 4000 μ_o and square cross-
sectional area of side length = 3 cm. The air gap length ℓ_g = 0.1 cm. Determine the
following:

1. The flux in the air gap if NI = 200 AT.
2. The mmf (NI) required to produce a flux density in the air gap of B = 0.5 Wb/m^2.

Solution

1. The equivalent circuit is shown in Figure 4.37b. We start by calculating the various
   reluctances $\mathcal{R}$ in the circuit,

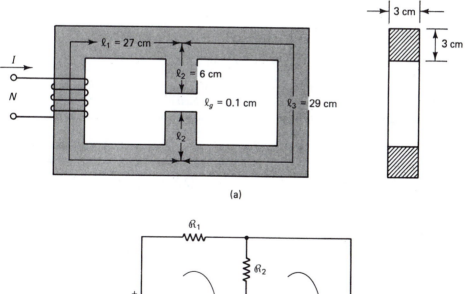

(a)

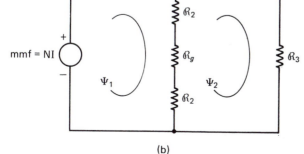

(b)

Figure 4.37 A ferromagnetic circuit with air gap and its equivalent circuit.

$$\mathcal{R}_1 = \frac{\ell_1}{\mu s} = \frac{27 \times 10^{-2}}{4000 \times 4\pi \times 10^{-7} \times 9 \times 10^{-4}} = 5.97 \times 10^4 \text{ H}^{-1}$$

$$\mathcal{R}_2 = \frac{\ell_2}{\mu s} = \frac{6 \times 10^{-2}}{4000 \times 4\pi \times 10^{-7} \times 9 \times 10^{-4}} = 1.326 \times 10^4 \text{ H}^{-1}$$

$$\mathcal{R}_3 = \frac{\ell_3}{\mu s} = 6.41 \times 10^4 \text{ H}^{-1}$$

and

$$\mathcal{R}_g = \frac{\ell_g}{\mu_o s_g}$$

To account for the fringing effect in the air gap, we assume that the cross-sectional area,

$$s_g = (\ell_s + \ell_g)^2$$

where ℓ_s is the side length of the square cross section of the ferromagnetic core. We note that we added a gap length to each linear dimension in the cross section, which is customarily used to account for the fringing effects.

$$s_g = (3.1)^2 = 9.61 \text{ cm}^2$$

The reluctance of the air gap is then

$$\mathcal{R}_g = \frac{0.1 \times 10^{-2}}{4\pi \times 10^{-7} \times 9.61 \times 10^{-4}} = 8.28 \times 10^5 \text{ H}^{-1}$$

It should be noted that although the air gap is only 0.1 cm long, its reluctance is more than an order of magnitude larger than the reluctances of any of the arms in the ferromagnetic core. This is due to the significantly lower permeability of air compared with that of ferromagnetic core.

Now we are ready to solve for the flux in the magnetic circuit given that $NI = 200$. By analogy to electric circuits, we write the following two loop equations:

$$NI = (\mathcal{R}_1 + 2\mathcal{R}_2 + \mathcal{R}_g)\psi_1 - (2\mathcal{R}_2 + \mathcal{R}_g)\psi_2$$

$$0 = -(2\mathcal{R}_2 + \mathcal{R}_g)\psi_1 + (2\mathcal{R}_2 + \mathcal{R}_3 + \mathcal{R}_g)\psi_2$$

Solving for ψ_1 and ψ_2, we obtain

$$\psi_1 = 1.075\psi_2, \qquad \psi_2 = 15.59 \times 10^{-4} \text{ Wb}$$

The total flux in the air gap ψ_g is then

$$\psi_g = \psi_1 - \psi_2 = 1.169 \times 10^{-4} \text{ Wb}$$

The flux density in the air gap is

$$B_g = \frac{\psi_g}{s_g} = 0.1217 \text{ Wb/m}^2$$

2. If we are given the flux density in the air gap $B_g = 0.500 \text{ Wb/m}^2$, we calculate the total flux

$$\psi_g = \psi_1 - \psi_2 = 0.500 \times 9.61 \times 10^{-4} = 4.81 \times 10^{-4} \text{ Wb}$$

From Kirchhoff's voltage law as used in part 1, we know that

$$\psi_1 = 1.075\,\psi_2$$

We then have

$$\psi_2 = 64.133 \times 10^{-4}\ \text{Wb}, \qquad \psi_1 = 68.943 \times 10^{-4}\ \text{Wb}$$

and the required mmf

$$NI = 91.422 \times 68.943 - 85.452 \times 64.133 = 822.6\ \text{AT}$$

which is 4.11 times the NI in part 1. This is clearly expected for a linear circuit. If the magnetic flux increases by a factor of $0.5/0.1217 = 4.11$, the required mmf should be increased by the same factor. The solution for part 2, however, was carried out in detail merely to illustrate the procedure.

There is one more comment to make regarding magnetic circuits. The core in real magnetic circuits is often made of ferromagnetic materials that have nonlinear **B-H** curves. A typical **B-H** curve for a ferromagnetic material, which is completely demagnetized, is shown in Figure 4.38. In the absence of any mmf, both **B** and **H** are zero. On the application of mmf, we supply an increase in **H**, and from Figure 4.38, we see that the flux density also increases but not linearly with the increase in **H**. The nonlinear increase of both **H** and **B** continues until the curve reaches saturation. At such a stage, we see only slight or no change in the resulting flux density with the increase in **H**. After reaching this saturation stage, we start decreasing **H** by gradually reducing the current in the magnetizing coil. We realize that the new **B-H** curve does not detract from the initial curve. Instead, we notice that there will be residual flux **B** even after the magnetizing current and, hence, **H** reaches zero. The fact that the **B-H** curve does not retract itself in the magnetization and demagnetization process is known as the "hys-

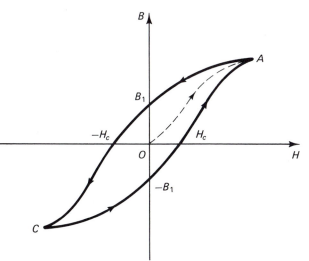

Figure 4.38 A typical initial magnetization and a hysteresis curve for ferromagnetic material. The dotted portion of the curve is for the initial magnetization, whereas the solid line is for the hysteresis loop. H_c is the coercive force, and B_1 is the remnant flux.

teresis" effect. Figure 4.38 shows that the magnetizing current and, hence, **H** has to be reversed to reduce the flux density to zero. The value of **H** required to bring the flux density to zero is known as the "coercive force" $\mathbf{H}_c$. If we continue to increase **H** in the reverse direction, the material gets magnetized in the reverse direction, and flux is induced in this reverse direction. From Figure 4.38, it is clear that once again saturation will be reached with the increase in **H**, and subsequent decrease of **H** to zero leaves the material permanently magnetized in the reverse direction with a remnant flux $(-\mathbf{B}_1)$. The hysteresis loop is closed with the repeated increase of **H** in the original direction. It should be noted that although the curve OA represents the initial magnetization of a completely demagnetized material, repeated magnetization and demagnetization follows the $A(-H_c)C(H_c)$ curve.

Of particular interest to us is how to use these nonlinear magnetization curves to solve magnetic circuits. The situation is clearly similar to the solution of electric circuits that include nonlinear resistive elements, such as diodes, thermometers, and so on. In general, iterative procedures that provide successive approximations of ψ's and B's are quite successful in solving these problems. We will conclude this section by solving a simple example that uses an actual **B-H** curve of ferromagnetic material.

EXAMPLE 4.24

In the magnetic circuit shown in Figure 4.39a, determine the current I in the 200-turn coil that is required to produce a magnetic flux density of 0.1 Wb/m² in the air gap. The ferromagnetic core has the initial magnetization curve shown in Figure 4.39b.

Solution

This is actually one of the simplest nonlinear magnetic circuits. From Ampere's law, the required mmf (NI) is equal to the integral (or the sum) of the **H** along a closed contour, hence,

$$\mathrm{NI} = \oint_c \mathbf{H} \cdot d\ell = \sum_{i=1}^{k} H_i \ell_i$$

where H_i is the magnetic field intensities in the various sections of the magnetic circuit, and ℓ_i is the mean path in each of these sections. k is the number of sections. In our case, $k = 2$, $\ell_1 = 30$ cm, and $\ell_2 = \ell_g = 0.15$ cm. The required flux density in the air gap is $B_g = 0.1$ Wb/m². H_g is then

$$H_g = \frac{B_g}{\mu_o} = \frac{0.1}{4\pi \times 10^{-7}} = 7.958 \times 10^4 \text{ A/m}$$

Considering the fringing effect in the air gap, the total air gap flux is given by

$$\psi_g = 0.1 (3.15)^2 \times 10^{-4} = 0.9923 \times 10^{-4} \text{ Wb}$$

In calculating ψ_g, we added the air gap length to each of the linear dimensions of the cross-sectional area of the magnetic core. Assuming that the core flux, $\psi_c = \psi_g$, the flux density in the magnetic core is given by

$$B_c = \frac{0.9923 \times 10^{-4}}{9 \times 10^{-4}} = 0.11 \text{ Wb/m}^2$$

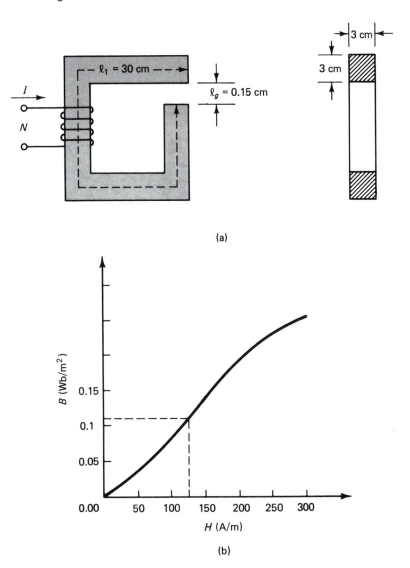

(a)

(b)

Figure 4.39   (a) The nonlinear magnetic circuit and (b) the initial magnetization curve.

From the **B-H** curve given in Figure 4.39b, the value of H_c required to produce B_c is $H_c = 127$.

The total mmf is then

$$NI = 7.958 \times 10^4 \times 0.15 \times 10^{-2} + 127 \times 30 \times 10^{-2}$$

$$= 119.37 + 38.1$$

$$= 157.47 \text{ AT}$$

The required current I is then

$$I = \frac{NI}{200} = 0.787 \text{ A}$$

It is interesting to note that the mmf $(H_g \ell_g)$ is much larger than $H_c \ell_c$. This can also be verified by the large value of reluctance of the air-gap region. This is similar to having the largest voltage drop across the largest resistor in a series resistance circuit.

4.12 SELF-INDUCTANCE AND MUTUAL INDUCTANCE

Inductance in magnetic circuits is like capacitance in electric ones. It is a geometrical property (in linear materials) that describes the ability of the physical arrangement or conductor configuration to store magnetic energy. Alternatively, it describes the amount of magnetic flux a particular conductor configuration is capable of producing for a unit current passing in it. To clarify these statements, as well as the magnetic energy storage concepts to be described in the next section, it is helpful to refer back to Faraday's experiment, which we described in chapter 1. A schematic of Faraday's experiment is shown in Figure 4.40 for convenient reference. Actually, what Figure 4.40 shows is two magnetically coupled loops. When a current I_1 is circulating in the transmitting loop on the left-hand side, magnetic flux of density **B** will be generated, as shown in Figure 4.40. Part of this flux will intersect the second or the receiving loop on the right-hand side. We may try to determine the total flux linking the transmitting loop owing to its own current I_1. This flux is denoted by ψ_{11}, where the subscript 11 is used to emphasize the fact that this is the flux linking loop 1 owing to its own current I_1. ψ_{11} is given by

$$\psi_{11} = \int_{s_1} \mathbf{B} \cdot d\mathbf{s}_1 \text{ Wb} \tag{4.114}$$

We define the "self-inductance" L_{11} as the ratio of the flux linking the first (transmitting) loop owing to its own current I_1, hence,

$$L_{11} = \frac{\psi_{11}}{I_1}$$

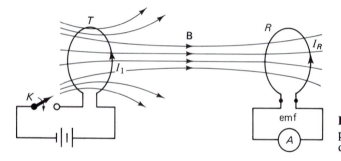

Figure 4.40 Schematic of the experimental apparatus used in Faraday's experiment.

If loop 1 consists of N_1 turns, we may approximate the total flux linking the N_1 turn loop to be $N_1 \psi_{11}$. The self-inductance in this case is given by

$$L_{11} = \frac{N_1 \psi_{11}}{I_1} \tag{4.115}$$

In SI units, the inductance is in henry (H). Furthermore, from Figure 4.40, it is clear that some of the magnetic flux generated by I_1 links the second (receiving) loop. The total flux linking the second loop is given by

$$\psi_{12} = \int_{s_2} \mathbf{B} \cdot d\mathbf{s}_2 \tag{4.116}$$

The subscript 12 emphasizes the fact that ψ is the flux linking the second loop owing to the current in the first one I_1. We define the "mutual inductance" L_{12} as the ratio of the flux linking loop 2 to the current I_1 producing it, hence,

$$L_{12} = \frac{\psi_{12}}{I_1} \tag{4.117}$$

If the receiving loop is made up of N_2 turns, the total flux linking the loop is the $N_2 \psi_{12}$, and the mutual inductance is then

$$L_{12} = \frac{N_2 \psi_{12}}{I_1} \tag{4.118}$$

It is worth noting that ψ_{11} in equation 4.114 is different from ψ_{12} in equation 4.116. This may be due to the fact that the areas s_1 and s_2 of the transmitting and receiving loops are different, their orientation with respect to the magnetic flux (dot product) is different, and also because not all the flux produced by the transmitting loop is linking the receiving one, as clearly illustrated in Figure 4.40. The following examples will illustrate the use of equations 4.115 and 4.117 to calculate self-inductances and mutual inductances.

EXAMPLE 4.25

Consider the closely wound infinitely long solenoid of radius a as shown in Figure 4.41. The windings are such that we have N_1 turns per distance ℓ along the solenoid.

1. Calculate the self-inductance per unit length L_{11}.
2. If a second solenoid of N_2 turns per distance ℓ and radius b is placed coaxially with the first one, calculate the mutual inductance per unit length L_{12}.

Solution

1. From our discussion of example 1.28, we learned that for the closely wound infinitely long solenoid, the magnetic field intensity obtained from Ampere's law is given by

$$\mathbf{H} = \frac{N_1 I}{\ell} \mathbf{a}_z$$

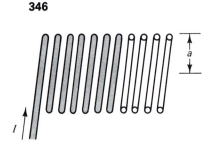

(a) Closely wound, infinitely long solenoid
of N_1 turns per distance ℓ

(b) Two coaxially placed closely wound infinitely
long solenoids. The outer one carries current I

Figure 4.41   (a) Self-inductances and (b) mutual inductances.

In this example, we assume air to be the core medium, hence,

$$\mathbf{B} = \mu_o \frac{N_1 I}{\ell} \mathbf{a}_z$$

The flux density (under the specified assumptions) is uniform within the solenoid. The magnetic flux linking one turn is therefore

$$\psi_m = \mu_o \frac{N_1 I}{\ell} s$$

where s is the cross-sectional area given by $s = \pi a^2$. Because we have N_1 turns in each length ℓ of the solenoid, the total flux linking N_1 turns is

$$\psi_m = \mu_o \frac{s}{\ell} N_1^2 I$$

The self-inductance L_{11} for a length ℓ of the solenoid is therefore

$$L_{11} = \frac{\psi_m}{I} = \mu_o \frac{s}{\ell} N_1^2$$

The self-inductance per unit length L_{11},

$$L_{11} = \frac{L}{\ell} = \mu_o \frac{s}{\ell^2} N_1^2 \text{ (H)}$$

In the preceding calculations, we clearly neglected the flux leakage outside the solenoidal windings by assuming a closely wound solenoid.

2. To calculate the mutual inductance, we first determine the flux linking the second solenoid, assuming the same flux density **B** resulting from the first one. Defining ψ_{12} as the flux linking N_2 turns (in a length ℓ) in the second solenoid, we have

$$\psi_{12} = \left(\mu_o \frac{N_1 I}{\ell} s_2 \right) N_2$$

where s_2 is the cross-sectional area of the interior solenoid, $s_2 = \pi b^2$. The mutual inductance per unit length,

$$L_{12} = \frac{1}{\ell} \left(\mu_o \frac{N_1 N_2}{\ell} s_2 \right) = \mu_o \frac{s_2}{\ell^2} N_1 N_2 \text{ (H)}$$

◆◆

EXAMPLE 4.26

Consider the long straight conductor of radius a and carrying a total current I. The conductor is placed coaxially with a toroidal core of magnetic materials $\mu = \mu_o \mu_r$, as shown in Figure 4.42. An N turn coil is wound around the toroid.

1. Calculate the mutual inductance between the straight conductor and the toroidal windings.
2. Determine the (internal) self-inductance of the straight wire.

Solution

Because of the cylindrical symmetry, we identify the magnetic field intensity and the magnetic flux to be in the azimuthal ϕ direction. Establishing the Amperian contour within the straight cylindrical conductor, we obtain

$$\mathbf{H} = \frac{I\left(\dfrac{\pi\rho^2}{\pi a^2}\right)\mathbf{a}_\phi}{2\pi\rho} = \frac{I}{2\pi a^2}\rho\,\mathbf{a}_\phi \qquad \rho \le a$$

This result was also obtained in example 1.28 of chapter 1.

Establishing the contour outside the straight conductor, we obtain

$$\mathbf{H} = \frac{I}{2\pi\rho}\,\mathbf{a}_\phi \qquad \rho \ge a$$

1. We may now calculate the mutual inductance L_{12}. The flux density in the toroidal core is given by

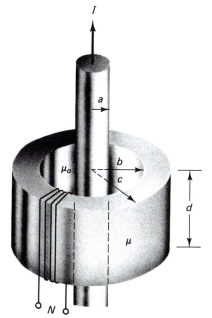

Figure 4.42 Coaxial arrangement of a straight conductor and a toroid.

$$\mathbf{B} = \frac{\mu I}{2\pi\rho} \mathbf{a_\phi}$$

and the flux linking the toroid is

$$\psi_m = \int_s \mathbf{B} \cdot d\mathbf{s} = \int_{z=0}^{d} \int_{\rho=b}^{c} \frac{\mu I}{2\pi\rho} \mathbf{a_\phi} \cdot d\rho \, dz \, \mathbf{a_\phi}$$

$$= \frac{\mu I d}{2\pi} \ell n \frac{c}{b}$$

The total flux linking the N turns of the toroidal windings is

$$\psi_{12} = N\psi_m = \frac{\mu N I d}{2\pi} \ell n \frac{c}{b}$$

The mutual inductance is then

$$L_{12} = \frac{\psi_{12}}{I} = \frac{\mu N d}{2\pi} \ell n \frac{c}{b} \text{ (H)}$$

2. The calculation of the internal self-inductance may be done more easily in terms of the energy stored in the conductor, as will be explained in the next section. The concept of flux linkage may, however, still be used if carefully modified to account for the distributed nature of the current and, hence, the associated flux within the straight cylindrical conductor.

   If we consider a cross section of radius ρ of the straight conductor, the total current in this section is

$$I \frac{\pi\rho^2}{\pi a^2} = I \left(\frac{\rho^2}{a^2} \right)$$

The differential flux per unit length $d\psi_m$ that links this portion of the current is

$$d\psi_m = \mathbf{B} \cdot d\mathbf{s} = \frac{\mu I}{2\pi a^2} \rho \, \mathbf{a_\phi} \cdot d\rho \, dz \, \mathbf{a_\phi}$$

It is clear that both the current and the differential flux depend on the radius ρ of the cross section under consideration. Based on stored magnetic energy consideration, it will be shown in the next section that

$$L_{11} = \frac{1}{I^2} \int_s i(\ell') \, d\psi_m$$

where the preceding integration is carried over the cross section of the conductor. The internal self-inductance is hence given by

$$L_{11} = \frac{1}{I^2} \int_{z=0}^{\ell} \int_{\rho=0}^{a} I \left(\frac{\rho^2}{a^2} \right) \frac{\mu I}{2\pi a^2} \rho \, d\rho \, dz$$

$$= \frac{\mu \ell}{8\pi} \text{ (H)}$$

and the inductance per unit length is then

$$L_{11} = \frac{\mu}{8\pi} \text{ (H/m)}$$

This inductance is known as the internal self-inductance, because it includes contribution from the internal flux inside the straight wire. As we know, there is magnetic flux external to the straight wire given by

$$\mathbf{B} = \frac{\mu I}{2\pi\rho}\, \mathbf{a}_\phi$$

This external flux results in what is known as the external inductance. Both the internal and external components of the self-inductance will be calculated in the following example.

EXAMPLE 4.27

Consider the coaxial cable of an inner conductor of radius a and an outer conductor of radius b, as shown in Figure 4.43. Calculate the inductance per unit length of this coaxial transmission line.

Solution

If we assume a current I in the inner conductor, the return current will be $(-I)$ in the outer conductor. At lower frequencies, including dc, the current in the inner conductor may be assumed uniformly distributed within the cross-sectional area πa^2. From Ampere's law, $\mathbf{H}$ inside and outside the inner conductor are given (as in the previous example) by

$$\mathbf{H} = \frac{I}{2\pi a^2}\rho\, \mathbf{a}_\phi \qquad 0 < \rho \le a$$

$$\mathbf{H} = \frac{I}{2\pi\rho}\, \mathbf{a}_\phi \qquad a \le \rho \le b$$

In the previous example, we dealt with the inner self-inductance, which results from the magnetic flux within the inner conductor. This resulted in a self-inductance per unit length given by

$$L_{\text{int}} = \frac{\mu}{8\pi}\ \ (\text{H/m})$$

The magnetic flux of density $\mu_o I/2\pi\rho\, \mathbf{a}_\phi$ between the inner and outer conductors $a \le \rho \le b$, conversely, contribute to the inductance that is known as the external induc-

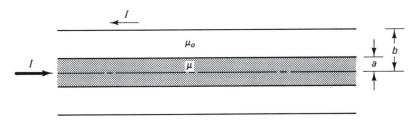

Figure 4.43   A sectional geometry of the coaxial transmission line.

tance. To evaluate this contribution, we note that the total flux per unit length ψ_{ex} between the inner and outer conductors is given by

$$\psi_{ex} = \int_{z=0}^{1} \int_{\rho=a}^{b} \frac{\mu I}{2\pi\rho} \, \mathbf{a}_\phi \cdot d\rho \, dz \, \mathbf{a}_\phi$$

$$= \frac{\mu I}{2\pi} \, \ell n \frac{b}{a} \text{ Wb}$$

This external flux clearly links the total current I, and the external self-inductance is, hence,

$$L_{ex} = \frac{\psi_{ex}}{I} = \frac{\mu}{2\pi} \, \ell n \frac{b}{a}$$

The total inductance per unit length of a coaxial transmission line is then

$$L = L_{int} + L_{ex} = \frac{\mu}{8\pi} + \frac{\mu}{2\pi} \, \ell n \frac{b}{a} \text{ (H/m)}$$

4.13 MAGNETIC ENERGY

In electrostatics, we obtained expressions for the energy stored in static electric fields. These expressions were derived based on the amount of work required to assemble a system of charges from infinity to their final locations within the desired charge distribution. A procedure analogous to this in the magnetic energy case is not possible simply because single magnetic charges do not exist and, if just introduced for the sake of obtaining the desired magnetic energy expression, the mathematical development will also require the introduction of a scalar magnetic potential, a concept that is of limited use otherwise. A magnetic energy expression can be easily introduced based on time-varying field considerations, which is described next.

Consider, once again, the setup of Faraday's experiment shown in Figure 4.40. We discussed in chapter 1 and once again in the previous section that closing the switch in the transmitting loop will result in an induced emf (v_2) in the receiving loop. Clearly, there should be some work done to help increase the current $I_1(t)$ in the transmitting loop from its zero initial value to its final value, for example, I_1. This power P, which is the rate of increase of energy dW/dt, is given by

$$P = v_1 i_1$$

where v_1 is the emf induced to counter the increase in current in the transmitting loop. The work required is then

$$W = \int v_1 i_1 \, dt \qquad (4.119)$$

According to Faraday's law, the induced emf v_1 is related to the total flux linking the transmitting loop by

$$v_1 = -\frac{d}{dt} \int_{s_1} \mathbf{B} \cdot d\mathbf{s} = -\frac{d\psi_1}{dt}$$

According to the definition of the inductance for linear media,

$$L_1 = \frac{\psi_1}{i_1} \quad \text{or} \quad \psi_1 = i_1 L_1$$

v_1 is hence given by

$$v_1 = L_1 \frac{di_1}{dt}$$

The minus sign is omitted because it just indicates the polarity of the induced voltage. It is implicitly considered by noting that work needs to be done to establish the current i_1.

Substituting v_1 in the work expression of equation 4.119, we obtain

$$W = \int_0^{I_1} L_1 \frac{di_1}{dt} i_1 \, dt = \frac{1}{2} L_1 I_1^2 \tag{4.120}$$

An alternative expression may be obtained by substituting $\psi_1 = L_1 I_1$, hence,

$$W = \frac{1}{2} \psi_1 I_1 \tag{4.121}$$

This amount of work done results in an increase in the magnetic energy stored in the circuit. Hence, the energy in the transmitting inductor,

$$W_m = \frac{1}{2} L I^2 = \frac{1}{2} \psi_m I \tag{4.122}$$

It is also often helpful to obtain an expression for the magnetic energy stored (equation 4.122) in terms of the **H** and **B** magnetic field quantities. Recalling that the total flux ψ_m is given by

$$\psi_m = \int_s \mathbf{B} \cdot d\mathbf{s} = \int_s \nabla \times \mathbf{A} \cdot d\mathbf{s} \tag{4.123}$$

Then using Stokes's theorem, we reduced equation 4.123 to a line integral around a closed contour surrounding s, hence,

$$\psi_m = \oint_c \mathbf{A} \cdot d\boldsymbol{\ell}$$

The magnetic energy stored in equation 4.122 is then given by

$$W_m = \frac{1}{2} \oint_c \mathbf{A} \cdot I \, d\boldsymbol{\ell} \tag{4.124}$$

Replacing $I \, d\boldsymbol{\ell}$, the linear current, by a volume current density **J**, we obtain

$$I \, d\boldsymbol{\ell} = \mathbf{J} \, dv$$

and equation 4.124 reduces to

$$W_m = \frac{1}{2} \int_v \mathbf{A} \cdot \mathbf{J} \, dv \tag{4.125}$$

where v is the volume containing the current density **J**. Equation 4.125 is not quite the desired expression for W_m in terms of the field quantities. Relating the current density **J** to the magnetic field intensity **H** through Ampere's law,

$$\nabla \times \mathbf{H} = \mathbf{J}$$

Equation 4.125 becomes

$$W_m = \frac{1}{2} \int_v \mathbf{A} \cdot \nabla \times \mathbf{H}\, dv \tag{4.126}$$

Making use of the vector identity,

$$\nabla \cdot (\mathbf{A} \times \mathbf{H}) = \mathbf{H} \cdot (\nabla \times \mathbf{A}) - \mathbf{A} \cdot (\nabla \times \mathbf{H})$$

Hence,

$$\mathbf{A} \cdot \nabla \times \mathbf{H} = \mathbf{H} \cdot (\nabla \times \mathbf{A}) - \nabla \cdot (\mathbf{A} \times \mathbf{H})$$
$$= \mathbf{H} \cdot \mathbf{B} - \nabla \cdot (\mathbf{A} \times \mathbf{H}) \tag{4.127}$$

Substituting equation 4.127 into equation 4.126, we obtain

$$W_m = \frac{1}{2} \int_v \mathbf{H} \cdot \mathbf{B}\, dv - \frac{1}{2} \int_v \nabla \cdot (\mathbf{A} \times \mathbf{H})\, dv$$
$$= \frac{1}{2} \int_v \mathbf{H} \cdot \mathbf{B}\, dv - \frac{1}{2} \oint_s \mathbf{A} \times \mathbf{H} \cdot d\mathbf{s} \tag{4.128}$$

The last form in equation 4.128 was obtained using the divergence theorem.

For a localized current distribution, **A** outside the current volume decreases as $1/r$, whereas **H** decreases as $1/r^2$. The surface area s is proportional to r^2, and, if the integration is carried out over an infinitely large surface (s could be any surface outside the current distribution), the second term in equation 4.128 will be zero. Equation 4.128, hence, reduces to

$$W_m = \frac{1}{2} \int_v \mathbf{H} \cdot \mathbf{B}\, dv \tag{4.129}$$

which is the desired expression for the magnetic energy stored in terms of field quantities. Recalling the magnetic energy stored expression in terms of the inductance of equation 4.122, and comparing it with equation 4.129, we obtain a new expression for the self-inductance,

$$W_m = \frac{1}{2} LI^2 = \frac{1}{2} \int_v \mathbf{H} \cdot \mathbf{B}\, dv$$

hence,

$$L = \frac{1}{I^2} \int_v \mathbf{H} \cdot \mathbf{B}\, dv \tag{4.130}$$

Equation 4.130 simply expresses the inductance in terms of the magnetic energy stored rather than the flux linkage.

EXAMPLE 4.28

Determine the internal self-inductance per unit length of cylindrical conductor of radius a, as shown in Figure 4.44. Compare the result with the value obtained in example 4.26.

Solution

If we assume a uniformly distributed current I in the cross-sectional area of the cylindrical conductor, the magnetic field intensity within the conductor is given by (see example 4.26)

$$\mathbf{H} = \frac{I}{2\pi a^2}\rho\,\mathbf{a}_\phi$$

If the conductor is made of linear material of permeability μ, the magnetic flux density within the conductor is

$$\mathbf{B} = \frac{\mu I}{2\pi a^2}\rho\,\mathbf{a}_\phi$$

From equation 4.130, the internal self-inductance per unit length is then

$$L_{\text{int}} = \frac{1}{I^2}\int_v \frac{I^2\mu}{(2\pi a^2)^2}\rho^2\,(\rho\,d\rho\,d\phi\,dz)$$

$$= \frac{\mu}{(2\pi a^2)^2}\int_{z=0}^{1}\int_{\rho=0}^{a}\int_{\phi=0}^{2\pi}\rho^3\,d\rho\,d\phi\,dz$$

$$= \frac{\mu}{8\pi}\qquad\text{(H/m)}$$

which is the same result obtained in example 4.26. It is just easier to calculate the internal self-inductance, using the energy stored expression rather than the flux linkage one. The flux linkage formulas are generally useful for problems dealing with filamentary currents. For thick conductors carrying current distributions, determination of the flux linking a specific portion of the current is involved, and it is often simpler to use energy-stored related expressions to calculate the inductance. The following example is a typical one of such cases.

Figure 4.44   Cylindrical conductor of radius a.

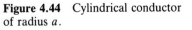

EXAMPLE 4.29

Consider a thick cylindrical tube of inner radius a and outer radius b, as shown in Figure 4.45. Calculate the internal self-inductance per unit length.

Solution

If we assume the total current in the cylindrical tube to be I, and that the current is uniformly distributed throughout the cross section of the tube, the current density will then be

$$\frac{I}{\pi(b^2 - a^2)} \text{ A/m}^2$$

From Ampere's law,

$$\oint_c \mathbf{H} \cdot d\boldsymbol{\ell} = \int_s \mathbf{J} \cdot d\mathbf{s}$$

and establishing the contour c of radius ρ somewhere between a and b, we obtain

$$H_\phi \cdot 2\pi\rho = \frac{I}{\pi(b^2 - a^2)} \pi(\rho^2 - a^2)$$

or

$$\mathbf{H} = \frac{I}{2\pi} \frac{(\rho^2 - a^2)}{\rho(b^2 - a^2)} \mathbf{a}_\phi$$

The magnetic flux density **B** is then

$$\mathbf{B} = \frac{\mu I}{2\pi(b^2 - a^2)} \frac{(\rho^2 - a^2)}{\rho} \mathbf{a}_\phi$$

where μ is the permeability of the tube material. The internal self-inductance per unit length is then

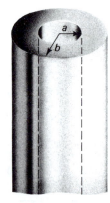

Figure 4.45 Cylindrical tube of inner and outer radii a and b, respectively.

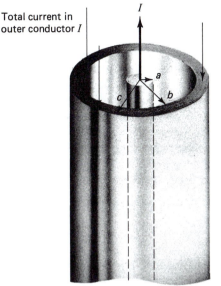

Total current in
outer conductor I

Figure 4.46 Coaxial transmission
line with thick outer conductor
walls.

$$L = \frac{1}{I^2} \int_v \mathbf{H} \cdot \mathbf{B} \, dv$$

$$= \frac{1}{I^2} \int_{z=0}^{1} \int_{\rho=a}^{b} \int_{\phi=0}^{2\pi} \frac{\mu I^2}{(2\pi)^2} \frac{(\rho^2 - a^2)^2}{(b^2 - a^2)^2 \rho^2} \, \rho \, d\rho \, d\phi \, dz$$

$$= \frac{\mu}{(2\pi)^2} \frac{2\pi}{(b^2 - a^2)^2} \int_{\rho=a}^{b} \left(\rho - \frac{a^2}{\rho} \right)^2 \rho \, d\rho$$

$$= \frac{\mu}{2\pi} \frac{1}{(b^2 - a^2)^2} \int_a^b \left(\rho^3 - 2a^2 \rho + \frac{a^4}{\rho} \right) d\rho$$

$$= \frac{\mu}{2\pi(b^2 - a^2)^2} \left[a^4 \ell n \frac{b}{a} - a^2(b^2 - a^2) + \frac{1}{4}(b^4 - a^4) \right] \qquad \text{(H/m)}$$

It is worth noting that if we have a coaxial transmission line with thick outer conductor, such as that shown in Figure 4.46, the result of example 4.28 gives the internal inductance of the center conductor, whereas the result of this example may be used to obtain the internal self-inductance of the outer conductor. The result from example 4.27 gives the external inductance L_{ex} as a result of the flux between the inner and outer conductors. At higher frequencies and if the transmission line is made of sufficiently highly conducting material so that we may assume zero magnetic field penetration in the center and outer conductors, the two internal self-inductances will be zero, and the total transmission line inductance will be due to the magnetic flux between the inner and outer conductors.

SUMMARY

The subject of this chapter is a particularly interesting one. In many textbooks, the material covered in this chapter is used to lead the way for an introductory course in electromagnetics. The reason is simply the separate treatment of the static electric and magnetic field problems. It is shown that for static fields, the electric field may be calculated separately in terms of the static charge sources, whereas the magnetic field may be calculated in terms of the current sources and their distributions.

The separability of the electric and magnetic fields is only one aspect of the many interesting features of these fields. Other interesting features include the following:

Electrostatic Potential and Its Solution Procedure

This concept is possible based on the conservative property of the static electric field

$$\oint_c \mathbf{E} \cdot d\ell = 0$$

This makes it possible to express the electric field $\mathbf{E}$ in terms of a scalar quantity known as the electric potential function Φ by

$$\mathbf{E} = -\nabla\Phi$$

This, in turn, significantly simplifies analysis of electrostatic problems. Typically, the scalar electric potential function, Φ, is related to the charge distribution source; once quantified, Φ may be used to determine the vector electric field $\mathbf{E}$. Examples 4.3 and 4.4 are given to illustrate this procedure.

Furthermore, by combining the two Maxwell's equations for electrostatic fields, we obtain the Poisson's equation

$$\nabla^2\Phi = -\frac{\rho_v}{\epsilon}$$

This is a partial differential equation commonly used to relate the electric potential, Φ, to the charge density ρ_v. For simple geometries and charge density distribution ρ_v, this differential equation may be solved anlaytically using procedures such as the separation of variables. Examples 4.12 and 4.13 illustrate such a procedure. For more general (complex) geometries and charge density distributions ρ_v, numerical techniques are used to solve Poisson's equation. Laplace's equation is a special case of Poisson's equation for $\rho_v = 0$, hence,

$$\nabla^2\Phi = 0$$

One such numerical solution of Poisson's or Laplace's equations is the finite-difference method, which is based on the central difference representation of the differential operators. Examples 4.14 and 4.15 illustrate such a solution procedure.

Alternatively, the electric potential Φ may be related to the electric charge by an integral equation of the form

$$\Phi(\mathbf{r}) = \frac{1}{4\pi\epsilon_o} \int_s \frac{\rho_s(\mathbf{r}')\,ds'}{R}$$

where $R = |\mathbf{r} - \mathbf{r}'|$, $\mathbf{r}$ and $\mathbf{r}'$ are the coordinates of the observation and source points, respectively. For a given charge distribution, an analytical or possibly numerical integration may be performed to determine the potential $\Phi(\mathbf{r})$ at a specified $\mathbf{r}$; consequently, the electric field $\mathbf{E}$ may be obtained from $\mathbf{E} = -\nabla\Phi$. In many instances, however, the potential Φ is known at specified boundaries, and it is desired to determine the charge distribution $\rho_s(\mathbf{r}')$, so that the electric potential and the electric field may be calculated everywhere. This is a typical integral equation formulation in which the unknown quantity (charge density in this case) is part of the integrand. A popular solution procedure for this kind of problem is the method of moments. This solution procedure is illustrated by examples 4.16 and 4.17.

Magnetic Vector Potential

Similar to the case of electrostatic fields where the electric potential is used as an auxiliary function to help solve electrostatic problems, magnetic field problems may be solved using the magnetic vector potential auxiliary function $\mathbf{A}$. There are, however, some differences between $\mathbf{A}$ and Φ. First, $\mathbf{A}$ is a vector, Φ is a scalar, and the other is related to the fact that unlike Φ, the use of $\mathbf{A}$ is not limited to static magnetic fields. This is because the introduction of $\mathbf{A}$ is based on $\nabla \cdot \mathbf{B} = 0$, which is valid for both static and time-varying fields. The similarity between $\mathbf{A}$ and Φ stems from the fact that both are directly related to their respective current and charge sources. Therefore, similar to the electrostatic case, it is easier to relate $\mathbf{A}$ to the current source and then solve for the field quantities such as $\mathbf{B}$ and $\mathbf{H}$ from $\mathbf{A}$. For static magnetic fields, $\mathbf{A}$ is related to the current density $\mathbf{J}$ by

$$\mathbf{A}(\mathbf{r}) = \frac{\mu_o}{4\pi} \int_v \frac{\mathbf{J}(\mathbf{r}')\,dv}{R}$$

where $R = |\mathbf{r} - \mathbf{r}'|$ is the distance between the source $\mathbf{r}'$ and observation $\mathbf{r}$ points. Examples 4.18 to 4.20 illustrate the procedure for solving magnetostatic problems using the magnetic vector potential $\mathbf{A}$.

Capacitance, Inductance, and Magnetic Circuits

Calculations of capacitance and inductance were presented as interesting applications of electrostatic and magnetostatic problems. They represent geometrical properties that describe the ability of a system of conductors to store the electric or magnetic energies, respectively. For a given arrangement of a system of conductors and the potential difference between them, the capacitance is calculated as

$$C = \frac{Q}{V}$$

where Q is the induced charge on the positively or negatively charged conductors. For a given potential difference on a set of conductors, Q may be calculated analytically for simple geometries or numerically using techniques such as the finite-difference method or the method of moments. Examples 4.14 to 4.17 illustrate such calculations.

For a given arrangement of conductors and the currents flowing in them, the mutual inductance is defined as

$$L_{12} = \frac{\psi_{12}}{I_1}$$

where ψ_{12} is the magnetic flux linking a second conductor because of a current I_1 flow in a first set of conductors. The self-inductance L_{11} is defined as

$$L_{11} = \frac{\psi_{11}}{I_1}$$

where ψ_{11} is the magnetic flux linking the same conductor carrying the current I_1. For a given current I_1 in a system of conductors, the flux linking the same conductors ψ_{11} or linking a second set of conductors ψ_{12} may be calculated from

$$\psi_m = \int_s \mathbf{B} \cdot d\mathbf{s}$$

and the self-inductance or mutual inductance is then determined.

Finally, a simple analysis procedure for magnetic circuits was introduced based on an analogy with electric circuits. Table 4.4 summarizes analogous quantities and examples 4.21 to 4.24 illustrate the solution procedure.

PROBLEMS

1. A conducting sphere of radius a is charged with a charge Q_a. Consider placing this sphere inside a spherical shell of radius $b > a$. Determine the potential on the spherical shell in the following cases:
   (a) It is uncharged.
   (b) It is charged with a charge Q_b.

2. Consider a circular loop of radius a. If the loop is uniformly charged with a total charge Q, determine the following:
   (a) Electrostatic potential at a point along the axis normal to the plane of the loop.
   (b) Electric field intensity at the same point in part a.
   (c) Using Coulomb's law, and integrating along the contour of the loop, determine the electric-field intensity at a point on the loop axis and normal to its plane. Compare your result with part b.

3. A very long straight conductor is charged with a linear charge density ρ_ℓ C/m.
   (a) Determine the electrostatic potential Φ at a point a distance ρ from the conductor.
   (b) Use Φ in part a to calculate the vector electric field intensity at that point.
   (c) Use Coulomb's law and use symmetry considerations to calculate the electric field intensity at the point described in part a.
   (d) Construct a suitable Gaussian surface around the charged conductor and use Gauss's law to determine the electric field intensity at the point described in part a.
   (e) Compare the obtained results for the parts b to d.

4. Consider the linear quadrapole shown in Figure P4.4. It basically consists of two dipoles superposed along the z axis. Determine the potential at P at a far distance r (i.e., $r \gg d$) from the charges and the electric field at that distance point. Use the approximations

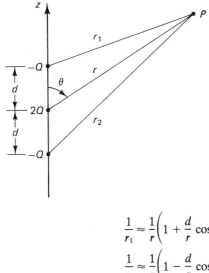

Figure P4.4 The geometry of a linear quadrapole.

$$\frac{1}{r_1} \approx \frac{1}{r}\left(1 + \frac{d}{r}\cos\theta\right)$$

$$\frac{1}{r_2} \approx \frac{1}{r}\left(1 - \frac{d}{r}\cos\theta\right)$$

to simplify your calculations.

5. Following a procedure similar to the one illustrated in examples 4.7 and 4.8, determine the capacitance of two concentric spherical conductors of radii a and b when a spherical dielectric shell of thickness d is placed concentrically between the conductors as shown in Figure P4.5a. The dielectric shell has an inner radius c and dielectric constant $\epsilon = \epsilon_o \epsilon_r$. Show that the total capacitance is the sum of three series capacitances each with a homogeneous dielectric layer as shown in Figure P4.5b.

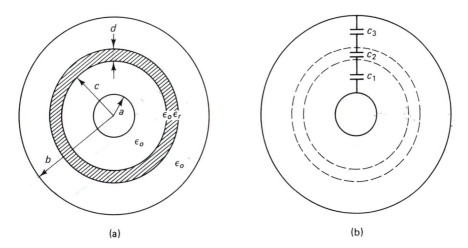

(a)   (b)

Figure P4.5 Geometry of spherical capacitor filled with multilayer dielectric.

6. Consider a spherical volume charge of radius R and a uniform charge density ρ_v C/m^3.
   (a) Use Gauss's law to determine the electric field inside $r < R$ and outside $r > R$ the spherical volume charge.

(b) Show that the total energy stored (inside and outside the spherical charge) is given by $W = (3/20) Q^2/\pi\epsilon_o R$, where $Q = 4/3\pi R^3 \rho_v$.

7. Show that the electrostatic energy for the case of a total charge Q distributed on a spherical surface of radius R is given by $W = Q^2/8\pi\epsilon_o R$. (*Hint:* The electric field inside the sphere $r < R$ is equal to zero in this case.)

8. In a capacitor, all the charge resides on the conducting electrodes as a surface charge.
   (a) Use this fact and equation 4.30 for the electrostatic energy stored to show that the energy stored in a capacitor is given by

$$W = \frac{1}{2} QV$$

   where Q is the magnitude of the total charge on each conducting electrode, and V is the potential difference between the electrodes.
   (b) In a capacitor in which the charge and voltage are linearly related—hence, $Q = CV$ where C is the capacitance—show that

$$W = \frac{1}{2}\frac{Q^2}{C}$$

   (c) A battery charges a parallel plate capacitor to a potential difference V. The battery is then disconnected, and the spacing between the plates is increased by a factor of 3. Determine the factor by which the potential energy is increased.

9. The two infinite conductors shown in Figure P4.9 form a wedge region between $\phi = \phi_1$ and $\phi = \phi_2$. The conductor at $\phi = \phi_1$ is kept at a potential $\Phi = 0$, whereas the other conductor at $\phi = \phi_2$ is at potential $\Phi = 100$ V. The charge density in the wedged region between the two conductors is zero.

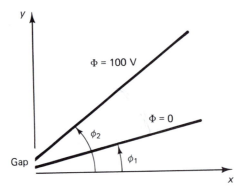

Figure P4.9 Two infinite conductor planes maintain at a potential difference of 100 V.

   (a) Solve Laplace's equation in cylindrical coordinates to determine the potential distribution in the wedged region.
   (b) Determine the electric field in the region $\phi_1 \leq \phi \leq \phi_2$.
   (c) Calculate the charge density on each conducting plate.

10. The two infinite conducting cones $\theta = \theta_1$ and $\theta = \theta_2$ are maintained at the two potentials $\Phi_1 = 100$ V and $\Phi_2 = 0$, respectively, as shown in Figure P4.10.
    (a) Use Laplace's equation in the spherical coordinates to solve for the potential variation between the two cones.
    (b) Calculate the electric field vector in the region between the two cones and the charge density on each conductor.

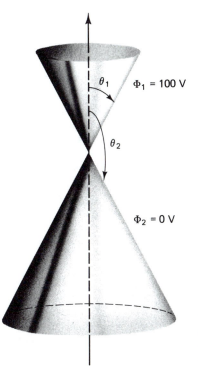

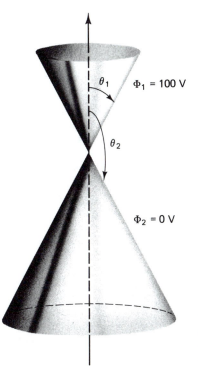

θ_1

$\Phi_1 = 100$ V

θ_2

$\Phi_2 = 0$ V

Figure P4.10 Two infinite conductor cones maintained at two different potentials $\Phi_1 = 100$ V and
$\Phi_2 = 0$ V.

11. Consider the parallel plates capacitor shown in Figure P4.11. The region between the parallel
    plates is filled with a nonuniform charge distribution of density $\rho(y) = \sigma_o y$, where σ_o is a
    constant. Solve Poisson's equation in the region between the parallel plates to show that the
    potential distribution $\Phi(y)$ is given by

$$\Phi(y) = \frac{V}{d}y + \frac{\sigma_o}{6\epsilon_o}(yd^2 - y^3)$$

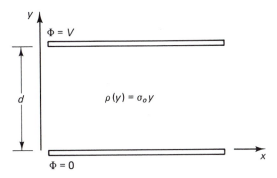

y

$\Phi = V$

d

$\rho(y) = \sigma_o y$

x

$\Phi = 0$

Figure P4.11 Parallel plate capacitor with a nonuniform charge distribution between the plates.

12. Use the expression of the potential distribution in problem 11 to obtain an expression for
    the electric field between the two plates. Show that the charge density at the lower plate is
    given by

$$\rho_s = -\epsilon_o\left(\frac{V}{d} + \frac{\sigma_o}{6\epsilon_o}d^2\right)$$

whereas the charge density at the upper plate is given by

$$\rho_s = \epsilon_o \left(\frac{V}{d} - \frac{\sigma_o}{3\epsilon_o} d^2 \right)$$

The capacitance is defined as $C = Q/V$ and because in this case there are two different values of Q on the lower and upper plates for the same potential difference V, there is no unique value for the capacitance C under these circumstances.

13. Consider the coaxial line shown in Figure P4.13. If the region between the center and outer conductors is filled with a charge density $\rho_s(\rho) = \sigma_o/\rho$, where σ_o is constant, determine the following:
    (a) Potential distribution between the inner and outer conductors.
    (b) Electric field intensity in the region between the conductors.
    (c) Charge density on the surfaces of the two conductors, $\rho = a$ and $\rho = b$. Can you define a unique value of the capacitance in this case? Why?

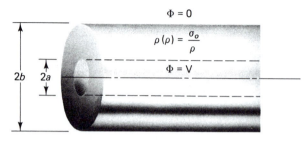

Figure P4.13 Coaxial line with a nonuniform charge distribution between the two conductors.

14. Use the finite-difference method to solve for the potential distribution in the bounded geometry shown in Figure P4.14.

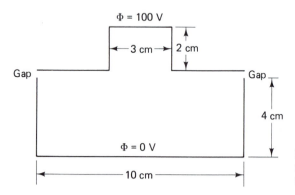

Figure P4.14 Geometry of a system of conductors kept at fixed potentials.

15. Use the finite-difference method to calculate the capacitance per unit length for the microstrip line geometry shown in Figure P4.15. The potential may be assumed zero at the boundaries indicated by the dotted line.

16. Consider the coplanar wave guide shown in Figure P4.16. The center conductor is maintained at a potential of 5 V, and the two side conductors were grounded. For the given geometry, the potential may be assumed zero at the boundaries indicated by the dotted lines. Use the finite-difference method and the finite-difference representation at dielectric interfaces to determine the potential distribution in this coplanar geometry.

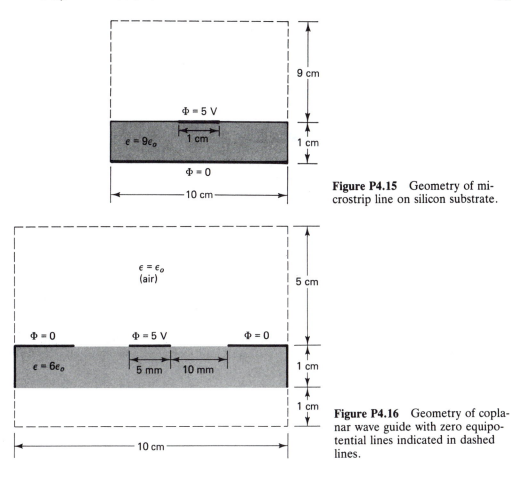

Figure P4.15 Geometry of microstrip line on silicon substrate.

Figure P4.16 Geometry of coplanar wave guide with zero equipotential lines indicated in dashed lines.

17. Use the method of moments solution procedure (equation 4.65) to calculate the charge distribution on the two cylindrical conductors shown in Figure P4.17. Make your calculations for the following three cases:
    (a) $\Phi_1 = \Phi_2$.
    (b) $\Phi_1 = 2\Phi_2$.
    (c) $\Phi_1 = 1/2\,\Phi_2$.

    Compare the charge distributions for these three cases.

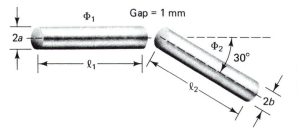

Figure P4.17 Geometry of two conductors maintained at two different potentials. $\ell_1 = 5$ cm, $\ell_2 = 100$ cm, $a = 2$ mm, and $b = 5$ mm.

18. Use the method of moments program developed for problem 17 to calculate the capacitance of two parallel conductors of radius $a = 1$ mm and separated by a distance $d = 5$ cm. The two cylindrical conductors are of equal length $\ell = 0.5$ m and displaced by a distance x as shown in Figure P4.18. Make your calculations as a function of $0.1\ell \leq x \leq \ell$.

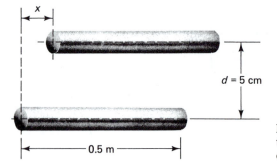

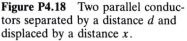

Figure P4.18 Two parallel conductors separated by a distance d and displaced by a distance x.

19. It is often simple to calculate the capacitance of a parallel plate capacitor, particularly if the separation distance is much smaller than the area of the plates. Sometimes it is of interest to calculate the capacitance if the plates are displaced by a distance x as shown in Figure P4.19. In this case, numerical techniques would be handy, and calculations may be made as a function of x. Use the method of moments and write a computer program according to equation 4.73 to calculate the capacitance of the parallel plate capacitor shown in Figure P4.19. Make your calculations as a function of x and check the accuracy of your results for the case when $x = 0$ and $d \gg \ell$.

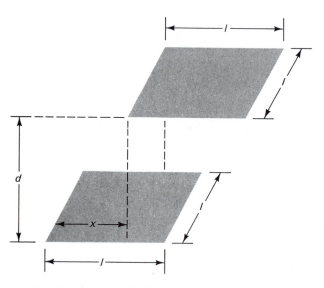

Figure P4.19 Parallel plate capacitor with the plates displaced by a distance x. $\ell = 1$ m, $0.05 < d \leq \ell$, $0 \leq x \leq \ell$.

20. Use the method of moments program developed for solving problem 19 to calculate the capacitance of a parallel plate capacitor for the case in which the areas of the plates are not equal. Assume one plate has a square area ℓ^2, whereas the other has a rectangular area $\ell_1 \ell_2$, where $\ell_1 = \ell$ and the value of ℓ_2 may vary $0.2\ell \leq \ell_2 \leq \ell$. Make all calculations for $d = 0.5\ell$.

21. In cylindrical geometries such as the junction of two coaxial transmission lines shown in

Figure P4.21, accurate results of the potential distribution and junction capacitance may be obtained using the finite-difference method in cylindrical coordinates. Starting from Laplace's equation in the cylindrical coordinates system, show that for a symmetrical potential in ϕ the finite-difference representation at point P (e.g., in the Nth row) in terms of the potentials at its four immediate neighbors in the cylindrical coordinates is given by (see Figure P4.21b)

$$\Phi_A + \Phi_B + \left(1 - \frac{h}{2\rho}\right)\Phi_C + \left(1 + \frac{h}{2\rho}\right)\Phi_D - 4\Phi_P = 0$$

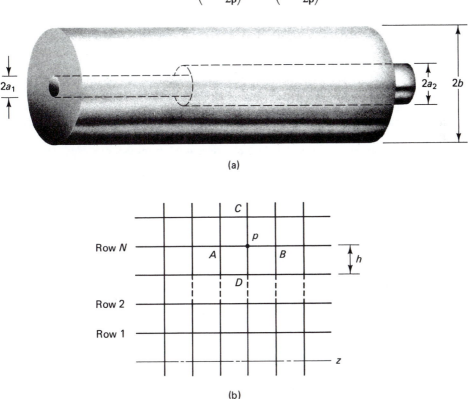

(a)

(b)

Figure P4.21(b)   Five-point arrangement of finite difference equation in cylindrical coordinates.

22. Write a computer program that uses the finite-difference representation of Laplace's equations in the cylindrical coordinate system to calculate the capacitance per unit length of the coaxial transmission line shown in Figure P4.22. To limit the computation domain, select length $\ell = 1$ m of the coaxial line and apply the boundary condition $\partial\Phi/\partial n = 0$, (i.e., Φ at nodes just before the boundary equal to Φ at nodes just after the boundary) on the A and B boundaries. Select values of a and b and compare numerical results with analytical values of example 4.8 assuming $\epsilon_1 = \epsilon_2 = \epsilon_o$.

23. To calculate the junction capacitance for the coaxial geometry of Figure P4.21a, follow the following steps:

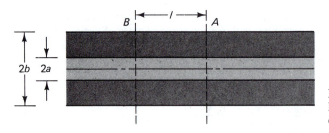

Figure P4.22   Coaxial transmission line and limits of the computation domain.

(a) Use the computer program of problem 22 to calculate the capacitance per unit length for a coaxial line of inner and outer diameters $2a_1$ and $2b$, respectively.

(b) Repeat part a for a coaxial line of inner and outer diameters $2a_2$ and $2b$, respectively.

(c) Select a section of the coaxial junction of length $\ell = 2$ m and use the finite-difference program of problem 22 to calculate the total capacitance of this section. Select the dimensions such that $\ell \geq 10b$. This is important because the boundary condition $\partial\Phi/\partial n = 0$ is valid at the end of the computation domain only far away from the junction. Why?

(d) Subtract the capacitances calculated from parts a and b from the total capacitance of part c to obtain the junction capacitance.

(e) Check your results by comparing them with those in Table P4.23.

TABLE P4.23   DISCONTINUITY CAPACITANCE IN COAXIAL LINE

| α | $T = 1$ | $\alpha = \dfrac{b - a_2}{b - a_1}$, $T = \dfrac{b}{a_1}$ | | | |
		$T = 3$	$T = 6$	$T = 11$	$T = \infty$
0.1	0.10864	0.11072	0.11308	0.11524	0.13633
0.2	0.06977	0.07209	0.7434	0.07642	0.09630
0.3	0.04779	0.04975	0.05184	0.05379	0.07298
0.4	0.03291	0.03456	0.03643	0.03821	0.05643
0.5	0.02212	0.02344	0.02504	0.02661	0.04355
0.6	0.01408	0.01507	0.01635	0.01765	0.03298
0.7	0.00810	0.00877	0.00969	0.01069	0.02399
0.8	0.00382	0.00420	0.00475	0.00539	0.01612
0.9	0.00110	0.00123	0.00143	0.00170	0.00895
1.0	0.00000	0.00000	0.00000	0.00000	0.00000

24. At the interface between two magnetic materials shown in Figure P4.24, a surface current density $\mathbf{J}_s = 0.1\,\mathbf{a}_y$ is flowing. The magnetic field intensity $\mathbf{H}_2$ in region 2 is given by

$$\mathbf{H}_2 = 3\,\mathbf{a}_x + 9\,\mathbf{a}_z$$

Determine the magnetic flux densities $\mathbf{B}_1$ and $\mathbf{B}_2$ in regions 1 and 2, respectively.

25. Consider the problem of determining the magnetic vector potential $\mathbf{A}$ inside and outside an infinite circular cylindrical solenoid of radius a. The solenoid has N turns per unit length and the current in the winding is I.

(a) Use the curl relation between $\mathbf{A}$ and the magnetic flux density $\mathbf{B} = \nabla \times \mathbf{A}$ and Stokes's theorem to show that

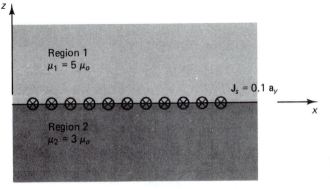

Figure P4.24 Surface boundary between two magnetic media.

$$\oint_c \mathbf{A} \cdot d\boldsymbol{\ell} = \int_s \mathbf{B} \cdot d\mathbf{s}$$

where s is the area encircled by c.

(b) Based on symmetry considerations, select suitable contours for $\mathbf{A}$ inside and outside the solenoid to show that

$$\mathbf{A} = \frac{\mu_o NI\rho}{2}\,\mathbf{a}_\phi \qquad \text{for } \rho < a$$

$$\mathbf{A} = \frac{\mu_o NIa^2}{2\rho}\,\mathbf{a}_\phi \qquad \text{for } \rho > a$$

26. Consider the magnetic circuit shown in Figure P4.26. Calculate the flux density in each of the three legs assuming that μ_r of the silicon steel under the given operating conditions to be $\mu_r = 4 \times 10^3$.

27. In the magnetic circuit shown in Figure P4.27a, it is desired to produce a magnetic flux density $B = 1 \text{ Wb/m}^2$ in the air gap. Determine the required ampere turns to produce the desired flux density. Use the **B-H** curve given in Figure P4.27b.

28. As in the case of a coaxial transmission line, it is often desirable to calculate the inductance per unit length of a parallel wire transmission line such as the one shown in Figure P4.28.

(a) Show that the magnetic flux produced by the lower conductor and linking a section of length ℓ of the upper conductor is given by

$$\Psi_m = \frac{\ell \mu_o I}{2\pi} \ell n \, d/a \qquad (\text{use } d \gg a)$$

(b) Show that by calculating the total flux linking the two conductors, the inductance per unit length is given by

$$L = \frac{\mu_o}{\pi} \ell n \, d/a$$

29. Use the result of equation 4.99a for the magnetic flux at a far point from a circular current loop to determine approximately the mutual inductance between two thin coaxial circular rings of radii a and b. Assume that the distance d between the two rings is much larger than a and b.

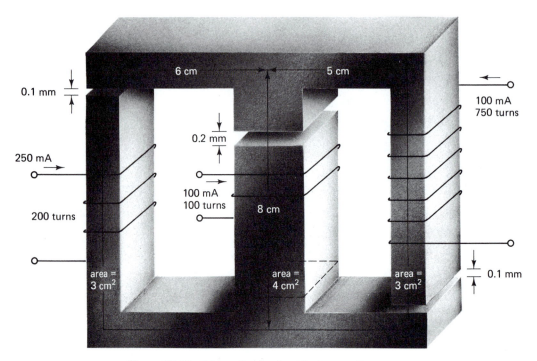

Figure P4.26   Magnetic circuit with three coil sources.

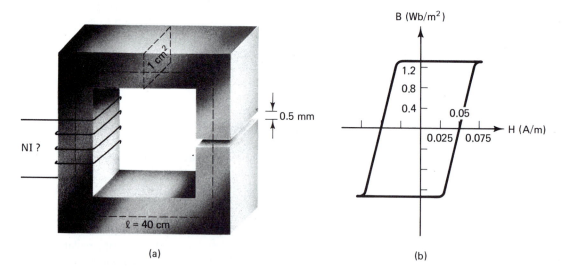

Figure P4.27   Magnetic circuit and the **B-H** curve of the core magnetic materials.

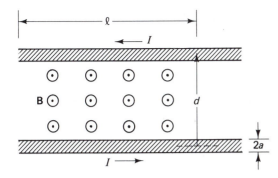

Figure P4.28 Parallel wire transmission line.

30. Determine the mutual inductance between an infinitely long conductor carrying a current I and a rectangular loop of side lengths a and b. The leading edge of the loop is placed at a distance d from the wire as shown in Figure P4.30.

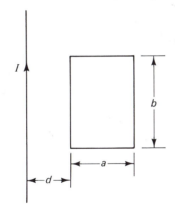

Figure P4.30 Mutual coupling between a current-carrying conductor and a rectangular loop.

31. Consider the N-turn solenoid shown in Figure P4.31. The ferromagnetic core consists of two identical halves.
   (a) If the solenoid is carrying a current I, calculate the magnetostatic energy stored assuming an infinitely long solenoid.
   (b) Determine the mechanical force required to separate the two halves of the ferromagnetic core by a small distance $d\ell$.

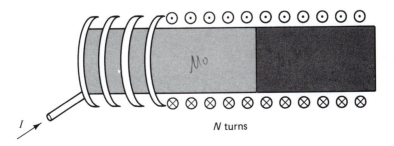

Figure P4.31 An N-turn solenoid with two halves ferromagnetic core.

32. Consider the horseshoe electromagnet A, and the horseshoe-shaped iron bar B shown in Figure P4.32. From equation 4.129 and if we assume uniform magnetic flux density $\mathbf{B} = \mu_o\mathbf{H}$ in the air gap, show that the total magnetic energy stored in the air gap is given by

$$W_m = 2\left(\frac{\mu_o}{2} H^2 sx\right)$$

where s is the cross-sectional area of the gap, and x is the gap length. The factor of 2 is included because of the presence of two air gaps. Use the law of conservation of energy—that is, the change in magnetic energy equals the mechanical work done—to show that the magnetic force is given by

$$F_x = \mu_o H^2 s \ (\text{N})$$

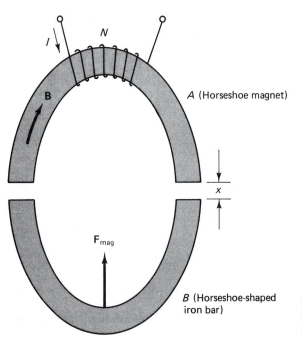

A (Horseshoe magnet)

B (Horseshoe-shaped iron bar)

Figure P4.32 Magnetic force between an electromagnet and a horseshoe-shaped iron bar.

CHAPTER 5

NORMAL-INCIDENCE PLANE WAVE REFLECTION AND TRANSMISSION AT PLANE BOUNDARIES

1 INTRODUCTION

In our previous discussion of the plane wave propagation in free space (chapter 2) and in conductive medium (chapter 3), we assumed the propagation medium to be either free space (air) or filled with a single material medium of constant properties throughout the entire propagation space. In this chapter, we shall examine the reflection and transmission properties of plane waves when incident on boundaries between regions of different electrical properties. To introduce these types of reflection problems, let us consider a plane wave propagation in a conductive medium of electrical properties ϵ_1, μ_1, and σ_1. This wave is incident on a three-dimensional dielectric object of properties ϵ_2, μ_2, and σ_2. As a result of the presence of this object in the path of the wave, the wave will be scattered in all directions around the object. To explain this scattering process, we may note that, as a result of the interaction of the incident fields with the three-dimensional object, polarization currents and charges will be induced inside and at the surface of the object. These new polarization and magnetization sources will in turn radiate what we call the scattered fields. The total field (electric or magnetic) at any point in space is the sum of the originally incident fields plus the scattered ones generated as a result of the interaction of the incident wave with the dielectric object.

With this brief introduction emphasizing the need to generate new waves if there is a discontinuity in the propagation path, let us examine the general procedure for determining the amplitudes and propagation properties of these scattered waves. Because these new waves are generated by the induced polarization and magnetization currents and polarization charge, their field expressions should satisfy Maxwell's equations. As we know, Maxwell's equations are the general mathematical relations between the fields and their sources, and expressions for the scattered fields should, therefore, satisfy Maxwell's equations. The originally incident fields, together with the scattered fields, should also satisfy the boundary conditions on the surface of the object. From chapter 3, we learned that these boundary conditions are general mathematical relations that should be satisfied at interfaces separating different media. In summary, therefore, the general solution procedure of these types of scattering problems includes first assuming expressions for the scattered fields that should satisfy Maxwell's equations and then determining the amplitudes of these fields by requiring that they, together with the originally incident ones, satisfy the boundary conditions on the surface of the object.

In this chapter, we are not going to solve scattering problems from three-dimensional objects. This would require solving the wave equation in the spherical coordinate system, which is far beyond the scope of this text. Instead we are going to limit our discussion to the case of one-dimensional problems in which we have normal-incident plane waves approaching on infinite plane interfaces separating two or more different media. We will start with the case in which we have a single infinite plane interface separating two different conductive media and then move on to generalize the solution procedure to the case in which we have multiple interfaces separating different materials. The very important case of reflection from a perfectly conducting plane will be described as a special case of the single-interface problem. Also, because of the difficulty in using the boundary-condition solution procedure to solve for the reflection and transmission at multiple interfaces, we will introduce a systematic solution method for solving these problems. The graphical implementation of the latter method on a Smith chart will also be described. We will then conclude the chapter by describing some practical applications that use the reflection and transmission ideas developed in the earlier sections of this chapter.

5.2 NORMAL INCIDENCE PLANE WAVE REFLECTION AND TRANSMISSION AT PLANE BOUNDARY BETWEEN TWO CONDUCTIVE MEDIA

Consider a plane wave that is propagating along the positive z axis with its electric field oriented in the x direction. This wave is normally incident on an infinite interface separating two conductive media as shown in Figure 5.1. Region 1 ($z < 0$) has the

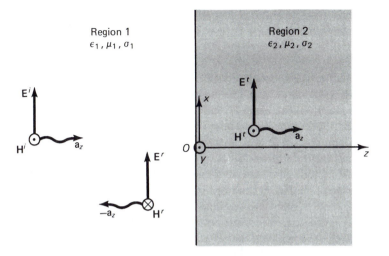

Figure 5.1   The geometry of the positive z propagating plane wave, which is normally incident on a plane interface between regions 1 and 2.

electrical properties ϵ_1, μ_1^*, and σ_1, whereas region 2 ($z > 0$) has the properties ϵ_2, μ_2, and σ_2. The electric and magnetic fields associated with the incident wave are given by

$$\hat{E}_x^i = \hat{E}_{m1}^+ e^{-\hat{\gamma}_1 z}$$

$$\hat{H}_y^i = \frac{\hat{E}_{m1}^+}{\hat{\eta}_1} e^{-\hat{\gamma}_1 z} \tag{5.1}$$

where $\hat{\gamma}_1 = \alpha_1 + j\beta_1$ is the propagation constant in region 1, whereas $\hat{\eta}_1$ is the wave impedance in the same region. Expressions for $\hat{\gamma}_1$ and $\hat{\eta}_1$ in terms of the properties of medium 1 are given by equations 3.76 and 3.79 in chapter 3. Clearly, some of the energy associated with the incident wave will be transmitted across the boundary surface $z = 0$ into region 2, thus providing a transmitted wave moving in the positive z direction in medium 2. The electric and magnetic fields associated with the transmitted wave are given by

$$\hat{E}_x^t = \hat{E}_{m2}^+ e^{-\hat{\gamma}_2 z}$$

$$\hat{H}_y^t = \frac{\hat{E}_{m2}^+}{\hat{\eta}_2} e^{-\hat{\gamma}_2 z} \tag{5.2}$$

In this case, $\hat{\gamma}_2$ and $\hat{\eta}_2$ are the propagation constant and the wave impedance in region 2, which are generally different from $\hat{\gamma}_1$ and $\hat{\eta}_1$.

Based on the solution of the wave equation that we developed in chapter 3, equations 5.1 and 5.2 clearly satisfy Maxwell's equations. To determine the unknown amplitude of the transmitted wave $\hat{E}_{m2}^+$ we simply require that these solutions satisfy the boundary condition at the interface $z = 0$, which separates the two media. Because both $\hat{E}_x^t$ and $\hat{E}_x^i$ are tangential to the interface, the boundary conditions require that these fields must be equal at $z = 0$. Equating the two electric fields in 5.1 and 5.2 and setting $z = 0$, we obtain

$$\hat{E}_{m1}^+ = \hat{E}_{m2}^+ \tag{5.3}$$

The magnetic fields associated with both the incident and transmitted waves are also tangential to the interface. The boundary conditions require that these fields must also be equal at $z = 0$. Equating $\hat{H}_y^t$ to $\hat{H}_y^i$ at $z = 0$ and noting that $\hat{E}_{m1}^+ = \hat{E}_{m2}^+$ from equation 5.3, we conclude that it is impossible to satisfy the magnetic field boundary condition as long as $\hat{\eta}_1 \neq \hat{\eta}_2$. We are therefore unable to satisfy both the electric and magnetic fields boundary conditions by using the incident and transmitted waves alone. Instead, we should include in our analysis a reflected wave in region 1 that travels away from the interface—that is, in the negative z direction. To justify the need for including the reflected wave in our analysis, besides the mathematical requirement of satisfying the boundary condition, we should recall that only a portion of the energy associated with the incident wave will be transmitted to region 2. Because of the adjustment the incident fields have to go through before they can cross the boundary—for example, the ratio between $\hat{E}_x^i / \hat{H}_y^i = \hat{\eta}_1$, which is different from $\hat{\eta}_2$, and hence not all the incident $\hat{E}_x^i$ and $\hat{H}_y^i$ may cross the boundary—there will be some fields left behind, and these fields constitute the reflected wave. This simple physical explanation may help us justify the need for including the reflected wave in our analysis. From the solution of the wave equation in chapter 3, we noted that the electric and magnetic fields associated with a wave traveling in the negative z direction (reflected wave) are given by

$$\hat{E}_x^r = \hat{E}_{m1}^- e^{\hat{\gamma}_1 z}$$
$$\hat{H}_y^r = -\frac{\hat{E}_{m1}^-}{\hat{\eta}_1} e^{\hat{\gamma}_1 z} \tag{5.4}$$

In equation 5.4 we introduced the complex amplitude of the reflected electric field $\hat{E}_{m1}^-$ as another unknown quantity in addition to $\hat{E}_{m2}^+$. We should also note that because the reflected wave is traveling in the negative z direction, the electric and magnetic fields in equation 5.4 were related by

$$\frac{\hat{E}_x^r}{\hat{H}_y^r} = -\hat{\eta}_1$$

so that the Poynting vector $\mathbf{E} \times \mathbf{H}$ would be in the $-\mathbf{a}_z$ direction.

With the introduction of the reflected wave, let us now try to satisfy the boundary condition. For the tangential electric field, we have at $z = 0$,

$$\left(\hat{E}_x^i + \hat{E}_x^r\right)\big|_{z=0} = \hat{E}_x^t\big|_{z=0}$$

hence,

$$\hat{E}_{m1}^+ + \hat{E}_{m1}^- = \hat{E}_{m2}^+ \tag{5.5}$$

Similarly, enforcing the continuity of the tangential magnetic field at $z = 0$, we have

$$\left(\hat{H}_y^i + \hat{H}_y^r\right)\big|_{z=0} = \hat{H}_y^t\big|_{z=0}$$

hence,

$$\frac{\hat{E}_{m1}^+}{\hat{\eta}_1} - \frac{\hat{E}_{m1}^-}{\hat{\eta}_1} = \frac{\hat{E}_{m2}^+}{\hat{\eta}_2} \tag{5.6}$$

To solve for $\hat{E}_{m2}^+$, we multiply equation 5.6 by $\hat{\eta}_1$ and add the resulting equation to equation 5.5. Carrying out this process, we find

$$2\hat{E}_{m1}^+ = \hat{E}_{m2}^+\left(1 + \frac{\hat{\eta}_1}{\hat{\eta}_2}\right)$$

or

$$\frac{\hat{E}_{m2}^+}{\hat{E}_{m1}^+} = \frac{2\hat{\eta}_2}{\hat{\eta}_1 + \hat{\eta}_2} \tag{5.7}$$

This ratio of the amplitudes of the transmitted to the incident fields is known as the "transmission coefficient" and is denoted by $\hat{T}$. From equation 5.7, $\hat{T}$ is given by

$$\boxed{\hat{T} = \frac{2\hat{\eta}_2}{\hat{\eta}_1 + \hat{\eta}_2}} \tag{5.8}$$

To solve for the amplitude of the reflected wave, we multiply equation 5.6 by $\hat{\eta}_2$ and subtract the resulting equation from equation 5.5. Carrying out this process, we find

$$\hat{E}_{m1}^+\left(1 - \frac{\hat{\eta}_2}{\hat{\eta}_1}\right) + \hat{E}_{m1}^-\left(1 + \frac{\hat{\eta}_2}{\hat{\eta}_1}\right) = 0$$

or

$$\hat{E}_{m1}^- = \hat{E}_{m1}^+\left(\frac{\hat{\eta}_2 - \hat{\eta}_1}{\hat{\eta}_2 + \hat{\eta}_1}\right) \tag{5.9}$$

The ratio of the amplitudes of the reflected and incident electric fields is called the reflection coefficient $\hat{\Gamma}$, which is given by

$$\boxed{\hat{\Gamma} = \frac{\hat{E}_{m1}^-}{\hat{E}_{m1}^+} = \frac{\hat{\eta}_2 - \hat{\eta}_1}{\hat{\eta}_2 + \hat{\eta}_1}} \tag{5.10}$$

From equations 5.8 and 5.10, we may note that the reflection and transmission coefficients are related by

$$1 + \hat{\Gamma} = \hat{T}$$

Let us now use these relations to solve some examples.

EXAMPLE 5.1

The electric field associated with a uniform plane wave propagating in air is given by

$$\mathbf{E}^i = 1000 \cos(10^8 \, \pi t - \beta_o z) \, \mathbf{a}_x \text{ V/m}$$

If this wave is normally incident on a glass medium of the following electrical properties ($\epsilon_g = 5\epsilon_o$, $\mu_g = \mu_o$, and $\sigma_g = 0$), determine the following:

1. β_o in air and β_g in glass.
2. Reflection $\hat{\Gamma}$ and transmission $\hat{T}$ coefficients.
3. Amplitudes of reflected and transmitted electric and magnetic fields.

Solution

1. Because $\sigma = 0$ in both media,

$$\beta_o = \omega\sqrt{\mu_o \epsilon_o} = \frac{\pi \times 10^8}{c} = \frac{\pi \times 10^8}{3 \times 10^8} = 1.05 \text{ rad/m}$$

$$\beta_g = \omega\sqrt{\mu_o \epsilon_o \epsilon_r} = \beta_o\sqrt{\epsilon_r} = 2.35 \text{ rad/m}$$

2.
$$\eta_1 \text{ in air} = \sqrt{\frac{\mu_o}{\epsilon_o}} = 120\pi \ \Omega$$

$$\eta_2 \text{ in glass} = \sqrt{\frac{\mu_o}{\epsilon_o \epsilon_r}} = \frac{120\pi}{\sqrt{5}} = 168.6 \ \Omega$$

$$\Gamma = \frac{\eta_2 - \eta_1}{\eta_2 + \eta_1} = -0.38$$

$$T = \frac{2\eta_2}{\eta_1 + \eta_2} = 0.62$$

3. The amplitude of the reflected electric field is

$$\hat{E}_{m1}^- = \Gamma \hat{E}_{m1}^+ = -0.38 \times 1000 = -380 \text{ V/m}$$

The amplitude of the transmitted electric field is

$$\hat{E}_{m2}^+ = T \hat{E}_{m1}^+ = 0.62 \times 1000 = 620 \text{ V/m}$$

The amplitude of the reflected magnetic field is simply

$$-\frac{\hat{E}_{m1}^-}{\eta_1} = -\frac{(-380)}{1200\pi} = 1.01 \text{ A/m}$$

whereas the amplitude of the transmitted magnetic field is given by

$$\frac{\hat{E}_{m2}^+}{\eta_2} = 3.68 \text{ A/m}$$

◆◆

EXAMPLE 5.2

A uniform plane wave is normally incident on the interface (at $z = 0$) separating two media. The incident wave is propagating in free space (region 1) at a frequency of 1 GHz and with an amplitude of the electric field $\hat{E}_{m1}^+ = 100e^{j0^\circ}$ V/m. The other medium (region 2) is conductive with $\sigma_2 = 25$ S/m, $\epsilon_{r_2} = 2$, and $\mu_{r_2} = 1$. Obtain real-time (time-domain) expressions for the incident, reflected, and transmitted electric and magnetic fields.

Solution

To simplify the calculations, let us calculate first the ratio $\sigma/\omega\epsilon$ for the conductive medium.

$$\frac{\sigma}{\omega\epsilon} = \frac{25}{2\pi \times 10^9 \times 2\epsilon_o} = 225$$

The medium in region 2 may therefore be considered a good conductor. Under the approximation $\sigma/\omega\epsilon \gg 1$, the propagation constant and the wave impedance in region 2 are given by

$$\hat{\eta}_2 = \sqrt{\frac{\omega\mu}{\sigma}}\underline{/45^\circ} = 17.77\underline{/45^\circ}\ \Omega$$

$$\alpha_2 = \beta_2 = \sqrt{\frac{\omega\mu\sigma}{2}} = 314.2$$

The wave impedance and the propagation parameters in air (region 1), conversely, are given by

$$\eta_1 = 120\pi, \qquad \beta_1 = \omega\sqrt{\mu_o\,\epsilon_o} = 20.9, \qquad \alpha_1 = 0$$

The complex reflection and transmission coefficients are, hence,

$$\hat{\Gamma} = \frac{\hat{\eta}_2 - \eta_1}{\hat{\eta}_2 + \eta_1} = 0.935\underline{/176.15^\circ}$$

$$\hat{T} = \frac{2\hat{\eta}_2}{\hat{\eta}_2 + \eta_1} = 0.091\underline{/43.15^\circ}$$

To check the accuracy of the calculation, we may want to show that our numbers confirm the mathematical relation $1 + \hat{\Gamma} = \hat{T}$.

After obtaining numerical values for the wave impedances, propagation constants, and the reflection and transmission coefficients, we may now proceed with obtaining time-domain forms for the various fields.

$$\mathbf{E}^i(z,t) = E_{m1}^+ \cos(\omega t - \beta_o z)\,\mathbf{a}_x$$
$$= 100\cos(2\pi \times 10^9 t - 20.9z)\,\mathbf{a}_x\ \text{V/m}$$

$$\mathbf{H}^i(z,t) = \frac{100}{120\pi}\cos(2\pi \times 10^9 t - 20.9z)\,\mathbf{a}_y$$
$$= 0.27\cos(2\pi \times 10^9 t - 20.9z)\,\mathbf{a}_y\ \text{A/m}$$

If we express $\hat{\Gamma} = |\hat{\Gamma}|e^{j\theta_r}$ and $\hat{T} = |\hat{T}|e^{j\theta_t}$, we obtain

$$\mathbf{E}^r(z,t) = |\hat{\Gamma}|E_{m1}^+ \cos(\omega t + \beta_o z + \theta_r)\,\mathbf{a}_x$$
$$= 93.5\cos(2\pi \times 10^9 t + 20.9z + 176.15^\circ)\,\mathbf{a}_x\ \text{V/m}$$

and

$$\mathbf{H}^r(z,t) = -\frac{93.5}{120\pi}\cos(2\pi \times 10^9 t + 20.9z + 176.15^\circ)\,\mathbf{a}_y$$
$$= -0.25\cos(2\pi \times 10^9 t + 20.9z + 176.15^\circ)\,\mathbf{a}_y\ \text{A/m}$$

The plus sign in front of the $\beta_o z$ term and the minus sign in the ratio between the amplitudes of the electric and magnetic fields are typical for a reflected wave propagating in the negative z direction.

The transmitted electric and magnetic fields are given by

$$\mathbf{E}^t(z,t) = |\hat{T}|E_{m1}^+ e^{-\alpha_2 z} \cos(\omega t - \beta_2 z + \theta_t)\,\mathbf{a}_x$$

$$= 9.1e^{-\alpha_2 z} \cos(2\pi \times 10^9 t - 314.2z + 43.15°)\,\mathbf{a}_x \text{ V/m}$$

$$\mathbf{H}^t(z,t) = \frac{9.1}{|\hat{\eta}_2|}e^{-\alpha_2 z} \cos(2\pi \times 10^9 t - 314.2z + 43.15° - \theta_\eta)\,\mathbf{a}_y$$

where θ_η is the angle of the complex wave impedance in region 2

$$\mathbf{H}^t(z,t) = 0.51\,e^{-\alpha_2 z} \cos(2\pi \times 10^9 t - 314.2z + 43.15° - 45°)\,\mathbf{a}_y$$

$$= 0.51e^{-314.2z} \cos(2\pi \times 10^9 t - 314.2z - 1.85°)\,\mathbf{a}_y \text{ A/m}$$

In the expressions of the transmitted fields, we should remember to use β_2 and $\hat{\eta}_2$ of region 2. Otherwise, the expressions for the incident and transmitted fields are basically similar, with the exception of the attenuation factor, because these two waves are traveling along the positive z direction.

5.3 NORMAL-INCIDENCE PLANE-WAVE REFLECTION AT PERFECTLY CONDUCTING PLANE

This is just an important special case of the general analysis presented in the last section. We assume that region 2 is a perfect conductor $\sigma_2 \to \infty$, and the wave impedance in this region would then be

$$\hat{\eta}_2 = \sqrt{\frac{\mu_2}{\epsilon_2 - j\dfrac{\sigma_2}{\omega}}} = 0 \qquad \text{as } \sigma_2 \to \infty \qquad (5.11)$$

To simplify the standing wave analysis to be described next, we shall further assume that region 1 is a perfect dielectric $\sigma_1 = 0$.

Substituting equation 5.11 in the reflection and transmission coefficient expressions in equations 5.8 and 5.10, we obtain

$$\hat{T} = 0, \qquad \hat{\Gamma} = -1$$

The zero value of the transmission coefficient simply means that the amplitude of the transmitted field in region 2—that is, $\hat{E}_{m2}^+ = 0$. This can be explained in terms of the depth of penetration parameter, which is zero in a perfectly conducting region. In other words, there would be no transmitted wave in the perfectly conducting region because of the inability of time-varying fields to penetrate media with conductivities approaching ∞. With the absence of a transmitted wave, the incident and reflected fields in region 1 will be the only present ones in our special case. For $\hat{\Gamma} = -1$, the amplitude of the reflected wave $\hat{E}_{m1}^- = -\hat{E}_{m1}^+$. The reflected wave thus is equal in amplitude and is opposite in phase to the incident wave. In other words, all the incident energy is reflected back by the perfect conductor. The incident and reflected fields also combine with their equal magnitudes and opposite phases to satisfy the boundary condition at the surface of the perfect conductor. This can be illustrated by examining the expression for the total electric field $\mathbf{E}^{tot}(z)$ in region 1, which is assumed to be a perfect dielectric (i.e., $\alpha_1 = 0$)

$$\hat{\mathbf{E}}^{tot}(z) = \hat{\mathbf{E}}^i(z) + \hat{\mathbf{E}}^r(z) = \hat{E}_{m1}^+ e^{-j\beta_1 z} \mathbf{a}_x + \hat{E}_{m1}^- e^{j\beta_1 z} \mathbf{a}_x$$

Substituting $\hat{E}_{m1}^- = -\hat{E}_{m1}^+$, we obtain

$$\hat{\mathbf{E}}^{tot}(z) = \hat{E}_{m1}^+ (e^{-j\beta_1 z} - e^{j\beta_1 z}) \mathbf{a}_x$$
$$= -2j\hat{E}_{m1}^+ \sin\beta_1 z \, \mathbf{a}_x \tag{5.12}$$

From equation 5.12, it is clear that the total electric field is zero at the perfectly conducting surface ($z = 0$), which satisfies the required boundary condition.

To study the propagation characteristics of the compound wave in front of the perfect conductor, we need to obtain the real-time form of the electric field. We routinely multiply the complex form of the field in equation 5.12 by $e^{j\omega t}$ and take the real part of the resulting expression, hence,

$$\mathbf{E}^{tot}(z, t) = Re[e^{j\omega t} \hat{\mathbf{E}}^{tot}(z)]$$
$$= 2E_{m1}^+ \sin(\beta_1 z) \sin \omega t \, \mathbf{a}_x \tag{5.13}$$

In equation (5.13) the amplitude of the electric field was assumed real E_{m1}^+. It is our objective to show next that this total field in region 1 is not a traveling wave, although it was obtained by combining two traveling waves of the same frequency, equal amplitudes, and are propagating in opposite directions. To show this, let us sketch the variation of the total electric field in equation 5.13 as a function of z at various time intervals. Figure 5.2 shows these variations from which we may make the following observations:

1. The amplitude of the total electric field is always zero at the surface of the perfect conductor. This simply indicates that the total field satisfies the boundary condition at all times.
2. The maximum amplitude of the total electric field is double that of the incident wave. This maximum amplitude occurs at the specific locations ($z = \lambda/4, 3\lambda/4$, etc.) and at specific times ($\omega t = \pi/2, 3\pi/2$, etc.) at which both the incident and reflected waves constructively interfere.
3. There are locations in front of the perfect conductor ($z = \lambda/2, \lambda, 3\lambda/2$, etc.) at which the total electric field is always zero. These are the locations at which the incident and reflected fields go through a destructive interference process for all values of ωt. These locations are known as the null locations of the total electric field.
4. The null locations as well as the locations at which the constructive interferences occur do not change with time (i.e., as a function of ωt). All that actually changes

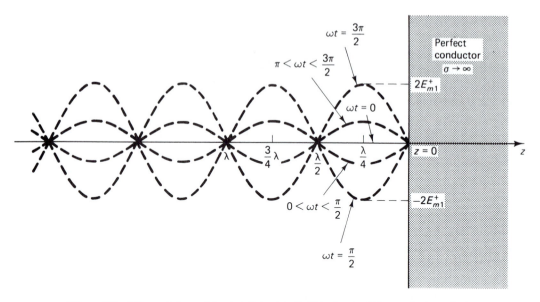

Figure 5.2   The variation of the total electric field in front of the perfect conductor as a function of z and at various time intervals ωt.

with time is the amplitude of the total field at the nonnull locations. This is why the wave resulting from the interference of the incident and reflected waves is called "standing wave" or nonpropagating wave.

We should also emphasize the difference between the electric field expressions for the traveling and standing waves. For a traveling wave, the electric field is given by

$$\mathbf{E}(z,t) = E_{m1}^{+} \cos(\omega t - \beta_1 z) \, \mathbf{a}_x$$

where the term $(\omega t - \beta_1 z)$ or $\omega(t - z/v_1)$ emphasizes the coupling between the location as a function of time of a specific point (constant phase) propagating along the wave. The constant phase term $(t - z/v_1)$ indicates that with the increase in t, z should also increase to maintain a constant value of $(t - z/v_1)$, which characterizes a specific point on the wave. This simply means that a wave with an electric field expression which includes $\cos(\omega t - \beta_1 z)$ is a propagating wave in the positive z direction. From equation 5.13, conversely, the time t and location z variables are uncoupled. In other words, the electric field distribution as a function of z in front of the perfect conductor follows a $\sin(\beta_1 z)$ variation, with the locations of the field nulls being those values of z at which $\sin(\beta_1 z) = 0$. The effect of the time term $\sin(\omega t)$ is simply to modify the amplitude of the field so as to vary as a function of time at the nonzero field locations as shown in Figure 5.2.

The permanent locations of the electric field nulls are determined by finding the values of $\beta_1 z$, which would make the value of the field zero. Thus, from equation 5.12, it may be seen that

$$\hat{E}^{tot}(z) = 0 \text{ at } \beta_1 z = n\pi \qquad (n = 0, \pm 1, \pm 2, \dots)$$

Hence,

$$\frac{2\pi}{\lambda_1} z = n\pi$$

or

$$z = n\frac{\lambda_1}{2} \tag{5.14}$$

where λ_1 is the wavelength in region 1. Equation 5.14 shows that $\hat{\mathbf{E}}^{tot}(z)$ is zero at the boundary $z = 0$ and at every half wavelength distance away from the boundary in region 1 as shown in Figure 5.2.

Let us also obtain an expression for the total magnetic field,

$$\hat{\mathbf{H}}^{tot}(z) = \hat{\mathbf{H}}^i(z) + \hat{\mathbf{H}}^r(z) = \left(\frac{E_{m1}^+}{\eta_1} e^{-j\beta_1 z} - \frac{E_{m1}^-}{\eta_1} e^{j\beta_1 z}\right) \mathbf{a}_y$$

The minus sign in the reflected magnetic field expression is simply because for a negative z-propagating wave the amplitude of the reflected magnetic field is related to that of the reflected electric field by $(-\eta_1)$. Substituting $E_{m1}^- = -E_{m1}^+$, we obtain

$$\hat{\mathbf{H}}^{tot}(z) = \frac{E_{m1}^+}{\eta_1}\left(e^{-j\beta_1 z} + e^{j\beta_1 z}\right) \mathbf{a}_y$$

$$= 2\frac{E_{m1}^+}{\eta_1} \cos\beta_1 z\ \mathbf{a}_y \tag{5.15}$$

The time-domain form of the magnetic field expression is obtained from equation 5.15 as

$$\mathbf{H}^{tot}(z,t) = 2\frac{E_{m1}^+}{\eta_1} \cos\beta_1 z\ \cos\omega t\ \mathbf{a}_y \tag{5.16}$$

This is also a standing wave as shown in Figure 5.3, with the maximum amplitude of the magnetic field occurring at the perfect conductor interface $(z = 0)$ where the total electric field is zero. The locations of the nulls in the magnetic field are at the values of z at which $\cos\beta_1 z = 0$, hence,

$$\beta_1 z = \text{odd number of } \frac{\pi}{2} = (2m + 1)\frac{\pi}{2} \qquad (m = 0, \pm 1, \pm 2, \dots)$$

or

$$z = (2m + 1)\frac{\lambda_1}{4} \tag{5.17}$$

The magnetic field distribution in front of a perfectly conducting boundary is shown in Figure 5.3, where it is clear that its first null occurs at $z = \lambda_1/4$, which is the location of the maximum electric field (see Figure 5.2). Comparing equation 5.13 with equation 5.16 also shows that the electric and magnetic fields of a standing wave are 90° out of time phase. This is simply because equation 5.13 contains a $\sin(\omega t)$ term, whereas 5.16

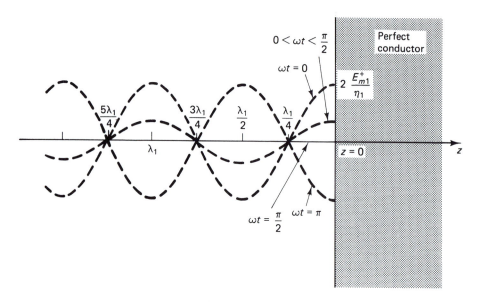

Figure 5.3 The magnetic field distribution in front of a perfect conductor as a function of time.

includes a $\cos(\omega t)$ time variation. The fact that these fields are 90° out of time phase results in a zero average power being transmitted in either direction by the standing wave. This can be illustrated by using the complex forms of the fields to calculate the time-average Poynting vector $\mathbf{P}_{av}(z)$

$$\mathbf{P}_{av}(z) = \frac{1}{2} Re\big[\hat{\mathbf{E}}(z) \times \hat{\mathbf{H}}^*(z)\big]$$

$$= \frac{1}{2} Re\left[-2j E_{m1}^+ \sin \beta_1 z\, \mathbf{a}_x \times 2\frac{E_{m1}^+}{\eta_1} \cos \beta_1 z\, \mathbf{a}_y\right] = 0 \qquad (5.18)$$

The zero value of $\mathbf{P}_{av}(z)$ is obtained because the result of the vector product of $\hat{\mathbf{E}}(z) \times \hat{\mathbf{H}}^*(z)$ is an imaginary number. This zero value of average power transmitted by this wave is yet another reason for calling the total wave in front of a perfect conductor a "standing wave."

Exercise

Use the time-domain forms of the electric and magnetic fields of a standing wave to show that by integrating the instantaneous Poynting vector over a complete period, the time-average Poynting vector is also zero, thus confirming the result in equation 5.18.

EXAMPLE 5.3

A plane wave of amplitude $\hat{E}_{m1}^+ = 100e^{j0°}$ and frequency = 150 MHz is propagating in a medium of $\mu_1 = \mu_o$ and of a wave impedance $\eta_1 = 100\ \Omega$. If this wave is reflected by a perfectly conducting plane boundary, determine the following:

1. The location of the first two consecutive nulls of the total electric field in front of the perfect conductor.
2. The location of the first null of the total magnetic field.

Also obtain time-domain expressions for the total electric and magnetic fields and determine the amplitude of the magnetic field at the surface of the conductor and at distance $z = -2$ m from it.

Solution

To determine the locations of the nulls as well as the required expressions for the electric and magnetic fields, we need the properties of the propagation medium. Because η_1 is given as a real number, it is clear that the medium is nonconductive—that is, $\sigma_1 = 0$. Substituting σ_1 and $\mu_1 = \mu_o$ in the expression of η_1, we obtain

$$\eta_1 = \sqrt{\frac{\mu_1}{\epsilon_1 - j\frac{\sigma_1}{\omega}}} = \sqrt{\frac{\mu_o}{\epsilon_o \epsilon_{r_1}}} = \frac{120\pi}{\sqrt{\epsilon_{r_1}}} = 100$$

$$\therefore \epsilon_{r_1} = 14.21$$

The propagation constant β_1 is then

$$\beta_1 = \omega\sqrt{\mu_o \epsilon_o \epsilon_{r_1}} = 11.84 \text{ rad/m}$$

$$\lambda_1 = \frac{2\pi}{\beta_1} = 0.53 \text{ m}$$

1. Locations of the first two nulls of the electric field are thus

$$z = 0, \frac{\lambda_1}{2}$$

$$= 0, 0.27 \text{ m}$$

2. Location of the first null of the magnetic field is

$$z = \frac{\lambda_1}{4} = 0.135 \text{ m}$$

The time-domain expressions of the electric and magnetic fields are obtained by substituting the appropriate constants in equations 5.13 and 5.16, hence,

$$\mathbf{E}(z,t) = 200 \sin(11.84z) \sin(9.4 \times 10^8 t)\mathbf{a}_x \text{ V/m}$$

$$\mathbf{H}(z,t) = 2 \cos(11.84z) \cos(9.4 \times 10^8 t)\mathbf{a}_y \text{ A/m}$$

The amplitude of the magnetic field at the surface of the conductor $z = 0$ is

$$|\mathbf{H}(0,t)| = 2 \cos(9.4 \times 10^8 t) \text{ A/m}$$

At $z = -2m$,

$$|\mathbf{H}(-2,t)| = 2 \cos(-23.68) \cos(9.4 \times 10^8 t) = 1.83 \cos(9.4 \times 10^8 t) \text{ A/m}$$

EXAMPLE 5.4

A uniform plane wave is propagating in a lossless medium and is normally incident on a plane perfect conductor at $f = 400$ MHz. If the measured distance between any two successive zeros of the total electric field in front of the conductor is 12.5 cm, determine the following:

1. The relative permittivity of the lossless medium, assuming $\mu = \mu_o$.
2. The shortest distance from the conductor at which the total magnetic field is zero.
3. If the amplitude of the incident electric field $\hat{E}_{m1}^+ = 120e^{j0°}$ V/m, calculate the magnitude of the magnetic field at a distance $z = 0.8$ m from the surface of the conductor, and also find the magnitude and direction of the induced surface current.

Solution

1. The distance between successive zeros is $\lambda_1/2$ where λ_1 is the wavelength in the medium. Hence, from the given information we note that,

$$\lambda_1 = 0.25 \text{ m}$$

β_1 for a lossless medium is given by $\beta_1 = \omega\sqrt{\mu_o \epsilon_o \epsilon_r}$. Relating β_1 to λ_1, we obtain

$$\lambda_1 = \frac{2\pi}{\beta_1} = \frac{2\pi}{\omega\sqrt{\mu_o \epsilon_o \epsilon_r}} = 0.25$$

Therefore the dielectric constant of the medium ϵ_r is given by

$$\epsilon_r = 9$$

2. The magnetic field is zero at distance $z = -\lambda_1/4 = -0.0625$ m.

3. $H_y(z,t) = \dfrac{2E_{m1}^+}{\eta_1} \cos \beta_1 z \, \cos \omega t$

η_1 for a lossless medium is given by

$$\eta_1 = \sqrt{\frac{\mu_o}{\epsilon_o \epsilon_r}} = \frac{120\pi}{\sqrt{9}} = 40\pi$$

$$\therefore H_y(-0.8, t) = \frac{2 \times 120}{40\pi} \cos(-6.4\pi) \cos \omega t$$

$$= 1.91 \cos(-6.4\pi) \cos \omega t \text{ A/m}$$

The current induced on the surface of the perfect conductor may be obtained from the boundary condition on the tangential component of the magnetic field. Hence,

$$\mathbf{n} \times (\mathbf{H}_1 - \mathbf{H}_2) = \mathbf{J}_s$$

Because the magnetic field inside the perfect conductor $\mathbf{H}_2$ is zero, we have

$$\mathbf{n} \times \mathbf{H}_1 = \mathbf{J}_s$$

or

$$\mathbf{J}_s = -\mathbf{a}_z \times \frac{2E_{m1}^+}{\eta_1} \cos \beta_1 z \, \cos \omega t \, \mathbf{a}_y \Big|_{z=0}$$

$$= \frac{2E_{m1}^+}{\eta_1} \cos \omega t \, \mathbf{a}_x = 1.91 \, \cos \omega t \, \mathbf{a}_x \text{ A/m}$$

<hr>

5.4 REFLECTION AND TRANSMISSION AT MULTIPLE INTERFACES

Thus far in our study we considered the reflection and transmission at one interface separating two media. In this section, we shall generalize the procedure to solve for the reflections and transmissions at multiple interfaces separating several different media. To do this, we need first to obtain expressions for the electric and magnetic fields in each of these regions. This can be accomplished by examining the development of the multiple reflections and transmissions at the interfaces separating say the three media shown in Figure 5.4a. The development of these reflections are shown in Figure 5.4b as a function of time. Initially, there is an incident wave ($\mathbf{E}^i$) at the interface between regions 1 and 2. At the interface between regions 1 and 2 this wave is partially reflected $\mathbf{E}_1^{-(1)}$ and partially transmitted $\mathbf{E}_2^{+(1)}$ to region 2. The amplitudes of the reflected and transmitted waves are determined from the reflection and transmission coefficients at the interface between regions 1 and 2. The subscript in the electric field notation (e.g., $[\mathbf{E}_2^{+(1)}]$) indicates the region's number, whereas the superscript plus or minus signs indicate whether the wave is transmitted (positive z propagation) or reflected (negative z propagation), respectively. The superscript 1, 2, and so on, conversely, is included to indicate the number of the reflection or transmission process. It should also be emphasized that although these plane waves are actually normally incident at the multiple interfaces, the reflection process in Figure 5.4b is shown at an angle with respect to the interface simply because this figure illustrates the development of these multiple reflections as a function of *time*. Hence, the travel time of a typical wave $\mathbf{E}_2^{+(1)}$ between two interfaces is measured by the vertical distance between points A and B at these two interfaces.

As the multiple reflection process continues between the three interfaces there would be an infinite number of waves in each region. The *steady-state* expressions for the electric fields in each of the three regions is obtained by summing all the fields that resulted from the multiple reflection process. These steady-state expressions of the field are hence given by the following:

Region 1

In this region at steady state, we have the incident wave propagating in the positive z direction and an infinite number of reflected waves $E_1^{-(1)}, E_1^{-(2)}, \ldots$, and so on propagating in the negative z direction. The total electric field in region 1 is hence given by

$$\hat{\mathbf{E}}_1(z) = \hat{\mathbf{E}}^i + \sum_{N=1}^{\infty} \hat{\mathbf{E}}_1^{-(N)} \tag{5.19}$$

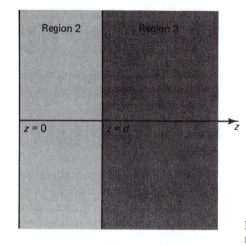

(a)

Figure 5.4a A plane wave normally incident on two interfaces separating three different media.

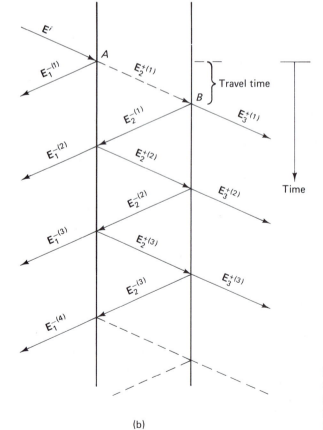

(b)

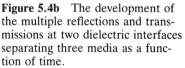

Figure 5.4b The development of the multiple reflections and transmissions at two dielectric interfaces separating three media as a function of time.

The expression for the incident electric field is $\hat{\mathbf{E}}^i = \hat{E}_{m1}^+ e^{-\hat{\gamma}_1 z} \mathbf{a}_x$, whereas the expression for each of the reflected waves is given by

$$\hat{\mathbf{E}}_1^{-(N)} = \hat{E}_{m1}^{-(N)} e^{\hat{\gamma}_1 z} \mathbf{a}_x \qquad (5.20)$$

where $\hat{E}_{m1}^{-(N)}$ is the amplitude of the (Nth) reflected wave. It should be emphasized that all the reflected waves are propagating in the negative z direction and hence the $e^{\hat{\gamma}_1 z}$ term is common to all the field expressions for these waves. Substituting equation 5.20 in equation 5.19, we obtain

$$
\begin{aligned}
\hat{\mathbf{E}}_1(z) &= \hat{E}_{m1}^+ e^{-\hat{\gamma}_1 z} \mathbf{a}_x + \sum_{N=1}^{\infty} \hat{E}_{m1}^{-(N)} e^{\hat{\gamma}_1 z} \mathbf{a}_x \\
&= \hat{E}_{m1}^+ e^{-\hat{\gamma}_1 z} \mathbf{a}_x + e^{\hat{\gamma}_1 z} \sum_{N=1}^{\infty} \hat{E}_{m1}^{-(N)} \mathbf{a}_x
\end{aligned}
\qquad (5.21)
$$

The summation $\sum_{N=1}^{\infty} \hat{E}_{m1}^{-(N)}$ over the complex amplitudes of the infinite number of the reflected waves is just another complex number $\hat{E}_{m1}^-$ representing the steady-state amplitude of the overall reflected wave. Equation 5.21, hence, reduces to

$$\hat{\mathbf{E}}_1(z) = \hat{E}_{m1}^+ e^{-\hat{\gamma}_1 z} \mathbf{a}_x + \hat{E}_{m1}^- e^{\hat{\gamma}_1 z} \mathbf{a}_x \qquad (5.22)$$

Equation 5.22 represents an expression for the overall fields in region 1 at steady state, which means after a period sufficiently long for the development of an infinitely large number of reflections.

Region 2

In this region, we also recognize from Figure 5.4b that at *steady state* there would be an infinite number of waves propagating in the positive z direction and an infinite number of other waves propagating in the negative z direction. The steady-state expression of the electric field is then

$$
\begin{aligned}
\hat{\mathbf{E}}_x(z) &= \sum_{N=1}^{\infty} \hat{\mathbf{E}}_2^{+(N)} + \sum_{N=1}^{\infty} \hat{\mathbf{E}}_2^{-(N)} \\
&= \sum_{N=1}^{\infty} \hat{E}_{m2}^{+(N)} e^{-\hat{\gamma}_2 z} \mathbf{a}_x + \sum_{N=1}^{\infty} \hat{E}_{m2}^{-(N)} e^{\hat{\gamma}_2 z} \mathbf{a}_x \\
&= e^{-\hat{\gamma}_2 z} \sum_{N=1}^{\infty} \hat{E}_{m2}^{+(N)} \mathbf{a}_x + e^{\hat{\gamma}_2 z} \sum_{N=1}^{\infty} \hat{E}_{m2}^{-(N)} \mathbf{a}_x
\end{aligned}
$$

The sum over the amplitudes of the waves propagating in the positive z direction $\sum_{N=1}^{\infty} \hat{E}_{m2}^{+(N)}$ is a complex number $\hat{E}_{m2}^+$, which represents the steady-state amplitude of the wave propagating in the positive z direction. Similarly, the steady-state amplitude of the wave propagating in the negative z direction is

$$\hat{E}_{m2}^- = \sum_{N=1}^{\infty} \hat{E}_{m2}^{-(N)}$$

A general expression for the steady-state fields in region 2 is, hence, given by

$$\hat{\mathbf{E}}_2(z) = \hat{E}_{m2}^+ e^{-\hat{\gamma}_2 z} \mathbf{a}_x + \hat{E}_{m2}^- e^{\hat{\gamma}_2 z} \mathbf{a}_x \qquad (5.23)$$

Region 3

From Figure 5.4b, it is clear that in region 3 all the transmitted waves in this region are propagating in the positive z direction. The steady-state expression for the electric field in region 3 is, hence,

$$\hat{E}_3(z) = \sum_{N=1}^{\infty} \hat{E}_{m3}^{+(N)} e^{-\hat{\gamma}_3 z} \mathbf{a}_x$$

$$= e^{-\hat{\gamma}_3 z} \sum_{N=1}^{\infty} \hat{E}_{m3}^{+(N)} \mathbf{a}_x \qquad (5.24)$$

$$= \hat{E}_{m3}^{+} e^{-\hat{\gamma}_3 z} \mathbf{a}_x$$

where

$$\hat{E}_{m3}^{+} = \sum_{N=1}^{\infty} \hat{E}_{m3}^{+(N)}$$

is the steady-state amplitude of the transmitted wave in region 3. A summary of the expressions of the electric and magnetic fields in the three regions is given in Figure 5.5.

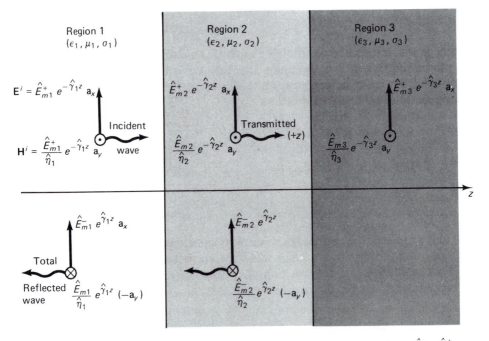

Figure 5.5 Steady-state expressions of the total fields in the three regions. $\hat{E}_{m1}^{-}$, $\hat{E}_{m2}^{+}$, $\hat{E}_{m2}^{-}$, and $\hat{E}_{m3}^{+}$ are unknown amplitudes of the fields that are to be determined from the boundary conditions at $z = 0$ and $z = d$.

After obtaining steady-state expressions for the electric fields in each of the three regions in Figure 5.4a, it is now possible to solve for the unknown amplitudes of the reflected and transmitted waves in each region in terms of a known amplitude of the incident electric field $\hat{E}_{m1}^{+}$. These unknown amplitudes can simply be obtained by enforcing the boundary conditions at the interfaces $z = 0$ and $z = d$.

At $z = 0$, the tangential electric fields in regions 1 and 2 are continuous, hence,

$$\hat{\mathbf{E}}_1(z)\big|_{z=0} = \hat{\mathbf{E}}_2(z)\big|_{z=0} \tag{5.25}$$

Substituting equations 5.22 and 5.23 in equation 5.25, we obtain

$$\hat{E}_{m1}^{+} + \hat{E}_{m1}^{-} = \hat{E}_{m2}^{+} + \hat{E}_{m2}^{-} \tag{5.26}$$

Also, by enforcing the continuity of the tangential component of the magnetic field at the interface $z = 0$, we obtain

$$\left(\frac{\hat{E}_{m1}^{+}}{\hat{\eta}_1}e^{-\hat{\gamma}_1 z} - \frac{\hat{E}_{m1}^{-}}{\hat{\eta}_1}e^{\hat{\gamma}_1 z}\right)\Bigg|_{z=0}\mathbf{a}_y = \left(\frac{\hat{E}_{m2}^{+}}{\hat{\eta}_2}e^{-\hat{\gamma}_2 z} - \frac{\hat{E}_{m2}^{-}}{\hat{\eta}_2}e^{\hat{\gamma}_2 z}\right)\Bigg|_{z=0}\mathbf{a}_y \tag{5.27}$$

or

$$\frac{\hat{E}_{m1}^{+}}{\hat{\eta}_1} - \frac{\hat{E}_{m1}^{-}}{\hat{\eta}_1} = \frac{\hat{E}_{m2}^{+}}{\hat{\eta}_2} - \frac{\hat{E}_{m2}^{-}}{\hat{\eta}_2} \tag{5.28}$$

In equation 5.27, the amplitude of the magnetic field propagating in the positive z direction is obtained by dividing $\hat{E}_{m1}^{+}$ by $\hat{\eta}_1$, whereas the amplitude of the magnetic field propagating in the negative z direction is obtained from $-(\hat{E}_{m1}^{-}/\hat{\eta}_1)$. The minus sign in the latter case is to maintain the Poynting vector in the negative z direction.

We similarly apply the boundary conditions for the tangential electric and magnetic fields in regions 2 and 3 at the interface $z = d$. Hence,

$$\hat{E}_{m2}^{+}e^{-\hat{\gamma}_2 d} + \hat{E}_{m2}^{-}e^{\hat{\gamma}_2 d} = \hat{E}_{m3}^{+}e^{-\hat{\gamma}_3 d} \tag{5.29}$$

$$\frac{\hat{E}_{m2}^{+}}{\hat{\eta}_2}e^{-\hat{\gamma}_2 d} - \frac{\hat{E}_{m2}^{-}}{\hat{\eta}_2}e^{\hat{\gamma}_2 d} = \frac{\hat{E}_{m3}^{+}}{\hat{\eta}_3}e^{-\hat{\gamma}_3 d} \tag{5.30}$$

Equations 5.26, 5.28, 5.29, and 5.30 are four equations in the four unknowns $\hat{E}_{m1}^{-}$, $\hat{E}_{m2}^{+}$, $\hat{E}_{m2}^{-}$, and $\hat{E}_{m3}^{+}$. These four equations may, of course, be solved using one of the well-known methods for solving simultaneous equations that we learned in the core mathematics courses.

Although we are not particularly interested in solving these equations, we would like to emphasize that every additional region that we introduce between regions 1 and 3 adds two unknowns to our system of equations, for the amplitudes of the positive z and negative z traveling waves. The boundary-value approach of obtaining the unknown amplitudes of the fields in each region would then become increasingly difficult to use as the number of regions continues to increase. In the following section, we shall therefore describe a simpler systematic procedure for solving the multiple-interface problem. This systematic procedure also lends itself to computer programming, which makes it quite easy to solve for any number of interfaces.

5.5 REFLECTION COEFFICIENT AND TOTAL FIELD IMPEDANCE SOLUTION PROCEDURE

In the previous section we discussed the boundary-value approach for solving the problem of reflections at multiple dielectric interfaces. We learned that the wave propagation in any intermediate region may be expressed in terms of two waves, one propagating in the positive z and the other in the negative z direction. The amplitudes of these waves are naturally unknown and should be determined by satisfying the boundary conditions at the interfaces between the various regions. We also indicated that such a solution procedure would be tedious as the number of interfaces increases because of the increase in the number of equations that need to be solved simultaneously.

In this section we shall describe a systematic procedure for solving the problem of reflection and refraction at multiple dielectric interfaces. To start with let us introduce a new parameter called "the total field impedance," $\hat{Z}(z)$, at any location z within any one of the dielectric media shown in Figure 5.6. The total field impedance is defined as the ratio between the total electric field and the total magnetic field at the location z within any of the dielectric regions. Hence,

$$\hat{Z}(z) = \frac{\hat{E}_x(z)}{\hat{H}_y(z)} \tag{5.31}$$

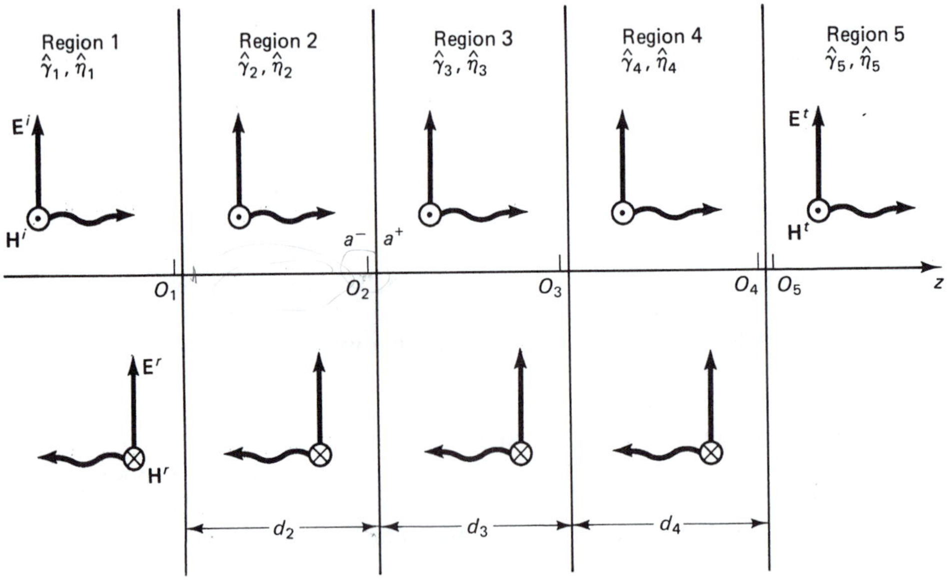

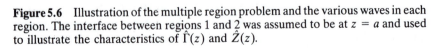

Figure 5.6 Illustration of the multiple region problem and the various waves in each region. The interface between regions 1 and 2 was assumed to be at $z = a$ and used to illustrate the characteristics of $\hat{\Gamma}(z)$ and $\hat{Z}(z)$.

Substituting the general expression in equation 5.22 for the total electric field in any region that contains reflections and remembering that the amplitude of the magnetic field component propagating in the positive z direction is related to the amplitude of the electric field by $\hat{\eta}$, whereas the amplitude of the magnetic field component propagating in the negative z direction is related to the corresponding component of the electric field by $-\hat{\eta}$, we obtain

$$\hat{Z}(z) = \frac{\hat{E}_m^+ e^{-\hat{\gamma}z} + \hat{E}_m^- e^{\hat{\gamma}z}}{\dfrac{\hat{E}_m^+}{\hat{\eta}} e^{-\hat{\gamma}z} - \dfrac{\hat{E}_m^-}{\hat{\eta}} e^{\hat{\gamma}z}}$$

$$= \hat{\eta} \frac{1 + \dfrac{\hat{E}_m^-}{\hat{E}_m^+} e^{2\hat{\gamma}z}}{1 - \dfrac{\hat{E}_m^-}{\hat{E}_m^+} e^{2\hat{\gamma}z}} \tag{5.32}$$

The ratio of the reflected to the incident waves $\hat{E}_m^-/\hat{E}_m^+ e^{2\hat{\gamma}z}$ at an arbitrary location z in any region is defined as the reflection coefficient at that location, $\hat{\Gamma}(z)$, hence,

$$\hat{\Gamma}(z) = \frac{\hat{E}_m^-}{\hat{E}_m^+} e^{2\hat{\gamma}z} \tag{5.33}$$

Substituting equation 5.33 in equation 5.32, the expression for the total field impedance at an arbitrary location z within one specific region reduces to

$$\hat{Z}(z) = \hat{\eta} \frac{[1 + \hat{\Gamma}(z)]}{[1 - \hat{\Gamma}(z)]} \tag{5.34}$$

A few points are worth emphasizing at this stage. The first is related to the difference between the total field impedance and the characteristic impedance of the medium. The characteristic impedance $\hat{\eta}$ is defined as the ratio between the electric and magnetic fields of a plane wave propagating in the positive z direction. For a plane wave propagating in the negative z direction this ratio becomes $-\hat{\eta}$, as explained in chapter 3. In the multiple regions problem, however, we indicated in the earlier sections of this chapter that the expressions of the fields in each region should include positive z and negative z traveling waves of different amplitudes $\hat{E}_m^+$ and $\hat{E}_m^-$, respectively. The "total field impedance" is then the ratio between these total electric and magnetic fields in a specific region. In the absence of a reflected wave in any one of the multiple regions—for example, region 4 in Figure 5.6—it is clear from equation 5.33 that the reflection coefficient $\hat{\Gamma}(z)$ would be zero and the total field impedance $\hat{Z}(z)$ in equation 5.34 will consequently be equal to the characteristic wave impedance $\hat{\eta}$. The other point to be noted is related to the difference between the reflection coefficient at a location z within one of the multiple media $\hat{\Gamma}(z)$, and the reflection coefficient at the interface $\hat{\Gamma}$ defined in equation 5.10. Clearly $\hat{\Gamma}$ is calculated specifically at the interface between two media and at the origin of the coordinate system for this case—that is, $z = 0$. $\hat{\Gamma}(z)$, conversely, is calculated at an arbitrary location z within any one of the multiple regions shown in Figure 5.6. In other words, $\hat{\Gamma}(z)$ in any one region of Figure 5.6 should not be substituted by the simplified expression of equation 5.10. A more general expression for $\hat{\Gamma}(z)$ and $\hat{Z}(z)$ at an arbitrary location z within any of the multiple regions of Figure 5.6 will be given in later sections of this chapter.

Another relation that we shall find useful in developing our systematic solution procedure is an expression for the reflection coefficient $\hat{\Gamma}(z)$ in terms of the total field impedance $\hat{Z}(z)$. From equation 5.34 it is quite straightforward to show that

$$\hat{\Gamma}(z) = \frac{\hat{Z}(z) - \hat{\eta}}{\hat{Z}(z) + \hat{\eta}} \tag{5.35}$$

Let us now discuss some interesting properties of the introduced quantities $\hat{Z}(z)$ and $\hat{\Gamma}(z)$.

1. The most important characteristic of $\hat{Z}(z)$ is related to its continuity at an interface separating two different media. We learned from the boundary conditions described in chapter 3 that the tangential components of the electric fields on both sides of an interface between two different media are continuous (equation 3.56 of chapter 3). Also in the absence of any external surface current density at the interface separating the two media, the tangential component of the magnetic field is continuous as given in equation 3.57 of chapter 3. Hence, if we have an interface at $z = a$ separating two media where the electric and magnetic fields are given by

$$\hat{E}_1(z) \text{ and } \hat{H}_1(z) \quad \text{(Region 1)}$$

$$\hat{E}_2(z) \text{ and } \hat{H}_2(z) \quad \text{(Region 2)}$$

then the boundary conditions at $z = a$ require that

$$\hat{E}_{t1}(a^-) = \hat{E}_{t2}(a^+) \tag{5.36}$$

$$\hat{H}_{t1}(a^-) = \hat{H}_{t2}(a^+) \tag{5.37}$$

where t indicates the tangential components of $\hat{E}$ and $\hat{H}$. In the boundary-value problems of interest to us we have normal incident plane wave reflection and transmission. The total electric and magnetic fields are then all tangential to the interfaces separating the various dielectric media. In other words for applications of interest to us we may replace the tangential components in equations 5.36 and 5.37 by the total fields. Hence, dividing equation 5.36 by 5.37 and noting that

$$\frac{\hat{E}_1(a^-)}{\hat{H}_1(a^-)} = \hat{Z}(a^-), \quad \text{and} \quad \frac{\hat{E}_2(a^+)}{\hat{H}_2(a^+)} = \hat{Z}(a^+)$$

we obtain

$$\hat{Z}_1(a^-) = \hat{Z}_2(a^+) \tag{5.38}$$

Therefore, enforcing the continuity of the tangential components of the electric and magnetic fields across an interface separating two media is equivalent to satisfying the boundary condition on the total field impedance as given by equation 5.38. The importance of this characteristic of $\hat{Z}(z)$ will be further clarified when we describe the systematic procedure for solving the reflection and transmission at multiple dielectric interfaces.

2. The reflection coefficient $\hat{\Gamma}(z)$: The following are two important characteristics of the reflection coefficient $\hat{\Gamma}(z)$ as applied to the systematic solution procedure to be developed next.

(a) First, the calculation of the reflection coefficient $\hat{\Gamma}(z')$ at z' from its value $\hat{\Gamma}(z)$ at z, where both z' and z are within a specific one of the regions shown in Figure 5.6. To obtain this relationship between $\hat{\Gamma}(z)$ and $\hat{\Gamma}(z')$, let us consider equation 5.33, which provides us with the value of $\hat{\Gamma}(z)$ within any one region in Figure 5.6. For region 2, hence, we have

$$\hat{\Gamma}_2(z) = \frac{\hat{E}_{m2}^-}{\hat{E}_{m2}^+} e^{2\hat{\gamma}_2 z} \tag{5.39}$$

In equation 5.39, we substituted the specific characteristic parameters of region 2 such as $\hat{E}_{m2}^+$, $\hat{E}_{m2}^-$, and $\hat{\gamma}_2$. At any other location—for example, z' within region 2—$\hat{\Gamma}_2(z')$ is given by

$$\hat{\Gamma}_2(z') = \frac{\hat{E}_{m2}^-}{\hat{E}_{m2}^+} e^{2\hat{\gamma}_2 z'} \tag{5.40}$$

From equations 5.39 and 5.40, it is clear that $\hat{\Gamma}_2(z')$ may be expressed in terms of $\hat{\Gamma}_2(z)$ by

$$\hat{\Gamma}_2(z') = e^{2\hat{\gamma}_2(z' - z)} \hat{\Gamma}_2(z) \tag{5.41}$$

Equation 5.41 presents an important relation that allows us to calculate the reflection coefficient at z' in terms of its value at z. It should be emphasized, however, that z' and z should be within the same region, say region 2.

(b) The other important characteristic of $\hat{\Gamma}(z)$ is related to its discontinuity at the interfaces between the different media. For example, if we assume that the interface between regions 2 and 3 in Figure 5.6 is at $z = a$, $\hat{\Gamma}(a^-)$ in region 2 just before the interface (i.e., at $z = a^-$) is given by

$$\hat{\Gamma}(a^-) = \frac{\hat{Z}(a^-) - \hat{\eta}_2}{\hat{Z}(a^-) + \hat{\eta}_2} \tag{5.42}$$

where the characteristic impedance of region 2 was, of course, assumed to be $\hat{\eta}_2$. At $z = a^+$ just after the interface, conversely, the characteristic impedance of region 3 is $\hat{\eta}_3$, and, hence, $\hat{\Gamma}(a^+)$ is given by

$$\hat{\Gamma}(a^+) = \frac{\hat{Z}(a^+) - \hat{\eta}_3}{\hat{Z}(a^+) + \hat{\eta}_3} \tag{5.43}$$

Because the total field impedance is continuous across the interfaces between the various regions—that is, $\hat{Z}(a^+) = \hat{Z}(a^-)$, as described in part 1—the reflection coefficients in equations 5.42 and 5.43 are, hence, discontinuous across the interface because of the different values of the characteristic impedances $\hat{\eta}_2$ and $\hat{\eta}_3$. In other words, the reflection coefficient is not an appropriate quantity to use for relating field quantities across an interface between two different media. Instead, the total field impedance $\hat{Z}(z)$ should be used to make transitions across interfaces between different dielectric media.

Let us now use the reflection coefficient $\hat{\Gamma}(z)$ and the total field impedance $\hat{Z}(z)$ quantities to develop the promised systematic solution procedure for solving the reflection and transmission at multiple interfaces. We start with the last region, which is region 5 in Figure 5.6. Because there is no interface beyond region 5, there will be no reflected wave in this region. The total field impedance in region 5 $\hat{Z}_5(O_5)$ is hence equal to the characteristic impedance $\hat{\eta}_5$ in this region, as can easily be verified from equation 5.34 by setting $\hat{\Gamma}(z) = 0$. Because our ultimate objective is to determine the amplitudes of the positive z and negative z traveling waves in each region in terms of the amplitude of the incident wave, we need to proceed from region 5, which is the easiest to handle, all the way to region 1 where the information on the incident wave is given. To do this, we shall use the new quantities that we introduced as $\hat{Z}(z)$ and $\hat{\Gamma}(z)$ several times, as will be explained next.

First, to cross the interface between regions 5 and 4, in our way to region 1, we need to satisfy the boundary conditions on the tangential components of the electric and magnetic fields. As stated earlier, these boundary conditions may be equivalently satisfied just by enforcing the continuity of the total field impedance at the interface, hence,

$$\hat{Z}_5(O_5) = \hat{Z}_4(O_4)$$

where O_5 and O_4 are two separate origins for regions 5 and 4, respectively. We shall use a separate origin for each of the regions to avoid an unnecessary and tedious process of carrying a phase constant throughout the calculations. In general, we shall place the origin of each region at the end of the region, except for the semi-infinite region 5, where the origin will be located at its beginning. To help us continue the process of moving from region 5 to region 1, we now use equation 5.35 to calculate $\hat{\Gamma}(O_4)$ in terms of the value of the total field impedance at that point $\hat{Z}_4(O_4)$. We then use equation 5.41 to calculate $\hat{\Gamma}(-d_4)$, where d_4 is the thickness in meters of region 4, in terms of the value of $\hat{\Gamma}_4(O_4)$ at the origin.

$$\hat{\Gamma}_4(-d_4) = \hat{\Gamma}_4(O_4)\, e^{2\hat{\gamma}_4(-d_4 - 0)}$$

The total field impedance $\hat{Z}_4(-d_4)$ is then calculated in terms of $\hat{\Gamma}_4(-d_4)$ as

$$\hat{Z}_4(-d_4) = \hat{\eta}_4 \frac{1 + \hat{\Gamma}_4(-d_4)}{1 - \hat{\Gamma}_4(-d_4)}$$

It should be noted that we calculated $\hat{\Gamma}_4(O_4)$ from $\hat{Z}_4(O_4)$ first and then calculated $\hat{Z}_4(-d_4)$ from $\hat{\Gamma}_4(-d_4)$ simply because we do not have an expression for calculating $\hat{Z}_4(-d_4)$ directly from $\hat{Z}_4(O_4)$. Also, the reason for calculating $\hat{Z}_4(-d_4)$ back from $\hat{\Gamma}_4(-d_4)$ rather than continuing our calculation in terms of $\hat{\Gamma}_4(-d_4)$ is because (as described in part b) the reflection coefficient quantities are not suitable for crossing boundaries between the various media. Crossing the boundaries requires satisfying the boundary conditions that may be achieved *only* by satisfying the continuity of the total field impedance $\hat{Z}$ at the interfaces. The process of matching the total impedance at the interfaces and using equation 5.41 to calculate the reflection coefficients $\hat{\Gamma}(-d)$ from their values at the origin $\hat{\Gamma}(O)$ continues until we obtain the values of the total field impedance $\hat{Z}_1(O_1)$ and the reflection coefficient $\hat{\Gamma}_1(O_1)$ in region 1. The amplitude of the reflected wave in region 1 is then calculated using $\hat{\Gamma}_1(O_1)$ and the amplitude of the incident wave. In other words from the definition of $\hat{\Gamma}(z)$ in equation 5.33, we have

$$\hat{E}_{m1}^- = [\hat{\Gamma}_1(z)\,\hat{E}_{m1}^+/e^{2\hat{\gamma}_1 z}]$$

Evaluating this expression at the origin O_1—that is, at $z = 0$—we obtain

$$\hat{E}_{m1}^- = \hat{\Gamma}(O_1)\,\hat{E}_{m1}^+ \qquad (5.44)$$

The total electric field in region 1 is then given by

$$\hat{E}_{x1}^{tot}(z) = \hat{E}_{m1}^+ e^{-\hat{\gamma}_1 z} + \hat{E}_{m1}^- e^{\hat{\gamma}_1 z}$$

where $\hat{E}_{m1}^-$ is obtained from equation 5.44. The total magnetic field in region 1 is given by

$$\hat{H}_{y1}(z) = \frac{\hat{E}_{m1}^+}{\hat{\eta}_1} e^{-\hat{\gamma}_1 z} - \frac{\hat{E}_{m1}^-}{\hat{\eta}_1} e^{\hat{\gamma}_1 z} \qquad (5.45)$$

The remaining part of the problem is now to calculate the amplitudes of the electric and magnetic fields in the other regions. This can be achieved by matching the boundary conditions at interfaces where the field is known on one side and is unknown on the other. For example, at the interface between regions 1 and 2, if we enforce the continuity of the electric field we obtain

$$\hat{E}_{x1}^{tot}(z)\big|_{\text{at } O_1} = \hat{E}_{x2}^{tot}(z)\big|_{z=-d_2}$$

$$= \hat{E}_{m2}^+ e^{-\hat{\gamma}_2 z} + \hat{E}_{m2}^- e^{\hat{\gamma}_2 z}\big|_{z=-d_2}$$

$$= \left\{ \hat{E}_{m2}^+ e^{-\hat{\gamma}_2 z}\left[1 + \frac{\hat{E}_{m2}^-}{\hat{E}_{m2}^+} e^{2\hat{\gamma}_2 z}\right] \right\}_{z=-d_2} \qquad (5.46)$$

$$\hat{E}_{m1}^+\left[1 + \hat{\Gamma}_1(O_1)\right] = \hat{E}_{m2}^+ e^{\hat{\gamma}_2 d}\left[1 + \hat{\Gamma}_2(-d_2)\right]$$

The amplitude $\hat{E}_{m2}^+$ is, hence,

$$\hat{E}_{m2}^+ = \frac{\hat{E}_{m1}^+\left[1 + \hat{\Gamma}_1(O_1)\right]}{e^{\hat{\gamma}_2 d_2}\left[1 + \hat{\Gamma}_2(-d_2)\right]} \qquad (5.47)$$

It should be emphasized that the reflection coefficients were calculated in the first part of the problem when we were relating the total impedances and the reflection coefficients in the various regions to that in region 1.

After calculating the amplitude of the electric field in region 2, $\hat{E}_{m2}^+$, we can now calculate the electric field at any point z within this region using the equation

$$\hat{E}_{x2}(-z) = \hat{E}_{m2}^+ e^{\hat{\gamma}_2 z}\left[1 + \hat{\Gamma}_2(-z)\right]$$

Of particular interest to us is the value of $\hat{E}_{x2}(-z)$ at the origin O_2, which is located at the interface between regions 2 and 3. At this interface, we enforce the continuity of the electric fields in regions 2 and 3 to calculate the amplitude of the electric field in region 3. Hence,

$$\hat{E}_{x2}(O_2) = \hat{E}_{m2}^+[1 + \hat{\Gamma}_2(O_2)]$$
$$= \hat{E}_{x3}(-d_3) = \hat{E}_{m3}^+ e^{+\hat{\gamma}_3 d_3}[1 + \hat{\Gamma}_3(-d_3)] \tag{5.48}$$

Because $\hat{E}_{m2}^+$, $\hat{\Gamma}_2(O_2)$, and $\hat{\Gamma}_3(-d_3)$ are known from earlier calculations, we use equation 5.48 to calculate the amplitude of the electric field $\hat{E}_{m3}^+$ in region 3. This procedure continues until we determine the amplitudes of the electric fields in all regions. The magnetic field expressions may then be obtained using expressions similar to equation 5.45 in each region. Let us illustrate the systematic procedure further by solving an example.

EXAMPLE 5.5

A uniform plane wave propagating in free space is normally incident on the multiple dielectric media shown in Figure 5.7. If the x-polarized electric field associated with this wave is given by

$$\mathbf{E}(z,t) = 50 \cos(2\pi \times 10^8 t - 2.094z)\,\mathbf{a}_x$$

determine the following:

1. Magnitude of electric field transmitted to region 2.
2. Time-average power density in region 1.

Solution

1. The solution procedure is summarized as follows:
   (a) From the given frequency of propagation and the electrical properties of each region, calculate the characteristic parameters $\hat{\gamma}$ and $\hat{\eta}$ in each region. From the given electric field expression, these parameters in region 1 are given by

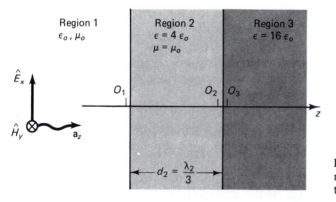

Figure 5.7   A plane wave is normally incident on the interfaces between three dielectric regions.

$$f = \frac{\omega}{2\pi} = \frac{2\pi \times 10^8}{2\pi} = 100 \text{ MHz}$$

$$\beta_1 = 2.094 \text{ rad/m}$$

$$\eta_1 = \sqrt{\frac{\mu_o}{\epsilon_o}} = 120\pi \ \Omega$$

and the phasor expression for the electric field is

$$\hat{\mathbf{E}}(z) = 50e^{-j2.094z} \ \mathbf{a}_x$$

In region 2,

$$\eta_2 = \sqrt{\frac{\mu_o}{4\epsilon_o}} = 60\pi \ \Omega$$

$$\alpha_2 = 0$$

because region 2 is lossless—that is, $\sigma_2 = 0$

$$\beta_2 = \omega\sqrt{\mu_o(4\epsilon_o)} = 2\omega\sqrt{\mu_o\,\epsilon_o} = 2\beta_1$$

$$= 4.188 \text{ rad/m}$$

In region 3,

$$\eta_3 = \sqrt{\frac{\mu_o}{16\epsilon_o}} = 30\pi \ \Omega$$

$$\alpha_3 = 0$$

$$\beta_3 = 4\beta_1 = 8.376 \text{ rad/m}$$

(b) To avoid a cumbersome process of carrying phase constants throughout the calculations, we set a separate origin in each region. We routinely locate origins at the end of each region, except in region 3 where we set its origin at the beginning as shown in Figure 5.7.

(c) We start the calculations from the simplest region—that is, region 3—which is free from any reflections.

$$\hat{Z}_3(O_3) = \eta_3 \frac{1 + \hat{\Gamma}_3(O_3)}{1 - \hat{\Gamma}_3(O_3)}$$

Because $\hat{\Gamma}_3(O_3) = 0$, the total field impedance $\hat{Z}_3(O_3)$ is, hence, equal to the characteristic impedance η_3

$$\hat{Z}_3(O_3) = \eta_3 = 30\pi \ \Omega$$

(d) Apply the continuity of the total field impedance at the interface between regions 2 and 3. Hence,

$$\hat{Z}_2(O_2) = \hat{Z}_3(O_3) = \eta_3 = 30\pi \ \Omega$$

(e) To relate the various parameters of the three regions to those of the incident plane wave, we need to continue moving toward region 1 ultimately to calculate $\hat{\Gamma}_1(O_1)$ in this region. To do so, we need to calculate $\hat{Z}_2(-d_2)$ from its value at the origin $\hat{Z}_2(O_2)$. Because we do not have an expression for relating $\hat{Z}_2(-d_2)$ to $\hat{Z}_2(O_2)$, we employ the reflection coefficient parameter. Hence,

$$\hat{\Gamma}_2(O_2) = \frac{\hat{Z}_2(O_2) - \eta_2}{\hat{Z}_2(O_2) + \eta_2} = \frac{30\pi - 60\pi}{30\pi + 60\pi} = -\frac{1}{3}$$

From equation 5.41, we have

$$\hat{\Gamma}_2(-d_2) = \hat{\Gamma}_2(O_2)\, e^{2\hat{\gamma}_2(-d_2 - 0)}$$

$$\hat{\Gamma}_2\left(-\frac{\lambda_2}{3}\right) = \left(-\frac{1}{3}\right) e^{-j4\pi/3} = \frac{1}{3} e^{-j\pi/3}$$

The desired value of $\hat{Z}_2(-d_2)$ is then

$$\hat{Z}_2(-d_2) = \eta_2 \frac{1 + \hat{\Gamma}_2(-d_2)}{1 - \hat{\Gamma}_2(-d_2)}$$

$$= 60\pi \frac{1 + \dfrac{1}{3} e^{-j\pi/3}}{1 - \dfrac{1}{3} e^{-j\pi/3}}$$

$$= 60\pi \frac{1.17 - j0.29}{0.83 + j0.29}$$

$$= 60\pi(1.15 - j0.75)$$

(f) From the continuity of the total field impedance at the interface between regions 1 and 2, we have

$$\hat{Z}_1(O_1) = \hat{Z}_2(-d_2)$$

$$= 60\pi(1.15 - j0.75)$$

The reflection coefficient in region 1 is then

$$\hat{\Gamma}_1(O_1) = \frac{\hat{Z}_1(O_1) - \eta_1}{\hat{Z}_1(O_1) + \eta_1}$$

$$= \frac{-160.23 - j141.37}{593.77 - j141.37} = 0.35 e^{j234.8°}$$

(g) The amplitude of the reflected wave in region 1 can now be calculated from $\hat{\Gamma}_1(O_1)$ and using equation 5.44

$$\frac{\hat{E}_{m1}^-}{\hat{E}_{m1}^+} = \hat{\Gamma}_1(O_1)$$

$$\hat{E}_{m1}^- = 50 \times \left(0.35 e^{j234.8°}\right)$$

$$= 17.5 e^{j234.8°} \text{ V/m}$$

(h) The total electric field in region 1 is hence

$$\hat{E}_{x1} = \hat{E}_{m1}^+ e^{-j\beta_1 z}\left[1 + \hat{\Gamma}_1(z)\right]$$

At $z = 0$, we have

$$\hat{E}_{x1}(0) = 50\left[1 + 0.35e^{j234.8°}\right]$$

Because of the continuity of the electric field at the interface between regions 1 and 2, we have

$$\hat{E}_{x1}(0) = \hat{E}_{x2}(-d_2)$$
$$= \hat{E}_{m2}^{+} e^{-j\beta_2(-d_2)}\left[1 + \hat{\Gamma}_2(-d_2)\right]$$

Hence,

$$\hat{E}_{m2}^{+} = \frac{50\left[1 + 0.35e^{j234.8°}\right]}{e^{j2\pi/\lambda_2(\lambda_2/3)}\left[1 + \dfrac{1}{3}e^{-j\pi/3}\right]}$$

$$= \frac{50\left[1 + 0.35e^{j234.8°}\right]}{e^{j2\pi/3} + \dfrac{1}{3}e^{j\pi/3}}$$

$$= 35.17e^{-j125.81°} \text{ V/m}$$

2. The time average power density in region 1 is given by

$$\mathbf{P}_{av1} = \frac{1}{2}Re\left(\hat{\mathbf{E}}_1 \times \hat{\mathbf{H}}_1^*\right)$$

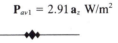

$$= \frac{1}{2}Re\left(\hat{E}_{m1}^{+}e^{-j\beta_1 z}\left[1 + \hat{\Gamma}_1(z)\right]\mathbf{a}_x \times \frac{\hat{E}_{m1}^{+*}}{\eta_1}e^{j\beta_1 z}\left[1 - \hat{\Gamma}_1^*(z)\right]\mathbf{a}_y\right)$$

$$= \frac{1}{2}\frac{|\hat{E}_{m1}^{+}|^2}{\eta_1}(1 - |\hat{\Gamma}_1(z)|^2)\,\mathbf{a}_z$$

where we have substituted $\hat{\mathbf{H}} = \hat{E}_{m1}/\eta_1\, e^{-j\beta_1 z}\left[1 - \hat{\Gamma}_1(z)\right]\mathbf{a}_y$ for the magnetic field in region 1 as given by equation 5.45. Substituting $|\hat{E}_{m1}^{+}| = 50$, $\eta_1 = 120\pi$, and $|\hat{\Gamma}_1(z)| = 0.35$ in the average power expression, we obtain

$$\mathbf{P}_{av1} = 2.91\,\mathbf{a}_z \text{ W/m}^2$$

Exercise

Write a computer program for solving the positive z and negative z traveling waves in a multiple region problem. Choose the number of layers to be N, and assume the characteristic parameters to be ϵ_i, μ_i, and σ_i in the ith region. The frequency and the thicknesses of the various layers should be included as part of the input data to the program. The general expressions given in chapter 3 for the characteristic impedance, attenuation constant, and the phase constant in each medium should be included as a part of the program. The output data should include the reflection coefficients and the total field impedances at the various interfaces as well as the amplitudes of the fields in the various regions.

5.6 GRAPHICAL SOLUTION PROCEDURE USING THE SMITH CHART

In the systematic procedure for solving wave propagation in multiple regions that we described in the previous section, we may have noticed that it is difficult to develop physical insight for what is happening in a specific problem by only looking at the calculated numbers that are often complex. Therefore, even with the availability of digital computers and the possibility of writing a computer program to make the calculations, it is desirable to solve these types of problems using a graphical approach. The graphical method that we shall describe in this section is based on the Smith Chart,[*] 1939, which is widely accepted and frequently used by the microwave engineering community. The basic idea of this chart is to provide a graphical way to find the total field impedance value from a given reflection coefficient, or vice versa. It also enables us to find the reflection coefficient $\hat{\Gamma}(z')$ at a given location z' from its known value $\hat{\Gamma}(z)$ at another location z within the same region. In other words, the Smith chart just helps us make the various calculations involved in the systematic solution procedure of the multiregion wave propagation problem graphically and, hence, more conveniently.

To start with, let us distinguish between impedance and reflection coefficient charts. Both the impedance $\hat{Z}$ and the reflection coefficient $\hat{\Gamma}$ are expressed in terms of complex numbers

$$\hat{Z} = R + jX \qquad \hat{\Gamma} = \Gamma_r + j\Gamma_i$$

where R and X are the resistive and reactive parts of the complex impedance, whereas Γ_r and Γ_i are the real and imaginary parts of the complex reflection coefficient. Figure 5.8 shows the polar and the rectangular representations of the two complex quantities $\hat{Z}$ and $\hat{\Gamma}$.

Although the representations of the reflection coefficient and the impedance look similar on their complex planes in Figure 5.8, there is actually a distinct difference between these two planes. For representing all possible values of impedances, the complex impedance plane extends from $-\infty$ to ∞ in both the directions of the real and imaginary parts of the complex impedance. This is obviously a disadvantage in making any graphical representation on this type of chart. In the reflection coefficient chart, conversely, all the possible values of the complex reflection coefficient are contained within a circle of radius equal to unity as shown in Figure 5.9. This is simply because the magnitude of the reflection coefficient never exceeds a unity. P. H. Smith was the first to report the advantages of using the complex reflection coefficient plane ($\hat{\Gamma}$ plane) to represent reflection coefficients and complex impedances graphically. The main advantage of using the $\hat{\Gamma}$ plane is that all the possible values of $\hat{\Gamma}$ and consequently $\hat{Z}$ will be within a circle of unit radius. The question now is: Although representing values of complex reflections are easy and clear from Figure 5.9, how are the complex

---

[*]P. H. Smith, "Transmission-Line Calculator," *Electronics,* January 1939. Also see P. H. Smith, "An Improved Transmission-Line Calculator," *Electronics,* January 1944.

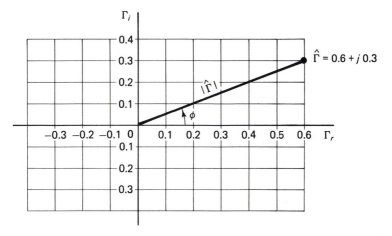

(a)

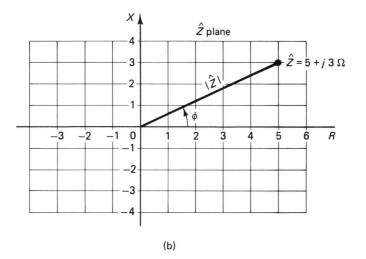

(b)

Figure 5.8   The reflection coefficient (a) and the impedance (b) planes.

impedance values going to be represented on the $\hat{\Gamma}$ plane? The answer to this question actually forms the mathematical basis of the Smith chart.

To start with, let us consider an expression that relates the complex impedance values to those of the reflection coefficient

$$\hat{Z}(z) = \hat{\eta}\frac{1 + \hat{\Gamma}(z)}{1 - \hat{\Gamma}(z)}$$

We defined the normalized total field impedance $\hat{z}_n(z)$ as the value of the total field impedance $\hat{Z}(z)$ divided by the characteristic impedance of the medium, hence,

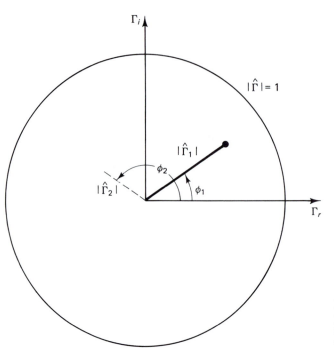

Figure 5.9 Reflection coefficient plane. All the possible reflection coefficient values are contained within a circle of radius $|\hat{\Gamma}| = 1$.

$$\hat{z}_n(z) = \frac{\hat{Z}(z)}{\hat{\eta}} = \frac{1 + \hat{\Gamma}(z)}{1 - \hat{\Gamma}(z)}$$

If we express the normalized field impedance and the reflection coefficient in terms of their real and imaginary parts, we obtain

$$r + jx = \frac{1 + \Gamma_r + j\Gamma_i}{1 - (\Gamma_r + j\Gamma_i)} \tag{5.49}$$

where $\hat{z}_n(z) = r + jx$, and $\hat{\Gamma}(z) = \Gamma_r + j\Gamma_i$. To separate the real and imaginary parts of equation 5.49 we multiply the right-hand side of this equation by the complex conjugate of the denominator

$$\frac{(1 - \Gamma_r) + j\Gamma_i}{(1 - \Gamma_r) + j\Gamma_i}$$

The real and imaginary parts are then given by

$$r = \frac{1 - \Gamma_r^2 - \Gamma_i^2}{(1 - \Gamma_r)^2 + \Gamma_i^2} \tag{5.50}$$

$$x = \frac{2\Gamma_i}{(1 - \Gamma_r)^2 + \Gamma_i^2} \tag{5.51}$$

To plot values of normalized impedances on the reflection coefficient diagram (Smith chart), we need to identify constant resistance and constant reactance curves on the chart. This way if we have a normalized impedance value of $2 + j3$, it would be plotted as the point of intersection of the $r = 2$ and $x = 3$ curves on the Smith chart. Our objective is, hence, to identify the constant r and constant x curves on the Smith chart. After a few lines of elementary algebra, equations 5.50 and 5.51 may be written in forms that easily display the nature of the r = constant and x = constant curves on the reflection coefficient (Γ_r, Γ_i) chart. These forms are given by

$$\left(\Gamma_r - \frac{r}{r+1}\right)^2 + \Gamma_i^2 = \left(\frac{1}{1+r}\right)^2 \tag{5.52}$$

$$(\Gamma_r - 1)^2 + \left(\Gamma_i - \frac{1}{x}\right)^2 = \left(\frac{1}{x}\right)^2 \tag{5.53}$$

Comparing these equations with that of a circle

$$(x - x_o)^2 + (y - y_o)^2 = a^2$$

of radius a and an origin located at (x_o, y_o) as shown in Figure 5.10, it is clear that both these equations describe families of circles with the following characteristics:

1. *Constant resistance circles.* From equation 5.52 it is clear that by setting r = constant we obtain a family of circles of radii $1/(r + 1)$ and origins at $[r/(r + 1), 0]$. Specifying values of r provides us with the set of circles shown in Figure 5.11. These are called the constant resistance circles.
2. *Constant reactance circles.* Similarly, by setting x = constant (rather than r = constant), it is clear that equation 5.53 would provide a family of circles of radii $(1/x)$ and origins at $(1, 1/x)$. Specifying values of x we obtain the set of circles shown in Figure 5.12, which are called the constant reactance circles.

The two sets of curves are shown in Figure 5.13, where it is now clear that if given a normalized value of an impedance $\hat{z}_n = \hat{Z}/\hat{\eta} = r + jx$, it would be quite straightforward to locate it on the Smith chart as the intersection of the appropriate r and x circles. Naturally, unlimited curves for all values of r and x would clutter up the chart; hence, these constant r and x circles are given only for a limited number of values of r and x, and in our solutions we may have to use interpolation as necessary. After plotting the normalized impedance values on the Smith chart, we may wonder how the correspond-

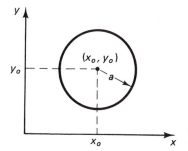

Figure 5.10 The circle of radius a and origin at (x_o, y_o), which is described by the equation $(x - x_o)^2 + (y - y_o)^2 = a^2$.

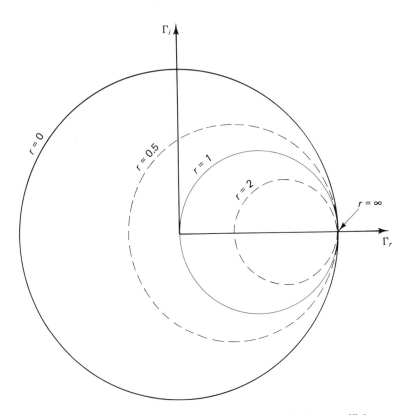

Figure 5.11 Constant r circles are shown on the reflection coefficient plane (Γ_r, Γ_i). The origins of these circles are at $\Gamma_r = r/(r + 1)$, and $\Gamma_i = 0$. The radii of these circles are $1/(r + 1)$.

ing reflection coefficient values may be obtained graphically in the absence of the constant reflection coefficient circles. By simply drawing a line between the plotted normalized impedance value and the origin of the chart and measuring its length (radial distance) we may obtain the magnitude of the reflection coefficient. The absolute value of the measured distance (e.g., in centimeters) should, of course, be normalized so that the radius of the Smith chart (i.e., $|\hat{\Gamma}| = 1$ circle) should be unity. A simple scale such as the graduate line shown below Figure 5.13 may help us directly transform absolute distance measurement (e.g., in centimeters) to values of $|\hat{\Gamma}|$. Regarding the angle of the reflection coefficient, it may be obtained simply by extending the line joining the normalized impedance and the origin to the outer circumference of the chart, and reading the angle of $\hat{\Gamma}$ in the counterclockwise direction as shown in Figure 5.13. This as well as the inverse process of obtaining the normalized impedance value from a given complex reflection coefficient are illustrated in the following two examples.

EXAMPLE 5.6

Given the normalized impedance value $\hat{z}_n = 2 + j3$, use the Smith chart to determine the reflection coefficient $\hat{\Gamma}$.

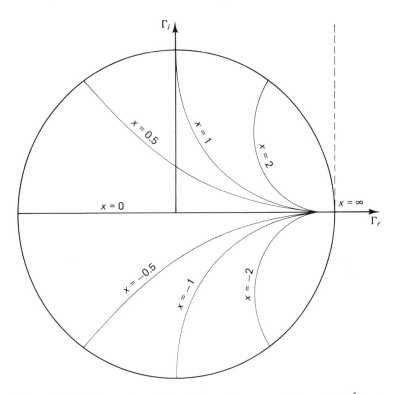

Figure 5.12   The portions of the circles of constant x lying within $|\hat{\Gamma}| = 1$. The origins of these circles are at $\Gamma_r = 1$, $\Gamma_i = 1/x$ (*dotted line*), and their radii have the values of $1/x$.

Solution

In the Smith chart shown in Figure 5.14 we first identified the $r = 2$ circle and the $x = 3$ curve. These two curves intersect at A where $\hat{z}_{nA} = 2 + j3$. To determine the reflection coefficient we join the line OA and measure its length with a divider or compass. According to the scale below the chart, the length OA corresponds to a reflection coefficient of magnitude $|\hat{\Gamma}| = 0.75$. By extending OA to the outer rim of the chart, it is easy to read the angle of the reflection coefficient $\phi = 26°$. The complex reflection coefficient is then $\hat{\Gamma} = 0.75e^{j26°}$.

EXAMPLE 5.7

If the magnitude and the phase of the reflection coefficient are given by $|\hat{\Gamma}| = 1/3$ and $\phi = 90°$, determine the corresponding value of the normalized impedance $\hat{z}_n = r + jx$, graphically using the Smith chart.

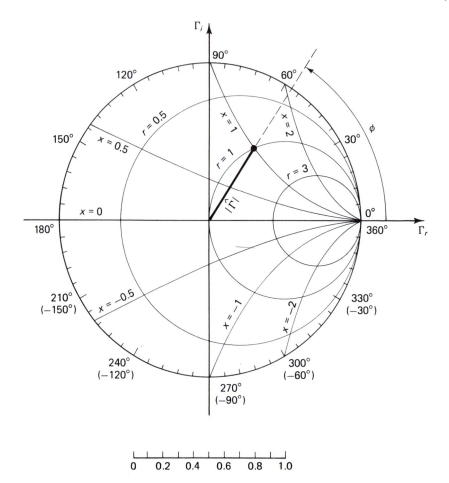

Figure 5.13 The two sets of curves r = constant and x = constance on the reflection coefficient plane. The magnitude of $\hat{\Gamma}$ is measured on the separate graduated scale, and ϕ is measured in the counterclockwise direction.

Solution

The normalized impedance value can be calculated from the reflection coefficient using equation 5.34. Hence,

$$\hat{z}_n = \frac{\hat{Z}}{\hat{\eta}} = \frac{1 + \hat{\Gamma}}{1 - \hat{\Gamma}}$$

Alternatively, we can use the graphical procedure and the Smith chart to calculate $\hat{z}_n$ from the given values of $\hat{\Gamma}$. In Figure 5.15 we use the scale on the bottom of the Smith chart to determine the radius of the circle that corresponds to the magnitude of the reflection coefficient $|\hat{\Gamma}| = 0.33$. We then draw a $|\hat{\Gamma}| = 1/3$ circle on the Smith chart. On the outer rim of the chart we determine the location at which $\phi = 90°$. On joining this location with the origin of the chart, it intersects the $|\hat{\Gamma}| = 1/3$ circle at point A. At point A the complex reflection coefficient is then $\hat{\Gamma}|_{\text{at } A} = 1/3\, e^{j90°}$. We determine the normalized impedance at

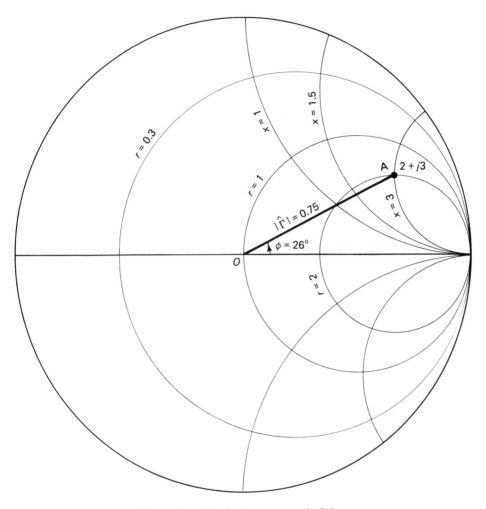

Figure 5.14 Solution to example 5.6.

A simply by reading the values on the $r =$ constant and $x =$ constant curves that are passing through A. From Figure 5.15, it can be seen that the desired value of $\hat{z}|_{\text{at } A} = 0.8 + j0.6$.

The complete Smith chart shown in Figure 5.16 contains more information than what we actually discussed thus far. From Figure 5.16 you may note more scales around the circumference of the chart and two arrows indicating rotations toward the source and toward the generator. From equation 5.41, which relates the reflection coefficient at z' to its value at another location z, we have

$$\hat{\Gamma}(z') = \hat{\Gamma}(z)\, e^{2\hat{\gamma}(z'-z)}$$
$$= \hat{\Gamma}(z)\, e^{2\alpha(z'-z)}\, e^{2j\beta(z'-z)} \qquad (5.54)$$

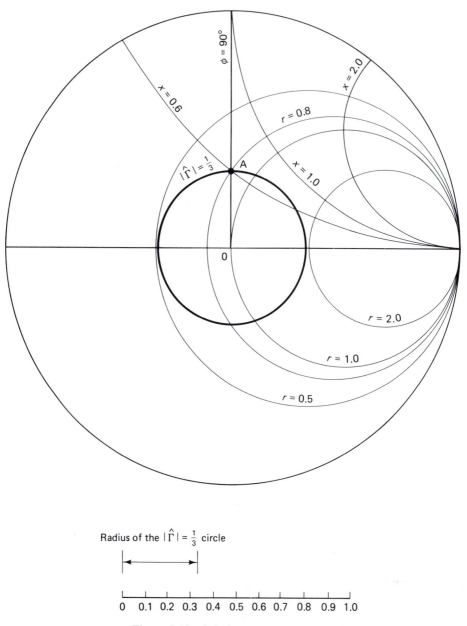

Radius of the $|\hat{\Gamma}| = \frac{1}{3}$ circle

0   0.1   0.2   0.3   0.4   0.5   0.6   0.7   0.8   0.9   1.0

Figure 5.15   Solution to example 5.7.

From the geometry of the multidielectric region problem of Figure 5.6, it is clear that if $z' > z$ it means that $\hat{\Gamma}(z')$ of equation 5.54 is being calculated at a distance z' that is further away from the incident plane wave and toward the load. From equation 5.54 it may be seen that when $z' > z$, the phase factor $e^{2j\beta(z'-z)}$ would be positive, thus changing the angle of $\hat{\Gamma}(z)$ by a positive amount. Therefore, rotating the reflection

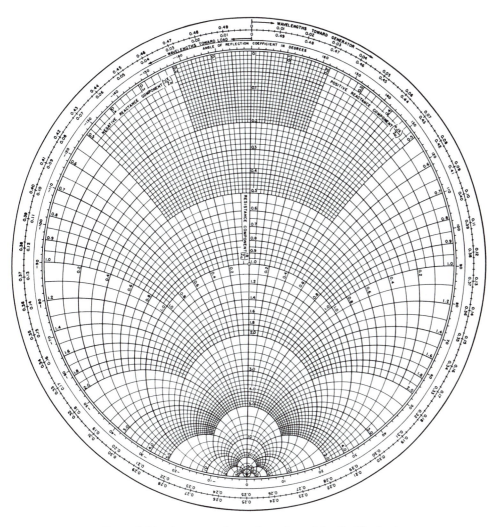

Figure 5.16   A commonly used version of the Smith chart.

coefficient in the counterclockwise direction, which is the direction of increasing ϕ, means z' is further toward the load than z. This is precisely why the arrow indicating rotation toward the load on the Smith chart points in the counterclockwise direction. If $z' < z$, conversely, the phase factor $e^{2j\beta(z'-z)}$ would be negative, thus resulting in a phase of $\hat{\Gamma}(z')$, which is smaller than that of $\hat{\Gamma}(z)$. Because $z' < z$ means moving closer to the incident plane wave (see Figure 5.6), rotation toward the generator (or toward the source) on the Smith chart is indicated by an arrow that points in the clockwise direction as shown in Figure 5.16. The final observation regarding the Smith chart is related to the scale on its circumference. To illustrate the usefulness of these scales, let us calculate the reflection coefficient $\hat{\Gamma}(z')$ from its value $\hat{\Gamma}(z)$, which is a distance $z' - z = \ell$. Hence,

$$\hat{\Gamma}(z') = \hat{\Gamma}(z)\, e^{2\alpha(\ell)}\, e^{2j\beta(\ell)}$$

$$= \hat{\Gamma}(z)\, e^{2\alpha\ell}\, e^{2j(2\pi/\lambda)\ell} = \hat{\Gamma}(z)\, e^{2\alpha\ell}\, e^{j4\pi\ell/\lambda} \tag{5.55}$$

Equation 5.55 shows that the magnitude of $\hat{\Gamma}(z)$ should be changed by the factor $e^{2\alpha\ell}$, which is to be calculated numerically outside the Smith chart. The change in the phase factor $e^{4\pi j\ell/\lambda}$, conversely, may be read directly on the Smith chart if the electrical distance ℓ/λ is specified. The two scales on the circumference of the chart are the ℓ/λ values, one for measuring the electrical distance toward the generator and the other for measuring the distance toward the load. It should be emphasized that one complete rotation around the chart corresponds to a distance $\ell/\lambda = 0.5$. This is because for $\ell/\lambda = 0.5$ the corresponding change in the phase of the reflection coefficient is

$$e^{4\pi j(\ell/\lambda)} = e^{j4\pi(0.5)} = e^{j2\pi}$$

In other words, one lap around the Smith chart corresponds to change in distance by one-half wavelength. This concludes our description of the Smith chart. We shall illustrate next its use in the graphical solution of the reflection and transmission problems at multidielectric interfaces.

EXAMPLE 5.8

In a lossless dielectric medium where the characteristic impedance is $\eta = 40\pi\ \Omega$, the total field impedance at a specific location was measured to be $\hat{Z} = 10\pi + j24\pi\ \Omega$. Use the graphical procedure and the Smith chart to determine the total field impedance at distance $\ell = 0.1\lambda$ toward the generator from the location where it was measured.

Solution

To use the Smith chart we first calculate the normalized value of the measured total field impedance

$$\hat{z}_n = \frac{\hat{Z}}{\eta} = 0.25 + j0.6$$

We then plot the value of $\hat{z}_n$ on the Smith chart as point A as shown in Figure 5.17. We do not actually need to calculate the reflection coefficient at location A. We realize, however, that because the dielectric medium is lossless the attenuation constant α in this medium is zero. In other words, the reflection coefficient at the new location which is 0.1λ toward the generator away from A is related to the reflection coefficient at A by only the following change in phase:

$$\hat{\Gamma}|_{\ell = 0.1\lambda} = \hat{\Gamma}|_A\, e^{2j(2\pi\ell/\lambda)}$$

$$= \hat{\Gamma}|_A\, e^{j0.4\pi}$$

Instead of making this calculation we can just extend the OA line to the scale on the circumference of the chart where will read 0.089 toward the generator (*twg*) and move a distance 0.1λ (i.e., to 0.189*twg*) on the same scale to point B as shown in Figure 5.17. It should be emphasized that we moved from A to B on a constant reflection coefficient circle because the dielectric medium is lossless and the attenuation constant α is zero. At point B we read the normalized impedance to be

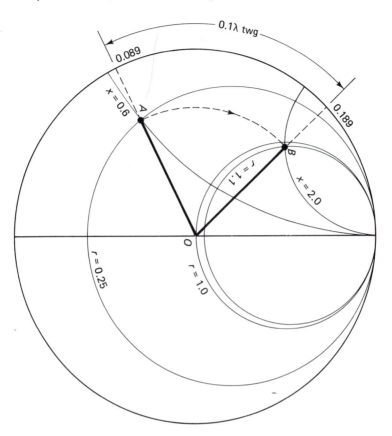

Figure 5.17   Solution to example 5.8.

$$\hat{z}\big|_{\text{at } B} = 1.1 + j2.0$$

The total field impedance at B that is a distance 0.1λ away from A is, hence,

$$\hat{Z}\big|_{\text{at } B} = 40\pi(1.1 + j2)$$

$$= 44\pi + j80 \ \Omega$$

◆◆◆

EXAMPLE 5.9

Consider the interface between two dielectric media shown in Figure 5.18. Region 1 is a lossless dielectric medium of characteristic impedance $\eta_1 = 50 \ \Omega$, whereas region 2 is a lossy dielectric medium of characteristic impedance $\hat{\eta}_2 = 100 + j50 \ \Omega$. Determine the total field impedance in region 1 and at point B a distance 0.2λ away from the interface.

Solution

The solution procedure may be summarized as follows:

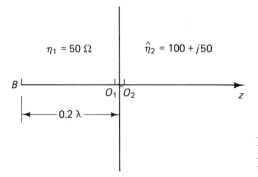

B

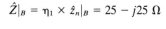

Figure 5.18 Geometry of example
5.9. Point B is 0.2λ away from interface.

1. Because there is no reflection in region 2, the total field impedance in this region
   should be equal to the characteristic impedance of the medium $\hat{\eta}_2$. Hence,

   $$\hat{Z}_2(O_2) = \hat{\eta}_2 = 100 + j50 \ \Omega$$

2. From the continuity of the total field impedance at the interface between regions 1
   and 2, we have

   $$\hat{Z}_1(O_1) = \hat{Z}_2(O_2) = 100 + j50 \ \Omega$$

3. To plot $\hat{Z}_1(O_1)$ on the Smith chart we need to normalize it first with respect to η_1,
   hence,

   $$\hat{z}_{n1}(O_1) = \frac{100 + j50}{50} = 2 + j1$$

   We then plot $\hat{z}_{n1}(O_1)$ on the Smith chart point A as shown in Figure 5.19.

4. We obtain the reflection coefficient at B by rotating (i.e., changing the phase of) the
   reflection coefficient at A, a distance 0.2λ toward the generator as shown in Figure 5.19. Once again we emphasize that there is no change in the magnitude of the
   reflection coefficient from A to B because region 1 is lossless and the attenuation
   constant α is zero.

5. From the Smith chart the value of the normalized impedance at B is

   $$\hat{z}_n|_B = 0.5 - j0.5$$

   The total field impedance at B is then

   $$\hat{Z}|_B = \eta_1 \times \hat{z}_n|_B = 25 - j25 \ \Omega$$

   ◆◆

EXAMPLE 5.10

Use the Smith chart to calculate the reflection coefficient at point A in the multidielectric
regions problem shown in Figure 5.20.

Solution

The step-by-step solution procedure is as follows:

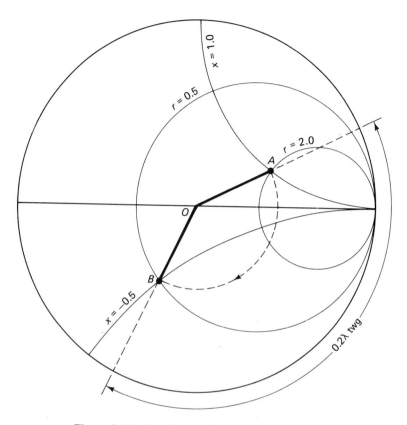

Figure 5.19   Graphical solution to example 5.9.

1. Locate separate origins in the three regions, and calculate the characteristic impedance of each.

$$\eta_3 = \sqrt{\frac{\mu_o}{9\epsilon_o}} = 40\pi$$

$$\eta_2 = 120\pi \qquad \text{and} \qquad \eta_1 = 60\pi$$

2. Start the solution with region 3 because there is no reflection in this region.

$$Z_3(O_3) = \eta_3 = 40\pi \ \Omega$$

3. From the continuity of the total field impedance at the interface between regions 3 and 2, we have

$$Z_2(O_2) = Z_3(O_3) = 40\pi \ \Omega$$

4. To plot $Z_2(O_2)$ on the Smith chart we need to calculate the normalized impedance value first. Hence,

$$\hat{z}_{n2}(O_2) = \frac{Z_2(O_2)}{\eta_2} = \frac{40\pi}{120\pi} = 1/3 + j0$$

We then plot $\hat{z}_{n2}(O_2)$ on the Smith chart (point P_1) as shown in Figure 5.21.

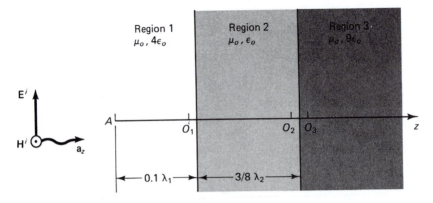

Figure 5.20 A plane wave is normally incident on the interfaces between three dielectric regions.

5. To obtain $\hat{z}_{n2}(-d_2) = \hat{z}_{n2}(-3/8\lambda_2)$ from $\hat{z}_{n2}(O_2)$ (point P_1), we simply rotate the value of $\hat{z}_{n2}(O_2)$ on a constant reflection coefficient circle a distance $3/8\lambda_2$ toward the generator. No adjustment in the magnitude of the reflection coefficient is necessary because region 2 is lossless. Also an easy way to achieve the required rotation is to use the outer scale on the rim of the chart. For example, we notice that P_1 is located at the zero of the "toward the generator" scale on the circumference of the chart. Hence, a rotation of $3/8\lambda_2 = 0.375\lambda_2$ corresponds to the 0.375 location (note the arrow) on the same scale. $\hat{z}_{n2}(-3/8\lambda_2)$ is then shown as point P_2 in Figure 5.21.

$$\hat{z}_{n2}(-3/8\lambda_2) = 0.6 - j0.8$$

6. Before we can apply the continuity of the total field impedance across the boundary between regions 1 and 2, we need to denormalize the value of the total field impedance $\hat{z}_{n2}(-3/8\lambda_2)$

$$\hat{Z}_2(-3/8\lambda_2) = \eta_2 \times \hat{z}_{n2}(-3/8\lambda_2)$$

$$= 120\pi(0.6 - j0.8)$$

$$= 226.2 - j301.6 \ \Omega$$

7. Applying the continuity of the total field impedance at the interface between regions 1 and 2 we, hence, obtain

$$\hat{Z}_1(O_1) = \hat{Z}_2(-3/8\lambda_2)$$

$$= 226.2 - j301.6$$

8. To determine the value of the reflection coefficient at O_1—that is, $\hat{\Gamma}_1(O_1)$, from the total field impedance at that point $\hat{Z}_1(O_1)$—we need to normalize $\hat{Z}_1(O_1)$ and then plot it on the Smith chart. Hence,

$$\hat{z}_{n1}(O_1) = \frac{\hat{Z}_1(O_1)}{\eta_1} = \frac{226.2 - j301.6}{60\pi}$$

$$= 1.2 - j1.6$$

The value of $\hat{z}_{n1}(O_1)$ is plotted as point P_3 in Figure 5.21.

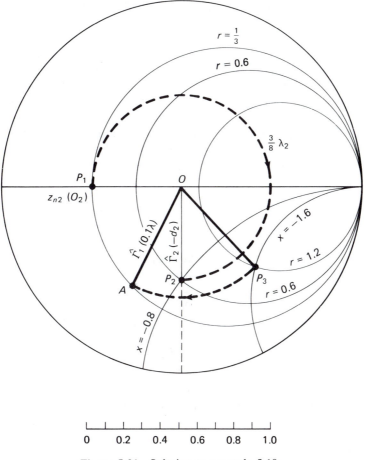

Figure 5.21   Solution to example 5.10.

9. The reflection coefficient at point A is then obtained from the value of $\hat{\Gamma}_1(O_1)$ at P_3 simply by rotating P_3 on a constant reflection coefficient circle a distance $0.1\lambda_1$ toward the generator.

10. The magnitude of the reflection coefficient is obtained by measuring the distance OA and determining $|\hat{\Gamma}_1(-0.1\lambda_1)|$ from the reflection coefficient scale on the side of the chart. The phase angle of $\hat{\Gamma}_1(-0.1\lambda_1)$ is shown in Figure 5.21. Hence,

$$\hat{\Gamma}_1(-0.1\lambda_1) = 0.6e^{j241.5°}$$

You may want to check the accuracy of this answer by solving the problem analytically, using the equations described in the previous section.

5.7 QUARTER- AND HALF-WAVELENGTH TRANSFORMERS

In the previous sections we focused on developing solution procedures for the problem of normal incidence plane wave reflection and transmission at multidielectric regions. In this section we wish to describe some practical applications that use some of the ideas we discussed in the previous sections. For example, let us assume that we wish to transmit electromagnetic energy efficiently across an interface between two media that have significantly different electrical properties. Because of the difference in the properties of the two media, we expect to have a large reflection of waves at the interface between the two media. Referring to Figure 5.22, suppose we want to transmit a plane wave across the interface between air and water. The characteristic wave impedance in air is

$$\eta_1 = \sqrt{\frac{\mu_o}{\epsilon_o}} = 120\pi \ \Omega$$

while η_2 in water (assuming $\sigma = 0$) is given by

$$\eta_2 = \sqrt{\frac{\mu_o}{64\epsilon_o}} = 15\pi \ \Omega$$

The reflection coefficient at the interface is given by equation 5.10, hence,

$$\Gamma = \frac{\eta_2 - \eta_1}{\eta_2 + \eta_1} = 0.778$$

A reflection coefficient of 0.778 does not promise an efficient transmission of the electromagnetic energy across the interface, and instead most of the incident energy is expected to be reflected back to region 1. If we want to eliminate the reflected wave in region 1, we may think of introducing a slab of dielectric material (intermediate layer) between our two original media (air and water). If we carefully choose the thickness and the dielectric properties of the intermediate layer, it may be possible to eliminate

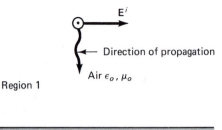

Region 1

Region 2

Figure 5.22 Plane wave reflection at air-water interface.

the reflected wave in region 1. We shall soon see that this is actually the idea of the quarter-wave transformer where the thickness of the dielectric slab is chosen to be $\lambda_m/4$, where λ_m is the wavelength inside the slab, and the dielectric properties of the slab is chosen such that the characteristic impedance η_m of the intermediate layer is the geometrical mean of the intrinsic impedances of the two other media on its sides—that is,

$$\eta_m = \sqrt{\eta_1 \eta_2}$$

In another application we may seek to introduce a dielectric slab in a medium without causing any reflections of waves traveling through that medium. In other words, in certain applications we may want to introduce what is known as a transparent dielectric window that is invisible to the waves propagating through the window. A special case of a dielectric window is a dome-shaped dielectric shell which is used to enclose microwave antennas for weather protection (e.g., snow, etc.) These transparent windows are known (in this case) as radomes. We shall shortly see that by making the thickness of the dielectric panel $d = \lambda_m/2$ we will be able to achieve the transparent property of the dielectric window.

   In these as well as many other applications that involve the process of reflection and transmission from multidielectric regions including, for example, the design of antireflection coatings for lenses, the geometry of the problem is illustrated in Figure 5.23. In all cases it is desired to eliminate the reflected wave in medium 1 by appropriately choosing the thickness, d_m, and the properties ϵ_m and μ_m of the intermediate layer. To eliminate the reflection in region 1, we should choose the parameters of the intermediate layer so that the total field impedance $\hat{Z}_m(-d_m) = \hat{Z}_1(O_1)$ is equal to η_1. Let us derive an expression for the total field impedance $\hat{Z}_m$ at any point $-z$ within the intermediate layer so that we would be able to determine values of d_m and ϵ_m that would satisfy the impedance matching condition $\hat{Z}_m(-d_m) = \eta_1$. Based on our previous discussions the total electric and magnetic fields in the intermediate layer are given by

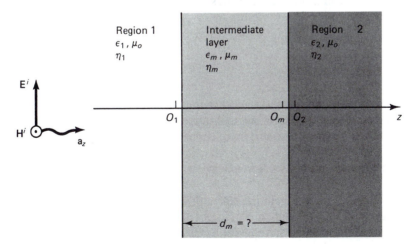

Figure 5.23   The geometry of the quarter-wave matching section (intermediate region) laced between regions 1 and 2.

$$\hat{\mathbf{E}}_m(z) = \hat{E}^+_{mm} e^{-j\beta_m z}(1 + \hat{\Gamma}_m e^{2j\beta_m z}) \mathbf{a}_x \tag{5.56}$$

$$\hat{\mathbf{H}}_m(z) = \frac{\hat{E}_{mm}}{\eta_m} e^{-j\beta_m z}(1 - \hat{\Gamma}_m e^{2j\beta_m z}) \mathbf{a}_y \tag{5.57}$$

where $\hat{E}^+_{mm}$ is the amplitude of the electric field, β_m is the phase constant ($\alpha_m = 0$ in a lossless medium), and $\hat{\Gamma}_m$ is the reflection coefficient at the interface between the intermediate layer and region 2. η_m is real because the intermediate region is assumed lossless, $\sigma_m = 0$. $\hat{\Gamma}_m$ at the interface (O_m) is clearly different from $\hat{\Gamma}_m(z)$ at an arbitrary location within the intermediate layer. These two reflection coefficients are related by

$$\hat{\Gamma}_m(z) = \frac{\hat{E}^+_{mm}}{\hat{E}_{mm}} e^{2j\beta_m z} = \hat{\Gamma}_m e^{2j\beta_m z}$$

Dividing equation 5.56 by equation 5.57 we obtain,

$$\hat{Z}_m(z) = \eta_m \left(\frac{1 + \hat{\Gamma}_m e^{2j\beta_m z}}{1 - \hat{\Gamma}_m e^{2j\beta_m z}} \right) \tag{5.58}$$

$\hat{\Gamma}_m$ at the interface, conversely, is given by

$$\hat{\Gamma}_m = \frac{\eta_2 - \eta_m}{\eta_2 + \eta_m} \tag{5.59}$$

Substituting equation 5.59 in the total field impedance expression of equation 5.58, we obtain

$$\hat{Z}_m(z) = \eta_m \frac{(\eta_2 + \eta_m) + (\eta_2 - \eta_m) e^{2j\beta_m z}}{(\eta_2 + \eta_m) - (\eta_2 - \eta_m) e^{2j\beta_m z}} \tag{5.60}$$

Multiplying both the numerator and denominator of equation 5.60 by $e^{-j\beta_m z}$, and separating terms containing η_m from those containing η_2, we obtain

$$\begin{aligned}
\hat{Z}_m(z) &= \eta_m \frac{\eta_2(e^{j\beta_m z} + e^{-j\beta_m z}) - \eta_m(e^{j\beta_m z} - e^{-j\beta_m z})}{\eta_m(e^{j\beta_m z} + e^{-j\beta_m z}) - \eta_2(e^{j\beta_m z} - e^{-j\beta_m z})} \\
&= \eta_m \frac{\eta_2 \cos(\beta_m z) - j\eta_m \sin(\beta_m z)}{\eta_m \cos(\beta_m z) - j\eta_2 \sin(\beta_m z)}
\end{aligned} \tag{5.61}$$

Equation 5.61 can be used to determine the total field impedance at any point within the intermediate dielectric layer. We shall use it next to determine the design parameters of the quarter- and half-wavelength transformers.

5.7.1 Quarter-wave Matching Section

As indicated earlier to eliminate the reflected wave in medium 1, $\hat{Z}_m(-d_m)$ should be made equal to η_1. From equation 5.61, if we choose $d_m = \lambda_m/4$—that is, a quarter of a wavelength in the intermediate layer, $\beta_m d_m$ would be

$$\beta_m d_m = \frac{2\pi}{\lambda_m} d_m = \frac{2\pi}{\lambda_m} \frac{\lambda_m}{4} = \frac{\pi}{2}$$

Substituting $\beta_m d_m = \pi/2$ in equation 5.61, we obtain

$$\hat{Z}_m\left(-\frac{\lambda_m}{4}\right) = \eta_m \frac{0 - j\eta_m}{0 - j\eta_2} = \frac{\eta_m^2}{\eta_2} \tag{5.62}$$

If $\hat{Z}_m(-\lambda_m/4) = \eta_1$, there will be no reflection in region 1. Substituting $\hat{Z}_m(-\lambda_m/4) = \eta_1$ in equation 5.62, we obtain

$$\eta_1 = \frac{\eta_m^2}{\eta_2}, \qquad \text{or} \qquad \eta_m = \sqrt{\eta_1 \eta_2}$$

This means that to achieve zero reflection in region 1 we choose the characteristic impedance of the intermediate region to be the geometrical mean of the intrinsic impedances on both sides. We also choose the thickness of the intermediate layer d_m to be $\lambda_m/4$, where λ_m is the wavelength in the intermediate region.

EXAMPLE 5.11

It is desired to minimize the reflections at the air-water interface shown in Figure 5.24a by using a quarter-wavelength dielectric slab of lossless dielectric material as shown in Figure 5.24b. Use the information in Table 5.1 to choose a suitable dielectric material and also determine the thickness of the slab at 10 GHz.

Solution

To eliminate the reflection in air, we choose the material of the quarter-wave matching section such that

$\eta_m = \sqrt{\eta_1 \eta_2}$

η_1 = characteristic wave impedance in free space = 120π

η_2 = characteristic wave impedance in water ($\epsilon_r = 64$) = $\sqrt{\dfrac{\mu_o}{64\epsilon_o}} = 15\pi$

$\eta_m = \sqrt{(120\pi)(15\pi)} = 30\sqrt{2}\pi$

$\qquad = \sqrt{\dfrac{\mu_o}{\epsilon_{rm}\epsilon_o}}$ (For lossless dielectric medium of $\mu = \mu_o$ and $\epsilon_m = \epsilon_o \epsilon_{rm}$)

TABLE 5.1   RELATIVE DIELECTRIC CONSTANTS OF SOME LOSSLESS DIELECTRIC MATERIALS

Material	Dielectric constant
Water (varies with frequency)	64–81
Air	1
Oil	2.3
Glass	6
Mica	6
Polystyrene	2.6
Flint glass	9.3
Glycerin	50

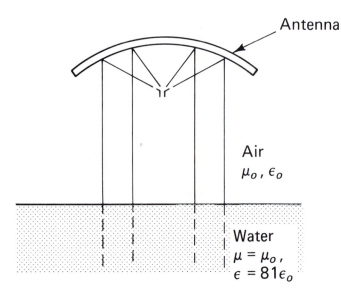

(a)

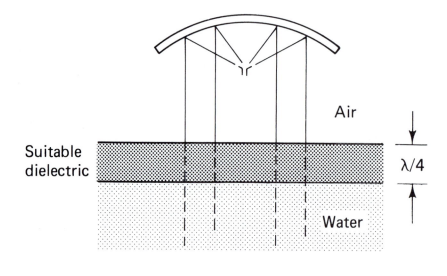

(b)

Figure 5.24   The quarter-wave transformer arrangement to minimize the reflection of the antenna radiation at the air-water interface.

The relative dielectric constant of the intermediate layer is hence,

$$\epsilon_{rm} = 8$$

The most suitable dielectric material available in Table 5.1 is the Flint glass with $\epsilon_r = 9.3$. The thickness of the slab is given by

$$d = \frac{\lambda_m}{4} = \frac{2\pi}{4\beta_m} = \frac{2\pi}{4\omega\sqrt{\mu_o\,\epsilon_o\,\epsilon_{rm}}}$$

$$= 0.25 \text{ cm}$$

It should be noted that because we did not select a material of $\epsilon_{rm} = 8$ the reflection coefficient will not be zero. Changing the frequency or the dielectric constant of the intermediate layer changes the reflection coefficient. Example 5.13 illustrates this point.

5.7.2 Half-wave Matching Section

In this case we want to eliminate the reflected wave in medium 1, but with the restriction that medium 1 and medium 2 are the same—that is, $\eta_1 = \eta_2$. In this case, if we require that the dielectric slab be a half wavelength in thickness—that is, $d_m = \lambda_m/2$, then $\beta_m\,d_m = 2\pi/\lambda_m(\lambda_m/2) = \pi$. Substituting $\beta_m\,d_m = \pi$ in equation 5.61, we obtain

$$\hat{Z}_m\left(-\frac{\lambda_m}{2}\right) = \eta_m\frac{\eta_2}{\eta_m} = \eta_2 = \eta_1$$

Hence, by choosing the thickness of the intermediate layer to be a half-wavelength in the medium, $\hat{Z}_m(-\lambda_m/2)$ would be equal to η_1, and the reflection coefficient in region 1 will, therefore, be zero. It should be emphasized that in the half-wavelength matching section case the design is independent on the specific dielectric properties of the intermediate layer. The matching section should, however, be made of lossless dielectric material to eliminate any energy losses resulting from attenuation.

EXAMPLE 5.12

A radome is to be constructed of Teflon, which is a lossless dielectric material of $\epsilon_r = 2.1$. Determine the thickness of the radome wall at 10 GHz.

Solution

For radomes the minimum thickness of the wall should be $\lambda_m/2$. Hence,

$$d = \frac{\lambda_m}{2} = \frac{2\pi}{2\beta_m} = \frac{2\pi}{2\omega\sqrt{\mu_o\,\epsilon_o\,\epsilon_r}}$$

$$= \frac{c}{2f\sqrt{\epsilon_r}} = \frac{3 \times 10^8}{2 \times 10^{10}\sqrt{2.1}} = 1.04 \text{ cm}$$

EXAMPLE 5.13

In designing coatings for optical equipment, one wishes to find the thickness and type of a transparent dielectric coating to use on glass to reduce the amount of reflected glare.

1. Given an assumed infinitely thick plate of glass (having $\epsilon_g = 6.6\epsilon_o$, and $\mu_g = \mu_o$) as shown in Figure 5.25, select reasonable values for the thickness d and the dielectric constant ϵ of a dielectric coating on this glass to completely prevent the reflection of light normally incident from air to the glass. The wavelength of the incident light in air and at which the reflection is desired to be zero is $\lambda = 6000$ Å.

2. Calculate the reflection coefficient in air if the wavelength of the incident light is changed slightly to $\lambda = 6120$ Å in air.

Solution

1. Because the properties of the two media to be matched are different (air and glass), we may use a quarter wave transformer matching section.

$$\eta_m = \sqrt{\eta_1\,\eta_g}$$

where

$$\eta_1 = 120\pi\ \Omega, \qquad \text{while} \qquad \eta_g = \sqrt{\frac{\mu_o}{6.6\epsilon_o}} = \frac{120\pi}{2.57}\ \Omega$$

$$\eta_m = \frac{120\pi}{\sqrt{2.57}} = 74.85\pi\ \Omega$$

$$= \sqrt{\frac{\mu_o}{\epsilon_o\,\epsilon_m}} = \frac{120\pi}{\sqrt{\epsilon_m}}$$

The dielectric constant of the coating material is therefore $\epsilon_m = 2.57$ (Polystyrene). The thickness of the coating d is given by

$$d = \frac{\lambda_m}{4} = \frac{2\pi}{4\beta_m} = \frac{2\pi}{4\omega\sqrt{\epsilon_o\,\epsilon_m\,\mu_o}} = \frac{c}{4f\sqrt{\epsilon_m}} = \frac{\lambda\ \text{in air}}{4\sqrt{\epsilon_m}}$$

$$= 935.7\ \text{Å}$$

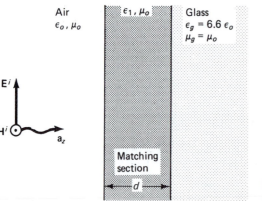

Air
ϵ_o, μ_o

ϵ_1, μ_o

Glass
$\epsilon_g = 6.6\,\epsilon_o$
$\mu_g = \mu_o$

E^i

H^i
a_z

Matching section
d

Figure 5.25 Illustrating the geometry of example 5.13.

2. After applying the coating, the wavelength of the incident light was changed by 2 percent from 6000 to 6120 Å. The thickness of the transformer material is no longer a quarter of a wavelength; hence, there will be reflections in air. To determine the reflection coefficient, we will follow the systematic procedure described in the earlier sections and also use the Smith chart. The origins in the various regions appropriate to the systematic solution procedure are shown in Figure 5.26.

$$\eta_1 = 120\pi, \qquad \eta_2 = 74.85\pi, \qquad \eta_3 = 46.7\pi$$

We start with region 3 where there is no reflection

$$\hat{Z}_3(O_3) = \eta_3 = 46.7\pi \ \Omega$$

From the continuity of the total field impedance at the interface between region 2 (coating) and region 3, we have

$$\hat{z}_{2n}(O_2) = \frac{\hat{Z}_2(O_2)}{\eta_2} = \frac{46.7\pi}{74.85\pi} = 0.62$$

$\hat{z}_{2n}(O_2)$ is plotted as point A on the Smith chart of Figure 5.27. At the new wavelength $\lambda_o = 6120$ Å, the wavelength in the coating material is given by

$$\lambda_m = \frac{\lambda_o}{\sqrt{\epsilon_m}} = \frac{6120 \times 10^{-10}}{\sqrt{2.57}} = 3817.6 \times 10^{-10} \ \text{m}$$

The electrical thickness of the coating in this case is

$$\frac{d}{\lambda_m} = \frac{935.7 \times 10^{-10}}{3817.6 \times 10^{-10}} = 0.245$$

which is slightly less than the quarter-wavelength thickness at the original wavelength. $\hat{z}_{2n}(-d_m)$ can now be obtained from $\hat{z}_{2n}(O_2)$ on the Smith chart by rotating point A a distance 0.245 toward the generator. The value of $\hat{z}_{2n}(-d_m)$ is shown as point B on the chart of Figure 5.27.

$$\hat{z}_{2n}(-d_m) = 1.61 + j0.025$$

The total field impedance at $z = d_m$ is then

$$\hat{Z}_2(-d_m) = \eta_2 \times \hat{z}_{2n}(-d_m)$$
$$= (120.5)\pi + j(3.)\pi \ \Omega$$

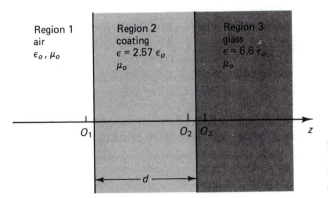

Region 1
air
ϵ_o, μ_o

Region 2
coating
$\epsilon = 2.57 \ \epsilon_o$
μ_o

Region 3
glass
$\epsilon = 6.6 \ \epsilon_o$
μ_o

O_1 O_2 O_3 z

d

Figure 5.26 The geometry of example 5.13 presented in a suitable way for the systematic solution procedure.

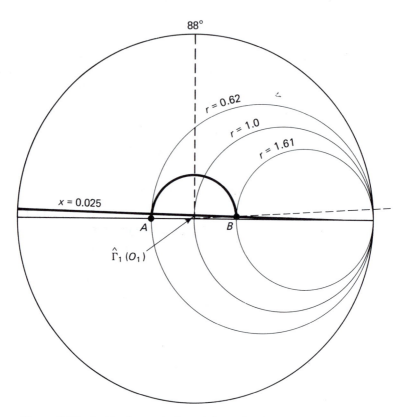

Figure 5.27 Smith chart solution of the reflection by the optical coating problem in example 5.13.

From the continuity of the total field impedance across the boundary between regions 1 and 2, we obtain

$$\hat{Z}_1(O_1) = \hat{Z}_2(-d_m) = 120.5\pi + j3.0\pi$$

$$\hat{z}_{1n}(O_1) = \frac{\hat{Z}_1(O_1)}{120\pi} = 1.00 + j0.025$$

It should be noted that if perfect matching is achieved, $\hat{z}_{1n}(O_1)$ would have been $1.0 + j0$, which is the origin on the reflection coefficient chart, thus indicating zero reflection. The slight change in the frequency therefore resulted in a slight mismatch as shown by point B on the chart of Figure 5.27. From Figure 5.27

$$\hat{\Gamma}_1(O_1) = .015e^{j88°}$$

which is still a negligible reflection in front of the coating material, of course, because it resulted from only a slight change in frequency.

SUMMARY

In this chapter we studied the reflection and transmission of plane waves normally incident on plane interfaces between different media.

Single Interface between Two Media

For the case of a single interface between two media, we obtained expressions for the reflection $\hat{\Gamma}$ and transmission $\hat{T}$ coefficients. These coefficients are given by

$$\hat{\Gamma} = \frac{\hat{E}_{m1}^-}{\hat{E}_{m1}^+} = \frac{\hat{\eta}_2 - \hat{\eta}_1}{\hat{\eta}_2 + \hat{\eta}_1}, \qquad \hat{T} = \frac{\hat{E}_{m2}^+}{\hat{E}_{m1}^+} = \frac{2\hat{\eta}_2}{\hat{\eta}_1 + \hat{\eta}_2}$$

where $\hat{E}_{m1}^+$ is the amplitude of the incident electric field while $\hat{E}_{m1}^-$ and $\hat{E}_{m2}^+$ are the amplitudes of the reflected and transmitted electric fields, respectively.

Also for this case of a single interface, we considered the special case of a total reflection from a perfectly conducting plane ($\hat{\Gamma} = -1$). We also defined and examined the propagation characteristics of standing waves. Standing waves result from interference between two waves of the same frequency, of equal magnitude, and are propagating in opposite directions. Typical characteristics of standing waves include zero time-average power, fixed locations of nulls, and a $\lambda/2$ distance between two successive nulls in the electric and magnetic fields patterns. The location of the first nulls in the electric field pattern is at the conducting plane and at a distance $\lambda/2$ from the perfectly conducting plane, whereas the first null location in the magnetic field pattern is $\lambda/4$ away from the conducting plane. Examples 5.3 and 5.4 were given to illustrate characteristics of standing waves.

Reflection and Transmission at Multiple Interfaces

For multiple interfaces we described the boundary value approach and the systematic solution procedure. The boundary value approach involves expressing the electric and magnetic fields in each region in form of positive z and negative z traveling waves with unknown amplitudes. For the last region, there is no reflection and only positive z traveling waves are needed in the electric and magnetic field expressions. The unknown amplitudes of the positive z and negative z traveling fields are determined by applying the boundary conditions. This basically involves enforcing the continuity of the tangential electric and magnetic fields across the interfaces. The systematic solution procedure, conversely, uses the total field impedance $\hat{Z}(z)$ and reflection coefficient to determine the amplitudes of the electric and magnetic fields in all regions.

It is shown that through consecutive enforcement of the continuity of the total field impedance at interfaces (i.e., $\hat{Z}(a^+) = \hat{Z}(a^-)$ at an interface located at $z = a$) and calculating the reflection coefficient at one interface from its known value at another location within the same medium (i.e., $\hat{\Gamma}_i(-d_i) = \hat{\Gamma}_i(O_i)\, e^{2\hat{\gamma}_i(-d_i)}$ in the ith medium of thickness d_i and propagation constant $\hat{\gamma}_i$) it is possible to calculate the amplitudes of fields in all regions. It is important to remember that (1) enforcing continuity of total field impedance is equivalent to applying the boundary conditions for both the electric and magnetic fields, and (2) the reflection coefficient cannot be used to cross interfaces between media. It can be used to relate reflection coefficients at different locations within the same medium. Example 5.5 illustrates this solution procedure.

Smith Chart

The Smith chart was also introduced to provide graphical implementation of the systematic solution procedure. It is a chart drawn on the reflection coefficient plane and provides an attractive graphical procedure that may be used to determine the following:

1. "Normalized" total field impedance from a known reflection coefficient.
2. Reflection coefficient at one location from its known value at another location within the same medium.
3. Reflection coefficient at a point from a known value of the normalized total field impedance at this point.

We should remember that in addition to the normalized resistance and normalized reactance circles clearly indicated on the chart, concentric circles with origins at the center of the chart represent constant reflection coefficient circles and that the complete circle around the chart represents $\lambda/2$ rotation of the reflection coefficient. Three scales are available on the rim of the chart; one to help us determine the phase angle of the reflection coefficient, whereas the others are intended to help us rotate the reflection coefficient a distance d/λ toward (*twg*) or away (*twl*) from the source. Additional calculations are required to account for the attenuation in the medium (if any). Examples 5.6 to 5.10 illustrate various applications of the Smith chart.

Quarter- and Half-wavelength Transformers

This section was included to provide applications of some of the ideas developed in this chapter. The quarter- and half-wavelength transformers are intermediate media introduced to eliminate reflections at interfaces. If we have an interface between two media of intrinsic impedances η_1 and η_2, the intrinsic impedance of the transformer is $\eta_m = \sqrt{\eta_1 \eta_2}$ and its thickness is $d = \lambda_m/4$ where λ_m is the wavelength calculated in the transformer medium. In other words, from the impedance equation we determine the dielectric properties of the transformer medium, and these properties are then used to calculate λ_m and thickness d of the medium.

In the half-wavelength transformer case, $\eta_1 = \eta_2$, and it is required to design a "transparent" window to electromagnetic waves such as the radome in antenna applications. It is shown that if the thickness of the transformer is selected to be $\lambda_m/2$, the reflection coefficient will be zero. The type of transformer material in this case is immaterial, but once selected it affects the thickness of the transformer because it enters in the calculation of λ_m. Examples 5.11 to 5.13 illustrate design procedures for quarter- and half-wavelength transformers.

These transformers are narrow-band ones and provide zero reflection at only the frequencies at which the thickness is designed. Also these transformers may be too thick at lower frequencies where λ is large. For broadband impedance matching, several cascaded sections of quarter-wave transformers may be used. This design procedure, however, is beyond the scope of this book.

PROBLEMS

1. A uniform plane wave traveling in free space is normally incident on the surface of a perfect conductor. If the *total* electric field is zero at a distance of 1 m away from the surface of the perfect conductor, determine the lowest possible frequency of the incident wave.

2. The complex forms of the total electric and magnetic fields in front of a perfectly conducting plane are given by

$$\hat{\mathbf{E}}^{tot} = -2jE_m^+ \sin \beta z \, \mathbf{a}_x$$

$$\hat{\mathbf{H}}^{tot} = \frac{2E_m^+}{\eta} \cos \beta z \, \mathbf{a}_y$$

(a) Obtain the real-time forms of these fields.
(b) Obtain the instantaneous Poynting vector of this wave.
(c) Determine the time-average Poynting vector using the time-domain forms of the fields.
(d) Determine the time-average Poynting vector using the complex forms of the fields.
(e) Show, based on the result obtained in parts c and d, that these fields represent a standing wave.

3. A uniform plane wave in lossless medium is normally incident on a plane perfect conductor at 150 MHz as shown in Figure P5.3. If the characteristic impedance of the lossless medium is $\eta_1 = 80 \, \Omega$ and $\mu = \mu_o$, determine the following:
(a) The shortest distance from the conductor surface at which the total electric field is zero. (Neglect the zero electric field at the interface.)
(b) The shortest distance from the conductor surface at which the total magnetic field is zero.
(c) If the amplitude of the incident electric field is $100e^{j0°}$, find the amplitude of the magnetic field at the surface of the conductor and at distance $z = -2$ m.

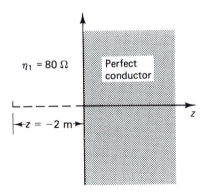

Figure P5.3   Geometry of problem 3.

4. A uniform plane wave in lossless medium is normally incident on a plane perfect conductor at 800 MHz as shown in Figure P5.4. If the measured distance between any two successive zeros of the total electric field in front of the conductor is 6.25 cm, find the following:

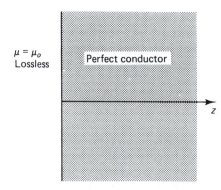

$\mu = \mu_o$
Lossless

Perfect conductor

z

Figure P5.4   Geometry of problem 4.

(a) The relative permittivity of the lossless medium, assuming $\mu = \mu_o$.
(b) The shortest distance from the conductor at which the total magnetic field is zero.
(c) If the amplitude of the incident electric field is $\hat{E}_{m1}^+ = 220e^{j0°}$ V/m, calculate the magnitude of the magnetic field at a distance $z = 0.4$ m from the surface of the conductor, and find the magnitude and direction of the induced surface current.

5. A transmitter consisting of a $\lambda_o/2$ antenna is placed in air (wavelength $= \lambda_o$) far above the surface of the sea. A receiver consisting of a $\lambda_w/2$ antenna is placed deep in the sea water (wavelength in sea water $= \lambda_w$). The incident plane wave on the sea water surface has an electric field $\mathbf{E}^i = E_o \cos(\omega t - \beta_o z)\mathbf{a}_x$ as shown in Figure P5.5. The electrical parameters of sea water are also given in Figure P5.5. The operating frequency is 20 kHz.
(a) Determine the lengths in meters of both antennas.
(b) Determine the depth d (in meters) at which the receiver is to be located if the signal received is 0.01 percent of the amplitude E_o of the wave in air.

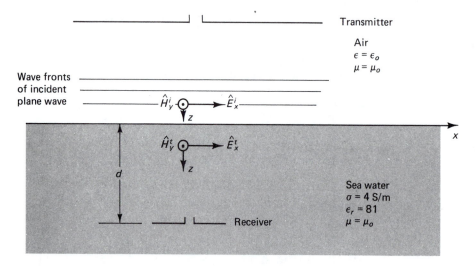

Figure P5.5   Communication system arrangement. The transmitter is placed in free space (air) where $\lambda = \lambda_o$. The receiver is placed in sea water, where $\lambda = \lambda_w$. The length of the transmitter is $\lambda_w/2$.

6. A uniform plane wave propagating in medium 1 ($z < 0$) is normally incident on medium 2 ($z > 0$). Medium 1 has the electrical characteristics $\epsilon_{r1} = 8$, $\mu_{r1} = 2$, $\sigma_1 = 0$, whereas medium 2 has $\epsilon_{r2} = 2$, $\mu_{r2} = 2$, $\sigma_2 = 0$. The electric field associated with the incident wave is given by $\hat{E} = 250e^{-j\beta_1 z}\mathbf{a}_x$. If the frequency of the incident wave is 2.5 GHz and the coordinate origin ($z = 0$) is located at the boundary surface between media 1 and 2, determine the following:
   (a) Reflection coefficient in region 1, $\hat{\Gamma}$.
   (b) Amplitude of reflected electric field $\hat{E}_{m1}^{-}$.
   (c) Amplitude of transmitted electric field $\hat{E}_{m2}^{+}$.
   (d) Amplitude of transmitted magnetic field $\hat{H}_{m2}^{+}$.
   (e) Total impedance at $z = -2.25$ cm.

7. (a) What is a standing wave? Describe reasons for causing it.
   (b) A 200-MHz uniform plane wave is traveling in free space in the positive z direction. At $z = 0$, the wave strikes normal to the surface a large block of material having $\epsilon_r = 4$, $\mu_r = 9$, $\sigma = 5$ S/m. If the incident magnetic field intensity vector is given by

$$\mathbf{H}^i(z,t) = \cos(\omega t - \beta_o z)\,\mathbf{a}_y \text{ A/m}$$

   (i) Write complete time-domain expressions for the incident, reflected, and transmitted electric field vectors.
   (ii) Determine the time-average power density associated with the transmitted wave.

8. A transmitter located in free space (ϵ_o, μ_o) is transmitting a positive z propagating plane wave with an x-directed electric field as shown in Figure P5.8. The transmitter is communicating with a submarine operating under sea water, which has the following electric parameters:

$$\epsilon = 81\epsilon_o, \qquad \mu = \mu_o, \qquad \sigma = 4 \text{ S/m}$$

The amplitude of the incident electric field is $\hat{E}_{m1}^{+} = 100e^{j0^\circ}$ V/m. The minimum detectable electric field by the submarine is 1 μV/m, and the frequency of operation is 20 kHz. Determine the largest possible depth (away from the surface of the air-water interface) for the submarine operation while maintaining a working communication system (can you neglect the conduction current?).

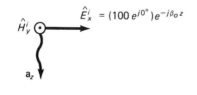

$$\hat{E}_x^i = (100\,e^{j0^\circ})e^{-j\beta_o z}$$

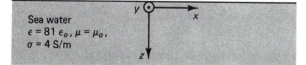

Sea water
$\epsilon = 81\,\epsilon_o,\ \mu = \mu_o,$
$\sigma = 4$ S/m

Figure P5.8 Geometry of problem 8.

9. A cruise missile is transmitting a 3-GHz signal, which is being received by a ground station as shown in Figure P5.9. If we assume that the earth has the following properties $\mu = \mu_o$, $\epsilon = 4\epsilon_o$, and $\sigma = 10^{-4}$ S/m and that the amplitude of the incident electric field is $\hat{E}_{m1}^{+} = 100e^{j0^\circ}$ V/m, determine the following:

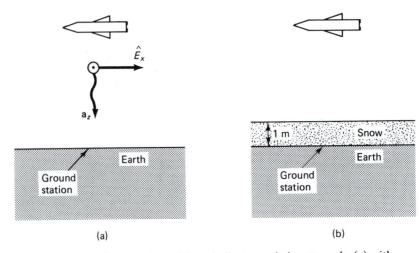

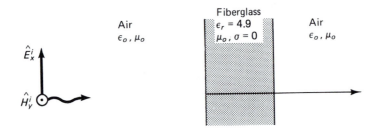

Figure P5.9 The geometry of the missile transmission to earth, (a) without snow and (b) with a layer of snow.

    (a) The amplitude of the *total* field received by the ground station. (*Hint*: Take the reflected field from the earth into account.)

    (b) Repeat part a for the case in which the ground station was covered by a 1-m-high layer of hard-packed snow. Use $\epsilon' = 1.5\epsilon_o$ and $\sigma/\omega\epsilon' = \sigma/\omega\epsilon_o\,\epsilon_r = 9 \times 10^{-4}$ for the hard-packed snow.

10. A 10-GHz uniform plane wave is propagating in the positive z direction and is incident from free space to a fiberglass ($\epsilon_r = 4.9, \mu_o, \sigma = 0$) radome as shown in Figure P5.10.

    (a) Determine the thickness of the radome so that there will be no reflections.

    (b) If the amplitude of the incident electric field $\hat{E}_{m1}^{+} = 100e^{j0°}$ and if we maintain the same physical thickness (in meters) of the fiberglass radome as determined in part a, use a Smith chart to determine the amplitude of the transmitted electric field in the fiberglass when the frequency is decreased by 10 percent.

Figure P5.10 The geometry of the fiberglass radome.

11. For the multiplane dielectric boundaries shown in Figure P5.11, calculate the reflection coefficient at point A in region 1.

12. Repeat the calculations in problem 11 if a lossy slab of $\mu = \mu_o$, $\epsilon = 2\epsilon_o$, and $\sigma/\omega\epsilon = 4$ is inserted in the free space of region 2. Use a frequency of 10 GHz.

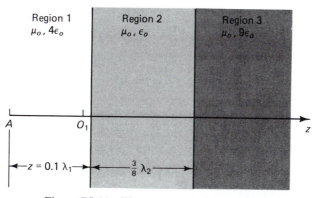

Figure P5.11   The geometry of problem 11.

13. A plane wave is normally incident on region 2, which is a lossy dielectric separating regions 1 and 3 as shown in Figure P5.13. Region 3 is a plane perfect conductor. If the frequency of the wave is 90 MHz and the amplitude of the incident electric field is $\hat{E}_{m1}^{+} = 10e^{j0°}$ V/m, determine the following:

(a) The reflection coefficient in region 1, just in front of the interface between regions 1 and 2.

(b) The amplitude of the reflected electric field in region 1 at the same point as in part a.

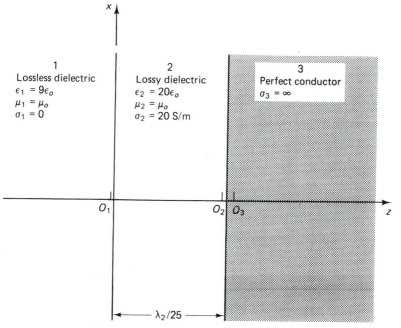

Figure P5.13   Geometry of problem 13.

14. It is desired to obtain zero reflection of a normally incident plane wave of frequency 1 MHz, at the interface of an infinitely thick plate of plexiglass (lossless dielectric, $\mu = \mu_o$) and air,

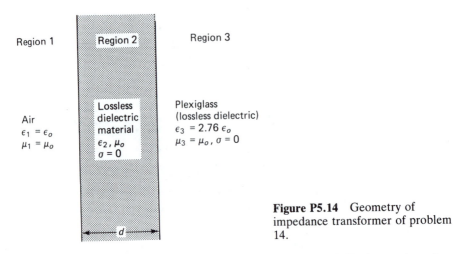

Figure P5.14 Geometry of impedance transformer of problem 14.

by using a slab of dielectric material as shown in Figure P5.14. Design an impedance transformer that will produce zero reflection in region 1. (*Hint:* Determine the thickness and the dielectric constant ϵ_2 of the transformer material.)

15. A uniform plane wave propagating in free space is normally incident on the multiple dielectric interfaces shown in Figure P5.15. The x-polarized electric field associated with this wave is given by

$$\mathbf{E}(z,t) = 50 \cos(2\pi \times 10^8 t - \beta_1 z)\, \mathbf{a}_x$$

where β_1 is the phase constant in region 1.
(a) If the thickness of the dielectric slab in region 2 is $d_2 = \lambda_2/3$, calculate the following:
    (i)   The magnitude of the electric field transmitted to region 2.
    (ii) The time-average power density in region 1.
(b) Show that if the thickness of the dielectric slab is changed to $d_2 = \lambda_2/4$, the reflection coefficient in region 1 would be zero. Also find the time-average power density in this case.

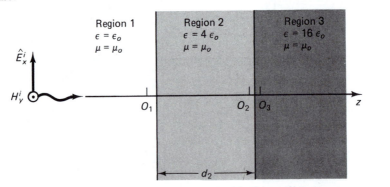

Figure P5.15 A plane wave normally incident on the interfaces between three dielectric regions.

16. A 500-MHz uniform plane wave traveling in a lossless medium with $\eta_1 = 60\pi\ \Omega$, $\mu_1 = \mu_o$ strikes normal to the surface a large block of copper ($\mu_2 = \mu_o$, $\epsilon_2 = \epsilon_o$, $\sigma_2 = 5.8 \times 10^7$ S/m).

Assume that the surface of the copper block lies in the x-y plane, the wave is propagating in the positive z direction, and the copper is a perfect conductor.

(a) Write real-time expressions for the incident and reflected electric and magnetic field vectors. Assume that the magnitude of the incident electric field is 1 V/m, and that it is oriented in the x direction.

(b) Determine the shortest distance from the copper surface at which the total electric field is zero.

(c) Determine the magnitude and direction of the induced surface current density.

(d) Calculate real-time average Poynting vector in the lossless region in front of the copper block.

17. (a) Explain what is meant by a "standing wave." Explain the reasons for its occurrence.

(b) The electric field $\mathbf{E}(z,t) = 24 \sin 10^9 t \sin 5z \, \mathbf{a}_x$ V/m is associated with a wave that is present in a lossless medium in front of a perfectly conducting plane.

    (i) Use the given electric field equation to explain the propagation nature of this wave. If $\mu = \mu_o$, determine the dielectric constant ϵ_r of the medium.

    (ii) Obtain a time-domain expression for the magnetic field intensity associated with this wave.

    (iii) Obtain an expression for the time-average Poynting vector associated with this wave. Use the obtained result to emphasize the wave propagation characteristic you indicated in part i above.

    (iv) Determine the locations (values of z in meters) of the zero values of the *electric* and *magnetic* fields in front of the perfect conductor.

18. (a) Explain what causes reflections at interfaces between different materials.

(b) In the electromagnetic hyperthermia applications, it is desired to heat cancerous tumors using electromagnetic radiation.

    (i) Assuming a plane wave normally incident on the tissue surface at 1 GHz, as shown in Figure P5.18a, determine the reflection coefficient. The electrical parameters of tissue at 1 GHz are

$$\mu = \mu_o, \qquad \epsilon_r = 50, \qquad \sigma = 3 \text{ S/m}$$

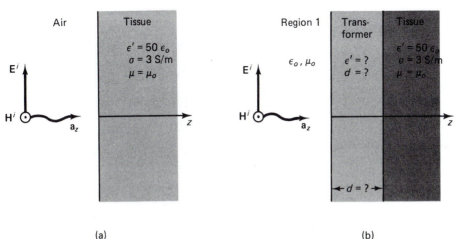

(a)   (b)

Figure P5.18   Heating tissue using electromagnetic radiation for cancer treatment using hyperthermia.

(ii) To minimize the large reflection coefficient calculated in part i, it is desired to design a quarter-wave transformer to be placed between air and the tissue to be heated, as shown in Figure P5.18b. To make the transformer out of lossless dielectric material, it was decided to base the transformer design on the *magnitude* of *the intrinsic wave impedance in tissue*. Determine the dielectric constant and thickness (in meters) of the transformer.

(iii) After designing the transformer in part ii, which is based on the assumption that $\hat{\eta}$ of tissue is real and equal to the *magnitude* of $\hat{\eta}$ in tissue, it is desired to calculate the reflection coefficient in region 1 (air). Calculate this reflection coefficient in air (region 1).

19. Using the Smith chart, show that introducing a slab of lossless dielectric normal to the propagation direction of a positive z propagating plane wave, as shown in Figure P5.19, will cause no reflection if the thickness of the slab is equal to $d_2 = \lambda_2/2$. The dielectric constant of the lossless dielectric slab may be chosen arbitrarily.

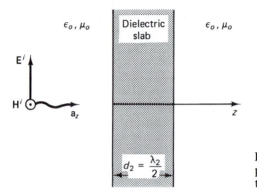

Figure P5.19 A $\lambda_2/2$ dielectric slab placed perpendicular to the direction of propagation.

20. Figure P5.20 shows three dielectric regions normally oriented to the direction of propagation of a positive z-propagating plane wave. The frequency of the incident wave is $f = 100$ MHz,

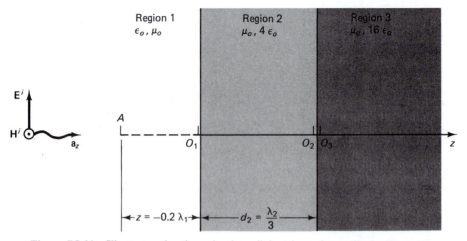

Figure P5.20 Illustrates the three lossless dielectric regions, the incident plane wave, and the point A at which the magnetic field is to be calculated.

and the amplitude of the electric field is $\hat{E}_{m1}^{+} = 100e^{j0°}$. Use the Smith chart to determine the following:

(a) Magnetic field at point A that is a distance $z = -0.2\lambda_1$ in region 1 as shown in Figure P5.20.

(b) Time-average power density in region 1.

(c) Amplitude of electric field intensity in region 2.

(d) Using the Smith chart, show that if the thickness of region 2, d_2, is equal to $\lambda_2/4$, the reflection coefficient in region 1 will be zero. Use a different Smith chart to solve this part of the question.

21. In discussing the reflection of plane waves from a perfectly conducting plane, it was shown that a totally reflected wave combines with the incident wave to form a standing wave. No traveling-wave component remains when there is perfect reflection; hence, no energy can be transported, and the Poynting vector is zero. When we have dielectric interface, there is only partial reflection. In this case the reflected wave combines with an equally strong part of the incident to form a standing wave. The remainder continues to be a traveling wave that carries energy and has a finite Poynting vector. For a plane wave normally incident on a dielectric interface as shown in Figure P5.21, show that the total electric field in region 1 is given by

$$\hat{E}_x(z) = \left(1 + \hat{\Gamma}\right) \hat{E}_m^i e^{-j\beta z} + \hat{\Gamma}\left(2j\,\hat{E}_m^i\,\sin\beta z\right)$$

$$= \left(1 + \hat{\Gamma}\right)(\text{traveling wave}) + \hat{\Gamma}\,(\text{standing wave})$$

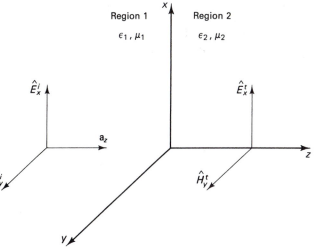

Figure P5.21   Plane wave incident on a plane interface between two media.

CHAPTER 6

OBLIQUE INCIDENCE PLANE WAVE REFLECTION AND TRANSMISSION

6.1 PLANE WAVE PROPAGATION AT ARBITRARY ANGLE

Thus far we considered plane wave propagation along the z direction. The mathematical treatment of this special case is simple, and therefore we used it as a start to examine the basic propagation characteristics of plane waves. In many boundary value problems, however, plane waves are not normally incident, as discussed in chapter 5. For this reason, we shall consider the general problem of uniform plane wave propagation along a specified axis that is oriented in an arbitrary manner relative to a rectangular coordinate system. This is most conveniently done in terms of the direction cosines of the normal to the plane of the wave. By definition of a uniform plane wave, the equiphase surfaces are planes perpendicular to the direction of propagation. Thus, in the expression

$$\hat{\mathbf{E}}(z) = \hat{E}_m e^{-j\beta z} \mathbf{a}_x \qquad (6.1)$$

for a wave traveling in the positive z direction, the planes of constant phase are given by the equation

$$\beta z = \text{constant}$$

In other words, for a plane wave propagating along the positive z axis, equation 6.1 states that each $z = \text{constant}$ plane represents an equiphase surface where there is no spatial variation in the electric or magnetic fields—that is, $\partial/\partial x = 0 = \partial/\partial y$ for a uniform plane wave.

Now, for a plane wave traveling in some arbitrary direction—for example, $\boldsymbol{\beta}$ direction, it is necessary to replace z with an expression that, when put equal to constant, gives the equiphase surfaces. The equation of an equiphase plane in this case is given by

$$\boldsymbol{\beta} \cdot \mathbf{r} = \text{constant} = \beta \, \mathbf{n}_\beta \cdot \mathbf{r}$$

where $\mathbf{r}$ is the radial vector from the origin to any point on the plane, and $\boldsymbol{\beta}$ is a vector normal to the plane, as shown in Figure 6.1.

In Figure 6.1, the plane perpendicular to the vector $\boldsymbol{\beta}$ is seen from its side, thus appearing as a line P-W. The dot product $\mathbf{n}_\beta \cdot \mathbf{r}$ is the projection of the radial vector $\mathbf{r}$ along the normal to the plane, and it is apparent that this will have the constant value OM for all points on the plane. Therefore, the equation $\boldsymbol{\beta} \cdot \mathbf{r} = \text{constant}$ is a characteristic property of a plane perpendicular to the direction of propagation $\boldsymbol{\beta}$. In the familiar rectangular coordinate system, the equation for the equiphase plane becomes

$$\boldsymbol{\beta} \cdot \mathbf{r} = \beta_x x + \beta_y y + \beta_z z$$
$$= \beta(\cos\theta_x x + \cos\theta_y y + \cos\theta_z z) = \text{constant}$$

where the radial vector $\mathbf{r}$ is equal to $\mathbf{r} = x \, \mathbf{a}_x + y \, \mathbf{a}_y + z \, \mathbf{a}_z$, $\boldsymbol{\beta} = \beta_x \, \mathbf{a}_x + \beta_y \, \mathbf{a}_y + \beta_z \, \mathbf{a}_z$, and θ_x, θ_y, and θ_z are the angles the vector $\boldsymbol{\beta}$ makes with x, y, and z axes, respectively.

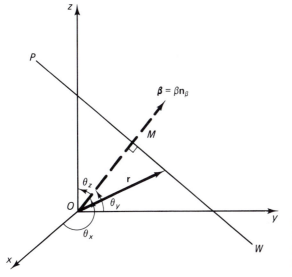

Figure 6.1   Plane wave front P-W with a propagation constant $\boldsymbol{\beta}$ making the angles θ_x, θ_y, and θ_z with the x, y, and z axes.

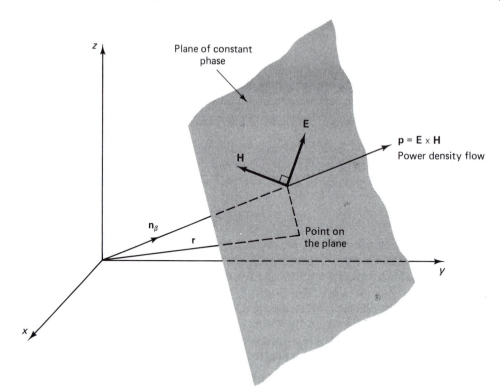

Figure 6.2 Plane wave front propagating in the $\mathbf{n}_\beta$ direction. The electric **E** and magnetic **H** fields are perpendicular to each other and in the plane wave front perpendicular to the direction of propagation $\mathbf{n}_\beta$. $\mathbf{n}_\beta$ is in the same direction as the power density flow **P**.

Now, because we have a transverse electromagnetic (TEM) wave, **H** is perpendicular to **E**, and both **E** and **H** are perpendicular to the direction of propagation $\boldsymbol{\beta}$. This allows us to write expressions for the $\hat{\mathbf{E}}$ and $\hat{\mathbf{H}}$ fields as

$$\hat{\mathbf{E}} = \hat{\mathbf{E}}_m e^{-j\boldsymbol{\beta} \cdot \mathbf{r}}$$

$$\hat{\mathbf{H}} = \frac{\mathbf{n}_\beta \times \hat{\mathbf{E}}}{\eta} \tag{6.2}$$

where $\mathbf{n}_\beta$ is a unit vector along $\boldsymbol{\beta}$ and η is the wave impedance in the propagation medium. The orthogonal relations between the electric and magnetic fields and the direction of propagation are illustrated in Figure 6.2.

EXAMPLE 6.1

The vector amplitude of an electric field associated with a plane wave that propagates in the negative z direction in free space is given by

$$\hat{\mathbf{E}}_m = 2\,\mathbf{a}_x + 3\,\mathbf{a}_y \text{ V/m}$$

Find the magnetic field strength.

Solution

The direction of propagation $\mathbf{n}_\beta$ is $-\mathbf{a}_z$. The vector amplitude of the magnetic field is then given by

$$\hat{\mathbf{H}}_m = \frac{\mathbf{n}_\beta \times \hat{\mathbf{E}}_m}{\eta} = \frac{1}{\eta_o} \begin{vmatrix} \mathbf{a}_x & \mathbf{a}_y & \mathbf{a}_z \\ 0 & 0 & -1 \\ 2 & 3 & 0 \end{vmatrix}$$

$$= \frac{1}{377}(3\,\mathbf{a}_x - 2\,\mathbf{a}_y) \text{ A/m}$$

———◆◆◆———

EXAMPLE 6.2

The phasor electric field expression in a plane wave is given by

$$\hat{\mathbf{E}} = \left[\mathbf{a}_x + \hat{E}_y\,\mathbf{a}_y + (2 + j5)\,\mathbf{a}_z\right]e^{-j2.3(-0.6x\,+\,0.8y)} \tag{6.3}$$

Find the following:

1. $\hat{E}_y$.
2. Vector magnetic field, assuming $\mu = \mu_o$ and $\epsilon = \epsilon_o$.
3. Frequency and wavelength of this wave.
4. Equation of surface of constant phase.

Solution

1. The general expression for a uniform plane wave propagating in an arbitrary direction is given by

$$\hat{\mathbf{E}} = \hat{\mathbf{E}}_m e^{-j\boldsymbol{\beta}\cdot\mathbf{r}}$$

   where the amplitude vector $\hat{\mathbf{E}}_m$, in general, has components in the x, y, and z directions. Comparing equation 6.3 with the general field equation for the plane wave propagating in an arbitrary direction, we obtain

$$\boldsymbol{\beta}\cdot\mathbf{r} = \beta_x x + \beta_y y + \beta_z z = \beta(\mathbf{n}_\beta\cdot\mathbf{r})$$

$$= \beta(\cos\theta_x x + \cos\theta_y y + \cos\theta_z z)$$

$$= 2.3(-0.6x + 0.8y + 0)$$

   Hence, a unit vector in the direction of propagation $\mathbf{n}_\beta$ is given by $\mathbf{n}_\beta = -0.6\,\mathbf{a}_x + 0.8\,\mathbf{a}_y$. Because the electric field $\hat{\mathbf{E}}$ must be perpendicular to the direction of propagation $\mathbf{n}_\beta$, it must satisfy the following relations:

$$\mathbf{n}_\beta\cdot\hat{\mathbf{E}} = 0$$

   Therefore,

$$(-0.6\,\mathbf{a}_x + 0.8\,\mathbf{a}_y)\cdot\left[\mathbf{a}_x + \hat{E}_y\,\mathbf{a}_y + (2 + j5)\,\mathbf{a}_z\right] = 0$$

   or

$$-0.6 + 0.8\hat{E}_y = 0$$

Hence, $\hat{E}_y = 0.75$. The electric field is given by

$$\hat{\mathbf{E}} = [\mathbf{a}_x + 0.75\,\mathbf{a}_y + (2 - j5)\,\mathbf{a}_z]\,e^{-j2.3(-0.6x\,+\,0.8y)}$$

2. The vector magnetic field $\hat{\mathbf{H}}$ is given by

$$\hat{\mathbf{H}} = \frac{1}{\eta}\,\mathbf{n}_\beta \times \hat{\mathbf{E}} = \frac{1}{377}\begin{vmatrix} \mathbf{a}_x & \mathbf{a}_y & \mathbf{a}_z \\ -0.6 & 0.8 & 0 \\ 1 & 0.75 & (2+j5) \end{vmatrix}$$

so that

$$\hat{H}_x = \frac{0.8(2+j5)}{377} = (4.24 + j10.6) \times 10^{-3}$$

$$\hat{H}_y = \frac{0.6(2+j5)}{377} = (3.18 - j7.95) \times 10^{-3}$$

$$\hat{H}_z = -\frac{0.6 \times 0.75 + 0.8}{377} = -3.31 \times 10^{-3}$$

The vector magnetic field is then given by

$$\hat{\mathbf{H}} = (\hat{H}_x\,\mathbf{a}_x + \hat{H}_y\,\mathbf{a}_y + \hat{H}_z\,\mathbf{a}_z)\,e^{-j2.3(-0.6x\,+\,0.8y)}$$

3. The wavelength λ is given by

$$\lambda = \frac{2\pi}{\beta} = \frac{2\pi}{2.3} = 2.73 \text{ m}$$

and the frequency

$$f = \frac{c}{\lambda} = \frac{3 \times 10^8}{2.73} = 0.11 \text{ GHz}$$

4. The equation of the surface of constant phase is

$$\mathbf{n}_\beta \cdot \mathbf{r} = -0.6x + 0.8y = \text{constant} \tag{6.4}$$

The general expression of this equation in terms of the direction cosines is given by

$$\mathbf{n}_\beta \cdot \mathbf{r} = (\cos\theta_x\,x + \cos\theta_y\,y + \cos\theta_z\,z) = \text{constant}$$

Comparison between equation 6.4 and the general expression shows that the plane given in equation 6.4 has no z dependence and, hence, defines a plane parallel to the z axis. In other words, equation 6.4 can be obtained by substituting $\theta_z = \pi/2$ in the general expression of the equiphase plane.

6.2 REFLECTION BY PERFECT CONDUCTOR—ARBITRARY ANGLE OF INCIDENCE

In this section we are interested in examining the reflection characteristics of a plane wave of an *arbitrary* incident angle on a perfect conductor. Considerable simplification in the analysis can be achieved by decomposing the general problem into two special cases: One in which the **E** field is polarized in the plane formed by the normal to the reflecting surface and the direction $\boldsymbol{\beta}_i$ of the incident wave, and the other in which the **E** field is perpendicular to the plane of incidence. The plane formed by the normal to the reflecting surface and the direction of propagation $\boldsymbol{\beta}$ is known as the plane of incidence. The general case can be considered as a superposition of the two cases; one in which **E** is parallel to the plane of incidence, while in the other **E** is pependicular to the plane of incidence.

6.2.1 E Field Parallel to Plane of Incidence

Figure 6.3 shows an incident wave polarized with the **E** field in the plane of incidence and the power flow in the direction of $\boldsymbol{\beta}_i$, which is at an angle θ_i with respect to the normal to the conducting surface. Because the Poynting vector gives the direction of propagation, $\boldsymbol{\beta}_i$, **E**, and **H** fields should be arranged such that $\boldsymbol{\beta}_i$ is in the same direction as $\mathbf{E}^i \times \mathbf{H}^i$ at any time. Hence, for the direction of the electric field shown in Figure 6.3, the direction of the magnetic field should be out of the plane of the paper—that is, $\mathbf{H} = \hat{H}_y \, \mathbf{a}_y$. There is no transmitted field within the perfect conductor, but there is a reflected field with power flow at an angle θ_r with respect to the normal to the interface. Taking the magnetic field to be still in the y direction, the direction of the electric field, therefore, will be as shown in Figure 6.3 so as to maintain the power density flow $\mathbf{E}^r \times \mathbf{H}^r$ in the same direction as $\boldsymbol{\beta}_r$. The total electric field in free space can be written as the sum of the incident and reflected fields, that is,

$$\hat{\mathbf{E}} = \hat{\mathbf{E}}^i + \hat{\mathbf{E}}^r = \hat{\mathbf{E}}_m^i e^{-j\boldsymbol{\beta}_i \cdot \mathbf{r}} + \hat{\mathbf{E}}_m^r e^{-j\boldsymbol{\beta}_r \cdot \mathbf{r}} \tag{6.5}$$

where

$$\boldsymbol{\beta}_i \cdot \mathbf{r} = \beta(\cos \theta_i \, \mathbf{a}_z + \sin \theta_i \, \mathbf{a}_x) \cdot (x \, \mathbf{a}_x + y \, \mathbf{a}_y + z \, \mathbf{a}_z)$$
$$= \beta(x \sin \theta_i + z \cos \theta_i) \tag{6.6}$$

$$\boldsymbol{\beta}_r \cdot \mathbf{r} = \beta(x \sin \theta_r - z \cos \theta_r) \tag{6.7}$$

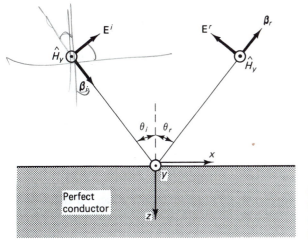

Figure 6.3   Reflection at a perfectly conducting interface for the case when the electric field is in the plane of incidence.

The total electric field is seen to have x and z components:

$$\hat{E}_x(x,z) = \hat{E}_m^i \cos\theta_i e^{-j\boldsymbol{\beta}_i \cdot \mathbf{r}} - \hat{E}_m^r \cos\theta_r e^{-j\boldsymbol{\beta}_r \cdot \mathbf{r}}$$

and

$$\hat{E}_z(x,z) = -\hat{E}_m^i \sin\theta_i e^{-j\boldsymbol{\beta}_i \cdot \mathbf{r}} - \hat{E}_m^r \sin\theta_r e^{-j\boldsymbol{\beta}_r \cdot \mathbf{r}}$$

The relationship between the incident and reflected amplitudes is provided by the boundary conditions that, for a perfect conductor, state that the total tangential E field at the surface must be zero. Hence,

$$\hat{E}_x\big|_{\text{at } z=0} = \hat{E}_m^i \cos\theta_i e^{-j\boldsymbol{\beta}_i \cdot \mathbf{r}} - \hat{E}_m^r \cos\theta_r e^{-j\boldsymbol{\beta}_r \cdot \mathbf{r}} = 0$$
$$= \hat{E}_m^i \cos\theta_i e^{-j\beta x \sin\theta_i} - \hat{E}_{-m}^r \cos\theta_r e^{-j\beta x \sin\theta_r} = 0 \qquad (6.8)$$

For equation 6.8 to be zero at all values of x along the surface of the conducting plane, the phase terms, which are the exponents, must be equal to each other; thus,

$$\theta_i = \theta_r \qquad (6.9)$$

that is, the angle of incidence equals the angle of reflection. Equation 6.9 is known as Snell's law of reflection. Substituting equation 6.9 in equation 6.8, we obtain

$$\hat{E}_m^i = \hat{E}_m^r \qquad (6.10)$$

Therefore, the total electric field in free space is given by

$$\hat{\mathbf{E}}(x,z) = \hat{E}_x(x,z)\,\mathbf{a}_x + \hat{E}_z(x,z)\,\mathbf{a}_z$$
$$= \hat{E}_m^i \cos\theta_i e^{-j\beta x \sin\theta_i}\big(e^{-j\beta z \cos\theta_i} - e^{j\beta z \cos\theta_i}\big)\,\mathbf{a}_x$$
$$\quad -\hat{E}_m^i \sin\theta_i e^{-j\beta x \sin\theta_i}\big(e^{-j\beta z \cos\theta_i} + e^{j\beta z \cos\theta_i}\big)\,\mathbf{a}_z$$
$$= -2j\,\hat{E}_m^i \cos\theta_i \,\sin(\beta z \,\cos\theta_i)\,e^{-j\beta x \sin\theta_i}\,\mathbf{a}_x \qquad (6.11)$$
$$\quad -2\hat{E}_m^i \sin\theta_i \,\cos(\beta z \,\cos\theta_i)\,e^{-j\beta x \sin\theta_i}\,\mathbf{a}_z$$
$$= 2\hat{E}_m^i\big[-j \cos\theta_i \,\sin(\beta z \,\cos\theta_i)\,\mathbf{a}_x$$
$$\quad -\sin\theta_i \,\cos(\beta z \,\cos\theta_i)\,\mathbf{a}_z\big]e^{-j\beta x \sin\theta_i}$$

From equation 6.11 and after we recover the time-domain form of the total electric field $\mathbf{E}(\mathbf{r},t) = Re\big(\hat{\mathbf{E}}(\mathbf{r})\,e^{j\omega t}\big)$, we observe that the variation of the total field with the x variable indicates that there is a traveling wave in the x direction with a phase constant $\beta_x = \beta \sin\theta_i$, but in the z direction the field forms a standing wave, as can be seen by examining the field variation with the z variable. This is expected to have standing wave normal to the conducting plane and traveling wave tangential to it.

The total magnetic field is given by

$$\hat{\mathbf{H}}(x,z) = \hat{H}_y(x,z)\,\mathbf{a}_y = \big[\hat{H}_y^i(x,z)\,\mathbf{a}_y + \hat{H}_y^r(x,z)\,\mathbf{a}_y\big]$$

Using the relation $\mathbf{H} = (\mathbf{n}_\beta \times \mathbf{E})/\eta$ for each of the incident and reflected fields and employing the expressions for the x and z components of the incident and reflected electric fields, we obtain

$$\hat{\mathbf{H}}^i = \frac{\mathbf{n}_{\beta i}}{\eta} \times \hat{\mathbf{E}}^i$$

$$= \frac{1}{\eta} \begin{vmatrix} \mathbf{a}_x & \mathbf{a}_y & \mathbf{a}_z \\ \sin\theta_i & 0 & \cos\theta_i \\ \hat{E}_m^i \cos\theta_i\, e^{-j\beta(\sin\theta_i x\, +\, \cos\theta_i z)} & 0 & -\hat{E}_m^i \sin\theta_i\, e^{-j\beta(\sin\theta_i x\, +\, \cos\theta_i z)} \end{vmatrix}$$

From the solution of this determinant, we find that the only nonzero component of $\hat{\mathbf{H}}^i$ is the $\mathbf{a}_y$ component and is given by

$$\hat{\mathbf{H}}^i = \frac{1}{\eta}\mathbf{a}_y\big[\hat{E}_m^i \cos^2\theta_i\, e^{-j\beta(\sin\theta_i x\, +\, \cos\theta_i z)}$$

$$+ \hat{E}_m^i \sin^2\theta_i\, e^{-j\beta(\sin\theta_i x\, +\, \cos\theta_i z)}\big]$$

$$= \frac{\hat{E}_m^i}{\eta}\, e^{-j\beta(\sin\theta_i x\, +\, \cos\theta_i z)}\, \mathbf{a}_y$$

Similarly, it can be shown that the reflected magnetic field is given by

$$\hat{\mathbf{H}}^r = \frac{\hat{E}_m^i}{\eta}\, e^{-j\beta(\sin\theta_i x\, -\, \cos\theta_i z)}\, \mathbf{a}_y$$

The total magnetic field $\hat{\mathbf{H}}(x, z)$ is then,

$$\hat{\mathbf{H}}(x, z) = \mathbf{a}_y \frac{2\hat{E}_m^i}{\eta} \cos(\beta z\, \cos\theta_i)\, e^{-j\beta x\, \sin\theta_i}$$

The average power flow parallel to the conducting surface is given by

$$\mathbf{P}_{ave}(x, z) = \frac{1}{2} Re\big[\hat{\mathbf{E}} \times \hat{\mathbf{H}}^*\big]$$

$$= \frac{1}{2} Re \begin{vmatrix} \mathbf{a}_x & \mathbf{a}_y & \mathbf{a}_z \\ \hat{E}_x & 0 & \hat{E}_z \\ 0 & \hat{H}_y^* & 0 \end{vmatrix}$$

From the cross product, it is apparent that we have two components: one in the x direction and the other in the z direction,

$$\mathbf{P}_{ave} = \frac{1}{2} Re\big[-\hat{E}_z \hat{H}_y^*\, \mathbf{a}_x + \hat{E}_x \hat{H}_y^*\, \mathbf{a}_z\big]$$

From the expressions of the various components of the electric and magnetic fields, it can be seen that in the expression of $\hat{E}_x \hat{H}_y^*$, the exponential terms will cancel out, and the overall product will be an imaginary number because of the (j) factor in the expression of $\hat{E}_x$. The expression for $\mathbf{P}_{ave}(x, z)$, hence, reduces to

$$\mathbf{P}_{ave}(x, z) = \frac{1}{2} Re\big[-\hat{E}_z \hat{H}_y^*\big]\mathbf{a}_x$$

$$= \frac{2|\hat{E}_m^i|^2}{\eta} \sin\theta_i\, \cos^2[\beta z\, \cos\theta_i]\, \mathbf{a}_x$$

Thus, for glancing incidence ($\theta_i \rightarrow 90°$), $\mathbf{P}_{ave} = (2(\hat{E}_m^i)^2)/\eta\, \mathbf{a}_x$ and the power flow is maximum. Conversely, for normal incidence ($\theta_i = 0$), power flow in the x direction is zero, that is, $\mathbf{P}_{x, ave} = 0$.

Average power flow perpendicular to the conducting surface is zero, because the average Poynting vector is zero in that direction, that is,

$$\mathbf{P}_{z, ave} = \frac{1}{2} Re(\hat{E}_x \hat{H}_y^*) = 0$$

This is because $\hat{E}_x$ is multiplied by j and, hence, $\hat{E}_x$ and $\hat{H}_y$ are out of phase by 90°. Therefore, while a traveling-wave pattern occurs in the x direction, because the incident and reflected waves travel in the *same* direction, a standing-wave pattern is observed in the z direction, because the incident and the reflected waves travel in the opposite directions. Another interesting parameter to study is the location of zeros (nodes) of the $\hat{E}_x$ field, which may be obtained by letting $\sin(\beta z \cos \theta_i) = 0$—that is, at distance z from the conducting plane given by

$$\beta z \cos \theta_i = n\pi$$

or

$$z = n\frac{\lambda}{2 \cos \theta_i} \qquad n = 0, 1, 2, \ldots$$

This field distribution is similar to the standing-wave pattern discussed in the normal incidence case, except that the zeros occur at distances larger than integer multiples of $\lambda/2$. Obviously, for normal incidence, $\theta_i = 0$, $\cos \theta_i = 1$, and the positions of the zeros are the same as those discussed before in chapter 5. Actually, even for the case of oblique incidence, the locations of standing-wave nodes are $\lambda/2$ apart along the direction of propagation. The wavelength measured along the z axis, however, is greater than the wavelength of the incident waves along the direction of propagation. The relation between these wavelengths is $\lambda_z = \lambda/(\cos \theta_i)$ as shown in Figure 6.4.

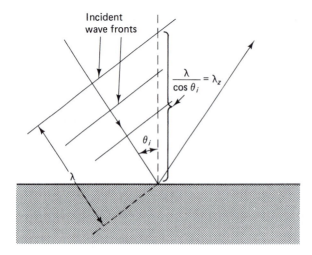

Incident
wave fronts

$\dfrac{\lambda}{\cos \theta_i} = \lambda_z$

θ_i

λ

Figure 6.4 The wavelength λ along the direction of propagation and the projected wavelength λ_z normal to the plane interface.

Therefore, while the planes of zero $\hat{E}_x$ field occur at multiples of $\lambda/2$ along the direction of propagation, they are located at integer multiples of $\lambda_z/2$ along the z axis and, hence, appear separated by larger distances. Also note, the standing-wave pattern associated with the $\hat{E}_z$ component, where it may be seen that there is no zero value of the electric field at $z = 0$. This is not in violation of any boundary condition in this case, because the $\hat{E}_z$ component is normal to the reflecting surface.

6.2.2 Electrical Field Normal to Plane of Incidence

In this case, the entire electric field is in the y direction (out of the paper) and the magnetic field has both x and z components, as shown in Figure 6.5.

The incident electric and magnetic fields are given by

$$\hat{\mathbf{E}}^i = \hat{E}_m^i e^{-j\boldsymbol{\beta}_i \cdot \mathbf{r}} \mathbf{a}_y$$

$$\hat{\mathbf{H}}^i = \frac{\mathbf{n}_{\beta_i} \times \hat{\mathbf{E}}^i}{\eta} = \frac{\hat{E}_m^i}{\eta}(-\cos\theta_i\, \mathbf{a}_x + \sin\theta_i\, \mathbf{a}_z) e^{-j\boldsymbol{\beta}_i \cdot \mathbf{r}}$$

where $\boldsymbol{\beta}_i \cdot \mathbf{r} = \beta(\sin\theta_i x + \cos\theta_i z)$. The reflected electric field is also assumed to be in the y direction so the magnetic field, which must be perpendicular to both $\mathbf{E}$ and the Poynting vector $\mathbf{P} = \mathbf{E} \times \mathbf{H}$, is in the direction shown in Figure 6.5.

$$\hat{\mathbf{E}}^r = \hat{E}_m^r e^{-j\boldsymbol{\beta}_r \cdot \mathbf{r}} \mathbf{a}_y$$

$$\hat{\mathbf{H}}^r = \frac{\mathbf{n}_{\beta_r} \times \hat{\mathbf{E}}^r}{\eta} = \frac{\hat{E}_m^r}{\eta}(\cos\theta_r\, \mathbf{a}_x + \sin\theta_r\, \mathbf{a}_z) e^{-j\boldsymbol{\beta}_r \cdot \mathbf{r}}$$

where $\boldsymbol{\beta}_r \cdot \mathbf{r} = \beta(\sin\theta_r x - \cos\theta_r z)$. Again, to determine the angle of reflection θ_r and the amplitude of the reflected electric field $\hat{E}_m^r$, we use the boundary conditions at $z = 0$. This includes zero values of the total tangential electric field $\mathbf{E}$ and the normal component of the magnetic field $\mathbf{H}$. (Note that we are not going to use the tangential component of $\mathbf{H}$ because of the unknown induced surface currents.)

Hence,

$$\hat{E}_y(x, z) = \hat{E}_y^i + \hat{E}_y^r = 0 \qquad \text{at } z = 0$$

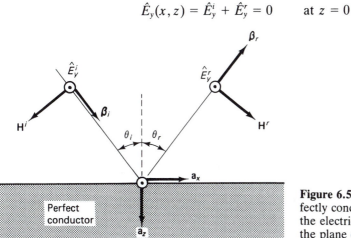

Figure 6.5 Reflection at a perfectly conducting interface when the electric field is perpendicular to the plane of incidence.

Therefore,

$$\hat{E}_y(x,0) = \hat{E}_m^i e^{-j\beta x \sin \theta_i} + \hat{E}_m^r e^{-j\beta x \sin \theta_r} = 0$$

and

$$\hat{H}_z(x,0) = \frac{1}{\eta} \hat{E}_m^i \sin \theta_i e^{-j\beta x \sin \theta_i} + \frac{1}{\eta} \hat{E}_m^r \sin \theta_r e^{-j\beta x \sin \theta_r} = 0$$

It should be noted that these two conditions will provide the same results for the unknowns θ_r and $\hat{E}_m^r$. These conditions must be true for every value of x along $z = 0$ plane, so that the phase factors must be equal. This simply leads to

$$\theta_r = \theta_i$$

and

$$\hat{E}_m^r = -\hat{E}_m^i$$

The negative sign indicates that the direction of the reflected electric field is opposite to that shown in Figure 6.5 (i.e., it should be into the paper). The total **E** field is then

$$\hat{E}_y(x,z) = \hat{E}_m^i e^{-j\beta x \sin \theta_i}\left(e^{-j\beta z \cos \theta_i} - e^{+j\beta z \cos \theta_i}\right)$$

$$= -2j\,\hat{E}_m^i[\sin(\beta z \cos \theta_i)]\, e^{-j\beta x \sin \theta_i}$$

The total $\hat{\mathbf{H}}$ field is given by

$$\hat{\mathbf{H}} = \hat{\mathbf{H}}^i + \hat{\mathbf{H}}^r = \left[\frac{\mathbf{n}_{\beta_i}}{\eta} \times \mathbf{a}_y \hat{E}_m^i e^{-j\beta_i \cdot \mathbf{r}}\right] - \left[\frac{\mathbf{n}_{\beta_r}}{\eta} \times \mathbf{a}_y \hat{E}_m^i e^{-j\beta_r \cdot \mathbf{r}}\right]$$

where the substitution $\hat{E}_m^r = -\hat{E}_m^i$ has been made. The direction vectors of the incident and reflected waves are given by

$$\mathbf{n}_{\beta_{i,r}} = \sin \theta_i\, \mathbf{a}_x \pm \cos \theta_i\, \mathbf{a}_z$$

and

$$\mathbf{n}_{\beta_{i,r}} \times \mathbf{a}_y = \sin \theta_i\, \mathbf{a}_z \mp \cos \theta_i\, \mathbf{a}_x$$

The components of the total magnetic field are

$$\hat{H}_x(x,z) = -\frac{2\hat{E}_m^i}{\eta} \cos \theta_i \cos(\beta z \cos \theta_i)\, e^{-j\beta x \sin \theta_i}$$

$$\hat{H}_z(x,z) = -\frac{2j}{\eta} \hat{E}_m^i \sin \theta_i \sin(\beta z \cos \theta_i)\, e^{-j\beta x \sin \theta_i}$$

where it can be seen that once again we have a standing wave in the z direction because the reflected and incident waves travel in opposite directions along the z axis. Also, the fields still travel in the x direction, and the only nonzero power flow is in the direction parallel to the interface. To illustrate this point, let us consider the average power density flow associated with this wave.

$$\mathbf{P}_{ave}(x,z) = \frac{1}{2} Re[\hat{\mathbf{E}} \times \hat{\mathbf{H}}^*]$$

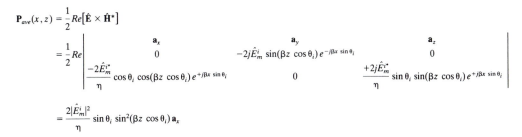

$$= \frac{2|\hat{E}_m^i|^2}{\eta} \sin\theta_i \sin^2(\beta z \cos\theta_i) \mathbf{a}_x$$

which indicates that the power flow is in the x direction.

EXAMPLE 6.3

Find the peak value of an induced surface current when a plane wave is incident at an angle on a large-plane, perfectly conducting sheet. The surface of the sheet is located at $z = 0$ and

$$\hat{\mathbf{E}}^i = 10 \cos\left(10^{10}t - \beta\frac{x}{\sqrt{2}} - \beta\frac{z}{\sqrt{2}}\right) \mathbf{a}_y \text{ V/m}$$

Solution

From the equation of the incident electric field, the propagation vector is given by

$$\boldsymbol{\beta} = \frac{\beta}{\sqrt{2}}\mathbf{a}_x + \frac{\beta}{\sqrt{2}}\mathbf{a}_z$$

$$= \beta(\sin 45° \mathbf{a}_x + \cos 45° \mathbf{a}_z), \qquad \text{that is, } \theta_i = 45°$$

Because the electric field is along the y direction—that is, perpendicular to the plane of incidence—we will use the equations given in section 6.2.2.

The sheet current $\hat{\mathbf{J}}$ (in ampere per meter) is determined by the total tangential magnetic field at the surface. From the boundary condition,

$$\hat{\mathbf{J}} = \mathbf{n} \times \hat{\mathbf{H}}$$

where the normal $\mathbf{n}$ to the surface for the geometry of Figure 6.5 is $\mathbf{n} = -\mathbf{a}_z$. The magnetic field in this case has two components:

$$\hat{H}_x = -\frac{2\hat{E}_m^i}{\eta_o} \cos\theta_i \cos(\beta z \cos\theta_i) e^{-j\beta x \sin\theta_i}$$

$$\hat{H}_z = -\frac{2j}{\eta_o} \hat{E}_m^i \sin\theta_i \sin(\beta z \cos\theta_i) e^{-j\beta x \sin\theta_i}$$

where η_o is the intrinsic impedance of free space. The surface current is then

$$\hat{\mathbf{J}}\big|_{\text{at } z = 0} = -\mathbf{a}_z \times \hat{\mathbf{H}} = \mathbf{a}_y \frac{2\hat{E}_m^i}{\eta_o} \cos\theta_i e^{-j\beta x \sin\theta_i}$$

and the peak value of the surface current at $z = 0$ is given by

$$\hat{J}|_{\text{peak value}} = \frac{2\hat{E}_m^i}{\eta_o} \cos \theta_i = \frac{2(10) \cos 45°}{377} = 3.75 \times 10^{-2} \text{ A/m}$$

---◆◆---

EXAMPLE 6.4

The electric field associated with a plane wave propagating in an arbitrary direction is given by

$$\hat{E} = (7.83 \, \mathbf{a}_x + 4 \, \mathbf{a}_y - 4.5 \, \mathbf{a}_z) e^{-j7(0.5x + 0.87z)}$$

If this wave is incident on a perfectly conducting plane oriented perpendicular to the z axis, find the following:

1. Reflected electric field.
2. Total electric field in region in front of the perfect conductor.
3. Total magnetic field.

Solution

Because a vector in the direction of propagation and a unit vector normal to the reflecting surface are contained in the x-z plane, we consider the x-z plane to be the plane of incidence as shown in Figure 6.6. The given electric field may, therefore, be decomposed into two components. The parallel polarization case in which the electric field is in the plane of incidence $\hat{E}_{\parallel}$ and the perpendicular polarization case in which the electric field is perpendicular to the plane of incidence $\hat{E}_{\perp}$. From the given equation of the electric field,

$$\hat{E}_{\parallel} = (7.83 \, \mathbf{a}_x - 4.5 \, \mathbf{a}_z) e^{-j7(0.5x + 0.87z)}$$

Comparing this with the equation of the electric field in the parallel polarization case, where the incident electric field is given by

$$\hat{E}^i = \hat{E}_m^i (\cos \theta_i \, \mathbf{a}_x - \sin \theta_i \, \mathbf{a}_z) e^{-j\beta(\sin \theta_i x + \cos \theta_i z)}$$

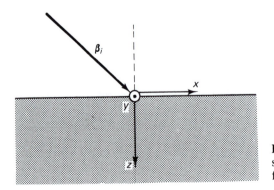

Figure 6.6 The propagation constant β_i associated with the electric field of example 6.4.

we observe that

$$\left.\begin{array}{l} \sin \theta_i = 0.5 \\ \cos \theta_i = 0.87 \end{array}\right\} \quad \text{that is, } \theta_i = 30°$$

The magnitude of the incident electric field $\hat{E}_m^i$ is therefore $= 7.83/0.87 = 9$ or $4.5/0.5 = 9$. Hence, the electric field associated with the parallel polarization case can be expressed in the form

$$\hat{\mathbf{E}}_\parallel^i = 9(0.87\,\mathbf{a}_x - 0.5\,\mathbf{a}_z)\,e^{-j7(0.5x + 0.87z)}$$

Based on the analysis of section 6.2.1, we have $\theta_r = 30°$, and the amplitude of the reflected electric field $\hat{E}_\parallel^r = \hat{E}_\parallel^i = 9$. Hence,

$$\hat{\mathbf{E}}^r = 9(-\cos 30°\,\mathbf{a}_x - \sin 30°\,\mathbf{a}_z)\,e^{-j7(\sin 30°x - \cos 30°z)}$$

We then treat the perpendicular polarization case where

$$\hat{\mathbf{E}}_\perp^i = 4\,\mathbf{a}_y\,e^{-j7(0.5x + 0.87z)}$$

Based on the analysis of section 6.2.2, it can be shown that

$$\hat{\mathbf{E}}_\perp^r = -4\,\mathbf{a}_y\,e^{-j7(0.5x - 0.87z)}$$

The total reflected electric field is then

$$\hat{\mathbf{E}}^r = (-7.83\,\mathbf{a}_x - 4\,\mathbf{a}_y - 4.5\,\mathbf{a}_z)\,e^{-j7(0.5x - 0.87z)}$$

Parts 2 and 3 can be easily obtained by following the analysis of section 6.2.
    For example, the magnetic field associated with the electric field in the parallel polarization case is given by

$$\hat{\mathbf{H}}_\parallel^i = \frac{9}{\eta}e^{-j7(0.5x + 0.87z)}\,\mathbf{a}_y$$

The reflected magnetic field intensity for this polarization is

$$\hat{\mathbf{H}}_\parallel^r = \frac{9}{\eta}e^{-j7(0.5x - 0.87z)}\,\mathbf{a}_y$$

For the perpendicular polarization case, the magnetic field has two components,

$$\hat{\mathbf{H}}_\perp^i = \frac{4}{\eta}(-\cos\theta_i\,\mathbf{a}_x + \sin\theta_i\,\mathbf{a}_z)e^{-j7(0.5x + 0.87z)}$$

$$= \left(-\frac{4}{\eta}\cos\theta_i\,\mathbf{a}_x + \frac{4}{\eta}\sin\theta_i\,\mathbf{a}_z\right)e^{-j7(0.5x + 0.87z)}$$

Because, for this case, $\hat{E}_\perp^r = -\hat{E}_\perp^i$,

$$\hat{\mathbf{H}}_\perp^r = -\frac{4}{\eta}(\cos 30°\,\mathbf{a}_x + \sin 30°\,\mathbf{a}_z)\,e^{-j7(0.5x - 0.87z)}$$

The total reflected magnetic field is then

$$\hat{\mathbf{H}}^r = \frac{1}{\eta}(-4\cos 30°\,\mathbf{a}_x + 9\,\mathbf{a}_y - 4\sin 30°\,\mathbf{a}_z)\,e^{-j7(0.5x - 0.87z)}$$

6.3 REFLECTION AND REFRACTION AT PLANE INTERFACE BETWEEN TWO MEDIA: OBLIQUE INCIDENCE

A plane wave incident at angle θ_i on a boundary between two media with electrical properties ϵ_1 and μ_1 in medium 1, and ϵ_2 and μ_2 in medium 2, will be partially transmitted into and partially reflected at the dielectric interface. The transmitted wave is refracted into the second medium—that is, its direction of propagation is different from that of the incident wave. The angles of reflection θ_r and refraction θ_t can be related to the angle of incidence θ_i with the aid of Figure 6.7, which shows two rays for each of the incident, reflected, and transmitted waves. (A ray is a line drawn normal to the equiphase surfaces and, therefore, this line is along the direction of propagation.)

Let us consider first the incident and reflected rays. In Figure 6.7 the incident ray 2 travels the distance CB, whereas the reflected ray 1 travels the distance AE. For both AC and BE to be the incident and reflected wave fronts (or planes of equiphase), it should take the incident ray the same time to cover distance CB as it takes the reflected ray to cover the distance AE. Because both the incident and reflected rays are in the same medium, their velocities will be the same, that is,

$$\frac{CB}{v_1} = \frac{AE}{v_1}$$

or $AB \sin \theta_i = AB \sin \theta_r$. Therefore,

$$\theta_i = \theta_r$$

As expected, the angle of reflection is equal to the angle of incidence.

To find the relationship between the angles of incidence θ_i and refraction θ_t, we

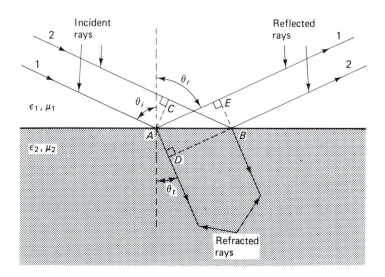

Figure 6.7   Reflection and refraction of an oblique incident plane wave at a plane interface between two media.

observe that it takes the incident ray the same time to cover distance CB as it takes the refracted ray to cover distance AD, that is,

$$\frac{CB}{v_1} = \frac{AD}{v_2}$$

where the magnitude of the velocity in medium 1 is $v_1 = 1/\sqrt{\mu_1 \epsilon_1}$ and its value is $v_2 = 1/\sqrt{\mu_2 \epsilon_2}$ in medium 2. From Figure 6.7 it may be seen that $CB = AB \sin \theta_i$, and $AD = AB \sin \theta_t$, hence,

$$\frac{CB}{AD} = \frac{\sin \theta_i}{\sin \theta_t} = \frac{v_1}{v_2} = \sqrt{\frac{\mu_2 \epsilon_2}{\mu_1 \epsilon_1}}$$

For most dielectrics $\mu_2 = \mu_1 = \mu_o$, therefore,

$$\frac{\sin \theta_i}{\sin \theta_t} = \left. \sqrt{\frac{\epsilon_2}{\epsilon_1}} \right|_{\mu_1 = \mu_2 = \mu_o} \tag{6.12}$$

This equation is known as Snell's law of refraction.

EXAMPLE 6.5

A plane wave is incident at an angle of 60° from air onto a half space of dielectric ($\mu_2 = \mu_o, \epsilon_2 = 6\epsilon_o$). Calculate the angle of refraction θ_t for the transmitted wave.

Solution

From Snell's law,

$$\frac{\sin \theta_i}{\sin \theta_t} = \sqrt{\frac{\epsilon_2}{\epsilon_1}}$$

$$\theta_i = 60°, \quad \text{and} \quad \frac{\epsilon_2}{\epsilon_1} = 6$$

$$\therefore \sin \theta_t = \sin 60°/\sqrt{6} = 0.354$$

$$\theta_t = 20.7°$$

As may be seen from the preceding discussion, the directions of the reflected and refracted waves can be determined from Snell's law. It remains, therefore, that we find the amplitudes of these waves in terms of the amplitude of the incident wave. As in the case of reflection by a perfect conductor, we can decompose the electric field of the incident wave into parallel and perpendicular components to the plane of incidence. Arbitrary polarization of the incident field is a superposition of two cases: One in which **E** is parallel to the plane of incidence, and one in which **E** is perpendicular to it. Recall that the plane of incidence is formed by the normal to the dielectric interface and the direction of incidence. Figure 6.8 illustrates the geometry of a plane wave incident on a dielectric interface and the relevance of the parallel and perpendicular polarization

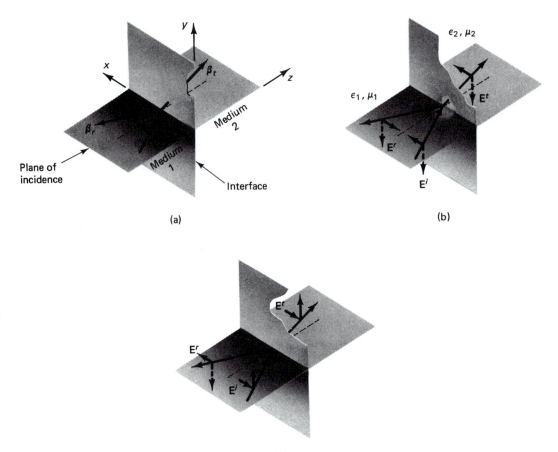

(a)

(b)

(c)

Figure 6.8   Schematic illustrating the geometry of (a) the plane of incidence, (b) the perpendicular, and (c) the parallel polarization cases.

components to the general formulation of the problem. In the following we will discuss these two polarization cases in detail.

6.3.1 Parallel Polarization Case —E is in Plane of Incidence

A plane wave with its electric field parallel to the plane of incidence is incident at angle θ_i on a dielectric interface between regions 1 and 2. Because the angles of reflection θ_r and refraction θ_t are known from Snell's laws, we need to determine the amplitudes of the reflected and transmitted waves. The unknown amplitudes of the reflected and transmitted electric fields $E_\parallel^r$ and $E_\parallel^t$ can be obtained by applying the boundary conditions at the dielectric interface. $E_\parallel^r$ and $E_\parallel^t$ will be used initially in the analysis instead of E_m^r and E_m^t to emphasize the fact that we are dealing with the parallel polarization case.

The tangential component of **H** should be continuous across the boundary. Hence,

$$\hat{H}^i_\| e^{-j\boldsymbol{\beta}_i \cdot \mathbf{r}} \mathbf{a}_y + \hat{H}^r_\| e^{-j\boldsymbol{\beta}_r \cdot \mathbf{r}} \mathbf{a}_y = \hat{H}^t_\| e^{-j\boldsymbol{\beta}_t \cdot \mathbf{r}} \mathbf{a}_y$$

Because the magnetic fields have only one component in the y direction, there is no need to carry the $\mathbf{a}_y$ vector indicating the use of the y component. Also recalling that this relation is valid at $z = 0$, we have

$$\hat{H}^i_\| e^{-j\beta_1(\sin\theta_i\, x)} + \hat{H}^r_\| e^{-j\beta_1(\sin\theta_r\, x)} = \hat{H}^t_\| e^{-j\beta_2(\sin\theta_t\, x)} \tag{6.13}$$

where β_1 and β_2 are the magnitudes of β in regions 1 and 2, respectively. For this relation to be valid at any value of x—that is, at any point on the interface, and recalling that $\theta_i = \theta_r$—we obtain

$$\beta_1 \sin\theta_i = \beta_2 \sin\theta_t$$

or

$$\frac{\sin\theta_i}{\sin\theta_t} = \frac{\beta_2}{\beta_1} = \frac{\dfrac{\omega}{v_2}}{\dfrac{\omega}{v_1}} = \frac{v_1}{v_2}$$

which is the same relation we obtained earlier from Snell's law. Substituting $\sin\theta_i/\sin\theta_t = v_1/v_2$ in equation 6.13, we obtain

$$\hat{H}^i_\| + \hat{H}^r_\| = \hat{H}^t_\| \qquad \text{at } z = 0 \tag{6.14}$$

Because **E** and **H** are related by η, equation 6.14 can be written in the form

$$\hat{E}^i_\| + \hat{E}^r_\| = \frac{\eta_1}{\eta_2} \hat{E}^t_\| \tag{6.15}$$

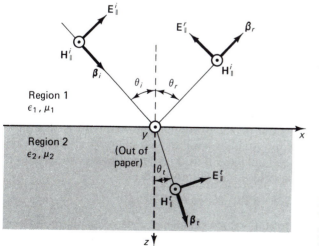

Figure 6.9 Reflection and refraction at a plane interface for the case when the electric field is parallel to the plane of incidence.

The tangential components of **E** must also be continuous across the boundary, hence,

$$\hat{E}_{\parallel}^i \cos \theta_i - \hat{E}_{\parallel}^r \cos \theta_r = \hat{E}_{\parallel}^t \cos \theta_t \qquad \text{at } z = 0 \qquad (6.16)$$

Remember that we once again used the amplitudes of the electric field in equation 6.16, because the exponential terms cancel out at $z = 0$ as required by Snell's law. Solving equations 6.15 and 6.16 for $E_{\parallel}^r$ and $\hat{E}_{\parallel}^t$, we obtain

$$\hat{E}_{\parallel}^r = \hat{E}_{\parallel}^i \frac{\eta_1 \cos \theta_i - \eta_2 \cos \theta_t}{\eta_1 \cos \theta_i + \eta_2 \cos \theta_t}$$

and

$$\hat{E}_{\parallel}^t = \hat{E}_{\parallel}^i \left(\frac{2\eta_2 \cos \theta_i}{\eta_1 \cos \theta_i + \eta_2 \cos \theta_t} \right) \qquad (6.17)$$

where we made use of the fact that $\theta_i = \theta_r$. If we define the reflection coefficient $\hat{\Gamma}_{\parallel} = -\hat{E}_{\parallel}^r / \hat{E}_{\parallel}^i$ and the transmission coefficient $\hat{\tau}_{\parallel} = \hat{E}_{\parallel}^t / \hat{E}_{\parallel}^i$, we obtain

$$\hat{\Gamma}_{\parallel} = -\frac{\hat{E}_{\parallel}^r}{\hat{E}_{\parallel}^i} = \frac{\eta_2 \cos \theta_t - \eta_1 \cos \theta_i}{\eta_2 \cos \theta_t + \eta_1 \cos \theta_i} = \left. \frac{\cos \theta_t - \sqrt{\epsilon_2/\epsilon_1} \cos \theta_i}{\cos \theta_t + \sqrt{\epsilon_2/\epsilon_1} \cos \theta_i} \right|_{\mu_1 = \mu_2 = \mu_o}$$

and

$$\hat{\tau}_{\parallel} = \frac{\hat{E}_{\parallel}^t}{\hat{E}_{\parallel}^i} = \frac{2\eta_2 \cos \theta_i}{\eta_2 \cos \theta_t + \eta_1 \cos \theta_i} = \left. \frac{2 \cos \theta_i}{\cos \theta_t + \sqrt{\epsilon_2/\epsilon_1} \cos \theta_i} \right|_{\mu_1 = \mu_2 = \mu_o}$$

The total electric field in region 1 is given by

$$\begin{aligned}
\hat{\mathbf{E}}_{\parallel}^{tot} &= \hat{\mathbf{E}}_{\parallel}^i + \hat{\mathbf{E}}_{\parallel}^r = \hat{E}_m^i (\cos \theta_i \, \mathbf{a}_x - \sin \theta_i \, \mathbf{a}_z) e^{-j\boldsymbol{\beta}_i \cdot \mathbf{r}} \\
&\quad + \hat{E}_m^r (-\cos \theta_r \, \mathbf{a}_x - \sin \theta_r \, \mathbf{a}_z) e^{-j\boldsymbol{\beta}_r \cdot \mathbf{r}} \\
&= \cos \theta_i \, \hat{E}_m^i e^{-j\beta x \sin \theta_i} (e^{-j\beta z \cos \theta_i} + \hat{\Gamma}_{\parallel} e^{j\beta z \cos \theta_i}) \, \mathbf{a}_x \qquad (6.18) \\
&\quad + \sin \theta_i \, \hat{E}_m^i e^{-j\beta x \sin \theta_i} (-e^{-j\beta z \cos \theta_i} + \hat{\Gamma}_{\parallel} e^{j\beta z \cos \theta_i}) \, \mathbf{a}_z
\end{aligned}$$

$$\underbrace{\qquad\qquad\qquad}_{\substack{\text{Traveling-wave} \\ \text{part}}} \underbrace{\qquad\qquad\qquad}_{\substack{\text{Standing plus} \\ \text{traveling waves}}}$$

where we have substituted $\boldsymbol{\beta}_i \cdot \mathbf{r}, \boldsymbol{\beta}_r \cdot \mathbf{r}$ from expressions derived before, and $\hat{E}_m^r / \hat{E}_m^i = -\hat{\Gamma}_{\parallel}$.

From equation 6.18, it is clear that we have a traveling-wave field in the x direction, and a traveling- as well as a standing-wave field in the z direction. The difference between this case and the perfectly conducting boundary is that $\hat{\Gamma}_{\parallel} \neq 1$ (remember that $\hat{\Gamma}_{\parallel} = -\hat{E}_{\parallel}^r / \hat{E}_{\parallel}^i$). Hence, the standing wave is only a part of the second term in equation 6.18. This can be explained if we rearrange the second term in the $\mathbf{a}_x$ component of the total electric field in the following form:

$$[(1 - \hat{\Gamma}_{\parallel}) e^{-j\beta z \cos \theta_i} + 2\hat{\Gamma}_{\parallel} \cos(\beta z \cos \theta_i)]$$

Hence, the preceding equation indicates that although a wave of amplitude $(1 - \hat{\Gamma}_{\parallel})$ is propagating along the z direction, another wave of amplitude $2\hat{\Gamma}_{\parallel}$ has the characteristics of a standing wave along the z axis. In other words, the characteristic of the wave along the z axis is neither a totally traveling wave or a totally standing wave, but, instead, a combination of both. If $\hat{\Gamma}_\| = 1$ (i.e., reflection from a perfect conductor), the amplitude of the traveling wave will be zero, and the wave characteristic along the z axis will be a totally standing wave. Conversely, if $\hat{\Gamma}_\| = 0$ (i.e., region 2 is a continuation of region 1), the amplitude of the standing wave will be zero, and we will have a totally traveling wave along the z direction.

The magnetic field in region 1 is given by

$$\hat{\mathbf{H}}_\|^{tot} = \hat{\mathbf{H}}_\|^i + \hat{\mathbf{H}}_\|^r = \hat{H}_m^i\, e^{-j\boldsymbol{\beta}_i \cdot \mathbf{r}}\, \mathbf{a}_y + \hat{H}_m^r\, e^{-j\boldsymbol{\beta}_r \cdot \mathbf{r}}\, \mathbf{a}_y$$

$$= \frac{\hat{E}_m^i}{\eta_1} e^{-j\beta x \sin\theta_i} \left(e^{-j\beta z \cos\theta_i} + \frac{\hat{E}_m^r}{\hat{E}_m^i} e^{j\beta z \cos\theta_i} \right) \mathbf{a}_y$$

$$= \frac{\hat{E}_m^i}{\eta_1} e^{-j\beta x \sin\theta_i} \left(e^{-j\beta z \cos\theta_i} - \hat{\Gamma}_\| e^{j\beta z \cos\theta_i} \right) \mathbf{a}_y$$

The transmitted fields in medium 2 are

$$\hat{\mathbf{E}}_\|^t = \hat{E}_m^t (\cos\theta_t\, \mathbf{a}_x - \sin\theta_t\, \mathbf{a}_z) e^{-j\boldsymbol{\beta}_t \cdot \mathbf{r}}$$

$$= \hat{\tau}_\| \hat{E}_m^i (\cos\theta_t\, \mathbf{a}_x - \sin\theta_t\, \mathbf{a}_z) e^{-j\boldsymbol{\beta}_t \cdot \mathbf{r}}$$

and

$$\hat{\mathbf{H}}_\|^t = \hat{H}_m^t\, \mathbf{a}_y\, e^{-j\boldsymbol{\beta}_t \cdot \mathbf{r}} = \frac{\hat{\tau}_\| \hat{E}_m^i}{\eta_2} e^{-j\boldsymbol{\beta}_t \cdot \mathbf{r}}\, \mathbf{a}_y$$

where $\boldsymbol{\beta}_t \cdot \mathbf{r} = \beta_2(x \sin\theta_t + z \cos\theta_t)$ and $\hat{E}_m^t / \hat{E}_m^i = \tau_\|$.

Brewster Angle.   From the expression of the reflection coefficient

$$\hat{\Gamma}_\| = \frac{\eta_2 \cos\theta_t - \eta_1 \cos\theta_i}{\eta_2 \cos\theta_t + \eta_1 \cos\theta_i}$$

it is clear that there is an angle of incidence at which $\hat{\Gamma}_\| = 0$. This angle may be obtained when

$$\eta_1 \cos\theta_i = \eta_2 \cos\theta_t$$

or

$$\cos\theta_i = \frac{\eta_2}{\eta_1} \cos\theta_t \tag{6.19}$$

This angle of incidence θ_i at which $\hat{\Gamma}_\| = 0$ is known as the Brewster angle. To obtain an expression for this angle in terms of the dielectric properties of media 1 and 2, we first consider Snell's law for the special case $\mu_1 = \mu_2 = \mu_o$, hence

$$\frac{\sin\theta_i}{\sin\theta_t} = \frac{v_1}{v_2} = \sqrt{\frac{\epsilon_2}{\epsilon_1}}\Bigg|_{\mu_1 = \mu_2 = \mu_o}$$

The condition $\mu_1 = \mu_2 = \mu_o$ is an important one because it is usually satisfied by the materials often used in optical applications. For this type of materials, equation 6.19 takes the form,

$$\cos \theta_i = \sqrt{\frac{\epsilon_1}{\epsilon_2}} \cos \theta_t \qquad (6.20)$$

Squaring both sides of equation 6.20 and using Snell's law for the special case of $\mu_1 = \mu_2 = \mu_o$, we obtain

$$\cos^2 \theta_i = \frac{\epsilon_1}{\epsilon_2} \cos^2 \theta_t = \frac{\epsilon_1}{\epsilon_2}(1 - \sin^2 \theta_t)$$

$$= \frac{\epsilon_1}{\epsilon_2}\left(1 - \frac{\epsilon_1}{\epsilon_2} \sin^2 \theta_i\right)$$

where the last substitution was made based on Snell's law of refraction. Hence,

$$1 - \sin^2 \theta_i = \frac{\epsilon_1}{\epsilon_2} - \frac{\epsilon_1^2}{\epsilon_2^2} \sin^2 \theta_i$$

$$\left(1 - \frac{\epsilon_1}{\epsilon_2}\right) = \sin^2 \theta_i \left(1 - \frac{\epsilon_1^2}{\epsilon_2^2}\right)$$

and

$$\sin^2 \theta_i = \frac{\epsilon_2}{\epsilon_2 + \epsilon_1} \qquad (6.21)$$

The Brewster angle of incidence is then given by

$$\sin \theta_i = \sqrt{\frac{\epsilon_2}{\epsilon_2 + \epsilon_1}} \qquad (6.22)$$

A very familiar expression for this specific value of θ_i can be obtained from equation 6.21,

$$1 - \cos^2 \theta_i = \frac{\epsilon_2}{\epsilon_2 + \epsilon_1}$$

or

$$\cos^2 \theta_i = 1 - \frac{\epsilon_2}{\epsilon_2 + \epsilon_1} = \frac{\epsilon_1}{\epsilon_2 + \epsilon_1}$$

Hence,

$$\cos \theta_i = \sqrt{\frac{\epsilon_1}{\epsilon_2 + \epsilon_1}} \qquad (6.23)$$

From equations 6.22 and 6.23, it is clear that

$$\tan \theta_i = \sqrt{\frac{\epsilon_2}{\epsilon_1}}$$

This specific angle of incidence θ_i is called the Brewster angle θ_β, hence

$$\theta_\beta = \tan^{-1}\sqrt{\frac{\epsilon_2}{\epsilon_1}}$$

Examples related to this Brewster angle will be given after we discuss the perpendicular polarization case.

6.3.2 Perpendicular Polarization Case — E Normal to Plane of Incidence

Figure 6.10 shows a perpendicularly polarized wave incident at an angle θ_i on dielectric medium 2. From Snell's law, the reflected wave will be at the same angle $\theta_r = \theta_i$, and the transmitted wave in medium 2 will be at an angle θ_t which can also be calculated using Snell's law. To determine the amplitude of the reflected and transmitted waves, we apply the continuity of the tangential components of **E** and **H** at the boundary.

Continuity of the tangential component of **H** along the x direction is given by

$$\hat{H}_\perp^i \cos\theta_i - \hat{H}_\perp^r \cos\theta_i = \hat{H}_\perp^t \cos\theta_t$$

Because **E** and **H** are related by η, one can write

$$\frac{\hat{E}_\perp^i}{\eta_1} \cos\theta_i - \frac{\hat{E}_\perp^r}{\eta_1} \cos\theta_i = \frac{\hat{E}_\perp^t}{\eta_2} \cos\theta_t \qquad (6.24)$$

The amplitudes of the electric fields $\hat{E}_m^i$, $\hat{E}_m^r$, and $\hat{E}_m^t$ were denoted by $\hat{E}_\perp^i$, $\hat{E}_\perp^r$, and $\hat{E}_\perp^t$ to emphasize the analysis of the perpendicular polarization case. Similarly, continuity of the tangential component of **E** along the y direction gives

$$\hat{E}_\perp^i + \hat{E}_\perp^r = \hat{E}_\perp^t \qquad \text{at } z = 0 \qquad (6.25)$$

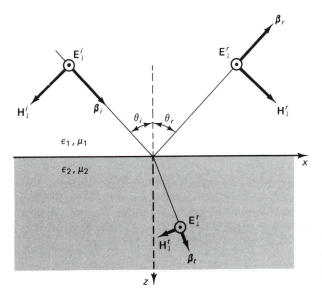

Figure 6.10 Reflection and refraction of plane waves at a plane interface between two media. The electric field is perpendicular to the plane of incidence (perpendicular polarization case).

It should be noted that in both equations 6.24 and 6.25, the exponential factors were canceled after substituting $z = 0$ and using Snell's laws. Equations 6.24 and 6.25 can then be solved for $\hat{E}^r_\perp$ and $\hat{E}^t_\perp$. It can be shown that

$$\hat{\Gamma}_\perp = \frac{\hat{E}^r_\perp}{\hat{E}^i_\perp} = \frac{\eta_2 \cos \theta_i - \eta_1 \cos \theta_t}{\eta_2 \cos \theta_i + \eta_1 \cos \theta_t}$$

and for nonmagnetic materials, $\mu_1 = \mu_2 = \mu_o$. Hence,

$$\hat{\Gamma}_\perp = \frac{\cos \theta_i - \sqrt{\epsilon_2/\epsilon_1} \cos \theta_t}{\cos \theta_i + \sqrt{\epsilon_2/\epsilon_1} \cos \theta_t}$$

The transmission coefficient (at $z = 0$) is given by

$$\hat{\tau}_\perp = \frac{\hat{E}^t_\perp}{\hat{E}^i_\perp} = \frac{2\eta_2 \cos \theta_i}{\eta_2 \cos \theta_i + \eta_1 \cos \theta_t}$$

and for nonmagnetic material, $\hat{\tau}_\perp$ is given by

$$\hat{\tau}_\perp = \frac{2 \cos \theta_i}{\cos \theta_i + \sqrt{\epsilon_2/\epsilon_1} \cos \theta_t}$$

6.4 COMPARISON BETWEEN REFLECTION COEFFICIENTS $\hat{\Gamma}_\parallel$ AND $\hat{\Gamma}_\perp$ FOR PARALLEL AND PERPENDICULAR POLARIZATIONS

The significant differences between the reflection coefficients for the parallel and perpendicular polarization cases $\hat{\Gamma}_\parallel$ and $\hat{\Gamma}_\perp$ will be illustrated in the following two examples.

EXAMPLE 6.6

Calculate the reflection coefficients as a function of the angle of incidence for parallel $(\hat{\Gamma}_\parallel)$ and perpendicularly $(\hat{\Gamma}_\perp)$ polarized waves incident from air $(\epsilon_r = 1)$ onto the following:

1. Water $(\epsilon_r = 81)$.
2. Paraffin $(\epsilon_r = 2)$.

Solution

In both cases, medium 2 can be assumed lossless dielectric and nonmagnetic material. Therefore the simplified forms of $\hat{\Gamma}_\parallel$ and $\hat{\Gamma}_\perp$ can be used. Figure 6.11 shows the calculated reflection coefficient for both polarizations for the two materials. The calculation procedure involves calculating θ_t first using Snell's law, and using its value to calculate $\hat{\Gamma}_\parallel$ and $\hat{\Gamma}_\perp$. For example, at $\theta_i = 0$, $\sin \theta_i / \sin \theta_t = \sqrt{\epsilon_2/\epsilon_1} = 0/\sin \theta_t$. Therefore,

$$\sin \theta_t = 0, \quad \text{or} \quad \theta_t = 0$$

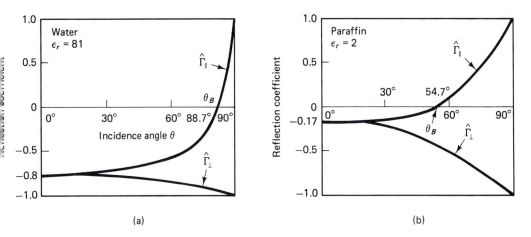

Figure 6.11   Reflection coefficient versus the angle of incidence θ_i for parallel and perpendicularly polarized waves. (a) Air-water interface. (b) Air-paraffin interface.

Then,

$$\hat{\Gamma}_{\parallel} = \frac{\cos \theta_t - \sqrt{\epsilon_2/\epsilon_1}\ \cos \theta_i}{\cos \theta_t + \sqrt{\epsilon_2/\epsilon_1}\ \cos \theta_i} = \frac{1 - \sqrt{2}}{1 + \sqrt{2}} \text{ for the paraffin case}$$

$$\hat{\Gamma}_{\parallel}\big|_{\text{paraffin at } \theta_i = 0} = -0.17$$

and so on.

From the reflection coefficients shown in Figure 6.11, it is clear that, for the perpendicular polarization case, there is no angle similar to the Brewster angle θ_B for which the reflected wave vanishes. The reason for this follows simply from Snell's law. Let us assume $\epsilon_2 > \epsilon_1$, then from Snell's law of reflection,

$$\frac{\sin \theta_i}{\sin \theta_t} = \sqrt{\frac{\epsilon_2}{\epsilon_1}} > 1 \qquad (\text{because } \epsilon_2 > \epsilon_1)$$

Then, $\sin \theta_t = (\sin \theta_i)/\sqrt{\epsilon_2/\epsilon_1}$, or θ_t should be less than θ_i.

Now from the expression of the reflection coefficient in the perpendicular polarization case,

$$\hat{\Gamma}_{\perp} = \frac{\cos \theta_i - \sqrt{\epsilon_2/\epsilon_1}\ \cos \theta_t}{\cos \theta_i + \sqrt{\epsilon_2/\epsilon_1}\ \cos \theta_t}$$

it may be seen that because $\sqrt{\epsilon_2/\epsilon_1} > 1$ and $\theta_t < \theta_i$, it is impossible to find a value of θ_i for which the numerator will be zero, i.e., $\cos \theta_i = \sqrt{\epsilon_2/\epsilon_1}\ \cos \theta_t$. Hence, the presence of an angle similar to the Brewster angle is not possible in the perpendicular polarization case. It should be noted, however, that such an angle can be found for the uncommon situation where $\epsilon_2 = \epsilon_1$ and $\mu_1 \neq \mu_2$. Such a case is of limited practical application and will not be discussed further.

EXAMPLE 6.7

1. Define what is meant by the Brewster angle.
2. Calculate the polarization angle (Brewster angle) for an air-water ($\epsilon_r = 81$) interface at which plane waves pass from the following:
   (a) Air into water.
   (b) Water into air.

Solution

1. *Brewster angle* is defined as the angle of incidence at which there will be no reflected wave. It occurs when the incident wave is polarized such that the **E** field is parallel to the plane of incidence.

2. (a) Air into water:

$$\epsilon_{r_1} = 1 \qquad \text{and} \qquad \epsilon_{r_2} = 81$$

The Brewster angle is then given by

$$\theta_B = \tan^{-1}\sqrt{\frac{\epsilon_2}{\epsilon_1}}$$

Therefore,

$$\theta_B = \tan^{-1}\sqrt{81} = 83.7°$$

(b) Water into air:

$$\epsilon_{r_1} = 81 \qquad \text{and} \qquad \epsilon_{r_2} = 1$$

Hence,

$$\theta_B = \tan^{-1}\sqrt{\frac{1}{81}} = 6.34°$$

To relate the Brewster angles in both cases, let us calculate the angle of refraction. From Snell's law,

$$\frac{\sin \theta_i}{\sin \theta_t} = \sqrt{\frac{\epsilon_2}{\epsilon_1}}$$

Therefore, in case a,

$$\frac{\sin \theta_B}{\sin \theta_t} = \sqrt{81}$$

Therefore,

$$\sin \theta_t = \frac{\sin 83.7}{9} = 0.11$$

or $\theta_t = 6.34°$, which is the same as the Brewster angle for case b. Also, the angle of refraction in case b is given by Snell's law as

$$\frac{\sin \theta_B}{\sin \theta_t} = \sqrt{\frac{\epsilon_o}{81\epsilon_o}} = \sqrt{\frac{1}{81}}$$

Therefore,

$$\sin \theta_t = \frac{\sin 6.34°}{\sqrt{\dfrac{1}{81}}} = 0.99$$

or $\theta_t = 83.7°$, which is the Brewster angle for case a.

————————◆◆————————

The preceding results simply indicate that when a monochromatic light is incident at an angle θ_B with respect to the normal of air-water interface, the angle of refraction θ_{t_1} is exactly the same as the Brewster angle when light is incident at the water-air interface (interface between regions 2 and 3). Hence, the incident monochromatic light will continue to suffer no reflections at the interface between regions 2 and 3 (see Figure 6.12). Similarly, the angle of refraction in the second case θ_{t_2} equals to the Brewster angle at the air-water interface. The light will, therefore, continue to suffer no reflection at the interface between regions 3 and 4, and so on. Such a phenomenon has an important application in optics, as will be discussed later.

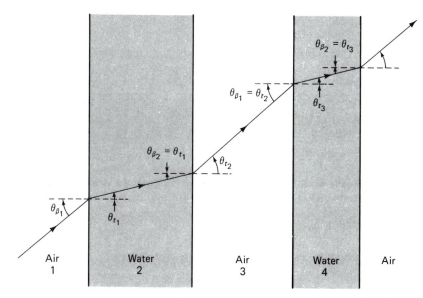

Figure 6.12 Monochromatic light refraction at multiple air-water interfaces. If the angle of incidence is equal to the Brewster angle, the alternate air-water parallel interfaces will cause no reflection.

6.5 TOTAL REFLECTION AT CRITICAL ANGLE OF INCIDENCE

In our previous discussion, we described the Brewster angle, which is the angle of incidence at which there is no reflection. It was also shown that for common dielectrics, this phenomenon (total transmission) exists only for the case in which the electric field is parallel to the plane of incidence (parallel polarization).

There is a second phenomenon that exists for both polarizations. Total reflection can occur at the interface between two dielectric media, and for a wave that is passing from a medium with a larger dielectric constant to a medium with smaller value of ϵ. From Snell's law of refraction, we have

$$\frac{\sin \theta_i}{\sin \theta_t} = \sqrt{\frac{\epsilon_2}{\epsilon_1}} \quad \text{or} \quad \sin \theta_t = \frac{\sin \theta_i}{\sqrt{\epsilon_2/\epsilon_1}} \qquad (6.26)$$

Hence, for a wave passing from a medium with a larger value of ϵ_1 to a medium of smaller dielectric constant $\epsilon_1 > \epsilon_2$, and from equation 6.26, it can be seen that $\theta_t > \theta_i$. In other words, when $\epsilon_1 > \epsilon_2$, a wave incident at an angle θ_i will pass into medium 2 at a larger angle θ_t. The critical angle of incidence θ_c is defined as that value of θ_i that makes $\theta_t = \pi/2$ (see Figure 6.13). Substituting $\theta_t = \pi/2$ in equation 6.26, we have

$$\sin \theta_c = \sqrt{\frac{\epsilon_2}{\epsilon_1}}, \quad \text{or} \quad \theta_c = \sin^{-1}\sqrt{\frac{\epsilon_2}{\epsilon_1}}$$

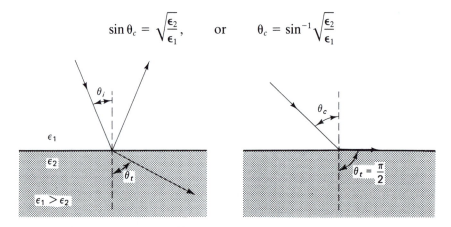

Figure 6.13 Illustrating the fact that $\theta_t > \theta_i$ if $\epsilon_1 > \epsilon_2$. The critical angle θ_c is defined as the value of θ_i at which $\theta_t = \pi/2$.

EXAMPLE 6.8

A uniform plane wave having the electric field given by

$$\mathbf{E}^i = 5[\cos \theta_i\, \mathbf{a}_x - \sin \theta_i\, \mathbf{a}_z] e^{-j10(\sin \theta_i x + \cos \theta_i z)}$$

is incident on the interface between a dielectric of permittivity $4\epsilon_o$ and the free space, as shown in Figure 6.14.

1. Find the angle of incidence θ_i so that there will be *no reflected wave* in region 1.

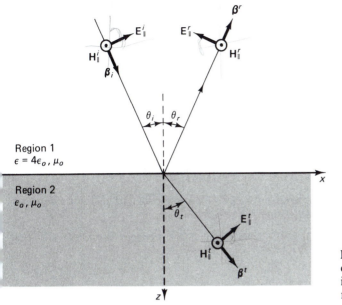

Figure 6.14 A uniform plane wave obliquely incident on a dielectric interface between region 1 and region 2.

2. Determine the transmitted field in region 2 at the angle of incidence calculated in part 1.

3. Find the angle of incidence θ_i so that there will be *no transmitted field* in region 2.

Solution

1. From the expression of the electric field and the schematic in Figure 6.14, it is clear that the incident electric field is parallel to the plane of incidence. In other words, in this example, we are basically dealing with the parallel polarization case.

    For this polarization, there is an angle of incidence that is called the Brewster angle at which there will be no reflected wave.

$$\theta_B = \tan^{-1}\sqrt{\frac{\epsilon_2}{\epsilon_1}} = \tan^{-1}\sqrt{\frac{1}{4}} = 26.56°$$

2. To obtain an expression for the transmitted electric field $\hat{E}_\parallel^t$, we need to find first the angle of refraction θ_t. From Snell's law,

$$\frac{\sin\theta_i}{\sin\theta_t} = \sqrt{\frac{\epsilon_2}{\epsilon_1}} = \sqrt{\frac{1}{4}}$$

$$\theta_t = 63.4°$$

The transmission coefficient for the parallel polarization case is also given by

$$\hat{\tau}_\parallel = \frac{2\eta_2\cos\theta_i}{\eta_2\cos\theta_t + \eta_1\cos\theta_i} = \frac{2\cos\theta_i}{\cos\theta_t + \sqrt{\dfrac{\epsilon_2}{\epsilon_1}}\cos\theta_i}$$

$$\approx 2$$

The expression for the transmitted electric field is, therefore,

$$\hat{\mathbf{E}}_{\parallel}^t = 5\hat{\tau}_{\parallel}[\cos\theta_t\,\mathbf{a}_x - \sin\theta_t\,\mathbf{a}_z]\,e^{-j\beta_2(\sin\theta_t\,x + \cos\theta_t\,z)}$$

To find a value for β_2, we use the relation between β_1 and β_2 and the given value of β_1, hence,

$$\frac{\beta_1}{\beta_2} = \frac{\omega\sqrt{\mu_1\,\epsilon_1}}{\omega\sqrt{\mu_2\,\epsilon_2}} = \sqrt{\frac{\epsilon_1}{\epsilon_2}}$$

$$\therefore\ \beta_2 = \beta_1\sqrt{\frac{\epsilon_2}{\epsilon_1}} = 10\sqrt{\frac{1}{4}} = 5$$

$$\therefore\ \hat{\mathbf{E}}_{\parallel}^t = 5(2)[0.448\,\mathbf{a}_x - 0.894\,\mathbf{a}_z]\,e^{-j5(0.894x + 0.448z)}$$

3. The angle of incidence θ_i at which there will be no transmitted field is the same for both the parallel and perpendicular polarization cases,

$$\theta_c = \sin^{-1}\sqrt{\frac{\epsilon_2}{\epsilon_1}} = \sin^{-1}\sqrt{\frac{1}{4}}$$

$$= 30°$$

◆◆◆

EXAMPLE 6.9

An unpolarized laser beam is incident on a dielectric interface between glass, $\epsilon_1 = 2.25\epsilon_o$, $\mu_1 = \mu_o$, and air $\epsilon_2 = \epsilon_o$, $\mu_2 = \mu_o$, as shown in Figure 6.15. The electric field associated with this wave is conveniently expressed in terms of a component parallel to the plane of incidence $\mathbf{E}_{\parallel}^i$ and another perpendicular to the plane of incidence $\mathbf{E}_{\perp}^i$. The total electric field is, therefore, given by

$$\hat{\mathbf{E}}^i = \hat{\mathbf{E}}_{\parallel}^i + \hat{\mathbf{E}}_{\perp}^i = [9\cos\theta_i\,\mathbf{a}_x + 4\,\mathbf{a}_y - 9\sin\theta_i\,\mathbf{a}_z]\,e^{-j7(x\,\sin\theta_i + z\,\cos\theta_i)}$$

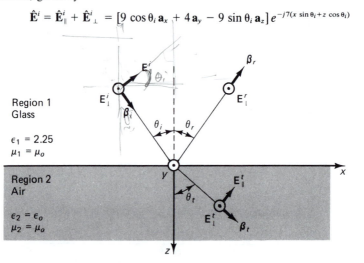

Figure 6.15   Unpolarized laser beam obliquely incident on a dielectric interface between glass and air.

1. Find the angle of incidence θ_i that would result in a reflected laser beam with the electric field totally polarized perpendicular to the plane of incidence.
2. For the angle of incidence θ_i calculated in part 1, obtain the following:
   (a) Expression for reflected electric field in region 1.
   (b) Expression for refracted (transmitted) electric field in region 2.
   (c) Time-average power density transmitted to region 2.

Solution

1. For the reflected laser beam to be totally polarized perpendicular to the plane of incidence, there has to be total transmission for the component of the electric field parallel to the plane of incidence. The angle of no reflection, or total transmission, for the parallel polarization case is known as the Brewster angle. This angle is given by

$$\theta_i = \theta_\beta = \tan^{-1}\sqrt{\frac{1}{2.25}} = 33.7°$$

By having the direction of incident to be at this angle $\theta_i = \theta_\beta$, the parallel polarization component of the field will be totally transmitted, and the only reflected field component in region 1 will be polarized perpendicular to the plane of incidence.

2. As indicated in part 1, the reflected electric field will be polarized only perpendicular to the plane of incidence.

(a)
$$\hat{\Gamma}_\perp = \frac{\eta_2 \cos\theta_i - \eta_1 \cos\theta_t}{\eta_2 \cos\theta_i + \eta_1 \cos\theta_t}$$

To find θ_t, we use Snell's law,

$$\frac{\sin\theta_i}{\sin\theta_t} = \sqrt{\frac{\epsilon_2}{\epsilon_1}} = \sqrt{\frac{1}{2.25}}$$

$$\theta_t = 56.3° \qquad \text{for } \theta_i = 33.7°$$

Hence,

$$\hat{\Gamma}_\perp = \frac{\cos\theta_i - \sqrt{\dfrac{\epsilon_2}{\epsilon_1}}\cos\theta_t}{\cos\theta_i + \sqrt{\dfrac{\epsilon_2}{\epsilon_1}}\cos\theta_t} = 0.385$$

$$\hat{E}^r = \hat{E}^r_\perp = \hat{\Gamma}_\perp \hat{E}^i_\perp \mathbf{a}_y\, e^{-j7(\sin 33.7°x - \cos 33.7°z)}$$

$$= 1.54\,\mathbf{a}_y\, e^{-j7(0.55x - 0.83z)}$$

(b) The transmitted electric field will contain the parallel and perpendicular polarization cases.

$$\frac{\beta_1}{\beta_2} = \sqrt{\frac{\epsilon_1}{\epsilon_2}} \therefore \beta_2 = 7\sqrt{\frac{\epsilon_2}{\epsilon_1}} = 4.67$$

$$\hat{\tau}_{\parallel} = \frac{2\eta_2 \cos\theta_i}{\eta_2 \cos\theta_t + \eta_1 \cos\theta_i} = \frac{2\cos 33.7°}{0.55 + \sqrt{\dfrac{1}{2.25}}(0.83)}$$

$$= 1.51$$

$$\hat{\tau}_{\perp} = \frac{2\cos 33.7°}{0.83 + \sqrt{\dfrac{1}{2.25}}\,0.55} = 1.38$$

$$\hat{\mathbf{E}}^t = \hat{\mathbf{E}}^t_{\parallel} + \hat{\mathbf{E}}^t_{\perp} = \{1.51(9)[\cos\theta_t\,\mathbf{a}_x - \sin\theta_t\,\mathbf{a}_z]$$
$$+ 1.38(4)\,\mathbf{a}_y\}e^{-j4.67(\sin\theta_t x + \cos\theta_t z)}$$

(c) The total transmitted magnetic field $\hat{\mathbf{H}}^t$ is given by

$$\hat{\mathbf{H}}^t = \frac{\boldsymbol{\eta}_{\beta_t} \times \hat{\mathbf{E}}^t}{\eta_2}$$

where the unit vector in the direction of the transmitted wave is given by

$$\boldsymbol{\eta}_{\beta_t} = \sin\theta_t\,\mathbf{a}_x + \cos\theta_t\,\mathbf{a}_z$$

and

$$\eta_2 = 120\pi$$

The average power transmitted to region 2 is then

$$\mathbf{P}_{ave2} = \frac{1}{2}Re(\hat{\mathbf{E}}^t \times \hat{\mathbf{H}}^{*t})$$

which can be easily obtained following the procedure repeatedly used in previous sections.

---◆◆◆---

6.6 ELECTROMAGNETIC SPECTRUM

As indicated earlier in this course, the spectrum of electromagnetic radiation extends from the long-wavelength radio waves to X rays and gamma rays at the shortest wavelength, as shown in Figure 6.16.

In the following section, we will discuss some electromagnetic based applications in the part of the spectrum extending from the near infrared through the visible to the ultraviolet. The visible region of the spectrum (where the human eye is sensitive) encompasses wavelengths from 400 nm to 700 nm. The properties of light waves can be discussed using the same mathematical formulation previously employed to examine the propagation characteristics of the lower frequency radio- and microwaves.

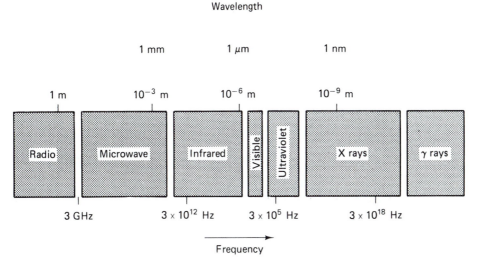

Figure 6.16   Electromagnetic spectrum from radio waves to X and γ rays.

6.7 APPLICATION TO OPTICS

In this section, we will discuss a few applications of some of the concepts described in this chapter. This includes control of polarization of incident waves using the concept of the Brewster angle, the role of the Brewster windows in light amplification by stimulated emission radiation (LASER) generation, and the use of the concept of angle of total reflection in optical fibers so widely used these days in communications.

6.7.1 Polarization by Reflection

Unpolarized light has both polarization cases we discussed earlier, namely, the parallel polarization, where the electric field is in the plane of incidence, and the perpendicular polarization, where the electric field is perpendicular to the plane of incidence. In several applications, it is desirable to separate these two polarizations. The principle of the Brewster angle of incidence, which is also called the polarization angle, can be used to separate the two orthogonal polarizations. To illustrate this application, consider an unpolarized light that is incident at the Brewster angle on a piece of glass with index of refraction $n = \sqrt{\epsilon_r} = 1.5$. The polarization, whose electric field is parallel to the plane of incidence, will be entirely transmitted (because it is incident at the Brewster angle), whereas the other polarization with the electric field perpendicular to the plane of incidence will be partially reflected and partially transmitted.

At the second interface (glass to air), we showed earlier (see example 6.7) that the angle of incidence will also be the Brewster angle for light incident from the glass side to free space. Therefore, again the polarization with **E** parallel to the plane of

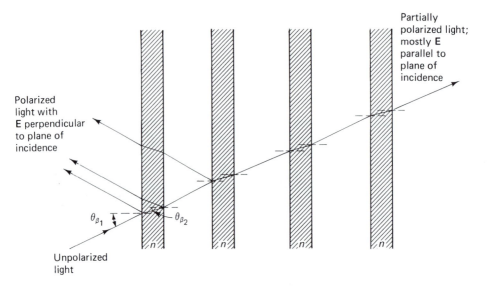

Figure 6.17   Light polarization by multiple reflections.

incidence will be entirely transmitted, and **E** perpendicular will be partially reflected and partially transmitted.

At the end of a long process such as that shown in Figure 6.17, the reflected wave will be entirely polarized, with the electric field perpendicular to the plane of incidence, whereas the transmitted waves will have both polarizations. The larger amplitude, however, will be for the polarization with the electric field parallel to the plane of incidence, because this polarization was entirely transmitted throughout the interfaces. With more glass elements, the transmitted light can be essentially completely polarized with the electric field parallel to the plane of incidence.

6.7.2 Brewster Windows or Brewster Cuts in LASER

Before discussing the reason for Brewster windows in a gas discharge laser or cutting the ends of a rod of a lasing material (ruby) at Brewster angles, let us summarize the steps leading to LASER action in three-level ruby laser material:

1. The laser material, in the shape of a long rod, is subjected to radiation from an extremely intense light source that causes interatomic transition from energy level 1 to level 3 as shown in Figure 6.18b.
2. If the nonradiative transition between level 3 and level 2 is sufficiently fast, the electrons in level 3 will transfer to level 2 rather than returning to level 1.
3. The population of electrons in level 2 now increases as a result of direct transition from level 1 to level 2 (during the radiation from light source) as well as the transfer from level 3 (see Figure 6.18c).

4. When the pumping action is sufficiently large and fast, the electron population at level 2 can be made larger than that at level 1. Radiation of light quanta at frequency f_{21} will occur when electrons make the transition from level 2 to level 1.

5. By placing reflectors (mirrors) at the end of the laser and forcing the radiation to be reflected back and forth to maintain the high-photon density, stimulated emission will increase and, as a result, a large photon density will build up (avalanche of photons).

6. The result is an intense light beam emerging from the end of the laser rod.

A schematic diagram illustrating the sequence of events occurring in laser action is shown in Figure 6.18.

With this brief background on the laser action, let us discuss the role of Brewster angle in polarizing the laser light. It is known that the output of many lasers is linearly polarized, with the ratio of the light polarized in one direction exceeding that polarized in the orthogonal direction by 1000 : 1. In most cases, this high degree of linear polarization is the result of a Brewster surface within the laser. Such a surface is commonly used in the construction of a laser because light must be transmitted out of the lasing medium with as little loss as possible.

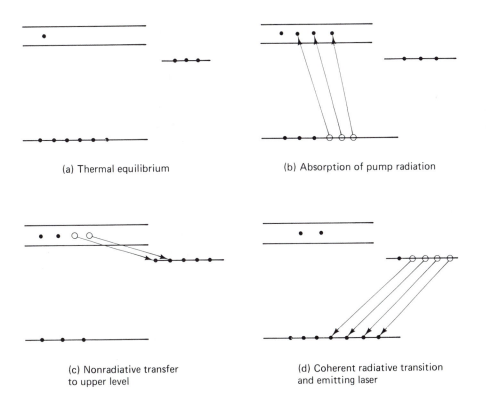

(a) Thermal equilibrium

(b) Absorption of pump radiation

(c) Nonradiative transfer
to upper level

(d) Coherent radiative transition
and emitting laser

Figure 6.18 Sequence of events occurring in laser action.

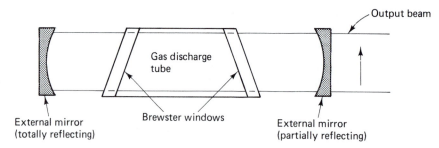

Figure 6.19 Schematic illustrating the use of Brewster windows in a gas discharge LASER.

Cutting the ends of a laser rod or mounting end windows in a gas laser tube at the *Brewster angle* assures that light of one polarization direction is transmitted out of the lasing medium to the reflecting mirrors and back into the lasing medium with no loss. These angle-cut surfaces and windows are called Brewster cuts and Brewster windows. For the light polarized perpendicular to the plane of incidence, there is a large loss at the Brewster surface because of the reflection out of the lasing medium. Usually this loss makes it impossible for light of this polarization to lase. The preferred polarization case lases, however, accounting for the high degree of polarization of the output, as shown in Figure 6.19.

6.7.3 Fiber Optics

Fiber optics is the branch of optics dealing with the transmission of light through small filamentary fibers (dielectric wave guides). The transmission of light through fibers is based on the phenomenon of total internal reflection, which occurs when light is obliquely incident on an interface between two media of different refractive indexes at an angle greater than the critical angle. Consider the dielectric fiber shown in Figure 6.20. Light is incident at an angle θ_i as shown, and it is required to determine the range of values of the index of refraction n so that internal reflections will occur for any value of θ_i.

From Snell's law of refraction, the relationship between θ_i and θ_t as the wave enters the fiber is

$$\frac{\sin \theta_i}{\sin \theta_t} = \sqrt{\frac{\epsilon_2}{\epsilon_1}} = n \tag{6.27}$$

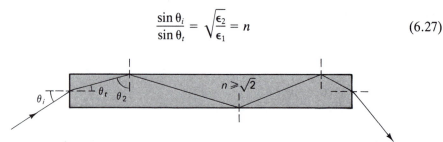

Figure 6.20 Schematic illustrating the principle of light propagation in optical fibers.

since $\epsilon_1 = \epsilon_o$. For θ_2 (see Figure 6.20) to be larger than the critical angle θ_c, we have

$$\sin \theta_2 = \cos \theta_t \geq \sin \theta_c \tag{6.28}$$

where for refraction from fiber to air $\sin \theta_c = 1/n$. Therefore, from equations 6.27 and 6.28, we obtain

$$\sin \theta_2 = \cos \theta_t = \sqrt{1 - \sin^2 \theta_t} = \sqrt{1 - \left(\frac{1}{n}\right)^2 \sin^2 \theta_i} \geq \frac{1}{n} \tag{6.29}$$

Equation 6.29, when solved for n, yields

$$n^2 \geq 1 + \sin^2 \theta_i \tag{6.30}$$

If this condition of equation 6.30 is to be met for grazing incident ($\theta_i = \pi/2$), all incident light will be passed by the fiber which requires that

$$n^2 \geq 2 \qquad \text{or} \qquad n \geq \sqrt{2}$$

Most types of glass have $n \approx 1.5$ so that this condition is easily met.

SUMMARY

This chapter starts with the derivation of expressions of electric and magnetic fields associated with a plane wave propagating in arbitrary direction $\mathbf{n_\beta}$. These fields are given by

$$\hat{\mathbf{E}} = (\hat{E}_x \mathbf{a}_x + \hat{E}_y \mathbf{a}_y + \hat{E}_z \mathbf{a}_z) e^{-j\beta(\mathbf{n_\beta} \cdot \mathbf{r})}$$

$$\hat{\mathbf{H}} = \frac{\mathbf{n_\beta} \times \hat{\mathbf{E}}}{\eta}$$

where $\mathbf{n_\beta} = \cos \theta_x \mathbf{a}_x + \cos \theta_y \mathbf{a}_y + \cos \theta_z \mathbf{a}_z$. $\cos \theta_x$, $\cos \theta_y$, and $\cos \theta_z$ are cosine directions of the unit vector $\mathbf{n_\beta}$, θ_x, θ_y, and θ_z are the angles the propagation constants β makes with respect to the x, y, and z axes.

Oblique reflection from a perfectly conducting plane was then discussed. Two polarization cases were considered: when the $\mathbf{E}$ field is parallel to the plane of incidence, and when the $\mathbf{E}$ field is perpendicular to the plane of incidence. Plane of incidence is defined as that which contains the direction of propagation and a unit vector perpendicular to the reflecting surface. Expressions for the reflected electric and magnetic fields in both parallel and perpendicular polarization cases were analyzed and examples 6.3 and 6.4 were used to illustrate the solution procedure. One of the more important observations is related to the fact that the total electric and magnetic field expressions have both traveling- and standing-wave characteristics. The obliquely incident wave has two components, one normal to the reflecting surface, whereas the other is tangential. The normally incident component, together with the reflected one, interfere and form standing waves. The tangentially incident component of the incident wave, however, continues to propagate—thus, the combined propagation characteristics associated with the resulting total fields. For example, for the parallel polarization case, the total electric field is given by

$$\hat{\mathbf{E}}(x, z) = 2 \hat{E}_m^i [-j \cos \theta_i \sin(\beta z \cos \theta_i) \mathbf{a}_x - \sin \theta_i \cos(\beta z \cos \theta_i) \mathbf{a}_z] e^{-j\beta x \sin \theta_i}$$

To examine the propagation characteristics, we multiply by $e^{j\omega t}$ and obtain the real part to derive the real-time expression of the electric field. On performing these steps, we end up with an x component of the electric field in the form

$$\hat{E}_x = 2 \hat{E}_m^i \cos \theta_i \sin(\beta z \cos \theta_i) \sin(\omega t - \beta x \sin \theta_i)$$

$\sin(\omega t - \beta x \sin \theta_i)$ represents a traveling-wave characteristic (a moving sine wave with time), whereas $\sin(\beta z \cos \theta_i)$ represents a standing-wave characteristic (no variation in the location of the sine wave with time). The fact that propagation is only tangential to the reflecting surface may be emphasized by noting that the expression of the time-average power density has only an x component tangential to the surface. The components of the incident and reflected waves perpendicular to the reflecting surface interfere and form a standing wave.

At an interface between two dielectric media, however, we noted that both reflections and transmission occur. The angle of reflection is equal to that of the incident $\theta_r = \theta_i$, whereas the angle of transmission is determined using Snell's law

$$\frac{\sin \theta_i}{\sin \theta_t} = \sqrt{\frac{\epsilon_2}{\epsilon_1}} \qquad (\mu_1 = \mu_2 = \mu_o)$$

Using this information, we obtained expressions for the reflection and transmission coefficients for both parallel and perpendicular polarization cases. For the parallel polarization case ($\mu_1 = \mu_2 = \mu_o$)

$$\hat{\Gamma}_\parallel = -\frac{\hat{E}_\parallel^r}{\hat{E}_\parallel^i} = \frac{\cos \theta_t - \sqrt{\epsilon_2/\epsilon_1} \cos \theta_i}{\cos \theta_t + \sqrt{\epsilon_2/\epsilon_1} \cos \theta_i}$$

$$\hat{\tau}_\parallel = \frac{\hat{E}_\parallel^t}{\hat{E}_\parallel^i} = \frac{2 \cos \theta_i}{\cos \theta_t + \sqrt{\epsilon_2/\epsilon_1} \cos \theta_i}$$

For the perpendicular polarization case ($\mu_1 = \mu_2 = \mu_o$)

$$\hat{\Gamma}_\perp = \frac{\cos \theta_i - \sqrt{\epsilon_2/\epsilon_1} \cos \theta_t}{\cos \theta_i + \sqrt{\epsilon_2/\epsilon_1} \cos \theta_t}$$

$$\hat{\tau}_\perp = \frac{2 \cos \theta_i}{\cos \theta_i + \sqrt{\epsilon_2/\epsilon_1} \cos \theta_t}$$

Differences between the reflection coefficients for both parallel and perpendicular polarization cases were illustrated in example 6.6.

One of the more important differences is the presence of a Brewster angle for the parallel polarization case. Brewster angle is the angle of incidence at which there is no reflection (i.e., $\hat{\Gamma}_\parallel = 0$). Another special angle of interest is the critical angle of incidence. This angle exists for both parallel and perpendicular polarizations and occurs when the wave is incident from a medium with larger dielectric constant, ϵ_1 to a medium with a smaller dielectric constant, ϵ_2. In Snell's law, the angle of incidence θ_i at which $\theta_t = \pi/2$ is the critical angle.

The chapter was concluded with the description of three interesting applications in optics including the polarization of light by multiple reflection (using the concept of the Brewster angle of incidence), the use of Brewster windows (in gas LASER) or Brewster cuts in LASER rods to produce polarized LASER, and the light guidance (propagation with minimal radiation) in optical fibers. The latter application is based on the concept of the critical angle of incidence at which total reflection occurs.

PROBLEMS

1. The vector magnetic field intensity associated with a plane wave is given by

$$\hat{\mathbf{H}} = [4\,\mathbf{a}_x + 3.1\,\mathbf{a}_y + (5 + j)\,\mathbf{a}_z]\,e^{-j(-2.28x + 3.04y)}$$

Find the following:
   (a) Wavelength and unit vector along the direction of propagation of this wave.
   (b) Vector electric field associated with this wave. Assume free-space propagation—that is, $\epsilon = \epsilon_o$ and $\mu = \mu_o$.

2. The electric field in a plane wave propagating in an arbitrary direction is given by

$$\hat{\mathbf{E}} = [2\,\mathbf{a}_x + 0.5\,\mathbf{a}_y + 5\,\mathbf{a}_z]\,e^{-j(\beta_x x + 1.8y - 2.1z)}$$

Find the following:
   (a) Direction of propagation of this plane wave.
   (b) Wavelength of this wave.
   (c) Vector magnetic field intensity, assuming $\mu = \mu_o$ and $\epsilon = \epsilon_o$.

3. The electric field vector associated with a plane wave propagating in free space is given by

$$\mathbf{E} = 5(\mathbf{a}_x + \sqrt{3}\,\mathbf{a}_y)\cos[6\pi \times 10^7 t - 0.05\pi(3x - \sqrt{3}y + 2z)]\ \text{V/m}$$

Find the following:
   (a) Frequency of propagation, the magnitude of the propagation constant $|\beta|$, and a unit vector in the direction of propagation.
   (b) Magnetic field vector associated with this plane wave.

4. The electric field vector of a uniform plane wave propagating in a perfect dielectric medium having $\epsilon = 9\epsilon_o$ and $\mu = \mu_o$ is given by

$$\mathbf{E} = 10(-\mathbf{a}_x - 2\sqrt{3}\,\mathbf{a}_y + \sqrt{3}\,\mathbf{a}_z)\cos[16\pi \times 10^6 t - 0.04\pi(\sqrt{3}x - 2y - 3z)]$$

Find the following:
   (a) Frequency.
   (b) Direction of propagation.
   (c) Wavelength in direction of propagation.

5. Given

$$\mathbf{E} = 10\cos[6\pi \times 10^7 t - 0.1\pi(y + \sqrt{3}z)]\,\mathbf{a}_x$$

   (a) Determine if the given $\mathbf{E}$ represents the electric field of a uniform plane wave propagating in free space.
   (b) If the answer to part a is yes, find the corresponding magnetic field vector $\mathbf{H}$.

6. Given

$$\mathbf{E} = (\mathbf{a}_x - \mathbf{a}_y - \sqrt{3}\,\mathbf{a}_z)\cos[15\pi \times 10^6 t - 0.05\pi(\sqrt{3}x + z)]$$

$$\mathbf{H} = \frac{1}{120\pi}(\mathbf{a}_x + 4\mathbf{a}_y - \sqrt{3}\,\mathbf{a}_z)\cos[15\pi \times 10^6 t - 0.05\pi(\sqrt{3}x + z)]$$

(a) Perform all the necessary tests to determine if these fields represent a uniform plane wave propagating in a perfect dielectric medium.

(b) Find the dielectric constant and the permeability of the medium.

7. The electric and magnetic fields of the composite wave resulting from the superposition of two uniform plane waves are given by

$$\mathbf{E} = E_{xo}\cos(\beta_x x)\cos(\omega t - \beta_z z)\,\mathbf{a}_x + E_{zo}\sin(\beta_x x)\sin(\omega t - \beta_z z)\,\mathbf{a}_z$$

$$\mathbf{H} = H_{yo}\cos(\beta_x x)\cos(\omega t - \beta_z z)\,\mathbf{a}_y$$

(a) Find the time-average Poynting vector.

(b) Discuss the nature of the composite wave.

8. The electric and magnetic fields of the composite wave resulting from the superposition of two uniform plane waves are given by

$$\hat{\mathbf{E}} = -2jE_o\sin(0.5\beta z)\,e^{-j0.87\beta x}\,\mathbf{a}_y$$

and

$$\hat{\mathbf{H}} = -H_o\cos(0.5\beta z)\,e^{-j0.87\beta x}\,\mathbf{a}_x - j1.7H_o\sin(0.5\beta z)\,e^{-j0.87\beta x}\,\mathbf{a}_z$$

(a) Find the *time-average* Poynting vector.

(b) Explain the propagation characteristics of this wave in the x and z directions.

9. A uniform plane wave having the electric field given by

$$\hat{\mathbf{E}}^i = 25(0.866\,\mathbf{a}_x - 0.5\,\mathbf{a}_z)\,e^{-j(5x + 8.66z)}$$

is incident on a dielectric interface between a medium of dielectric constant $\epsilon_1 = 4\epsilon_o$ and another of dielectric constant $\epsilon_2 = 10\epsilon_o$ as shown in Figure P6.9.

(a) Determine ω and write down the real-time form of the incident electric field.

(b) Obtain an expression for the transmitted electric field. (Your expression should in-

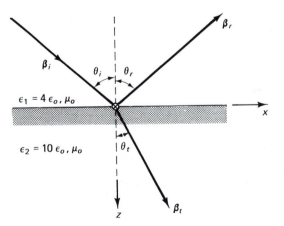

Figure P6.9 Plane wave parallel polarization reflection and refraction at a plane interface between two dielectrics.

clude the magnitude and polarization of the electric field as well as its direction of propagation.)

10. A planar dielectric slab is bounded by the planes $z = 0$ and $z = d$. The semi-infinite regions $z < 0$ and $z < d$ are free space (ϵ_o, μ_o). The parameters of the dielectric region $0 \le z \le d$ are $\mu_o, 2.1\epsilon_o$. A uniform plane wave that has an electric field parallel to the plane of incidence is incident on the interface $z = 0$ of the dielectric slab at the Brewster angle θ_B as shown in Figure P6.10.

   (a) Show that the incident wave is transmitted without any reflection at both the interfaces.
   (b) From first principles, prove that *regardless* of the type of dielectric in the slab region $0 \le z \le d$—that is, $\epsilon = \epsilon_o \epsilon_r$, $\mu = \mu_o$—the incident wave will be transmitted without any reflection at both the interfaces.

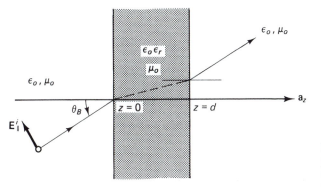

Figure P6.10   A dielectric slab subjected to a partially polarized incident plane wave.

11. A linearly polarized laser beam ($\lambda_1 = 0.6283 \times 10^{-6}$ m) is obliquely incident on the dielectric interface between air and glass $\epsilon_r = 6$, $\mu = \mu_o$. If the incident beam is oriented such that the incident electric field is parallel to the plane of incidence, as shown in Figure P6.11.

   (a) Find the Brewster angle of incidence.
   (b) Write down an expression (substitute numerical values) for the incident electric field when the angle of incidence is equal to the Brewster angle. Assume the amplitude of the incident electric field equals 5 V/m.
   (c) Determine the transmitted field in the glass region for the angle of incidence calculated in part a.
   (d) If the laser beam is incident from the glass to the air region, find the critical angle.

12. A uniform plane wave having an electric field given by

$$\hat{E}^i = [4\,a_x + 2\,a_y - 3\,a_z]e^{-j(6x+8z)}$$

is incident on the interface between free space and a dielectric of permittivity $4\epsilon_o$, as shown in Figure P6.12.

   (a) Determine the frequency and wavelength of this wave in free space.
   (b) Determine the angle of incidence (relative to a_z) and the angle of refraction.
   (c) Obtain an expression for the reflected wave in region 1.
   (d) Obtain an expression for the transmitted wave in region 2.

13. What is meant by the plane of incidence? Distinguish between the two different linear polarizations pertinent to the derivation of the reflection and transmission coefficients for oblique incidence on a dielectric interface.

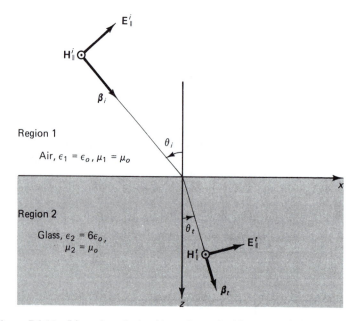

Figure P6.11   Linearly polarized laser beam incident on a dielectric interface between air and glass.

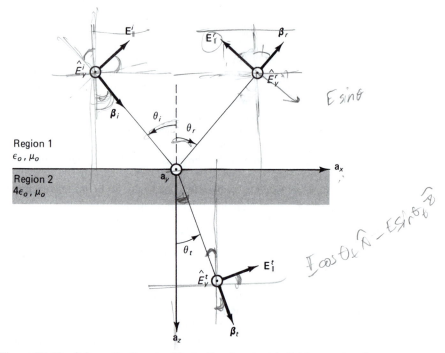

Figure P6.12   Schematic diagram illustrating the electric fields associated with the incident, reflected, and transmitted plane waves.

14. Briefly discuss the determination of the reflection and transmission coefficients for an obliquely incident wave on a dielectric interface.

15. What is the Brewster angle? What is the polarization of the reflected wave for an elliptically polarized wave incident on a dielectric interface at the Brewster angle?

16. Briefly discuss the differences between the reflection coefficients $\hat{\Gamma}_\parallel$ and $\hat{\Gamma}_\perp$ for the parallel and perpendicular polarizations.

17. A uniform plane wave having the electric field given by

$$\mathbf{E} = E_m \left(\frac{\sqrt{3}}{2} \mathbf{a}_x - \frac{1}{2} \mathbf{a}_z \right) \cos(6\pi \times 10^9 t - 10\pi(x + \sqrt{3}z))$$

is incident on the interface between free space and a dielectric of permittivity $1.5\epsilon_o$, as shown in Figure P6.17.
(a) Obtain the expression for the reflected field.
(b) Obtain the expression for the transmitted electric field.

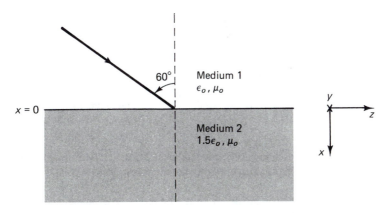

Figure P6.17   The geometry related to problem 17.

18. In many cases, the permeability of dielectric media equals that of free space. In this limit and with the aid of Snell's law, show that the reflection and transmission coefficients for waves obliquely incident on dielectric media are the following:
(a) For **E** perpendicular to the plane of incidence,

$$\hat{\Gamma}_\perp = \frac{\sin(\theta_t - \theta_i)}{\sin(\theta_i + \theta_t)}, \qquad \hat{\tau}_\perp = \frac{2 \cos\theta_i \sin\theta_t}{\sin(\theta_i + \theta_t)}$$

(b) For **E** parallel to the plane of incidence,

$$\hat{\Gamma}_\parallel = \frac{\tan(\theta_i - \theta_t)}{\tan(\theta_i + \theta_t)}, \qquad \hat{\tau}_\parallel = \frac{2 \cos\theta_i \sin\theta_t}{\sin(\theta_i + \theta_t) \cos(\theta_i - \theta_t)}$$

19. From Snell's law of refraction, we have

$$\frac{\sin\theta_i}{\sin\theta_t} = \frac{v_1}{v_2} = \sqrt{\frac{\mu_2 \epsilon_2}{\mu_1 \epsilon_1}} = \left. \sqrt{\frac{\epsilon_2}{\epsilon_1}} \right|_{\mu_1 = \mu_2 = \mu_o}$$

The ratio $v_1/v_2 = [(c/v_2)/(c/v_1)]$ and is known as the ratio of indexes of refraction. If medium 1 is free space, v_1/v_2 is $\sqrt{\epsilon_r \epsilon_o/\epsilon_o} = \sqrt{\epsilon_r} = n_2$, where the relative permittivity of medium 2 is $\epsilon_r = \epsilon_2/\epsilon_o$ and n_2 is the index of refraction in region 2. As we know, white light is composed of the entire visible spectrum. The index of refraction n for most materials is a weak function of wavelength λ, often described by Cauchy's equation:

$$n = A + \frac{B}{\lambda^2}$$

If a beam of white light is incident at 30° to a piece of glass with $A = 1.5$ and $B = 5 \times 10^{-15} \text{m}^2$ as shown in Figure P6.19, calculate the transmitted angles for the colors violet (400 nm), blue (450 nm), green (550 nm), yellow (600 nm), orange (650 nm), and red (700 nm). This separation of colors is called dispersion.

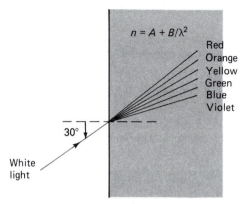

Figure P6.19 Incident white light on a material with frequency dependent index of refraction.

20. A straight-light pipe with refractive index n_1 has a dielectric coating with index n_2 added for protection. The light pipe is usually in free space, so that n_3 is typically unity as shown in Figure P6.20.
   (a) If light within the pipe is incident on the first interface at angle θ_1, what are the angles θ_2 and θ_3?
   (b) What value of θ_1 will make θ_2 just equal to the critical angle for total reflection at the second interface?
   (c) How does this value differ from the critical angle if the coating was not present so that n_1 was directly in contact with n_3?
   (d) If we require that total reflection occurs at the first interface, what is the allowed range of incident angle θ_1?

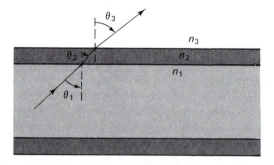

Figure P6.20 Light propagation in a solid dielectric rod of the index of refraction n_1 coated with another dielectric of index of refraction n_2.

Iskander P6.20

CHAPTER 7

TRANSMISSION LINES

INTRODUCTION

Low-frequency radiation cannot be directed efficiently using antennas. Therefore, to transport electromagnetic energy from one point to another, we usually use transmission lines. A transmission line is an electromagnetic guiding system consisting of two or more conductors embedded in a suitable dielectric system. The conductor shape generally takes on the form of a system of parallel or coaxial cylinders or parallel plates, as shown in Figure 7.1.

When the dielectric system is of a single material, as it is in most cases, the line is then said to be homogeneous. A uniform transmission line is a system that has identical cross-sectional configurations for all positions along its length. In this chapter we will examine the propagation characteristics of electromagnetic waves on transmission lines. Both transient and sinusoidal steady state analysis will be considered.

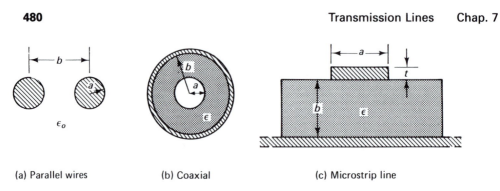

(a) Parallel wires   (b) Coaxial   (c) Microstrip line

Figure 7.1   Cross sections of transmission lines.

7.1 CHARACTERISTICS OF WAVE PROPAGATION IN TRANSMISSION LINES

To describe the fields configuration in a transmission line (called mode of propagation), let us consider the parallel wire transmission line shown in Figure 7.2. One conductor is at zero potential (ground), and the other conductor supports a potential V. The first conductor (1) carries a current I that is returned to the source by means of the ground conductor (2). When the voltage V is positive, the direction of the electric field is from the upper conductor to the lower conductor. From Ampere's law, it is easy to show that the magnetic fields surround the two wires. From Figure 7.2 it is clear that the magnetic and electric fields are transverse to the direction of propagation that is the direction of current flow. Therefore, the mode of propagation in low-frequency transmission lines is a transverse electromagnetic (TEM) mode. If the direction of propagation is taken along the positive z direction in the TEM mode of propagation, $E_z = 0 = B_z$.

To examine the propagation characteristics further, let us study Maxwell's equations. From Faraday's law, we have

$$\nabla \times \mathbf{E} = -\frac{\partial \mathbf{B}}{\partial t} = \begin{vmatrix} \mathbf{a}_x & \mathbf{a}_y & \mathbf{a}_z \\ \partial/\partial x & \partial/\partial y & \partial/\partial z \\ E_x & E_y & 0 \end{vmatrix} \tag{7.1}$$

From the x component of equation 7.1, we obtain

$$\frac{\partial E_y}{\partial z} = \mu \frac{\partial H_x}{\partial t} \tag{7.2}$$

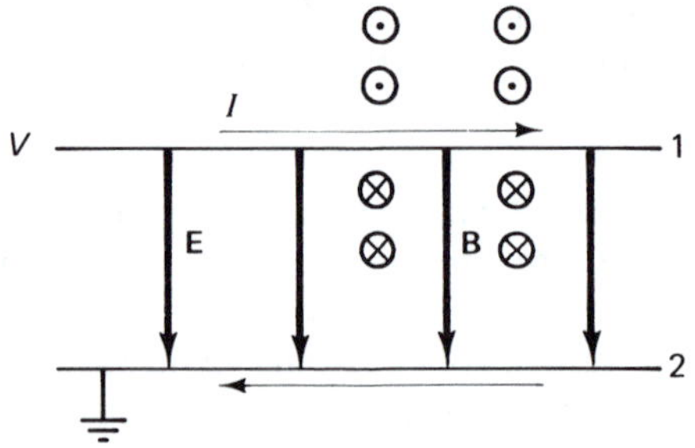

Figure 7.2   Electric and magnetic fields associated with a TEM mode in a two-conductor transmission line.

Also from Ampere's law, we have

$$\nabla \times \mathbf{H} = \epsilon \frac{\partial \mathbf{E}}{\partial t} = \begin{vmatrix} \mathbf{a}_x & \mathbf{a}_y & \mathbf{a}_z \\ \partial/\partial x & \partial/\partial y & \partial/\partial z \\ H_x & H_y & 0 \end{vmatrix} \tag{7.3}$$

The y component of equation 7.3 is

$$\frac{\partial H_x}{\partial z} = \epsilon \frac{\partial E_y}{\partial t} \tag{7.4}$$

Combining equations 7.2 and 7.4, we obtain

$$\frac{\partial^2 E_y}{\partial z^2} = \mu\epsilon \frac{\partial^2 E_y}{\partial t^2} \tag{7.5}$$

We saw earlier, in chapters 2 and 3 of the text, that equation 7.5 has a solution of the form

$$E_y(z,t) = E_y^+(z - ut) + E_y^-(z + ut) \tag{7.6}$$

where $u = 1/\sqrt{\mu\epsilon}$. The first term on the right-hand side of equation 7.6 represents a wave traveling along the positive z axis, whereas the second term represents a wave traveling along the negative z axis. In addition, we may also notice that the velocity of propagation u is independent of the geometry and the frequency. This simply means that the TEM mode propagates along the transmission line with a velocity that is independent of the frequency. In other words, the TEM mode of propagation has no cutoff frequency.

In summary, the mode (field configuration) of propagation along the transmission lines is the TEM mode with the following properties:

1. The electric and magnetic fields are perpendicular to the direction of propagation—that is, $E_z = 0 = H_z$.
2. It has a zero cutoff frequency.

In the following section, we will use the properties of the TEM mode of propagation to show that there is a unique relationship between the electric field and the voltage between the two transmission-line conductors, and also a unique relationship between the current and the magnetic field circulating the transmission-line conductors. To start with, let us consider Faraday's law in integral form,

$$\oint_c \mathbf{E} \cdot d\ell = -\frac{d}{dt} \int_s \mathbf{B} \cdot ds$$

If we take the contour of integration c in the transverse plane perpendicular to the direction of propagation, the total magnetic flux crossing the area s enclosed by c will be zero because $B_z = 0$. Hence,

$$\oint_c \mathbf{E} \cdot d\ell = 0 \tag{7.7}$$

Equation 7.7 simply indicates that the electric field associated with the TEM mode is conservative, which means that the line integral of **E** along any contour between the transmission-line conductors is constant and independent of the specific shape of the contour. In other words,

$$v(z,t) = \int_1^2 \mathbf{E} \cdot d\ell$$

which means that the voltage between the two conductors is uniquely related to the electric field, independent of the specific contour of integration between the conductors.

Similarly, from Ampere's law we have

$$\oint_c \mathbf{H} \cdot d\ell = i(z,t) + \frac{d}{dt} \int_s \epsilon \mathbf{E} \cdot d\mathbf{s}$$

where c is any contour in the transverse plane encircling conductor 1. From the properties of the TEM mode, however, we have $E_z = 0$. The total electric flux crossing the area s encircled by c—that is, $\int_s \epsilon \mathbf{E} \cdot d\mathbf{s}$—will, hence, be zero. Ampere's law, hence, reduces to

$$\oint_c \mathbf{H} \cdot d\ell = i(z,t)$$

which means that the magnetic field is uniquely related to the current in the transmission-line conductors. In our analysis of the propagation characteristics in a transmission line, we will therefore use the scalar voltage and current quantities $v(z,t)$ and $i(z,t)$ rather than the vector electric and magnetic field quantities. The voltage $v(z,t)$ and the current $i(z,t)$ along the transmission line, although dependent on the location z and time t at which they are calculated, they are uniquely related to the electric and magnetic fields, respectively, at the specific location z and time t.

7.2 DISTRIBUTED CIRCUIT REPRESENTATION OF TRANSMISSION LINES

To help us analyze the voltage and current characteristics along transmission lines, we will develop an equivalent distributed circuit representation of a section of a transmission line. A transmission line may be broken into an infinite number of elemental sections, and each such section may be modeled by a network of lumped parameters of resistance, inductance, conductance, and capacitance. Before proceeding to the derivation of the transmission-line equations, we should distinguish between the transmission line and the conventional network in an ordinary electric circuit. The geometrical dimensions of the circuit elements in an ordinary circuit network are so small in comparison with the wavelength that the elements may be reasonably considered to be lumped at a point. A transmission line is usually used in the radio frequency range (i.e., its length is comparable with the wavelength) so the validity of the equivalent circuit analysis requires that the circuit elements of the transmission line be distributed (rather

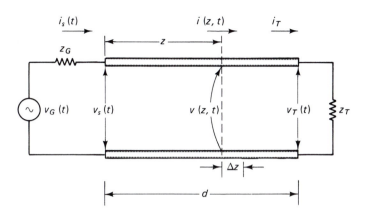

Figure 7.3 Voltage and current at an arbitrary point on a transmission line.

than lumped) throughout its length. In other words, whereas the ordinary electric circuit consists of lumped elements, the transmission line has to be treated as a distributed parameter network.

In our equivalent circuit representation, therefore, we shall select a small section of a uniform transmission line. This section should be sufficiently small compared with a wavelength so that equivalent circuit representation in terms of lumped elements would be valid. The distributed parameter network of the overall transmission line may be obtained by connecting in tandem a large number of the developed equivalent circuit for each section. With this in mind, let us proceed with the equivalent circuit analysis.

Consider the situation in which a uniform transmission line has a generator connected to its sending end and a load at its receiving end, as shown in Figure 7.3.

An approximate circuit representation of an incremental section Δz of the transmission line of Figure 7.3 is shown in Figure 7.4. The distributed parameters, on a per unit length basis, are a series resistance r, series inductance ℓ, shunt capacitance c, and shunt leakage conductance g. The resistance r is included in the equivalent circuit to account for the ohmic losses along the transmission line, whereas the conductance g

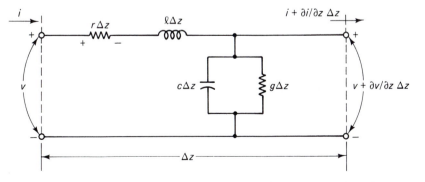

Figure 7.4 Approximate circuit representation of an incremental section Δz of a transmission line.

is introduced to account for the dielectric losses in the dielectric material between the conductors. The inductance and capacitance are included to describe the geometrical property of the transmission line and its relation to the ability of the line to store the electric energy for a given value of the applied voltage (capacitance) and to store magnetic energy for a given value of the input current (inductance). We will now use this equivalent circuit to describe the transmission characteristics of the voltage and the current along the transmission line. As a first-order approximation to the voltage drop across the Δz section of the line, we will assume a constant current through the section—that is, the series parameters need only to be considered. From Kirchhoff's voltage law we have

$$v - ir\Delta z - \ell\Delta z \frac{\partial i}{\partial t} - v - \frac{\partial v}{\partial z}\Delta z = 0$$

Therefore,

$$-\frac{\partial v}{\partial z}\Delta z = r\Delta zi + \ell\Delta z\frac{\partial i}{\partial t}$$

or

$$\boxed{-\frac{\partial v}{\partial z} = ri + \ell\frac{\partial i}{\partial t}} \tag{7.8}$$

Similarly, assuming a constant voltage across the section and using Kirchhoff's current law, we obtain

$$i - \left(i + \frac{\partial i}{\partial z}\Delta z\right) = g\Delta zv + c\Delta z\frac{\partial v}{\partial t}$$

Therefore,

$$-\frac{\partial i}{\partial z}\Delta z = g\Delta zv + c\Delta z\frac{\partial v}{\partial t}$$

or

$$\boxed{-\frac{\partial i}{\partial z} = gv + c\frac{\partial v}{\partial t}} \tag{7.9}$$

Equations 7.8 and 7.9 are known as the *telegrapher's equations*. Let us now consider an important special case of these transmission-line equations.

7.3 LOSSLESS LINE

A line where it is assumed that $r = 0$ and $g = 0$ is called a lossless line. This case is of significant practical importance because most transmission lines may be considered lossless. In this case the transmission line equations reduce to

$$-\frac{\partial v}{\partial z} = \ell \frac{\partial i}{\partial t} \tag{7.10}$$

and

$$-\frac{\partial i}{\partial z} = c \frac{\partial v}{\partial t} \tag{7.11}$$

A differentiation of equations 7.10 and 7.11 enables us to obtain a differential equation in terms of v or i alone. Thus the voltage equation is obtained by differentiating equation 7.10 with respect to z and equation 7.11 with respect to t, and eliminating $\partial^2 i/\partial z \partial t$ between the resulting two equations. The obtained voltage equation is then

$$\frac{\partial^2 v}{\partial z^2} = \ell c \frac{\partial^2 v}{\partial t^2} \tag{7.12}$$

Similarly, for the current, we have

$$\frac{\partial^2 i}{\partial z^2} = \ell c \frac{\partial^2 i}{\partial t^2} \tag{7.13}$$

For $u = 1/\sqrt{\ell c}$, equations 7.12 and 7.13 are

$$\frac{\partial^2 v}{\partial z^2} = \frac{1}{u^2} \frac{\partial^2 v}{\partial t^2} \tag{7.14}$$

and

$$\frac{\partial^2 i}{\partial z^2} = \frac{1}{u^2} \frac{\partial^2 i}{\partial t^2} \tag{7.15}$$

As indicated by equation 7.6, the solutions to equations 7.14 and 7.15 are propagating waves in positive and negative z directions at the speed u. Hence,

$$v(z,t) = v^+(z - ut) + v^-(z + ut)$$

and

$$i(z,t) = i^+(z - ut) + i^-(z + ut) \tag{7.16}$$

Each one of the solutions v^+, v^-, i^+, and i^- is a propagating wave—that is, a wave of voltage or current that shifts its position with time along the transmission line without change in form or magnitude (because we are considering lossless lines). The wave $v^+(z - ut)$, for example, shifts in the positive z direction, moving from the generator toward the load, and is called the *incident wave*. To illustrate this, let us consider the voltage wave form $v^+(z - ut)$ shown in Figure 7.5, as a function of distance z. Because the value of the function v^+ depends uniquely on its argument, it follows that the voltage at $z = z_1$ and $t = t_1$ (i.e., at point P) is equal to the voltage at $z = z_2$ and $t = t_2 > t_1$ (i.e., at the same point P but at an advanced time t_2) only if

$$z_1 - ut_1 = z_2 - ut_2$$

or

$$z_2 = z_1 + u(t_2 - t_1) \tag{7.17}$$

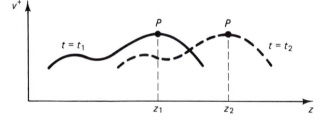

Figure 7.5 Incident voltage wave form along a section of transmission line at two specific instants of time.

From equation 7.17 it follows that the voltage at z_1 and t_1 has traveled to a point z_2 (which is larger than z_1) in time $(t_2 - t_1)$. Figure 7.5 shows a reference point on the wave, such as P, at two different instants of time. The corresponding locations of point P, z_1 and z_2, are determined such that the incident wave voltage has a constant argument; hence, the value of the voltage is the same at both locations. Similarly, it is easy to show that the wave $v^-(z + ut)$ propagates in the negative z direction with a velocity u. This wave is called the *reflected wave*.

In summary, the total voltage at any point on the transmission line is, in general, the sum of two propagating waves of voltage. One is an incident wave moving toward the receiving end, and the other is a reflected wave moving toward the sending end.

Characteristic Impedance of Lossless Lines.   The current i^+ associated with the incident voltage v^+ can be determined by substituting i^+ and v^+ in equation 7.11 and changing the variables of differentiation, z and t, to $(z - ut)$. Thus,

$$-\frac{\partial i^+}{\partial(z - ut)}\frac{\partial(z - ut)}{\partial z} = c\frac{\partial v^+}{\partial(z - ut)}\frac{\partial(z - ut)}{\partial t}$$

Therefore,

$$\frac{\partial i^+}{\partial(z - ut)} = cu\frac{\partial v^+}{\partial(z - ut)}$$

Integrating both sides with respect to $(z - ut)$, we obtain

$$i^+ = cuv^+ \tag{7.18}$$

which simply states that the propagating wave of current is related to the propagating wave of voltage through the constant cu.

$$cu = c\frac{1}{\sqrt{\ell c}} = \sqrt{\frac{c}{\ell}}$$

From equation 7.18, we have

$$\frac{v^+}{i^+} = \frac{1}{cu} = \sqrt{\frac{\ell}{c}} = Z_o$$

where Z_o is known as the characteristic impedance of the transmission line. *This characteristic impedance is analogous to the wave impedance η of a plane wave propagating in unbounded medium.*

Following a similar procedure, it can be shown that the reflected waves of voltage and currents are related through the characteristic impedance by $v^-/i^- = -Z_o$. The minus sign for the reflected wave simply indicates that the current in Figure 7.4 should be in the negative z direction. The total voltage and current on the lossless line are then

$$v = v^+ + v^- = Z_o(i^+ - i^-)$$

$$i = i^+ + i^- = Y_o(v^+ - v^-)$$

where Y_o is the characteristic admittance $Y_o = 1/Z_o$.

7.4 VOLTAGE REFLECTION COEFFICIENT

Because the voltage and the current along the transmission line are related by a specific ratio known as the characteristic impedance Z_o, any discontinuity along or at the end of the line that causes alteration of this characteristic ratio results in the presence of reflected voltages and currents along the transmission line. The reflection coefficient is defined as the ratio of the reflected voltage to the incident voltage:

$$\Gamma = \frac{v^-}{v^+}$$

Because $v^- = -Z_o i^-$ and $v^+ = Z_o i^+$, the reflection coefficient, in terms of the current waves, is

$$\Gamma = -\frac{i^-}{i^+}$$

In other words, Γ is defined as the voltage reflection coefficient and the current reflection coefficient (i.e., i^-/i^+) is simply $-\Gamma$. Let us now study reflections from discontinuities at the end of transmission lines. Specifically, we will treat the following two cases:

1. Resistive terminations
2. Arbitrary terminations

7.4.1 Reflections from Resistive Terminations

In Figure 7.6, we see a positively traveling voltage wave incident on a load resistor R_T at the end of the transmission line $z = d$. Because the termination resistance R_T is, in general, different from the characteristics impedance Z_o, there will be an impedance mismatch at the load; hence, reflected voltage and current waves will be introduced at the load location. In other words, the incident voltage and current that are related by the characteristic impedance of the transmission line cannot, in general, satisfy Ohm's law at the termination, the matter that results in the introduction of the reflected waves so that the total voltage and current may satisfy the required condition at the termination. These reflected waves travel back toward the source as v^- and i^- waves. At the terminal end of the transmission line, the boundary condition requires that

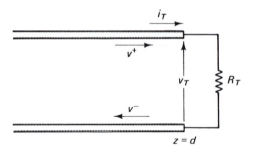

Figure 7.6 Lossless transmission line with resistive termination.

$$\frac{v_T}{i_T} = R_T$$

In terms of the propagating waves, we have

$$\frac{v_T^+ + v_T^-}{i_T^+ + i_T^-} = R_T$$

Because $i_T^+ = v_T^+/Z_o$ and $i_T^- = -v_T^-/Z_o$, therefore,

$$(R_T/Z_o) = \frac{1 + v_T^-/v_T^+}{1 - v_T^-/v_T^+} = \frac{1 + \Gamma_T}{1 - \Gamma_T}$$

Solving for Γ_T we obtain

$$\boxed{\Gamma_T = \frac{R_T - Z_o}{R_T + Z_o}}$$

Let us examine the reflection coefficient values for a number of specially important terminations.

Matched line.   When $R_T = Z_o$, the load voltage reflection coefficient $\Gamma_T = 0$ and $v_T^- = 0$. Therefore, there will be no reflected voltages and currents on the transmission line—that is, $v = v^+$ and $i = i^+$. At the line input, the generator therefore sees a resistance equal to the transmission line characteristic impedance as shown in Figure 7.7. The incident voltage and current at the sending end v_s^+ and i_s^+ are obtained, in terms of the generator voltage v_G and internal impedance R_G, from the relations

$$v_s^+ = \frac{v_G}{R_G + Z_o}Z_o \qquad \text{and} \qquad i_s^+ = \frac{v_G}{R_G + Z_o}$$

The voltage and current at the sending end, v_s^+ and i_s^+, will continue to travel through the transmission line without any change in the magnitude or shape.

Open-circuited line.   When $R_T = \infty$, the reflection coefficient is unity—that is, $\Gamma_T = 1$ and $v_T^- = v_T^+$. The incident voltage is totally reflected by the load and the terminal voltage is double the incident value, $v_T = 2v_T^+$. In this case the reflected current is $i_T^- = -i_T^+$ and hence the total current $i_T = 0$ as expected at an open-circuit termination.

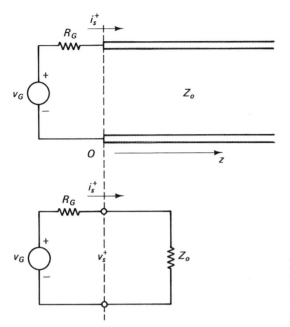

Figure 7.7 A voltage source is applied to the transmission line through a series resistor R_G. The voltage across the line v_s^+ is given by the voltage divider relation.

Short-circuited line.   When $R_T = 0$, the load reflection coefficient $\Gamma_T = -1$, so that $v_T^- = -v_T^+$ and $i_T^- = i_T^+$. The incident current is totally reflected by the load, and the terminal current is double the incident value—that is, $i_T = 2i_T^+$. The total voltage at the termination, conversely, is equal to zero, a value that is expected for a short-circuit termination.

7.4.2 Reflections from Arbitrary Terminations

For resistive terminations, we related the amplitudes of the reflected voltage to that of the incident through the use of the voltage reflection coefficient. Such a substitution was possible because the voltage across and current through a resistive termination are algebraically related by Ohm's law. For an arbitrary termination, however, the voltage and current at the termination are not related algebraically but instead by an integral or differential equation; hence, it is necessary to solve an integral or a differential equation that describes the circuit problem at the end of the line.

For an arbitrary termination Z_T at the end of a transmission line of length d (see Figure 7.8a) the voltage and current at $z = d$ are given as a function of time by

$$v_T(t) = v^+(d - ut) + v^-(d + ut) \tag{7.19}$$

$$i_T(t) = i^+(d - ut) + i^-(d + ut)$$
$$= Y_o[v^+(d - ut) - v^-(d + ut)] \tag{7.20}$$

Because the incident voltage wave v^+ is known, we shall solve for $v_T(t)$ or $i_T(t)$ by eliminating v^- between equations 7.19 and 7.20. Hence,

$$2v^+(d - ut) = v_T(t) + Z_o i_T(t) \tag{7.21}$$

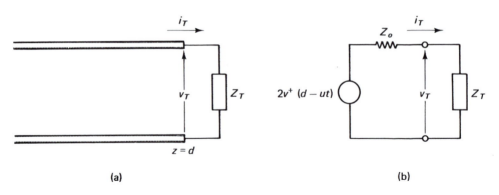

(a)   (b)

Figure 7.8 (a) A transmission line with an arbitrary termination at $z = d$; (b) equivalent circuit.

Equation 7.21 suggests the equivalent circuit shown in Figure 7.8(b). Because $v^+(d - ut)$ is known, once we calculate v_T or i_T from equation 7.21, which represents the equivalent circuit, the reflected voltage $v^-(d + ut)$ can be calculated from the relation

$$v^-(d + ut) = v_T(t) - v^+(d - ut)$$

The following example illustrates the solution procedure for an arbitrarily terminated transmission line.

EXAMPLE 7.1

A transmission line of characteristic impedance Z_o and length d is terminated by a capacitor load of capacitance C. Determine the total voltage and current at the termination. Explain physically the characteristics of the termination's voltage and current as a function of time.

Solution

The capacitive termination of the transmission line is shown in Figure 7.9. If the magnitude of the incident voltage at the termination is v_o, hence equation 7.21 reduces to

$$2v_o = v_T(t) + Z_o i_T(t) \tag{7.22}$$

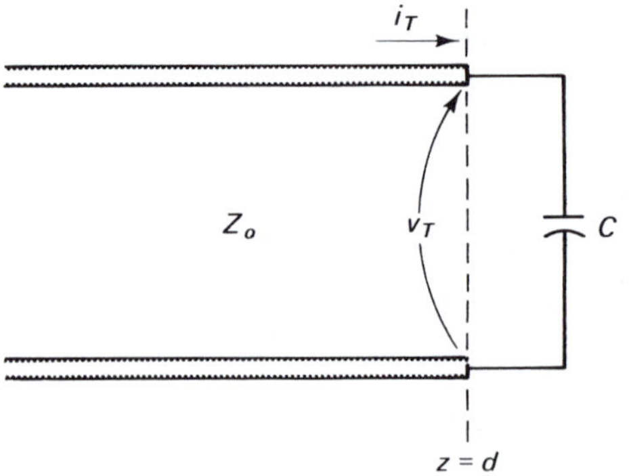

Figure 7.9 Capacitive termination of a transmission line of characteristic impedance Z_o.

v_T and i_T are the terminal voltage and current across the capacitor C. They are, therefore, related by

$$i_T(t) = C\frac{dv_T(t)}{dt}$$

Substituting this relation in equation 7.22, we obtain

$$2v_o = v_T(t) + Z_o C\frac{dv_T(t)}{dt} \tag{7.23}$$

In the introductory mathematics courses, we often faced first-order differential equations such as equation 7.23. We noted from these previous experiences that the solution for equation 7.23 consists of two parts:

1. **Particular integral part (forced response)** that is related to the characteristics of the input function, which is $2v_o$ in our case.
2. **Complementary function (transient or free response)** that is related to the characteristics of the equation independent of the forced function.

If we assume a *constant (dc) input voltage* on the transmission line, the particular integral part of the solution is obtained by letting $dv_T/dt = 0$, which is equivalent to waiting a long enough time until the transient part of the solution dies out. Hence, for dc input voltage on the transmission line, we have

$$2v_o = v_T \qquad \left(\text{as } t \to \infty \text{ and } \frac{dv_T}{dt} \to 0\right) \tag{7.24}$$

The transient part of the solution, conversely, is obtained by letting the forced function $2v_o = 0$. Hence,

$$\frac{dv_T}{dt} + \frac{1}{Z_o C}v_T = 0$$

or

$$v_T = Ae^{-t/Z_o C} \tag{7.25}$$

where A is a constant to be determined from the initial conditions.

The total solution for v_T is obtained by summing equations 7.24 and 7.25; hence,

$$v_T(t) = 2v_o + Ae^{-t/Z_o C} \tag{7.26}$$

If the capacitor is initially (i.e., at $t = 0$) uncharged, $v_T(0) = 0$. Substituting this condition in 7.26, we obtain

$$A = -2v_o$$

The final solution for $v_T(t)$ is given by

$$v_T(t) = 2v_o(1 - e^{-t/Z_o C}) \tag{7.27}$$

The termination current $i_T(t)$ is obtained from the relation

$$i_T(t) = C\frac{dv_T(t)}{dt} = \frac{2v_o}{Z_o}e^{-t/Z_o C} \tag{7.28}$$

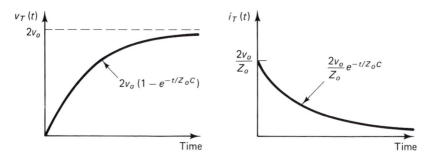

Figure 7.10 Plots of the voltage and the current across a capacitor terminating a transmission line (as given by equations 7.27 and 7.28) as a function of time.

Plots of the termination voltage $v_T(t)$ and currents $i_T(t)$ as a function of time is shown in Figure 7.10. The variation of $v_T(t)$ and $i_T(t)$ as a function of time may be explained physically in terms of the characteristics of the capacitor termination. At $t = 0$, we assumed the capacitor to be discharged. Because the capacitance has the characteristic of opposing any sudden change in voltage, the termination voltage $v_T (t = 0^+)$ just after the arrival of the incident voltage at the terminal capacitor will still be zero. In other words, the capacitor will act initially like a short circuit. For a short-circuit termination (see section 7.4), v_T $(t = 0) = v^+ + v^- = 0$, and $i_T (t = 0) = i^+ + i^- = 2i^+ = 2v_o/Z_o$. These initial values are in agreement with the values at $t = 0$ of Figure 7.10.

After a long time $(t \to \infty)$ from the initial arrival of the incident voltage and current to the terminal capacitor, the capacitor will be fully charged, and we will have an open-circuit termination. Under these circumstances (see section 7.4), $v_T(t \to \infty) = v^+ + v^- = 2v_o$ and $i_T (t \to \infty) = i^+ + i^- = 0$. These results also agree with the values on the plots of Figure 7.10 at $t \to \infty$. Like many physical phenomena, the voltage and current change from their initial values at $t = 0$ to the final values at $t \to \infty$ in an exponential fashion as shown in Figure 7.10.

7.5 TRANSIENTS ON TRANSMISSION LINE

Current and voltage waves travel on a transmission line at a speed u. Therefore, once generated, say at $z = 0$, they cannot reach any position $z = d$ along the transmission line until a time d/u later. Waves traveling in the positive z direction are described by the function $v^+(z - ut)$ and waves traveling in the negative z direction by $v^-(z + ut)$. At any time t and position z along the transmission line, the total voltage is equal to the sum of both the incident and reflected waves, whereas the current is proportional to their differences (i.e., the reflected current is in opposite direction of the incident current). Study of transients on transmission lines basically provides this type of information, which includes the variation in the voltage and the current along the transmission line as the multiple reflections develop with time at the various discontinuities. These discontinuities may be between the transmission line and the termina-

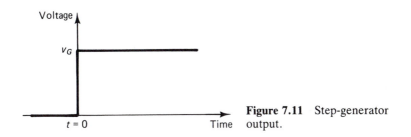

Figure 7.11   Step-generator output.

tion, the transmission line and the generator, or at a junction between different transmission lines connected between the generator and the load. Unlike our analysis in the previous sections where we were basically concerned with the voltage and the current at the termination, transients includes all reflections at the generator, termination, and at discontinuities in between, if any. To illustrate transients on transmission lines, we will consider first the step function excitation.

7.5.1 Step Excitation of Resistively Terminated Lossless Line

The step-generator output is shown in Figure 7.11. If such a generator is connected to a transmission line, the propagating wave excited on the line will have the same shape as the excitation function except for a time delay. For example, let us consider a dc battery of voltage v_G and series resistance R_G connected to a transmission line of characteristic impedance Z_o at $t = 0$ as shown in Figure 7.12a. At $z = 0$ and $t = 0$, the source has no knowledge of the line's length, load condition, or any discontinuities along the transmission line. Thus, the lossless transmission line looks to the source like a resistor of value Z_o, and there will be no reflected voltage v^- at $t = 0$. The incident voltage at the sending end is then determined by a voltage divider ratio as shown in Figure 7.12b, hence at $t = 0$ we have

$$v_s = v^+ = \frac{v_G}{R_G + Z_o} Z_o, \qquad i_s = i^+ = \frac{v_G}{R_G + Z_o}$$

The incident voltage v^+ travels down the line at speed u. If the load end is matched, there will be no further reflections at the load; hence, the steady state will

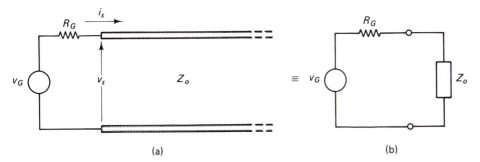

Figure 7.12   (a) Line input conditions at $t = 0$; (b) equivalent circuit at $t = 0$.

be reached after one transit time $T = d/u$, where T is the time required for a wave to propagate between the two ends of the transmission line. If the load is not matched, there will be a new reflected voltage wave v^- generated at $z = d$ and $t = T = d/u$. As the incident voltage v^+ continues to propagate in the positive z direction, the reflected voltage v^- propagates back toward the source. If the source end is matched, after one round trip $t = 2T = 2d/u$, there will be no further reflections, and the steady state will be reached after time $t = 2T$.

EXAMPLE 7.2

Consider the dc voltage of Figure 7.13a, which is switched onto a transmission line of characteristic impedance Z_o and terminated by a load resistance R_L. Discuss the process of multiple reflections for $R_L = 0$ (short circuit) and for $R_L = \infty$ (open-circuit termination), and determine the voltage and the current along the transmission line in both cases.

Solution

In Figure 7.13b it is clear that because the source resistance $R_G = Z_o$ and is, hence, matched to the transmission line, only half of the source voltage propagates down the line toward the load. If the transmission line is short circuited—that is, $R_L = 0$—the reflection coefficient at the load $\Gamma_T = -1$ so that $v^-/v^+ = -1$, and $i^-/i^+ = -\Gamma_T = 1$. In other words, the reflected voltage wave will cancel the incident wave and the reflected current wave will add, in phase, an equal value to the incident current wave as shown in Figure 7.13c. Because the source end is matched, no further reflections will occur at $z = 0$ with the arrival of the reflected voltage v^- and the reflected current i^- to the generator end, and the steady state is reached for $t \geq 2T$. Similarly, if the transmission line is open circuited, $R_L = \infty$, $\Gamma_T = 1$; hence, $v^-/v^+ = 1$ and $i^-/i^+ = -1$. The reflected voltage wave will therefore add to the incident, whereas the reflected and incident current waves will cancel each other, as shown in Figure 7.13d.

Let us consider next the case in which the source and the load ends are not matched. In this case, reflections will occur at both ends, and the voltage and current waves will continue bouncing back and forth forever. It should be noted, however, that because reflection coefficients are less than unity (i.e., $\Gamma_G < 1$ and $\Gamma_T < 1$)*, each successive reflection will be reduced in magnitude, and after a few round trips, the changes in v^+ and v^- become negligible, and the steady state is approximately reached.

7.6 REFLECTION DIAGRAM

For the case of resistive termination of transmission lines, the development of the multiple reflections and the consequent build-up of the voltage and current from their initial to final (steady-state) values can be easily understood by using the reflection diagrams. Because these reflection diagrams are applicable for both voltage and cur-

---

*Γ_G and Γ_T are the reflection coefficients at the generator and the termination ends, respectively.

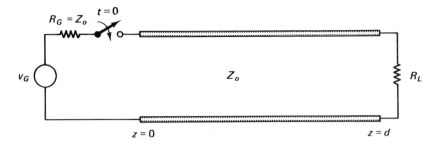

(a)

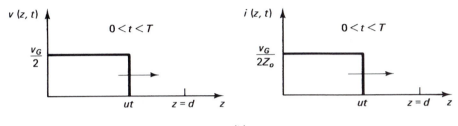

(b)

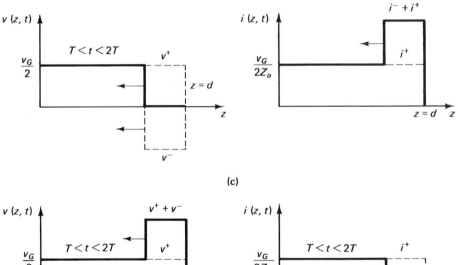

(c)

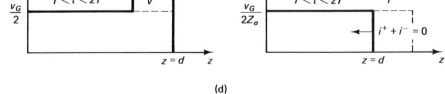

(d)

Figure 7.13a Transient analysis of a section of a transmission line terminated by a resistive load R_L. (b) Voltage and current distributions along a short-circuited transmission line for time $0 < t < T$. (c) Voltage and current distributions along the short-circuited transmission line for time $T < t < 2T$. (d) Voltage and current distributions along the open-circuit terminated transmission line for $T < t < 2T$.

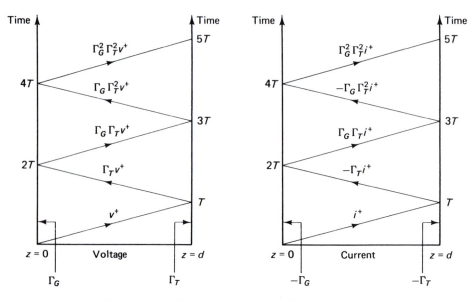

Figure 7.14 Voltage and current reflection diagram.

rent, it is usual to write "voltage" or "current," as the case may be, at the bottom of the reflection diagram. The abscissa of the reflection diagram denotes the length along the transmission line. In Figure 7.14, for example, the length extends from $z = 0$ to $z = d$. The reflection coefficients at the terminals have also been indicated at the appropriate positions along the abscissa as shown in Figure 7.14. The ordinate of the reflection diagram denotes the time required for the multiple reflections to develop along the transmission line. For example, the time required for an incident voltage or current to reach the termination at $z = d$ is $T = d/u$, where u is the velocity of propagation along the transmission line.

On the reflection diagram, the initial voltage along the transmission line (v^+) is marked at a point corresponding to the ordinate $t = 0$. In the time range $0 < t < T$, there is only v^+, which is the voltage wave traveling in the positive z direction and produced by the voltage generator at $z = 0$. At $z = d$ and $t = T$, an additional wave traveling in the negative z direction is generated. On the voltage reflection diagram, the reflected voltage is denoted by $\Gamma_T v^+$, whereas on the current reflection diagram, the reflected current is denoted by $-\Gamma_T i^+$. The negative sign in the current relations is simply included because the current reflection coefficient is the negative of the voltage reflection coefficient. The incremental voltages and currents corresponding to the successive reflections at both ends of the transmission line can thus be obtained in a similar manner.

There are two basic uses of the reflection diagram. These include the following:

1. Obtaining the voltage or current distribution along the transmission line at a given time.

2. Obtaining the voltage or current at any specified point on the transmission line as a function of time.

To illustrate these uses, let us consider the following example.

EXAMPLE 7.3

Consider a lossless transmission line of characteristic impedance $Z_o = 50$ ohms and extends from $z = 0$ to $z = 900$ m. The velocity of propagation along the transmission line is $u = 3 \times 10^8$ m/s. A battery of voltage 4 V and internal resistance $R_G = 150$ ohms is connected to the input terminal of the line ($z = 0$), whereas the output terminal is left open. Determine the following:

1. The voltage distribution along the line at time $t = 10 \times 10^{-6}$ s.
2. The time dependence of the voltage at $z = 600$ m, up to time $t = 12 \times 10^{-6}$ s.

Solution

The voltage reflection diagram is shown in Figure 7.15a. The reflection coefficient at the termination Γ_T and the generator Γ_G ends are given by

$$\Gamma_G = \frac{150 - 50}{150 + 50} = \frac{1}{2}, \qquad \text{and} \qquad \Gamma_T = \frac{1 - Z_o/R_T}{1 + Z_o/R_T} = 1$$

The time T required for a wave to travel the entire length of the transmission line is

$$T = \frac{d}{u} = \frac{900 \text{ m}}{3 \times 10^8} = 3 \times 10^{-6} \text{ s}$$

The reflection diagram in Figure 7.15a is drawn for total time $t = 4T = 12 \times 10^{-6}$ s. The incident voltage v^+ is determined from the initial conditions at the generator end—that is,

$$v^+ = \frac{v_G}{R_G + Z_o} Z_o = \frac{4 \times 50}{150 + 50} = 1 \text{ V}$$

1. To find the voltage distribution along the transmission line at $t = 10 \times 10^{-6}$ s, we draw a dashed line corresponding to the ordinate $t = 10 \times 10^{-6}$ s, and this intersects the locus straight line at an abscissa point corresponding to $z = 600$ m (see Figure 7.15a). This simply indicates that at $t = 10 \times 10^{-6}$ s there will be a voltage discontinuity located at $z = 600$ m. If the time is incremented by Δt—that is, from 10×10^{-6} to $(10 \times 10^{-6} + \Delta t)$, as shown in Figure 7.15a—it is clear that z changes from 600 m to $(600 - \Delta z)$, which indicates that the voltage discontinuity is traveling toward the generator end. The voltage distribution along the transmission line at $t = 10 \times 10^{-6}$ s can be obtained by adding the voltages that resulted from the multiple reflections (e.g., $\Gamma_T v^+$, $\Gamma_G \Gamma_T v^+$, etc.) with their appropriate signs to the incident voltage v^+. Of course, only the reflections that occurred at time t less than or equal to 10×10^{-6} s will be considered and counted in the addition process. From Figure 7.15a, we find that the total voltage along the transmission line from $z = 0$

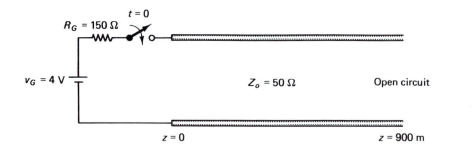

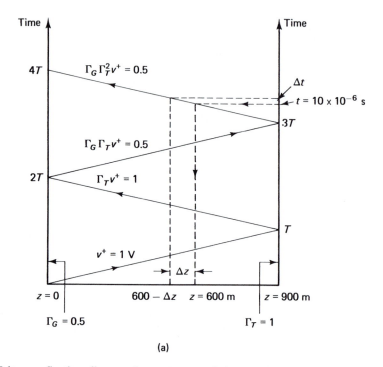

(a)

Figure 7.15a Voltage reflection diagram for an open-ended transmission line of length 900 m.

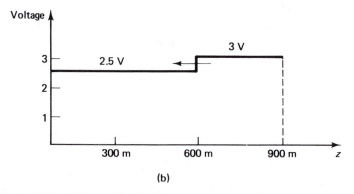

(b)

Figure 7.15b Voltage distribution at $t = 10 \times 10^{-6}$ s.

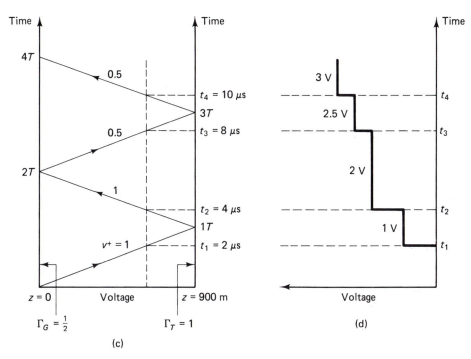

(c)

(d)

Figure 7.15 (c) Reflection diagram with specific emphasis on the voltage distribution at $z = 600$ m. (d) Voltage distribution at $z = 600$ m as a function of time.

to $z = 600$ m is 2.5 V and that from $z = 600$ m to $z = 900$ m is 3 V. Thus, we obtain the voltage distribution shown in Figure 7.15b.

2. To find the time dependence of the voltage at $z = 600$ m from an initial time $t = 0$ up to a time $t = 12 \times 10^{-6}$ s, we draw a dashed line corresponding to the abscissa point $z = 600$ m as shown in Figure 7.15c. This intersects the locus straight lines representing the multiple reflections at $t = t_1, t_2, t_3,$ and t_4, where for this example $t_1 = 2 \times 10^{-6}$ s, $t_2 = 4 \times 10^{-6}$ s, $t_3 = 8 \times 10^{-6}$ s, and $t_4 = 10 \times 10^{-6}$ s. The development of the voltage wave form as a function of time is shown in Figure 7.15d. The voltage at $z = 600$ m is zero from $t = 0$ to 2×10^{-6} s, when the voltage changes to 1 V with the arrival of the incident voltage. At each of the subsequent times $t_2, t_3,$ and t_4 indicated on Figure 7.15d, the voltage at $z = 600$ m changes by an amount indicated on the appropriate locus straight line representing the multiple reflection as shown in Figure 7.15c. In this manner we obtain the voltage at $z = 600$ m as a function of time up to $t = 12 \times 10^{-6}$ s.

7.7 TANDEM CONNECTION OF TRANSMISSION LINES

In addition to discontinuities at the generator and the termination locations on the transmission line, reflections may occur at junctions between various transmission lines. For example, consider two transmission lines of characteristic impedances Z_{o1} and Z_{o2} connected as shown in Figure 7.16.

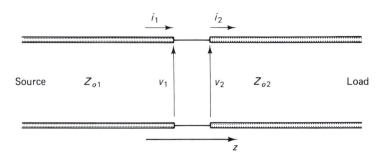

Figure 7.16 The junction of two transmission lines, source wave is incident from line 1.

The boundary condition at the junction of two transmission lines requires that the voltage and current be continuous across the junction—that is, $v_1 = v_2$ and $i_1 = i_2$. If the incident voltage wave is from line 1 (v_1^+), at the junction there will be a reflected wave v_1^- in line 1 and a transmitted wave v_2^+ in line 2. In this case, $v_1 = v_1^+ + v_1^-$, whereas $v_2 = v_2^+$. The voltage boundary condition is then satisfied when

$$v_1^+ + v_1^- = v_2^+$$

which may be rewritten as

$$1 + \frac{v_1^-}{v_1^+} = \frac{v_2^+}{v_1^+} \tag{7.29}$$

The ratio v_1^-/v_1^+ has been previously defined as the reflection coefficient Γ and will be denoted here as Γ_{11} to emphasize the fact that the reflection occurs at the end of line 1. Similarly, we define a transmission coefficient τ_{12} as

$$\tau_{12} = \frac{v_2^+}{v_1^+} \tag{7.30}$$

This coefficient gives the fraction of incident wave from line 1 that is transmitted through to the second line, hence, the coefficient τ_{12}. Equation 7.29 then becomes

$$\tau_{12} = 1 + \Gamma_{11} \tag{7.31}$$

An expression for Γ_{11} may be obtained by following a procedure similar to that of section 7.4.

The total voltage and current at the junction are given by

$$v_T = v^+ + v^-$$

$$i_T = i^+ + i^- = \frac{v^+}{Z_{o1}} - \frac{v^-}{Z_{o1}}$$

Z_{o2} acts as termination to transmission line 1. Therefore, $v_T/i_T = Z_{o2}$, and substituting expressions for v_T and i_T we obtain,

$$Z_{o2} = Z_{o1} \frac{v^+ + v^-}{v^+ - v^-}$$

or

$$Z_{o2} = Z_{o1}\frac{1 + \Gamma_{11}}{1 - \Gamma_{11}}$$

The expression for Γ_{11} is hence

$$\Gamma_{11} = \frac{Z_{o2} - Z_{o1}}{Z_{o2} + Z_{o1}}$$

From equation 7.31 the transmission coefficient is given by

$$\tau_{12} = \frac{2Z_{o2}}{Z_{o1} + Z_{o2}}$$

Let us next consider the situation where the source is moved to line 2, as shown in Figure 7.17.

In this case there will be an incident and a reflected wave in line 2 (v_2^+ and v_2^-), and only a transmitted wave in line 1 (v_1^+). Thus, the voltage boundary condition gives

$$v_1^+ = v_2^+ + v_2^-$$

or

$$\frac{v_1^+}{v_2^+} = 1 + \frac{v_2^-}{v_2^+} \tag{7.32}$$

If we define $v_2^-/v_2^+ = \Gamma_{22}$, the reflection coefficient in line 2, and $v_1^+/v_2^+ = \tau_{21}$, the transmission coefficient from line 2 to line 1, then we can rewrite equation 7.32 in the form

$$\tau_{21} = 1 + \Gamma_{22} \tag{7.33}$$

where

$$\Gamma_{22} = \frac{Z_{o1} - Z_{o2}}{Z_{o1} + Z_{o2}}$$

and from equation 7.33,

$$\tau_{21} = \frac{2Z_{o1}}{Z_{o1} + Z_{o2}}$$

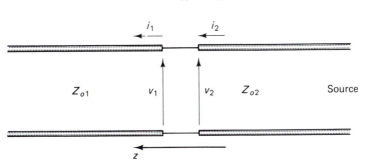

Figure 7.17   The junction of two transmission lines; source wave is incident from line 2.

The reflection diagram is a very useful and frequently used tool in determining the voltage and current at any point in a cascaded transmission-line system. To illustrate the development of the multiple reflections in this case, let us consider the following example.

EXAMPLE 7.4

Consider the tandem transmission line system shown in Figure 7.18. All the system parameters are shown in Figure 7.18. If the velocity of propagation in both transmission lines is the same—that is, $u_1 = u_2 = 3 \times 10^8$ m/s—determine the voltage distribution on the tandem system at time $t = 7.5 \times 10^{-8}$ s.

Solution

First we determine the reflection and transmission coefficients at the various junctions and terminations along the transmission line:

$$\Gamma_G = \frac{R_G - Z_{o1}}{R_G + Z_{o1}} = 0, \qquad \text{because } R_G = Z_{o1}$$

$$\Gamma_T = \frac{R_T - Z_{o2}}{R_T + Z_{o2}} = \frac{\frac{1}{3}Z_{o2} - Z_{o2}}{\frac{1}{3}Z_{o2} + Z_{o2}} = -\frac{1}{2}$$

$$\Gamma_{11} = \frac{Z_{o2} - Z_{o1}}{Z_{o2} + Z_{o1}} = -\frac{1}{2}$$

$$\tau_{12} = \frac{2Z_{o2}}{Z_{o1} + Z_{o2}} = \frac{1}{2}, \qquad \text{or} \qquad \tau_{12} = 1 + \Gamma_{11}$$

$$\Gamma_{22} = \frac{Z_{o1} - Z_{o2}}{Z_{o1} + Z_{o2}} = \frac{1}{2}$$

$$\tau_{21} = \frac{2Z_{o1}}{Z_{o1} + Z_{o2}} = \frac{3}{2}, \qquad \text{or} \qquad \tau_{21} = 1 + \Gamma_{22}$$

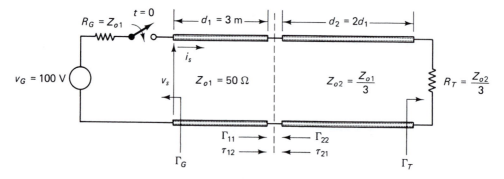

Figure 7.18 Tandem-line system.

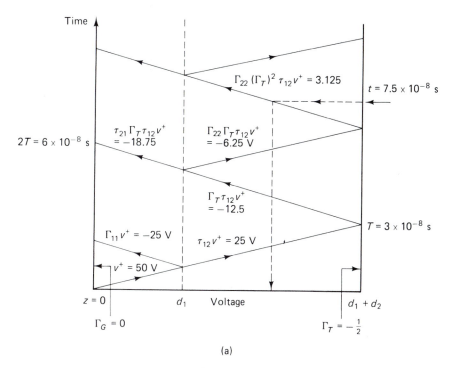

Figure 7.19a   Reflection diagram for a tandem connection of two transmission lines of the same velocity of propagation.

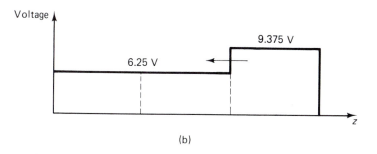

Figure 7.19b   Voltage distribution at $t = 7.5 \times 10^{-8}$ s.

The wave initially launched on line 1 by the step generator is given by

$$v_s = v^+ = \frac{v_G}{R_G + Z_{o1}} Z_{o1} = \frac{v_G}{2} = 50 \text{ V}$$

The reflection diagram pertinent to this problem is shown in Figure 7.19a. The resulting voltage distribution along the line is shown in Figure 7.19b.

EXAMPLE 7.5

Two transmission lines 1 and 2 of characteristic impedances $Z_{o1} = 50 \ \Omega$ and $Z_{o2} = 100 \ \Omega$, respectively, are connected in tandem as shown in Figure 7.20. At $t = 0$, transmission line 1 is connected at one side to a battery of 3 V and internal resistance $R_G = 150 \ \Omega$, and at the other side to the transmission line 2 through a series resistor $R_s = 100 \ \Omega$. The transmission line 2 is terminated by a load resistance $R_L = 100 \ \Omega$.

If the velocity of propagation along line 1 is $u_1 = 3 \times 10^8$ m/s, and the velocity of propagation along line 2 is $u_2 = 2 \times 10^8$ m/s, determine the following:

1. The voltage at the sending end for a period up to $t = 12 \times 10^{-6}$ s.
2. The voltage at the termination for a period up to $t = 9 \times 10^{-6}$ s.

Solution

We start the solution, like in all cases, by determining the reflection coefficients at all the discontinuities.

$\Gamma_G \equiv$ reflection coefficient between the source and transmission line 1

$$= \frac{150 - 50}{150 + 50} = \frac{1}{2}$$

$\Gamma_L \equiv$ load reflection coefficient $= 0$

$\Gamma_{11} \equiv$ reflection coefficient between lines 1 and 2

$$\Gamma_{11} = \frac{v_1^-}{v_1^+}$$

To obtain a relation between v_1^- and v_1^+ when a series or any other combination of resistive discontinuities is present at the junction between the transmission lines, we use our basic equations that relate the current to the voltage at the junction. In transmission line 1,

$$v_T = v_1^+ + v_1^-$$

$$i_T = i_1^+ + i_1^- = \frac{1}{Z_{o1}}(v_1^+ - v_1^-)$$

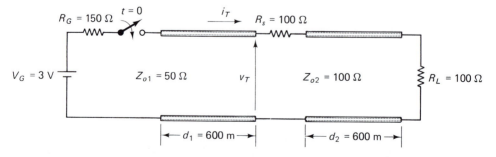

Figure 7.20 Tandem connection of two transmission lines through a series resistor R_s.

In Figure 7.20, however, v_T/i_T is simply $R_s + Z_{o2}$. Hence,

$$R_s + Z_{o2} = Z_{o1}\frac{1 + \Gamma_{11}}{1 - \Gamma_{11}}$$

The reflection coefficient Γ_{11} is then given by

$$\Gamma_{11} = \frac{(R_s + Z_{o2}) - Z_{o1}}{(R_s + Z_{o2}) + Z_{o1}} = 0.6$$

To determine $\tau_{12} = v_2^+/v_1^+$, we also need to relate v_2^+ to v_1^+:

$$v_1 = v_1^+ + v_1^- = v_1^+(1 + \Gamma_{11})$$

$$v_2^+ = \frac{v_1}{R_s + Z_{o2}}Z_{o2} = \frac{v_1^+(1 + \Gamma_{11})Z_{o2}}{R_s + Z_{o2}}$$

$$\therefore \tau_{12} = \frac{v_2^+}{v_1^+} = \frac{(1 + \Gamma_{11})Z_{o2}}{R_s + Z_{o2}} = 0.8$$

Looking from line 2 into line 1,

$$\Gamma_{22} = \frac{150 - 100}{100 + 150} = 0.2$$

Following a procedure similar to that we used in determining τ_{12}, we can determine τ_{21}:

$$\tau_{21} = \frac{v_1^+}{v_2^+} = \frac{(1 + \Gamma_{22})50}{150} = 0.4$$

The incident voltage

$$v_1^+ = \frac{v_G Z_{o1}}{Z_{o1} + R_G} = \frac{3}{4} \text{ V}$$

The travel times T_1 and T_2 in transmission lines 1 and 2 are given respectively by

$$T_1 = \frac{600}{3 \times 10^8} = 2 \times 10^{-6} \text{ s}$$

$$T_2 = \frac{600}{2 \times 10^8} = 3 \times 10^{-6} \text{ s}$$

With the identification of all these reflection and transmission parameters, we are now ready to construct the reflection diagram shown in Figure 7.21. It should be noted that the slopes of the locus straight lines are different in the two different transmission lines. This is simply because the velocities of propagation are different in the two lines. The slope of the locus straight line is basically $1/u$; hence, larger slopes will be in the region of smaller velocities of propagation.

1. The required voltage at the sending end is obtained by constructing a vertical line at $z = 0$ (i.e., at the sending end) and adding the incremental contributions from the multiple reflections as the time elapses. For example, for time t between 0 and 4×10^{-6} s, we have at the sending end only the incident voltage of 3/4 V. At $t = 4 \times 10^{-6}$ s, the voltage reflected from the series discontinuity arrives at the sending end, and it simultaneously generates a reflected voltage wave as a result of

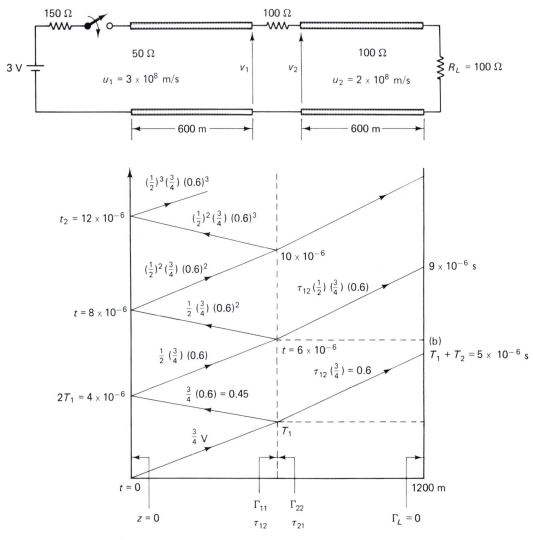

Figure 7.21 Bounce diagram for example 7.5. Tandem connection of two transmission lines of two different propagation velocities.

the mismatch between line 1 and the source. The total voltage at $t = 4 \times 10^{-6}$ is, hence,

$$\underbrace{\frac{3}{4}}_{v^+} \quad + \quad \underbrace{\Gamma_{11}\left(\frac{3}{4}\right)}_{\substack{\text{reflection from}\\\text{series}\\\text{discontinuity}}} \quad + \quad \underbrace{\Gamma_G\left[\Gamma_{11}\left(\frac{3}{4}\right)\right]}_{\substack{\text{reflection at the}\\\text{generator}}}$$

The obtained result for the voltage at the sending end as a function of time is shown in Figure 7.22.

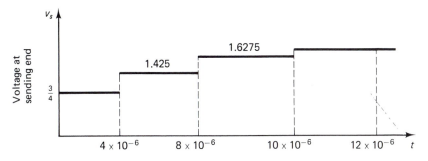

Figure 7.22   Voltage at the sending end as a function of time.

2. Similarly the voltage at the termination is obtained by adding the contributions from the multiple reflections at the load location. The obtained result is shown in Figure 7.23.

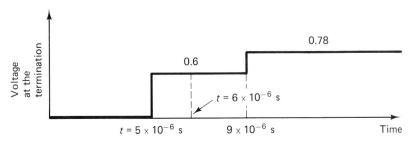

Figure 7.23   Voltage at the load as a function of time.

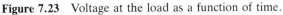

7.8 PULSE PROPAGATION ON TRANSMISSION LINES

Thus far we have considered only the propagation of dc transients on transmission lines. The same techniques can be used to study propagation of pulses on lines of finite lengths. To illustrate the solution procedure, let us consider the following example.

EXAMPLE 7.6

A transmission line of 50-ohm characteristic impedance and 600 m long is connected to a pulse generator that has an internal resistance of 150 ohms and produces a 40 volts, 1-μs pulse. The line is terminated by a load resistance of $R_T = 16.7$ ohms. If the velocity of propagation on the transmission line is 300 m/μs, determine the sending end voltage and current as a function of time.

Solution

The transmission line geometry is shown in Figure 7.24a. At $t = 0$, the incident voltage pulse is unaffected by the receiving end, and it begins to travel down the line as if the length

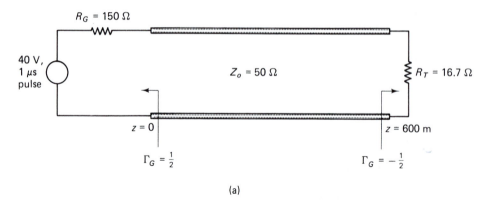

Figure 7.24a   Pulse excitation of a 50 Ω transmission line of length $d = 600$ m.

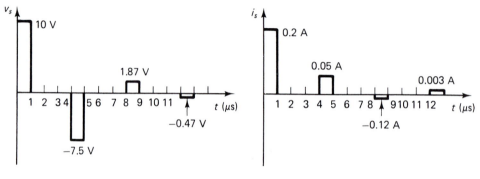

Figure 7.24b   Sending-end voltage and current as a function of time.

of the line is infinitely long. The 40-V pulse is initially divided between the generator resistance R_G and the characteristic impedance Z_o of the line to give a sending-end pulse of

$$v_s = v^+ = \frac{40}{150 + 50} \times 50 = 10 \text{ V}$$

The 10-V pulse travels to the load in 2 μs and is partially absorbed and partially reflected, with the reflection $\Gamma_R = -1/2$. The reflected pulse, which is -5 V, travels back toward the generator in 2 μs and is partially reflected by the mismatched resistance R_G. The reflection coefficient at the generator is $\Gamma_G = 1/2$, which results in a reflected pulse of -2.5 V. Because the initially incident wave lasts only for 1 μs, its value does not contribute to the total voltage at the sending end for times larger than 1 μs—that is, $v^+ = 0$ for $t > 1$ μs. Therefore, the total voltage at $t = 2T = 4$ μs is basically the sum of the voltage reflected from the load (-5 V) and the reflected pulse at the generator end (-2.5 V). The second reflected pulse of -2.5 V results from the reflection of the load reflected voltage (-5 V) at the generator with $\Gamma_G = 1/2$. Therefore $v^{tot}(t = 2T) = -5 - 2.5 = -7.5$ V. From this point on, the process repeats itself with the pulse being multiplied by $\Gamma_T = -1/2$ at the termination and $\Gamma_G = 1/2$ at the generator, as shown in Figure 7.24b.

Similar procedure can be used for the current pulse with the initial current being $i^+ = 40/(150 + 50) = 0.2$ A. It should be noted, however, that the current reflection coefficient at the load is $i_T^-/i_T^+ = -\Gamma_T = 1/2$ and is $i_s^-/i_s^+ = -\Gamma_G = -1/2$ at the generator. The sending-end current is also shown in Figure 7.24b.

In summary, the process of pulse multiple reflections along transmission lines may be treated using the reflection diagram in a fashion similar to that used in the case in which we have step-voltage excitation. The only difference between the two cases is simply in the addition process of the contributions of the various reflections at a specific location along the transmission line to determine the voltage or current distribution along the line. In the step-voltage-excitation case, all the contributions from the multiple reflections that occurred at times previous or equal to the instant of interest are added in determining the final voltage or current distribution along the transmission line. In the pulse excitation case, however, this may or may not be the case depending on the duration of the pulse. In general, all contributions from the various multiple reflections should be considered, and only pulses (incident or reflected) that are still "on" at the desired instant of time will contribute to the value of the voltage or the current distribution at that time.

7.9 TIME-DOMAIN REFLECTOMETER

Evaluation of microwave components and devices may be routinely achieved by measuring their reflection and transmission properties as a function of frequency. For complete characterization, however, measurement over a broad frequency band is often desired. This broadband information can be obtained by sweeping the frequency in the desired range or by applying a short rise-time voltage pulse, or a step voltage, to the device under test. Based on a simple Fourier transformation of the applied pulse or step voltage, it can be shown that such a waveform contains a broad frequency band and that in the ideal case in which the applied voltage is a delta function, the frequency band extends from zero to infinity. Of course, it is impossible practically to generate a delta function voltage pulse with a zero rise time. This is why time-domain analysis does not provide frequency domain information over unlimited band. Instead, it provides frequency domain information in a band that is determined by the rise time of the incident pulse or step function voltage. The shorter the rise time, the broader the frequency band will be. This withstanding, it is clear that the desired broadband information can either be measured using swept-frequency techniques or by measuring the response of the system under test to an input short rise-time pulse- or a step-voltage excitation. In other words, the transient response of a microwave component or device together with a simple analysis procedure involving the Fourier transform may be used for a complete and broadband characterization of the device instead of the routine point-by-point or swept frequency domain measurements.

Another advantage of the time-domain pulse-type characterization of microwave systems is that it facilitates separating, in time, the responses from various discontinuities along the transmission-line system. In the frequency domain measurements where

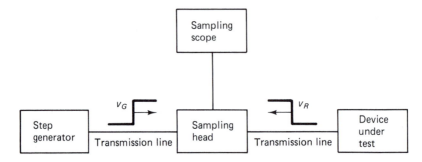

Figure 7.25 Basic components of the time-domain reflectometer.

sinusoidal voltages are often used, the composed reflection and transmission coefficients are measured, and it is quite difficult to decompose these measured values to the various components generated from the various discontinuities at different locations along the transmission line. In time-domain measurements, the contributions resulting from the various discontinuities along the transmission-line system are all separated in time and, hence, can be recognized separately. This is why time-domain or pulse measurement techniques have been used for many years to locate faults in cables and along telephone lines. Modern microwave automatic network analyzers provide gatting capabilities that facilitate isolation of reflections from various discontinuities and their subsequent analysis in the frequency domain.

A commercially available device that provides the capability of making pulse echo or time-domain measurements is known as the time domain reflectometer (TDR). A block diagram of this device is shown in Figure 7.25, where it can be seen that it consists of the following basic components:

1. A step-voltage generator that provides the step function input voltage to the transmission line.
2. A sampling head that includes a high-impedance probe to sample the voltage along the transmission line.
3. A sampling oscilloscope that is a broadband scope to display the probed voltage along the transmission line.
4. The device under test.

It should be noted that the sampling probe actually measures the total voltage $v_G + v_R$ along the transmission line. Therefore, to obtain the step-function response of the device under test, the incident step voltage should be subtracted from the total signal displayed on the scope.

The following are examples of some typical TDR displays and their interpretation.

7.9.1 Time-Domain Display of Resistive Terminations

Schematic diagram illustrating the operation principle of the TDR is shown in Figure 7.26. At $t = 0$, the step voltage is

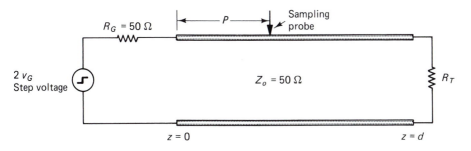

Figure 7.26 Schematic illustrating the function of the TDR when terminated by a resistance R_T.

$$v^+ = \frac{2v_G}{R_G + Z_o} Z_o = v_G \text{ V}$$

This incident voltage will be sampled by the high impedance probe at time $t = P/u$, where u is the velocity of propagation along the transmission line and P is the distance from probe to the generator. At $t = d/u$, this incident voltage v^+ will arrive at the termination. Assuming R_T is different from Z_o, there will be a reflected voltage v^- generated at $z = d$. The total voltage—that is, incident plus reflected—will be monitored by the sampling probe and displayed on the TDR scope. The time elapsed between the first monitoring of v^+ by the probe and the probing of the total voltage $v^+ + v^-$ is $2(d - p)/u$, which is basically the time required for a voltage wave to travel from the probe to the load and back to the probe.

At $t = 2d/u$, the reflected voltage will arrive at the generator, but because the generator's impedance R_G is equal to the characteristic impedance Z_o of the transmission line, no further reflections will occur, and steady state will be reached. As an example of the typical waveforms expected on the TDR, consider the case in which $R_T = 25 \ \Omega$. In this case the reflection coefficient is given by,

$$\Gamma = \frac{R_T - Z_o}{R_T + Z_o} = \frac{25 - 50}{25 + 50} = -\frac{1}{3}$$

The reflected voltage v^- is then

$$v^- = \Gamma v^+ = \Gamma v_G = -\frac{1}{3} v_G$$

The total voltage detected by the sampling probe is then $v^+ + v^- = 2/3v_G$. A representative TDR waveform is shown in Figure 7.27a and a typical photograph of the TDR scope display is shown in Figure 7.27b.

Other examples of possible TDR displays for a variety of resistive loads are shown in Figure 7.28.

7.9.2 Time-Domain Displays of Arbitrary Terminations

From the discussion in the previous section it is clear that resistive termination (R_T) of a lossless transmission line (Z_o real) results in a reflected voltage of the same shape as

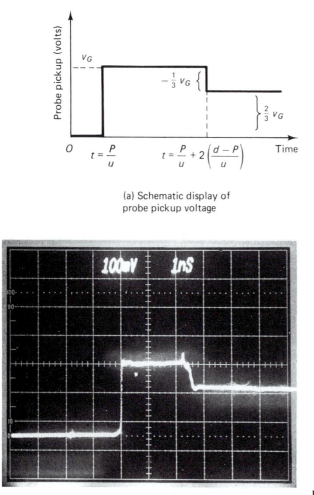

(a) Schematic display of
probe pickup voltage

(b) Photograph of TDR scope

Figure 7.27 TDR display for
$R_T = 25\ \Omega$.

the applied one. The magnitude and polarity (positive or negative) of the reflected voltage, however, are determined by the specific value of R_T relative to Z_o.

In cases in which the transmission line is terminated by arbitrary impedances, the reflected voltage wave forms are generally different from the shape of the applied voltage. For example, consider the case in which we have a lossy capacitor terminating a lossless transmission line of characteristic impedance Z_o. A schematic illustrating the operation of the TDR with the lossy capacitor termination is shown in Figure 7.29.

At $t = 0$, a step voltage $v^+ = 2v_G Z_o/R_G + Z_o = v_G$ will be applied to the transmission line. At $t = P/u$, the incident voltage will be sampled by the sampling probe and will also be displayed on the TDR scope. At $t = d/u$, the incident step will be applied to the capacitive termination, and a reflected voltage traveling back to the source will be generated at this point ($z = d$). To determine the reflected voltage, it

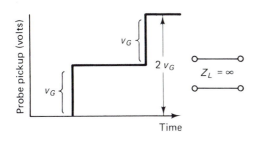

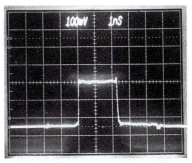

(a) Open-circuit termination $Z_L = \infty$ ($\Gamma = 1$)

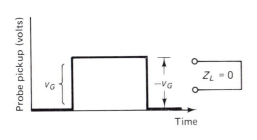

(b) Short-circuit termination $Z_L = 0$ ($\Gamma = -1$)

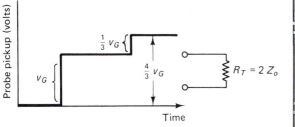

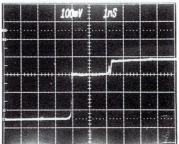

(c) Termination resistance R_T equals double characteristic impedance Z_o ($\Gamma = 1/3$).

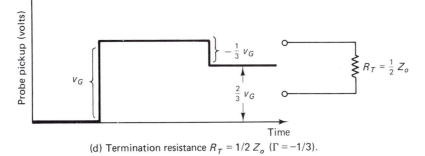

(d) Termination resistance $R_T = 1/2\ Z_o$ ($\Gamma = -1/3$).

Figure 7.28 TDR displays for various resistive terminations. (a) Open circuit termination $Z_L = \infty$ ($\Gamma = 1$), (b) Short circuit termination $Z_L = 0$ ($\Gamma = -1$), (c) Termination resistance $R_T = 2Z_o$ ($\Gamma = \frac{1}{3}$), and (d) Termination resistance $R_T = \frac{1}{2}Z_o$ ($\Gamma = -\frac{1}{3}$).

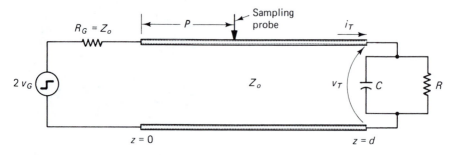

Figure 7.29 TDR terminated with a lossy capacitor, which is modeled as a parallel combination of R and C.

is necessary to use the arbitrary termination solution procedure described in section 7.4. It is shown that, at the termination, the incident voltage is related to the total voltage and current by

$$2v^+(d - ut) = v_T + Z_o i_T(t)$$

where t is measured from the time of arrival of v^+ at the termination. We will continue to develop our analysis based on this reference time t and will adjust it later to account for the reference $t = 0$ at the generator location. Substituting $v^+ = v_G$, and $i_T = C\,dv_T/dt + v_T/R$, which is the voltage current relation across the parallel load (i.e., $i_T = i_C + i_R$), we obtain

$$2v_G = \left(v_T + \frac{Z_o}{R}v_T\right) + Z_o C\frac{dv_T}{dt}$$

Once again the preceding equation is a first-order differential equation, and, for a dc input voltage (i.e., $v_G = $ constant), the solution of this equation will take the form $v_T = v_t + v_f$ where v_t is the transient part of the response obtained by suppressing the source term and solving for the resulting homogeneous differential equation, and v_f is the forced part of the response which for dc input voltage is obtained by letting $dv_T/dt = 0$. It can be shown that

$$v_f = \frac{2v_G}{1 + \dfrac{Z_o}{R}}, \qquad v_t = Ae^{\frac{-t}{\tau}}$$

where τ in this case is the time constant given by $\tau = Z_o RC/(Z_o + R)$. The complete solution is, hence, given by

$$v_T = \frac{2v_G}{1 + \dfrac{Z_o}{R}} + Ae^{\frac{-t}{\tau}}$$

To determine the constant of integration A, we assume the condition that the charge on the capacitor was initially zero. Capacitors oppose sudden change in voltage; hence, $v_T = 0$ at $t = 0$.

Substituting this initial condition in the total voltage equation we obtain

$$A = \frac{-2v_G}{1 + \dfrac{Z_o}{R}}$$

The total voltage is, hence,

$$v_T = \frac{2v_G}{1 + \dfrac{Z_o}{R}}(1 - e^{\frac{-t}{\tau}}), \qquad \tau = \frac{Z_o R}{Z_o + R}C$$

To explain the resulting TDR display, let us rearrange the total voltage response equation in the form

$$v_T = \frac{2R\,v_G}{R + Z_o}(1 - e^{\frac{-t}{\tau}})$$

$$= v_G\left(1 + \frac{R - Z_o}{R + Z_o}\right)(1 - e^{\frac{-t}{\tau}})$$

The TDR response for this lossy capacitor termination is shown in Figure 7.30. From Figure 7.30, the following observations should be noted:

1. The sampling probe will first monitor $v_T = v_G$ traveling down the transmission line.
2. After time Δt (see Figure 7.30), which is the time required for the incident voltage to reach the termination and for the reflected voltage to travel back to the sampling probe—that is, $\Delta t = 2(d - P)/u$, the total voltage—that is, $v^+ + v^-$ along the transmission line—will be sampled by the probe and displayed on the scope.
3. With the arrival of the incident voltage at the termination and because the initial voltage on the capacitor is assumed to be zero, the overall termination will act initially as a short circuit. The reflected voltage wave v^- will be equal to

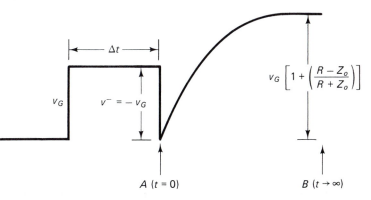

Figure 7.30   TDR display of a lossy capacitor termination.

$-v^+ = -v_G$, and the total voltage on the transmission line will, hence, be zero (point A) in Figure 7.30.

4. After a long time, the capacitor terminating the line will be fully charged and, hence, will act as an open circuit (i.e., current flow in the capacitor element = 0). The TDR termination will, hence, be basically the resistance R, and the reflected voltage in this case will be

$$v^- = v^+ \Gamma = v_G \frac{R - Z_o}{R + Z_o}$$

The total voltage along the line as picked up by the sampling probe will, hence, be

$$v_T = v_G + v_G \frac{R - Z_o}{R + Z_o} = v_G \left(1 + \frac{R - Z_o}{R + Z_o} \right)$$

which is indicated by the point B in Figure 7.30.

5. The values of the total voltages at points A and B can also be obtained from the final expression of the total voltage simply by substituting the appropriate times $t = 0$ for point A and $t \to \infty$ for point B.

7.10 SINUSOIDAL STEADY-STATE ANALYSIS OF TRANSMISSION LINES

In the previous sections we developed the transmission-line equations, and for the case of lossless transmission lines we discussed the transients analysis of voltages and currents along these lines. We also indicated that such transient information may be used for locating faults along the transmission lines and telephone cables as well as in characterizing microwave components and devices over a broad frequency band with the aid of Fourier transformation.

Most practical applications of transmission lines, however, involve sinusoidal excitations. The 60-Hz power lines, broadcast stations, and other antenna operations are examples of sinusoidal excitations of transmission lines. In the following sections we will discuss the analysis and the various applications of transmission lines under sinusoidal excitation. In addition to the sinusoidal steady-state analysis of the voltage, current, and the power along transmission lines, we will discuss the use of sections of lossless lines to achieve impedance matching of microwave devices such as antennas. The use of slotted lines to measure the voltage and current characteristics along lossless lines and to measure unknown input impedances of microwave components will also be described.

7.10.1 Sinusoidal Steady-State Solution of Transmission-Line Equations

In section 7.2 we derived the transmission-line equations without specifying the time variation of the voltage and the current along the line. We used these equations to study the step voltage and pulse excitation of lossless transmission lines. In the case of sinusoidal excitation the time variation of the input voltage and current is of the form $\cos(\omega t \pm \theta)$. We learned in our introductory circuits courses as well as in solving Maxwell's equations for time harmonic fields (chapters 2 and 3) that for sinusoidal excitation, and if we are interested in only the steady-state part of the solution, it is more convenient to use the phasor representation rather than the real-time forms of the quantities of interest such as the current and voltage along the line.

If we have a sinusoidal source—that is, $v_G(t) = v_G \cos \omega t$, the output voltage may be equivalently represented in the form

$$v_G(t) = Re(\hat{V}_G e^{j\omega t}) \tag{7.34}$$

Hence, we may proceed with the sinusoidal steady-state analysis by assuming an $e^{j\omega t}$ time variation and using the phasor representation of the source voltage $\hat{V}_G = V_G e^{j0^\circ}$ to obtain the voltage and current distributions along the transmission line. It should be emphasized, once again, however, that there are no physical sources with $e^{j\omega t}$ time variation. Instead, $e^{j\omega t}$ is only an assumed form that facilitates the use of the phasor concept and, hence, simplifies the analysis. The phasor quantities are only useful in simplifying the analysis, and to study the propagation characteristics, the time variable should be included using equations such as equation 7.34.

Let us now use the $e^{j\omega t}$ time variation and the phasor representation of the voltage and current to solve the transmission-line equations. The voltage and current at any point along the transmission line shown in Figure 7.31 are given by

$$\frac{-\partial}{\partial z}[v(z,t)] = r\,i(z,t) + \ell\,\frac{\partial i(z,t)}{\partial t} \tag{7.35}$$

$$\frac{-\partial}{\partial z}[i(z,t)] = g\,v(z,t) + c\,\frac{\partial v(z,t)}{\partial t} \tag{7.36}$$

For sinusoidal excitation, $v(z,t)$ and $i(z,t)$ may be expressed as

$$v(z,t) = Re[\hat{V}(z)\,e^{j\omega t}] \tag{7.37}$$

$$i(z,t) = Re[\hat{I}(z)\,e^{j\omega t}] \tag{7.38}$$

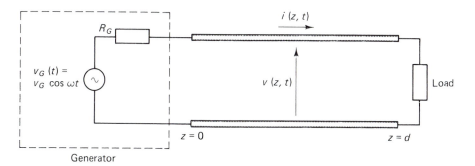

Figure 7.31   Sinusoidal excitation of a transmission line.

where $\hat{V}(z)$ and $\hat{I}(z)$ are the phasor voltage and current, respectively. Substituting equations 7.37 and 7.38 in equations 7.35 and 7.36, and eliminating $e^{j\omega t}$ from both sides of the equations, we obtain

$$-\frac{d\hat{V}(z)}{dz} = r\hat{I}(z) + j\omega\ell\hat{I}(z) \tag{7.39}$$

$$-\frac{d\hat{I}(z)}{dz} = g\hat{V}(z) + j\omega c\hat{V}(z) \tag{7.40}$$

The importance of assuming a source of time variation of the form $e^{j\omega t}$ is clear from equations 7.39 and 7.40 in which the derivative with respect to time has been eliminated. Differentiating equation 7.39 with respect to z and substituting $d\hat{I}(z)/dz$ from equation 7.40, we obtain

$$\frac{d^2\hat{V}(z)}{dz^2} - (r + j\omega\ell)(g + j\omega c)\hat{V}(z) = 0 \tag{7.41}$$

Similarly, by differentiating equation 7.40 with respect to z and substituting $d\hat{V}(z)/dz$ from equation 7.39, we obtain

$$\frac{d^2\hat{I}(z)}{dz^2} - (r + j\omega\ell)(g + j\omega c)\hat{I}(z) = 0 \tag{7.42}$$

Equations 7.41 and 7.42 are second-order differential equations only in the voltage and the current along the transmission line, respectively. Solutions of equations 7.41 and 7.42 are in the form

$$\hat{V}(z) = \hat{V}_m^+ e^{-\hat{\gamma}z} + \hat{V}_m^- e^{\hat{\gamma}z} \tag{7.43}$$

and

$$\hat{I}(z) = \hat{I}_m^+ e^{-\hat{\gamma}z} + \hat{I}_m^- e^{\hat{\gamma}z} \tag{7.44}$$

where $\hat{\gamma} = \sqrt{(r + j\omega\ell)(g + j\omega c)} = \alpha + j\beta$ is the propagation constant and $\hat{V}_m^+, \hat{V}_m^-, \hat{I}_m^+$, and $\hat{I}_m^-$, in general, are complex numbers representing the amplitudes of the voltage and the current along the transmission line. To examine the propagation characteristics of the voltage and current, we have to convert equations 7.43 and 7.44 first to time-domain expressions $v(z,t)$ and $i(z,t)$. Based on similar discussions in chapters 2 and 3, it is easy to show that

$$\begin{aligned} v(z,t) &= Re(\hat{V}(z)e^{j\omega t}) = Re[(\hat{V}_m^+ e^{-\alpha z}e^{-j\beta z} + \hat{V}_m^- e^{\alpha z}e^{j\beta z})e^{j\omega t}] \\ &= V_m^+ e^{-\alpha z}\cos(\omega t - \beta z + \theta^+) + V_m^- e^{\alpha z}\cos(\omega t + \beta z + \theta^-) \end{aligned} \tag{7.45}$$

where the complex amplitudes $\hat{V}_m^+$ and $\hat{V}_m^-$ were replaced by $\hat{V}_m^+ = V_m^+ e^{j\theta^+}$ and $\hat{V}_m^- = V_m^- e^{j\theta^-}$. It should be familiar by now that the first term in equation 7.45 represents a voltage wave traveling along the positive z direction, whereas the second term represents a voltage wave traveling along the negative z direction. Because the transmission line was assumed lossy (r and g are not zero), the voltage will attenuate with attenuation constant α as it travels down the line. Similarly, the real-time form of the current may be obtained from equation 7.44 as

$$i(z,t) = I_m^+ e^{-\alpha z} \cos(\omega t - \beta z + \phi^+) + I_m^- e^{\alpha z} \cos(\omega t + \beta z + \phi^-) \qquad (7.46)$$

where the complex amplitudes $\hat{I}_m^+$ and $\hat{I}_m^-$ were replaced by $\hat{I}_m^+ = I_m^+ e^{j\phi^+}$ and $\hat{I}_m^- = I_m^- e^{j\phi^-}$. To obtain the relationship between the voltage and current on a transmission line, let us consider the current and voltage waves traveling along the positive z direction from the generator to the load. In phasor form, these voltages and currents are given by

$$\hat{V}^+(z) = \hat{V}_m^+ e^{-\alpha z} e^{-j\beta z} \qquad (7.47a)$$

$$\hat{I}^+(z) = \hat{I}_m^+ e^{-\alpha z} e^{-j\beta z} \qquad (7.47b)$$

$\hat{V}_m^+$ and $\hat{I}_m^+$ are related by substituting equations 7.47a and 7.47b in the transmission-line equation 7.39, hence,

$$+(\alpha + j\beta)\,\hat{V}_m^+ e^{-(\alpha + j\beta)z} = (r + j\omega\ell)\,\hat{I}_m^+ e^{-(\alpha + j\beta)z}$$

$$\therefore \frac{\hat{V}_m^+}{\hat{I}_m^+} = \frac{(r + j\omega\ell)}{\hat{\gamma}} = \sqrt{\frac{r + j\omega\ell}{g + j\omega c}} = \hat{Z}_o$$

where $\hat{Z}_o$ is the complex characteristic impedance of the lossy transmission line.

Similarly, it may be shown that the complex amplitudes of the negative z propagating voltage and current signals are related by

$$\frac{\hat{V}_m^-}{\hat{I}_m^-} = -\hat{Z}_o$$

The negative sign simply indicates that for such a case, the current is propagating in the negative z direction—that is, from the load to the generator.

The phase velocity of propagation may be obtained by examining the velocity of a specific point (constant phase) in the wave. Hence, for a positive z traveling wave, the point of constant phase satisfies $\omega t - \beta z + \theta^+ = $ constant, and $\omega t + \beta z + \theta^- = $ constant, for a negative z traveling wave. By differentiating each of these constant phase equations, it can be shown that the phase velocity of the positive z traveling wave is $u_p = dz/dt = \omega/\beta$ and $u_p = -\omega/\beta$ for a negative z traveling wave.

In summary, the solution of the transmission-line equations for sinusoidal excitation showed that at steady state the voltage and the current along these transmission lines consist, in general, of two waves, one traveling along the positive z direction from the generator to the load and the other traveling along the negative z from the load to the generator. In the general case of lossy line, the propagation constant $\hat{\gamma}$ is a complex number and is given by $\hat{\gamma} = \sqrt{(r + j\omega\ell)(g + j\omega c)} = \alpha + j\beta$ where α and β are the attenuation and phase constants, respectively. In addition, the amplitudes of the voltage and the current in any one wave are related by the complex characteristic impedance $\hat{Z}_o$ of the transmission line. The phase velocity of propagation is given by $u_p = \pm\omega/\beta$, where the plus sign is for a wave traveling along positive z and the minus sign is for a wave traveling along negative z direction. The following are values of the characteristic impedances and propagation constants for special transmission lines of practical interest.

Lossless transmission line.   In this case, the ohmic losses parameter r and the dielectric losses owing to the conductance g are both assumed to be zero. Hence, the propagation constant from equation 7.44 reduces to $\hat{\gamma} = j\omega\sqrt{\ell c}$, which means

$$\alpha = 0, \quad \text{and} \quad \beta = \omega\sqrt{\ell c}$$

The characteristic impedance $\hat{Z}_o$ in this case is given by $Z_o = \sqrt{\ell/c}$. Z_o is a real number indicating that the voltage and the current along a lossless, infinitely long, transmission line are in phase.

The phase velocity is given by

$$u_p = \frac{\omega}{\beta} = \frac{1}{\sqrt{\ell c}}$$

which indicates that the velocity of propagation along the transmission lines is constant and independent of the frequency.

Distortionless transmission line.   To transmit a broadband signal on a transmission line, and to minimize distortion to the signal, it is necessary to satisfy the following conditions:

1. The phase velocity of the transmission line should be independent of frequency. In this way, the various frequency components of the broadband signal will travel with the same velocity, thus minimizing distortion to the broadband signal.

2. The attenuation constant should also be independent of the frequency. Thus, the various frequency components of the input broadband signal attenuate with equal amounts as they travel down the transmission line, thus minimizing the distortion in the transmitted signal.

The lossless transmission line described in the first case satisfies both of these conditions. A lossless transmission line, therefore, is also a distortionless line. In the following, we will show that the transmission line may be lossy but still distortionless if the following condition is satisfied

$$\frac{r}{\ell} = \frac{g}{c} \tag{7.48}$$

In this case, the propagation constant $\hat{\gamma}$ is given by

$$\hat{\gamma} = \alpha + j\beta = \sqrt{(r + j\omega\ell)(g + j\omega c)}$$

$$= \sqrt{c\ell}\sqrt{\left(\frac{r}{\ell} + j\omega\right)\left(\frac{g}{c} + j\omega\right)}$$

Because $r/\ell = g/c$, $\hat{\gamma}$ is, hence,

$$\hat{\gamma} = \sqrt{\ell c}\left(\frac{r}{\ell} + j\omega\right)$$

$$\hat{\gamma} = \alpha + j\beta = \sqrt{\frac{c}{\ell}}(r + j\omega\ell)$$

$$\therefore \alpha = r\sqrt{\frac{c}{\ell}}$$

(7.49a)

$$\beta = \omega\sqrt{\ell c}$$

(7.49b)

The phase velocity is given by

$$u_p = \frac{\omega}{\beta} = \frac{1}{\sqrt{\ell c}}$$

(7.50)

From equations 7.49 and 7.50, it is clear that α and u_p are independent of the frequency ω. Hence, a transmission line with parameter values satisfying equation 7.48 will be distortionless. The characteristic impedance of the distortionless transmission line is given, utilizing equation 7.48, by

$$\hat{Z}_o = \sqrt{\frac{r + j\omega\ell}{g + j\omega c}} = \sqrt{\frac{\ell}{c}}\sqrt{\frac{r/\ell + j\omega}{g/c + j\omega}} = \sqrt{\frac{\ell}{c}}$$

which is the same value for a lossless transmission line.

Low-loss transmission lines.   The distortionless property of a transmission line is not only limited to those lines that satisfy the ideal conditions indicated in the first and second cases, but it may also be extended in an approximate way, to low-loss lines. For low-loss transmission lines, the conditions $r \ll \omega\ell$ and $g \ll \omega c$ are satisfied, and the propagation constant may be approximated by

$$\hat{\gamma} = \alpha + j\beta = \sqrt{(r + j\omega\ell)(g + j\omega c)}$$

$$= j\omega\sqrt{\ell c}\sqrt{\left(1 + \frac{r}{j\omega\ell}\right)\left(1 + \frac{g}{j\omega c}\right)}$$

$$= j\omega\sqrt{\ell c}\left(1 + \frac{r}{j\omega\ell}\right)^{\frac{1}{2}}\left(1 + \frac{g}{j\omega c}\right)^{\frac{1}{2}}$$

Using the binomial expansion, we obtain

$$\hat{\gamma} = \alpha + j\beta \approx j\omega\sqrt{\ell c}\left(1 + \frac{r}{2j\omega\ell}\right)\left(1 + \frac{g}{2j\omega c}\right)$$

$$\approx j\omega\sqrt{\ell c}\left[1 + \frac{1}{j2\omega}\left(\frac{r}{\ell} + \frac{g}{c}\right)\right]$$

$$\therefore \beta = \omega\sqrt{\ell c}$$

and

$$\alpha \simeq \frac{1}{2}\left(r\sqrt{\frac{c}{\ell}} + g\sqrt{\frac{\ell}{c}}\right)$$

The phase velocity is, hence,

$$u_p = \frac{\omega}{\beta} \simeq \frac{1}{\sqrt{\ell c}}$$

The attenuation constant α and the phase velocity are thus approximately independent of the frequency, and the transmission line may be characterized as distortionless.

The characteristic impedance of the line in this case is given by

$$Z_o = \sqrt{\frac{r + j\omega\ell}{g + j\omega c}}$$

$$\simeq \sqrt{\frac{\ell}{c}}$$

because $r \ll \omega\ell$ and $g \ll \omega c$. Similar results may be obtained by expanding both the numerator and denominator in binomial expansions.

EXAMPLE 7.7

A coaxial cable has the following constants: $r = 32\ \Omega/\text{km}$, $\ell = 1.4\ \text{mH/km}$, $c = 88\ \text{nF/km}$, and g is negligible.

1. Show that this cable is a distortionless transmission line in the frequency range from 1 to 20 MHz.
2. Calculate its characteristic impedance and propagation constant at 20 MHz.

Solution

1. For this lossy cable to be distortionless, r should be much smaller than $\omega\ell$, even at the lowest frequency of interest.

$$\omega\ell \text{ at } 1 \text{ MHz} = 2\pi \times 10^6 \times 1.4 \times 10^{-3}$$

$$= 2.8\pi \times 10^3\ \Omega/\text{km}$$

$\omega\ell$ is thus much larger than r at any frequency in the band of interest, and the transmission line is distortionless.

2. The characteristic impedance for distortionless line Z_o is given by

$$Z_o = \sqrt{\frac{\ell}{c}} = 126.13\ \Omega$$

The phase constant $\beta = \omega\sqrt{\ell c} = 1394.81$ rad/km and the attenuation constant α for $g = 0$ is given by

$$\alpha = \frac{r}{2}\sqrt{\frac{c}{\ell}} = \frac{r}{2Z_o} = 0.21\ \text{Np/km}$$

7.11 REFLECTIONS ON TRANSMISSION LINES WITH SINUSOIDAL EXCITATION

In the previous section, it was shown that the incident voltage and current propagating along a transmission line are related by a specific ratio called the characteristic impedance, which depends on the line geometrical dimensions as well as the dielectric properties of the material between the transmission line conductors. A discontinuity along the transmission line or a termination that is different from the characteristic impedance, therefore, requires the alteration of this specific ratio between the voltage and the current, and, hence, results in reflected voltage and current waves. The magnitudes of these reflected waves depend on the disparity between the impedance of the discontinuity and the characteristic impedance of the transmission line. To account for these possible reflections, we use the general solution for the voltage and the current along transmission lines, which is given by

$$\hat{V}(z) = \hat{V}_m^+ e^{-\hat{\gamma}z} + \hat{V}_m^- e^{\hat{\gamma}z} \tag{7.51}$$

$$\hat{I}(z) = \frac{\hat{V}_m^+}{\hat{Z}_o} e^{-\hat{\gamma}z} - \frac{\hat{V}_m^-}{\hat{Z}_o} e^{\hat{\gamma}z} \tag{7.52}$$

Equations 7.51 and 7.52 are of the same form as the steady-state expressions for the electric and magnetic fields given in chapter 5. Figure 7.32 shows the analogy between waves of voltage and current in a cascaded line system, and the electric and magnetic fields associated with normally incident plane waves on a multilayered system. This analogy is useful because it allows us to use previously developed solution procedures to solve for reflections along transmission lines. For example, we may want to use the boundary value approach to solve for reflections in a tandem system of transmission lines. In this case, the current and voltage in each line of the tandem system are expressed in terms of the general forms given by equations 7.51 and 7.52 with unknown amplitudes $\hat{V}_m^+$ and $\hat{V}_m^-$ for the voltage and $\hat{I}_m^+$ and $\hat{I}_m^-$ for the current. These unknown amplitudes are then determined by enforcing the continuity of the voltage and current signals across the discontinuity. For example, at the junction between transmission lines 1 and 2 shown in Figure 7.33, it is clear that the current and the voltage at both sides of the junction are continuous. Hence,

$$\hat{V}(z^-) = \hat{V}(z^+) \tag{7.53a}$$

$$\hat{I}(z^-) = \hat{I}(z^+) \tag{7.53b}$$

where z^- and z^+ are the locations just before and after the discontinuity at z. Equations such as equation 7.53 may be applied at all the tandem junctions and used to determine the unknown amplitudes of the voltages and the currents in the various transmission lines in terms of the known applied voltage and current.

Alternatively, the general expressions of the current and voltage along the line may be expressed as

$$\hat{V}(z) = \hat{V}_m^+ e^{-\hat{\gamma}z}[1 + \hat{\Gamma}(z)] \tag{7.54a}$$

$$\hat{I}(z) = \frac{\hat{V}_m^+}{Z_o} e^{-\hat{\gamma}z}[1 - \hat{\Gamma}(z)] \tag{7.54b}$$

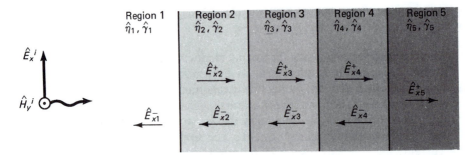

(a) Plane wave normal incidence on multiple regions

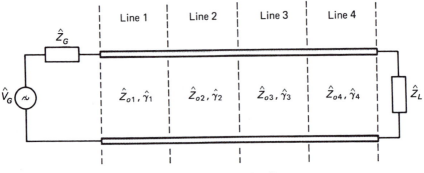

(b) Cascaded (tandem) transmission lines

Figure 7.32 Analogy between plane wave propagation in multiple regions and cascaded transmission lines.

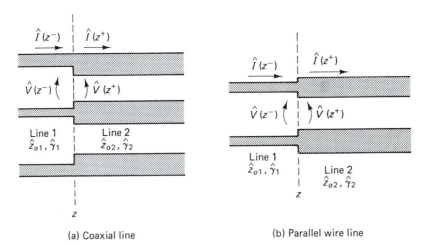

(a) Coaxial line   (b) Parallel wire line

Figure 7.33 Continuity of $\hat{V}$ and $\hat{I}$ at a junction separating two different transmission lines connected at z.

where $\hat{\Gamma}(z)$ is the complex reflection coefficient at any location z along a transmission line of characteristic impedance Z_o. $\hat{\Gamma}(z)$ is given by

$$\hat{\Gamma}(z) = \frac{\hat{V}_m^-}{\hat{V}_m^+} e^{2\hat{\gamma}z} \tag{7.55}$$

The ratio between the *total* voltage and current along the transmission line is also defined as the total line impedance $\hat{Z}(z)$, hence,

$$\hat{Z}(z) = \frac{\hat{V}(z)}{\hat{I}(z)} = \hat{Z}_o \frac{1 + \hat{\Gamma}(z)}{1 - \hat{\Gamma}(z)} \tag{7.56}$$

Equation 7.56 provides a useful relationship between the total line impedance $\hat{Z}(z)$ at some location z along the line, and the reflection coefficient $\hat{\Gamma}(z)$ at the same location. Alternatively, the reflection coefficient in equation 7.56 may be expressed in terms of the total line impedance as

$$\hat{\Gamma}(z) = \frac{\hat{Z}(z) - \hat{Z}_o}{\hat{Z}(z) + \hat{Z}_o}$$

The reflection coefficient at a location z' may be expressed in terms of its value $\hat{\Gamma}(z)$ at z, where z' and z are within the same transmission line section by

$$\hat{\Gamma}(z') = \hat{\Gamma}(z) e^{2\hat{\gamma}(z' - z)} \tag{7.57}$$

A summary of the similarities between the plane wave propagation in a multilayered region and the cascaded transmission line is given in Table 7.1. With the introduction of these appropriate definitions, let us now summarize the systematic procedure for solving a cascaded system of transmission lines such as that shown in Figure 7.34.

1. Calculate the characteristic impedances $\hat{Z}_{o1}, \hat{Z}_{o2}, \ldots$ and the propagation constants $\hat{\gamma}_1, \hat{\gamma}_2, \ldots$ as well as the electrical lengths $d_1/\lambda_1, d_2/\lambda_2, \ldots$ of all the lines. For example, for line 1:

$$\hat{Z}_{o1} = \sqrt{\frac{r_1 + j\omega\ell_1}{g_1 + j\omega c_1}}$$

$$\hat{\gamma}_1 = \sqrt{(r_1 + j\omega\ell_1)(g_1 + j\omega c_1)} = \alpha_1 + j\beta_1$$

and

$$\frac{d_1}{\lambda_1} = \frac{d_1}{2\pi/\beta_1} = \frac{d_1\beta_1}{2\pi}$$

$r_1, \ell_1, c_1,$ and g_1 are the characteristic parameters of transmission line 1.

2. To avoid carrying cumbersome phase coefficients throughout the calculations, we employ a separate origin in each transmission line. All origins are customarily located at the end of each line as shown in Figure 7.34. The location of a point along any one line is measured with respect to the origin of this line.

3. Start from the load where the ratio between the phasor voltage and current is known and is equal to the load impedance

$$\frac{\hat{V}_L}{\hat{I}_L} = \hat{Z}_L$$

At the junction between the load and the last transmission line (line N), the total voltage and current should be continuous. Instead of enforcing the continuity of the voltage and the current, we may enforce the continuity of the total impedance at the junction

$$\hat{V}_N(O_N) = \hat{V}_L$$
$$\hat{I}_N(O_N) = \hat{I}_L$$

Hence,

$$\hat{Z}_N(O_N) = \hat{Z}_L$$

It should be noted that at any junction located at z between two transmission lines, the following equation should be satisfied:

$$\hat{Z}_1(z^-) = \hat{Z}_2(z^+) \tag{7.58}$$

TABLE 7.1 SIMILARITIES BETWEEN ELECTRIC AND MAGNETIC FIELD EXPRESSIONS IN PLANE WAVE PROPAGATING IN MULTILAYERED REGION, AND VOLTAGE AND CURRENT EXPRESSIONS IN CASCADED TRANSMISSION LINES

Plane wave propagation in multilayered region	Cascaded transmission lines
1. Complex electric field expression in the ith region: $$\hat{E}_x(z) = \hat{E}_{mi}^+ e^{-\hat{\gamma}_i z}[1 + \hat{\Gamma}_i(z)]$$	1. Voltage expression in the ith transmission-line section: $$\hat{V}(z) = \hat{V}_{mi}^+ e^{-\hat{\gamma}_i z}[1 + \hat{\Gamma}_i(z)]$$
2. Complex magnetic field expression in the ith region: $$\hat{H}_y = \frac{\hat{E}_{mi}^+}{\hat{\eta}_i} e^{-\hat{\gamma}_i z}[1 - \hat{\Gamma}_i(z)]$$	2. Current expression in the ith transmission-line section: $$\hat{I}(z) = \frac{\hat{V}_{mi}^+}{\hat{Z}_{oi}} e^{-\hat{\gamma}_i z}[1 - \hat{\Gamma}_i(z)]$$
3. Total field impedance: $$\hat{Z}_i(z) = \frac{\hat{E}_x(z)}{\hat{H}_y(z)} = \hat{\eta}_i \frac{1 + \hat{\Gamma}_i(z)}{1 - \hat{\Gamma}_i(z)}$$	3. Total transmission-line impedance: $$\hat{Z}_i(z) = \frac{\hat{V}(z)}{\hat{I}(z)} = \hat{Z}_{oi} \frac{1 + \hat{\Gamma}_i(z)}{1 - \hat{\Gamma}_i(z)}$$
4. Boundary conditions at the interfaces between the ith and $i + 1$ regions: $$\hat{E}_{x_i}(z^-) = \hat{E}_{x_{i+1}}(z^+)$$ $$\hat{H}_{y_i}(z^-) = \hat{H}_{y_{i+1}}(z^+)$$ or equivalently $$\hat{Z}_i(z^-) = \hat{Z}_{i+1}(z^+)$$	4. Boundary conditions at the junction between the ith and $i + 1$ transmission-line sections $$\hat{V}_i(z^-) = \hat{V}_{i+1}(z^+)$$ $$\hat{I}_i(z^-) = \hat{I}_{i+1}(z^+)$$ or equivalently $$\hat{Z}_i(z^-) = \hat{Z}_{i+1}(z^+)$$

5. Other expressions that are common to both cases:

$$\hat{\Gamma}_i(z) = \frac{\hat{Z}_i(z) - \hat{\eta}_i}{\hat{Z}_i(z) + \hat{\eta}_i}$$

$$\hat{\Gamma}_i(z') = \hat{\Gamma}_i(z) e^{2\hat{\gamma}_i(z' - z)}$$

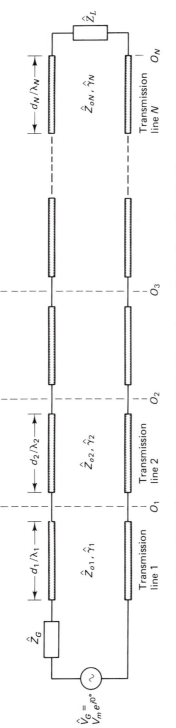

Figure 7.34 Tandem connection of N sections of transmission lines. Each line has a characteristic impedance $\hat{Z}_{oi}$, propagation constant $\hat{\gamma}_i$, and origin O_i.

This equation is an expression of the continuity of the voltage and the current across two transmission lines. It should, therefore, be used to make transitions from one line to another. It is to be emphasized, however, that because $\hat{\Gamma}_1(z^-)$ and $\hat{\Gamma}_2(z^+)$ are calculated in two different transmission lines on both sides of the junction—that is,

$$\hat{\Gamma}_1(z^-) = \frac{\hat{Z}_1(z^-) - \hat{Z}_{o1}}{\hat{Z}_1(z^-) + \hat{Z}_{o1}}$$

and

$$\hat{\Gamma}_2(z^+) = \frac{\hat{Z}_2(z^+) - \hat{Z}_{o2}}{\hat{Z}_2(z^+) + \hat{Z}_{o2}}$$

and because $\hat{Z}_1(z^-) = \hat{Z}_2(z^+)$ at the junction, hence,

$$\hat{\Gamma}_1(z^-) \neq \hat{\Gamma}_2(z^+)$$

The reflection coefficient, therefore, is an inadequate parameter to be used for crossing junctions between transmission lines. Instead, the total line impedance $\hat{Z}(z)$ should be used to make such a transition between various transmission lines because it automatically satisfies the continuity of the voltage and the current across the junction.

4. From step 3 we calculated $\hat{Z}_N(O_N) = \hat{Z}_L$. To relate the voltages and currents along the transmission line system to the known input conditions, we need to determine the input impedance as seen by the generator. In other words, the known impedance at the load should appropriately be transformed to the input conditions as seen by the generator. To do this, we first calculate the reflection coefficient at the load end of the transmission line

$$\hat{\Gamma}_N(O_N) = \frac{\hat{Z}_N(O_N) - \hat{Z}_{oN}}{\hat{Z}_N(O_N) + \hat{Z}_{oN}}$$

where $\hat{Z}_{oN}$ is the characteristic impedance of the Nth section of the transmission-line system. The reflection coefficient at the input end of transmission line N (i.e., at $z = -d_N$) is then calculated using equation

$$\hat{\Gamma}_N(-d_N) = \hat{\Gamma}_N(O_N)\, e^{2\hat{\gamma}_N(-d_N - 0)} \tag{7.59}$$

where $\hat{\gamma}_N$ is the propagation constant in the Nth transmission line.

5. With the calculation of $\hat{\Gamma}_N(-d_N)$ described in step 4, we are now ready to cross the junction between transmission lines N and $N-1$. As indicated in step 3, we have to use the total transmission-line impedance to cross such a junction. Hence,

$$\hat{Z}_N(-d_N) = \hat{Z}_{oN} \frac{1 + \hat{\Gamma}_N(-d_N)}{1 - \hat{\Gamma}_N(-d_N)}$$

and

$$\hat{Z}_{N-1}(O_{N-1}) = \hat{Z}_N(-d_N) \tag{7.60}$$

6. To calculate the input impedance of the tandem transmission-line system as seen by the generator, we repeat steps 3 and 4 for the other sections of transmission lines $N-1$ to 1. In other words, we relate the reflection coefficient at the load end of

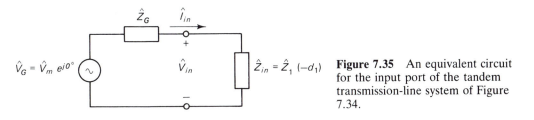

Figure 7.35 An equivalent circuit for the input port of the tandem transmission-line system of Figure 7.34.

each line to its input end value using equation 7.59 and cross the junction between various lines using the total line impedances as given by equation 7.60. Ultimately we will obtain a value of the input impedance of the tandem system $\hat{Z}_{in} = \hat{Z}_1(-d_1)$, which may be used in the equivalent circuit of the generator port shown in Figure 7.35. The input voltage, current, and the input average power to the tandem transmission-line system are then

$$\hat{V}_{in} = \frac{\hat{V}_G}{\hat{Z}_G + \hat{Z}_{in}} \hat{Z}_{in}$$

$$\hat{I}_{in} = \frac{\hat{V}_G}{\hat{Z}_G + \hat{Z}_{in}}$$

and the time-average input power is given by

$$P_{ave}^{in} = \frac{1}{2} Re(\hat{V}_{in} \hat{I}_{in}^*)$$

7. These input voltages and currents may then be used to calculate the voltage or the current distributions in any section of the tandem system. For example, to calculate the current distribution in transmission line 1, we use the general current expression evaluated at the input port—that is,

$$\hat{I}_{in} = \hat{I}_1(-d_1) = \hat{I}_{m1}^+ e^{-\hat{\gamma}_1(-d_1)}[1 - \hat{\Gamma}_1(-d_1)] \tag{7.61}$$

where $\hat{\Gamma}_1(-d_1)$ is known from the input impedance calculations (steps 1 to 6) and $\hat{I}_{m1}^+$ is an unknown amplitude of the current distribution in transmission line 1.

From equation 7.61, we determine $\hat{I}_{m1}^+$ as

$$\hat{I}_{m1}^+ = \frac{\hat{I}_{in}}{e^{\hat{\gamma}_1 d_1}[1 - \hat{\Gamma}_1(-d_1)]}$$

Knowledge of $\hat{I}_{m1}^+$ allows us to determine the current distribution at any location $(-z)$ along transmission line 1, in terms of the known reflection coefficient at the same location. Hence,

$$\hat{I}(-z) = \hat{I}_{m1}^+ e^{-\hat{\gamma}_1(-z)}[1 - \hat{\Gamma}_1(-z)] \tag{7.62}$$

Similarly, we determine the amplitude of the voltage distribution $\hat{V}_{m1}^+$ using the input conditions

$$\hat{V}_{m1}^+ = \frac{\hat{V}_{in}}{e^{-\hat{\gamma}_1(-d_1)}[1 + \hat{\Gamma}_1(-d_1)]}$$

and $\hat{V}(-z)$ at any location $-z$ is given by

$$\hat{V}(-z) = \hat{V}_{m1}^{+} e^{-\hat{\gamma}(-z)}[1 + \hat{\Gamma}_1(-z)] \tag{7.63}$$

To calculate the current and voltage distributions in transmission line 2, we use equations 7.62 and 7.63 to calculate $\hat{V}$ and $\hat{I}$ at O_1, $z = 0$, which is the location of the junction between transmission lines 1 and 2

$$\hat{V}_1(O_1) = \hat{V}_{m1}^{+}[1 + \hat{\Gamma}_1(O_1)]$$
$$\hat{I}_1(O_1) = \hat{I}_{m1}^{+}[1 - \hat{\Gamma}_1(O_1)]$$

From the continuity of the voltage and the current at the junction, we have

$$\hat{V}_2(-d_2) = \hat{V}_1(O_1) \tag{7.64a}$$
$$\hat{I}_2(-d_2) = \hat{I}_1(O_1) \tag{7.64b}$$

Therefore, the amplitudes of the voltage and the current in transmission line 2 may be determined from

$$\hat{V}_2(-d_2) = \hat{V}_{m2}^{+} e^{-\hat{\gamma}_2(-d_2)}[1 + \hat{\Gamma}_2(-d_2)]$$
$$\hat{I}_2(-d_2) = \hat{I}_{m2}^{+} e^{-\hat{\gamma}_2(-d_2)}[1 - \hat{\Gamma}_2(-d_2)]$$

Because $\hat{\Gamma}_2(-d_2)$ is known from our calculation of $\hat{Z}_{in}$, the amplitudes $\hat{V}_{m2}^{+}$ and $\hat{I}_{m2}^{+}$ are obtained from

$$\hat{V}_{m2}^{+} = \frac{\hat{V}_2(-d_2)}{e^{-\hat{\gamma}_2(-d_2)}[1 + \hat{\Gamma}_2(-d_2)]}$$

$$\hat{I}_{m2}^{+} = \frac{\hat{I}_2(-d_2)}{e^{-\hat{\gamma}_2(-d_2)}[1 - \hat{\Gamma}_2(-d_2)]}$$

Of course, $\hat{I}_{m2}^{+}$ and $\hat{V}_{m2}^{+}$ are related by $\hat{Z}_{o2}$, the characteristic impedance of the transmission line 2.

Depending on the number of transmission lines, the preceding procedure may be repeated as many times as needed to calculate the voltage and the current across the load. For our cascaded system shown in Figure 7.34,

$$\hat{V}_L = \hat{V}_{mN}^{+}[1 + \hat{\Gamma}_N(O_N)]$$

$$\hat{I}_L = \frac{\hat{V}_L}{\hat{Z}_L} = \hat{I}_{mN}^{+}[1 - \hat{\Gamma}_N(O_N)]$$

where $\hat{V}_{mN}^{+}$ and $\hat{I}_{mN}^{+}$ are the amplitudes of the voltage and current on transmission line N. $\hat{\Gamma}_N$ is the reflection coefficient in the Nth transmission line at the origin O_N.

The time-average power delivered to the input port of the transmission lines system is

$$P_{in} = \frac{1}{2} Re(\hat{V}_{in} \hat{I}_{in}^{*}) = \frac{1}{2} Re(\hat{I}_{in} \hat{I}_{in}^{*} \hat{Z}_{in})$$

$$= \frac{1}{2} |\hat{I}_{in}|^2 Re(\hat{Z}_{in})$$

The time-average power delivered to the load is

$$P_L = \frac{1}{2} Re(\hat{V}_L \hat{I}_L^*) = \frac{1}{2} |\hat{I}_L|^2 Re(\hat{Z}_L)$$

This systematic solution procedure for solving for tandem connection of transmission line will be illustrated by the following examples.

EXAMPLE 7.8

A 40-mile transmission line is connected to a generator of $\hat{V}_G = 100e^{j0°}$ V, $Z_G = 700\ \Omega$, and frequency of 1 kHz. On the other end, the transmission line is terminated by a $400 + j180\ \Omega$ load. If the characteristic impedance and the propagation constant of the transmission line are given by

$$\hat{Z}_o = 375e^{j12°}\ \Omega, \qquad \hat{\gamma} = 0.009\ (\text{Np/mile}) + j0.04\ (\text{rad/mile})$$

Determine the following:

1. Input impedance of transmission line.
2. Voltage and current at input port of transmission line.
3. Current at load and power delivered to load.

Solution

The transmission-line geometry is shown in Figure 7.36.

1. The step-by-step solution is as follows:
    (a) $\hat{Z}(0) = \hat{Z}_L = 400 + j180\ \Omega$

    (b) $\hat{\Gamma}(0) = \dfrac{\hat{Z}(0) - \hat{Z}_o}{\hat{Z}(0) + \hat{Z}_o}$ where $\hat{Z}_o$ is the characteristic impedance of the transmission

    line and $\hat{Z}(0)$ is the impedance at $z = 0$,

    $$\hat{\Gamma}(0) = 0.079 + j0.106 = 0.132\underline{/53.3°}$$

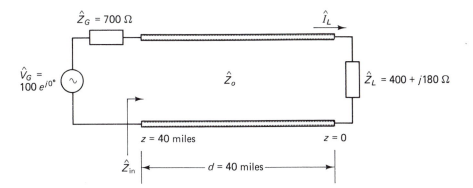

Figure 7.36 Transmission-line geometry of example 7.8.

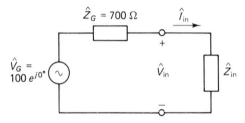

Figure 7.37 The equivalent circuit of the input port of the transmission line of example 7.8.

(c) Using equation 7.59, we obtain $\hat{\Gamma}(-d)$ from $\hat{\Gamma}(0)$ as

$$\hat{\Gamma}(-d) = \hat{\Gamma}(0)\, e^{-2\alpha d}\, e^{-2j\beta d}$$

Substituting $\alpha = 0.009$ Np/mile, $\beta = 0.04$ rad/mile and $d = 40$ miles, we obtain

$$\hat{\Gamma}(-d) = 0.064\underline{/-130°}$$

(d) $\hat{Z}(-d) = \hat{Z}_{in} = \hat{Z}_o \dfrac{1 + \hat{\Gamma}(-d)}{1 - \hat{\Gamma}(-d)} = 345.375\underline{/6.38°}$

(e) The equivalent circuit of the input port of the transmission line is shown in Figure 7.37.

2. From the equivalent circuit of input port we have,

$$\hat{I}_{in} = \frac{\hat{V}_G}{\hat{Z}_G + \hat{Z}_{in}} = 0.1\underline{/-1.9 \times 10^{-6°}} \approx 0.1 \text{ A}$$

$$\hat{V}_{in} = \hat{I}_{in}\hat{Z}_{in} = 34.5\underline{/6.38°}$$

3. The current and power at the load can be calculated from,

$$\hat{I}_{in} = \hat{I}(-d) = \hat{I}_m^+ e^{-\alpha(-d)} e^{-j\beta(-d)}[1 - \hat{\Gamma}(-d)]$$

$$\therefore \hat{I}_m^+ = 0.067\, e^{-j94.42°} \text{ A}$$

Knowledge of $\hat{I}_m^+$ allows us to calculate the current at any location along the transmission line including at $z = 0$. Hence,

$$\hat{I}(0) = \hat{I}_L = \hat{I}_m^+ e^{-\alpha(0)} e^{-j\beta(0)}[1 - \hat{\Gamma}(0)]$$

$$= 0.067\, e^{-j94.42°}[1 - (0.079 + j0.106)]$$

$$= 0.062\, e^{-j101.0°}$$

The power delivered to the load is given by

$$P_L = \frac{1}{2} Re(\hat{V}_L\, \hat{I}_L^*) = \frac{1}{2}|\hat{I}_L|^2\, Re(\hat{Z}_L)$$

$$= \frac{1}{2}(0.062)^2\, 400$$

$$= 0.769 \text{ W}$$

7.12 USE OF SMITH CHART

Some of the calculations involved in the reflection coefficient $\hat{\Gamma}(z)$ and the total transmission line impedance $\hat{Z}(z)$ solution procedure may be made graphically using the Smith chart. Basically the Smith chart may help us calculate the normalized impedance from a known value of the reflection coefficient and to calculate the reflection coefficient $\hat{\Gamma}(z)$ at one location from its known value $\hat{\Gamma}(z')$ somwhere else along the same transmission line. In other words, the various calculations included in steps 3 to 5 of the systematic solution procedure described in the previous section may be conveniently made using the Smith chart. The following example will illustrate the use of the Smith chart for making these calculations.

EXAMPLE 7.9

Two transmission lines are connected in tandem with a bad lossy connector. The transmission lines and the equivalent circuit of the connector are shown in Figure 7.38. The tandem system is connected to a load of impedance $\hat{Z}_L = 75 - j125 \ \Omega$ and to a generator of output voltage $\hat{V}_G = 100 \, e^{j0°}$ and an internal resistance equals to 180 Ω. Determine the power delivered to the load.

Solution

The step-by-step solution procedure may be indicated as follows:

1. From the continuity of the total impedance at the junction between transmission line 2 and the load, we have

$$\hat{Z}_2(O_2) = \hat{Z}_L = 75 - j125 \ \Omega$$

2. To move along transmission line 2 toward the generator, we need to obtain the reflection coefficient at O_2 and then use this value to obtain the reflection coefficient at $z = -0.4\lambda_2$, just before the connector. $\hat{\Gamma}_2(O_2)$ may be obtained using the Smith

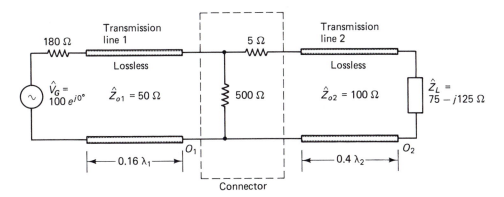

Figure 7.38   Tandem connection of two transmission lines through a lossy connector.

chart simply by calculating the normalized impedance $\hat{z}_{n2}(O_2)$ and plotting it on the Smith chart. Hence,

$$\hat{z}_{n2}(O_2) = \frac{\hat{Z}_2(O_2)}{\hat{Z}_{o2}} = 0.75 - j1.25$$

$\hat{z}_{n2}(O_2)$, when plotted on the Smith chart shown in Figure 7.39a, gives point A.

3. $\hat{\Gamma}_2(-0.4\lambda_2)$ is obtained graphically on the Smith chart by rotating $\hat{\Gamma}_2(O_2)$ a distance $0.4\lambda_2$ toward the generator (point B). It should be emphasized that if transmission line 2 is lossy, the magnitude of $\hat{\Gamma}_2(-0.4\lambda_2)$ would be different from that of $\hat{\Gamma}_2(O_2)$. From equation 7.57 it is shown that

$$\hat{\Gamma}_2(-0.4\lambda_2) = \hat{\Gamma}_2(O_2)\, e^{2\alpha_2(-0.4\lambda_2 - 0)}\, e^{2j\beta_2(-0.4\lambda_2 - 0)}$$

where α_2 and β_2 are the attenuation and phase constants of transmission line 2, respectively. Although the phase term $e^{2j\beta_2(-0.4\lambda_2)}$ may be routinely taken into account by the rotation on the Smith chart, the attenuation term $e^{2\alpha_2(-0.4\lambda_2)}$ should be calculated separately and the magnitude of $\hat{\Gamma}_2(-0.4\lambda_2)$ be modified accordingly.

4. From the Smith chart shown in Figure 7.39a we read the value of $\hat{z}_{n2}(-0.4\lambda_2)$ at point B

$$\hat{z}_{n2}(-0.4\lambda_2) = 3.8 + j0.65$$

The denormalized value of the total impedance is, hence,

$$\hat{Z}_2(-0.4\lambda_2) = \hat{z}_{n2}(-0.42\lambda_2)\,\hat{Z}_{o2} = 380 + j65 \ \Omega$$

5. To cross the junction between the transmission lines 1 and 2, the connector equivalent circuit should be combined with $\hat{Z}_2(-0.4\lambda_2)$ to give the value of the impedance loading transmission line 1. Hence,

$$\hat{Z}_1(O_1) = \frac{[(380 + j65) + 5] \times 500}{[(380 + j65) + 5] + 500} = 219 + j20.6 \ \Omega$$

6. To move a distance $0.16\lambda_1$ from O_1 to the input port of transmission line 1, we normalize $\hat{Z}_1(O_1)$ with respect to $\hat{Z}_{o1}$, plot the normalized $\hat{z}_{n1}(O_1)$ on the Smith chart, and then rotate the resulting reflection coefficient $\hat{\Gamma}_1(O_1)$ a distance $0.16\lambda_1$ toward the generator. Once again, because transmission line 1 is lossless no attenuation calculations need to be considered and $\hat{\Gamma}_1(-0.16\lambda_1)$ may be obtained from $\hat{\Gamma}_1(O_1)$ just through the rotation of a distance $0.16\lambda_1$ on the Smith chart. Hence,

$$\hat{z}_{n1}(O_1) = \frac{\hat{Z}_1(O_1)}{\hat{Z}_{o1}} = 4.38 + j0.41$$

Plot $\hat{z}_{n1}(O_1)$ on the Smith chart (point C in Figure 7.39a), and rotate the resulting reflection coefficient $\hat{\Gamma}_1(O_1)$ a distance $0.16\lambda_1$ toward the generator point (D). The normalized impedance at D is

$$\hat{z}_{n1}(-0.16\lambda_1) = 0.33 - j0.63$$

$\hat{Z}_{in}$ is, hence,

$$\hat{Z}_{in} = \hat{Z}_{o1}\,\hat{z}_{n1}(-0.16\lambda_1) = 16.5 - j31.5 \ \Omega$$

7. The equivalent circuit at the input port is shown in Figure 7.39b.

$$\hat{V}_{in} = \frac{100^{j0}\,\hat{Z}_{in}}{\hat{Z}_{in} + R_G} = 17.87\, e^{-j53.24} \ V$$

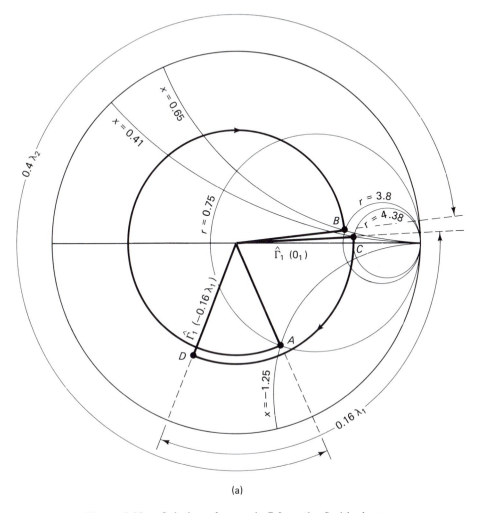

(a)

Figure 7.39a   Solution of example 7.9 on the Smith chart.

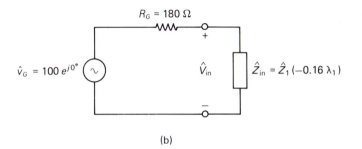

(b)

Figure 7.39b   The equivalent circuit at the input port of transmission line 1.

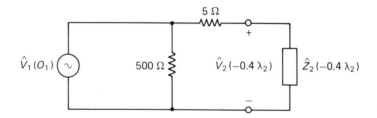

Figure 7.40   The equivalent circuit at the connector between transmission lines 1 and 2. $\hat{V}_1(O_1)$ is the voltage at the end of transmission line 1, and $\hat{V}_2(-0.4\lambda_2)$ is the input voltage to transmission line 2.

8. The voltage distribution in transmission line 1 is given by

$$\hat{V}_1(z) = \hat{V}_m^+ e^{-j\beta_1 z}[1 + \hat{\Gamma}_1(z)]$$

Because $\hat{V}_{in} = \hat{V}_1(-0.16\lambda_1)$, we use this known value of the voltage to determine the amplitude of the voltage distribution $\hat{V}_{m1}^+$

$$\hat{V}_{m1}^+ = \frac{\hat{V}_{in}}{e^{-j\beta_1(-0.16\lambda_1)}[1 + \hat{\Gamma}_1(-0.16\lambda_1)]} = 18.35\, e^{-j196°}\text{ V}$$

The voltage at the end of transmission line 1 is then obtained as

$$\hat{V}_1(O_1) = \hat{V}_{m1}^+[1 + \hat{\Gamma}_1(O_1)] = 29.91\, e^{j196.8°}\text{ V}$$

It should be noted that the values of $\hat{\Gamma}_1(O_1)$ and $\hat{\Gamma}_1(-0.16\lambda_1)$ are both known from the first part of the solution involving the calculation of $\hat{Z}_{in}$ using the Smith chart.

9. To determine the voltage transmitted to transmission line 2, we use the equivalent circuit shown in Figure 7.40. The input voltage to transmission line 2 denoted by $\hat{V}_2(-0.4\lambda_2)$ is given using a voltage divider as

$$\hat{V}_2(-0.4\lambda_2) = \frac{\hat{V}_1(O_1)\,\hat{Z}_2(-0.4\lambda_2)}{\hat{Z}_2(-0.4\lambda_2) + 5} = 29.53\, e^{j196.9°}\text{ V}$$

10. At this stage, we either determine the voltage at the load $\hat{V}_L$ following a procedure similar to that indicated in steps 7 and 8, and then determine the load power P_L from $\hat{V}_L$, or calculate the input power to transmission line 2 as

$$P_{in_2} = \frac{1}{2}Re[\hat{V}_2^*(-0.4\lambda_2)\,\hat{I}_2(-0.4\lambda_2)]$$

$$= \frac{1}{2}|\hat{V}_2(-0.4\lambda_2)|^2\, Re\frac{1}{\hat{Z}_2(-0.4\lambda_2)}$$

P_{in_2} is the same as P_L because transmission line 2 is lossless. In either case, we will find that

$$P_L = 1.1\text{ W}$$

EXAMPLE 7.10

Two transmission lines are connected in parallel, and the parallel combination is connected to a third transmission line as shown in Figure 7.41a. All the transmission lines have the same characteristic impedance of 50 Ω. Determine the minimum length of transmission line 3 (in fraction of a wavelength) so that the input impedance at the A-A port is real. Also find this value of the input impedance.

Solution

1. Place the three origins, O_1, O_2, and O_3 in the three transmission lines 1, 2, and 3, respectively.

2. $\hat{z}_{nd_1}(O_1) = \dfrac{50 + j75}{50} = 1 + j1.5$

   $\hat{z}_{nd_2}(O_2) = \dfrac{75 - j125}{50} = 1.5 - j2.5$

3. Plot $\hat{z}_{nd_1}$ point A and $\hat{z}_{nd_2}$ point B on the Smith chart as shown in Figure 7.41b.

4. Because transmission line 1 is lossless, $\hat{\Gamma}_1(-0.5\lambda_1)$ may be obtained from $\hat{\Gamma}_1(O_1)$ simply by rotating a distance $0.5\lambda_1$ toward the generator. A distance of $0.5\lambda_1$ corresponds to a complete rotation on the Smith chart and $\hat{z}_{n1}(-0.5\lambda_1) = 1.0 + j1.5$.

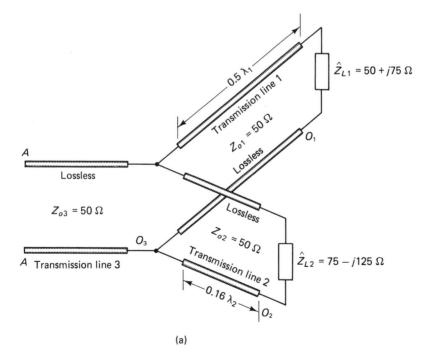

(a)

Figure 7.41a   Geometry of the problem in example 7.10.

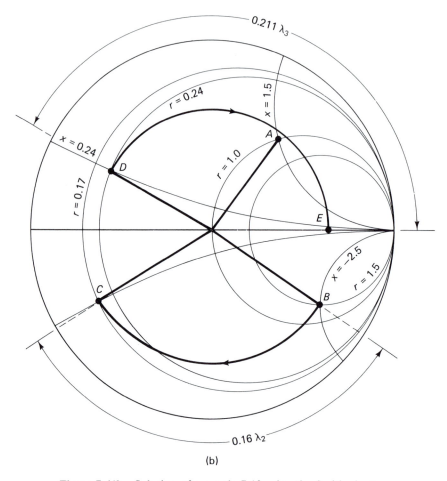

Figure 7.41b   Solution of example 7.10 using the Smith chart.

Similarly, $\hat{\Gamma}_2(-0.16\lambda_2)$ (point C) on the Smith chart of Figure 7.41b is obtained by rotating $\hat{\Gamma}_2(O_2)$ a distance $0.16\lambda_2$ toward the generator. $\hat{z}_{n2}(-0.16\lambda_2) = 0.17 - j0.27$.

5. The load impedance of transmission line 3 is the parallel combination of the denormalized impedances $\hat{Z}_1(-0.5\lambda_1)$ and $\hat{Z}_2(-0.16\lambda_2)$.

$$\hat{Z}_3(O_3) = \frac{(50 + j75)(8.5 - j13.5)}{(50 + j75) + (8.5 - j13.5)} = 12 + j12$$

6. Normalize $\hat{Z}_3(O_3)$ by dividing its value by $\hat{Z}_{o3} = 50\ \Omega$, we obtain

$$\hat{z}_{n3}(O_3) = 0.24 + j0.24$$

Plot $\hat{z}_{n3}(O_3)$ on the Smith chart, point D in Figure 7.41b.

7. It can be seen from the Smith chart that point D should be rotated toward the generator on the constant reflection coefficient circle until it intersects the zero

reactance line at E. The required length of transmission line 3 is simply the rotated distance from D to E. From the chart $\ell_3 = 0.211\lambda_3$.

8. The normalized impedance at E is $\hat{z}_{n3}(-d_3) = 4.4$. Therefore, the input impedance at the A-A port is

$$Z_{in} = \hat{Z}_3(-d_3) = (4.4)50 = 220 \ \Omega$$

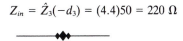

7.13 ANALYTICAL EXPRESSION OF TRANSMISSION-LINE IMPEDANCE

To examine some interesting features of the variation of the total impedance $\hat{Z}(z)$ along a transmission line of characteristic impedance $\hat{Z}_o$ when terminated by a load impedance $\hat{Z}_L$, which is, in general, different from $\hat{Z}_o$, we need to derive an analytical expression of the total transmission line impedance $\hat{Z}(z)$. Consider the transmission-line geometry shown in Figure 7.42. From equation 7.56, $\hat{Z}(-z)$ in terms of the coefficient $\hat{\Gamma}(-z)$ at the same location $(-z)$ is given by

$$\hat{Z}(-z) = \hat{Z}_o \frac{1 + \hat{\Gamma}(-z)}{1 - \hat{\Gamma}(-z)} \qquad (7.65)$$

$\hat{\Gamma}(-z)$ may be expressed in terms of $\hat{\Gamma}(0)$ at the origin as follows:

$$\hat{\Gamma}(-z) = \hat{\Gamma}(0)\, e^{2\hat{\gamma}(-z)} \qquad (7.66)$$

Substituting equation 7.65 into equation 7.66, we obtain

$$\hat{Z}(-z) = \hat{Z}_o \frac{1 + \hat{\Gamma}(0)\, e^{-2\hat{\gamma}z}}{1 - \hat{\Gamma}(0)\, e^{-2\hat{\gamma}z}} \qquad (7.67)$$

$\hat{\Gamma}(0)$ in equation 7.66 may be expressed in terms of $\hat{Z}_L$ and $\hat{Z}_o$ as

$$\hat{\Gamma}(0) = \frac{\hat{Z}_L - \hat{Z}_o}{\hat{Z}_L + \hat{Z}_o}$$

Hence,

$$
\begin{aligned}
\hat{Z}(-z) &= \hat{Z}_o \frac{1 + (\hat{Z}_L - \hat{Z}_o)/(\hat{Z}_L + \hat{Z}_o)\, e^{-2\hat{\gamma}z}}{1 - (\hat{Z}_L - \hat{Z}_o)/(\hat{Z}_L + \hat{Z}_o)\, e^{-2\hat{\gamma}z}} \\
&= \hat{Z}_o \frac{(\hat{Z}_L + \hat{Z}_o)\, e^{\hat{\gamma}z} + (\hat{Z}_L - \hat{Z}_o)\, e^{-\hat{\gamma}z}}{(\hat{Z}_L + \hat{Z}_o)\, e^{\hat{\gamma}z} - (\hat{Z}_L - \hat{Z}_o)\, e^{-\hat{\gamma}z}} \\
&= \hat{Z}_o \frac{\hat{Z}_L(e^{\hat{\gamma}z} + e^{-\hat{\gamma}z}) + \hat{Z}_o(e^{\hat{\gamma}z} - e^{-\hat{\gamma}z})}{\hat{Z}_o(e^{\hat{\gamma}z} + e^{-\hat{\gamma}z}) + \hat{Z}_L(e^{\hat{\gamma}z} - e^{-\hat{\gamma}z})} \\
&= \hat{Z}_o \frac{\hat{Z}_L \cosh(\hat{\gamma}z) + \hat{Z}_o \sinh(\hat{\gamma}z)}{\hat{Z}_o \cosh(\hat{\gamma}z) + \hat{Z}_L \sinh(\hat{\gamma}z)}
\end{aligned} \qquad (7.68)
$$

Equation 7.68 is the desired expression of the total transmission-line impedance $\hat{Z}(-z)$ at any point $(-z)$ along the line, in terms of its load impedance $\hat{Z}_L$ and the line char-

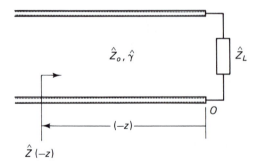

Figure 7.42 A transmission line of characteristic impedance $\hat{Z}_o$, and propagation constant $\hat{\gamma}$, terminated by a load impedance $\hat{Z}_L$. The total transmission-line impedance $\hat{Z}(-z)$ is evaluated at a distance $-z$ from the origin O.

acteristic parameters $\hat{Z}_o$ and $\hat{\gamma}$. It should be noted that equation 7.68 is also suitable for calculating the total impedance in any transmission line section in the tandem system of Figure 7.34 subject to replacing $\hat{Z}_L$ by the value of the total impedance $\hat{Z}$ loading that section (e.g., obtained using the systematic procedure described in the previous sections or any other means) and, of course, the appropriate parameters $\hat{\gamma}$ and $\hat{Z}_o$ of that section. In other words, instead of using reflection coefficients to relate $\hat{Z}$ to $\hat{Z}(-d)$ at the ends of each section of the tandem transmission lines, equation 7.68 may be alternatively used to relate these quantities. Many important special cases of equation 7.68, which will be extensively used in the remaining sections of this chapter, are related to the special case of the input impedance of a *lossless* section of transmission line terminated by an arbitrary impedance $\hat{Z}_L$. In this case Z_o is real and $\hat{\gamma} = j\beta$. Equation 7.68 reduces to

$$\hat{Z}(-z) = Z_o \frac{\hat{Z}_L \cosh(j\beta z) + Z_o \sinh(j\beta z)}{Z_o \sinh(j\beta z) + \hat{Z}_L \cosh(j\beta z)}$$

Utilizing the following relations from Appendix C,

$$\cosh(j\beta z) = \frac{e^{j\beta z} + e^{-j\beta z}}{2} = \cos(\beta z)$$

and

$$\sinh(j\beta z) = \frac{e^{j\beta z} - e^{-j\beta z}}{2} = j \sin(\beta z)$$

we obtain

$$\hat{Z}(-z) = Z_o \frac{\hat{Z}_L \cos(\beta z) + jZ_o \sin(\beta z)}{Z_o \cos(\beta z) + j\hat{Z}_L \sin(\beta z)} \tag{7.69}$$

For this special case of practical interest, we will examine the variation of $\hat{Z}(-z)$ along the transmission line (i.e., as a function of z) for a variety of load conditions $\hat{Z}_L$.

7.13.1 Short-Circuit Termination of Lossless Section of Transmission Line

Substituting $Z_L = 0$ in equation 7.69, we obtain

$$\hat{Z}(-z) = jZ_o \tan(\beta z) \tag{7.70}$$

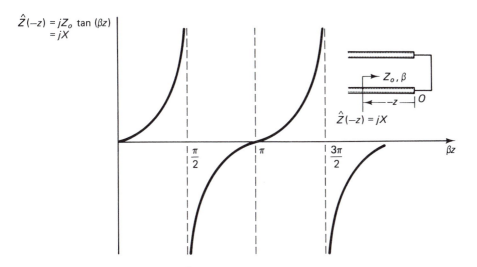

Figure 7.43 The variation of $\hat{Z}(-z)$ as a function of $(-z)$, the distance along the transmission line, for a short-circuit-terminated lossless line. Note that $Z(-z)$ is purely reactive, and its value varies from 0 to $\pm\infty$, depending on the location (values of z along the line).

Because the characteristic impedance Z_o is real for a lossless line, $\hat{Z}(-z)$ represents a reactive input impedance with zero real or resistive part. Furthermore, examining the variation of $\hat{Z}(-z)$ along the transmission line—that is, as a function of z—shows that $\hat{Z}(-z)$ may attain any value from $-\infty$ to ∞. Figure 7.43 illustrates such a variation.

To summarize our observations thus far, we note that $\hat{Z}(-z)$ is a purely reactive input impedance (zero resistive part), and that by varying the length $(-z)$ of the short-circuited transmission line, we may obtain any value of inductive reactance (positive X) from 0 to ∞, and also any value of capacitive reactance (negative X) from 0 to $-\infty$. Varying the length of the line between 0 to $\lambda/4$ (i.e., βz from 0 to $\pi/2$) would provide values of inductive reactance from 0 to ∞, and varying the length from $\lambda/4$ to $\lambda/2$ (i.e., βz from $\pi/2$ to π) would provide capacitive reactances from $-\infty$ to 0. Hence, varying the length of a lossless transmission line from 0 to $\lambda/2$ would provide all possible values of inductive and capacitive reactances from $X = 0$ to $X = \pm\infty$. This information will be used in discussing the stub matching in the following sections.

7.13.2 Open-Circuit Section of Lossless Transmission Line

Substituting the condition $Z_L = \infty$ in equation 7.69, we obtain

$$\hat{Z}(-z) = -jZ_o \cot(\beta z) \qquad (7.71)$$

which is, once again, a purely reactive input impedance. The variation of $\hat{Z}(-z)$ as a function of z is plotted in Figure 7.44, where it is clear that varying the length of an open-ended section of transmission line from 0 to $\lambda/2$ provides any value of reactive impedance from 0 to $\pm\infty$. Unlike the short-circuited section, however, the capacitive reactance may be obtained by varying the length from 0 to $\lambda/4$ and the inductive

$\hat{Z}(-z)$
$= -jZ_o \cot(\beta z)$
$= -jX$

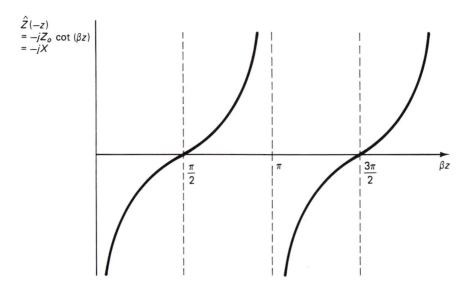

Figure 7.44   The input impedance of an open-ended section of lossless transmission line is purely reactive and varies from $X = -\infty$ to ∞ by changing the length of the line from 0 to $\lambda/2$.

reactance by varying the length from $\lambda/4$ to $\lambda/2$. An open-ended section of lossless line may, therefore, provide the same function as a short-circuited one. In practice, however, the use of a short-circuited section is preferable, because it contains the energy within the line and, hence, avoids problems owing to leakage radiation from the open end of the line.

7.13.3 Lossless Section of Transmission Line of Length $d = \lambda/4$

In the two previous special cases, we specified the load impedance $\hat{Z}_L$ and examined the variation of $\hat{Z}(-z)$ along the line or as we continuously change the length of the line. In the present case, we examine the relation between $\hat{Z}(-z)$ and an arbitrary value of $\hat{Z}_L$ for a specified length of the line $z = \lambda/4$—that is, a quarter-wavelength-long transmission line.

Substituting $z = \lambda/4$, or $\beta z = 2\pi/\lambda\,(\lambda/4) = \pi/2$ in equation 7.69, we obtain

$$\hat{Z}\left(-\frac{\lambda}{4}\right) = \frac{Z_o^2}{\hat{Z}_L} \tag{7.72}$$

Equation 7.71 may be rearranged in the form

$$\frac{\hat{Z}\left(-\dfrac{\lambda}{4}\right)}{Z_o} = \frac{1}{(\hat{Z}_L/Z_o)}$$

or

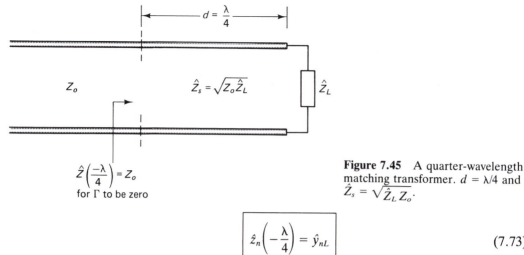

$$\hat{Z}\left(\frac{-\lambda}{4}\right) = Z_o$$
for Γ to be zero

Figure 7.45 A quarter-wavelength matching transformer. $d = \lambda/4$ and $\hat{Z}_s = \sqrt{\hat{Z}_L\, Z_o}$.

$$\hat{z}_n\left(-\frac{\lambda}{4}\right) = \hat{y}_{nL} \tag{7.73}$$

Equation 7.73 simply states that the normalized value (dimensionless) of the input impedance of a $\lambda/4$ long section of lossless transmission line is *numerically* equal to the normalized (no units) admittance of the load. This is actually an important observation, because we use such normalized impedance values on the Smith chart. If we are given the normalized impedance value on the chart and we need to obtain the normalized admittance value at that location, we simply rotate the normalized impedance point a distance $\lambda/4$ or equivalently $180°$ on the constant reflection coefficient circle on the Smith chart. Such a graphical procedure of obtaining $\hat{y}_n$ from $\hat{z}_n$ and vice versa is certainly useful when solving for transmission lines connected in parallel. Before we illustrate the usefulness of this procedure in solving transmission-line problems by solving examples, let us examine another application that is based on equation 7.72. Consider a load of an arbitrary value of impedance $\hat{Z}_L$, and it is desired to connect this load to a transmission line of characteristic impedance Z_o (which is generally different from $\hat{Z}_L$) under the condition that there will be no reflection in the transmission line. To achieve this we may connect $\hat{Z}_L$ to the transmission line of interest through a section of another transmission line of characteristic impedance $\hat{Z}_s$ and length d. If we choose the length of the connecting transmission line d to be $\lambda/4$, as shown in Figure 7.45, equation 7.72 states that the impedance of the connecting section $\hat{Z}_s$ should be

$$\hat{Z}_s^2 = \hat{Z}_L\,\hat{Z}\left(-\frac{\lambda}{4}\right) \tag{7.74}$$

For zero reflection in the transmission line, $\hat{Z}(-\lambda/4)$ should be equal to Z_o, and, hence,

$$\hat{Z}_s = \sqrt{\hat{Z}_L\, Z_o}$$

In other words, with the proper design of the connecting section such that $d = \lambda/4$ and $\hat{Z}_s = \sqrt{\hat{Z}_L\, Z_o}$, the load may be connected to the transmission line of interest Z_o without introducing any reflections. Such a procedure is known as the impedance matching using a quarter-wave transformer. This, as well as other matching procedures, will be described in the following section.

EXAMPLE 7.11

The impedance at some point along a lossless transmission line of characteristic impedance $Z_o = 50 \; \Omega$ is $\hat{Z}(-z) = 50 + j75 \; \Omega$. Determine the admittance at the same point.

Solution

The admittance may be obtained from the impedance value using the relation

$$\hat{Y} = \frac{1}{\hat{Z}} = \frac{1}{50 + j75} = 0.011 \, e^{-j56.3°} \text{ mhos} = 0.0061 - 0.00915 \text{ mhos}$$

Alternatively, we may use the Smith chart to determine $\hat{Y}$ graphically, as follows:

1. Obtain the normalized value of the impedance

$$\hat{z}_n(-z) = \frac{50 + j75}{50} = 1 + j1.5$$

2. Plot $\hat{z}_n(-z)$ on the Smith chart as shown in Figure 7.46 (point A).

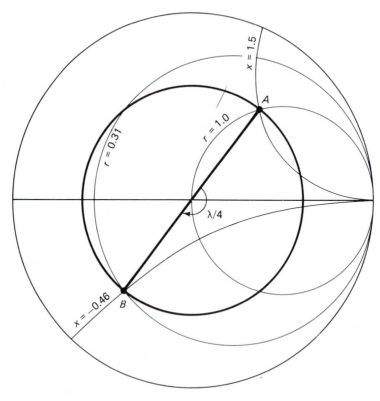

Figure 7.46 Use of the Smith chart to calculate a normalized admittance from a given value of normalized impedance.

3. Rotate A on a constant reflection coefficient circle an angle of 180° (which is equivalent to moving away from the impedance point a distance of $\lambda/4$).

4. The value of the normalized admittance at B is

$$\hat{y}_n = 0.31 - j0.46$$

5. The admittance $\hat{Y}(-z)$ is, hence,

$$\hat{Y}(-z) = \hat{y}_n \, Y_o = (0.31 - j0.46)0.02 = 0.0062 - j0.0092 \text{ mhos}$$

which is approximately the same value as obtained analytically.

As indicated earlier, this graphical procedure of obtaining $\hat{y}_n$ from $\hat{z}_n$ and vice versa may become useful in solving some transmission line problems involving parallel connection of lines.

———————◆◆◆———————

7.14 IMPEDANCE MATCHING OF LOSSLESS LINES

In many practical applications, transmission lines are often connected to loads of impedances different from the characteristic impedance of the line. We learned in the previous sections that impedance mismatches result in reflections. It is, therefore, desirable to devise means of eliminating these reflections along transmission lines, even if they were connected to loads of different impedances. The following are some of the more important reasons why the load reflections need to be eliminated or minimized:

1. *Protect generator*. Generators are often designed to deliver electromagnetic energy at specific frequencies and not to absorb or dissipate it. Excessive reflections, hence, result in damaging generators, particularly in high-power applications.

2. *Minimize distortion of transmitted signals*. Communication signals are often broadband. If there are reflections along the transmission line, these reflections are different at different frequencies. Hence, the different frequencies in the broadband communication signal would be reflected by different amounts, thus significantly increasing the signal distortion. An obvious way of overcoming this problem is to eliminate the reflection along the transmission line.

3. *Maximize power delivered to load*. If the main reason of using transmission lines is to transmit the electromagnetic energy from one location to another, the presence of reflections clearly causes loss of energy and a decrease in the transmission efficiency. The matching procedures to be described next use lossless networks; hence, by achieving the impedance matching, they basically improve the transmission efficiency.

The preceding discussion emphasizes the importance of impedance matching, and, in the following, we will describe some of the methods most commonly used to achieve such a matching.

7.14.1 Impedance Matching Using Quarter-Wavelength Transformers

In discussing the analytical expression of the impedance along lossless transmission lines in section 7.13, we indicated in the special third case that by inserting a $\lambda/4$ section of connecting transmission line between the load, which may be of arbitrary impedance, and the main transmission line of interest, impedance matching will be achieved if the impedance of the connecting section is chosen to be

$$\hat{Z}_s = \sqrt{Z_o \hat{Z}_L}$$

where $\hat{Z}_L$ and Z_o are the impedances of the load and the main transmission line, respectively, and $\hat{Z}_s$ is the impedance of the transformation (connecting) section, as shown in Figure 7.47. The main disadvantage of this impedance-matching procedure is the narrow band matching. This is because perfect impedance matching occurs at only one frequency at which the length of the matching section is $\lambda/4$. The procedure for achieving impedance matching at several frequencies may require replacing the transformer section at each frequency or connecting a few of these sections in tandem. In either case, the procedure may be tedious, certainly limited, and, hence, undesirable. Another limitation of the quarter-wavelength transformer impedance matching may be explained if we notice that if the load has a complex impedance, the matching section should also have a complex characteristic impedance. From the expression of the characteristic impedance

$$\hat{Z}_o = \sqrt{\frac{r + j\omega\ell}{g + j\omega c}}$$

it is clear that only lossy lines will have complex characteristic impedances. Because of the attenuation associated with these lossy lines, however, their use is undesirable in

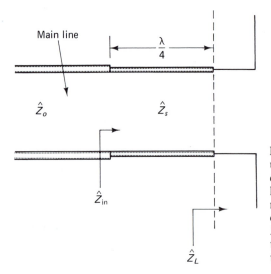

Figure 7.47 Matching procedure using quarter-wavelength transformers. $\hat{Z}_L$ is the input impedance of a load antenna that needs to be matched to a transmission line of characteristic impedance $\hat{Z}_o$. For $\hat{Z}_{in} = \hat{Z}_o$—that is, zero reflection in the main line—$\hat{Z}_s$ should be equal to $\sqrt{\hat{Z}_L \hat{Z}_o}$.

these matching applications. Hence, $\lambda/4$ transformers may *practically* be used for matching only real impedances. The use of some of the stub-matching methods, to be described next, is certainly more attractive when the load impedance is complex.

EXAMPLE 7.12

The dipole antenna of Figure 7.47 has an input impedance of 73 Ω at 400 MHz. It is desired to feed this antenna using a coaxial line of characteristic impedance $Z_o = 50\ \Omega$. Determine the impedance and the length of a quarter-wavelength transformer that may be used to achieve impedance matching between the antenna and the feed line at 400 MHz.

Solution

The characteristic impedance of the transformer section is

$$Z_s = \sqrt{50 \times 73} = 60.42\ \Omega$$

The length of the transformer is $\lambda/4$ at 400 MHz. If we use an air-filled line $\lambda = c/f =$ 75 cm, the transformer length is hence $d = 18.75$ cm.

EXAMPLE 7.13

Quarter-wavelength transformers are usually made of sections of lossless transmission lines. Describe a procedure to match a complex load $\hat{Z}_L$ to a line whose characteristic impedance is real and equal to Z_o.

Solution

One of the various ways to achieve matching in this case is to insert a section of lossless transmission line between the load and the matching section. From Figure 7.48, it can be seen that by inserting a section of lossless line, the input impedance $\hat{z}_{nin}$ may be obtained by rotating the normalized load impedance $\hat{z}_{nL}$ on a constant reflection coefficient circle. By adjusting the length of the lossless line d_R so that the input impedance would be one the real axis (i.e., $\hat{z}_{nin} = r + j0$), the quarter-wavelength transformer may then be inserted

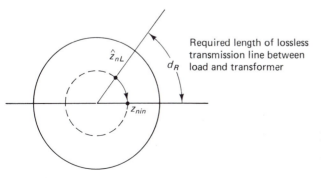

Figure 7.48   A procedure to use $\lambda/4$ lossless transformers when the load is complex.

at this location and impedance matching may be achieved. The characteristic impedance of the transformer line is

$$Z_s = \sqrt{Z_o Z_{in}}, \qquad \text{where} \qquad Z_{in} = z_{nin} \times Z_o$$

The length of the transformer will be routinely taken $d_x = \lambda/4$, where λ is the wavelength at the operating frequency.

7.14.2 Single-Stub Matching

The geometry of this matching procedure basically involves placing a short- or an open-ended section of lossless transmission line (stub) in series or parallel with the main line and at a specific distance from the load. As explained in section 7.13, by varying the length of these lossless sections (stubs) from 0 to $\lambda/2$, values of reactive impedances from 0 to $\pm\infty$ may be obtained. It is therefore, desired to determine the location at which these stubs should be introduced so that perfect matching may be achieved. Let us illustrate the matching procedure by considering the matching of a load impedance $\hat{Z}_L$, which is in general complex, to a transmission line of characteristic impedance Z_o. If the impedance matching is to be achieved using a single parallel stub, it is desirable to determine the location and the length of the stub. The geometry of the desired matching arrangement is shown in Figure 7.49.

To simplify the analysis, let us assume that the stub has the same characteristic impedance as the main line. In the matching arrangement of Figure 7.49, the following should be noted:

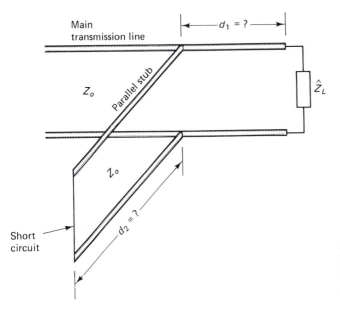

Figure 7.49 The impedance-matching arrangement using a single stub connected in parallel with the main transmission line.

1. By using a short-circuited section of a lossless transmission line as a stub, varying the length of the stub provides only reactive impedances $\hat{Z}_s = jX$.
2. By impedance matching, we mean that $\hat{Z}_{in}$, after introducing the stub, should be equal to Z_o; hence, the reflection coefficient in the main line after the stub will be zero.
3. Because we are dealing with the parallel connection of transmission lines, the Smith admittance chart will be used in the analysis. The Smith chart in Figure 7.50 shows the following various steps in the solution procedure.
   (a) Obtain the normalized load impedance $\hat{z}_{nL} = \hat{Z}_L/Z_o$ and plot it on the Smith chart (point A).
   (b) Obtain $\hat{y}_{nL}$, as suggested earlier, by rotating point A on a constant reflection coefficient circle 180° to point B.

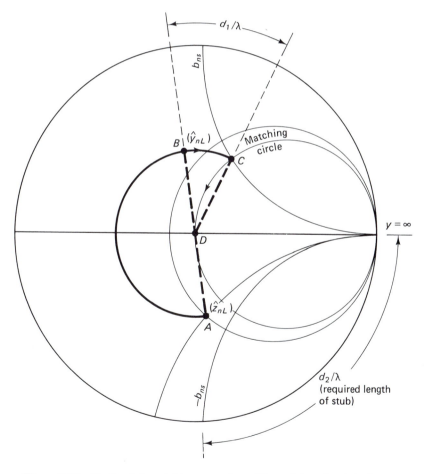

Figure 7.50   Single-stub matching procedure. The stub location is indicated as d_1/λ, and the stub length is shown as d_2/λ.

(c) To find a suitable location for the stub, we recall first that by achieving our matching objective, the normalized admittance after the stub should be at point D, which is the zero reflection coefficient point on the Smith chart. Because the admittance of a short-circuited section of lossless line is purely reactive (susceptance), however, the admittance just before the stub $\hat{y}_{nb}$ should be different from $\hat{y}_n$ at D by only a susceptance value. In other words, the locus of the admittance, just before the stub, is the circle $g = 1$ (constant conductance circle), which is also known as the *matching circle*. The required location of the stub should then be the length d_1/λ away from the load so that it would bring the load admittance $\hat{y}_{nL}$ to the matching circle at point C.

(d) The susceptance of the stub $\hat{y}_{ns}$ is the difference between the values of the admittances at D and C,

$$\hat{y}_{ns} = \hat{y}_{nD} - \hat{y}_{nC}$$

where $\hat{y}_{nD}$ and $\hat{y}_{nC}$ are the values of the normalized admittances at D and C, respectively.

(e) To determine the length of the short-circuited stub from its known value of input admittance $\hat{Y}_s = \hat{y}_{ns} Y_o$, where $Y_o = 1/Z_o$, the characteristic admittance of the stub, we use equation 7.70 for the impedance of the short-circuited section of transmission line

$$\hat{Y}_s = \frac{1}{jZ_o \tan(\beta d_2)} = \frac{1}{jZ_o \tan(2\pi d_2/\lambda)}$$

where d_2/λ is the required length of the stub in terms of λ. Alternatively, we may graphically use the Smith chart to determine the length of the stub. Remembering that $\hat{y}$ at the short-circuited end of the stub $= \infty$, it is desired to continue to move away from this load location (i.e., toward the generator) until the desired value of $\hat{y}_{ns}$ is obtained. The length of the stub d_2/λ is determined from the scale on the rim of the Smith chart between $\hat{y} = \infty$ and $\hat{y}_{ns}$, as shown in Figure 7.50. The following example will illustrate further this solution procedure.

EXAMPLE 7.14

An antenna of normalized input impedance equals to $0.4 - j0.4$. It is desired to match the antenna to the transmission line by placing a stub (purely reactive element having an adjustable, sliding short on its end) at a distance d_1 from the antenna. If the frequency is 3 GHz, find the design distance d_1 from the antenna to the stub, and the length d_2 of the stub for proper matching.

Solution

The solution shown on the Smith chart (Figure 7.51) is detailed as follows:

1. Plot the antenna normalized impedance $0.4 - j0.4$ (point A) and by rotating this point on $|\hat{\Gamma}| = $ constant circle, convert the antenna (load) impedance to admittance (point B). This normalized admittance value is $1.23 + j1.23$.

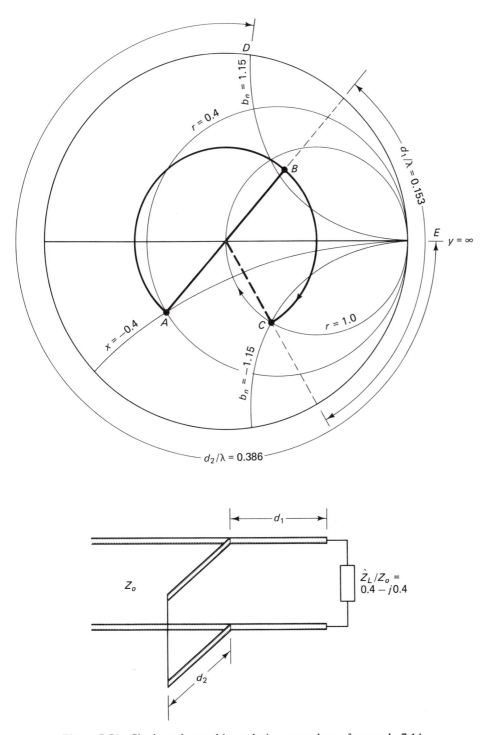

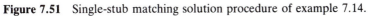

Figure 7.51 Single-stub matching solution procedure of example 7.14.

2. Move around the $|\hat{\Gamma}|$ = constant circle toward the generator to the constant conductance circle, which passes through the center of the chart (point C on the *matching circle*). At point C, the admittance of the line looking back toward the antenna is $1.0 - j1.15$. If a susceptance of $+j1.15$ is added in parallel at this point, the combination will equal a resistive component of 1.0, and the line will be perfectly matched from that point on, toward the generator. This means that the input admittance of the stub should be $+j1.15$ capacitive susceptance.

3. The distance d_1 can be easily computed from the phase angle between points B and C, and is found to be 0.153λ or, at $f = 3$ GHz, $d_1 = 1.53$ cm from the antenna to the matching stub. This distance may also be read directly from the rim scale on the Smith chart.

4. To determine the length of stub d_2, it is necessary to enter on the Smith chart the desired capacitive susceptance of $+j1.15$ (point D). From point D, move around the chart on the conductance $= 0$ circle (because we have a short-circuited stub, which is a purely reactive element; i.e., $g = 0$) until a short-circuited point is obtained at point E. The length of the stub is simply $d_2 = 0.386\lambda$ and at $f = 3$ GHz, $d_2 = 3.86$ cm.

---◆◆---

7.14.3 Double-Stub Matching

A serious practical limitation of the single-stub matching procedure is varying the location of the stub d_1 for each different load impedance. The double-stub matching method overcomes precisely this problem by changing the adjustable unknown variables from being the location and the length of the stub to the unknown lengths of two stubs located at fixed distances from each other and from the load. In the double-stub matching procedure, shown in Figure 7.52, therefore, the distance between the stubs is known, and it is required to determine the lengths d_1 and d_2 of the two stubs.

To help us understand the solution procedure, the following points should be noted:

1. The admittance just after the second stub $\hat{Y}_{B^+}$ should be equal to the characteristic admittance of the main line $Y_o = 1/Z_o$. This way the reflection coefficient will be zero, and the desired impedance matching would be achieved. The normalized admittance just after the second stub $\hat{y}_{nB^+}$ is, hence, equal to 1, which is located at the origin B^+ of the Smith chart shown in Figure 7.53.

2. The admittance of the short-circuited stub is purely susceptance. Hence, the normalized admittance $\hat{y}_{nB^-}$, just before the second stub should lie on the $g = 1$ circle, known as the *matching circle*. The specific location of $\hat{y}_{nB^-}$ on the matching circle is, however, unknown and depends on the specific value of the susceptance of the stub, which is to be determined.

3. The admittances $\hat{y}_{nB^-}$ and $\hat{y}_{nA^+}$ just before the second stub and just after the first stub, respectively, are separated by a section of length d (known) of *lossless* transmission line. The specific value of $\hat{y}_{nA^+}$ may, therefore, be obtained from $\hat{y}_{nB^-}$ by rotating $\hat{y}_{nB^-}$ a distance (d) toward the load (counterclockwise). Because the specific

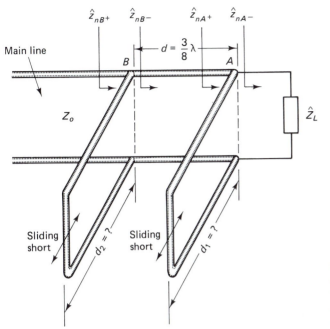

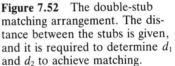

Figure 7.52 The double-stub matching arrangement. The distance between the stubs is given, and it is required to determine d_1 and d_2 to achieve matching.

value of $\hat{y}_{nB^-}$ is not known on the matching circle, however, we obtain a locus for $\hat{y}_{nA^+}$ by rotating the whole matching circle a distance d toward the load.

Graphically, this may be achieved by rotating the origin of the matching circle a distance d and drawing a circle of the same radius at the new origin. The new circle is the locus of $\hat{y}_{nA^+}$ and is known as the *rotated circle*.

4. The first stub is also a short-circuited section of transmission line and, hence, provides a susceptive value of admittance. The difference between $\hat{y}_{nA^-} = \hat{y}_{nL}$ and $\hat{y}_{nA^+}$ is simply the susceptance of the first stub. The specific value of $\hat{y}_{nA^+}$ may then be obtained from $\hat{y}_{nL}$ by moving along the *constant conductance*, g = *constant, circle* from $\hat{y}_{nL}$ until it intersects the rotated circle.

The susceptance for stub 1 is therefore chosen to alter the admittance from $\hat{y}_{nL}$ at point A^- (Figure 7.53) to $\hat{y}_{nA^+}$ at A^+ on the rotated circle. The corresponding point just to the right of stub 2 is obtained by rotating point A^+, on the rotated circle, a distance $d = 3\lambda/8$ toward the generator. This procedure will result in point B^- shown in Figure 7.53. The stub length d_2 is chosen to modify the admittance at B^- to $\hat{y}_{nB^+} = 1$, thus producing a matched transmission-line system. Stub lengths required to provide the necessary susceptances are found by moving around the chart perimeter g = 0 circle from the short-circuit position ($y_n = \infty$) to the desired normalized admittance value.

The step-by-step procedure for double-stub matching is summarized as follows:

1. Draw the rotated circle by rotating the matching circle g = 1 an angle $2\beta d$ toward the load, where d is the known distance between the two stubs.

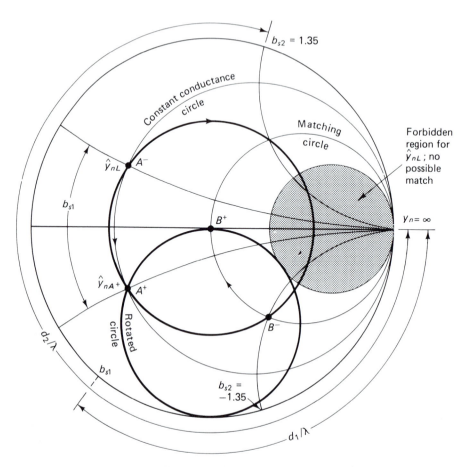

Figure 7.53 Double-stub matching solution procedure.

2. Locate the normalized load admittance $\hat{y}_{nL}$ on the Smith chart.
3. Follow the constant conductance locus $g = $ constant on the Smith chart to the point where it intersects the rotated circle. The admittance at this point is $\hat{y}_{nA^+} = \hat{y}_{nL} + jb_{s1} = g_L + jb_1$, where $b_{s1} = b_1 - b_L$, b_1 is the susceptance at $\hat{y}_{nA^+}$, and b_L is the load normalized susceptance (imaginary part of $\hat{y}_{nL}$).
4. Because b_1 and b_L are known, the required susceptance of the first stub can be determined and the length of the stub d_1 may hence be calculated.
5. From $\hat{y}_{nA^+}$, follow the $|\hat{\Gamma}| = $ constant circle a distance $2\beta d$ until it intersects the matching circle at $\hat{y}_{nB^-}$. The susceptance of the second stub is then chosen to obtain matched system immediately after the stub. If the normalized admittance immediately to the right of stub 2 is $\hat{y}_{nB^-} = 1 + jb_2$, the normalized susceptance of stub 2 should therefore be $b_{s2} = -b_2$. The length of the second stub is obtained using a procedure similar to that in step 4.

EXAMPLE 7.15

The layout for a double-stub tuner is shown in Figure 7.54. Determine the required lengths of stubs d_1 and d_2.

Solution

The step-by-step solution is as follows:

1. Draw the matching circle ($g = 1$ circle) and the rotated circle (0.25λ toward the load away from the matching circle) as shown in Figure 7.54.
2. Plot the load normalized admittance $\hat{y}_{nL}$ on the Smith chart (point A).
3. Move on a constant reflection coefficient circle a distance of 0.1λ toward the generator to point B. The normalized admittance $\hat{y}_{nB} = 0.6 - j0.685$.
4. At point B, we insert the first stub. Hence, the admittance just before and just after the stub should have the same conductance. We therefore move from B on the constant conductance line until we intersect the rotated circle at point C. The normalized admittance $\hat{y}_{nC} = 0.6 - j0.5$.
5. Because the admittance $\hat{y}_{nC}$ just after the first stub and the admittance, say $\hat{y}_{nD}$, just before the second stub are separated by a 0.25λ section of a transmission line, $\hat{y}_{nD}$ may be obtained by rotating $\hat{y}_{nC}$ on a constant reflection coefficient circle a distance 0.25λ toward the generator or until the constant $|\hat{\Gamma}|$ circle intersects the matching circle at D. The normalized admittance $\hat{y}_{nD} = 1 + j0.81$.
6. The susceptance of the second stub is required to change the admittance at D, $\hat{y}_{nD}$, to that at the matching point at the center of the Smith chart O.
7. From Figure 7.54, the normalized susceptance of the first stub is given by

$$\hat{y}_{ns_1} = \hat{y}_{nC} - \hat{y}_{nB} = j0.185$$

The normalized susceptance of the second stub is

$$\hat{y}_{ns_2} = 1.0 + j0 \text{ (at matching point } O) - \hat{y}_{nD} = -j0.81$$

8. The length of each stub is determined by rotating from the short-circuited end of each stub ($y_n = \infty$) along the rim of the Smith chart ($g = 0$ circle) toward the generator, sufficient distances (i.e., lengths of the stubs) so as to obtain the desired values of the admittances of these stubs. From Figure 7.54, it can be seen that the length of the first stub $d_1/\lambda = 0.278$, whereas the length of the second stub $d_2/\lambda = 0.142$.

7.15 VOLTAGE STANDING-WAVE RATIO (VSWR) ALONG TRANSMISSION LINES

In chapter 5 when we discussed reflections of plane waves, it was indicated that as a result of the interference between the incident and the reflected waves that are propagating in opposite directions and of the same frequency, there will be standing waves. An amplitude maxima occurs whenever there is a constructive interference in which

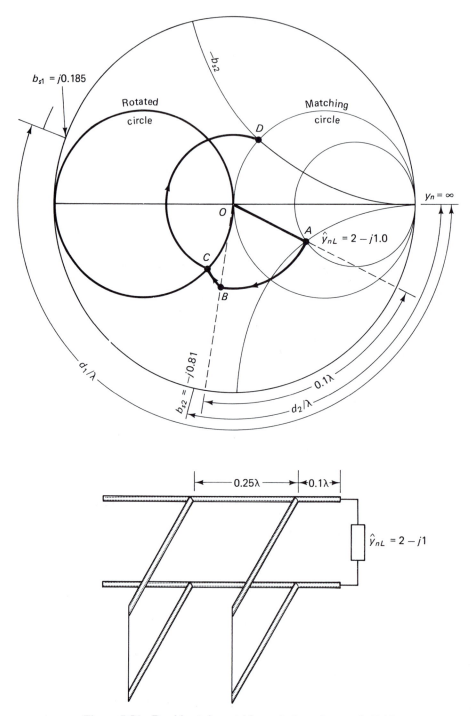

Figure 7.54 Double-stub matching solution of example 7.15.

the incident and reflected waves are in phase, and an amplitude minima occurs whenever there is a destructive interference at which the incident and the reflected waves are opposite in phase. Similar standing-wave phenomena occur for the voltage and current along the transmission lines if there are reflections along the lines. Consider, for example, the voltage along a lossless transmission line. Because we assumed the presence of reflections, the general expression for the voltage is given by

$$\hat{V}(z) = \hat{V}_m^+ e^{-j\beta z} + \hat{V}_m^- e^{j\beta z}$$
$$= \hat{V}_m^+ e^{-j\beta z} + \hat{V}_m^- e^{j\beta z} + \hat{V}_m^- e^{-j\beta z} - \hat{V}_m^- e^{-j\beta z} \qquad (7.75)$$

where the term $\hat{V}_m^- e^{-j\beta z}$ has been added and subtracted from the general expression of the voltage. Equation 7.75 may be rearranged in the form

$$\hat{V}(z) = \hat{V}_m^+ e^{-j\beta z}(1 + \hat{\Gamma}) + \hat{V}_m^-(e^{j\beta z} - e^{-j\beta z})$$
$$= \underbrace{\hat{V}_m^+(1 + \hat{\Gamma})e^{-j\beta z}}_{\substack{\text{Traveling} \\ \text{wave}}} + \underbrace{2j\hat{\Gamma}\,\hat{V}_m^+ \sin\beta z}_{\substack{\text{Standing} \\ \text{wave}}} \qquad (7.76)$$

where

$$\hat{\Gamma} = \frac{\hat{V}_m^-}{\hat{V}_m^+}$$

The first term in equation 7.76 represents a wave of amplitude $\hat{V}_m^+(1 + \hat{\Gamma})$ and traveling along the positive z direction, whereas the second term represents a standing wave of amplitude $2j\hat{\Gamma}\,\hat{V}_m^+$. The propagation properties of traveling and standing waves may be further clarified by recovering the real-time form of the voltage along the line as described in chapter 5. Hence, we may summarize by indicating that as a result of the interference between the incident and reflected voltages along transmission lines, the reflected voltage wave combines with a portion of equal amplitude of the incident wave resulting in a standing wave (second part of the right side of equation 7.76) and the remaining part of the incident voltage continued to propagate down the transmission line. To illustrate this interference process further, let us consider the following special cases:

1. $\hat{\Gamma} = 0$, no reflections along the line. In this case, the voltage along the line will be given from equation 7.76 by

$$\hat{V}(z) = \hat{V}_m^+ e^{-j\beta z}$$

or

$$v(z,t) = V_m^+ \cos(\omega t - \beta z + \phi^+)$$

which represents a voltage wave propagating along the positive z axis. V_m^+ and ϕ^+ are the magnitudes and phase of the complex amplitude

$$\hat{V}_m^+ = V_m^+ e^{j\phi^+}$$

2. $\hat{\Gamma} = -1$, reflection from a short-circuit termination of the line. In this case, the voltage from equation 7.76 will be given by

$$\hat{V}(z) = -2j\,\hat{V}_m^+ \sin \beta z$$

and the real-time form of the voltage is, hence,

$$v(z,t) = 2V_m^+ \sin \beta z \, \sin(\omega t + \phi^+)$$

which is, as described in chapter 5, a standing wave because the spatial coordinate z is independent of time t.

The special cases 1 and 2 represent the resulting voltage waves in some extreme situations, but, in general, we would have combinations of traveling and standing waves along the transmission line.

To study further the standing wave characteristics along transmission lines, let us consider the general expressions of the voltage and current along a reflective transmission line

$$\hat{V}(z) = \hat{V}_m^+ e^{-j\beta z} + \hat{V}_m^- e^{j\beta z}$$

$$\hat{I}(z) = \hat{I}_m^+ e^{-j\beta z} + \hat{I}_m^- e^{j\beta z}$$

These expressions may be rearranged in the form

$$\hat{V}(z) = \hat{V}_m^+ e^{-j\beta z}[1 + \hat{\Gamma}(z)] \tag{7.77a}$$

$$\hat{I}(z) = \hat{I}_m^+ e^{-j\beta z}[1 - \hat{\Gamma}(z)] \tag{7.77b}$$

where

$$\hat{\Gamma}(z) = \frac{\hat{V}_m^-}{\hat{V}_m^+} e^{2j\beta z} = -\frac{\hat{I}_m^-}{\hat{I}_m^+} e^{2j\beta z}$$

From equation 7.77a, it can be seen that as we move along a lossless transmission line (change values of z), the phase of $\hat{\Gamma}(z)$ will change resulting in a voltage maxima when $\hat{\Gamma}(z)$ is a positive real number and a voltage minima when $\hat{\Gamma}(z)$ is a negative real number. We also note that because of the negative sign in front of $\hat{\Gamma}(z)$ in equation 7.77b, current minima occurs at locations of voltage maxima, and current maxima occurs at the locations of voltage minima. To illustrate the occurrence of the voltage and current maxima and minima along transmission lines graphically using the Smith chart, let us consider the magnitudes of these quantities given in equation 7.77,

$$|\hat{V}(z)| = |\hat{V}_m^+||1 + \hat{\Gamma}(z)| \tag{7.78a}$$

$$|\hat{I}(z)| = |\hat{I}_m^+||1 - \hat{\Gamma}(z)| \tag{7.78b}$$

Because we are interested in the relative variations of the voltage and current, we develop normalized voltage and current quantities by dividing by their arbitrary magnitudes $|\hat{V}_m^+|$ and $|\hat{I}_m^+|$, hence,

$$|\hat{V}(z)|_n = |1 + \hat{\Gamma}(z)| \tag{7.79a}$$

$$|\hat{I}(z)|_n = |1 - \hat{\Gamma}(z)| \tag{7.79b}$$

where the subscript n stands for normalized. Figure 7.55 illustrates the normalized magnitudes of the voltage and current as given by equation 7.79 for a specified value of $\hat{\Gamma}(z)$ at a location z along the line. The normalized voltage is represented by the length $OA = |1 + \hat{\Gamma}(z)|$, and the normalized current is represented by the length $OB = |1 - \hat{\Gamma}(z)|$. The radius of the outer circle on the Smith chart OC, of course, has the normalized value of 1. This graphical representation is very helpful in examining the variation of the voltage and current along a transmission line because as we move along a lossless line (change values of z), $\hat{\Gamma}(z)$ rotates on a constant reflection coefficient circle, thus resulting in different values of the normalized voltage $|1 + \hat{\Gamma}(z)|$ and the normalized current $|1 - \hat{\Gamma}(z)|$. Figure 7.55 illustrates these variations in the normalized voltage and current along the transmission line as projected from the rotation of the reflection coefficient.

From Figure 7.55 we may make the following observations:

1. The voltage maxima V_{max} occurs at the location of I_{min} and vice versa.
2. V_{max} occurs when $\hat{\Gamma}(z)$ becomes a real positive number at A_{max} in Figure 7.55. At this location

$$V_{max} = 1 + |\hat{\Gamma}(z)|$$

because $\hat{\Gamma}(z)$ is a real positive number.
    Similarly,

$$V_{min} = 1 - |\hat{\Gamma}(z)|$$

at A_{min} where the reflection coefficient is a negative real number.

3. The larger the reflection coefficient, the larger will be the variation of the voltage and current along the transmission line. We define the Voltage Standing Wave Ratio (VSWR) as the ratio of V_{max} to V_{min}, hence,

$$\text{VSWR} = \frac{V_{max}}{V_{min}} = \frac{1 + |\hat{\Gamma}(z)|}{1 - |\hat{\Gamma}(z)|}$$

Because the $|\hat{\Gamma}(z)|$ varies between 0 and 1, VSWR takes the value 1 when $|\hat{\Gamma}(z)| = 0$, and a value of ∞ at $|\hat{\Gamma}(z)| = 1$. Hence,

$$0 \leq |\hat{\Gamma}| \leq 1,$$

whereas

$$1 \leq \text{VSWR} \leq \infty$$

4. The distance between two successive V_{max} or two successive V_{min} along the transmission line is $\lambda/2$ because the reflection coefficient should rotate a complete cycle on the Smith chart to generate these successive maxima or minima.
5. The distance between V_{max} and V_{min} or between I_{max} and I_{min} is $\lambda/4$. This can be seen from Figure 7.55 by noting that the distance between A_{max} and A_{min} corresponds to a $\lambda/4$ rotation on the Smith chart.

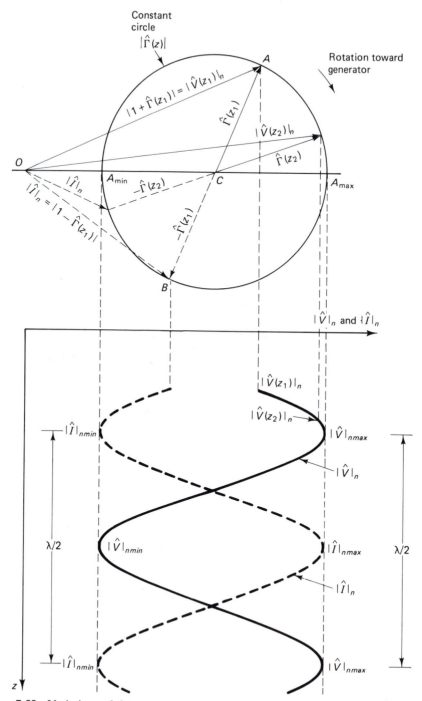

Figure 7.55 Variations of the normalized magnitudes of the voltage and current along a transmission line as obtained from the variation of $\hat{\Gamma}(z)$ with the rotation (z-variation) toward the generator.

EXAMPLE 7.16

In measuring the VSWR along a slotted transmission line, the distance between two successive minima was 15 cm, and the VSWR was 3.2. Determine the magnitude of the reflection coefficient and the operating frequency.

Solution

$$\text{VSWR} = \frac{1 + |\hat{\Gamma}(z)|}{1 - |\hat{\Gamma}(z)|}$$

$$\therefore |\hat{\Gamma}(z)| = \frac{\text{VSWR} - 1}{\text{VSWR} + 1} = \frac{2.2}{4.2}$$

$$= 0.52$$

The distance between two successive minima is $\lambda/2$. Hence,

$$\lambda/2 = 15 \text{ cm}$$

or

$$\lambda = 30 \text{ cm}$$

If we assume an air-filled line $\lambda = c/f$, the operating frequency is then

$$f = \frac{c}{\lambda} = \frac{3 \times 10^{10}}{30} = 1 \text{ GHz}$$

7.15.1 Representation of VSWR on Smith Chart

A convenient way of representing the VSWR on the Smith chart may be described by noting the variation of the reflection coefficient $\hat{\Gamma}(z)$ as a function of z along the line, as shown in Figure 7.56. The quantities $1 + |\hat{\Gamma}(z)|$ and $1 - |\hat{\Gamma}(z)|$ at V_{nmax} and V_{nmin}, respectively, are simply the distances OA and OB. Hence,

$$\text{VSWR} = \frac{1 + |\hat{\Gamma}(z)|}{1 - |\hat{\Gamma}(z)|} = \frac{OA}{OB}$$

At the point A, conversely, the normalized impedance $z_n = r + jx$ is real ($x = 0$) and the reflection coefficient is also a real positive number. At A,

$$z_n = r + j0 = \frac{1 + (\Gamma_r + j\Gamma_i)}{1 - (\Gamma_r + j\Gamma_i)}\Big|_{\Gamma_i = 0} = \frac{1 + \Gamma_r}{1 - \Gamma_r} = \frac{1 + |\hat{\Gamma}|}{1 - |\hat{\Gamma}|}$$

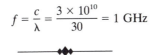

$$= \text{VSWR}$$

The VSWR circle may thus be drawn on the Smith chart by noting that it is the $r = $ constant circle that passes through A. The value of r on the circle is equal to the VSWR value.

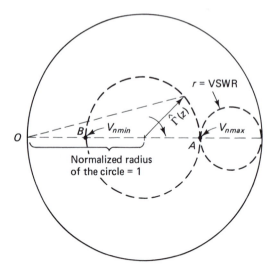

Figure 7.56 The variation (rotation) of $\hat{\Gamma}(z)$ along a lossless transmission line and the r = VSWR circle.

EXAMPLE 7.17

The VSWR measured along a slotted transmission line was 2.6. Use the Smith chart to determine the magnitude of the reflection coefficient and also plot the voltage variation along the line. Indicate the points of V_{max} and V_{min}. Compare the results with the case in which VSWR = ∞.

Solution

The VSWR numerically equals the value of the normalized resistance r on the Smith chart. The VSWR is, hence, represented by the r = constant = 2.6 circle as shown in Figure 7.57a. From Figure 7.57a, it can be seen that the magnitude of the reflection coefficient is given by the distance OA, which, when normalized according to the reflection coefficient scale, provides $|\hat{\Gamma}| = 0.38$. For the VSWR = ∞ case, the magnitude of the reflection coefficient $|\hat{\Gamma}(z)| = 1$. Therefore, at the locations of V_{nmax}, the incident and reflected voltages constructively interfere, resulting in a value of $V_{nmax} = 2$, whereas at the locations of V_{nmin}, two equal voltages destructively interfere (180° out of phase) and the resulting value of $V_{nmin} = 0$. Comparison between the voltage distribution along the transmission line for VSWR = 2.6 and VSWR = ∞ are shown in Figure 7.57b. It should also be noted from Figure 7.57b that the minima (particularly for larger values of VSWR) are much narrower than the voltage maxima, thus they are easier to locate in VSWR measurements.

7.16 USE OF VSWR MEASUREMENT TO DETERMINE UNKNOWN IMPEDANCES

Voltage standing-wave patterns including VSWR values and the locations of various voltage maxima and minima may be measured using an instrument known as slotted line. Figure 7.58 illustrates such a line where it can be seen that a slot is milled or cut lengthwise in the outer conductor of a rigid line (similar to a coaxial line) to allow for the penetration of a small probe into the space between the outer and inner conductors. The probe samples the magnitude of the **E** field between the conductors and, hence, measures the standing-wave pattern. The probe is movable along the line in a carriage and is, of course, connected to a detector and an instrument suitable for recording or displaying the voltage variation along the slotted transmission line. Positions of the probe may be also directly read on a scale attached to the side of the slot. The penetration of the probe into the slot should be kept to a minimum so as to minimize the distortion of the fields being measured. When such a slotted line is inserted between the source and the load, the standing-wave pattern may be examined by moving the carriage along the line and reading the detected output. It should be noted that the use

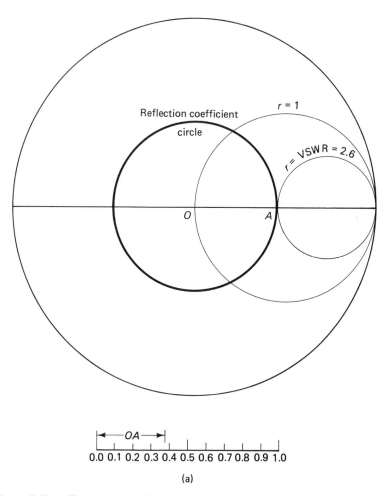

(a)

Figure 7.57a   The use of the Smith chart to determine reflection coefficient values from measured VSWR.

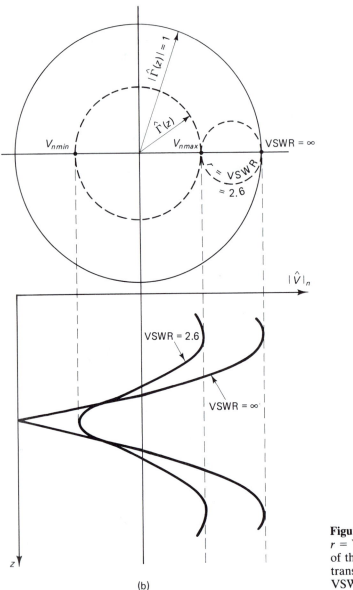

Figure 7.57b The VSWR circle
r = VSWR = 2.6, and comparison
of the voltage variation along the
transmission line when
VSWR = 2.6 and ∞.

of these slotted lines is limited to higher frequencies where the length of the line is at least $\lambda/2$ and, hence, may provide a complete standing-wave pattern that includes both voltage maxima and minima.

The main use of the slotted line is to determine the value of an unknown impedance when connected to its load side. The measurement procedure involves measuring the VSWR and the location of the first voltage minima away d_{min} from the load as shown in Figure 7.59. The operating frequency may also be determined by

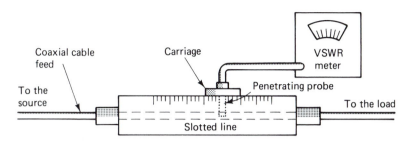

Figure 7.58   Geometry of a slotted line and its connection to a VSWR meter to measure the standing-wave pattern.

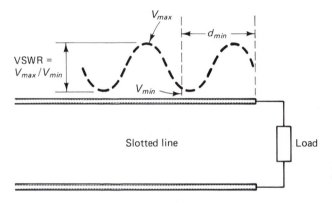

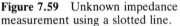

Figure 7.59   Unknown impedance measurement using a slotted line.

measuring the distance between two successive minima and equating the result by $\lambda/2$. It should be emphasized that in making these slotted-line measurements, we would rather deal with voltage minima than voltage maxima because the minima are narrower and, hence, better defined, and may be easily located on the scale of the slotted line. The following example illustrates the determination of unknown impedance from slotted-line measurements.

EXAMPLE 7.18

The VSWR on a 100 Ω transmission line is 3. The distance between successive voltage minima is 50 cm, and the distance from the load to the first minima is 20 cm. Find the reflection coefficient and the load impedance.

Solution

The solution procedure is illustrated in Figure 7.60. We first locate the r = VSWR circle. We then draw the VSWR circle with origin at the center of the Smith chart and a radius equal to the distance from the center to the point V_{max} as shown. The VSWR circle will then intercept the negative real axis of the chart at point M denoting the location of V_{min}.

Because the distance between two successive minima = $\lambda/2$ = 50 cm, hence λ = 100 cm. The distance from the load to the first minima is then 20/100 = 0.2λ. The

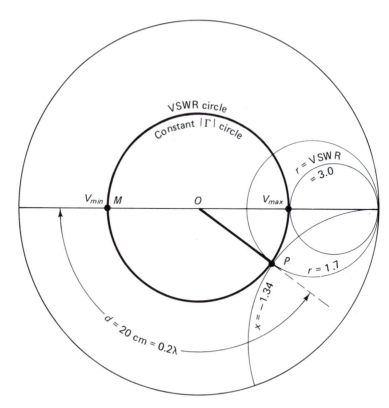

Figure 7.60   Solution of example 7.18.

normalized load impedance can then be obtained by rotating the point $M(V_{min})$ a distance 0.2λ back toward the load (counterclockwise). From the Smith chart, it can be seen that rotating point M a distance 0.2λ toward the load results in point P. At point P the normalized impedance is

$$\hat{z}_{nL} = 1.7 - j1.34$$

Therefore the load impedance $\hat{Z}_L = Z_o \times \hat{z}_{nL} = 170 - j134\ \Omega$. The magnitude of the reflection coefficient may be determined directly from the chart as the distance OV_{max}. From Figure 7.60, $|\hat{\Gamma}| = 0.5$.

EXAMPLE 7.19

A lossless transmission line of $Z_o = 200\ \Omega$ has an unknown load impedance $\hat{Z}_L$ and a standing wave ratio of 5. The first voltage minimum is 4 cm from the load, and the successive minima are 20 cm apart. We wish to match the line by placing a short-circuited stub in parallel with the load and a second stub 10 cm from the load. Find the required lengths of these stubs.

Solution

The first step in the solution is to find the load impedance $\hat{Z}_L$ from the given information. Because the distance between successive minima = 20 cm, λ = 40 cm. VSWR = 5, therefore, from the r = 5 circle shown in Figure 7.61, we determine V_{max} and, hence, the VSWR circle. Again this circle will intercept the negative real axis at $M(V_{min})$, which is known (measured to be 4 cm away from the load). The distance between V_{min} and the load is 4 cm or 4/40 = 0.1λ. Therefore, by rotating M (toward the load) a distance equal to 0.1λ, we locate the normalized load impedance at P. Because we will use parallel stub tuning, the normalized load admittance is required and can be obtained by rotating the impedance a distance $\lambda/4$ (180° on the Smith chart) as shown in Figure 7.61.

The procedure used to determine the stub lengths is the same one described in previous examples and is illustrated in Figure 7.62. It should be noted that the distance between the two stubs in 10/40 = $\lambda/4$. The following are some values of normalized admittances and the required lengths of the stubs.

$$\hat{y}_{nL} = 0.55 + j1.22 \qquad \text{(From Figure 7.61)}$$

$$\hat{y}_n(O) = 0.55 + j0.5 \qquad \text{(Obtained by moving } \hat{y}_{nL} \text{ on constant}$$
$$\text{conductance circle until it intersects the}$$
$$\text{rotated circle as shown in Figure 7.62)}$$

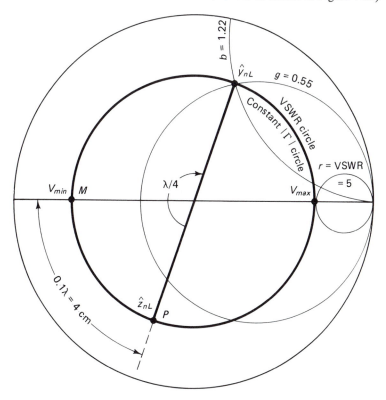

Figure 7.61   Use of VSWR measurements to determine the unknown load impedance of example 7.19.

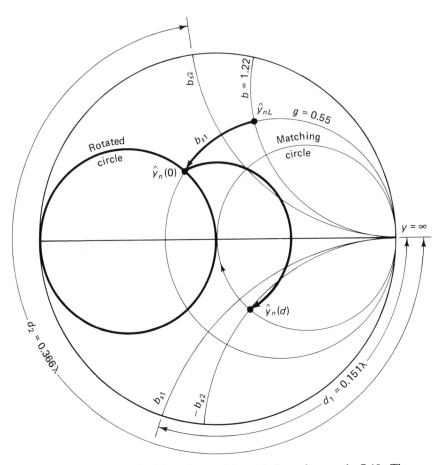

Figure 7.62 The double-stub matching solution of example 7.19. The rotated circle is obtained by rotating the matching circle a distance $\lambda/4$ toward the load.

$$\hat{y}_n(d) = 1 - j0.9 \qquad \text{(Obtained by rotating } \hat{y}_n(O) \text{ a distance}$$

(Obtained by rotating $\hat{y}_n(O)$ a distance $\lambda/4$ toward generator. $\hat{y}_n(d)$ is on the matching circle as shown in Figure 7.62)

$$b_{s1} = -j0.72, \text{ therefore, } d_1 = 0.151\lambda$$

$$b_{s2} = j0.9, \text{ therefore, } d_2 = 0.366\lambda$$

In many practical measurements, it is actually very difficult to determine accurately the location of the first voltage minima away from the load. The load may be connected to the slotted line with a long coaxial cable that is several wavelength long and determining $d_{min} + n\,\lambda/2$ may actually be practically difficult and, more important, inaccurate. To overcome this problem, the practical procedure of determining an unknown impedance involves the following:

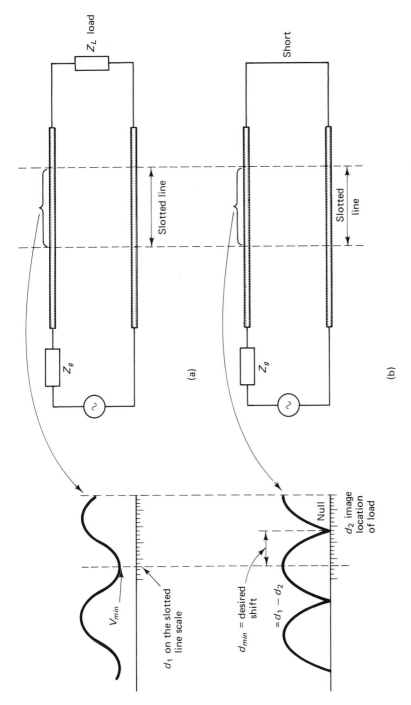

Figure 7.63 Slotted-line measurement procedure of an unknown impedance (a) the load connected to the slotted line; (b) short circuit connected to the slotted line. The desired shift in minima $d_{min} = d_1 - d_2$ toward the load.

569

1. With the load connected to the slotted line, determine VSWR and the location of the first minima that can be read on the scale of the slotted line.
2. Replace the unknown load with a short circuit (to have effectively a short circuit in the plane of the load) and determine the voltage nulls on the slotted line. Because the voltage is zero at the short-circuit termination, these nulls on the slotted line may be regarded as images ($n \lambda/2$ away) from the real location of the load.
3. We therefore determine d_{min} by measuring the difference between the locations of the first minima that can be read on the slotted line with the load and then with the short-circuit terminations.

The solution procedure is illustrated in Figure 7.63.

EXAMPLE 7.20

To measure the input impedance of an antenna, it is connected to a 50 Ω slotted transmission line. The measured VSWR was 3.6, and the location of the first minima was at the 8-cm mark on the scale of the slotted line. The antenna was then replaced by a short circuit, and the first null was located at the 4.8-cm mark along the scale. The distance between two successive nulls was also found to be 14 cm.

1. Determine the operating frequency and the value of the unknown impedance.
2. The double-stub arrangement shown in Figure 7.64 is to be used to match the antenna to a 100 Ω feed coaxial transmission line. Determine the lengths w_1 and w_2 of the two parallel stubs needed to achieve the impedance matching.

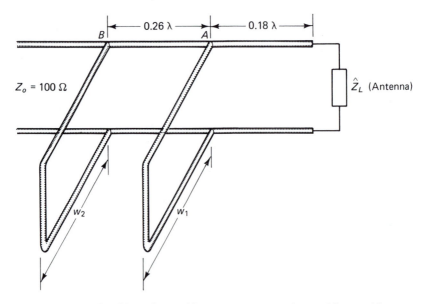

Figure 7.64 Double-stub matching arrangement to be used in matching the antenna $\hat{Z}_L$ to a 50 Ω coaxial feed line.

Solution

1. Figure 7.65 illustrates the first part of the solution. The distance between two successive nulls is 14 cm; hence, $\lambda = 2 \times 14 = 28$ cm. The shift in the minima positions between the load and the short circuit is $d_{min}/\lambda = (8 - 4.8)/28 = 0.114\lambda$. From Figure 7.65, $\hat{z}_{nL} = 0.46 - j0.76$; hence, $\hat{Z}_L = 23 - j38$ Ω. The operating frequency $f = c/\lambda = 3 \times 10^{10}/28 = 1.07$ GHz.

2. The double-stub matching procedure is shown in Figure 7.66. We first obtain the rotated circle 0.26λ toward the load away from the matching circle. $\hat{z}_{nL}$ is obtained by dividing $\hat{z}_{nL}$ of part 1 by 100 Ω, which is the characteristic impedance of the new transmission line.

$$\hat{z}_{nL} = 0.23 - j0.38$$

We then plot $\hat{z}_{nL}$ on the Smith chart and obtain $\hat{y}_{nL}$ by rotating $\hat{z}_{nL}$ 180° as shown in Figure 7.66. $\hat{y}_{nA^-}$ just before the first stub is obtained by rotating $\hat{y}_{nL}$ 0.18λ toward the generator. The 0.18λ is actually the distance from the load to the first stub. The stub 1 takes $\hat{y}_{nA^-}$ from its present value to $\hat{y}_{nA^+}$ on the rotated circle. $\hat{y}_{nA^-}$ and $\hat{y}_{nA^+}$

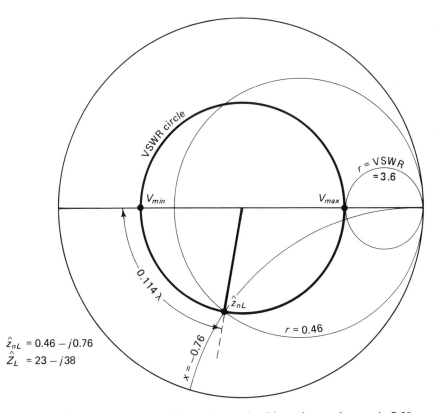

Figure 7.65 Determination of the unknown load impedance of example 7.20.

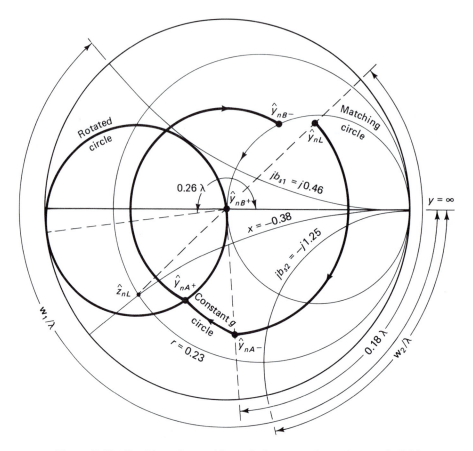

Figure 7.66 Double-stub matching solution procedure of example 7.20.

are on the same constant conductance circle. The susceptance and length of the first stub are then

$$jbs_1 = -j0.55 - (-j0.99) = j0.46$$

$$w_1 = (0.25 + 0.068)\lambda = 0.318\lambda$$

The normalized admittance $\hat{y}_{nB^-}$ just before the second stub is obtained by rotating $\hat{y}_{nA^+}$ from the rotated circle back to the corresponding point $(0.26\lambda$ away$)$ on the matching circle. The introduction of the second stub changes $\hat{y}_{nB^-}$ to $\hat{y}_{nB^+}$ at the center of the chart for a perfect match.

The susceptance and length of the second stub are

$$jbs_2 = -j1.25$$

$$w_2 = (0.357 - 0.25)\lambda = 0.107\lambda$$

SUMMARY

This chapter dealt with the transient and the steady-state analysis of two-conductor transmission lines. It is shown that the dominant electric and magnetic field configuration is the TEM mode, the transverse electromagnetic mode. In this mode, there are no components of the electric nor the magnetic fields along the direction of propagation. Also there is no cutoff frequency, which means that two-conductor transmission lines may be used to transmit electromagnetic energy at any frequency. At frequencies higher than 3 GHz, however, attenuation effects over long distances become serious, and the use of other means for transmitting electromagnetics energy is desirable. For the TEM mode, we identified unique relationships between the vector electric and magnetic fields, and the voltage and current on the transmission line. Hence, we used the scalar voltage and current quantities in our transient and steady-state analysis of transmission lines. Most important, we used an equivalent circuit model to develop the telegrapher's equations, the backbone of transmission-line analysis. These equations are given by

$$-\frac{\partial v}{\partial z} = ri + \ell\frac{\partial i}{\partial t}$$

$$-\frac{\partial i}{\partial z} = gv + c\frac{\partial v}{\partial t}$$

Simultaneous solution of these two equations provides the propagation characteristics of the voltage and current quantities on transmission lines.

Among the more important observations that resulted from this analysis is the fact that the voltage and current on a transmission line are related by a parameter known as the characteristic impedance Z_o

$$\frac{v^+}{i^+} = Z_o, \qquad \frac{v^-}{i^-} = -Z_o$$

Z_o is related to the transmission-line parameters r, ℓ, g, and c and the operating frequency. These parameters may be calculated based on the physical dimensions of the transmission line and the dielectric material between the conductors.

If the load, internal resistance of a generator, or a discontinuity along the transmission line is different from the characteristic impedance (Z_o), reflections occur. The amount of reflection is directly related to the mismatch between the discontinuity impedance and the characteristic impedance Z_o of the transmission line.

Transient Analysis

In this type of analysis we are basically concerned with the voltage and current propagation characteristics as a function of time and with the development of the multiple-reflection process as a result of discontinuities along or at either end of a transmission line. The following two cases were treated:

Resistive discontinuities and terminations.   For this case the reflection coefficient solution procedure is helpful. Voltage and current reflection coefficients are first calculated, and a reflection diagram is constructed to help visualize the development of multiple reflections. These reflection diagrams provide information on the voltage and current distributions along the transmission line at a specified time, and also show the development of multiple reflections at a specific location along the line. Reflection diagrams are useful for both step-function and pulse transient analyses of transmission lines. Observe that the current reflection coefficient has the opposite sign of the voltage reflection coefficient. Examples 7.2 to 7.6 illustrate the use of the reflection coefficient and reflection diagram solution procedure to analyze resistive discontinuities at the end and along lossless transmission lines.

Arbitrary termination.   The reflection coefficient solution procedure is inadequate in this case. Instead, an equation that relates the voltage and current at the load should be solved. This equation is given by

$$2v^+ = v_T + Z_o i_T$$

where v^+ is the incident voltage, and v_T and i_T are the voltage and current at the load (T). A differential equation may be developed by using the appropriate relationship between v_T and i_T at the load. For example, if we have an inductive load L, $v_T = L \, \partial i_T / \partial t$ and by making the appropriate substitution, the differential equation in the current i_T may be solved. If we have a capacitive load, conversely, $i_T = c \, \partial v_T / \partial t$ and a differential equation in the voltage v_T will be in hand. In both cases, once the inductor current or the capacitor voltage is solved, other quantities of interest may be obtained. An interesting application of some of the ideas in this section is the TDR described in section 7.9. In addition to the resistive terminations described in section 7.9, inductive and capacitive terminations may be analyzed, and the obtained results may be compared with the displayed response on the TDR screen.

Sinusoidal Steady-State Analysis

For this purpose we assumed the hypothetical $e^{j\omega t}$ time dependence and used the phasor voltage $\hat{V}$ and current $\hat{I}$ quantities to analyze the propagation characteristics. Two important observations were made. The first is related to the calculation of the characteristic impedance $\hat{Z}_o$ and the propagation constant $\hat{\gamma}$ of a transmission line

$$\hat{Z}_o = \sqrt{\frac{r + j\omega\ell}{g + j\omega c}}, \qquad \hat{\gamma} = \sqrt{(r + j\omega\ell)(g + j\omega c)}$$

The frequency dependence of these parameters is important because it causes signal distortion on transmission lines. For example, pulses in digital circuits may suffer significant distortion when propagated on transmission lines of strong frequency-dependent characteristics.

The other observation is related to the similarities between the analysis of tandem sections of transmission lines and of the normal incidence reflection and transmission at multiple-dielectric interfaces described in chapter 5. Table 7.1 summarizes these similarities, the matter that significantly simplifies carrying out sinusoidal analysis of cascaded transmission lines. The graphical implementation of the total impedance and the reflection coefficient solution procedure using the Smith chart is also possible in this case. Examples 7.7 to 7.10 were given to illustrate this solution procedure.

Short- and Open-Circuit Stubs and Transmission-Line Matching Techniques

Based on an expression of the input impedance of a lossless section of a transmission line (equation 7.69), it is shown that a short- or open-circuited section of transmission line may provide any capacitive or inductive susceptance values varying from ($-\infty$ to ∞). In other words, by adjusting the length of the open- or short-circuited section of lossless transmission line, we may obtain any desired value of susceptance. These sections of transmission lines are known as stubs and in practical application short-circuited stubs are more often used. This is to minimize potential radiation problems that may occur at the open end of the line. These stubs are often used as impedance-matching devices on transmission lines. Among the most commonly used procedures are the single- and double-stub matching. In the single-stub matching, both the location of the stub (away from the load) and the length of the stub are determined to achieve a perfect impedance match to the characteristic impedance of the transmission line. The location of the stub is selected such that the input impedance or admittance (for parallel connection of stubs) of the transmission line at this location is on the matching circle ($r = 1$ [series], $g = 1$ [parallel] circle). Then by adding the appropriate reactance or susceptance of the stub (by selecting the length of the stub), impedance matching may be achieved.

For the double-stub matching procedure, the locations of the stubs are fixed, and the lengths of the two stubs are adjusted to achieve impedance matching. Three instead of two stubs at fixed locations are needed to guarantee achieving impedance matching for any value of load impedance. Hence, with the double-stub matching procedure described here, impedance matching may or may not be achievable. Example 7.14 illustrates the single-stub and example 7.15 illustrates the double-stub matching procedures.

We concluded this chapter with a brief description of the VSWR and of how a VSWR measurement on a slotted transmission line may be used to determine an unknown load impedance. Examples 7.16 to 7.20 illustrate this solution procedure.

PROBLEMS

1. A small section of a transmission line may be represented by an equivalent circuit of either the L-type configuration given in the text or the Pi-type configuration shown in Figure P7.1.

   Obtain the same transmission line equations using the Pi-type alternative equivalent circuit.

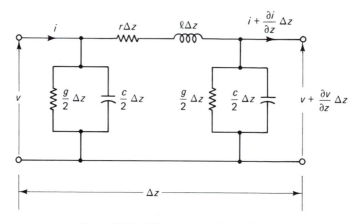

Figure P7.1 Pi-type equivalent circuit.

2. Consider the following voltage and current distributions:

$$v(z,t) = v_o \cos \beta(z - ut); \qquad i(z,t) = \frac{v_o}{Z_o} \cos \beta(z - ut),$$

where β is a constant, $u = 1/\sqrt{\ell c}$, and $Z_o = \sqrt{\ell/c}$. By direct substitution, verify that $v(z,t)$ and $i(z,t)$ satisfy the transmission-line equations 7.10 to 7.13.

3. Draw the equivalent circuit at the input of a transmission line for the case in which time equals zero. Give the equations for the input voltage and current step-function wave forms. Also, draw the equivalent circuit for the case in which time equals infinity.

4. Given the lossless transmission line shown in Figure P7.4, if it takes T seconds for a traveling wave to move from the sending end to the receiving end of the line:
   (a) Plot the voltage at the receiving end as a function of time up to $t = 9T$ seconds. Write the voltage value on the plot for each time interval.
   (b) Plot the sending-end current up to $t = 9T$ seconds and write the current values on the plot for each time interval.
   (c) What are the final values of the load voltage and current?

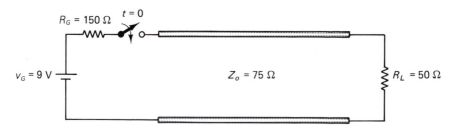

Figure P7.4 Transient analysis of resistively terminated transmission line.

5. For the transmission line shown in Figure P7.5, obtain equations describing the variation of the load current with time. Plot this current variation as a function of time and explain physically the reason for the occurrence of such variation.

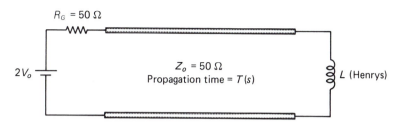

Figure P7.5 Time-domain analysis of a lossless transmission line terminated by an inductive load.

6. A transmission line of characteristic impedance $Z_o = 50\ \Omega$ and length $d = 900$ m is terminated by a load resistance of $200\ \Omega$. At distance $z = 600$ m, a resistive discontinuity is
introduced as shown in Figure P7.6, where $R_1 = R_2 = 50\ \Omega$ and $R_p = 100\ \Omega$. The line is then
connected at $t = 0$ to a battery of 3 V, which has an internal resistance of $100\ \Omega$. If the velocity
of the wave propagation along the transmission line is $u = 3 \times 10^8$ m/s, use the reflection
diagram to determine the following:
(a) Sending-end voltage versus time, up to time $t = 6 \times 10^{-6}$ s.
(b) Voltage distribution along the transmission line at time $t = 3.5 \times 10^{-6}$ s.

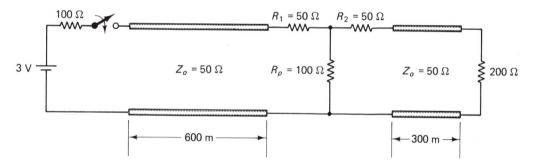

Figure P7.6 A step-function excitation of a $50\ \Omega$ transmission line with a resistive
discontinuity.

7. A transmission line of characteristic impedance $Z_o = 50\ \Omega$ and length $d = 900$ m is terminated by a load resistance of $200\ \Omega$. At distance $z = 450$ m, a resistive discontinuity is
introduced as shown in Figure P7.7, where $R_1 = 50\ \Omega$ and $R_p = 100\ \Omega$. The line is then

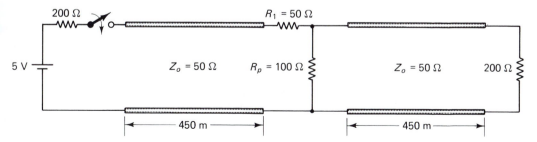

Figure P7.7 A step-function excitation of a $50\ \Omega$ transmission line with resistive
discontinuities.

connected at $t = 0$ to a battery of 5 V, which has an internal resistance of 200 Ω. If the velocity of the wave propagation along the transmission line is $u = 3 \times 10^8$ m/s, use the reflection diagram to determine the following:

  (a) Sending-end voltage versus time, up to time $t = 6 \times 10^{-6}$ s.

  (b) Voltage distribution along the transmission line at time $t = 7 \times 10^{-6}$ s.

8. Consider the assemblage of three transmission lines joined together as in Figure P7.8. The series resistance $R_s = 200 \; \Omega$ is introduced to account for the junction resistance. Determine the voltage at the sending end for a time period up to $t = 4.0\tau$. (Include in your analysis the reflections occurring at $t = 4\tau$.)

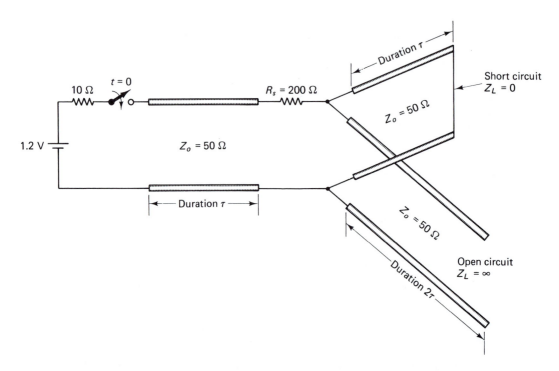

Figure P7.8 Transients on parallel connection of transmission lines.

9. Consider the tandem connection of the three transmission lines shown in Figure P7.9. The discontinuity between transmission line 1 and the generator is represented by the parallel resistance $R_{p1} = 300 \; \Omega$, whereas the discontinuity between transmission lines 1 and 2 is represented by the small series resistance $R_{s2} = 4 \; \Omega$. In addition, the discontinuity between transmission lines 2 and 3 is represented by a series resistance $R_{s3} = 2.5 \; \Omega$ and a parallel resistance $= 200 \; \Omega$. Transmission line 1 is fed by a pulse generator of internal resistance $R_G = 50 \; \Omega$, and the generated pulse has a magnitude $V_G = 8$ V and time duration of 1 μs. A load of resistance $R_L = 100 \; \Omega$ terminates transmission line 3.

  (a) Find the voltage at the sending and the load ends of the transmission lines system shown in Figure P7.9 as a function of time and up to time $t = 5.6 \; \mu$s.

  (b) Obtain the voltage distribution along the tandem connection of the transmission lines at time $t = 2.8 \; \mu$s.

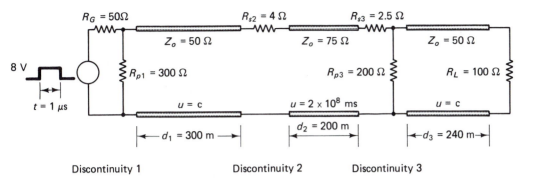

Figure P7.9   Tandem connection of three transmission lines.

10. In Figure P7.10, the battery voltage is 10 V and its internal resistance $R_G = 25\ \Omega$. The characteristic impedances and the propagation velocities of the tandem transmission lines are the following:

LINE 1	LINE 2
$Z_{o1} = 50\ \Omega$	$Z_{o2} = 75\ \Omega$
$u_{p1} = 2 \times 10^8$ m/s	$u_{p2} = 3 \times 10^8$ m/s
Length $= 400$ m	Length $= 900$ m

The load termination is $Z_L = 75\ \Omega$, and the switch is closed at $t = 0$.
(a) Use the reflection diagram to determine and plot the voltage at the load v_L as a function of time and up to $t = 14\ \mu$s.
(b) Obtain the current reflection and transmission coefficients at the various junctions, and use the current reflection diagram to determine the load current i_L as a function of time and up to $t = 14\ \mu$s.
(c) Determine the steady-state values of v_L and i_L.
(d) If the battery voltage is replaced by a pulse generator of the same internal resistance, where the pulse has a magnitude of 10 V and duration of 1 μs, determine the voltage distribution at the sending end as a function of time, up to $t = 12\ \mu$s.

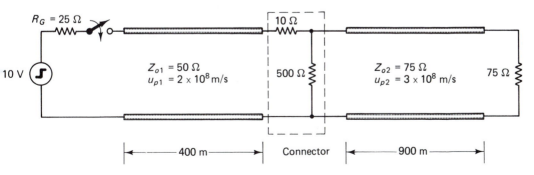

Figure P7.10   Transients on tandem transmission lines.

11. Two transmission lines are connected together by a faulty connector that introduced resistive discontinuity at the junction between the two transmission lines. Transmission line 1 has a characteristic impedance $Z_{o1} = 50\,\Omega$ and a velocity of propagation $u_1 = 3 \times 10^8$ m/s, whereas transmission line 2 has a characteristic impedance $Z_{o2} = 100\,\Omega$ and a velocity of propagation $u_2 = 2 \times 10^8$ m/s. Transmission line 1 has a length of 300 m, whereas transmission line 2 has a length of 150 m. The resistive discontinuity at the junction is represented by a series resistance $R_s = 2.5\,\Omega$ and a parallel resistance $R_p = 300\,\Omega$. Transmission line 2 is terminated by a load resistance $R_L = 150\,\Omega$. At $t = 0$, transmission line 1 is connected to a source of internal resistance $R_G = 50\,\Omega$ and source voltage $V_G = 10$ V. Use the reflection diagram to determine the following:
(a) The sending-end voltage as a function of time up to time $t = 4.5$ μs.
(b) The voltage distribution along the transmission line assemblage (i.e., both lines 1 and 2) at time $t = 4$ μs.

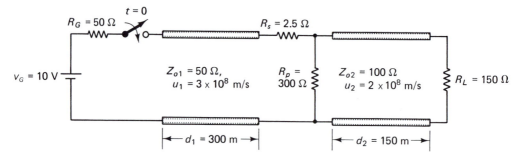

Figure P7.11   Transient response of tandem transmission lines.

12. A pulse generator of output voltage $V_G = 1$ V and pulse duration of 1 μs is connected to a transmission line of characteristic impedance $Z_o = 50\,\Omega$ and a velocity of propagation $u = 3 \times 10^8$ m/s. The internal resistance of the pulse generator is $R_G = 75\,\Omega$, and the resistance terminating the transmission line is $R_L = 150\,\Omega$. The length of the transmission line is $d = 900$ m. Use the reflection diagram to determine the following:
(a) The voltage at the sending end as a function of time for time t up to $t = 12$ μs.
(b) The voltage at the load terminating the transmission line as a function of time up to time $t = 10$ μs.

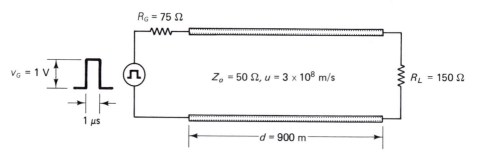

Figure P7.12   Pulse propagation along lossless transmission lines.

13. For each of the terminations given in Figure 7.28, determine the reflection coefficient, and the reflected and the total voltages along the transmission line. Verify the TDR displays shown in Figure 7.28.

14. Determine and plot the TDR display for the following loads (Figure P7.14):
  (a) Z_L is a series connection of R and L.
  (b) Z_L is a parallel connection of R and L.
  (c) Z_L is a series connection of R and C.

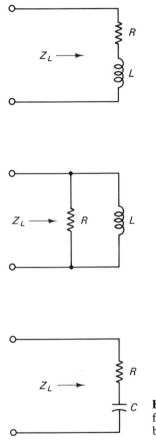

Figure P7.14   The various loads for which the TDR displays are to be determined.

15. (a) The TDR shown in Figure P7.15a is connected to the parallel junction of transmission lines 1 and 2 through the 50 Ω coaxial cable 3. Transmission line 1 is terminated by a short circuit, whereas transmission line 2 is terminated by a matched load of 50 Ω impedance. Also, the connector used for the parallel connection of the lines is lossy and has the equivalent circuit shown in Figure P7.15a. Determine the TDR response (total voltage picked up by the high impedance probe) as a function of time up to $5 = 7\tau$.
  (b) If the parallel combination of transmission lines in part a is replaced by the inductive load shown in Figure P7.15b, *illustrate graphically* and *explain physically* (do not use detailed mathematical derivation) the expected TDR response. The initial current in the inductor is assumed zero.

16. In an experiment to determine the dielectric properties of materials over a broad frequency band using the TDR, the equivalent circuit of the capacitance sample holder was found to be as shown in Figure P7.16. The sample holder with this equivalent circuit is connected to

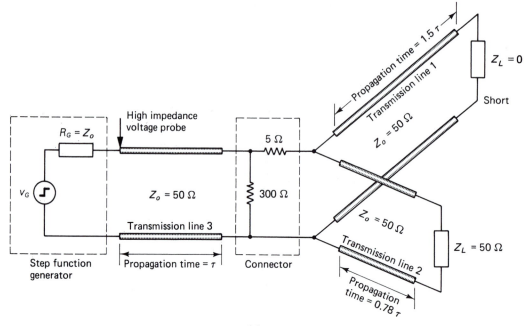

(a)

Figure P7.15a   TDR response of parallel combination of transmission lines.

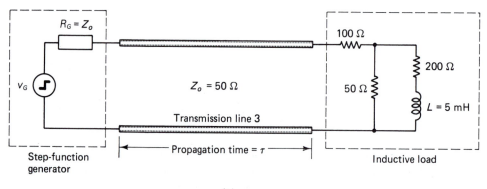

(b)

Figure P7.15b   TDR response of inductive load.

the TDR through a lossless section of transmission line of length d and characteristic impedance Z_o. The step-function generator of the TDR has an output voltage V_G and internal resistance $R_G = Z_o$. The velocity of propagation along the connecting section of transmission line is u in m/s. Determine the following:

(a)   The step-function response of the sample holder as measured by a voltage probe located at the sending-end P of the connecting transmission line.

(b) Illustrate graphically the variation of the probe voltage v_p in part a as a function of time, and explain physically the reason for such a variation.

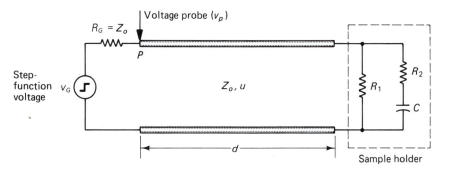

Figure P7.16   TDR method for measuring dielectric properties of materials.

17. An antenna of input impedance $\hat{Z}_L = 75 + j150 \ \Omega$ at 2 MHz is connected to a transmitter through a 100-m section of coaxial cable, which has the following distributed constants: $r = 153 \ \Omega/\text{km}$, $\ell = 1.4 \ \text{mH/km}$, $c = 88 \ \text{nF/km}$, and $g = 0.8 \ \mu\text{S/km}$. If the output voltage of the transmitter is $\hat{V}_G = 100e^{j0°} \ \text{V}$ and its internal impedance is 75 Ω as shown in Figure P7-17, determine the following:
    (a) The characteristic impedance $\hat{Z}_o$ and the propagation constant $\hat{\gamma} = \alpha + j\beta$ of the coaxial cable.
    (b) The input impedance $\hat{Z}_{in}$ of the cable when terminated by the antenna.
    (c) The average power delivered to the load antenna.

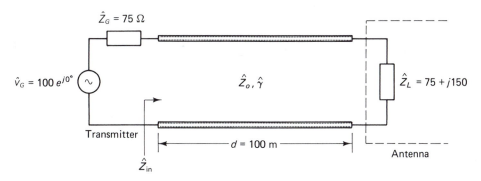

Figure P7.17   An antenna-transmitter system connected by a 100 Ω lossy coaxial cable.

18. Solve example 7.9 using the reflection coefficient and total transmission-line impedance systematic procedure but without the aid of the Smith chart. Compare the obtained results with those given in the example.

19. In the tandem connection of transmission lines shown in Figure P7.19, determine the following:
    (a) Input impedance as seen by the generator.
    (b) Input voltage and power to transmission line 1.
    (c) Power delivered to load terminating transmission line 2.

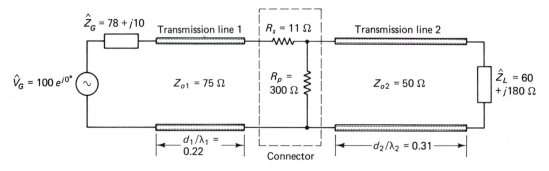

Figure P7.19   Tandem connection of two transmission lines with a lossy connector.

20. A transmission line with a normalized load admittance of $0.55 + j0.27$ is to be matched using a single stub (short circuited). Determine the minimum distance of the stub from the load and the required stub length.

21. An open-wire transmission line is constructed of two parallel wires spaced 3 cm apart, each of 1.29 mm radius. At 100 MHz, the line has these constants: $r = 600 \, \Omega/\text{km}$, $\ell = 1.3 \, \text{mH/km}$, $c = 8.8 \, \text{nF/km}$, and g is negligible.
    (a) Calculate the characteristic impedance $\hat{Z}_o$ and the propagation constant $\hat{\gamma}$ of the transmission line.
    (b) A 500-m section of this transmission line is terminated by an antenna of impedance $\hat{Z}_L = 50 - j217 \, \Omega$ at 100 MHz as shown in Figure P7.21. To test the antenna–transmission-line operation, a generator of voltage $\hat{V}_G = 100e^{j0°}$ V and internal resistance of $R_G = 125 \, \Omega$, is connected to the other end of the line as also shown in Figure P7.21. Determine the admittance $\hat{Y}$ that should be connected to the input terminal of the line so that the generator will see an input impedance Z_{in} that is matched to the generator internal resistance.
    (c) Under the matched conditions in part b, determine the power delivered to the load antenna.

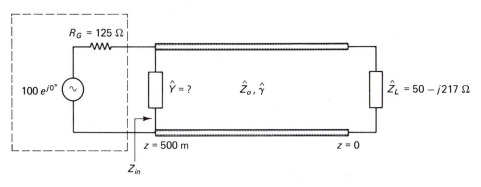

Figure P7.21   Geometry of lossy transmission line.

22. A lossless transmission line of $Z_o = 50$ ohms is terminated with an unknown impedance $\hat{Z}_L$. The standing-wave ratio on the line is 3. Successive voltage minima are 20 cm apart, and the first minimum is 5 cm from the load.
    (a) What is the terminating impedance?
    (b) Find the location and length of the short-circuited stub required to match the line.

23. A load impedance $\hat{Z}_L = 60 + j80\ \Omega$ terminates a 50-Ω transmission line. It is desired to use a double-stub tuner shown in Figure P7.23 with two short-circuited stubs spaced by a distance $d = \lambda/8$ to match $\hat{Z}_L$ to the transmission line. Determine the required lengths of the stubs.

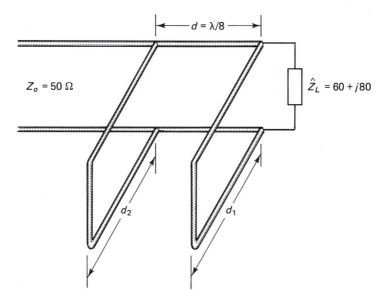

Figure P7.23   Double-stub tuner arrangement.

24. A single-stub tuner is to be designed to match a 75-Ω transmission line to an antenna load of impedance $\hat{Z}_L = 30 + j75\ \Omega$. The characteristic impedance of both the transmission line

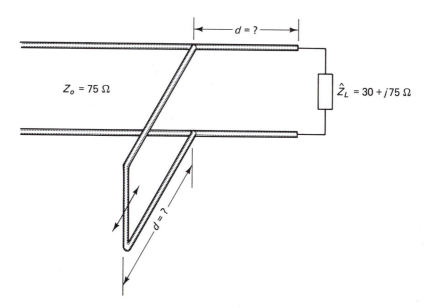

Figure P7.24   A single-stub matching arrangement.

and the short-circuited stub tuner is 75 Ω. The operating wavelength is λ = 1.5 m. Determine the distance from the load impedance to the tuning stub and the proper length of the stub. Examine the two possible locations of the stub and determine the length of the stub in each case.

25. A phased-array radar has ten antenna elements that are matched to their feed lines as shown in Figure P7.25. All transmission lines are 50 Ω coaxial lines. Find the location (d_1) and the length (d_2) of the stub tuner on the mainline that will maximize the power delivered to the antenna array. The operating frequency is 1 GHz.

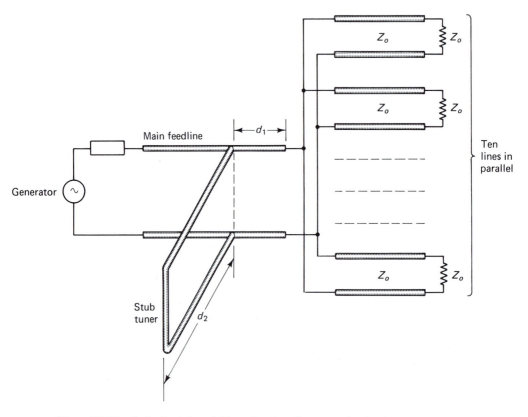

Figure P7.25   A single-stub matching of a phased-array radar that has ten elements.

26. A load impedance of $\hat{Z}_L = 100 + j50$ Ω is being matched to a lossless transmission line of characteristic impedance 300 Ω using two parallel stubs.
    (a) Show that if the stub arrangement is such that the spacing between the stubs is $3\lambda/8$, with one stub connected directly in parallel with the load, the matching is not possible.
    (b) If the double-stub matching arrangement is modified by adding a length of the 300 Ω transmission line between the load and the first stub:
        (i) Determine the minimum required additional line length d.
        (ii) Find the lengths of the short-circuited stub tuners required to achieve matching after introducing the length of the transmission line determined in part i.
27. The geometry of the double-stub tuner available in the laboratory is shown in Figure P7.27.

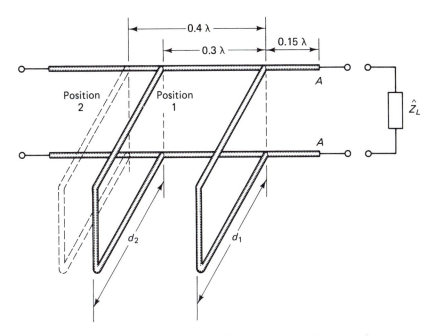

Figure P7.27   The double-stub matching device available in the laboratory. Determine which position, 1 or 2, you would select to locate the second stub to achieve matching.

The position of the first stub is fixed at 0.15λ from the A-A port, whereas the other stub may either be located at position 1, which is a distance 0.3λ from the first stub, or at position 2, which is a distance 0.4λ from the first stub.

It is desired to match an antenna of input impedance $\hat{Z}_L = 50 + j100$ Ω to a 50 Ω coaxial transmission line using this double-stub tuner. Determine which one of the two positions, 1 or 2, or both, is suitable for achieving the desired impedance matching. Also determine the lengths d_1 and d_2 of both stubs required to achieve this matching.

28. Two transmission lines are connected in parallel to a third one as shown in Figure P7.28. All the transmission lines have the same characteristic impedance of 50 Ω. Transmission line 1 is 0.4λ long and is terminated by a load impedance $\hat{Z}_{L1} = 50 + j75$ Ω. Transmission line 2 is 0.16λ long and is terminated by a load impedance $\hat{Z}_{L2} = 75 - j125$ Ω.
    (a) What are the standing wave ratios on transmission lines 1 and 2?
    (b) What is the minimum length of transmission line 3 (in fraction of a wavelength) so that the *input impedance* at section A-A is real? Also find the value of this input impedance.
    (c) If a transmitter of internal impedance equals 180 Ω and an output voltage $\hat{V}_G = 100e^{j0°}$ V is connected to the input section A-A of transmission line 3, find the value of the power absorbed by the load $\hat{Z}_{L1}$.

29. A lossless transmission line of characteristic impedance $Z_o = 50$ ohms is terminated in a load impedance that is to be determined. The VSWR on the transmission line is measured as 5.
    (a) Determine the load impedance if the distance between successive minima in the VSWR pattern is 0.2 m, and the distance from the load impedance to the first minimum is 0.08 m.

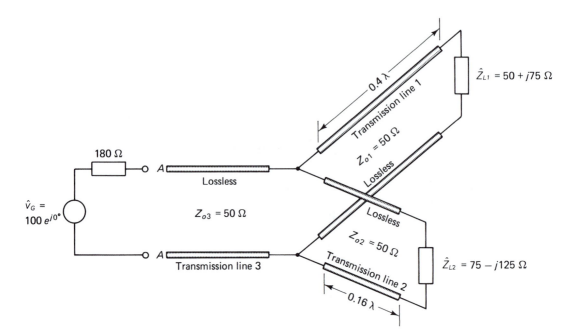

Transmission line 1

$\hat{Z}_{L1} = 50 + j75\ \Omega$

0.4 λ

$Z_{o1} = 50\ \Omega$

Lossless

180 Ω

A

Lossless

$Z_{o3} = 50\ \Omega$

Lossless

$Z_{o2} = 50\ \Omega$

$\hat{V}_G =$
$100\ e^{j0°}$

A

Transmission line 3

Transmission line 2

$\hat{Z}_{L2} = 75 - j125\ \Omega$

0.16 λ

Figure P7.28 Schematic diagram showing the parallel connection of the transmission line in problem 28.

(b) Repeat part a if a maximum of the voltage standing wave is found to occur at the termination of the transmission line.

30. (a) In an experiment to determine the unknown impedance of a microwave antenna, a slotted section of transmission line is first connected to a short-circuit load. In this case, adjacent voltage minima are found 23.6 and 35.4 cm away from the load. After replacing the short circuit with the antenna, the location of minima shifted to points 27.8 and 39.6 cm, respectively, away from the load with the VSWR value equal to 2.6. Assuming that the slotted transmission line is lossless of characteristic impedance $Z_o = 50\ \Omega$ and the velocity of propagation $u = c$, determine the following:

    (i) Operating frequency.

    (ii) Unknown impedance of microwave antenna.

(b) It is required to match the antenna in part a to a section of transmission line of characteristic impedance $Z_o = 50\ \Omega$. Show that the double-stub tuner available in the laboratory and shown in Figure P7.30 is *not suitable* to achieve the required matching.

(c) It is suggested that we insert a length d of a 50 Ω lossless transmission line between the antenna and the side A-A of the double stub tuner. Find the *minimum* length of such a transmission-line section and the lengths d_1 and d_2 of the stubs required to achieve matching.

31. (a) In an experiment to determine the unknown impedance of antenna, a 50-Ω, air-filled slotted section of transmission line is first connected to a short-circuit load. In this case, adjacent voltage minima are found at the 8-cm and 22-cm marks on the scale of the slotted line. After removing the short circuit and connecting the antenna, a VSWR value

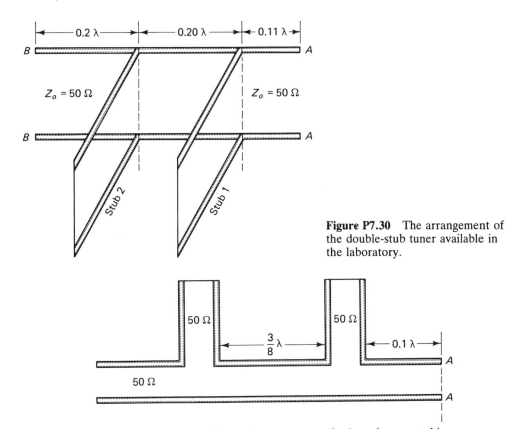

Figure P7.30 The arrangement of the double-stub tuner available in the laboratory.

Figure P7.31 Two series stubs arrangement for impedance matching.

of 3.6 was measured, and the location of the first minima was at the 12.8-cm mark along the same scale.

(i) Determine the operating frequency.

(ii) Find the unknown impedance of the antenna at the frequency in part (i).

(b) If it is required to match the antenna in part a to a section of transmission line of characteristic impedance $Z_o = 50 \ \Omega$, use the arrangement of the two series stubs shown in Figure P7.31. Also if the antenna is to be connected to the A-A port of the double stub arranged, find the reactances of these two stubs that are required to achieve matching.

(c) Find VSWRs between the antenna and the first stub, and between the two stubs.

32. An antenna has a measured input impedance of $\hat{Z}_L = 27.5 - j42.5 \ \Omega$ at $f = 400$ MHz. This antenna is to be fed from a 50-Ω coaxial line through the double-stub tuner shown in Figure P7.32. The characteristic impedance of each of the stubs is 50 Ω and is equal to the characteristic impedance of the section of the transmission line connecting them. Determine to which side of the double-stub tuner (side A-A or B-B) the antenna should be connected to achieve impedance matching. For the specific arrangement that you find adequate, find the following:

(a) Length d_1 and d_2 of two stubs required for matching.

(b) Value of VSWR between load and first stub, and VSWR between two stubs.

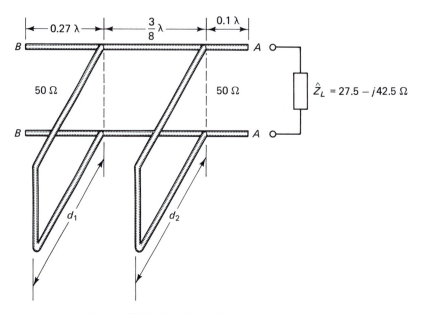

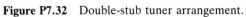

Figure P7.32   Double-stub tuner arrangement.

CHAPTER 8

WAVE GUIDES

8.1 INTRODUCTION

In the previous chapters we described the propagation characteristics of plane waves in unbounded medium, reflection and transmission of these waves at plane interfaces, and the mode of propagation as well as the fields configurations in two-conductor transmission lines. In two-conductor transmission lines we identified the TEM mode as the dominant field configuration. The electric and magnetic fields configurations in the TEM mode bear a close resemblance to those of a plane wave propagating in an unbounded medium. For example, it is shown that the electric and magnetic field components are perpendicular to each other and have no components along the direction of propagation. The absence of a cutoff frequency is yet another resemblance between the TEM mode in the two-conductor transmission line and the plane wave propagation in unbounded space.

In this chapter we describe the propagation characteristics in single conductor transmission lines often referred to as "wave guides." These wave guides are used in practice for high-frequency transmission. This is because, as we will see shortly, the

TEM mode cannot propagate in these single-conductor transmission lines. Only higher-order modes in the form of transverse electric (TE) and transverse magnetic (TM) modes can propagate in wave guides. These higher-order modes do have cutoff frequencies below which they will not propagate in the wave guide. Because the cutoff frequencies depend on the geometry and the dimensions of the wave guide, these wave guides are used only at higher frequencies to help keep their dimensions practical. In a way, the practical use of wave guides at higher frequencies complements the use of two-conductor transmission lines at lower frequencies. This is not because the dominant TEM mode in two-conductor transmission lines ceases to propagate at higher frequencies, but instead the transmission losses (attenuation) at these higher frequencies become excessive and impractical in two-conductor transmission lines. Single-conductor wave guides provide (whenever propagation is possible) larger cross-sectional area for the wave to propagate and, hence, less attenuation owing to ohmic losses in the conductors.

We start our analysis of wave guides by showing that the familiar TEM mode cannot propagate in a single-conductor wave guide. We then show that because the electromagnetic waves within the wave guide have to satisfy certain boundary conditions at the metallic walls of the wave guide, only a discrete number of modes or field configurations are possible and can propagate. We then analyze the various TE and TM modes in a rectangular wave guide. Various propagation characteristics such as cutoff frequencies, fields configurations, and attenuation resulting from ohmic losses on the wave-guide metallic walls are then described.

8.2 GUIDED MODES IN WAVE GUIDES

Let us start our discussion of the field configurations or modes within perfectly conducting cylindrical wave guides by showing that the TEM mode of propagation, which does not have a cutoff frequency nor electric or magnetic field components along the direction of propagation, is not possible in a single-conductor wave guide. Figure 8.1a shows the electric and magnetic field configurations associated with the TEM mode in two-conductor transmission line. According to Ampere's law, it can be seen that the transverse magnetic field **H** is supported by the axial flow of the conduction current I. In the single-conductor cylindrical wave guide shown in Figure 8.1b, conversely, the magnetic field **H** within the wave guide should be supported, according to Ampere's law, either by an axial flow of conduction current or an axial component of the electric field. In other words, if we briefly examine Ampere's law, we see that

$$\oint_c \mathbf{H} \cdot d\boldsymbol{\ell} = \underbrace{\int_s \mathbf{J} \cdot d\mathbf{s}}_{\substack{\text{True con-}\\\text{duction}\\\text{current}\\\text{similar to}\\\text{the case}\\\text{of a two-}\\\text{conductor}\\\text{transmis-}\\\text{sion line}}} \quad + \quad \underbrace{\frac{d}{dt}\int_s \epsilon_o \mathbf{E} \cdot d\mathbf{s}}_{\substack{\text{Displacement}\\\text{current that}\\\text{requires the}\\\text{presence of an}\\\text{axial component}\\\text{of the electrical}\\\text{field}}} \qquad (8.1)$$

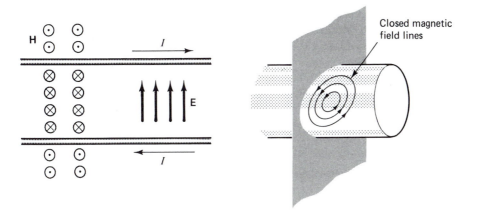

(a) Two conductor transmission line
with the **E** and **H** field configurations
in the TEM mode.

(b) TEM mode is not possible in a
single-conductor wave guide.

Figure 8.1 Illustration of the reason why the propagation of the TEM is
not possible in a cylindrical wave guide. The absence of the second conductor makes it necessary to have axial components of fields.

If we take the contour of integration c in the transverse plane perpendicular to the
direction of propagation, we note the following:

1. The first term on the right-hand side of equation 8.1 represents a true axial
   conduction-type current similar to the one that flows in the center conductor in
   a two-conductor transmission line.
2. The second term on the right-hand side of equation 8.1 represents the displacement current that requires the presence of an axial component of the electric field.

The absence of both a second conductor and the axial E field component in a
single-conductor wave guide make supporting a TEM mode not possible.

The next question would then be what type of modes (field configurations) should
we expect in a single-conductor wave guide? To answer this question and specifically
to identify the presence of cutoff frequencies in wave guides, let us consider a solution
of Faraday's and Ampere's equations.

Consider the electric and magnetic field vectors propagating along the z axis in
the cylindrical wave guide shown in Figure 8.1b.

$$\mathbf{E}(x,y,z,t) = \hat{\mathbf{E}}(x,y)\,e^{j\omega t \,\pm\, \hat{\gamma}z}$$

$$\mathbf{H}(x,y,z,t) = \hat{\mathbf{H}}(x,y)\,e^{j\omega t \,\pm\, \hat{\gamma}z}$$

where $\hat{\gamma}$ is the propagation constant, which is in general complex $\hat{\gamma} = \alpha + j\beta$. For time
harmonic fields, Faraday's law is given by

$$\nabla \times \hat{\mathbf{E}} = -j\omega\mu\hat{\mathbf{H}}$$

Writing this equation in terms of its various components for a positive z propagating wave $(e^{-\hat{\gamma}z})$, we obtain

$$\frac{\partial \hat{E}_z}{\partial y} + \hat{\gamma}\hat{E}_y = -j\omega\mu\,\hat{H}_x \tag{8.2a}$$

$$\frac{\partial \hat{E}_z}{\partial x} + \hat{\gamma}\hat{E}_x = j\omega\mu\,\hat{H}_y \tag{8.2b}$$

$$\frac{\partial \hat{E}_y}{\partial x} - \frac{\partial \hat{E}_x}{\partial y} = -j\omega\mu\,\hat{H}_z \tag{8.2c}$$

Similarly from the time harmonic expression of Ampere's law we obtain

$$\nabla \times \hat{\mathbf{H}} = j\omega\epsilon\,\hat{\mathbf{E}},$$

hence,

$$\frac{\partial \hat{H}_z}{\partial y} + \hat{\gamma}\hat{H}_y = j\omega\epsilon\,\hat{E}_x \tag{8.3a}$$

$$\frac{\partial \hat{H}_z}{\partial x} + \hat{\gamma}\hat{H}_x = -j\omega\epsilon\,\hat{E}_y \tag{8.3b}$$

$$\frac{\partial \hat{H}_y}{\partial x} - \frac{\partial \hat{H}_x}{\partial y} = j\omega\epsilon\,\hat{E}_z \tag{8.3c}$$

If we substitute $\hat{E}_y$ from equation 8.3b into equation 8.2a, $\hat{E}_x$ from equation 8.3a into equation 8.2b, $\hat{H}_x$ from equation 8.2a into equation 8.3b, and $\hat{H}_y$ from equation 8.2b into equation 8.3a, we obtain the following expressions for the transverse components of the electric and magnetic fields:

$$\hat{E}_y = \frac{1}{\hat{\gamma}^2 + \omega^2\mu\epsilon}\left(j\omega\mu\frac{\partial \hat{H}_z}{\partial x} - \hat{\gamma}\frac{\partial \hat{E}_z}{\partial y}\right)$$

$$\hat{E}_x = -\frac{1}{\hat{\gamma}^2 + \omega^2\mu\epsilon}\left(\hat{\gamma}\frac{\partial \hat{E}_z}{\partial x} + j\omega\mu\frac{\partial \hat{H}_z}{\partial y}\right)$$

$$\hat{H}_y = -\frac{1}{\hat{\gamma}^2 + \omega^2\mu\epsilon}\left(j\omega\epsilon\frac{\partial \hat{E}_z}{\partial x} + \hat{\gamma}\frac{\partial \hat{H}_z}{\partial y}\right) \tag{8.4}$$

$$\hat{H}_x = \frac{1}{\hat{\gamma}^2 + \omega^2\mu\epsilon}\left(j\omega\epsilon\frac{\partial \hat{E}_z}{\partial y} - \hat{\gamma}\frac{\partial \hat{H}_z}{\partial x}\right)$$

Equation 8.4 simply shows the following:

1. All the transverse electric and magnetic field components are expressed in terms of the axial components $\hat{E}_z$ and $\hat{H}_z$. Therefore, in our solution for the various modes (field configurations) in wave guides we need only to find the z-directed components of the electric and magnetic fields.
2. If $\hat{E}_z = 0 = \hat{H}_z$, all the other components of the electric and magnetic fields would also be zero. It should be noted, however, that for $\hat{E}_z = \hat{H}_z = 0$, (i.e., TEM wave) a nontrivial solution may still exist if the denominator $\hat{\gamma}^2 + \omega^2\mu\epsilon = 0$. In this case,

the propagation constant $\hat{\gamma} = j\omega\sqrt{\mu\epsilon}$, which is identical to the propagation constant of TEM waves in two-conductor transmission lines and to the propagation constant of a plane wave in an unbounded medium. Because the TEM mode is not possible in the single-conductor wave guides, neither this mode nor any other field configuration that has both $\hat{H}_z = \hat{E}_z = 0$ can propagate in single-conductor wave guides.

3. In solving equation 8.4 for the transverse field components in terms of the axial components $\hat{E}_z$ and $\hat{H}_z$, we may use one of the following procedures:

    (a) Assume the simultaneous presence of both $\hat{E}_z$ and $\hat{H}_z$, and solve for the transverse components using equation 8.4. This type of wave is known as the hybrid wave.

    (b) We may assume the presence of only the axial component of the electric field (i.e., $\hat{E}_z \neq 0$, whereas $\hat{H}_z = 0$) and solve for the transverse components under this condition using equation 8.4. Clearly, in this case all the magnetic field components will be transverse to the direction of propagation. The mode of propagation associated with such fields structure is, hence, called the transverse magnetic (TM) mode.

    (c) We may also assume the presence of only the axial magnetic field component (i.e., $\hat{H}_z \neq 0$, whereas $\hat{E}_z = 0$). The mode of propagation in this case does not have an electric field component along the direction of propagation and, hence, is called the transverse electric (TE) mode.

Clearly, the general solution in part a, which assumes the simultaneous presence of both $\hat{H}_z$ and $\hat{E}_z$ can be obtained from the special-case solutions in parts b and c using the superposition. Hence, to simplify our wave-guide analysis procedure, we will solve for the TM ($\hat{H}_z = 0$) and TE ($\hat{E}_z = 0$) modes and the general solution (if needed) can be obtained using the superposition.

To solve for the z components of the electric and magnetic fields, we once again use Faraday's and Ampere's laws

$$\nabla \times \hat{\mathbf{E}} = -j\omega\mu\hat{\mathbf{H}}, \qquad \nabla \times \hat{\mathbf{H}} = j\omega\epsilon\hat{\mathbf{E}}$$

Taking the curl of both sides of Faraday's law and using Gauss's law to substitute $\nabla \cdot \hat{\mathbf{E}} = 0$, we obtain

$$(\nabla^2 + \omega^2\mu\epsilon)\hat{\mathbf{E}} = 0$$

which is the homogeneous wave equation for the time harmonic electric field. Limiting the preceding expression to the desired $\hat{E}_z$ component of the electric field, we obtain

$$(\nabla^2 + \omega^2\mu\epsilon)\hat{E}_z = 0$$

Similarly, the wave equation for the axial component of the magnetic field is given by

$$(\nabla^2 + \omega^2\mu\epsilon)\hat{H}_z = 0$$

Expressing the Laplacian in terms of its three components

$$\nabla^2 = \frac{\partial^2}{\partial x^2} + \frac{\partial^2}{\partial y^2} + \frac{\partial^2}{\partial z^2}$$

and using the fact that we assumed a wave propagating along the axial z direction of the wave guide—that is, assumed $e^{\pm\hat{\gamma}z}$ for the z variation of the fields—we obtain

$$\left[\boldsymbol{\nabla}_t^2 + \left(\omega^2\mu\epsilon + \hat{\gamma}^2\right)\right]\big|_{\hat{H}_z}^{\hat{E}_z} = 0 \tag{8.5}$$

where $\boldsymbol{\nabla}_t^2$ stands for the Laplacian in terms of the transverse x and y coordinates.

As indicated earlier, to simplify our wave-guide analysis procedure we will solve equation 8.5 for $\hat{E}_z$ assuming $\hat{H}_z = 0$—that is, TM modes—and then solve equation 8.5, one more time, for $\hat{H}_z$ assuming $\hat{E}_z = 0$—that is, TE modes. The general solution for the field configurations propagating in wave guides may be obtained from the superposition of the TM and TE modes. In the following sections we will develop detailed solutions of equation 8.5 for TM and TE modes propagating in rectangular wave guides. These rectangular wave guides are the most commonly used wave guides in practice and fortunately the easiest to analyze.

8.3 TM MODES IN RECTANGULAR WAVE GUIDES

Consider the rectangular wave guide shown in Figure 8.2. The broader side of the rectangular cross section of the hollow metallic cylinder is of length a, whereas the length of the shorter side is b. The broader side is placed along the x axis as shown in Figure 8.2. To solve for the TM modes ($\hat{H}_z = 0$) in this wave guide, we will solve for $\hat{E}_z$ in equation 8.5 using the separation of variables. This solution procedure involves expressing the electric field $\hat{E}_z$ as the product of three functions, each of which is a function of only one of the coordinate variables. Hence,

$$\hat{E}_z(x,y,z) = X(x)Y(y)e^{-\hat{\gamma}z} \tag{8.6}$$

Substituting equation 8.6 in the Helmholtz equation 8.5 for the axial electric $\hat{E}_z$, we obtain

$$\frac{1}{X}\frac{d^2X}{dx^2} + \frac{1}{Y}\frac{d^2Y}{dy^2} + \left(\hat{\gamma}^2 + \omega^2\mu\epsilon\right) = 0 \tag{8.7}$$

Because X is a function of x only and Y is a function of y only, each term on the left side of equation 8.7 must be a constant. If this is not the case, equation 8.7 could not

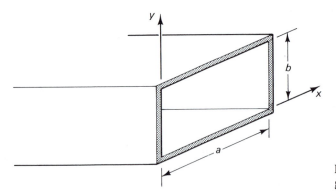

Figure 8.2   Geometry of a rectangular wave guide.

be satisfied for all values of x and y. Equating each of the x- and y-dependent functions to a constant, we obtain:

$$\frac{1}{X}\frac{d^2 X}{dx^2} = -M^2 \tag{8.8}$$

$$\frac{1}{Y}\frac{d^2 Y}{dy^2} = -N^2 \tag{8.9}$$

The sign of these constants and the fact that each is in the form of a square of a number is arbitrarily chosen to simplify the solution. Substituting equation 8.8 and equation 8.9 in equation 8.7, we obtain

$$-M^2 - N^2 + \left(\hat{\gamma}^2 + \omega^2 \mu\epsilon\right) = 0 \tag{8.10}$$

Solution to equation 8.8 is given by

$$X = A \sin Mx + B \cos Mx$$

Solution of equation 8.9 is given by

$$Y = C \sin Ny + D \cos Ny$$

Hence, the complete solution for the axial component of the phasor electric field $\hat{E}_z$ is

$$\hat{E}_z = (A \sin Mx + B \cos Mx)(C \sin Ny + D \cos Ny)\,e^{-\hat{\gamma}z} \tag{8.11}$$

To determine the unknown constants A, B, C, D, M, and N we require that the expression for $\hat{E}_z$ in equation 8.11 satisfies the boundary conditions on the metallic walls of the wave guide. These boundary conditions require that the tangential electric field must be zero at the walls of the guide. Hence, from Figure 8.2, we have

$$
\begin{aligned}
\hat{E}_z &= 0 \text{ at } x = 0 \\
\hat{E}_z &= 0 \text{ at } x = a \\
\hat{E}_z &= 0 \text{ at } y = 0 \\
\hat{E}_z &= 0 \text{ at } y = b
\end{aligned}
\tag{8.12}
$$

Applying the boundary conditions

$$\hat{E}_z\big|_{x=0} = 0 \qquad \text{and} \qquad \hat{E}_z\big|_{y=0} = 0$$

we find that the coefficients B and D should be zero. Hence, the remainder can be written as

$$\hat{E}_z = A' \sin Mx \sin Ny\, e^{-\hat{\gamma}z} \tag{8.13}$$

where $A' = AC$, which is an arbitrary amplitude of the electric field to be determined based on the amount of the input power to the wave guide. The two other remaining unknowns M and N in equation 8.13 should be obtained by satisfying the boundary conditions at $x = a$ and $y = b$.

Applying the boundary condition $\hat{E}_z|_{x=a} = 0$, we obtain

$$A' \sin Ma \sin Ny = 0 \tag{8.14}$$

Because A' cannot be made zero (trivial solution), equation 8.14 may, hence, be satisfied if Ma is chosen such that

$$Ma = m\pi, \qquad m = 1, 2, 3, \ldots$$

or the constant M is given by

$$M = \frac{m\pi}{a}$$

Similarly, by satisfying the boundary condition $\hat{E}_z|_{y=b} = 0$, we obtain

$$N = \frac{n\pi}{b}, \qquad n = 1, 2, 3, \ldots$$

Equation 8.13 for the axial component of the electric field is, hence, given by

$$\boxed{\hat{E}_z = A' \sin\left(\frac{m\pi}{a}x\right) \sin\left(\frac{n\pi}{b}y\right) e^{-\hat{\gamma}z}} \qquad \begin{array}{l} m = 1, 2, 3, \ldots \\ \\ n = 1, 2, 3, \ldots \end{array} \tag{8.15}$$

From equation 8.15, it is clear that there is an infinite set of the integers m and n that may be chosen. For each pair of m and n there is a corresponding field structure within the wave guide designated as TM_{mn} mode; TM mode because of the absence of the axial component of the magnetic field $\hat{H}_z = 0$. The subscript mn, conversely, indicates the mode number. The lowest order TM mode is the TM_{11} mode. Before we describe the field configurations associated with the various modes, however, let us discuss the impact of having this discrete number of modes on the propagation characteristics of the TM modes. Substituting $M = m\pi/a$ and $N = n\pi/b$ in the propagation constant equation 8.10, we obtain

$$-\left(\frac{m\pi}{a}\right)^2 - \left(\frac{n\pi}{b}\right)^2 + (\omega^2 \mu\epsilon + \hat{\gamma}^2) = 0 \tag{8.16}$$

The propagation constant $\hat{\gamma}$ is, hence, given by

$$\hat{\gamma} = \sqrt{\left(\frac{m\pi}{a}\right)^2 + \left(\frac{n\pi}{b}\right)^2 - \omega^2 \mu\epsilon} \tag{8.17}$$

From our study of the plane wave propagation in conductive medium we indicated that for the wave propagation to occur, $\hat{\gamma}$ must be an imaginary number $\hat{\gamma} = j\beta$ (or at least have a nonzero imaginary part). This can be further explained by obtaining the real-time form of the electric field $\hat{E}_z$. If $\hat{\gamma}$ is imaginary,

$$\hat{\gamma} = j\beta_{mn} = j\sqrt{\omega^2 \mu\epsilon - \left(\frac{m\pi}{a}\right)^2 - \left(\frac{n\pi}{b}\right)^2} \tag{8.18}$$

Then multiplying the $\hat{E}_z$ by $e^{j\omega t}$ and taking the real part to obtain the time-domain form of E_z we obtain

$$E_z(x,y,z,t) = A' \sin\left(\frac{m\pi}{a}x\right) \sin\left(\frac{n\pi}{b}y\right) Re(e^{-\hat{\gamma}z} e^{j\omega t})$$

$$= A' \sin\left(\frac{m\pi}{a}x\right) \sin\left(\frac{n\pi}{b}y\right) \cos(\omega t - \beta_{mn} z)$$

which represents a cosine wave traveling in the positive z direction. We also realize from equation 8.18 that $\hat{\gamma}$ will be purely imaginary to provide a traveling wave propagation characteristic if

$$\omega^2 \mu\epsilon > \left[\left(\frac{m\pi}{a}\right)^2 + \left(\frac{n\pi}{b}\right)^2\right] \tag{8.19}$$

If, conversely, $\omega^2 \mu\epsilon < (m\pi/a)^2 + (n\pi/b)^2$, $\hat{\gamma}$ in equation 8.17 will be real—that is, $\hat{\gamma} = \alpha_{mn}$, where α_{mn} is a real number. Obtaining the time-domain expression for the electric field will include the term $Re(e^{-\alpha_{mn}z} e^{j\omega t}) = e^{-\alpha_{mn}z} \cos \omega t$, which indicates a nonpropagating mode, and instead an attenuating wave with an attenuation factor $e^{-\alpha_{mn}z}$ along the z direction. These nonpropagating modes are called "evanescent" modes. Thus, to have propagation within the wave guide, we require that the guide dimensions a and b, and the frequency of excitation ω satisfy equation 8.19. The cutoff frequency that is defined as the lowest possible excitation frequency for a wave propagation to occur along the guide axis is obtained by substituting $\hat{\gamma} = 0$ in equation 8.16. Hence,

$$\omega_c \sqrt{\mu\epsilon} = \sqrt{\left(\frac{m\pi}{a}\right)^2 + \left(\frac{n\pi}{b}\right)^2}$$

$$\boxed{f_{c,mn} = \frac{1}{2\pi\sqrt{\mu\epsilon}} \sqrt{\left(\frac{m\pi}{a}\right)^2 + \left(\frac{n\pi}{b}\right)^2}} \tag{8.20}$$

At frequencies $f > f_{c,mn}$, the propagation constant $\hat{\gamma}$ will be purely imaginary and is called, in this case, the phase constant β_{mn}. This phase constant is given from equation 8.18 by

$$\beta_{mn} = \sqrt{\omega^2 \mu\epsilon - \left(\frac{m\pi}{a}\right)^2 - \left(\frac{n\pi}{b}\right)^2}$$

In terms of the cutoff frequency $f_{c,mn}$, the phase constant β_{mn} may be given by

$$\boxed{\beta_{mn} = \omega\sqrt{\mu\epsilon} \sqrt{1 - \left(\frac{f_{c,mn}}{f}\right)^2}} \tag{8.21}$$

The final expression for the axial electric field $\hat{E}_z$ for a propagating mode is, hence, given by

$$\hat{E}_z = A' \sin\left(\frac{m\pi}{a}x\right) \sin\left(\frac{n\pi}{b}y\right) e^{-j\beta_{mn}z} \tag{8.22}$$

The remaining field components may be obtained by substituting $\hat{E}_z$ given in equation 8.22 in equation 8.4 and noting the fact that $\hat{H}_z = 0$ for TM modes. Once again we should note that the lowest-order TM mode is the TM_{11} mode because setting either m or n equal to zero in equation 8.22 renders a trivial solution with all zero field components.

The wavelength λ_{mn} of the TM_{mn} mode in the wave guide is defined as the distance in the z direction of propagation required for a phase change of 2π (rad). Hence,

$$\lambda_{mn} \beta_{mn} = 2\pi,$$

or

$$\boxed{\lambda_{mn} = \frac{2\pi}{\beta_{mn}} = \frac{\lambda}{\sqrt{1 - \left(\frac{f_{c,mn}}{f}\right)^2}}} \tag{8.23}$$

where $\lambda = 2\pi/\omega\sqrt{\mu\epsilon} = 2\pi/\beta_o$ is the wavelength of a uniform plane wave propagating in the same medium (ϵ, μ) with a phase constant $\beta_o = \omega\sqrt{\mu\epsilon}$.

From equation 8.23, it can be seen that the wave guide wavelength λ_{mn} is longer than the wavelength λ of a plane wave propagating at the same frequency. This apparently longer guide wavelength may be explained by using the interpretation of the wave-guide modes as the superposition of plane waves bouncing back and forth between the guide conducting boundaries at various angles. Because the mathematical development of such analogy is rather involved for the general TM_{mn} mode of propagation, it suffices at this point to mention that it is possible to show that wave-guide modes may be viewed as the superposition of uniform plane waves propagating and bouncing off the metallic boundaries at various angles. Because these bouncing waves are propagating at an angle with respect to the true propagation direction along the axis of the wave guide, the measured wavelength along the waveguide axis λ_{mn} is related to that along the direction of propagation of the bouncing plane waves λ through a geometrical projection process that results in this apparently longer wavelength. Such a process may be further visualized by considering the physical example of waves approaching the seashore at some angle with respect to the shore line. As the wave strikes the shore, the wave crest appears to be separated by a distance along the shore longer than the actual wavelength along the actual direction of propagation that is perpendicular to the crest. The interpretation of wave-guide modes as the superposition of plane waves is illustrated in example 8.3.

Similar arguments may be used in explaining the fact that the phase velocity of a propagating mode is actually larger than the velocity of light.

The phase velocity of the TM_{mn} mode is given by

$$v_{pmn} = \frac{\omega}{\beta_{mn}}$$

Substituting β_{mn} from equation 8.21, we obtain

$$v_{pmn} = \frac{1/\sqrt{\mu\epsilon}}{\sqrt{1 - \left(\frac{f_{c,mn}}{f}\right)^2}} = \frac{v_p}{\sqrt{1 - \left(\frac{f_{c,mn}}{f}\right)^2}} \tag{8.24}$$

which shows that for a propagating mode, $f > f_{c,mn}$ and $v_{pmn} > v_p$, which is the speed of light in the material filling the wave guide. According to the arguments presented earlier, v_{pmn} is just the projection of the phase velocity of the bouncing plane waves on the direction of propagation along the axis of the wave guide. Moreover, it should be noted that v_{pmn} is just the velocity of the constant phase fronts of the wave, which is different from the velocity of energy transmission.

The final observation regarding the analogous quantities in wave guides and uniform plane waves propagating in unbounded space is related to the intrinsic wave impedance of the mode. Using equation 8.22 for the axial electric field expression to obtain the transverse components in equation 8.4, for TM modes ($\hat{H}_z = 0$), we obtain

$$\hat{E}_z = A' \sin\left(\frac{m\pi}{a}x\right) \sin\left(\frac{n\pi}{b}y\right) e^{-j\beta_{mn} z} \tag{8.25a}$$

$$\hat{E}_y = \frac{-j\beta_{mn}\left(\frac{n\pi}{b}\right)}{\left(\frac{m\pi}{a}\right)^2 + \left(\frac{n\pi}{b}\right)^2} A' \sin\left(\frac{m\pi}{a}x\right) \cos\left(\frac{n\pi}{b}y\right) e^{-j\beta_{mn} z} \tag{8.25b}$$

$$\hat{E}_x = \frac{-j\beta_{mn}\left(\frac{m\pi}{a}\right)}{\left(\frac{m\pi}{a}\right)^2 + \left(\frac{n\pi}{b}\right)^2} A' \cos\left(\frac{m\pi}{a}x\right) \sin\left(\frac{n\pi}{b}y\right) e^{-j\beta_{mn} z} \tag{8.25c}$$

$$\hat{H}_y = \frac{-j\omega\epsilon\left(\frac{m\pi}{a}\right)}{\left(\frac{m\pi}{a}\right)^2 + \left(\frac{n\pi}{b}\right)^2} A' \cos\left(\frac{m\pi}{a}x\right) \sin\left(\frac{n\pi}{b}y\right) e^{-j\beta_{mn} z} \tag{8.25d}$$

$$\hat{H}_x = \frac{j\omega\epsilon\left(\frac{n\pi}{b}\right)}{\left(\frac{m\pi}{a}\right)^2 + \left(\frac{n\pi}{b}\right)^2} A' \sin\left(\frac{m\pi}{a}x\right) \cos\left(\frac{n\pi}{b}y\right) e^{-j\beta_{mn} z} \tag{8.25e}$$

where equation 8.10 was used to replace the coefficients $\hat{\gamma}^2 + \omega^2\mu\epsilon$ in equation 8.4 by $(m\pi/a)^2 + (n\pi/b)^2$.

From the transverse field expressions in equation 8.25, it is clear that

$$\frac{\hat{E}_x}{\hat{H}_y} = \frac{\beta_{mn}}{\omega\epsilon}$$

Substituting β_{mn} from equation 8.21, we obtain

$$\frac{\hat{E}_x}{\hat{H}_y} = \sqrt{\frac{\mu}{\epsilon}}\sqrt{1 - \left(\frac{f_{c,mn}}{f}\right)^2} \tag{8.26}$$

Similarly, the two other transverse electric and magnetic field components $\hat{E}_y$ and $\hat{H}_x$ are related by

$$\frac{\hat{E}_y}{\hat{H}_x} = -\frac{\beta_{mn}}{\omega\epsilon} = -\sqrt{\frac{\mu}{\epsilon}}\sqrt{1 - \left(\frac{f_{c,mn}}{f}\right)^2} \tag{8.27}$$

Equations 8.26 and 8.27 suggest that the transverse field components are related by the same coefficient that we may define as the intrinsic wave impedance of the TM_{mn} mode, that is $\eta_{\text{TM}_{mn}}$,

$$\boxed{\eta_{\text{TM}_{mn}} = \sqrt{\frac{\mu}{\epsilon}}\sqrt{1 - \left(\frac{f_{c,mn}}{f}\right)^2}}$$

$$\eta_{\text{TM}_{mn}} = \eta_o \sqrt{1 - \left(\frac{f_{c,mn}}{f}\right)^2} \tag{8.28}$$

where $\eta_o = \sqrt{\mu/\epsilon}$ is the intrinsic wave impedance of a plane wave propagating in an unbounded medium of constitutive parameters μ and ϵ.

Thus, the transverse fields are related by

$$\hat{E}_x = \eta_{\text{TM}_{mn}}\hat{H}_y \tag{8.29a}$$

$$\hat{E}_y = -\eta_{\text{TM}_{mn}}\hat{H}_x = \eta_{\text{TM}_{mn}}(-\hat{H}_x) \tag{8.29b}$$

where the negative sign relating $\hat{E}_y$ to $\hat{H}_x$ is consistent with the direction of the power flow of the TM modes propagating in the positive z direction. As indicated in chapter 3, the direction of the power flow is provided by the Poynting vector $\mathbf{E} \times \mathbf{H}$. Hence, equation 8.29a shows that $\hat{E}_x$ and $\hat{H}_y$ fields are associated with a positive z-propagating wave, whereas equation 8.29b shows that $\hat{E}_y$ and $(-\hat{H}_x)$ or alternatively $(-\hat{E}_y)$ and $\hat{H}_x$ are associated with the positive z-traveling wave. Also from equation 8.28, it can be seen that for $f > f_{c,mn}$, the intrinsic wave impedance increases with the increase in frequency and asymptotically approaches η_o as shown in Figure 8.3. At $f = f_{c,mn}$, $\eta_{\text{TM}_{mn}}$

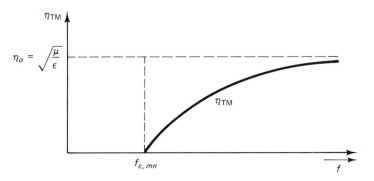

Figure 8.3 Wave impedance of TM modes.

is zero and the wave guide is effectively short circuited. As described earlier, at frequencies $f \leqslant f_{c,mn}$ it is not possible to propagate fields in this particular TM_{mn} mode in the wave guide.

EXAMPLE 8.1

An air-filled rectangular wave guide has 2″ × 4″ inside dimensions. Determine the cutoff frequency of the TM_{11} mode in this guide. Also calculate the phase constant β_{11} and the intrinsic wave impedance at an operating frequency f that is 30 percent higher than the cutoff frequency.

Solution

The wave guide dimensions are

$$a = 4'' = 0.1016 \text{ m}$$

$$b = 0.0508 \text{ m}$$

The cutoff frequency for the TM_{11} mode is

$$f_{c_{11}} = \frac{1}{2\pi\sqrt{\mu_o\,\epsilon_o}}\sqrt{\left(\frac{\pi}{a}\right)^2 + \left(\frac{\pi}{b}\right)^2}$$

$$= \frac{3 \times 10^8}{2}\sqrt{\left(\frac{1}{a}\right)^2 + \left(\frac{1}{b}\right)^2} = 3.3 \text{ GHz}$$

$$\beta_{11} = \omega\sqrt{\mu_o\,\epsilon_o}\sqrt{1 - \left(\frac{f_{c_{11}}}{f}\right)^2}$$

$$= \frac{2\pi \times 1.3 \times 3.3 \times 10^9}{3 \times 10^8}\sqrt{1 - \left(\frac{f_{c_{11}}}{1.3f_{c_{11}}}\right)^2}$$

$$= 57.41 \text{ rad/m}$$

$$\eta_{TM_{11}} = \sqrt{\frac{\mu_o}{\epsilon_o}}\sqrt{1 - \left(\frac{f_{c_{11}}}{1.3f_{c_{11}}}\right)^2} = 120\pi\sqrt{1 - \left(\frac{1}{1.3}\right)^2}$$

$$= 240.89 \ \Omega$$

8.4 TE MODES IN RECTANGULAR WAVE GUIDES

For TE modes, the axial component of the electric field $\hat{E}_z = 0$ and, hence, all the transverse electric and magnetic field components in equation 8.4 may be obtained in terms of the solution for $\hat{H}_z$ using the homogeneous wave equation 8.5. Following a separation of variable solution procedures similar to that used in the TM modes case, the general expression for $\hat{H}_z$ will be

$$\hat{H}_z = (A \sin Mx + B \cos Mx)(C \sin Ny + D \cos Ny)e^{-\hat{\gamma}z} \tag{8.30}$$

Equation 8.30 assumes a mode propagating along the positive z direction. To determine the unknown constants M, N, A, B, C, and D we need to enforce the boundary conditions. Unlike the TM modes case, the boundary conditions for $\hat{H}_z$ in the TE modes case are rather involved and certainly less straightforward. A simpler way to determine the unknown constants in equation 8.30 is to obtain the transverse electric field components from equation 8.30 and determine the unknown coefficients by enforcing zero tangential electric field values at the metallic boundaries of the wave guide.

From equation 8.4, the electric field components for TE mode ($\hat{E}_z = 0$) are given by

$$\hat{E}_y = \frac{j\omega\mu}{\hat{\gamma}^2 + \omega^2\mu\epsilon} \frac{\partial\hat{H}_z}{\partial x} \tag{8.31a}$$

$$\hat{E}_x = \frac{-j\omega\mu}{\hat{\gamma}^2 + \omega^2\mu\epsilon} \frac{\partial\hat{H}_z}{\partial y} \tag{8.31b}$$

Hence, the boundary conditions that should be satisfied in this case are

$$\hat{E}_y|_{x=0} = 0 \tag{8.32a}$$

$$\hat{E}_y|_{x=a} = 0 \tag{8.32b}$$

$$\hat{E}_x|_{y=0} = 0 \tag{8.32c}$$

$$\hat{E}_x|_{y=b} = 0 \tag{8.32d}$$

Substituting equation 8.30 in equation 8.31 and enforcing the conditions in equation 8.32a and 8.32c, we obtain

$$\hat{H}_z = A' \cos Mx \cos Ny\, e^{-\hat{\gamma}z} \tag{8.33}$$

where $A' = BD$ is a constant related to the input power to the wave guide. Enforcing equation 8.32b and remembering that A' cannot be zero (trivial solution), we find that the following condition should be satisfied:

$$\sin(Ma) = 0, \quad \text{hence} \quad Ma = m\pi \quad \text{or} \quad M = \frac{m\pi}{a}, \quad m = 0, 1, 2$$

Similarly, the condition necessary to satisfy equation 8.32d is

$$N = \frac{n\pi}{b}, \quad n = 0, 1, 2, \ldots$$

Substituting these values of M and N in equation 8.33, we obtain

$$\boxed{\hat{H}_z = A' \cos\left(\frac{m\pi}{a}x\right) \cos\left(\frac{n\pi}{b}y\right) e^{-\hat{\gamma}_{mn}z}}$$

$$m = 0, 1, 2, \ldots$$

$$n = 0, 1, 2, \ldots$$

$$\tag{8.34}$$

It should be noted that unlike the TM case the $m = 0$, and $n = 0$ values are allowed in equation 8.34.

The propagation constant $\hat{\gamma}_{mn}$ is given from equation 8.17 by

$$\hat{\gamma}_{mn} = \sqrt{\left(\frac{m\pi}{a}\right)^2 + \left(\frac{n\pi}{b}\right)^2 - \omega^2 \mu\epsilon} \tag{8.35}$$

Once again, true propagation $\hat{\gamma}_{mn} = j\beta_{mn}$ occurs when

$$\omega^2 \mu\epsilon > \left[\left(\frac{m\pi}{a}\right)^2 + \left(\frac{n\pi}{b}\right)^2\right]$$

while evanescent or non-propagating modes occur when

$$\omega^2 \mu\epsilon < \left[\left(\frac{m\pi}{a}\right)^2 + \left(\frac{n\pi}{b}\right)^2\right]$$

The cutoff frequency for a TE_{mn} mode is at $\hat{\gamma}_{mn} = 0$, and is given by

$$\boxed{f_{c,mn} = \frac{1}{2\pi\sqrt{\mu\epsilon}}\sqrt{\left(\frac{m\pi}{a}\right)^2 + \left(\frac{n\pi}{b}\right)^2}} \tag{8.36}$$

which is the same as that given by equation 8.20 for the TM_{mn} modes case. An important difference between the cutoff frequencies in the TM_{mn} and TE_{mn} modes arises from the fact that for TE modes $m = 0$ or $n = 0$ are allowed. This, however, is not the case for the TM modes because either $m = 0$ or $n = 0$ would provide a trivial solution. It should be noted that although $m = 0$ and $n = 0$ is a mathematically allowed solution in equation 8.34, the transverse field components associated with the $\hat{H}_z = $ constant that result from substituting $m = 0$ and $n = 0$ are all zero in this case. This does not support a physical mode and hence, the first physically possible solution for the TE_{mn} modes is for $m = 0$ and $n = 1$ or $n = 0$ and $m = 1$, depending on the cross-sectional dimensions of the wave guide. For example, for a wave guide with $b < a$, the lowest-order mode is the TE_{10}. For this mode the cutoff frequency is

$$f_{c_{10}} = \frac{1}{2\pi\sqrt{\mu\epsilon}}\sqrt{\left(\frac{\pi}{a}\right)^2} = \frac{v_p}{2a} \tag{8.37}$$

For the TE_{01} mode (i.e., $m = 0, n = 1$), conversely, the cutoff frequency is given by

$$f_{c_{01}} = \frac{1}{2\pi\sqrt{\mu\epsilon}}\sqrt{0 + \left(\frac{\pi}{b}\right)^2} = \frac{v_p}{2b}$$

because we assumed $b < a$, $f_{c_{01}} > f_{c_{10}}$ and the lowest order mode in this case ($b < a$) is the TE_{10} mode.

The expression for the intrinsic wave impedance $\eta_{\text{TE}_{mn}}$ for TE_{mn} modes is different from $\eta_{\text{TM}_{mn}}$ of the TM_{mn} modes. Substituting equation 8.34 in equation 8.4 and remembering that $\hat{E}_z = 0$ for the TE modes, it is rather straightforward to show that the ratio between the transverse electric and magnetic field components is given by

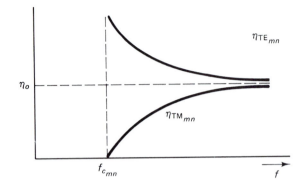

Figure 8.4 Comparison between the variations of the η_{TE} and η_{TM} as a function of frequency.

$$\eta_{TEmn} = \frac{\hat{E}_x}{\hat{H}_y} = \frac{\omega\mu}{\beta_{mn}} = \frac{\sqrt{\frac{\mu}{\epsilon}}}{\sqrt{1 - \left(\frac{f_{cmn}}{f}\right)^2}} = \frac{\eta_o}{\sqrt{1 - \left(\frac{f_{cmn}}{f}\right)^2}}\,\Omega$$

and

$$\boxed{\frac{\hat{E}_y}{\hat{H}_x} = \frac{-\omega\mu}{\beta_{mn}} = \frac{-\eta_o}{\sqrt{1 - \left(\frac{f_{cmn}}{f}\right)^2}} = -\eta_{TEmn}}$$ (8.38)

where η_o is the intrinsic impedance of a uniform plane wave in an unbounded medium.

There is an important difference between the intrinsic wave impedances of the TM modes given in equation 8.28 and that of the TE modes given in equation 8.38. η_{TM} in equation 8.28 is zero at the cutoff frequency $f = f_c$ and increases with the increase in frequency until it reaches its asymptotic value η_o at frequencies $f \gg f_c$. η_{TE}, conversely, approaches infinity at $f = f_c$ and decreases with the increase in frequency approaching its asymptotic value $\eta_{TE} = \eta_o$ at $f \gg f_c$. Figure 8.4 illustrates these variations of the η_{TE} and η_{TM} as a function of frequency.

EXAMPLE 8.2

Determine the cutoff frequencies of several of the lowest-order TE and TM modes in a rectangular wave guide with the following side ratio:

1. $b/a = 1/2$
2. $b/a = 1$

Solution

The expressions for the cutoff frequencies for both TE_{mn} and TM_{mn} modes are the same and given by

$$f_{c,mn} = \frac{1}{2\pi\sqrt{\mu\epsilon}}\sqrt{\left(\frac{m\pi}{a}\right)^2 + \left(\frac{n\pi}{b}\right)^2}$$

As indicated earlier the lowest possible TM mode is TM_{11} for which $m = 1$ and $n = 1$, whereas the lowest-order TE mode is TE_{10} ($m = 1, n = 0$) or TE_{01} ($m = 0, n = 1$) depending on the side ratio b/a of the wave guide.

1. For $b/a = 1/2$, the cutoff frequencies for the lowest-order TE modes are

$$f_{c10} = \frac{c}{2a}$$

where

$$c = \frac{1}{\sqrt{\mu_o \epsilon_o}}$$

equals the velocity of light in air

$$f_{c01} = \frac{c}{2b} = 2f_{c10}$$

because $b = 1/2\ a$.

$$f_{c11} = \frac{c}{2}\sqrt{\left(\frac{1}{a}\right)^2 + \left(\frac{1}{b}\right)^2} = \frac{c}{2a} \times 2.235 = 2.235 f_{c10}$$

$$f_{c20} = \frac{c}{2}\sqrt{\left(\frac{2}{a}\right) + 0} = 2f_{c10} = f_{c01}$$

$$f_{c02} = \frac{c}{2}\sqrt{0 + \left(\frac{2}{b}\right)^2} = 2f_{c10} = 4f_{c10}$$

f_{c11} is the cutoff frequency for the lowest possible TM mode.

Figure 8.5 shows the relative cutoff frequencies of the lower-order modes in a rectangular wave guide with $b/a = 1/2$.

2. For a wave guide with a side ratio $b/a = 1$ (i.e., square wave guide) the cutoff frequencies of several lower-order TE and TM modes are shown in Figure 8.6.

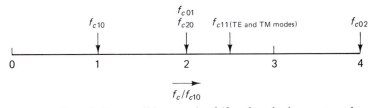

Figure 8.5 The relative cutoff frequencies f_c/f_{c10} of modes in a rectangular wave guide with $b/a = 1/2$.

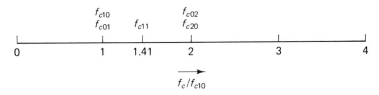

Figure 8.6 Relative cutoff frequencies of a square wave guide with $b/a = 1$.

From Figures 8.5 and 8.6, it is clear that different modes may have the same cutoff frequency. For example, in Figure 8.5, we see that $f_{c_{01}}$ for the TE_{01} mode has the same cutoff frequency as the TE_{20} mode. The field distributions for the TE_{01} ($m = 0, n = 1$) and TE_{20} ($m = 2, n = 0$) are however different even though the cutoff frequencies are the same. Modes that have different field distributions, but the same cutoff frequency are called *degenerate modes*.

EXAMPLE 8.3

Show that the field solution of the TE_{10} mode in a rectangular wave guide of cross-sectional dimensions $a(m)$ and $b(m)$ can be interpreted as a superposition of two plane waves propagating at an angle, and bouncing back and forth between the wave-guide walls.

Solution

The phasor expression for the axial component of the magnetic field of the TE_{10} mode is given by

$$\hat{H}_z = H_o \cos\left(\frac{\pi}{a}x\right)e^{-j\beta_{10}z}$$

where H_o is an arbitrary amplitude and

$$\beta_{10} = \sqrt{\omega^2 \mu\epsilon - \left(\frac{\pi}{a}\right)^2}$$

The phasor magnetic field $\hat{H}_z$ may be alternatively written as

$$\hat{H}_z = H_o\left(\frac{e^{j\frac{\pi}{a}x} + e^{-j\frac{\pi}{a}x}}{2}\right)e^{-j\beta_{10}z}$$

$$= \frac{H_o}{2}\left[e^{-j(-\frac{\pi}{a}x + \beta_{10}z)} + e^{-j(\frac{\pi}{a}x + \beta_{10}z)}\right] \tag{8.39}$$

Comparing the preceding expression with the oblique incident plane wave propagation described in chapter 6, we identify two plane waves. One wave is propagating in the negative x and positive z direction (first term), whereas the other propagates in the positive x and positive z direction. To clarify this, let us recall the general expression for a plane wave propagating in the $\boldsymbol{\beta} = \beta\,\mathbf{n}_\beta$ where $\mathbf{n}_\beta$ is a unit vector perpendicular to the wave front and is in the direction of propagation. An expression for the z component of the magnetic field associated with this wave may be given by

$$\hat{H}_z = \hat{H}_m e^{-j\boldsymbol{\beta}\cdot\mathbf{r}} = \hat{H}_m e^{-j(\beta_x x + \beta_y y + \beta_z z)} \tag{8.40a}$$

$$\hat{H}_z = \hat{H}_m e^{-j\beta(\cos\theta_x x + \cos\theta_y y + \cos\theta_z z)} \tag{8.40b}$$

where θ_x, θ_y, and θ_z are the angles that the direction of propagation $\mathbf{n}_\beta$ makes with the x, y, and z axes, respectively. It should be noted that the total magnetic field vector $\mathbf{H}$ and the propagation constant $\boldsymbol{\beta}$ should be oriented such that $\mathbf{H}\cdot\boldsymbol{\beta} = 0$. Comparing equations 8.39 and 8.40, it can be seen that the first term in equation 8.39 represents a plane wave

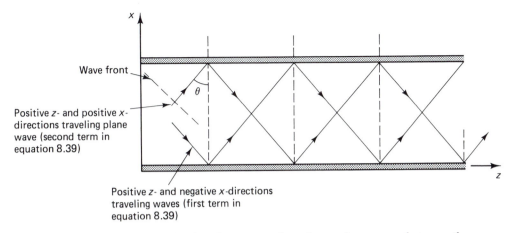

Figure 8.7 Diagram showing the propagation of two plane waves between the wave-guide walls.

propagating in the x-z plane with the direction of propagation in the positive z and negative x directions, whereas the second term represents another wave propagating in the positive z and positive x directions. These two waves and their bouncing back and forth between the conducting walls of the wave guide are shown in Figure 8.7.

Let us consider further the positive z-, positive x-propagating plane wave. From Figure 8.7 and equation 8.39, it is clear that

$$\beta \cos \theta = \frac{\pi}{a} \qquad (x \text{ component of propagation constant}) \qquad (8.41a)$$

$$\beta \sin \theta = \beta_{10} \qquad (z \text{ component of propagation constant}) \qquad (8.41b)$$

Solving equation 8.41 for β_{10}, we obtain

$$\beta_{10} = \sqrt{\beta^2 - \left(\frac{\pi}{a}\right)^2} = \sqrt{\omega^2 \mu \epsilon - \left(\frac{\pi}{a}\right)^2} \qquad (8.42)$$

where β, the propagation constant of the plane wave, has been replaced by $\beta = \omega \sqrt{\mu \epsilon}$. Equation 8.42 provides an expression for β_{10}, which is identical to that obtained from the wave-guide theory. Solving equation 8.41a for θ, we obtain

$$\cos \theta = \frac{\pi}{\beta a} = \frac{\lambda}{2a} \qquad (8.43)$$

Equation 8.43 shows that physical values of θ, hence, true mode of propagation, are possible as long as $\lambda < 2a$. For the TE_{10} mode under consideration $f_{c_{10}} = c/2a$ and $\lambda_{c_{10}} = 2a$. Hence, equation 8.43 simply states that physical propagation is possible for $\lambda < \lambda_{c_{10}}$ or alternatively for $f > f_{c_{10}}$. At cutoff, $\lambda = \lambda_{c_{10}} = 2a$ and the angle θ in equation 8.43 will be zero. The bouncing of the plane waves off the conducting walls in this case is shown in Figure 8.8, where it is clear that the lateral bounce of the waves does not result in any forward progression (i.e., no propagation).

As indicated earlier and illustrated in this example, the visualization of the various wave-guide modes as superposition of plane waves propagating at various angles is very

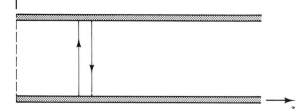

Figure 8.8 The lateral bounce of plane waves at cutoff.

helpful in developing a physical insight of these modes and also helps explain many relations between the wave-guide and plane wave parameters. This includes the longer wave-guide wavelength λ_g as compared with that of a plane wave of the same frequency (equation 8.23) and the fact that the wave guide phase velocity is larger than the speed of light in the material filling the wave guide (equation 8.24).

8.5 FIELD CONFIGURATIONS IN WAVE GUIDES

Besides the cutoff frequencies, propagation constants, and the intrinsic wave impedances of the various TE and TM modes in wave guides, it is important to study the electric and magnetic field configurations associated with the various wave-guide modes. For appropriate design of excitation and reception probe systems in wave guides as well as for many other microwave circuits and slot-antenna designs, it is important to know the field configurations of the various modes. To sketch the electric and magnetic field lines of the different modes we start by obtaining the instantaneous (time-domain) expressions of the field components of the mode of interest. Electric and magnetic field lines are then sketched in a cross-section plane across or along the wave guide to illustrate the spatial variation of these fields.

Let us illustrate the procedure by sketching the field configurations associated with the dominant mode (i.e., the mode with the lowest cutoff frequency) in a rectangular wave guide. The phasor expressions of the various field components of the TE_{10} mode are

$$\hat{H}_z = H_o \cos\left(\frac{\pi}{a}x\right)e^{-j\beta_{10}z} \tag{8.44a}$$

$$\hat{E}_y = \frac{-j\omega\mu a}{\pi} H_o \sin\left(\frac{\pi}{a}x\right)e^{-j\beta_{10}z} \tag{8.44b}$$

$$\hat{H}_x = \frac{j\beta_{10}a}{\pi} H_o \sin\left(\frac{\pi}{a}x\right)e^{-j\beta_{10}z} \tag{8.44c}$$

$$\hat{E}_x = 0, \qquad \hat{H}_y = 0, \qquad \hat{E}_z = 0$$

Assuming a real amplitude H_o, the instantaneous (time-domain) forms of these fields are given by

$$H_z(x,z,t) = H_o \cos\left(\frac{\pi}{a}x\right) \cos(\omega t - \beta_{10}z) \qquad (8.45a)$$

$$E_y(x,z,t) = \frac{\omega\mu a}{\pi} H_o \sin\left(\frac{\pi}{a}x\right) \sin(\omega t - \beta_{10}z) \qquad (8.45b)$$

$$H_x(x,z,t) = \frac{-\beta_{10}a}{\pi} H_o \sin\left(\frac{\pi}{a}x\right) \sin(\omega t - \beta_{10}z) \qquad (8.45c)$$

The spatial variations of these instantaneous fields with the (x,y,z) coordinates are sketched next. First, the variation of the electric and magnetic fields in a plane across the wave guide is shown in Figure 8.9. From equation 8.45b, it is clear that the electric field component E_y is independent of y and varies as a sine function with x. Figure 8.9 shows the flux representation of a y-directed electric field with its maximum value at the center (i.e., $x = a/2$) of the wave guide. In the cross section in the x-y plane we can also illustrate the spatial variation of the H_x component of the magnetic field. H_x is independent of y, and, hence, the magnetic field lines are equally spaced in the y direction. The sine variation of H_x with x is difficult to illustrate in Figure 8.9 with the exception of showing zero values of H_x at the wave-guide walls at $x = 0$ and $x = a$.

To illustrate further the spatial variation of the fields associated with the TE$_{10}$ mode, let us sketch these variations in a longitudinal cross section in the y-z plane. Such variations are shown in Figure 8.10. The electric field lines are shown with arrows in the positive or negative y direction depending on their z variation at time $t = 0$. From equation 8.45b, it is clear that at $t = 0$, E_y varies as $-\sin\beta_{10}z$ along the z direction. This results in electric field lines pointing in the negative y direction in the first half of the $\sin\beta_{10}z$ cycle and pointing in the positive y direction in the second half as shown in Figure 8.10. In illustrating the variations of the various magnetic field components in the y-z plane, conversely, we assume that the y-z plane shown in Figure 8.10 is taken at the plane $x = a/2$. At this plane $H_z = 0$ as can be verified from equation 8.45a, and the only magnetic field component that needs to be sketched in the cross section of Figure 8.10 is H_x. From equation 8.45c, we may see that H_x at $t = 0$ is x directed and varies as $\sin\beta_{10}z$ along the z direction. The dart notation is used in Figure 8.10 to illustrate the magnetic field out $\odot$ or into $\otimes$ the plane of the paper. Finally, the spatial variation of the various field components in the x-z plane is shown in Figure 8.11. The y-directed electric field component is shown into or out of the plane of the paper using

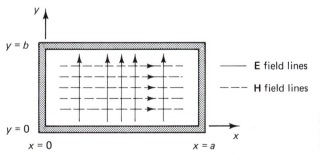

E field lines

H field lines

Figure 8.9 Sketch of the electric and magnetic fields in a plane across the wave guide.

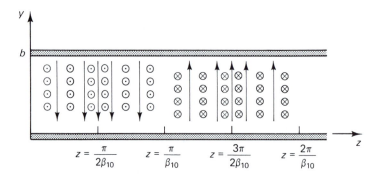

Figure 8.10 Sketch of the electric and magnetic field lines in the y-z plane.

the dart notation. From equation 8.45b it may be seen that the magnitude of E_y varies with x as $\sin \pi/a\,x$. Such a variation is illustrated in Figure 8.11 by including more electric field lines around the $x = a/2$ area of the cross section. This electric field component also varies as $-\sin \beta_{10} z$ with z at $t = 0$. This is also illustrated in Figure 8.11 by including more electric field lines around the $z = \pi/2\beta_{10}$ and $z = 3\pi/2\beta_{10}$ areas of the cross section.

Regarding the graphical illustration of the magnetic field equation 8.45 shows that there are two components H_z and H_x in the x-z plane. Setting $t = 0$ in equations 8.45a and 8.45c, we notice that H_z varies as $\cos(\beta_{10} z)$ with z, whereas H_x varies as $\sin(\beta_{10} z)$. This is why the magnetic field lines are in the z direction at $z = 0$ (i.e., there is only H_z component of the magnetic field) and in the x direction at $z = \pi/2\beta_{10}$. Why? Also to illustrate the x dependence of the magnetic field lines, these lines were drawn closer to each other at $x = 0$ and $x = a$ to emphasize the larger magnitude of H_z (H_z is proportional to $\cos(\pi x/a)$ at these locations. Similarly, to illustrate the increase in the magnitude of H_x at $z = \pi/2\beta_{10}$ and $z = 3\pi/2\beta_{10}$ (H_x is proportional to $\sin(\beta_{10} z)$), a larger number of the magnetic field lines were drawn at these locations. It should be noted

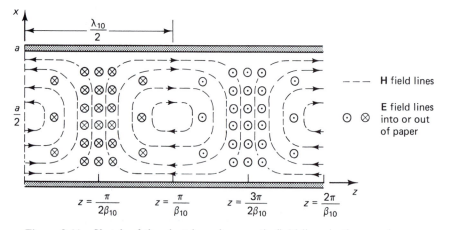

Figure 8.11 Sketch of the electric and magnetic field lines in the x-z plane.

that the physical considerations of the variation of the various field components in equation 8.45 may lead to the graphical illustration of Figure 8.11. The rigorous procedure for sketching the magnetic field lines, however, is based on using the slope of these lines at $t = 0$, which is governed by the following equation

$$\frac{dx}{dz} \text{ (in the } x\text{-}z \text{ plane)} = \frac{H_x}{H_z} = \frac{\beta_{10} a}{\pi} \tan\left(\frac{\pi}{a} x\right) \tan(\beta_{10} z)$$

which can be used to draw the **H** field lines either directly or after integration to obtain the equation for the magnetic field lines. A summary of the electric and magnetic field lines in different TE and TM modes in a rectangular wave guide is given in Table 8.1.*

8.6 EXCITATION OF VARIOUS MODES IN WAVE GUIDES

The process of efficiently exciting the various modes in a rectangular wave guide at the feed end and receiving the transmitted waves at the receiving end is not a simple one. The design of the excitation system involves the following:

1. Obtaining a field configuration that closely resembles those of the desired mode.
2. The design of an appropriate circuit of impedance that matches the feed system to minimize energy losses resulting from reflections. The process of designing a feed system that provides a field configuration appropriate to the desired mode clearly starts with a detailed sketch of the electric and magnetic field lines. The following guidelines should also be observed:
   (a) Introduce a probe or a wire monopole antenna near a maximum of the electric field of the desired mode. The probe should also be oriented along the direction of the electric field. Such an excitation procedure is shown in Figure 8.12a where a single coaxial feed monopole (probe) is used to excite the TE_{10} mode in a rectangular wave guide. From Figures 8.9 and 8.12a, we note the probe is oriented parallel to the electric field lines and at the location of maximum. In case of more than one electric field maximum, multiple probes should be used with an appropriate phase arrangement between the currents feeding the various probes. Figure 8.12b illustrates such a procedure that is used to excite the TE_{20} mode in a rectangular wave guide. The currents feeding these two probes are out of phase by 180° to provide the desired full sine wave variation of the electric field along the broader dimension of the waveguide.
   (b) If current loops are used as excitation probes, these loops should be oriented in a plane normal to the magnetic field and at the location of its maximum. An example of such excitation procedure is shown in Figure 8.13 where a current loop was used to excite the TE_{10} mode in a rectangular wave guide. In case of modes requiring more than one current loop, the phase relation between the currents in these loops should be observed.

---

* S. Ramo, J. R. Whinnery, and T. Van Duzer, *Fields and Waves in Communication Electronics* (New York: John Wiley & Sons, Inc., 2nd ed., 1984), p. 414. Reprinted with permission.

TABLE 8.1 SUMMARY OF WAVE TYPES FOR RECTANGULAR GUIDES*

*Electric field lines are shown solid and magnetic field lines are dashed. (From S. Ramo, J. R. Whinnery, and T. Van Duzer, *Fields and Waves in Communication Electronics*, New York: John Wiley & Sons, Inc., 2nd ed., 1984, p. 414. Reprinted with permission).

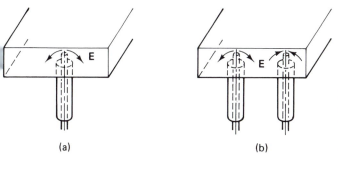

(a)   (b)

Figure 8.12   Excitation of wave guide modes using monopole antennas. (a) Excitation of TE_{10} mode and (b) excitation of TE_{20} mode.

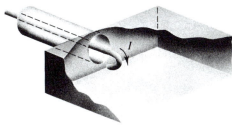

Figure 8.13   Excitation of wave-guide modes using current loops as excitation probes.

It should be noted that these excitation sources do not generally excite one mode but instead provide the most favorable conditions for a specific mode to be excited. All other modes that may be excited accidentally should attenuate as they propagate down the wave guide. If the wave-guide dimensions are large enough to support more than one mode at the excitation frequency, all the allowed modes will propagate in the wave guide. Also, it was indicated earlier in this section that providing the appropriate probe system is not sufficient to design an efficient excitation or receiving system. The design of a circuit that matches the wave guide to the feed system is also an important factor. For example, in a simple single probe system, as shown in Figure 8.12a, for exciting the TE_{10} mode, the length of the probe inside the wave guide, its diameter and location from the end of the guide (which is often shorted by placing a conductor at the end) are all important design parameters. In general these parameters should be adjusted appropriately to provide a good impedance match between the wave guide and the feed system.[†] Figure 8.14 shows the calculated input impedance of a coaxial probe placed in a rectangular wave guide. Contours of the real, R = constant, and imaginary, $X = 0$, parts of the input impedance of the transition $Z_{in} = R + jX$ is shown as function of the location $\beta_{10}\ell$ and electrical length $2\pi d/\lambda_o$ of the probe for an X-band wave guide at $f = 10\,\text{GHz}$, $\lambda_o = 3.14\,\text{cm}$, $a = 2.29\,\text{cm}$, $b = 1.02\,\text{cm}$, and probe radius $r = 0.05a$. Similar graphs for other wave guides are available in the reference cited in the footnote.

[†]R. E. Collin, *Field Theory of Guided Waves* (New York: McGraw-Hill Book Company, 1960), p. 271.

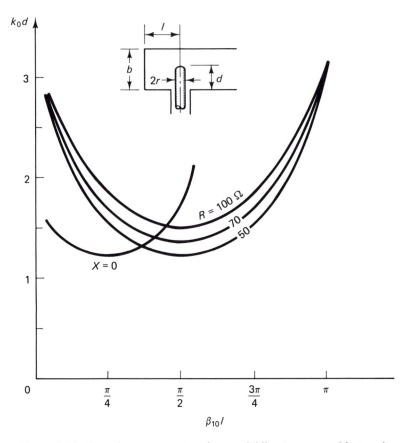

Figure 8.14 Impedance parameters for coaxial line-to-wave-guide transition. $X = 0$ and $R =$ constant contours in $k_0 d - \beta_{10} \ell$ plane for $a = 2.29$, $b = 1.02$ cm, $r = 0.05a$, and $\lambda_o = 3.14$ cm. (*From* R. E. Collin, *Field Theory of Guided Waves* [New York: McGraw-Hill Book Company, 1960], p. 271; reprinted with permission.)

8.7 ENERGY FLOW AND ATTENUATION IN RECTANGULAR WAVE GUIDES

Thus far we discussed the various propagation characteristics as well as the field components associated with TE and TM modes in rectangular wave guides. In this section we will describe the power flow in wave guides and the power attenuation as a result of losses in the dielectric material filling the guide as well as the finite conductivity of its metallic walls. The power density flow along the axis of a wave guide may be calculated in terms of the time-average Poynting vector. It is shown in equation 3.90 of chapter 3 that the time-average Poynting vector $\mathbf{P}_{av}$ is given in terms of the phasor components of the electric and magnetic field by

$$\mathbf{P}_{av} = \frac{1}{2} Re(\hat{\mathbf{E}} \times \hat{\mathbf{H}}^*)$$

Expressing both $\hat{\mathbf{E}}$ and $\hat{\mathbf{H}}$ in terms of their various components and assuming a power flow along the z axis of the wave guide, we obtain

$$\mathbf{P}_{av} = \frac{1}{2} Re[\hat{E}_x \hat{H}_y^* - \hat{H}_x^* \hat{E}_y] \mathbf{a}_z \qquad (8.46)$$

From our wave-guide analysis we noted that

$$\frac{\hat{E}_x}{\hat{H}_y} = \eta_{mn}, \qquad \text{and} \qquad \frac{\hat{E}_y}{\hat{H}_x} = -\eta_{mn}$$

where η_{mn} stands for $\eta_{TM_{mn}}$ for TM modes or $\eta_{TE_{mn}}$ for TE modes. Substituting these intrinsic impedance relations in equation 8.46, we obtain

$$\boxed{\mathbf{P}_{av} = \frac{1}{2} \frac{|\hat{E}_x|^2 + |\hat{E}_y|^2}{\eta_{mn}} \mathbf{a}_z} \qquad (8.47)$$

Equation 8.47 provides an expression for the power density transmitted in the positive z direction. To obtain the total average power transmitted across the cross section of the guide, we simply integrate equation 8.47 over the rectangular cross section. Hence,

$$\mathbf{P}_{av}^t = \int_{y=0}^{y=b} \int_{x=0}^{x=a} \mathbf{P}_{av} \cdot dx\, dy\, \mathbf{a}_z$$

$$= \frac{1}{2} \int_{y=0}^{y=b} \int_{x=0}^{x=a} \frac{|\hat{E}_x|^2 + |\hat{E}_y|^2}{\eta_{mn}} dx\, dy \qquad (8.48)$$

Numerical calculations of equations 8.47 and 8.48 for specific wave-guide modes will be illustrated by examples at the end of this section.

Of significant importance in many practical applications, however, is the attenuation of the power flow resulting from the following:

1. Losses in the dielectric materials filling the wave guide.
2. Losses as a result of the finite conductivity of the guide metallic walls.

Assuming that the dielectric losses are sufficiently small so that no appreciable change in the field patterns occur, the attenuation constant resulting from the dielectric losses may be calculated by replacing the real ϵ in equation 8.35 by $\epsilon^* = \epsilon' - j\epsilon'' = \epsilon - j\sigma/\omega$. Making this substitution, we obtain

$$\hat{\gamma} = \sqrt{\left(\frac{m\pi}{a}\right)^2 + \left(\frac{n\pi}{b}\right)^2 - \omega^2 \mu\epsilon\left(1 - j\frac{\sigma}{\omega\epsilon}\right)}$$

$$= j\left[\omega^2 \mu\epsilon\left(1 - j\frac{\sigma}{\omega\epsilon}\right) - \left(\frac{m\pi}{a}\right)^2 - \left(\frac{n\pi}{b}\right)^2\right]^{1/2} \qquad (8.49)$$

$$= j\sqrt{\left[\omega^2 \mu\epsilon - \left(\frac{m\pi}{a}\right)^2 - \left(\frac{n\pi}{b}\right)^2\right]\left\{1 - j\sigma\omega\mu\left[\omega^2 \mu\epsilon - \left(\frac{m\pi}{a}\right)^2 - \left(\frac{n\pi}{b}\right)^2\right]^{-1}\right\}^{1/2}}$$

Assuming

$$\frac{\sigma\omega\mu}{\omega^2\mu\epsilon - \left(\frac{m\pi}{a}\right)^2 - \left(\frac{n\pi}{b}\right)^2} \ll 1$$

and using the binomial expansion to approximate the expression between the outside brackets in equation 8.49 by its two first terms, we obtain

$$\hat{\gamma} = j\sqrt{\omega^2\mu\epsilon - \left(\frac{m\pi}{a}\right)^2 - \left(\frac{n\pi}{b}\right)^2}\left\{1 - \frac{j\sigma\omega\mu}{2}\left[\omega^2\mu\epsilon - \left(\frac{m\pi}{a}\right)^2 - \left(\frac{n\pi}{b}\right)^2\right]^{-1}\right\} \qquad (8.50)$$

The complex propagation constant in equation 8.50 may be expressed in terms of its real α_{dmn} and imaginary β_{mn} components as

$$\hat{\gamma} = \alpha_{d_{mn}} + j\beta_{mn} = \frac{\sigma\omega\mu}{2}\left[\omega^2\mu\epsilon - \left(\frac{m\pi}{a}\right)^2 - \left(\frac{n\pi}{b}\right)^2\right]^{-1/2}$$

$$+ j\sqrt{\omega^2\mu\epsilon - \left(\frac{m\pi}{a}\right)^2 - \left(\frac{n\pi}{b}\right)^2} \qquad (8.51)$$

$$= \frac{\frac{\sigma}{2}\sqrt{\mu/\epsilon}}{\sqrt{1 - \left(\frac{f_{cmn}}{f}\right)^2}} + j\omega\sqrt{\mu\epsilon}\sqrt{1 - \left(\frac{f_{cmn}}{f}\right)^2}$$

where f_{cmn} is given by equation 8.36.

The attenuation constant resulting from the dielectric losses α_{dmn} is, hence, given by

$$\alpha_{dmn} = \frac{\sigma\eta}{2\sqrt{1 - \left(\frac{f_{cmn}}{f}\right)^2}} \qquad (8.52)$$

where $\eta = \sqrt{\mu/\epsilon}$ is the intrinsic impedance of a plane wave in the medium filling the guide. The second term in equation 8.51 is identical to equation 8.21 and describes the phase constant of TM or TE modes in a wave guide filled with lossless dielectric. Retention of higher-order terms in the binomial expansion provides a correction to the phase term as a result of the losses in the dielectric. Such a correction may be important in considering the dispersive properties of wave guides.

We next focus our attention on the power loss or attenuation owing to the finite conductivity of the guide metallic walls. In other words, this type of attenuation is due to the ohmic losses that results from the current flow in the metallic walls of the guide. These kinds of losses were accounted for in the transmission-line theory (chapter 7) by including the resistive element r (Ω/m) in the low-frequency equivalent circuit of a section of the transmission line.

The calculation of the ohmic losses in the wave-guide case is much more involved than the simplified approach used in the transmission-line theory. It starts with defining the power loss per unit length P_L as the rate of decrease of the time-average power $P(z)$ with distance along the wave guide. Hence,

$$-\frac{\partial P(z)}{\partial z} = P_L$$

If we assume an exponentially decaying time-average power transmission along the wave guide—that is,

$$P(z) = P_o e^{-2\alpha z}$$

where α is the attenuation constant; hence,

$$P_L = 2\alpha P(z)$$

or

$$\alpha = \frac{P_L}{2P(z)} \text{ Np/m} \tag{8.53}$$

Calculation of the attenuation constant in equation 8.53, hence, requires calculating the time-average power transmission through the cross section of the wave guide $P(z)$, and the time-average power loss resulting from the finite conductivity of the wave-guide walls P_L. Calculation of $P(z)$ is based on integrating the time average Poynting vector over the wave-guide cross section as described earlier in this section; hence,

$$P(z) = \frac{1}{2} \int_{y=0}^{y=b} \int_{x=0}^{x=a} \frac{|\hat{E}_x|^2 + |\hat{E}_y|^2}{\eta_{mn}} dx\, dy$$

For example, in the TE_{10} mode, we have

$$\hat{E}_x = 0, \qquad \hat{E}_y = -j\frac{\omega\mu a}{\pi} H_o \sin\left(\frac{\pi}{a}x\right) e^{-j\beta_{10}z}$$

Hence,

$$P(z) = \frac{\omega^2 \mu^2 a^2 H_o^2}{2\pi^2 \eta_{TE_{10}}} \int_0^b \int_0^a \sin^2\left(\frac{\pi}{a}x\right) dx\, dy$$

Substituting $\eta_{TE_{10}} = \omega\mu/\beta_{10}$, we obtain

$$P(z) = \frac{\omega\mu\beta_{10}}{2}\left(\frac{a}{\pi}\right)^2 H_o^2 b \cdot \frac{1}{2}a$$

$$= \frac{\omega\mu\beta_{10} a^3 b}{(2\pi)^2} H_o^2 \tag{8.54}$$

Calculation of P_L, conversely, requires knowledge of the current flowing on the wave-guide walls as well as the resistance of these walls per unit length. Carrying out this calculation for an arbitrary TM_{mn} or TE_{mn} mode is tedious. Hence, to illustrate the procedure we will carry out the analysis for the TE_{10} mode. For this mode, the current density on the four walls of the rectangular wave guide (i.e., $x = 0, x = a, y = 0,$

$y = b$) are obtained from the magnetic field components tangential to these conductors. In general the magnetic field boundary condition requires that

$$\mathbf{n} \times \mathbf{H} = \mathbf{J}_s \qquad (8.55)$$

where $\mathbf{n}$ is an inward unit vector normal to the wave-guide wall. Applying equation 8.55
at the $x = 0$ wall, we obtain

$$\mathbf{a}_x \times (\hat{H}_x \mathbf{a}_x + \hat{H}_z \mathbf{a}_z) = \mathbf{J}_s \qquad (x = 0)$$

hence,

$$-\hat{H}_z|_{x = 0}\mathbf{a}_y = \mathbf{J}_s|_{(x = 0)}$$

or

$$\hat{\mathbf{J}}_s(x = 0) = -H_o\,\mathbf{a}_y \qquad (8.56a)$$

Similarly, at the $x = a$ wall we have

$$\hat{\mathbf{J}}_s(x = a) = -H_o\,\mathbf{a}_y \qquad (8.56b)$$

For the induced currents on the broader sides of the wave guide $y = 0$ and $y = b$ we
follow a similar procedure to obtain
    At $y = 0$,

$$\mathbf{a}_y \times (\hat{H}_x \mathbf{a}_x + \hat{H}_z \mathbf{a}_z) = \hat{\mathbf{J}}_s \qquad (y = 0)$$

Hence,

$$\hat{\mathbf{J}}_s(y = 0) = -\hat{H}_x|_{y = 0}\mathbf{a}_z + \hat{H}_z|_{y = 0}\mathbf{a}_x$$
$$= H_o\,\cos\!\left(\frac{\pi}{a}x\right)\mathbf{a}_x - \frac{\beta_{10}\,a}{\pi}H_o\,\sin\!\left(\frac{\pi}{a}x\right)\mathbf{a}_z \qquad (8.56c)$$

and

$$-\hat{\mathbf{J}}_s(y = 0) = \hat{\mathbf{J}}_s(y = b) \qquad (8.56d)$$

The total power loss is the sum of all the ohmic losses in the wave-guide walls. Hence,

$$P_L = P_L|_{x = 0} + P_L|_{x = a} + P_L|_{y = 0} + P_L|_{y = b}$$

and

$$P_L|_{x = 0} = P_L|_{x = a} = \frac{1}{2}Re(\hat{V}\,\hat{I}^*)$$
$$= \frac{1}{2}|\hat{I}|^2\,R_s \qquad (8.57)$$

where R_s is the resistance per unit length of the wave-guide walls.

$$P_L|_{x = 0} = \frac{1}{2}\int_{y = 0}^{y = b}|\hat{\mathbf{J}}_s(x = 0)|^2\,R_s\,dy$$
$$= \frac{b}{2}H_o^2\,R_s \qquad (8.58)$$

The power loss on the other two sides of the wave guide is given by

$$P_L|_{y=0} = P_L|_{y=b} = \frac{1}{2}\int_{x=0}^{x=a} |\hat{\mathbf{J}}_s(y=0)|^2 R_s\, dx$$

Noting that $\hat{\mathbf{J}}_s(y=0)$ has two components as shown in equation 8.56c, we obtain

$$P_L|_{y=0} = \frac{a}{4}H_o^2 R_s + \frac{\beta_{10}^2 a^3}{4\pi^2}H_o^2 R_s$$

The total ohmic losses are then

$$P_L = 2\left(\frac{b}{2}H_o^2 R_s\right) + 2\left(\frac{a}{4}\right)H_o^2 R_s\left[1 + \left(\frac{\beta_{10}a}{\pi}\right)^2\right]$$

$$= \left\{b + \frac{a}{2}\left[1 + \left(\frac{\beta_{10}a}{\pi}\right)^2\right]\right\}H_o^2 R_s \tag{8.59}$$

Equation 8.59 may be put in the more familiar form by noting that

$$\beta_{10} = \sqrt{\omega^2\,\mu\epsilon - \left(\frac{\pi}{a}\right)^2} = \omega\sqrt{\mu\epsilon}\,\sqrt{1 - \left(\frac{f_{c_{10}}}{f}\right)^2}$$

Squaring the first two terms, we may obtain

$$\left(\frac{\beta_{10}a}{\pi}\right)^2 + 1 = \omega^2\,\mu\epsilon\left(\frac{a}{\pi}\right)^2 \tag{8.60}$$

Squaring the last two terms and rearranging the various terms, we obtain

$$\left(\frac{f}{f_{c_{10}}}\right)^2 = \omega^2\,\mu\epsilon\left(\frac{a}{\pi}\right)^2 \tag{8.61}$$

Comparing equations 8.60 and 8.61, we have

$$\left(\frac{\beta_{10}a}{\pi}\right)^2 + 1 = \left(\frac{f}{f_{c_{10}}}\right)^2 \tag{8.62}$$

Substituting equation 8.62 in equation 8.59, we obtain

$$P_L = \left\{b + \frac{a}{2}\left(\frac{f}{f_{c_{10}}}\right)^2\right\}H_o^2 R_s \tag{8.63}$$

From equations 8.53, 8.54, and 8.63, we calculate the attenuation constant due to ohmic losses as

$$\alpha = \frac{[1 + 2b/a(f_c/f)^2]\,R_s}{b\eta\sqrt{1 - (f_c/f)^2}} \tag{8.64}$$

The simplified relation in equation 8.64 was obtained by using the expression for β_{10} and equation 8.61. The resistance of the conducting walls R_s in equation 8.64 may be obtained as a limiting case of the expression for the intrinsic impedance of a wave propagating in a conductive medium. From equation 3.79 of chapter 3 we have

$$\eta = \sqrt{\frac{\mu}{\epsilon - j\dfrac{\sigma}{\omega}}} = R_s + jX_s$$

(8.65)

$$= \sqrt{\mu/\epsilon}\sqrt{\frac{1}{1 - j\dfrac{\sigma}{\omega\epsilon}}}$$

For a highly conducting wall—that is, $\sigma/\omega\epsilon \gg 1$—we obtain

$$R_s + jX_s = \sqrt{\frac{\mu}{\epsilon}}\sqrt{\frac{\omega\epsilon}{\sigma}}\sqrt{j} = \sqrt{\frac{\omega\mu}{\sigma}}\frac{(1 + j)}{\sqrt{2}} = \sqrt{\frac{\pi f \mu}{\sigma}}(1 + j)$$

R_s is, hence, given by

$$R_s = \sqrt{\frac{\pi f \mu}{\sigma}}$$

where μ and σ are the permeability and conductivity of the conducting material of the wave-guide walls, respectively. More complicated but similar expressions for α for the general TE_{mn} and TM_{mn} modes in a rectangular wave guide are given elsewhere.[‡] Figure 8.15 shows attenuation α versus frequency for the dominant TE mode (i.e., TE_{10}) and the dominant TM mode (i.e., TM_{11}) in a rectangular wave guide made of copper. The attenuation results are shown for various values of the guide side ratio b/a. Figure 8.15 clearly illustrates the very rapid decrease of the attenuation with the frequency increase from the cutoff. After a minima is reached, the attenuation continues to increase with the increase in frequency. The presence of a frequency, close to the cutoff frequency, at which the attenuation is minimum, is significant in designing and operating wave guides. Furthermore, the increase of the attenuation with the decrease in the side ratio b/a is also important. We noted from equation 8.37 and equation 8.44 that neither the propagation constants of the TE_{10} mode (i.e., $\beta_{10}, f_{c_{10}}, \eta_{TE_{10}}$) nor the electric and magnetic field equations depend on the cross-sectional dimension b. Hence, choosing the dimension b of the wave guide when operating in the TE_{10} relies on other factors including minimizing the attenuation owing to the wall losses. From Figure 8.15 it is clear that least attenuation values are obtained for a square wave guide (i.e., $b/a = 1$). Such a choice, however, results in degenerate mode because TE_{10} and TE_{01} will have the same cutoff frequency. To avoid such an undesirable situation of having two different field configurations of the same cutoff frequency, b/a should be made <1. A good choice of the side ratio b/a is to make $b/a \leq 1/2$. This makes the next higher-order mode be the TE_{20} mode and not any other intermediate mode. In other words, making $b/a = 1/2$ makes the cutoff frequency of the second wave-guide mode (next to the dominant TE_{10} mode) TE_{20} and TE_{10}. With such a choice of $b/a = 1/2$, we achieve the largest possible spacing between the cutoff of the TE_{10} and the next higher-order mode.

[‡] S. Ramo, J. R. Whinnery, and T. V. Duzer, *Fields and Waves in Communication Electronics* (New York: John Wiley & Sons, 1984), Chapter 8.

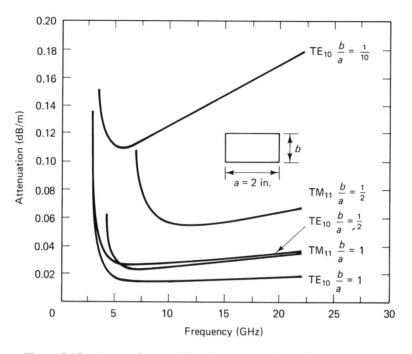

Figure 8.15   Attenuation resulting from copper losses in rectangular wave guides of fixed width (*From* S. Ramo, J. R. Whinnery, T. V. Duzer, *Fields and Waves in Communication Electronics,* page 417, New York: John Wiley & Sons, 1984. Reprinted with permission).

Figure 8.15, however, indicates that this will be at the expense of increasing the attenuation due to ohmic losses. Also, making $b/a < 1/2$ limits the power-handling capacity of the wave guide because of the possible danger of voltage breakdown. In practice $b/a = 1/2$ is a common choice for the ratio of the cross-sectional dimensions of the wave guide.

EXAMPLE 8.4

A rectangular wave guide of dimensions $a = 5.08$ cm and $b = 2.54$ cm. If the frequency of operation when the wave guide is air filled is 5 GHz.

1. Determine all the possible propagating modes in the wave guides.
2. Repeat part 1 assuming that the wave guide is filled with a dielectric material of $\epsilon_r = 4$.

Solution

1. The cutoff frequencies of the various modes are determined from

$$f_c = \frac{1}{2\pi\sqrt{\mu\epsilon}}\sqrt{\left(\frac{m\pi}{a}\right)^2 + \left(\frac{n\pi}{b}\right)^2}$$

for air-filled wave guide

$$f_c = \frac{c}{2\pi}\sqrt{\left(\frac{m\pi}{a}\right)^2 + \left(\frac{n\pi}{b}\right)^2}$$

where c is the velocity of light in air $c = 3 \times 10^8$ m/s, and m and n are the subscripts of the particular TE or TM mode.

$$f_{c\text{TE}_{10}} = \frac{3 \times 10^8}{2} \cdot \frac{1}{a} = 2.953 \text{ GHz}$$

$$f_{c\text{TE}_{01}} = \frac{3 \times 10^8}{2} \cdot \frac{1}{b} = 5.906 \text{ GHz}$$

$$f_{c\text{TE}_{11}} = \frac{3 \times 10^8}{2}\sqrt{\left(\frac{1}{a}\right)^2 + \left(\frac{1}{b}\right)^2} = 6.6 \text{ GHz}$$

$$f_{c\text{TM}_{11}} = \frac{3 \times 10^8}{2}\sqrt{\left(\frac{1}{a}\right)^2 + \left(\frac{1}{b}\right)^2} = 6.6 \text{ GHz}$$

Therefore, we conclude from the calculation of these various cutoff frequencies that the operating frequency of 5 GHz is higher than only the cutoff frequency of the TE_{10} mode. Therefore, only the dominant mode TE_{10} will propagate in the air-filled wave guide.

2. If the wave guide is filled with a dielectric material $\epsilon_r \doteq 4$, conversely, the cutoff frequencies are given by

$$f_c = \frac{1}{2\pi\sqrt{\mu_o \, \epsilon_o \, \epsilon_r}}\sqrt{\left(\frac{m\pi}{a}\right)^2 + \left(\frac{n\pi}{b}\right)^2}$$

$$= \frac{c}{2\sqrt{\epsilon_r}}\sqrt{\left(\frac{m}{a}\right)^2 + \left(\frac{n}{b}\right)^2}$$

$f_{c\text{TE}_{10}}$ in this case is given by

$$f_{c\text{TE}_{10}} = \frac{c}{2\sqrt{4}}\frac{1}{a} = \frac{f_{c\text{TE}_{01}} \text{ (in air)}}{2} = 1.477 \text{ GHz}$$

$$f_{c\text{TE}_{01}} = \frac{f_{c\text{TE}_{01}} \text{ (in air)}}{2} = 2.95 \text{ GHz}$$

Also,

$$f_{c\text{TE}_{11}} \text{ (in dielectric)} = 0.5 f_{c\text{TE}_{11}} \text{ (in air)} = 3.30 \text{ GHz}$$

$$f_{c\text{TM}_{11}} \text{ (in dielectric)} = 3.3 \text{ GHz}$$

$$f_{c\text{TE}_{20}} = \frac{c}{2\sqrt{4}} \cdot \frac{2}{a} = 2.95 \text{ GHz}$$

$$f_{c\text{TE}_{02}} = \frac{c}{2\sqrt{4}}\frac{2}{b} = 5.9 \text{ GHz}$$

From the preceding calculations it is clear that filling the wave guide with dielectric materials (i.e., dielectric loading) results in lowering the cutoff frequencies by a factor of

$\sqrt{\epsilon_r}$. For given wave-guide dimensions this means the possible excitation of many higher-order modes in the wave guide, which is, in most practical applications, an undesirable feature. To avoid the excitation of these higher-order modes, we may decide to operate this wave guide at a frequency that is lower than the operating frequency in air by a factor of $\sqrt{\epsilon_r}$. In other words, the dielectric-filled (loaded) wave guide may be operated at a frequency $f = 5 \text{ GHz}/\sqrt{\epsilon_r} = 2.5 \text{ GHz}$. In this case, only the dominant TE_{10} mode will propagate. Dielectric loading of wave guides, therefore, provides a valuable procedure for lowering the cutoff frequencies of the modes without having to increase the wave-guide dimensions. Lossless dielectric should, however, be used to minimize the energy losses in the wave guide. An example of these dielectric loaded wave guides include rectangular wave guides filled with deionized water ($\epsilon_r = 80$) and used in electromagnetic hyperthermia for cancer treatment.

EXAMPLE 8.5

Similar to the transmission line measurements, slotted sections of wave guide are often used to measure VSWR in wave guides and input impedances of terminating loads. In rectangular wave guides these slotted sections are made by milling an axial narrow slot at the center of the broader dimension of the wave guide as shown in Figure 8.16. A small probe penetrates the slot and samples the E field within the wave guide; this measures the VSWR pattern. The probe is movable along the wave guide in a carriage, and its position may be directly read on a scale attached to the side of the slot. Let us examine the use of this slotted wave guide in determining the operating frequency and the unknown impedance of a load terminating the wave guide. Assuming that in a rectangular wave guide ($a = 2.54$ cm, $b = 1.27$ cm), the VSWR pattern has a distance between two successive minima of 3 cm and a VSWR value of 3. When the load connected to the end of the slotted wave guide was replaced by a short circuit, the location of the first minima away from the load shifted by a distance of 1.3 cm.

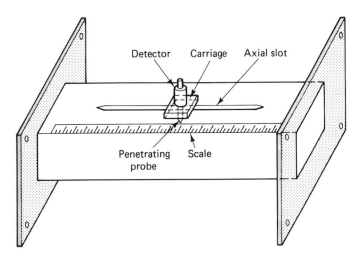

Figure 8.16   Slotted section of rectangular wave guide.

1. Determine the operating frequency in the wave guide.
2. Use the Smith chart to calculate the impedance of the unknown load terminating the wave guide.

Solution

With the exception of replacing the slotted transmission line with the slotted wave guide, the solution procedure for this example follows closely the method of solution described in section 7.16 on transmission-line measurements.

1. The distance between two successive minima is equal to $\lambda_g/2$, where λ_g is the wave-guide wavelength. Hence,

$$\lambda_g = 2 \times 0.03 = 0.06 \text{ m}$$

The cutoff frequency for the TE_{10} mode in the given wave guide is

$$f_{c\,TE_{10}} = \frac{1}{2\pi\sqrt{\mu_o\,\epsilon_o}}\frac{\pi}{a}$$

$$= \frac{3 \times 10^8}{2 \times 0.0254} = 5.9 \text{ GHz}$$

Knowledge of the wave-guide wavelength λ_g and the cutoff frequency of the mode allows us to calculate the operating frequency in the wave guide. From equation 8.23,

$$\lambda_g = \frac{\lambda}{\sqrt{1 - \frac{f_c^2}{f^2}}} = \frac{c}{\sqrt{f^2 - f_c^2}}$$

where c is the velocity of light. Hence,

$$0.06 = \frac{3 \times 10^8}{\sqrt{f^2 - (5.9 \times 10^9)^2}}$$

or

$$f = \sqrt{\frac{(3 \times 10^8)^2}{(0.06)^2} + (f_c)^2}$$

$$= 7.73 \text{ GHz}$$

2. In the Smith chart shown in Figure 8.17, we identify the VSWR circle as the normalized resistance $r = 3$ circle (see section 7.15 in chapter 7). The locations of V_{max} and V_{min} are indicated on the chart. Because the location of V_{min} with the load replaced by short circuit is a true image of the load location (within a multiple of $\lambda_g/2$ that corresponds to complete rotation on the Smith chart), the true value of the load impedance is determined by moving toward the load on the Smith chart an electrical distance corresponding to the shift in minima between the short circuit and the load. The rotated electrical distance is $0.013/0.06 = 0.22\lambda_g$. This solution is very similar to that of example 7.18 of chapter 7. $\hat{z}_{nL}$ is obtained after rotating V_{min} a distance of $0.22\lambda_g$ toward the load as shown in Figure 8.17. The obtained value of

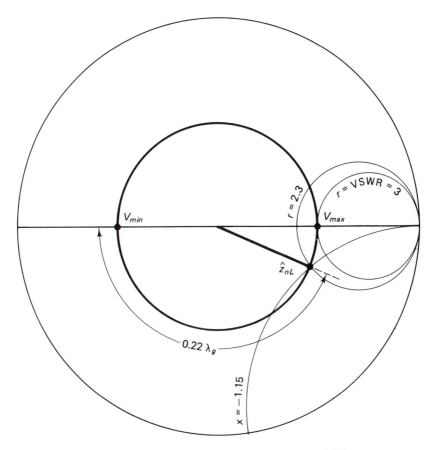

Figure 8.17   Determination of load impedance using VSWR measurements on slotted wave guide.

the normalized load impedance can be read directly on the Smith chart and from Figure 8.17 its value is

$$\hat{z}_{nL} = 2.3 - j1.15$$

The true value in ohms of the load impedance is obtained by multiplying $\hat{z}_{nL}$ by intrinsic impedance of the wave guide calculated at the operating frequency

$$\eta_{\text{TE}} = \frac{\eta}{\sqrt{1 - \left(\frac{f_c}{f}\right)^2}} = \frac{120\pi}{\sqrt{1 - \left(\frac{5.9}{7.73}\right)^2}}$$

$$= 583.5 \ \Omega$$

The load impedance is, hence,

$$\hat{Z}_L = \hat{z}_{nL} \cdot \eta_{\text{TE}} = 1342.2 - j671.1 \ \Omega$$

EXAMPLE 8.6

An air-filled wave guide has the inside dimensions of 7.62 cm × 3.81 cm and is operating in the TE_{10} mode. In an attempt to determine the operating frequency, the wave-guide wavelength λ_g was measured using a slotted wave guide.

1. If the measured distance between two successive minima was 0.1 m, determine the operating frequency.
2. At the operating frequency determined in part 1, and if the magnitude of the total electric field is 1000 V/m, calculate the total power transmitted down the wave guide.

Solution

1. The distance between two successive minima is $\lambda_g/2 = 0.1$ m.

$$\therefore \lambda_g = 0.2 \text{ m}$$

$$\lambda_g = \frac{\lambda}{\sqrt{1 - \left(\frac{f_c}{f}\right)^2}} = \frac{c}{\sqrt{(f)^2 - (f_c)^2}}$$

$$f = \sqrt{\left(\frac{c}{\lambda_g}\right)^2 + (f_c)^2} = \sqrt{\left(\frac{c}{\lambda_g}\right)^2 + \left(\frac{c}{2a}\right)^2} = 2.475 \text{ GHz}$$

2. The average power density is given by

$$\mathbf{P}_{av} = \frac{1}{2}\frac{|\hat{E}_y|^2}{\eta_{TE_{10}}} = \frac{1}{2}\frac{(1000)^2}{\eta_{TE_{10}}} \sin^2\left(\frac{\pi}{a}x\right)\mathbf{a}_z$$

$$\eta_{TE_{10}} = \frac{\eta_o}{\sqrt{1 - \left(\frac{f_c}{f}\right)^2}} = \frac{120\pi}{\sqrt{1 - \left(\frac{c}{2af}\right)^2}} = 622.0 \ \Omega$$

where f_c was substituted by $f_c = c/2a$

$$\mathbf{P}_{av} = \frac{1}{2}\frac{(1000)^2}{622} \sin^2\left(\frac{\pi}{a}x\right)\mathbf{a}_z$$

The total average power $\mathbf{P}_{av}^{tot}$ is then

$$P_{av}^{tot} = \int_0^a \int_0^b \mathbf{P}_{av} \cdot dx \, dy \, \mathbf{a}_z = \frac{1}{2}\frac{(1000)^2}{622} b \int_0^a \frac{1}{2}\left(1 - \sin\frac{2\pi}{a}x\right) dx$$

$$= \frac{1}{4}ba\frac{(1000)^2}{622}$$

$$= 1.167 \text{ W}$$

EXAMPLE 8.7

It is desired to design a rectangular wave guide that propagates only the lowest-order mode TE_{10} and operates at 3 GHz. If the wave guide is air filled, determine the wave guide dimensions (a and b) such that the desired operating frequency is 27 percent above the cutoff frequency of the TE_{10} mode. Although the dimension b of the wave guide is often taken $b = a/2$, for attenuation considerations and also to improve the wave-guide power-handling capacity, it is desired to design a wave guide with the largest possible value of b. Furthermore, to avoid mode degeneracy, b should be chosen such that, $b < a$, but $b > a/2$. Determine the dimension b so that the operating frequency of 3 GHz is 30 percent below the cutoff frequency of the next higher-order mode.

Solution

First, to determine the dimension a, we note that $f = 3$ GHz is 27 percent higher than $f_{c\,TE_{10}}$. Hence,

$$3 \times 10^9 = 1.27 f_{c\,TE_{10}} = 1.27 \frac{c}{2a}$$

where c is the velocity of light in air for an air-filled wave guide.

$$a = \frac{1.27}{20} = 6.35 \text{ cm}$$

The determination of b is based on the attenuation owing to wall losses and power-handling capacity of the wave guide, on one hand, and on maintaining a single mode of propagation on the other. Both the attenuation and the power-handling capacity require the use of the largest possible value of $b \leq a$ (see Figure 8.15), while keeping a single mode of operation requires $b \leq a/2$. Because the attenuation and power-handling capacity were more important considerations in this case, we will select a value of b that is larger than $a/2$. It should be chosen such that the operating frequency f is 30 percent lower than the next higher-order mode. The next higher-order mode may be one of the following:

$$f_{c\,TE_{01}} = \frac{c}{2b}, \qquad f_{c\,TE_{20}} = \frac{c}{a}$$

or

$$f_{c\,TE_{11}} = \frac{c}{2\pi} \sqrt{\left(\frac{\pi}{a}\right)^2 + \left(\frac{\pi}{b}\right)^2} = \frac{c}{2} \sqrt{\left(\frac{1}{a}\right)^2 + \left(\frac{1}{b}\right)^2}$$

for $a/2 < b < a$, the next higher-order mode will be TE_{01}. Hence,

$$(1.3) \times 3 \text{ GHz} = f_{c\,TE_{01}} = \frac{c}{2b}$$

or

$$b = \frac{3 \times 10^8}{(1.3) \times 2 \times 3 \times 10^9} = 3.85 \text{ cm}$$

The wave-guide dimensions are then

$$a = 6.35 \text{ cm}$$

$$b = 3.85 \text{ cm}$$

————— ◆◆◆ —————

SUMMARY

This chapter starts with a brief analysis that shows that the familiar TEM mode cannot propagate in a single-conductor wave guide. Instead, a possible propagating mode should have either an axial electric, $\hat{E}_z$, an axial magnetic, $\hat{H}_z$, or both axial field components along the direction of propagation. In the former case, modes are labeled "transverse magnetic" or TM, whereas in the second case, modes are known as "transverse electric" or TE modes. Detailed expansions of Maxwell's equations (equation 8.4) further show that all transverse electric and magnetic field components may be expressed in terms of the axial ones. Therefore, TE modes in rectangular wave guides were analyzed in terms of $\hat{H}_z$, whereas the TM modes were analyzed in terms of $\hat{E}_z$. Assuming propagation along the z direction $e^{\pm \hat{\gamma} z}$, the wave equation for the axial electric $\hat{E}_z$ or magnetic $\hat{H}_z$ fields is given by

$$\left[\frac{\partial^2}{\partial x^2} + \frac{\partial^2}{\partial y^2} + (\omega^2 \mu \epsilon + \hat{\gamma}^2) \right] \begin{matrix} \hat{E}_z \\ \hat{H}_z \end{matrix} = 0$$

We then used separation of variables to solve the preceding wave equation for rectangular wave guides. Separate solutions for the TM ($\hat{E}_z$) and TE ($\hat{H}_z$) cases were obtained, and the following is a summary of the obtained results.

TM Modes

1. The axial component of the electric field $\hat{E}_z$ that satisfies the wave equation and the boundary conditions is given by

$$\hat{E}_z = A' \sin\left(\frac{m\pi}{a} x\right) \sin\left(\frac{n\pi}{b} y\right) e^{\pm \hat{\gamma} z} \qquad \begin{matrix} m = 1, 2, 3, \ldots \\ n = 1, 2, 3, \ldots \end{matrix}$$

a and b are the cross-sectional dimensions of the wave guide in the x and y directions, respectively, and A' is an amplitude constant that depends on the input power to the wave guide. As mentioned earlier, all other transverse field components may be obtained from $\hat{E}_z$ using equation 8.4.

2. The cutoff frequency defined as the lowest possible excitation frequency for a specific mode, for example, the cutoff frequency for TM_{mn} mode, is given by

$$f_{c,mn} = \frac{1}{2\pi \sqrt{\mu \epsilon}} \sqrt{\left(\frac{m\pi}{a}\right)^2 + \left(\frac{n\pi}{b}\right)^2}$$

From this equation, it may be seen that the larger the dimensions of the wave guide, the lower the cutoff frequency for a specific mode.

3. For a propagating mode (for example, the TM$_{mn}$ mode), the propagation constant, wave-guide wavelength, and the intrinsic wave impedance are given by

$$\hat{\gamma} = j\beta_{mn} = j\omega\sqrt{\mu\epsilon}\sqrt{1 - \left(\frac{f_{c,mn}}{f}\right)^2}$$

$$\lambda_{mn} = \frac{\lambda}{\sqrt{1 - \left(\frac{f_{c,mn}}{f}\right)^2}}$$

$$\eta_{mn} = \sqrt{\frac{\mu}{\epsilon}}\sqrt{1 - \left(\frac{f_{c,mn}}{f}\right)^2}$$

where μ, ϵ are the electrical properties of the medium filling the wave guide, and $\lambda = 2\pi/\omega\sqrt{\mu\epsilon}$ is the wavelength in the same medium.

4. From the preceding equations the following observations may be made:
   (a) At frequencies f lower than the cutoff frequency $f < f_{c,mn}$, the propagation constant is real, which means that the mode will be highly attenuated rather than truly propagating as with an imaginary propagation (phase) constant $e^{-j\beta_{mn}z}$. Attenuating modes in wave guides excited below the cutoff frequency are known as evanescent modes.
   (b) The wave guide wavelength λ_{mn} for propagating modes $f > f_{c,mn}$ is longer than the plane wave wavelength λ at the same frequency. This was explained in terms of the fact that wave-guide modes (field configurations) may be considered as a superposition of plane waves bouncing off the wave-guide walls at an angle. Therefore, the wavelength of the bounding plane waves along their direction of propagation is shorter than the wavelength along the axis of the wave guide that is the direction of propagation and the direction of the energy flow in the wave guide.
   (c) The intrinsic impedance at $f = f_{c,mn}$ is zero and is imaginary at frequencies f lower than the cutoff frequency. This shows that the wave guide is short-circuited for its fields at cutoff, and the imaginary ratio between the electric and magnetic fields below cutoff means zero time-average power and zero energy flow in the wave guide.

TE Modes

1. The axial component of the magnetic field associated with the TE$_{mn}$ mode is given by

$$\hat{H}_z = A'\cos\left(\frac{m\pi}{a}x\right)\cos\left(\frac{n\pi}{b}y\right)e^{\pm j\beta_{mn}z}$$

$$m = 0, 1, 2, \ldots$$
$$n = 0, 1, 2, \ldots$$
$$m = 0, n = 0 \text{ is not a}$$
$$\text{physical mode}$$

2. The cutoff frequency for the TE_{mn} mode is

$$f_{c,mn} = \frac{1}{2\pi\sqrt{\mu\epsilon}}\sqrt{\left(\frac{m\pi}{a}\right)^2 + \left(\frac{n\pi}{b}\right)^2}$$

3. The intrinsic wave impedance for a propagating TE_{mn} mode is

$$\eta_{mn} = \frac{\sqrt{\frac{\mu}{\epsilon}}}{\sqrt{1 - \left(\frac{f_{c,mn}}{f}\right)^2}}$$

Unlike the TM case, the intrinsic impedance is open circuit at $f = f_{c,mn}$. Expressions for the propagation constant and cutoff wavelength are the same as in the TM case. The only difference is that modes with either $m = 0$ or $n = 0$ are possible in the TE case and not possible for TM modes. Mathematically, this is because either $m = 0$ or $n = 0$ will make $\hat{E}_z = 0$, and no modes will be possible.

Other topics covered in this chapter include the drawing of the electric and magnetic field configurations associated with the different modes, and the current distribution on the walls of a wave guide. Excitation of wave guides was also discussed where it is shown that for effective excitation of a specific mode in a wave guide, feed probes should be located at positions of maximum electric field in the desired mode. The probe should also be oriented along the direction of the electric field.

Energy flow in a wave guide is also considered, and it is shown that the time-average power density in a wave guide supporting either TM_{mn} or TE_{mn} modes is given by

$$\mathbf{P}_{av} = \frac{1}{2}\frac{|\hat{E}_x|^2 + |\hat{E}_y|^2}{\eta_{mn}}\mathbf{a}_z$$

where $\hat{E}_x$ and $\hat{E}_y$ are the transverse field components associated with the mode. η_{mn} stands for either $\eta_{\text{TM}_{mn}}$ or $\eta_{\text{TE}_{mn}}$.

Finally, it is shown that wave-guide problems, including calculation of input impedance of a section of a wave guide terminated with an arbitrary load (for example, an antenna) and the measurement of a load impedance using slotted wave guides, may be analyzed using solution procedures similar to those used in the transmission-line analysis described in chapter 7. Wave-guide wavelength should be used in this case, and graphical solutions using Smith charts are also possible. Examples 8.5 to 8.7 illustrate such solutions.

PROBLEMS

1. A commonly used procedure for reducing the cutoff frequency of a wave guide of given dimensions is to load (fill) the wave guide dielectrically with materials of high-dielectric constant. For an X-band wave guide with $a = 2.3$ cm and $b = 1.2$ cm, determine the cutoff frequency of the lowest few TE and TM modes if the wave guide is filled with the following materials:

(a) Air.
(b) Glass dielectric with $\epsilon = 4\epsilon_o$.
(c) Deionized water with $\epsilon = 81\epsilon_o$.

2. Using the procedure described in section 8.5, obtain expressions for the electric field lines for a TE_{11} mode in a rectangular wave guide. Plot a few electric field lines and compare your results with those given in Table 8.1.

3. Derive an expression for the magnetic field lines in the transverse plane of a TM_{11} mode. Plot a few lines and compare your results with those given in Table 8.1.

4. Following the procedure described in problems 2 and 3, write a computer program that calculates and plots the electric and magnetic field lines in the transverse and longitudinal planes of a rectangular wave guide. Input to your program should include the type of mode—that is, TM or TE excitation—and two integers for the m and n mode numbers. For plotting, you may color code the intensity of the electric and magnetic field lines to illustrate high- and low-intensity regions.

5. From the expression of the axial magnetic field component $\hat{H}_z$ given in equation 8.34, derive expressions for all the electric and magnetic field components associated with the TE_{mn} mode in a rectangular wave guide.

6. In a rectangular wave guide, the time-domain expressions for the transverse electric field component E_x is given by

$$E_x = A \cos\left(\frac{\pi}{a}x\right) \sin\left(\frac{2\pi y}{b}\right) \sin(7\pi \times 10^{10}t - \beta z)$$

Determine the following:
(a) Operating mode.
(b) Frequency of operation.
(c) Propagation constant β if the wave-guide dimensions are $a = 2.3$ cm and $b = 1.2$ cm. Assume the wave guide is filled with air.
(d) Cutoff frequency and wave impedance.

7. The x component of the magnetic field intensity associated with a TM mode is given by

$$H_x = 7 \sin\left(\frac{3\pi}{a}x\right) \cos\left(\frac{\pi y}{b}\right) \sin(\omega t - \beta z)$$

(a) Determine the operating wave-guide mode.
(b) If $a = 2.3$ cm and $b = 1.2$ cm, calculate the cutoff frequency of this mode assuming an air-filled wave guide.
(c) Calculate ω so that the operating frequency is 25 percent higher than the cutoff frequency.
(d) Calculate the propagation constant β at the frequency in part c.

8. An air-filled rectangular wave guide is operating in the TM_{21} mode at a frequency f that is 27 percent above the cutoff frequency. The wave-guide dimensions are $a = 11.5$ cm and $b = 6$ cm.
(a) Calculate the cutoff and operating frequencies.
(b) Calculate the phase constant β and the intrinsic wave impedance.
(c) Write complete expressions (substitute numerical values) for the electric and magnetic fields associated with the mode of operation.

9. The cutoff frequency of the TM_{12} mode in an air-filled rectangular wave guide is 10 GHz, whereas the cutoff frequency of the TM_{21} mode is 6 GHz.

(a) Calculate the a and b dimensions of this wave guide.

(b) Calculate the cutoff frequency of the TM_{11} mode.

(c) Calculate the cutoff frequency of the TM_{11} mode if the wave guide is filled with teflon material of $\epsilon = 2.1\epsilon_o$.

10. The cutoff frequency of the TE_{10} mode is 10 GHz, whereas the cutoff frequency of the TE_{03} mode is 60 GHz. The wave guide is filled with a glass dielectric of $\epsilon = 4\epsilon_o$.

(a) Calculate the a and b dimensions of the wave guide.

(b) Calculate the cutoff frequency of the TE_{10} mode in air.

(c) For an air-filled wave guide of the same dimensions as calculated in part a, calculate the number of possible modes if the excitation frequency is 65 GHz.

11. Design a rectangular wave guide operating in the TE_{10} mode. For minimum attenuation consideration, select $a = 2b$. Determine the dimensions such that the middle frequency of the operating band for only the TE_{10} mode is 9 GHz.

12. Determine the surface currents on the walls of a rectangular wave guide supporting the lowest TM mode. Sketch the variations of these surface currents on the broad top wall ($y = b$) of the wave guide.

13. A rectangular wave guide is operating in the TE_{10} mode. The wave-guide dimensions are $a = 7.62$ cm and $b = 4$ cm, and the total power transmitted is 1.2 W at an operating frequency of 30 percent higher than the cutoff frequency.

(a) Calculate the cutoff frequency, operating frequency, and the intrinsic wave impedance, assuming an air-filled guide.

(b) Determine the amplitude of the electric field associated with the given transmitted power.

(c) Write complete expressions for the electric and magnetic fields associated with this mode. Substitute numerical values for all the constants.

14. A slotted wave guide of dimensions $a = 7.6$ cm and $b = 3.8$ is used to measure the wave guide wavelength and the operating frequency. If the distance between two successive minima is 9 cm, and assuming TE_{10} mode of operation in air, calculate the following:

(a) Wave-guide wavelength λ_g, cutoff frequency, and operating frequency.

(b) Propagation constant and intrinsic impedance.

15. The slotted wave guide of problem 14 is terminated by a horn antenna, and the measured VSWR is 2.1. When the horn antenna is replaced by a short circuit, a 3-cm shift towards the load is observed in the position of the first minimum. Calculate the input impedance of the horn antenna.

16. A rectangular wave guide is loaded with a dielectric material of $\epsilon_r^* = 50 - j12$. The high-dielectric constant $\epsilon' = 50$ was desired to reduce the cutoff frequency, whereas the dielectric losses $\epsilon'' = 12$ could not be avoided. The wave guide dimensions are $a = 2.54$ cm and $b = 1.3$ cm.

(a) Because ϵ_r^* is complex in this case, the cutoff frequency in equation 8.36 is expected to be complex. Calculate a real value of the cutoff frequency of the TE_{10} mode based on the real part of ϵ_r^*.

(b) Based on an operating frequency 10 percent higher than the cutoff frequency calculated in part a, calculate the attenuation per unit length as a result of the unavoidable dielectric losses.

(c) At what distance along the guide is the transmitted power reduced to 50 percent of its initial value?

17. Assume an air-filled rectangular wave-guide made of copper of conductivity $\sigma = 5.8 \times 10^7$

S/m, with dimensions $a = 7.6$ cm and $b = 4$ cm, and a TE_{10} mode of operation at a frequency $f = 1.1f_c$. Calculate the attenuation constant owing to ohmic losses in the wave-guide walls. Also calculate the distance along the axis at which the power is reduced to 95 percent of its initial value. Comment on the importance of the ohmic losses in this case.

18. Assume an X-band air-filled rectangular wave guide of dimensions $a = 2.3$ cm and $b = 1.2$ cm. The operating mode is the TE_{10} mode. Calculate the maximum power that can be transmitted in the wave guide if the breakdown electric field in air is 2×10^6 V/m and a minimum 25 percent safety factor is required. The operating frequency is 30 percent higher than the cutoff frequency for the TE_{10} mode.

19. In example 8.3, it was shown that the propagation of various modes in wave guides may be interpreted in terms of a zigzag motion of plane waves bouncing back and forth between the wave-guide walls. As a result, we may identify three types of velocities.

   (a) The velocity of propagation in the medium filling the wave guide. This velocity is denoted by v in Figure P8.19 and is actually the velocity of propagation of plane waves at an angle θ. $v = 1/\sqrt{\mu\epsilon}$.

   (b) The phase velocity, which is the velocity of propagation of constant phase planes as projected along the guide axis. This is

$$v_p = \frac{v}{\sqrt{1 - \left(\frac{f_c}{f}\right)^2}}$$

   (c) The third velocity v_g is known as the group velocity and is the velocity with which the multiple-reflected waves actually travel along the axis of the wave guide. The group velocity v_g is, hence, the velocity of the energy propagation down the guide. v_g is given using equation 8.21 by

$$v_g = \frac{\partial\omega}{\partial\beta} = v\cos\theta = v\sqrt{1 - \left(\frac{f_c}{f}\right)^2}$$

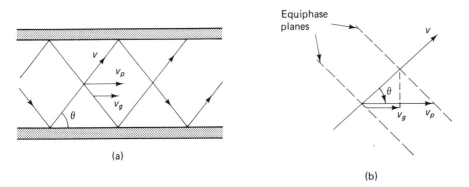

(a)

(b)

Figure P8.19 (a) Wave-guide modes may be interpreted as superpositions of plane waves bouncing at an angle θ. (b) Relationship by v, v_p, and v_g. v is the velocity of propagation along the direction of the bouncing waves, v_p is the projected velocity of equiphase planes along the wave-guide axis, and v_g is the projection of v along the wave-guide axis.

The group velocity in an air-filled wave guide is, hence, less than the velocity of light, whereas the phase velocity is larger. For a rectangular wavelength of dimensions $a = 7$ cm and $b = 4$ cm and supporting TE_{10} mode, calculate v, v_p, and v_g at $f = 1.3f_c$ and for the following cases:

  (i) Air-filled wave guide.
  (ii) Glass-filled wave guide $\epsilon = 4\epsilon_o$.
  (iii) Repeat the above two calculations for $f = 1.1f_c$ and $f = 1.9f_c$.
  (iv) Comment on your results regarding the relative values of v, v_p, and v_g in air, glass, and as a function of frequency.

20. A rectangular wave guide of dimensions $a = 5$ cm and $b = 2.5$ cm is operating in the TE_{10} mode at a frequency midway between the fundamental TE_{10} and the next higher-order mode.
  (a) Calculate the cutoff and operating frequencies for the two cases of an air-filled wave guide and a glass-filled wave guide of $\epsilon_r^* = 4 - j0.1$. (*Hint:* Use the assumption $\epsilon''/\epsilon' \ll 1$ to calculate a real number for f_c.)
  (b) Calculate the attenuation constant as a result of the dielectric losses in the glass material.
  (c) If the wave-guide walls are made of copper of $\sigma = 5.8 \times 10^7$ S/m, calculate the attenuation constant as a result of ohmic losses in the wave-guide walls for both air- and glass-filled wave guides.
  (d) Compare values of the dielectric and ohmic attenuation constants for the glass-filled wave-guide case.

CHAPTER 9

ANTENNAS

9.1 INTRODUCTION

In the previous chapters we discussed transmission lines and wave guides as possible means for transmitting electromagnetic energy. Antennas are devices that are used for transmitting and receiving electromagnetic energy in the form of radio waves. These antennas are used whenever it is uneconomical, impractical, or even impossible to use transmission lines or wave guides. Examples of these applications that require the use of antennas include aircraft communication, communication between ships, broadcasting, and the various other advanced communication systems that include satellites. Besides these communication applications, antennas are also used in other industrial and medical applications that may include geophysical exploration, and the electromagnetic induced hyperthermia for cancer treatment.

In all these applications antennas are designed to transmit and receive radio waves efficiently. A successful antenna design is one that achieves good impedance matching to the feed transmission line so as to maximize the available power for radiation and is also one that achieves the best possible compromise between the various constraints imposed on the desired radiation pattern. Optimizing the radiation pattern may include maximizing the radiation in one direction and suppressing it in others.

If achieving a specific radiation pattern is difficult or impossible using a single antenna, antenna engineers often resort to designing arrays of simple antennas. By adjusting the amplitude and phase of the feed voltages to the various elements in the array, as well as the geometrical arrangement of these elements, it is often possible to achieve the desired radiation characteristics. An array design is complicated by the mutual interaction between the various elements in the array. Compromises are often made to simplify the array designs.

Although antenna shapes may be countless, there are several basic types of antennas. This includes "wire antennas" that are often used in radio stations, portable radio receivers, automobiles, and on televisions. They may take the shape of a simple straight wire, loop, helix, or spiral. A wire antenna is shown in Figure 9.1. At higher frequencies where wave guides are used as feed structures, "aperture antennas" are often used. This may include the electromagnetic horn shown in Figure 9.1 where the dimensions of the feed wave guide are linearly flared to improve the directivity of the radiation pattern. An antenna array may be formed by combining several of the basic elements shown in Figure 9.1. "Reflector antennas" and "lens antennas" are examples of the many engineering procedures that are used to collimate, focus, and effectively direct the electromagnetic energy radiated by antennas. Frequency-independent antennas are designed and used to provide equal performance over a broad frequency band. Examples of these as well as other types of antennas are shown in Figure 9.1.

Figure 9.1 Examples of various types of antennas.

In this chapter we briefly discuss the physical aspects for radiation from antennas. Simple types of antennas will then be analyzed. To identify the basic radiation characteristics we start by analyzing the radiation fields from an infinitesimally small current element known as the oscillating electric dipole. Analysis of linear wire antennas then follows, and the radiation characteristics of a linear antenna array are described. Mutual effects between array elements are analyzed, and the effect of ground on the radiation characteristics of antennas will be examined.

9.2 PHYSICAL ASPECTS OF RADIATION

Before we describe geometries of commonly used antennas and the mathematical basis for their operation and design, it is important that we understand the physical aspects of the radiation mechanism and, more specifically, the process that describes the detachment of the fields from the antenna. Through our study of the time-varying fields in the previous chapters, we know that time-varying electric and magnetic fields generate each other; hence, once these fields are separated from the antenna they can sustain themselves and actually propagate in the surrounding medium. So, although the propagation process is fairly well understood, the detachment process of these fields from the antenna still needs some explanation.

To explain this field separation process, let us consider the transient propagation characteristic of a sinusoidal voltage wave along a two-conductor transmission line shown in Figure 9.2a. This line is considered here to be the feed structure for the antenna as shown in Figure 9.2b. The electric field distribution associated with the sinusoidal voltage across the two-wire transmission line is shown in Figure 9.2a, where it can be seen that the electric field lines change direction every half-wavelength ($\lambda/2$) as the voltage changes in phase. The distribution of the electric field lines along the flared (antenna) section of the transmission line at the end of a quarter of a period $T/4$ is shown in Figure 9.2c. After this period, the applied voltage reaches maximum at the feed point and, hence, the electric field on the flared section reaches its maximum, which is represented by three electric field lines as shown in Figure 9.2c. As the applied voltage continues $T/4 < t < T/2$ to propagate down the transmission line toward the open end, the voltage at the feed point will start to decrease from its peak value to zero at the end of the first half-period. This is equivalent to neutralizing the charge distribution on the flared (antenna) section with charges of the opposite polarity, which results in establishing electric field lines of the opposite direction as shown in Figure 9.2d. In other words, the increase in the applied voltage during the first quarter of the sinusoidal cycle is represented graphically in Figure 9.2c with the increase in the number of the electric field lines to a maximum of three. The decrease in the applied voltage at the feed during the period $T/4 < t < T/2$ is represented by an additional electric field line in the opposite direction as shown in Figure 9.2d. After a half-period $t = T/2$, the voltage at feed point will be zero, and equal numbers of oppositely directed electric field lines will be formed on the flared section (antenna) as shown in Figure 9.2e. Also at $t = T/2$ the electric field lines will be extended to a distance of a half-wavelength away from the plane of the flared antenna wires. The equal and

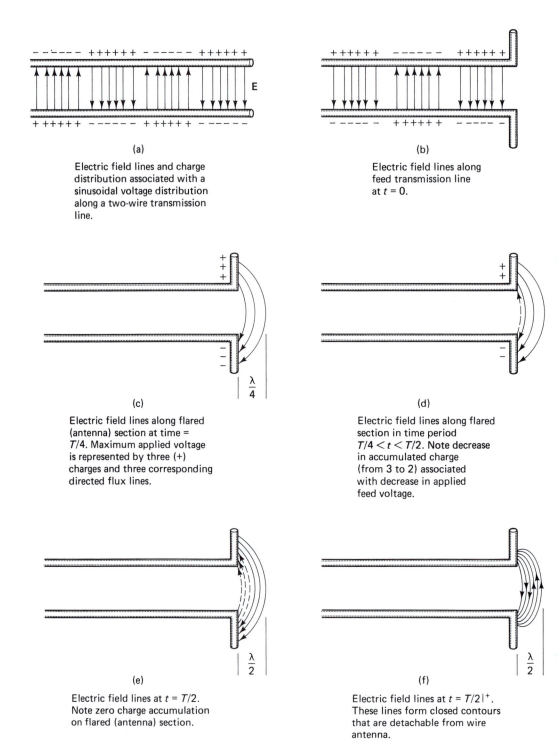

(a)

Electric field lines and charge
distribution associated with a
sinusoidal voltage distribution
along a two-wire transmission
line.

(b)

Electric field lines along
feed transmission line
at $t = 0$.

(c)

Electric field lines along flared
(antenna) section at time =
$T/4$. Maximum applied voltage
is represented by three (+)
charges and three corresponding
directed flux lines.

(d)

Electric field lines along flared
section in time period
$T/4 < t < T/2$. Note decrease
in accumulated charge
(from 3 to 2) associated
with decrease in applied
feed voltage.

(e)

Electric field lines at $t = T/2$.
Note zero charge accumulation
on flared (antenna) section.

(f)

Electric field lines at $t = T/2|^+$.
These lines form closed contours
that are detachable from wire
antenna.

Figure 9.2 Graphical illustration of the physical aspects related to the separation
of the electric fields from wire antennas.

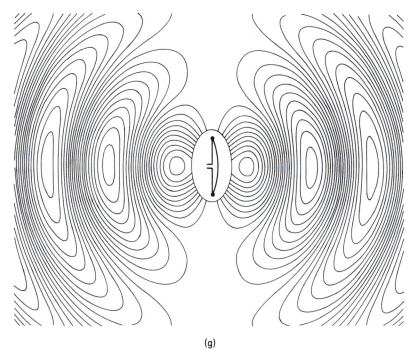

(g)

Electric field lines illustrating continuous separation
and propagation of fields away from wire antennas.

Figure 9.2   *Continued*.

oppositely directed electric field lines shown in Figure 9.2f shortly after $(t = T^+/2)$ form closed lines. With the instantaneously zero voltage at $t = T/2$, zero voltage at the open end of the transmission line, these closed electric field lines will be free to separate and, hence, propagate away from the wire structure. The continuous detachment of the closed groups of the electric field lines will form the basis for the wave propagation away from the antenna as shown in Figure 9.2g. Once again these time-varying electric and magnetic fields once separated will reproduce each other and, hence, will be able to propagate away from the antenna.

9.3 RADIATION FROM SHORT ALTERNATING CURRENT ELEMENT

To illustrate the mathematical basis for determining the radiated electric and magnetic fields from wire antennas as well as the physical properties of these fields near and far away from the antenna, let us consider the simplest possible radiating element that is an infinitesimal current element of length $d\ell$ and carrying an alternating current $I \cos \omega t$. Clearly, having a uniform current distribution of magnitude I throughout the element is unrealistic because currents should be zero at both open ends of the wire.

Such a case, which is needed to simplify the mathematical treatment of the field quantities, can, however, be justified by considering the current element as a building block in a more complicated long-wire antenna structure. Analysis of the radiation characteristics will be based on the magnetic vector potential.

9.3.1 Magnetic Vector Potential

The geometry of the current element is illustrated in Figure 9.3. The objective is to calculate the electric and magnetic fields radiated from this current element. To facilitate such calculations, it is mathematically simpler to relate the current distribution on the antenna to the auxiliary function known as the vector magnetic potential **A** (see chapter 4) and then use **A** to calculate the desired electric and magnetic fields. To illustrate this, let us recall that $\nabla \cdot \mathbf{B} = 0$ from Gauss's law of the magnetic field. Hence, **B** can be expressed as

$$\mathbf{B} = \nabla \times \mathbf{A} \tag{9.1}$$

where **A** is the magnetic vector potential function. From vector algebra it can be shown that $\nabla \cdot \nabla \times \mathbf{A} = 0$ always. To obtain a differential equation that relates **A** to the current distribution along the antenna, we use Faraday's law for time harmonic fields

$$\nabla \times \mathbf{E} = -j\omega\mu_o \mathbf{H} = -j\omega\nabla \times \mathbf{A}$$

or

$$\nabla \times (\mathbf{E} + j\omega\mathbf{A}) = 0$$

The function $(\mathbf{E} + j\omega\mathbf{A})$ with a zero curl can be expressed as the gradient of a scalar function; hence,

$$\mathbf{E} + j\omega\mathbf{A} = -\nabla\Phi \tag{9.2}$$

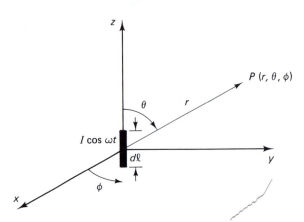

Figure 9.3 Schematic illustrating the geometry of a short $(d\ell)$ current element $I \cos \omega t$. The point $P(r, \theta, \phi)$ is the observation point at which the radiation fields are to be determined.

where Φ is a scalar potential function. To obtain the desired differential equation, we use Ampere's law,

$$\nabla \times \mathbf{H} = \mathbf{J} + j\omega\epsilon_o \mathbf{E} \qquad (9.3)$$

Substituting equations 9.1 and 9.2 in equation 9.3, we obtain

$$\nabla \times \nabla \times \mathbf{A} = \mu_o \mathbf{J} + j\omega\epsilon_o \mu_o (-j\omega\mathbf{A} - \nabla\Phi)$$

$$= \mu_o \mathbf{J} + \omega^2 \mu_o \epsilon \mathbf{A} - j\omega\epsilon_o \mu_o \nabla\Phi$$

Using the vector identity $\nabla \times \nabla \times \mathbf{A} = \nabla(\nabla \cdot \mathbf{A}) - \nabla^2 \mathbf{A}$, we obtain

$$\nabla^2 \mathbf{A} + \omega^2 \mu_o \epsilon_o \mathbf{A} = -\mu_o \mathbf{J} + \nabla(\nabla \cdot \mathbf{A} + j\omega\mu_o \epsilon_o \Phi) \qquad (9.4)$$

To simplify equation 9.4, we choose $\nabla \cdot \mathbf{A}$ such that

$$\nabla \cdot \mathbf{A} = -j\omega\mu_o \epsilon_o \Phi \qquad (9.5)$$

This is known as Lorentz's condition and is possible because of the fact that we only specified the $\nabla \times \mathbf{A}$ in equation 9.1 and are free to choose $\nabla \cdot \mathbf{A}$ for a complete characterization of the vector quantity $\mathbf{A}$. Substituting equation 9.5 in equation 9.4, the desired equation for $\mathbf{A}$ becomes

$$\nabla^2 \mathbf{A} + \beta_o^2 \mathbf{A} = -\mu_o \mathbf{J} \qquad (9.6)$$

where $\beta_o = \omega\sqrt{\mu_o \epsilon_o}$ is the free space wave number. A similar equation may be obtained for the scalar potential function Φ. From Gauss's law, we have

$$\nabla \cdot \epsilon \mathbf{E} = \rho_v$$

Substituting $\mathbf{E}$ from equation 9.2 and using Lorentz's condition in equation 9.5, we obtain

$$\nabla \cdot \epsilon_o (-\nabla\Phi - j\omega\mathbf{A}) = -\epsilon_o \nabla^2 \Phi - j\omega\epsilon_o \nabla \cdot \mathbf{A}$$

$$= -\epsilon_o \nabla^2 \Phi + \epsilon_o \beta_o^2 \Phi = \rho_v$$

or

$$\nabla^2 \Phi + \beta_o^2 \Phi = -\frac{\rho_v}{\epsilon_o} \qquad (9.7)$$

Equation 9.7 can be solved to determine the scalar potential Φ in terms of the charge ρ_v. This charge, however, is not an independent quantity and in fact is related to the current $\mathbf{J}$ through the continuity equation. It is therefore not necessary to solve for the scalar potential function Φ and instead determine it from $\mathbf{A}$ through the Lorentz's condition (equation 9.5). Based on the preceding discussion, it is clear that in our radiation problem we are interested in solving equation 9.6 for $\mathbf{A}$ in terms of the current distribution. The radiation field quantities can then be determined from equation 9.1 and Maxwell's equations, as is illustrated next.

For the infinitesimal current source of Figure 9.3, the current is z directed—that is, $\mathbf{J} = J_z \mathbf{a}_z$—hence, equation 9.6 reduces to

$$\nabla^2 A_z + \beta_o^2 A_z = -\mu_o J_z \qquad (9.8)$$

Based on symmetry considerations, it can be shown that A_z is independent of the polar angle θ and the azimuth angle ϕ. A_z will be a function of the radial distance r only as shown in Figure 9.3.

Writing the Laplacian operator ∇^2 in the spherical coordinate system and maintaining only the derivatives with respect to r, we obtain

$$\frac{1}{r^2}\frac{d}{dr}\left(r^2\frac{dA_z}{dr}\right) + \beta_o^2 A_z = 0 \tag{9.9}$$

In equation 9.9, the current J_z term was substituted by zero at the observation point P. If we make the substitution $\psi = rA_z$, we obtain

$$\frac{dA_z}{dr} = \frac{1}{r}\frac{d\psi}{dr} - \frac{1}{r^2}\psi$$

Equation 9.9 then reduces to

$$\frac{d^2\psi}{dr^2} + \beta_o^2\psi = 0 \tag{9.10}$$

The general solution for equation 9.10 may be in the form

$$\psi = C_1 e^{-j\beta_o r} + C_2 e^{j\beta_o r}$$

and consequently

$$A_z = C_1\frac{e^{-j\beta_o r}}{r} + C_2\frac{e^{j\beta_o r}}{r} \tag{9.11}$$

Considering the $e^{j\omega t}$ time-harmonic variation, we have to choose the first part $C_1 e^{-j\beta_o r}/r$ as our proper solution because it has the outgoing traveling wave characteristic. The solution for A_z is, hence,

$$A_z = C_1\frac{e^{-j\beta_o r}}{r} \tag{9.12}$$

To determine the unknown coefficient C_1 we derive an expression for the magnetic field $\mathbf{H} = \nabla \times \mathbf{A}$ and compare the result with Biot-Savart's law for the case of direct current. To carry out the curl operation in the spherical coordinate system, we express A_z in terms of its spherical components to obtain

$$A_r = A_z \cos\theta = C_1 \cos\theta\frac{e^{-j\beta_o r}}{r}$$

and

$$A_\theta = -A_z \sin\theta = -C_1 \sin\theta\frac{e^{-j\beta_o r}}{r}$$

The magnetic field intensity $\mathbf{H}$ is now given by

$$\mathbf{H} = \frac{1}{\mu_o} \nabla \times \mathbf{A} = \frac{1}{\mu_o} \begin{vmatrix} \dfrac{\mathbf{a}_r}{r^2 \sin\theta} & \dfrac{\mathbf{a}_\theta}{r \sin\theta} & \dfrac{\mathbf{a}_\phi}{r} \\ \dfrac{\partial}{\partial r} & \dfrac{\partial}{\partial\theta} & \dfrac{\partial}{\partial\phi} \\ A_r & rA_\theta & 0 \end{vmatrix}$$

$$= \frac{1}{\mu_o} - \left((-j\beta_o) C_1 \sin\theta\, e^{-j\beta_o r} - C_1 \sin\theta \frac{e^{-j\beta_o r}}{r} \right) \frac{\mathbf{a}_\phi}{r} \qquad (9.13)$$

$$\mathbf{H} = \frac{C_1}{\mu_o} \sin\theta\, e^{-j\beta_o r} \left(\frac{j\beta_o}{r} + \frac{1}{r^2} \right) \mathbf{a}_\phi$$

Comparing the $1/r^2$ term with Biot-Savart's law for the magnetic field associated with direct currents, we obtain

$$C_1 = \frac{\mu_o}{4\pi} I d\ell$$

and equation 9.13 reduces to

$$\mathbf{H} = \frac{1}{4\pi} I d\ell\, \sin\theta\, e^{-j\beta_o r} \left(\frac{j\beta_o}{r} + \frac{1}{r^2} \right) \mathbf{a}_\phi \qquad (9.14)$$

From Ampere's law the electric field is given by

$$\mathbf{E} = \frac{1}{j\omega\epsilon_o} \nabla \times \mathbf{H}$$

$$= \frac{j\eta_o I d\ell}{2\pi\beta_o} \cos\theta \left(\frac{j\beta_o}{r^2} + \frac{1}{r^3} \right) e^{-j\beta_o r} \mathbf{a}_r \qquad (9.15)$$

$$- \frac{j\eta_o I d\ell}{4\pi\beta_o} \sin\theta \left(-\frac{\beta_o^2}{r} + \frac{j\beta_o}{r^2} + \frac{1}{r^3} \right) e^{-j\beta_o r} \mathbf{a}_\theta$$

where $\eta_o = \sqrt{\mu_o/\epsilon_o}$. In the following, we briefly describe the physical meaning associated with each term in equations 9.14 and 9.15.

9.3.2 Discussion of Radiation Fields

To begin with, it is rather surprising to see that an infinitesimal current element carrying alternating current of constant amplitude would produce fields with such complicated expressions as those given in equations 9.14 and 9.15. Soon we will see that there is a physical significance for each term, which will help us not only accept them but also appreciate their presence. We will also learn that some of these terms, as well as one of the electric field components, will be omitted when we discuss the far field radiation from the current element. Because the far fields are of prime interest in our antenna studies, the fields in the near zone of the antenna will not be considered in our future analysis. This will significantly reduce the mathematical effort required in future analysis.

With this in mind let us begin our analysis of the various terms associated with the magnetic field intensity. From equation 9.14, we note that the magnetic field H_ϕ has two components: one is proportional to $1/r^2$, and the other varies as $1/r$. The second term in equation 9.14 that is proportional to $1/r^2$ is similar to that obtained from Biot-Savart's law. This $1/r^2$ term is known as the induction term, because it relates the current on the radiating element to a magnetic field component that is present even if direct (not time varying) currents are present. This induction term certainly contributes to the magnetic energy stored in the neighborhood of the electric dipole. The other term in equation 9.14 is proportional to $1/r$ and is known as the radiation term. At larger distances from the radiating dipole, the contribution from the $1/r$ term to the value of the magnetic field will be larger than that due to the $1/r^2$ term. Hence, the $1/r$ radiation term is actually related to the outflow of the electromagnetic energy away from the antenna. Based on the calculation of the time-average power density flow away from the source that is described next, it will be shown that the time-average contribution of the $1/r^2$ term is zero. This physically means that the $1/r^2$ term makes only a transient contribution to the energy stored in the immediate neighborhood of the current element. The nonzero contribution from the $1/r$ term to the time-average Poynting vector describes the electromagnetic energy flow away from the current element. It should be noted that the $1/r$ term was not included in Biot-Savart's law for direct currents; hence, its appearance in equation 9.14 resulted from the assumed oscillating (time-varying) current on the radiating element. From equation 9.14, the distance r at which the induction and radiation fields are equal is given by

$$\frac{\beta_o}{r} = \frac{1}{r^2}$$

or

$$r = \frac{1}{\beta_o} = \frac{\lambda}{2\pi} \tag{9.16}$$

At distances shorter than $\lambda/2\pi$ the magnetic field will be dominated by induction fields that contribute to the energy storage in the "near field" zone of the radiating element, whereas at distances greater than $\lambda/2\pi$ the magnetic field will be dominated by the $1/r$ term that contributes to "radiation" and the actual energy propagation away from the source.

Let us further emphasize these ideas and also introduce new ones by discussing the characteristics of the various terms in electric field expression of equation 9.15. From equation 9.15, we notice that in addition to the $1/r$ and $1/r^2$ terms described earlier there is a new term proportional to $1/r^3$ that will be denoted as the electrostatic field term. The reason for relating the $1/r^3$ term to electrostatic fields may be clarified by briefly considering the physical realization of an infinitesimal current element with constant current distribution on it. If such a current element physically exists, the current at the end points should be zero, and a triangular rather than uniform current distribution should be assumed, as shown in Figure 9.4b. Assuming a constant current

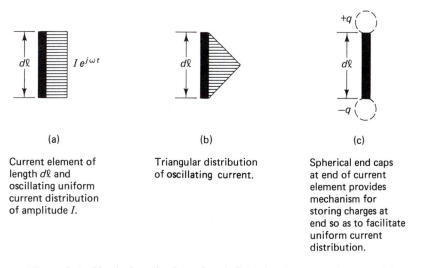

(a)

Current element of
length $d\ell$ and
oscillating uniform
current distribution
of amplitude I.

(b)

Triangular distribution
of oscillating current.

(c)

Spherical end caps
at end of current
element provides
mechanism for
storing charges at
end so as to facilitate
uniform current
distribution.

Figure 9.4   Physical realization of an infinitesimal current element with constant current distribution.

distribution on the antenna, therefore, requires a mechanism for storing electric charges at both ends of the current element so that the continuity equation may be satisfied. In other words, the nonzero flow of current at the end of the antenna requires a structure such as a circular disk or a spherical cap at the end of the antenna to help store the charge accumulated as a result of the current flow at the end. Such an arrangement is illustrated in Figure 9.4c. The charge accumulation q at the end spherical caps is related to current flow through the continuity equation

$$\frac{\partial q}{\partial t} = I e^{j\omega t}$$

Integrating with respect to time, we obtain

$$q = \frac{I}{j\omega} e^{j\omega t}$$

Next we calculate the electric potential Φ at an observation point P resulting from the two charges $+q$ and $-q$ at the end caps. From Figure 9.5, we obtain

$$\Phi_P = \frac{q}{4\pi\epsilon_o r_1} - \frac{q}{4\pi\epsilon_o r_2}$$

$$= \frac{q}{4\pi\epsilon_o \left(r - \dfrac{d\ell}{2} \cos\theta \right)} - \frac{q}{4\pi\epsilon_o \left(r + \dfrac{d\ell}{2} \cos\theta \right)}$$

$$= \frac{q}{4\pi\epsilon_o} \left(\frac{d\ell \cos\theta}{r^2 - \left(\dfrac{d\ell}{2} \cos\theta \right)^2} \right)$$

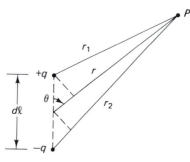

Figure 9.5 Electric potential at P resulting from the dipole charges $+q$ and $-q$.

If the observation point is located at a distance r much larger than the length $d\ell$ of the current element (i.e., $r^2 \gg (d\ell/2 \cos \theta)^2$), we obtain

$$\Phi_p = \frac{q}{4\pi\epsilon_o} \frac{d\ell \cos \theta}{r^2} \tag{9.17}$$

The electric field at P may be obtained from the potential expression (equation 9.17) through the relation

$$\mathbf{E}_p = -\boldsymbol{\nabla} \Phi_p = -\frac{\partial \Phi_p}{\partial r} \mathbf{a}_r - \frac{1}{r} \frac{\partial \Phi_p}{\partial \theta} \mathbf{a}_\theta - \frac{1}{r \sin \theta} \frac{\partial \Phi_p}{\partial \phi} \mathbf{a}_\phi.$$

Hence,

$$E_r = \frac{2q}{4\pi\epsilon_o} \frac{d\ell \cos \theta}{r^3} \tag{9.18a}$$

$$E_\theta = \frac{q}{4\pi\epsilon_o} \frac{d\ell \cos \theta}{r^3} \tag{9.18b}$$

and

$$E_\phi = 0$$

From equation 9.18, it can be seen that both electric field components have $1/r^3$ distance dependence. As a matter of fact, if we substitute $q = I/j\omega\, e^{j\omega t}$, the electric field components in equation 9.18 would reduce to exactly the same $1/r^3$ terms in the electric field expression in equation 9.15.

We summarize by indicating that the electric and magnetic field expressions for a short electric dipole include near and far field terms. The near fields include the $1/r^2$ and $1/r^3$, which contribute to the energy storage in the neighborhood of the antenna. The $1/r^2$ terms correspond to the induction fields produced by the current flow in the dipole element, whereas the $1/r^3$ terms are related to the electrostatic fields produced by the charge accumulated at the ends of the current elements to maintain uniform current distribution. The far or radiation fields, conversely, are related to the $1/r$ terms that dominate the field values at distance $r > \lambda/2\pi$. To illustrate further the radiative nature of the $1/r$ fields, let us calculate the time-average Poynting vector $\mathbf{P}_{ave}$, which is related to the power density flow away from the source.

$$\mathbf{P}_{ave} = \frac{1}{2} Re(\mathbf{E} \times \mathbf{H}^*) \tag{9.19}$$

From equations 9.14 and 9.15, we obtain

$$\mathbf{P}_{ave} = -\frac{1}{2}\frac{(Id\ell)^2}{2(2\pi)^2}\sin\theta\,\cos\theta\,Re\left\{\frac{j\eta_o}{\beta_o}\left(\frac{j\beta_o}{r^2}+\frac{1}{r^3}\right)\left(\frac{-j\beta_o}{r}+\frac{1}{r^2}\right)\right\}\mathbf{a}_\theta$$

$$+\frac{1}{2}\frac{(Id\ell)^2}{(4\pi)^2}\sin^2\theta\,Re\left\{\left(\frac{-j\eta_o}{\beta_o}\right)\left(\frac{-\beta_o^2}{r}+\frac{j\beta_o}{r^2}+\frac{1}{r^3}\right)\right. \tag{9.20}$$

$$\left.\times\left(\frac{-j\beta_o}{r}+\frac{1}{r^2}\right)\right\}\mathbf{a}_r$$

The θ component of $\mathbf{P}_{ave}$ is zero because this term does not include a real part as required by equation 9.19. Hence,

$$\mathbf{P}_{ave} = \frac{(Id\ell)^2}{2(4\pi)^2}\sin^2\theta\,\frac{\eta_o\,\beta_o^2}{r^2}\mathbf{a}_r$$

$$= \frac{\eta_o}{2}\left(\frac{\beta_o\,Id\ell\,\sin\theta}{4\pi r}\right)^2\mathbf{a}_r \tag{9.21}$$

Equation 9.21 shows that the power density flows in the outward $\mathbf{a}_r$ direction away from the current element and that the $1/r$ components of the electric and magnetic fields are the only terms that contribute to such a flow of power. Hence, we conclude that the $1/r$ terms in the field expressions are the dominant far field components, and that they are the only contributor to the actual flow of the electromagnetic power away from the current element. Other terms in the electric and magnetic field expressions such as the $1/r^2$ and $1/r^3$ terms are responsible for quantifying the magnetic and electric energies stored in the neighborhood of the current element.

The far zone fields are essential to quantifying and characterizing the radiation characteristics of antennas. Near fields do not contribute to the radiated power and are generally not of great interest in quantifying radiation characteristics of antennas. Near fields, however, represent energy storage in the space immediately surrounding the antenna. This certainly impacts the input impedance calculation of the antenna. Thus, apart from impedance calculations, near zone fields are neglected. From this point on, we will focus on the calculations of far fields of antennas and explain the role the length of an antenna plays in changing its radiation characteristics. In the following we will emphasize some of the more important properties of the far field radiation.

9.3.3 Properties of Far Zone Radiation Fields

When r is large compared with the wavelength λ, the only important terms in the electric and magnetic field expressions are those that vary as $1/r$. From equations 9.14 and 9.15, the far zone fields are given by

$$\mathbf{E} = j\eta_o\,\beta_o(Id\ell)\,\sin\theta\frac{e^{-j\beta_o r}}{4\pi r}\mathbf{a}_\theta \tag{9.22a}$$

$$\mathbf{H} = j\beta_o(Id\ell)\,\sin\theta\frac{e^{-j\beta_o r}}{4\pi r}\mathbf{a}_\phi \tag{9.22b}$$

As noted in the previous section the time-average power or the radiated power prop-agates away in the **a**$_r$ direction. Based on these observations one may make the following conclusions:

1. The far zone electric **a**$_\theta$ and magnetic **a**$_\phi$ fields are perpendicular to each other, and both are perpendicular to the outward direction of propagation **a**$_r$.
2. The ratio between the far zone electric and magnetic fields is

$$\frac{E_\theta}{H_\phi} = \eta_o = \sqrt{\frac{\mu_o}{\epsilon_o}}$$

which is the characteristic impedance of free space.

The preceding two characteristics of the far zone radiation are general for any antenna. With the exception that the magnitudes of these fields decrease as $1/r$ and also vary as $\sin\theta$, they have the same properties as those of a plane wave propagating in free space of characteristic impedance $\eta_o = \sqrt{\mu_o/\epsilon_o}$. The fact that the fields vary as $1/r$, and the power density decreases as $1/r^2$ is expected because of the spreading out of the fields as they propagate radially outward from the radiating element. The $\sin\theta$ variation of E_θ and H_ϕ means that the radiated fields are not spherically symmetric around the antenna. In the plane wave case where uniform wave fronts and constant amplitudes are assumed as the wave continues to propagate were made possible by an assumed infinitely large source. Physical antennas are clearly limited in size; hence, spreading out of the fields as well as variation in their amplitudes as a function of θ are expected. In the following section we introduce some parameters often used to characterize radiation from antennas.

9.4 BASIC ANTENNA PARAMETERS

Clearly it is not sufficient to derive complicated expressions for the various field components radiated from an antenna. To help evaluate the performance of these antennas and clearly understand their radiation characteristics, it is necessary to use graphical representation of the fields in a standard fashion and also introduce parame-ters that would help quantify the efficiencies and facilitate quantitative comparison between antennas. The following are some examples of these antenna parameters.

9.4.1 Radiation Pattern

The radiation pattern is a graphical representation of the relative distribution of the radiated power as a function of the space coordinates. In practice the radiation pattern is determined using the far field radiation from the antenna. It is basically a three-dimensional plot that illustrates the relative distribution of the radiated power as a function of the coordinate angles θ and ϕ and at a fixed distance r from the antenna. To illustrate the usefulness of this parameter let us consider the radiated power from the infinitesimal electric dipole (current element) as given in equation 9.21. From

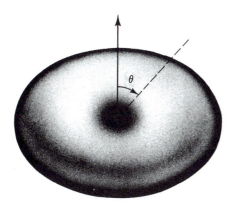

Figure 9.6 The donut-shaped radiation pattern $P(\theta, \phi) = \sin^2 \theta$ of an infinitesimal electric dipole.

equation 9.21, we observe that the radiated power density at a fixed value of r varies as $\sin^2 \theta$ and is independent of ϕ (i.e., there is no ϕ variation). A three-dimensional plot of the variation of the relative distribution (i.e., disregard the multiplicative constants in equation 9.21) of the radiated power as a function of θ and ϕ (i.e., for a fixed r) is shown in Figure 9.6. Because equation 9.21 is independent of ϕ, the resulting radiation pattern is a doughnut-shaped one as illustrated in Figure 9.6. It is sometimes difficult, however, to visualize complicated three-dimensional plots of the radiation patterns. Instead, it is common practice to show two-dimensional cross sections; however, these cross sections must be carefully selected to emphasize the important features of the three-dimensional plots. Two of the most commonly used two-dimensional views are the E- and H-plane patterns. The E-plane pattern is a section of the three-dimensional pattern in which the electric field lies. This section also contains the maximum value of the electric field. From equation 9.15, we observe that the electric field at a far distance (i.e., the $1/r$ component) is in the $\mathbf{a}_\theta$ direction. Hence, the two-dimensional E-plane pattern should be a vertical section containing $\mathbf{a}_\theta$ in Figure 9.6. From Figure 9.7a and equation 9.21, it is clear that the radiated power distribution in a plane containing the electric field varies according to $\sin^2 \theta$ and results in a "figure 8" pattern as shown in Figure 9.7b. The two-dimensional E-plane pattern shown in Figure 9.7b clearly does not contain all the main features of the three-dimensional plot of Figure 9.7a. This is why more than one two-dimensional sectional view is often required. Of particular interest is the H-plane pattern that is the plane of the magnetic field and also includes its maximum value. Such a sectional view is shown in Figure 9.8a and the resulting H-plane pattern for an infinitesimal current element is shown in Figure 9.8b. The x-y plane was chosen to be the H-plane pattern because the magnetic field given by equation 9.14 is totally in the $\mathbf{a}_\phi$ direction in this case. Because the radiation power is independent of ϕ, however, a circular (i.e., constant) magnitude of the radiated power is shown in Figure 9.8b.

From the preceding discussion it is clear that the qualitative description of the radiation patterns (i.e., Figure 9.7, the term uniform, etc.) is not sufficient for quantitative comparison between various antennas. Three-dimensional or multisection two-dimensional visualization of the radiated power from an antenna is important, but in

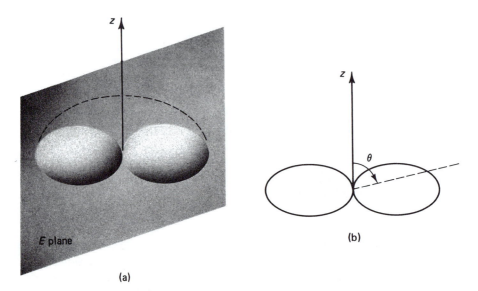

(a)

(b)

Figure 9.7   Two-dimensional view of the radiation pattern. (a) The cross section is taken in a plane in which the E field lies, i.e. an E-plane pattern. (b) The resulting "figure 8" E-plane pattern.

addition means for quantifying these patterns are needed. Antenna engineers use several parameters to quantify radiation patterns, and the following are just examples:

   Half-power beam width.   This is the angular width between points at which the radiated power density is one-half its maximum value. For example, in our electric

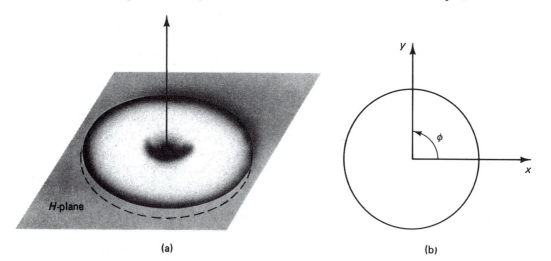

(a)   (b)

Figure 9.8   The H-plane two-dimensional sectional view of the radiation pattern. (a) The cross section in the plane of the H-field (x-y plane). (b) The resulting uniform H-plane pattern.

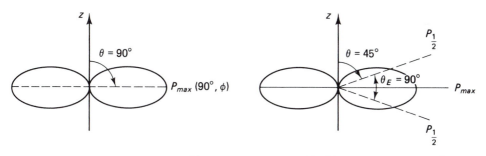

Figure 9.9 *E*-plane pattern. (a) P_{max} occurs at $\theta = 90°$. (b) $P_{1/2}$ occurs at $\theta = 45°$, thus resulting in $\theta_E = 90°$.

dipole case the half-power beam width in the *E*-plane pattern θ_E is 90°, whereas the half-power beam width for the *H*-plane pattern θ_H does not exist because of the uniformity of the pattern in this plane. The calculation procedure for half-power beam width in a specific plane is as follows:

E-Plane Pattern.   From equation 9.21, the radiated power density $P(\theta, \phi)$ for a constant *r* is proportional to $\sin^2 \theta$. Hence,

$$P(\theta, \phi) = K \sin^2 \theta$$

where *K* is a constant.

The maximum value of $P(\theta, \phi)$ is attained at $\theta = 90°$ as shown in Figure 9.9a. The half-power points $P_{1/2}$, conversely, occur at $\theta = 45°$, where $\sin \theta = 1/\sqrt{2}$. The angle between the half-power points is, therefore, 90° as shown in Figure 9.9b.

H-Plane Pattern.   The *H*-plane pattern of Figure 9.8 for an electric dipole does not show half-power points because the pattern is constant in the *H* plane. For other antennas, however, the *H*-plane pattern may vary with ϕ, thus resulting in a true value of θ_H. In this case, the calculation procedure for θ_H follows that of θ_E with the vertical angle θ replaced by the azimuthal angle ϕ.

First null beam width.   This is defined as the angular width between directions at which the radiated power is zero. Once again as an example let us consider the *E*- and *H*-plane patterns of an electric dipole. From Figure 9.10, it may be seen that the first null beam width is $\theta_N = 180°$ for the *E*-plane pattern and does not exist for the uniformly distributed radiated power in the *H*-plane pattern. For the *E*-plane pattern $P(\theta, \phi) = K \sin^2 \theta$. The radiated power *P* is zero for $\theta_1 = 0°$ and $\theta_2 = 180°$. The first null beam width is, hence, $\theta_N = \theta_2 - \theta_1 = 180°$, as shown in Figure 9.10a.

9.4.2 Directivity and Gain

To help us understand the gain and directivity parameters, it is important to start by defining a hypothetical source known as the isotropic radiator. An isotropic antenna is a fictitious radiator that radiates equally in all directions. It is often used as a reference source for comparing the performances of the various antennas. We also need to define

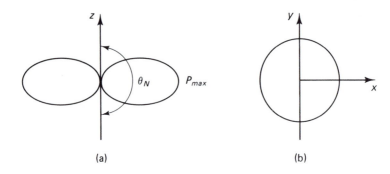

Figure 9.10 The first null beam width for (a) *E*-plane and (b) *H*-plane patterns.

the "radiation intensity." The radiation intensity is simply the radiated power density multiplied by the square of the distance. This multiplication provides the power radiated per unit solid angle, which is the definition of the radiation intensity $I(\theta, \phi)$. Hence,

$$I(\theta, \phi) = r^2 P_{ave}(\theta, \phi) \qquad \text{W/unit solid angle}$$

where P_{ave} is the radiated power density in W/m². With the preceding definitions we may now introduce terms such as *directive gain*, *directivity*, and *antenna gain*.

The directive gain in a given direction (i.e., specific values of θ and ϕ) is defined as ratio of the radiation intensity in this direction to the radiation intensity of an isotropic source that radiates equally in all directions. The isotropic source and the antenna are assumed to radiate equal amounts of power. The directive gain D is, hence,

$$D(\theta, \phi) = \frac{I(\theta, \phi)}{P_{tot}/4\pi}$$

where I is the radiation intensity, and P_{tot} is the total power radiated obtained by integrating P_{ave} over the solid angle. P_{tot} is assumed equal for both the antenna and the reference isotropic source.

The directivity, D_o, conversely, is defined as the ratio between the maximum value of radiation intensity to the radiation intensity of an isotropic source.

$$D_o = \frac{I_{max}}{P_{tot}/4\pi}$$

where I_{max} is the radiation intensity evaluated in the direction of its maximum.

The gain of an antenna is defined in a manner similar to that of the directive gain, except that the input power rather than the radiated power is used for the reference isotropic source. Hence,

$$G(\theta, \phi) = \frac{I(\theta, \phi)}{\text{Input power}/4\pi}$$

The antenna gain, therefore, in addition to providing information on the ability of the antenna to direct the radiated power (relative to the reference isotropic source), includes information on the efficiency of the antenna that is the ratio of the radiated power to the input power. The gain is therefore related to the directive gain by the antenna efficiency parameter η—that is,

$$G(\theta, \phi) = \eta\, D(\theta, \phi)$$

The maximum gain that is simply the value of $G(\theta, \phi)$ evaluated in the direction (i.e., values of θ, ϕ) of its maximum is often known in practice as the gain of the antenna.

Let us illustrate these antenna parameters by solving the following example.

EXAMPLE 9.1

The far field power density radiated from an antenna is given by

$$\mathbf{P}_{ave} = K\frac{\cos\theta}{r^2}\mathbf{a}_r, \text{ W/m}^2 \qquad 0 \le \theta \le \frac{\pi}{2} \qquad \text{and} \qquad 0 \le \phi \le 2\pi$$

where K is a constant. The far electric and magnetic fields have θ and ϕ components, respectively. Determine the following:

1. Half-power beam width in both E and H planes.
2. First null beam width in E and H planes.
3. Directive gain $D(\theta, \phi)$.
4. Directivity D_o.
5. Gain, given that antenna efficiency is 85 percent.

Solution

1. The E- and H-plane radiation patterns are shown in Figure 9.11. The half-power directions in the E-plane pattern occur at $\theta = 60°$ since $\cos 60° = 1/2$. θ_E is, hence, given by

$$\theta_E = 120°$$

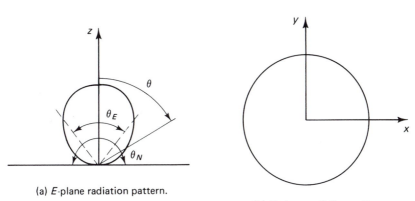

(a) E-plane radiation pattern.

(b) H-plane radiation pattern.

Figure 9.11 E- and H-plane radiation patterns. Also illustrated is the half-power beam width θ_E and the first null beam width θ_N.

Because $\mathbf{P}_{ave}$ is independent of ϕ, we conclude that the antenna radiates uniformly in the H plane and there is no half-power beam width θ_H.

2. Similarly, it may be seen from Figure 9.11 that the first null beam width in the E plane is given by $\theta_N = 180°$ and that there is no similar angle in the H plane.

3. To calculate the directive gain we need to evaluate the total radiated power from the antenna.

$$P_{tot} = \int_{\phi = 0}^{2\pi} \int_{\theta = 0}^{\pi/2} \mathbf{P}_{ave} \cdot d\mathbf{s}$$

where $d\mathbf{s} = r^2 \sin\theta \, d\theta \, d\phi \, \mathbf{a}_r$ is the element of area over a spherical surface surrounding the antenna

$$P_{tot} = \int_{\phi = 0}^{2\pi} \int_{\theta = 0}^{\pi/2} K \cos\theta \sin\theta \, d\theta \, d\phi$$

$$P_{tot} = 2\pi K \int_{\theta = 0}^{\pi/2} \frac{1}{2} \sin(2\theta) \, d\theta$$

$$= \pi K \quad \text{W}$$

$$D(\theta, \phi) = \frac{I(\theta, \phi)}{P_{tot}/4\pi} = \frac{K \cos\theta}{\pi K/4\pi} = 4 \cos\theta$$

4. The directivity

$$D_o = \frac{I(\theta, \phi)|_{max}}{P_{tot}/4\pi} = 4$$

where $I(\theta, \phi)|_{max}$ was evaluated at $\theta = 0$. This result shows that the antenna is four times more directive than a reference isotropic source radiating the same amount of power.

5. By the gain we simply mean the maximum gain that is often used in practice.

$$G_o = \eta D_o = 3.2$$

9.4.3 Radiation Resistance

Another antenna parameter that is useful in quantifying the ability of antennas to radiate power is the "radiation resistance," R_a. R_a is defined as the equivalent resistance that would dissipate the same amount of total power as that radiated by an antenna when the same current flows in both. For example, let us consider the electric dipole case. The total power radiated by the source can be obtained by integrating the power density expression in equation 9.21 over a spherical surface surrounding the source. Hence,

$$P_{tot} = \int \mathbf{P}_{ave} \cdot d\mathbf{s}$$

$$= \frac{\eta_o \beta_o^2}{32\pi^2} I^2 \, d\ell^2 \int_{\phi = 0}^{2\pi} \int_{\theta = 0}^{\pi} \frac{\sin^2\theta}{r^2} \mathbf{a}_r \cdot r^2 \sin\theta \, d\theta \, d\phi \, \mathbf{a}_r \qquad (9.23)$$

$$= \frac{\eta_o \beta_o^2 (I d\ell)^2}{12\pi}$$

If the same current at the antenna terminal I is applied to a resistance R_a the total power P_R dissipated in the resistor will be

$$P_R = \frac{1}{2} I^2 R_a \tag{9.24}$$

Equating equations 9.23 and 9.24, the equivalent value of the resistance R_a that would dissipate the same amount of power (i.e., $P_R = P_{tot}$) is given by

$$R_a = \frac{\eta_o \beta_o^2 (d\ell)^2}{6\pi} \ \Omega$$

EXAMPLE 9.2

Calculate the total power radiated from an infinitesimal electric dipole of $d\ell/\lambda = 0.02$. Use the obtained value to calculate the directive gain and the directivity of this source.

Solution

The total radiated power calculated from equation 9.23 is given by

$$P_{tot} = \frac{\eta_o I^2 (2\pi)^2 \left(\dfrac{d\ell}{\lambda}\right)^2}{12\pi}$$

$$= 39.44 I^2 \left(\frac{d\ell}{\lambda}\right)^2 = 0.0158 I^2 \ \text{W}$$

$D(\theta, \phi)$ for very short electric dipole is obtained from

$$D(\theta, \phi) = \frac{I(\theta, \phi)}{P_{tot}/4\pi}$$

From equations 9.21 and 9.23, we obtain

$$D(\theta, \phi) = 1.5 \sin^2 \theta$$

The directivity D_o is then $D_o = 1.5$ where θ was substituted by $\pi/2$ at which maximum radiation occurs.

EXAMPLE 9.3

Consider an antenna used for an amateur radio station. This antenna is 10 m long and operates at 600 kHz. Calculate the radiation resistance of this antenna and discuss the obtained result.

Solution

At 600 kHz, the wavelength $\lambda_o = 3 \times 10^8/600 \times 10^3 = 500$ m, the electrical length of the antenna ℓ/λ_o is then

$$\ell/\lambda_o = \frac{10}{500} = 0.02$$

Because the length of this antenna is a very small fraction of the wavelength, its radiation resistance may be approximated by that of an electric dipole. Hence, from section 9.4.3, we obtain

$$R_a = \frac{\eta_o}{6\pi}(2\pi)^2\left(\frac{d\ell}{\lambda_o}\right)^2$$

$$= 80\pi^2(0.02)^2 \approx 0.31 \ \Omega$$

The small value of the radiation resistance clearly indicates the small amount of power radiated by this antenna. Hence this antenna is a very inefficient radiator. This is actually a general observation which applies to antennas of lengths representing a very small fraction of the wavelength at the operating frequency. We will see in the following sections that the value of the radiation resistance may be significantly increased by making the antenna length comparable to the wavelength.

9.5 LINEAR WIRE ANTENNAS

The infinitesimal electric dipole described in the previous section actually represents a linear wire antenna of a very short electrical length, $\ell/\lambda \ll 1$. Such a short length enables us to make simplifying assumptions regarding the current distribution on the antenna (i.e., assumed constant current); this, in turn, simplified the mathematical analysis of the resulting fields. We discovered, however, that this current element is a very inefficient radiator with a very small radiation resistance. Hence, there is a need to improve the radiation efficiency, for example, by increasing the length of the antenna. We also found that near fields of an antenna are related only to the energy storage in the immediate neighborhood of the antenna. Because we are now interested in the far field radiation characteristics, there is no need to calculate the complicated expressions for the near fields. This will significantly simplify the mathematical treatment of linear wire antennas. With this in mind, let us analyze the far field radiation characteristics of these antennas.

The first question we need to answer, however, is related to the appropriate current distribution that we may assume along the antenna. It has been found numerically (e.g., using the method of moments) and confirmed experimentally that the current distribution on thin wire linear antennas is approximately sinusoidal. This rather important observation may be clarified by considering the parallel wire transmission line shown in Figure 9.12a. As described in chapter 7, the sinusoidal steady-state current distribution on an infinitely long transmission line is sinusoidal. For the open-ended section of a transmission line shown in Figure 9.12b, however, the voltage reflection coefficient $\hat{\Gamma}$ is equal to one, thus resulting in a standing-wave current distribution with zero current at the open ends, as shown in Figure 9.12c. In all transmission-line systems it is assumed that the separation distance between the two

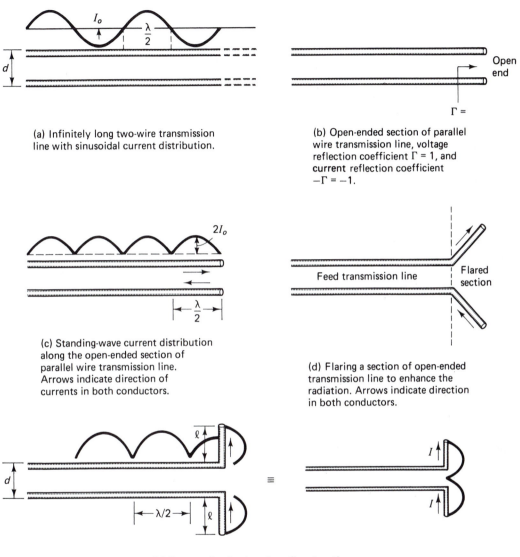

(a) Infinitely long two-wire transmission line with sinusoidal current distribution.

(b) Open-ended section of parallel wire transmission line, voltage reflection coefficient $\Gamma = 1$, and current reflection coefficient $-\Gamma = -1$.

(c) Standing-wave current distribution along the open-ended section of parallel wire transmission line. Arrows indicate direction of currents in both conductors.

(d) Flaring a section of open-ended transmission line to enhance the radiation. Arrows indicate direction in both conductors.

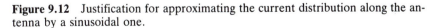

(e) Current distribution along flared section of parallel wire transmission line. Current in both sections will be in phase as long as $\ell/\lambda < 1/2$.

Figure 9.12   Justification for approximating the current distribution along the antenna by a sinusoidal one.

wires is sufficiently small—that is, $d/\lambda \ll 1$—so that the external radiation from these wave-guide structures may be neglected. Hence, for both systems shown in Figures 9.12a and c, the radiation fields may be assumed negligible. In an attempt to enhance the radiation fields, one may try to flare the open ends of the transmission line as shown in Figures 9.12d and e. This is actually equivalent to the creation of a thin wire antenna fed by a parallel wire transmission line. It is clear from Figures 9.12c and e, that as long as the length of the flared section is less than $\lambda/2$, the direction of the current in the upper and lower sections will be the same, thus enhancing the radiation. Besides the fact that flaring sections of the transmission line results in a structure with enhanced radiation fields, Figure 9.12d justifies the approximation that will be implemented in our antenna analysis; namely, the current distribution on a thin wire linear antenna may be assumed sinusoidal. We use such an approximation next to analyze the radiation fields of a linear thin wire antenna.

Consider the linear wire antenna shown in Figure 9.13. The sinusoidal current distribution may be assumed as

$$I(z') = \frac{I_o}{\sin \beta_o \ell} \sin \beta_o(\ell - |z'|) \quad -\ell < z' < \ell \tag{9.25}$$

where $\beta_o = \omega\sqrt{\mu_o \epsilon_o}$ is the propagation constant in air. It should be noted that the absolute value of z' is used in equation 9.25 so as to maintain symmetrical current distribution on both arms of the antenna as explained in Figure 9.12. I_o is divided by the normalization factor $\sin \beta_o \ell$ so as to maintain a sinusoidal current of amplitude I_o along the antenna. In other words, I_o is the driving point current at the feed point of the antenna.

As indicated in previous sections, the current distribution can be conveniently related to the magnetic vector potential function **A**, from which the radiation electric and magnetic fields may be calculated.

The magnetic vector potential $\mathbf{A} = A_z\,\mathbf{a}_z$ is given in terms of the current distribution by using (9.2), substituting C_1 from page 645, and integrating over the length of the antenna, hence;

$$A_z(r, \theta) = \frac{\mu_o}{4\pi} \int_{-\ell}^{\ell} \frac{I_o}{\sin \beta_o \ell} \sin \beta_o(\ell - |z'|) \frac{e^{-j\beta_o R}}{R} dz' \tag{9.26}$$

where R is the distance from the radiating element dz' along the antenna to the far field point. At an observation point sufficiently far from the antenna $r \gg \ell$, the distance

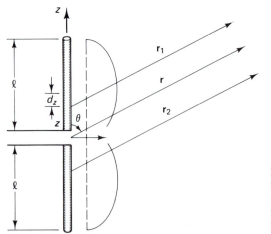

Figure 9.13 Schematic illustrating the sinusoidal current distribution and the geometry of the far field point from a linear wire antenna of total length equals 2ℓ.

$r = r_1$ and $r = r_2$ for points along the upper and lower halves of the antenna may be approximated by

$$r_1 = r - z' \cos \theta$$

$r_2 = r + z' \cos \theta = r - z' \cos \theta$ with the substitution of z' with a negative quantity along the lower section. Substituting these approximations in equation 9.26, we obtain

$$A_z(r, \theta) = \frac{\mu_o I_o}{4\pi \sin \beta_o \ell} \int_0^\ell \sin \beta_o(\ell - z') \frac{e^{-j\beta_o(r - z' \cos \theta)}}{r - z' \cos \theta} dz'$$
$$+ \int_{-\ell}^0 \sin \beta_o(\ell + z') \frac{e^{-j\beta_o(r - z' \cos \theta)}}{r - z' \cos \theta} dz' \tag{9.27}$$

At far distances from the antenna $r \gg z' \cos \theta$, equation 9.27 reduces to

$$A_z(r, \theta) = \frac{\mu_o I_o}{4\pi \sin \beta_o \ell} \frac{e^{-j\beta_o r}}{r} \left[\int_0^\ell \sin \beta_o(\ell - z') e^{j\beta_o z' \cos \theta} dz' \right.$$
$$\left. + \int_{-\ell}^0 \sin \beta_o (\ell + z') e^{j\beta_o z' \cos \theta} dz' \right] \tag{9.28}$$

It should be noted that in equation 9.28 the effect of the far field approximation on the amplitude (i.e., $1/R$) was neglected by assuming $1/r_1 = 1/r_2$ equal to the constant value $1/r$. The effect of such an approximation on the phase difference, or the path length phase delay, however, was considered by maintaining the $e^{-j\beta_o z' \cos \theta}$ factors.

Both integrals in equation 9.28 are of the form*

$$\int e^{ax} \sin(c + bx) dx = \frac{e^{ax}}{a^2 + b^2} [a \sin(c + bx) - b \cos(c + bx)]$$

Substituting the appropriate a, b, and c quantities and the integration limits in each portion of equation 9.28, we obtain

$$A_z(r, \theta) = \frac{\mu_o I_o}{2\pi \sin \beta_o \ell} \frac{e^{-j\beta_o r}}{r\beta_o} \left[\frac{\cos(\beta_o \ell \cos \theta) - \cos(\beta_o \ell)}{\sin^2 \theta} \right] \tag{9.29}$$

The magnetic and electric fields intensities may be obtained from equation 9.29 by expressing A_z in terms of its spherical components $A_r = A_z \cos \theta$, and $A_\theta = -A_z \sin \theta$ and using the vector differential operations

$$\mathbf{H} = \frac{1}{\mu_o} \nabla \times \mathbf{A}, \qquad \mathbf{E} = \frac{1}{j\omega\epsilon_o} \nabla \times \mathbf{H}$$

Maintaining the far field $1/r$ components of the electric and magnetic fields, we obtain

$$\mathbf{H} = \mathbf{H}_\phi \mathbf{a}_\phi = \frac{j\beta_o}{\mu_o} A_z \sin \theta \, \mathbf{a}_\phi$$
$$= \frac{j I_o}{2\pi \sin \beta_o \ell} \frac{e^{-j\beta_o r}}{r} \left[\frac{\cos(\beta_o \ell \cos \theta) - \cos(\beta_o \ell)}{\sin \theta} \right] \mathbf{a}_\phi \tag{9.30}$$

*J. D. Kraus, *Antennas* (New York: McGraw-Hill Book Company, 1950).

$$\mathbf{E} = \sqrt{\frac{\mu_o}{\epsilon_o}} H_\phi(\mathbf{a_\theta})$$

$$= j60 \frac{I_o}{\sin \beta_o \ell} \frac{e^{-j\beta_o r}}{r} \left[\frac{\cos(\beta_o \ell \cos \theta) - \cos(\beta_o \ell)}{\sin \theta} \right] \mathbf{a_\theta}$$

(9.31)

As expected the magnitudes of the far electric and magnetic field quantities are related by the intrinsic wave impedance $\eta_o = \sqrt{\mu_o/\epsilon_o}$. The directions of these fields, however, are mutually orthogonal ($\mathbf{E} = E_\theta \mathbf{a_\theta}$ and $\mathbf{H} = H_\phi \mathbf{a_\phi}$) and are perpendicular to the direction of the wave propagation $\mathbf{a_r}$. In the following sections the radiation characteristics of linear wire antennas are described in more detail.

9.5.1 Radiation Pattern

The time-average power density may be obtained from equations 9.30 and 9.31. Hence,

$$\mathbf{P}_{ave} = \frac{1}{2} Re(\mathbf{E} \times \mathbf{H}^*)$$

$$= \frac{1}{8\pi^2} \frac{|I_o|^2}{\sin^2 \beta_o \ell} \frac{\eta_o}{r^2} \left[\frac{\cos(\beta_o \ell \cos \theta) - \cos(\beta_o \ell)}{\sin \theta} \right]^2 \mathbf{a_r}$$

(9.32)

The factor $|F(\theta)|^2$ where

$$F(\theta) = \frac{\cos(\beta_o \ell \cos \theta) - \cos(\beta_o \ell)}{\sin \theta}$$

actually describes the relative radiation pattern variation as a function of θ. Detailed plots of $|F(\theta)|^2$ are shown in Figure 9.14 for linear antennas of various electrical lengths $2\ell/\lambda$. These patterns may be calculated by writing a simple computer program based on the $|F(\theta)|^2$ term of equation 9.32. From Figure 9.14, it may be seen that for $\ell/\lambda \leq 1/2$ the radiation pattern includes only one major lobe symmetrically placed around the axis of the antenna. The half-power beam widths of these patterns with a single main lobe are, however, different with a tendency to decrease with the increase in the electrical length of the antenna. Table 9.1 summarizes the beam width of linear antennas with ℓ/λ varying from $\ell/\lambda \ll 1$ to $\ell/\lambda = 1/2$. The decrease in the beam width with the increase in the antenna's electrical length simply indicates an increase in the ability of the antenna to direct the radiated energy. This characteristic is further clarified by calculating the directive gain as is illustrated next.

Before carrying out such gain calculations, however, let us emphasize another important characteristic of the radiation pattern. Figures 9.14b and c show radiation patterns of antennas of total length 2ℓ larger than λ. It is clear that there is now more than one main lobe for these longer antennas and that the relative intensity of the radiation along these lobes changes with the increase in the electrical length of the antenna. For example, although the lobe with maximum intensity was centered at $\theta = 90°$ for $\ell/\lambda = 5/8$, such a maxima was located at $\theta = 40°$, and $\theta = 140°$ for $\ell/\lambda = 3/4$. With further increase in the electrical length of the antenna, larger numbers of lobes begin to appear in the radiation pattern. Hence, we conclude that with the increase in

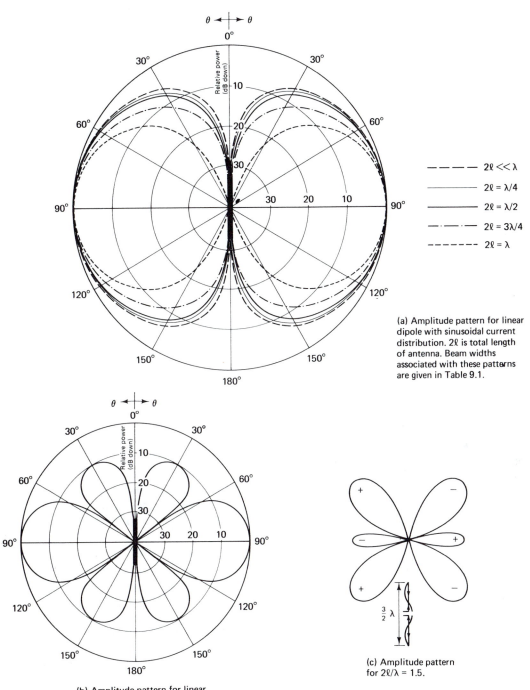

(a) Amplitude pattern for linear dipole with sinusoidal current distribution. 2ℓ is total length of antenna. Beam widths associated with these patterns are given in Table 9.1.

$$2\ell \ll \lambda$$
$$2\ell = \lambda/4$$
$$2\ell = \lambda/2$$
$$2\ell = 3\lambda/4$$
$$2\ell = \lambda$$

(b) Amplitude pattern for linear dipole of length $2\ell/\lambda = 1.25$.

(c) Amplitude pattern for $2\ell/\lambda = 1.5$.

Figure 9.14 Radiation patterns of linear dipole antennas of various electrical lengths $2\ell/\lambda$. (C. A. Balanis, *Antenna Theory—Analysis and Design*, Harper & Row, 1982, pp. 121, 122. Reprinted with permission by John Wiley & Sons).

TABLE 9.1   BEAM WIDTH OF LINEAR ANTENNAS WITH ONE MAIN LOBE IN RADIATION PATTERN

Half Length	Current distribution	Half-power beam width
$\ell/\lambda \ll 1$	Approximately triangular	90°
$\ell/\lambda = 1/8$		87°
$\ell/\lambda = 1/4$		78°
$\ell/\lambda = 3/8$		64°
$\ell/\lambda = 1/2$		47.8°

the antenna half length beyond $\ell/\lambda = 1/2$, the antenna will generally have multiple lobes in its radiation pattern and consequently lose its directional properties.

9.5.2 Directivity and Directive Gain

For linear antennas of electrical length $2\ell/\lambda \le 1$, the maximum radiation occurs at $\theta = 90°$ with a tendency of decreasing the beam width with the increase in the antenna length. Narrower beam width, while maintaining single radiation lobe, means an increase in the antenna length $0 < 2\ell/\lambda < 1$.

The evaluation of the directivity as defined in section 9.4.2 requires the calculation of the total power radiated from the antenna.

$$P_{tot} = \int_s \mathbf{P}_{ave} \cdot d\mathbf{s} \tag{9.33}$$

where s is a spherical surface surrounding the antenna. Substituting equation 9.32 in equation 9.33, we obtain

$$P_{tot} = \int_0^{2\pi} \int_0^{\pi} P_{ave} \, \mathbf{a}_r \cdot r^2 \sin\theta \, d\theta \, d\phi \, \mathbf{a}_r$$

Because P_{ave} is independent of ϕ, hence,

$$P_{tot} = \eta_o \frac{|I_o|^2}{4\pi \sin^2 \beta_o \ell} \int_0^{\pi} \frac{[\cos(\beta_o \ell \cos\theta) - \cos(\beta_o \ell)]^2}{\sin\theta} d\theta \tag{9.34}$$

This integral can be evaluated in terms of sine and cosine integrals $C_i(x)$ and $S_i(x)$, respectively. The resulting expression for the total radiated power is given by

$$P_{tot} = \frac{\eta_o}{4\pi} \frac{|I_o|^2}{\sin^2 \beta_o \ell} \left\{ \gamma + \ell n(2\beta_o \ell) - C_i(2\beta_o \ell) \right.$$
$$+ \frac{1}{2} \sin(2\beta_o \ell)[S_i(4\beta_o \ell) - 2S_i(2\beta_o \ell)] \tag{9.35}$$
$$\left. + \frac{1}{2} \cos(2\beta_o \ell)[\gamma + \ell n(\beta_o \ell) + C_i(4\beta_o \ell) - 2C_i(2\beta_o \ell)] \right\}$$

where $\gamma = 0.5772$ is Euler's constant and C_i and S_i are given by

$$S_i(x) = \int_0^x \frac{\sin u}{u} du$$

$$C_i(x) = \int_\infty^x \frac{\cos u}{u} du$$

the values of these integrals are tabulated in Appendix E. Following definition of the directive gain in section 9.4.2, we obtain

$$D(\theta, \phi) = \frac{\left[\dfrac{\cos(\beta_o \ell \cos \theta) - \cos(\beta_o \ell)}{\sin \theta} \right]^2}{P_{tot}/4\pi} \cdot \frac{|I_o|^2 \eta_o}{8\pi^2 \sin^2 \beta_o \ell}$$

The directivity D_o is plotted in Figure 9.15 as a function of the antenna electrical length $2\ell/\lambda$.

For a half wavelength antenna $2\ell/\lambda = 1/2$, P_{tot} is evaluated as

$$P_{tot} = 36.565|I_o|^2 \tag{9.36}$$

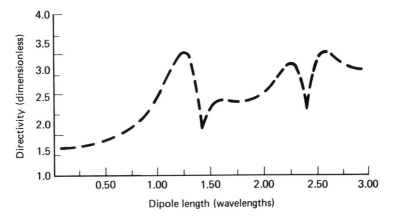

Figure 9.15   Variation of directivity of linear dipoles with the electrical length $2\ell/\lambda$ of the antenna. (C. A. Balanis, *Antenna Theory—Analysis and Design*, Harper & Row, 1982, p. 125. Reprinted with permission by John Wiley & Sons).

The directive gain is then

$$D(\theta, \phi) = 1.64 \left[\frac{\cos\left(\dfrac{\pi}{2} \cos\theta\right)}{\sin\theta} \right]^2$$

According to Figure 9.14, the maximum radiation occurs at $\theta = \pi/2$. The directivity of a half-wavelength antenna is, hence,

$$D_o = 1.64$$

which represent only a small increase over the directivity of an infinitesimal electric dipole that has a value $D_o = 1.5$, as shown in example 9.2. This value of directivity D_o can be obtained from Figure 9.15.

9.5.3 Radiation Resistance

Once again the expression for the total radiated power from a linear wire antenna (equation 9.35) may be used to calculate the radiation resistance. If a sinusoidal current of amplitude I_o flows in a resistance R_a, the total power dissipated is then $1/2\,|I_o|^2\,R_a$. Equating this expression to equation 9.35, we obtain the radiation resistance R_a as

$$
\begin{aligned}
R_a = \frac{\eta_o}{2\pi \sin^2\beta_o \ell} &\left\{ \gamma + \ell n(2\beta_o \ell) - C_i(2\beta_o \ell) \right. \\
&+ \frac{1}{2} \sin(2\beta_o \ell)[S_i(4\beta_o \ell) - 2S_i(2\beta_o \ell)] \\
&\left. + \frac{1}{2} \cos(2\beta_o \ell)[\gamma + \ell n(\beta_o \ell) + C_i(4\beta_o \ell) - 2C_i(2\beta_o \ell)] \right\}
\end{aligned}
$$

A plot of R_a as a function of the electric length $2\ell/\lambda$ of the antenna is shown in Figure 9.16. Using the value of the total radiated power from a half-wavelength dipole given in equation 9.36,

$$P_{tot} = 36.565|I_o|^2 = \frac{1}{2}|I_o|^2 R_a$$

$$R_a = 73.13\ \Omega$$

which is significantly higher than that for an infinitesimal dipole (a fraction of an ohm) as illustrated in example 9.3. This value of R_a may also be obtained from Figure 9.16.

It is important to conclude at this point that increasing the length of the linear antenna from that of an infinitesimal dipole $\ell/\lambda \ll 1$ to that of a half-wavelength antenna $2\ell/\lambda = 1/2$ does not significantly impact the directivity of the antenna (changes from 1.5 to 1.64) and, hence, does not considerably change the ability of the antenna to concentrate the radiated power in a given direction. The value of the radiation resistance R_a, however, shows remarkable change from a fraction of an ohm for $\ell/\lambda \ll 1$ to 73 Ω for $2\ell/\lambda = 0.5$. This shows a great increase in the amount of radiated power

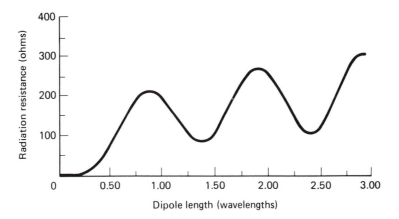

Figure 9.16   Radiation resistance of linear dipole as a function of electrical length $2\ell/\lambda$. (C. A. Balanis, *Antenna Theory—Analysis and Design,* Harper & Row, 1982, p. 125. Reprinted with permission by John Wiley & Sons).

from the antenna. Such an increase in the radiation resistance also makes it somewhat easier to feed the antenna from a 50-Ω transmission line without suffering a significant amount of reflection. Feeding an antenna with a radiation resistance of a fraction of an ohm is, conversely, practically not advisable because of the significant reflections. In other words, the infinitesimal dipole antenna is just an elementary source of academic interest because of its simplicity, but it is not used in practice because of its inefficiency and the difficulty of feeding it.

9.6 ANTENNA ARRAYS

We recognize from studying the radiation characteristics of a simple linear wire antenna that the radiation pattern in all cases is isotropic in the ϕ direction. This simply means that the radiated power is equally distributed in this direction and that the only gain may result from controlling the radiation pattern along the elevation angle in the θ direction. This is why it was not surprising to see that the directivity changed only from 1.5 for a very short electric dipole to 1.64 for a half-wavelength linear antenna. In many communication systems, it is necessary to improve the gain of the transmitting and receiving antennas. This helps establish much longer-range communication systems at reasonable amounts of input power. For such a system, an assembly of radiating elements may be used to produce a highly directional radiation pattern. An arrangement of several radiating elements is known as an antenna array. By selecting the appropriate type of radiating elements and adjusting their geometrical configuration, including the spacing between them, the desired radiation pattern may be produced. Other degrees of freedom that may be used to manipulate and adjust the overall radiation pattern of an array system may include changing the relative amplitudes and the phases of the currents feeding the various elements. Let us illustrate the role played by these adjustable parameters in changing the radiation pattern of the antenna array by solving the following examples. These examples deal with simple geometrical arrangements of isotropic sources (e.g., the φ pattern of linear antennas) so that the overall array pattern may be constructed on physical rather than complicated mathematical basis. Each example examines the effect of varying one specific parameter while maintaining the rest constant.

EXAMPLE 9.4

Consider two isotropic sources fed with identical currents both in amplitude and phase. Based on physical basis, determine the overall radiation patterns for the following two different spacings between the sources:

  1. Separation distance $d = \lambda/2$.
  2. Distance $d = \lambda$.

Solution

  1. Consider first the two isotropic sources separated by a distance $d = \lambda/2$ as shown in Figure 9.17a. At a point A located along the line connecting the two sources, the

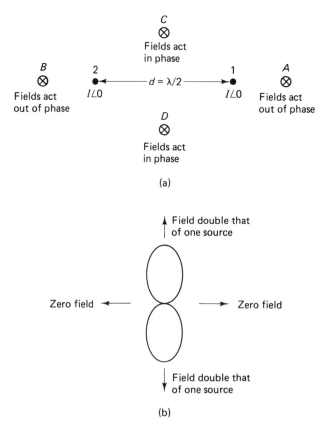

(a)

(b)

Figure 9.17 (a) The geometrical arrangement of the two equal in amplitude and in phase sources as well as the total fields at the far zone points A, B, C, and D. (b) The resulting radiation pattern of the array of two sources.

far fields will cancel; hence, the total field at A is zero. This is because the fields at A from source 2 have to travel an additional distance equal to a half-wavelength (the separation distance between sources) before interfering with the far field radiation from source 1. The currents in these sources are equal in amplitude and phase; hence, the half-wavelength separation distance will cause a 180° phase difference between the fields arriving at A and, consequently, the complete cancellation of the fields at this point. Similar arguments may be used to show another complete field cancellation at point B along the line joining the two sources.

At point C, which is at a far field distance along the perpendicular bisector of the line joining the two sources, the fields from the two sources will travel equal distances to C. Because the currents in the two sources are equal in amplitude and phase, the fields at C will be equal in amplitude and in phase and because of the constructive interference the total field at C will be double that of one source. A similar argument may be used to justify the constructive interference of the fields at D; hence, the resulting radiation pattern of these two sources is shown in Figure 9.17b. This resulting radiation pattern may also be justified based on simple calculation of the far field of the two sources. Taking the coordinate origin midway between the two sources as shown in Figure 9.18, the total far field is given by

$$E_\theta = A_1 \frac{e^{-j\beta_o r_1}}{r_1} + A_1 \frac{e^{-j\beta_o r_2}}{r_2}$$

where the constant A_1 was taken the same, because both sources were assumed identical. Substituting $r_1 = r - d/2 \cos\phi$ and $r_2 = r + d/2 \cos\phi$ and because in the far field $r \gg d/2 \cos\phi$ and as far as the amplitude is concerned $1/r_1 = 1/r_2 = 1/r$, we obtain

$$\begin{aligned}
E_\theta &= A_1 \frac{e^{-j\beta_o r}}{r} [e^{+j\beta_o d/2 \cos\phi} + e^{-j\beta_o d/2 \cos\phi}] \\
&= 2A_1 \frac{e^{-j\beta_o r}}{r} \cos\left(\beta_o \frac{d}{2} \cos\phi\right)
\end{aligned} \tag{9.37}$$

The quantity $\cos(\beta_o d/2 \cos\phi)$ introduces the variation in the array radiation pattern as a result of the constructive and destructive interference between the two sources. This quantity is known as the array factory (AF) and should help us explain the resulting radiation pattern of Figure 9.17b. Hence,

$$AF = \cos\left(\beta_o \frac{d}{2} \cos\phi\right) \tag{9.38}$$

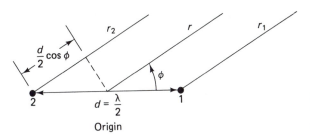

Figure 9.18 Schematic illustrating the procedure for calculating the array factory of two sources.

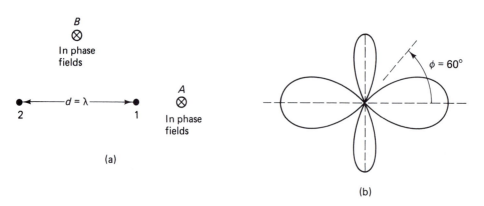

Figure 9.19 (a) Geometrical arrangement of an array of two identical sources separated by a distance $d = \lambda$. (b) The resulting radiation pattern.

In our case $d = \lambda/2$, hence, $\beta_o d/2 = 2\pi/\lambda \cdot \lambda/4 = \pi/2$. At $\phi = 0$ (i.e., along the line joining the two sources), $AF = \cos(\pi/2) = 0$, thus resulting in a null in the radiation pattern and at $\phi = \pi/2$, $AF = \cos(\pi/2 \cdot 0) = 1$, thus resulting in a maxima along the perpendicular bisector of the line joining the two sources. The calculation of the AF thus justifies the pattern calculated by inspection in Figure 9.17.

2. In this case the two identical sources are separated by a distance d equal to one wavelength. The geometry of this two-element array is shown in Figure 9.19a. By inspection, the fields should add along the line joining the two sources (e.g., point A), because the *additional* distance that will be traveled by source 2 is equal to λ (i.e., full 360°), which means that the fields from the two sources will still be in phase. Along the perpendicular line bisecting the line joining the two sources (e.g., point B), the fields from the two sources are in-phase and add. To determine the locations of nulls (if any) in the radiation pattern, let us calculate the AF of these two sources.

   Using equation 9.38 and substituting $d = \lambda$ in this case, we obtain

$$AF = \cos\left(\frac{2\pi}{\lambda} \frac{\lambda}{2} \cos\phi\right) = \cos(\pi \cos\phi) \qquad (9.39)$$

From equation 9.39, it may be seen that $|AF| = 1$ for both $\phi = 0$ and $\phi = 90°$, as indicated by inspection earlier. At $\phi = 60°$, however, $\cos\phi = 1/2$, and the $AF = \cos(\pi/2) = 0$. Hence, the resulting radiation pattern of the array will have nulls at $\phi = \pm60°$ and $\phi = \pm120°$, as shown in Figure 9.19b. It is clear from this example that the separation distance between array elements (two elements in this case) play a significant role in determining the radiation pattern of the antenna. The pattern in Figure 9.18b is certainly different from that in Figure 9.19b, and such a change was achieved simply by changing the separation distance between the two sources. In the following example, the effect of varying the phases of the feed currents to sources on the radiation pattern of the array is examined.

EXAMPLE 9.5

Consider the two isotropic sources shown in Figure 9.20. The currents feeding these sources are assumed to be of equal amplitude, whereas the separation distance d is maintained at a quarter of a wavelength long at the operating frequency. Determine the radiation pattern of the two-element array under the following conditions of the phase difference between the sources:

1. Source 1 leads source 2 by 90°.
2. Source 1 leads source 2 by 180°.
3. Source 1 is in phase with source 2.

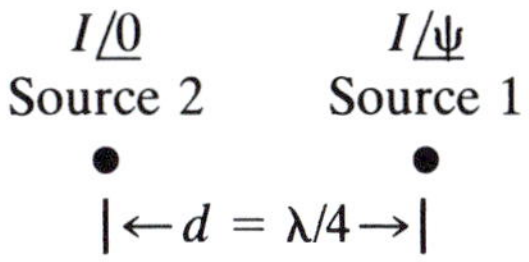

$$I\underline{/0} \qquad I\underline{/\psi}$$
Source 2   Source 1
•   •
$$|\leftarrow d = \lambda/4 \rightarrow|$$

Figure 9.20 Geometry of two isotropic sources separated by a distance $d = \lambda/4$. The phase angle ψ varies for the three conditions indicated in example 9.5.

Solution

The geometry of the two sources is shown in Figure 9.20. The resulting radiation patterns are shown in Figure 9.21 for the three conditions 1, 2, and 3 of the phase difference between the sources. Obtaining the radiation patterns in Figure 9.21 by inspection is quite straightforward following arguments similar to those used in the previous example. For example, in case 1 the cancellation of the fields at $\phi = 0°$ and the constructive interference at $\phi = 180°$ can be justified based on the separation distance and phase difference between the two sources. In this case, the phase difference $\psi = 90°$ and because the fields from source 2 have to travel an additional distance $d = \lambda/4$ to reach an observation point along the direction $\phi = 0°$, the fields from the two sources will be out of phase and, hence, cancel at $\phi = 0°$. For an observation point at $\phi = 180°$, conversely, the phase lag of 90° resulting from the additional travel distance of fields from source 1 will compensate for the 90° phase lead of the source, and the far fields of the two sources will constructively interfere at the observation point. This explains the pattern shown in Figure 9.21a. For the pattern in Figure 9.21c it is shown that the fields from the two sources do not cancel anywhere in space.

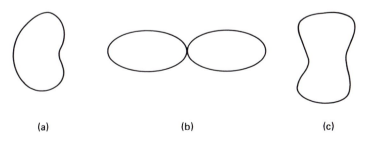

(a)   (b)   (c)

Figure 9.21 Resulting radiation patterns for three conditions of phase difference between two elements of array. (a) $\psi = 90°$, (b) $\psi = 180°$, and (c) $\psi = 0°$.

They are, however, totally in phase at $\phi = 90°$ and suffer a 90° phase difference for observation points along $\phi = 0°$. The resulting radiation pattern is, hence, stronger along $\phi = 90°$ than at $\phi = 0°$. To check whether or not the radiation pattern in Figure 9.21c has any field nulls, we examine the AF given in equation 9.38.

$$AF = \cos\left(\beta_o \frac{d}{2} \cos\phi\right)$$

Substituting $d = \lambda/4$,

$$AF = \cos\left(\frac{2\pi}{\lambda}\frac{\lambda}{8}\cos\phi\right) = \cos\left(\frac{\pi}{4}\cos\phi\right)$$

The AF has no nulls for any value of ϕ from 0 to 2π. This equation also shows that the AF is maximum at $\phi = \pi/2$, and that its value at $\phi = 0°$ is 0.707 of its maximum value. Such a characteristic is also illustrated in Figure 9.21c.

Similar arguments may be used to describe the pattern of Figure 9.21b. In this case 2, the two sources are out of phase by 180° and hence the constructive interference is at $\phi = 0$ and 180° and the destructive interference is at $\phi = 90°$ and 270°.

9.6.1 Linear Array of N Elements

Thus far, the physical interpretation of the array principle was illustrated by solving by inspection some examples of array arrangements. These examples illustrated the effect of varying the separation distance as well as the phase of the feed current on the resulting radiation pattern of the array. In the following, the radiation pattern of a linear array of N elements is analyzed.

Consider the N-element array shown in Figure 9.22. It consists of N-center-fed wire antennas of different lengths $\ell_1, \ell_2, \ldots, \ell_N$, and fed by different currents (in magnitude and phase) $I_1 e^{-j\psi_1}, I_2 e^{-j\psi_2}, \ldots, I_N e^{-j\psi_N}$. The first element is assumed to be located at the origin of the coordinate system, whereas the separation distances between the various elements were assumed different $d_1, d_2, \ldots, d_{N-1}$. The array elements were all aligned along the z axis in Figure 9.22.

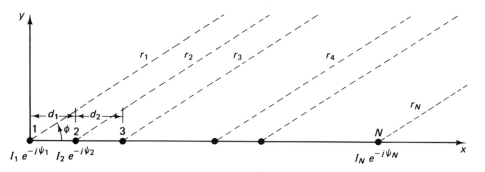

Figure 9.22 Linear array of N elements. The array elements are center-fed wire antennas oriented along the z axis.

The electric field from each element at an observation point placed in the far zone is given from equation 9.31 by

$$E_{\theta i} = \frac{j60I_i e^{-j\psi_i}}{\sin\beta_o \ell_i} \frac{e^{-j\beta_o r_i}}{r_i} \left[\frac{\cos(\beta_o \ell_i \cos\theta) - \cos(\beta_o \ell_i)}{\sin\theta} \right]$$

The total electric field from all the array elements is, hence,

$$E_{\theta}^{tot} = \frac{j60I_1 e^{-j\psi_1}}{\sin\beta_o \ell_1} \frac{e^{-j\beta_o r_1}}{r_1} \left[\frac{\cos(\beta_o \ell_1 \cos\theta) - \cos(\beta_o \ell_1)}{\sin\theta} \right]$$

$$+ \frac{j60I_2 e^{-j\psi_2}}{\sin\beta_o \ell_2} \frac{e^{-j\beta_o r_2}}{r_2} \left[\frac{\cos(\beta_o \ell_2 \cos\theta) - \cos(\beta_o \ell_2)}{\sin\theta} \right] \qquad (9.40)$$

$$+ \cdots + \cdots$$

$$+ \frac{j60I_N e^{-j\psi_N}}{\sin\beta_o \ell_N} \frac{e^{-j\beta_o r_N}}{r_N} \left[\frac{\cos(\beta_o \ell_N \cos\theta) - \cos(\beta_o \ell_N)}{\sin\theta} \right]$$

The electric field expression in equation 9.40 is quite general, and without specification of the lengths of the various elements, the separation distances between them, as well as the magnitudes and phases of the currents feeding the various elements of the array, it is difficult to describe the radiation characteristics of the array. Hence, despite the fact that arrays of such general geometrical and feeding characteristics do exist, not only in a linear but also in a variety of two-dimensional arrangements—for example, rectangular and circular arrays—it is necessary for us to implement some simplifying assumptions to help develop and understand some basic radiation characteristics of the linear array. The following are some realistic assumptions often used in practice:

1. The array elements are assumed identical (i.e., $\ell_1 = \ell_2 = \cdots = \ell_N = \ell$). The factor

$$F_i(\theta) = \left[\frac{\cos(\beta_o \ell_i \cos\theta) - \cos(\beta_o \ell_i)}{\sin\theta} \right]$$

which describes the variation of the radiation fields as a function of θ will, hence, be identical for all elements.

2. The magnitudes of the feed currents to the various array elements are also assumed constant and equal to I—that is, $I_1 = I_2 = \cdots = I_N = I$.

3. The phase of the feed currents to the array elements are assumed to be different by a constant progressive phase shift ψ between neighboring elements.

4. The separation distance between the various elements is constant—that is, $d_1 = d_2 = d_3 = \cdots d_{N-1} = d$.

Under these assumptions, as well as the usual far field approximation in which the amplitude fluctuations resulting from the various values of $1/r_i$ are neglected—that is,

$$\frac{1}{r_1} = \frac{1}{r_2} = \cdots = \frac{1}{r_N} = \frac{1}{r}$$

and only the effect of these various values of r_i on phase are accounted for, we obtain

$$E_\theta^{tot} = \frac{j60Ie^{-j\psi_o}}{\sin\beta_o\ell}\frac{e^{-j\beta_o r}}{r}\left[\frac{\cos(\beta_o\ell\cos\theta) - \cos(\beta_o\ell)}{\sin\theta}\right]$$
$$\{e^{-j\beta_o(r_1 - r)} + e^{-j\beta_o(r_2 - r)}e^{-j\psi} + e^{-j\beta_o(r_3 - r)}e^{-j2\psi}$$
$$+ \cdots + e^{-j\beta_o(r_N - r)}e^{-j(N-1)\psi}\} \tag{9.41}$$

where ψ_o is the phase of the current feeding the first element, ψ is the progressive constant phase difference between the various elements, and $r_1 = r$ is the distance to the far field observation point from the coordinate origin. Under the assumption of equally spaced array elements, we obtain

$$r_2 - r = -d\cos\phi \qquad , \qquad r_3 - r = -2d\cos\phi$$
$$\cdots \qquad\qquad , \qquad r_N - r = -(N-1)d\cos\phi$$

Equation 9.41, hence, reduces to

$$E_\theta^{tot} = \frac{j60Ie^{-j\psi_o}}{\sin\beta_o\ell}\frac{e^{-j\beta_o r}}{r}\left[\frac{\cos(\beta_o\ell\cos\theta) - \cos(\beta_o\ell)}{\sin\theta}\right]$$
$$\times (1 + e^{j(\beta_o d\cos\phi - \psi)} + e^{j2(\beta_o d\cos\phi - \psi)}$$
$$+ \cdots + e^{j(N-1)(\beta_o d\cos\phi - \psi)}) \tag{9.42}$$

Equation 9.42 may be divided into two parts. The first describes the far electric field radiated from each center-fed element in the antenna array, whereas the other

$$1 + e^{j(\beta_o d\cos\phi - \psi)} + \cdots + e^{j(N-1)(\beta_o d\cos\phi - \psi)}$$

represents the modification in the radiation fields as a result of the interference between the various elements of the array. The second part is known as the AF, and the principle demonstrated by equation 9.42 is known as the *principle of pattern multiplication*. It simply states that the total field pattern of an array of nonisotropic but similar sources is the product of the individual source pattern, and the pattern of an array of isotropic sources having the same relative amplitude and phase as the elements of the array.

$$E_\vartheta^{tot} = E_\theta \text{ (Each individual element)} \times \text{AF} \tag{9.43}$$

Because the radiation characteristics of linear wire antennas was examined in detail in the previous sections, attention is focused here on examining the various characteristics of the array factor AF. It should be emphasized first, however, that the radiation fields from linear wire antennas were all independent of ϕ and, hence, isotropic in the ϕ direction. The AF, conversely, is a function of ϕ and will, hence, result in focusing and controlling the radiation pattern in the ϕ direction. From equations 9.42 and 9.43, the AF is given by

$$\text{AF} = 1 + e^{j(\beta_o d\cos\phi - \psi)} + e^{j2(\beta_o d\cos\phi - \psi)} + \cdots$$
$$+ e^{j(N-1)(\beta_o d\cos\phi - \psi)}$$

This series of expressions is actually a geometric progression summation and may be expressed in the following closed form:

$$AF = \frac{1 - e^{jNx}}{1 - e^{jx}} = e^{j\left(\frac{N-1}{2}\right)x}\left[\frac{e^{-j\frac{N}{2}x} - e^{j\frac{N}{2}x}}{e^{-j\frac{x}{2}} - e^{j\frac{x}{2}}}\right]$$

$$= e^{j\left(\frac{N-1}{2}\right)x}\frac{\sin\left(\frac{N}{2}\right)x}{\sin\left(\frac{x}{2}\right)} \tag{9.44}$$

where $x = \beta_o d \cos\phi - \psi$.

Neglecting the phase factor

$$e^{j\left(\frac{N-1}{2}\right)x}$$

in equation 9.44, which actually does not contribute to the radiation pattern of the array, we obtain

$$|AF| = \frac{\sin N\frac{x}{2}}{\sin\frac{x}{2}} = \frac{\sin\frac{N}{2}(\beta_o d \cos\phi - \psi)}{\sin\frac{1}{2}(\beta_o d \cos\phi - \psi)} \tag{9.45}$$

Equation 9.45 gives the general expression of the array factor for a uniform linear array. The N elements of the uniform array can be arranged such that total constructive interference would occur in a given direction. The condition under which the array factor will attain its maximum value is $x = 0$. In this case, the maximum value of $|AF| = N$. From equation 9.45, the normalized $|AF|_n$ is, hence, given by

$$|AF|_n = \frac{1}{N}\frac{\sin\frac{N}{2}(\beta_o d \cos\phi - \psi)}{\sin\frac{1}{2}(\beta_o d \cos\phi - \psi)} \tag{9.46}$$

Figure 9.23 illustrates the variation of $|AF|_n$ with x for various values of N (i.e., the array size). The $|AF|_n$ in Figure 9.23 is shown only for values of $0 \le x \le \pi$. This is because $|AF|_n$ is symmetric about π and also periodic in 2π. The proof for the first property will be left as exercise, whereas the periodic property of the AF may be proved by noting

$$AF(x) = \sum_{n=0}^{N-1} e^{jnx} \tag{9.47}$$

Substituting $x + 2\pi$ for x in equation 9.47, we obtain

$$AF(\chi + 2\pi) = \sum_{n=0}^{N-1} e^{jn(\chi + 2\pi)}$$

$$= \sum_{n=0}^{N-1} e^{jn\chi} e^{jn(2\pi)}$$

$$= \sum_{n=0}^{N-1} e^{jn\chi} = AF(\chi)$$

The following properties can also be seen from Figure 9.23:

1. The main lobe of the $|AF|_n$ narrows with the increase in N. This simply emphasizes an increase in the ability of the array to focus the radiation with the increase in the number of elements.
2. As N increases, there are more side lobes.
3. The side lobe peaks decrease with increasing N.

In the following, some of the more interesting properties of the radiation patterns of N-element arrays are examined.

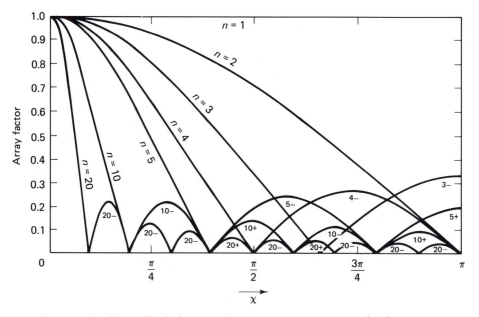

Figure 9.23 Normalized (universal) curves of array factor $|AF|_n$ versus χ. (From J. D. Kraus, *Antennas*, McGraw-Hill, New York, page 78, 1950. Reprinted with permission).

9.6.2 Radiation Pattern of N-Element Array

The radiation pattern of the array may be obtained by examining the variation of $|AF|_n$ with ϕ, as given in equation 9.46. From equation 9.46, it is clear that the nonlinear relationship between $|AF|_n$ and ϕ does not lend itself to a simple analytical procedure for fully determining the radiation pattern as a function of the various antenna parameters. For example, from equation 9.46, to find the nulls of the array, we set

$$\sin\frac{N}{2}(\beta_o d \cos\phi - \psi) = 0$$

Thus,

$$\frac{N}{2}(\beta_o d \cos\phi - \psi) = \pm n\pi \qquad \begin{array}{l} n = 1, 2, 3, \ldots \\ \\ n \neq N, 2N, 3N, \ldots \end{array}$$

and

$$\phi_n = \cos^{-1}\left[\frac{1}{\beta_o d}\left(\psi \pm \frac{2n\pi}{N}\right)\right] \qquad (9.48a)$$

It should be noted that the values of $n = N, 2N, 3N, \ldots$, were excluded from the preceding relation because at these values of n the array factor $|AF|_n$ attains its maximum values. At these values of n the array factor reduces to the form $\sin 0/\sin 0$ which attains the maximum value N after differentiating both numerator and denominator with respect to argument. The maximum values of $|AF|_n$ thus occur when

$$\frac{1}{2}(\beta_o d \cos\phi - \psi) = \pm m\pi \qquad m = 0, 1, 2, \ldots$$

$$\phi_m = \cos^{-1}\left[\frac{1}{\beta_o d}(\psi \pm 2m\pi)\right] \qquad (9.48b)$$

Besides obtaining relations for ϕ at which the AF attains interesting properties such as equation 9.48, it is rather difficult to see the variation of the array factor as a function of ϕ. It is therefore of great interest to develop a graphical procedure that may actually be used to obtain the complete radiation pattern of the array for a wide variety of array excitations or geometrical arrangements. Basically, this graphical procedure helps us relate $|AF|_n$ to ϕ directly and without the need to perform analytically nonlinear transformation that relates χ to ϕ.

The graphical procedure may be summarized as follows:

1. For the desired number of array elements N, select the appropriate curve for $|AF|_n$ from Figure 9.23.
2. To extend the chosen curve for values of χ beyond those available in Figure 9.23, we use the analytical properties of the array factor including its symmetry around $\chi = \pi$ and its periodic property every 2π. For an array of five elements, the extended drawing of the array factor is shown in Figure 9.24.

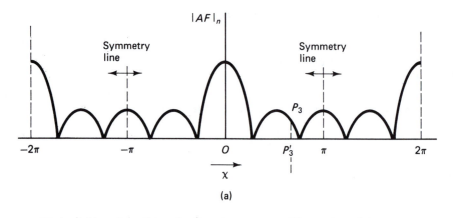

Figure 9.24a Array factor for five-element array. The portion of the curve $0 = \chi \leq \pi$ is given from Figure 9.23. The rest is constructed based on symmetry considerations and using the periodic property of $|AF|_n$.

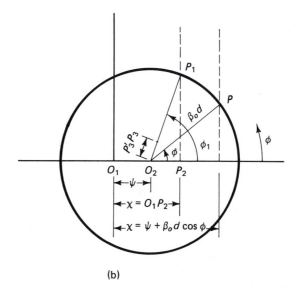

(b)

Figure 9.24b Graphical relationship between the angle ϕ of the radiation pattern and the variable χ of the normalized array factor. The value of the array factor at ϕ_1 is $P_3' P_3$. $P_3' P_3$ is measured from Figure 9.24a and plotted on $O_2 P_1$.

3. To help derive the relation between $|AF|_n$ and the angle ϕ, graphically, we extend the vertical axis of the AF graph downward to a sufficiently far-located horizontal line, which will be used to represent the variation of ϕ. The distance OO_1 (between top and lower figures) should be sufficiently large to draw a circle of radius $= \pi$.

4. Locate an origin O_2 along the horizontal line. The distance $O_2 O_1 = \psi$, the progressive phase difference between the elements of the uniform array. The origin O_2 will be to the right of O_1 for positive values of ψ, and to the left of O_1 for negative values of ψ.

5. From the origin O_2, draw a circle of radius $R = \beta_o d$, the electrical distance between the array elements. For any observation point P that makes an angle ϕ, the projection of P along the horizontal line as measured from the origin O_1, is given by

$$\text{Projection of } P = \beta_o d \cos \phi + \psi = \chi$$

which is the variable on the horizontal axis in the normalized AF curve of Figure 9.24a.

6. Hence, to relate the value of $|AF|_n$ to the angle ϕ of the radiation pattern, we determine (graphically) the value of χ that corresponds to the desired ϕ using the projection procedure described in step 5 and use the determined value of χ finally to evaluate $|AF|_n$ using Figure 9.24a.

   For example, let us assume that we need to calculate the relative magnitude of the radiation pattern at angle ϕ_1. We draw this angle from the horizontal line of Figure 9.24b and the origin of the circle O_2. At this value of ϕ_1, we determine χ as the distance $\chi = O_1 P_2 = \psi + \beta_o d \cos \phi_1$. For this value of χ, and because the origins O of Figure 9.24a and O_1 of Figure 9.24b are aligned, we extend the vertical line $P_2 P_1$ until it intersects the $|AF|_n$ pattern at point P_3, as shown in Figure 9.24a.

   The relative value of the radiation pattern at ϕ_1 is measured by $P_3' P_3$. Such a value is then transferred on the line $O_2 P_1$ indicating the relative amplitude of the radiation pattern at ϕ_1, the angle that makes $O_2 P_1$.

7. The procedure is repeated for several values of ϕ until a clear picture of the radiation pattern for values of $0 \leq \phi \leq \pi$ is obtained. This range of $0 \leq \phi \leq \pi$ is known as the "visible region," because it provides (together with the rotational symmetry of the radiation pattern about the line of the array) a complete structure of the radiation pattern. This rather powerful graphical procedure for determining the radiation pattern of a uniform array is further illustrated next by solving several examples.

EXAMPLE 9.6

In a broadside array, the separation distance d between the array elements and the progressive phase difference ψ are chosen such that the radiation pattern of the array would have a main lobe perpendicular to the array line—that is, at $\phi = 90°$. To determine ψ and d required for broadside array with beam maximum at $\phi = 90°$, we use equation 9.48, hence,

$$\frac{1}{2}\left(\beta_o d \cos \frac{\pi}{2} - \psi\right) = \pm m\pi, \qquad m = 0, 1, 2, \ldots$$

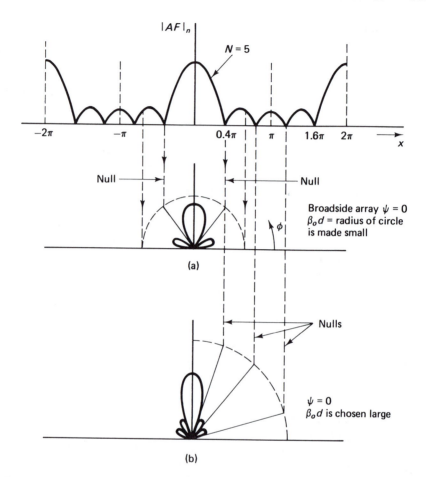

Figure 9.25   Radiation pattern of broadside array (i.e., main beam is at $\phi = 90°$).

For $m = 0$

$$\psi = 0$$

In other words, keeping the elements of the uniform array in phase ($\psi = 0$) will result in a broadside array with radiation pattern maximum at $\phi = 90°$. Figure 9.25 shows two constructed radiation patterns for two five-element broadside arrays ($\psi = 0$) of two different separation distances $\beta_o d$ between the elements of each array. It is clear from Figure 9.25 that by keeping the elements of the array in phase ($\psi = 0$), the radiation pattern will always have maxima at $\phi = 90°$. It is also shown that radiation patterns with sharper beams (Figure 9.25b) may be obtained by increasing the separation distance $\beta_o d$ between the array elements.

Such an increase in the interdistance $\beta_o d$ is also accompanied by an increase in the number of the side lobes, as shown in Figure 9.25b.

EXAMPLE 9.7

The end-fire array has a radiation pattern with main beam along the array line—that is, at $\phi = 0$ or π. To obtain a beam maxima at $\phi = 0, \pi$, we use equation 9.48 to obtain

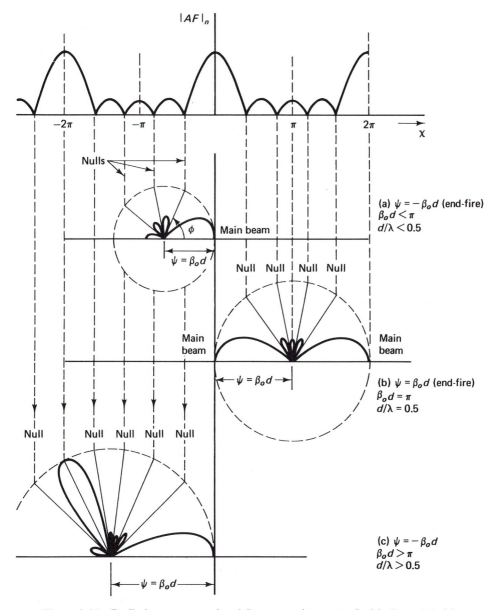

Figure 9.26   Radiation patterns of end-fire arrays ($\psi = \pm\beta_o d$). (a) $d/\lambda < 0.5$, (b) $d/\lambda = 0.5$, and (c) $d/\lambda > 0.5$.

$$\frac{1}{2}\left[\beta_o\, d\, \cos\!\left(\begin{matrix}0\\ \pi\end{matrix}\right) - \psi\right] = \pm m\pi, \qquad m = 0, 1, 2, \ldots$$

For $m = 0$

$$\psi = \pm \beta_o\, d$$

The progressive phase difference between the various elements should, hence, be equal to the electrical distance between the elements. The plus sign is for the main beam at $\phi = 0$, and the minus sign is for the main beam at $\phi = \pi$. Figure 9.26 shows three radiation patterns for three different end-fire arrays with three different values of the interdistance $\beta_o\, d$ between the array elements. Figure 9.26a has $\beta_o\, d < \pi$, which means $d/\lambda < 0.5$, whereas Figure 9.26b is for $\beta_o\, d = \pi$, which corresponds to an interdistance $d = \lambda/2$. Figure 9.26c illustrates the fact that more than one major radiation beam may result if d exceeds a half-wavelength. Hence, we conclude that although larger values of $\beta_o\, d$ may be required to sharpen (narrow) the main radiation beam, the matter that results in a higher antenna gain, there is a limit on how large $\beta_o\, d$ may be made. This limit is set to avoid the generation of multiple main beams in the radiation pattern.

EXAMPLE 9.8

In this example, we intend to illustrate that by changing the progressive phase difference between the elements of a uniform array we can vary the direction (ϕ) at which the maximum radiation beam occurs. Figure 9.27 illustrates a principle known as "electronic beam scanning" for an array of five elements. It is shown that by changing the progressive phase difference between the currents feeding the various elements ψ, the direction of the main beam would change. In Figure 9.27 two cases were considered; $\psi = 0$ and $\psi = 0.4\pi$, while $\beta_o\, d$ was kept constant at 1.2π.

Thus far the discussion was limited to the calculation of the radiation pattern of uniform array for which the magnitudes of the feed currents were assumed constant, whereas their phases were assumed to change by a progressive constant value ψ between neighboring elements. The solution procedure for determining the AF, however, is general and may be used to calculate the radiation patterns of different kinds of arrays. Furthermore, the graphical use of the normalized $|AF|_n$ for these general arrays to determine the radiation patterns are also general, and may be used successfully to design and optimize the radiation patterns of these arrays. This generalization is illustrated by solving the following example.

EXAMPLE 9.9

Consider a five-element nonuniform array. For simplicity, we will assume that we still have a constant value of a progressive phase difference between elements ψ. The amplitudes of the currents feeding these elements, however, are different and are assumed to have the following ratio $I_1 : I_2 : I_3 : I_4 : I_5$, which is equal to $1 : 3 : 5 : 3 : 1$.

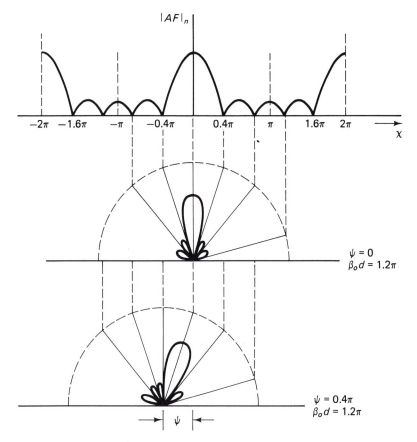

Figure 9.27 Principle of "electronic scanning." Changing the progressive phase between the elements of a uniform array results in changing the direction ϕ of the main beam.

1. Determine the AF of this nonuniform array.
2. Draw a curve for the $|AF|_n$ in this case and use it to determine the radiation pattern of such an array for the special case $d = \lambda/4$ and $\psi = 0$.
3. Use the results of part 2 to determine the half-power beam width and the beam width between the first nulls of the radiation pattern.

Solution

1. To derive an expression for the AF of an array with different values of currents feeding the various elements, we use equation 9.40. Under the same assumptions used in the case of the far field of a uniform array with the exception that $I_1 \neq I_2 \neq I_3 \neq I_4 \neq I_5$, equation 9.40 reduces to

$$E_\theta^{tot} = \frac{j60\,e^{-j\psi_o}}{\sin\beta_o\,\ell}\,\frac{e^{-j\beta_o\,r}}{r}\left[\frac{\cos(\beta_o\,\ell\,\cos\theta)-\cos(\beta_o\,\ell)}{\sin\theta}\right]\!\left[I_1 + I_2\,e^{j(\beta_o\,d\,\cos\phi-\psi)}\right.$$

$$\left. + I_3\,e^{j2(\beta_o\,d\,\cos\phi-\psi)} + \cdots + I_N\,e^{j(N-1)(\beta_o\,d\,\cos\phi-\psi)}\right] \tag{9.49}$$

For five-element array, the AF, is, hence,

$$\mathrm{AF} = I_1 + I_2\,e^{jx} + I_3\,e^{j2x} + I_4\,e^{j3x} + I_5\,e^{j4x}$$

where $\chi = \beta_o\,d\,\cos\phi - \psi$.

The maximum value of this array factor occurs when the radiation from all the five elements constructively interfere, resulting in $\mathrm{AF}|_{max} = I_1 + I_2 + I_3 + I_4 + I_5$.

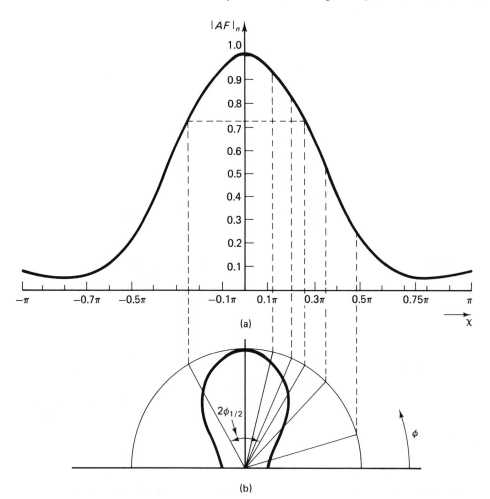

Figure 9.28   The normalized $|\mathrm{AF}|_n$ and the radiation pattern of a nonuniform array. (a) $|\mathrm{AF}|_n$ versus χ. (b) Radiation pattern for $d = \lambda/4$ and $\psi = 0$. $2\phi_{1/2}$ is the half-power beam width.

The normalized magnitude of the array factor is then

$$|\text{AF}|_n = \frac{1}{13}|1 + 3e^{jx} + 5e^{j2x} + 3e^{j3x} + e^{j4x}|$$

Multiplying by e^{-j2x}, taking the magnitude, and rearranging the terms, we obtain

$$|\text{AF}|_n = \frac{1}{13}|(e^{j2x} + e^{-j2x}) + 3(e^{jx} + e^{-jx}) + 5|$$

$$= \frac{1}{13}|5 + 2\cos 2\chi + 6\cos \chi| \qquad (9.50)$$

2. A curve illustrating the variation of $|\text{AF}|_n$ as a function of $-\pi \le \chi \le \pi$ is shown in Figure 9.28a. Up to this point, specific array parameters such as the distance d and ψ were not specified. For the special case of $d = \lambda/4$, (i.e., $\beta_o d = \pi/2$) and $\psi = 0$, the radiation pattern is shown in Figure 9.28b.
3. The half-power and first null beam widths are determined in Figure 9.28 with the aid of the curve for the normalized radiation pattern of Figure 9.28a and the resulting radiation pattern of Figure 9.28b.

9.6.3 Self- and Mutual Impedance

An important parameter in designing an antenna array—for example, in using equation 9.40—is the knowledge of the amplitude and the phase of the currents feeding the various elements. If these elements are separated by a sufficiently large electrical distance, the mutual interactions between them may be neglected, and the current distribution in each will be related to the applied voltage and the self- or driving-point impedance of each antenna. This driving-point impedance is simply the equivalent terminal impedance $\hat{Z}_{in}$ seen by the transmission-line feeding the antenna, as shown in Figure 9.29a. Knowledge of the value of this input impedance is important, not only in determining the current distribution—and, hence, the radiation fields of the antenna—but also for designing the proper matching network between the antenna and the feed transmission line. If the interaction between the array elements cannot be neglected, conversely, the antenna array may be represented by an N-port network with the mutual impedances between the various ports representing the degree of interaction between the elements. For example, in the two-element array shown in Figure 9.29b, the two-port (four terminals) network may be used to provide the following current-voltage relations:

$$\hat{V}_1 = \hat{Z}_{11}\hat{I}_1 + \hat{Z}_{12}\hat{I}_2 \qquad (9.51a)$$

$$\hat{V}_2 = \hat{Z}_{21}\hat{I}_1 + \hat{Z}_{22}\hat{I}_2 \qquad (9.51b)$$

where the $\hat{Z}$'s are the elements of the impedance matrix,

$$[\hat{Z}] = \begin{bmatrix} \hat{Z}_{11} & \hat{Z}_{12} \\ \hat{Z}_{21} & \hat{Z}_{22} \end{bmatrix}$$

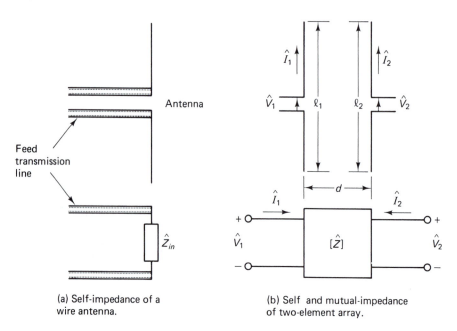

(a) Self-impedance of a   (b) Self and mutual-impedance
    wire antenna.   of two-element array.

Figure 9.29   Illustration of self- and mutual impedances of antenna array.

and they actually represent the proportionality factors between the terminal currents and voltages of the two antennas. These impedance elements are functions of the lengths ℓ_1 and ℓ_2 of the antennas, as well as of the separation distance d between them. From equation 9.51, it can be seen that the input or driving-point impedance of the first antenna $\hat{Z}_{in1}$ is given by

$$\hat{Z}_{in1} = \frac{\hat{V}_1}{\hat{I}_1} = \hat{Z}_{11} + \hat{Z}_{12}\frac{\hat{I}_2}{\hat{I}_1} \tag{9.52}$$

with the presence of the mutual interaction; hence, the input impedance of the antenna depends not only on its value $\hat{Z}_{11}$ when radiating in unbounded medium (i.e., $I_2 = 0$) but also on the ratio $\hat{I}_2/\hat{I}_1$, which describes the ratio between the current induced on the second antenna $\hat{I}_2$ caused by current $\hat{I}_1$ on the first, transferred to the input terminal of the first antenna by the function $\hat{Z}_{12}$. $\hat{Z}_{12}$ describes and quantifies the magnitude of the interaction between the two antennas. Based on the preceding discussion, we may see that

$$\hat{Z}_{in2}\big|_{I_1 = 0} = Z_{22} \qquad (\hat{I}_1 = 0 \text{ means antenna 2 is radiating in unbounded medium})$$

$$\hat{Z}_{in2} = \hat{Z}_{22} + \hat{Z}_{21}\frac{\hat{I}_1}{\hat{I}_2} \qquad (\text{In the presence of the mutual interaction})$$

For reciprocal networks $\hat{Z}_{12} = \hat{Z}_{21}$ and the $[\hat{Z}]$ matrix of Figure 9.29b may be equivalently represented by the T network shown in Figure 9.30.

When we calculate the current in an antenna, it is the driving-point impedance $\hat{Z}_{in}$ that must be used. Also, for impedance-matching purposes, the driving-point impedance must be used for designing the matching network between the antenna and

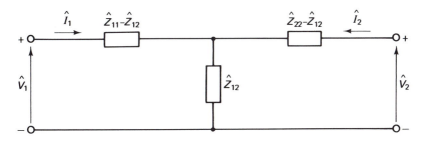

Figure 9.30 Equivalent circuit representing reciprocal mutual interaction between two antenna elements.

the feed transmission line. In other words, in designing and characterizing the radiation performance of an antenna array, the mutual interaction between the array elements must be considered. Clearly, the importance of accounting for the mutual interaction varies depending on the geometrical arrangement including the separation distance and the orientation of the array elements. It may be negligible for sufficiently separated array elements and could be substantial for closely spaced array elements. In an antenna such as the Yagi-Uda array, which includes "parasitic elements" with no voltage excitations, the induced currents on the parasitic elements are due to only mutual interactions and are known to play a dominant role in the overall performance of the antenna.

The variation of the input self-impedance of linear wire antennas as a function of their electrical length is shown in Figure 9.31.

Figures 9.32 and 9.33 show the variation of the mutual impedance between two linear antennas. From these figures, it can be seen that the side-by-side coupling is much stronger than that of the collinear arrangement. As a matter of fact, the coupling in the latter case may be neglected for a separation distance $s \geq \lambda/4$. This may be explained in terms of the negligible radiation from the ends of the antennas versus the presence of the mutual coupling in the direction of maximum radiation for the side-by-side coupling case.

The dependence of the mutual impedance in the side-by-side arrangement on the antenna length is shown in Figure 9.33. It is generally observed that the mutual coupling increases with the increase in the antenna length.

EXAMPLE 9.10

Calculate the driving-point (input) impedances of two antennas when arranged in the side-by-side geometry. Each antenna is a half-wavelength long and the two elements are separated by a distance $d = 2\lambda$. Compare your results with the case in which $d = 0.2\lambda$.

Solution

For each of the individual elements radiating in unbounded medium, the self-impedance from Figure 9.31 is given by

$$\hat{Z}_{11} = 73 + j42 \ \Omega$$

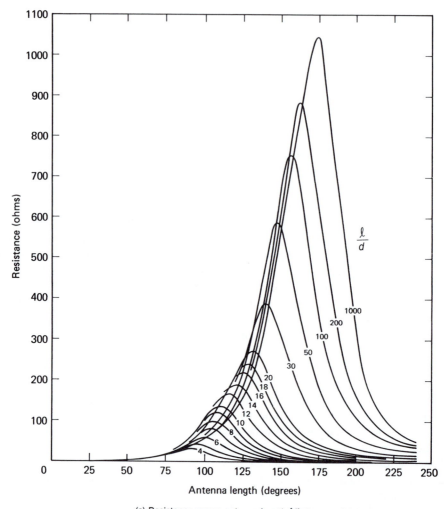

(a) Resistance versus antenna length ℓ/λ (monopole) in degrees.

Figure 9.31 Resistance and reactance of linear antenna versus antenna length in degrees. d is the diameter of the cylindrical antenna. (*From* H. Jasik, *Antenna Engineering Handbook* [New York: McGraw-Hill Book Company, 1961], pp. 3–4 and 3–5; reprinted with permission.)

At $d = 2\lambda$, the mutual impedance in the side-by-side arrangement is given from Figure 9.32 by

$$\hat{Z}_{12} = 0 + j10 \ \Omega$$

The driving-point impedance is given by

$$\hat{Z}_{in} = \frac{\hat{V}_1}{\hat{I}_1} = \hat{Z}_{11} + \hat{Z}_{12}\frac{\hat{I}_2}{\hat{I}_1}$$

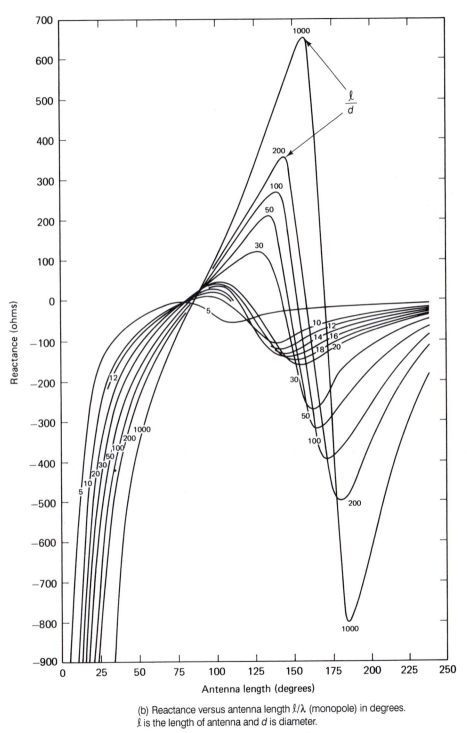

(b) Reactance versus antenna length ℓ/λ (monopole) in degrees.
ℓ is the length of antenna and d is diameter.

Figure 9.31 *Continued.*

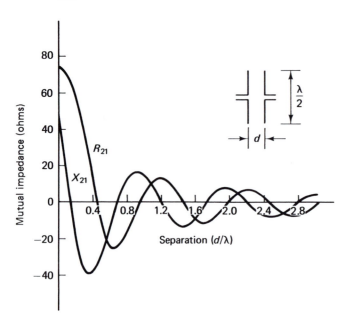

(a) Side-by-side arrangement.

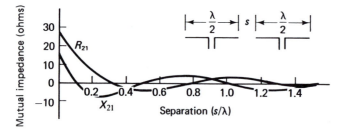

(b) Collinear arrangement.

Figure 9.32 Mutual impedance of side-by-side and collinear arrangements of $\lambda/2$ dipoles. (*From* W. L. Weeks, *Antenna Engineering* [New York: McGraw-Hill Book Company, 1968], p. 189; reprinted with permission.)

For equal currents in both elements,

$$\hat{Z}_{in} = \hat{Z}_{11} + \hat{Z}_{12} = 73 + j52 \ \Omega \qquad (9.53a)$$

At $d = 0.20\lambda$,

$$\hat{Z}_{12} = 59 - j15$$

which is significantly larger than that calculated for $d = 2\lambda$. In this case, and if equal currents are assumed in both elements, we obtain

$$\hat{Z}_{in} = 72 + j42 + (59 - j15)$$
$$= 132 + j27 \ \Omega \qquad (9.53b)$$

From this example, it is clear that the mutual coupling between the antennas may or may not have an impact on the input impedance of the antennas. From the impedance matching

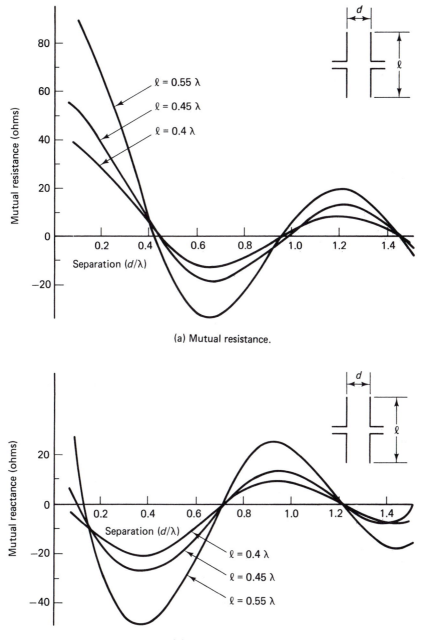

(a) Mutual resistance.

(b) Mutual reactance.

Figure 9.33 Mutual impedance of side-by-side coupled linear antennas of various lengths. The length ℓ was chosen around the $\lambda/2$ dipole described in Figure 9.32. (*From* W. L. Weeks, *Antenna Engineering* [New York: McGraw-Hill Book Company, 1968], p. 190; reprinted with permission.)

viewpoint alone, the antenna in the second case will have a significantly different matching network than that for the first case of $d = 2\lambda$. It is also to be noted that equal currents were assumed in the above calculations. A procedure for calculating these currents while accounting for the mutual coupling is described in the following section.

9.6.4 Radiation Characteristics of N-Element Array Including Mutual Effects

In the preceding sections, we discussed the radiation characteristics of N-element array, assuming that the magnitude and phase of the current in each element as well as the array geometries are known. In particular, equation 9.40, which may be written in the following compact form:

$$E_\theta^{tot} = j60 \frac{e^{-j\beta_o r_o}}{r_o} \sum_{i=1}^{N} \frac{I_i e^{-j\psi_i}}{\sin \beta_o \ell_i} e^{-j\beta_o \cdot (\mathbf{r}_i - \mathbf{r}_o)} F_i(\theta)$$

where

$$F_i(\theta) = \left[\frac{\cos(\beta_o \ell_i \cos \theta) - \cos(\beta_o \ell_i)}{\sin \theta} \right] \tag{9.54}$$

provides an expression for the total electric field of an N-element array, all of different lengths $2\ell_i, i = 1, \ldots, N$ and all are assumed to be fed with different currents $I_i e^{-j\psi_i}, i = 1, \ldots, N$. In writing equation 9.54, the far field approximation was made, and r_i in the denominator was replaced by r_o while maintaining the appropriate $\beta_o \cdot (\mathbf{r} - \mathbf{r}_o)$ phase for each of the terms. This phase term may be written as

$$\beta_o \cdot (\mathbf{r}_i - \mathbf{r}_o) = \beta_o[(x_i - x_o) \cos \phi \sin \theta + (y_i - y_o) \sin \phi \sin \theta$$
$$+ (z_i - z_o) \cos \theta] \tag{9.55}$$
$$= \beta_o[x_i' \cos \phi \sin \theta + y_i' \sin \theta \sin \phi + z_i' \cos \theta]$$

where (x_i', y_i', z_i') is the location of the driving (center) point of the ith element, and (x_i, y_i, z_i) is the coordinate of the observation point from an arbitrarily chosen original (x_o, y_o, z_o). In equation 9.55, it is clear that in calculating the phase difference between the various elements, it is important to account only for the relative locations (x_i', y_i', z_i') of the array elements. The time-average radiated power for this N-element array is given by

$$\mathbf{P}_{ave} = \frac{1}{2} R_e(\hat{\mathbf{E}} \times \hat{\mathbf{H}}^*) = \frac{15}{\pi r_o^2} \sum_{i=1}^{N} \sum_{j=1}^{N} \frac{I_i I_j F_i(\theta) F_j(\theta)}{\sin(\beta_o \ell_i) \sin(\beta_o \ell_j)}$$
$$\times \cos(\alpha_i - \alpha_j + \psi_i - \psi_j)\mathbf{a}_r \tag{9.56}$$

where α_i denotes $\beta_o(x_i' \cos \phi \sin \theta + y_i' \sin \phi \sin \theta + z_i' \cos \theta)$. A computer program to calculate equation 9.56 may be written and used to calculate the radiation patterns of a wide variety of antenna arrays that use parallel array elements.

A key step in the calculation of equation 9.56, however, is that the magnitude and the phase of the currents $I_i e^{-j\psi_i}$ feeding the various elements are assumed to be known. In an antenna array where the mutual interaction is strong, the current $I_i e^{-j\psi_i}$ may be significantly different from that expected based on the knowledge of the input voltages to each element as provided by the array feed structure. In other words, in the presence of a strong mutual coupling between the array elements, the input impedance of each element will be different from its own self-impedance; hence, the resulting current distribution will depend on the input voltage as well as on the arrangement of the array elements. Therefore, in the determination of the complex current $\hat{I}_i = I_i e^{-j\psi_i}$, input to each element of a closely spaced array, the mutual effects must be considered. To this end, we go back to equation 9.51 and rewrite it below for the N-element array,

$$\hat{V}_1 = \hat{I}_1 \hat{Z}_{11} + \hat{I}_2 \hat{Z}_{12} + \hat{I}_3 \hat{Z}_{13} + \cdots + \hat{I}_N \hat{Z}_{1N}$$
$$\hat{V}_2 = \hat{I}_1 \hat{Z}_{21} + \hat{I}_2 \hat{Z}_{22} + \cdots + \hat{I}_N \hat{Z}_{2N} \qquad (9.57)$$
$$- -$$
$$\hat{V}_N = \hat{I}_1 \hat{Z}_{N1} + \hat{I}_2 \hat{Z}_{N2} + \cdots + \hat{I}_N \hat{Z}_{NN}$$

Knowledge of the feed voltages $\hat{V}_i, i = 1, \ldots, N$ (magnitude and phase) and the mutual, as well as the self-impedances of the array elements, would facilitate the calculation of the input currents through the simultaneous solution of equation 9.57. These currents clearly take the mutual coupling effects into account and, hence, when substituted in equation 9.56 should provide the appropriate radiation pattern of an N-element array. It should be noted that the input voltages $\hat{V}_i, i = 1, \ldots, N$ in equation 9.57 are controlled by an appropriate design of the feed system of the array. Often, the voltage amplitudes are controlled through the use of voltage dividers and attenuators in the feed lines of the various elements. The phase differences, conversely, are usually introduced by including delay lines (transmission lines of various lengths) in the feed sections or electronic phase shifters. The self- and mutual impedances in equation 9.57 are obtained either using Figures 9.31 to 9.33 or alternatively using tabulated and graphical data available in the literature.[‡]

Exercise 9.1.   Write a computer program to calculate the average radiated power using equation 9.56. Input to the program should be the lengths of the various array elements $2\ell_i, i = 1, \ldots, N$, the amplitudes of the currents feeding the various elements $\hat{I}_i, i = 1, \ldots, N$, and the coordinates (x_i, y_i, z_i) of the center or the feed points of the elements of the array. Check the accuracy of the developed computer code by comparing the results you obtain with those reported in the literature.[‡]

Exercise 9.2.   Write or obtain a computer code that you may use to solve a system of equations such as those displayed in equation 9.57. For a given input voltage to the various array elements $\hat{V}_i, i = 1, \ldots, N$, and the elements of the impedance matrix $[\hat{Z}]$,

$$[\hat{Z}] = \begin{bmatrix} \hat{Z}_{11} \hat{Z}_{12} \cdots \hat{Z}_{1N} \\ \hat{Z}_{21} \quad \cdots \hat{Z}_{2N} \\ - - - - - - - - - - - \\ \hat{Z}_{N1} \quad \cdots \hat{Z}_{NN} \end{bmatrix}$$

[‡] C. A. Balanis, *Antenna Theory: Analysis and Design*, (New York: Harper & Row, 1982) chapter 7.

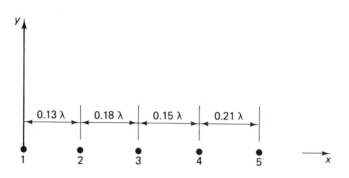

Figure 9.34 Geometry of array in exercise 9.3.

The program should provide the complex current amplitudes $\hat{I}_i, i = 1, \ldots, N$ as output. You may choose either direct methods for solving the system of equations, such as the Gauss's elimination method, or iterative procedures, such as Gauss-Seidal's method.[§] The program should be suitable for handling complex numbers.

Exercise 9.3. Consider the five-element linear array shown in Figure 9.34. The lengths of the five elements $2\ell_i, i = 1, \ldots, 5$, are all equal to 0.5λ. The radii a_i, $i = 1, \ldots, 5$ of the various elements are all chosen such that $\ell_i/a_i = 200$. For the array geometry shown in Figure 9.34, determine the following:

1. The self-impedances $\hat{Z}_{ii}, i = 1, \ldots, 5$ and mutual impedances $\hat{Z}_{ij}, i = 1, \ldots, 5$, $j = 1, \ldots, 5$ of the array using Figures 9.31 to 9.33.
2. Assuming that we arranged for a feed system to provide the following voltages at the driving points of the various elements,

$$\hat{V}_1 = 25\,e^{-j\pi/2}$$
$$\hat{V}_2 = 50\,e^{-j\pi/4}$$
$$\hat{V}_3 = 100\,e^{j0} \text{ (Reference)}$$
$$\hat{V}_4 = 50\,e^{-j\pi/4}$$
$$\hat{V}_5 = 25\,e^{-j\pi/2}$$

Use the computer code you developed in exercise 9.2, together with the impedance matrix developed in part 1, to calculate the driving-point currents, $\hat{I}_i, i = 1, \ldots, 5$.
3. Use the computer code you developed in exercise 9.1 to calculate the radiation pattern of this array.

9.6.5 Array with Parasitic Elements

The design of a feed system that would accurately provide the required driving-point voltages at the terminals of the various array elements is not a simple engineering task.

[§] M. J. Maron, *Numerical Analysis: A Practical Approach* (New York: Macmillan, 1982).

As a matter of fact, sometimes the array design is deliberately altered to simplify the design of the feed system. Some large arrays require truly creative engineering designs. For this reason, it is certainly desirable to design antenna arrays with parasitic elements. In this case, the currents "induced" on the nonexcited or "parasitic" elements of the array result in modifying the radiation characteristics of the active or "excited" elements. Such an arrangement has the clear advantage of simplifying the feed system because only a few elements in the array are excited. The degree by which the parasitic elements modify the radiation characteristics, including the input impedance and the radiation pattern of the array, clearly depends on the spacing between and the lengths of the various elements. To illustrate such effects, let us consider a two-element array with one excited and one parasitic element, as shown in Figure 9.35. To obtain strong mutual coupling and, hence, significant induced current on the parasitic element, there is often a tendency to make the separation distance electrically short, $0.03 < d/\lambda < 0.15$. For the specific case of $d/\lambda = 0.04$ and for array elements of equal length $L_1/\lambda = L_2/\lambda = 0.5$, the resulting radiation pattern is shown in Figure 9.35b. Because of the closeness of the parasitic element to the driven dipole, and also because the two elements are at resonance, the induced current on the parasitic is almost equal in magnitude and opposite in phase to that on the excited element. A two-element array with two equal but 180° out-of-phase currents should have an end-fire radiation pattern such as that shown in Figure 9.35b. In particular, the radiation pattern has a null along the perpendicular bisector of the line joining the two elements. If the length of the parasitic is slightly increased, say by 5 percent, but still keeping the same separation distance at $d/\lambda = 0.04$, the radiation pattern will be as shown in Figure 9.35c. In this case, the parasitic element will be off resonance, which results in a reduction in the magnitude of the induced current relative to that on the excited element. Furthermore, examination of the radiation pattern shows that the parasitic elements act as a reflector, thus increasing the directive gain of the array. Reducing the length of the parasitic element, conversely, say by 5 percent, results in essentially the same change in the radiation pattern, with the exception that the parasitic element acts as director rather than reflector (in this case as shown in Figure 9.35d). Examination of the results of Figures 9.35c and 9.35d shows that further increase in the gain of the antenna array may be achieved by combining the reflector, the director, and the driven dipole in a three-element array arrangement. Such an array is shown in Figure 9.36. It is known as the Yagi-Uda array[||] which is often used in television receivers. The following is a summary of some of the characteristics of the Yagi-Uda array:

Gain.   Experimental and detailed numerical examination of the maximum gain that may be achieved using this array arrangement shows that an increase in the number of reflectors does not appreciably impact the array gain. Increasing the number of directors, conversely, was found to increase the array gain significantly. Excessive increase in the number of directors, however, has a diminishing effect because of the reduction in the magnitude of the induced currents on the parasitic elements located further away. Furthermore, increasing the number of elements results in excessively large and, hence, impractical antennas. The number of parasitic elements is often kept

[||] W. L. Weeks, *Antenna Engineering* (New York: McGraw-Hill Book Company, 1968).

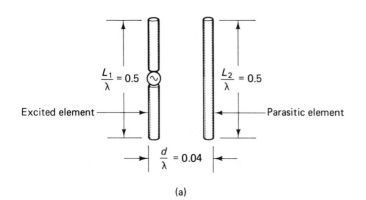

(a)

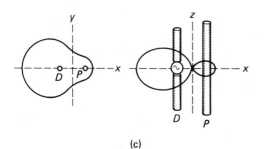

(b)

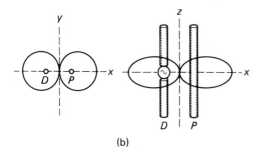

(c)

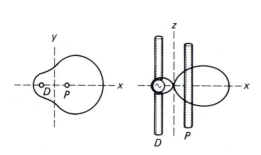

(d)

Figure 9.35 Radiation pattern of an array of two linear antennas, one of which is a parasitic element. (a) Geometry of too closely spaced elements of array. (b) Resulting radiation pattern with two elements of equal length. (c) Radiation pattern with parasitic element 5 percent longer (reflector). (d) Radiation pattern with parasitic element 5 percent shorter. (*From* A. B. Bailey, *TV and Other Receiving Antennas* [New York: John Francis Rider, 1950]).

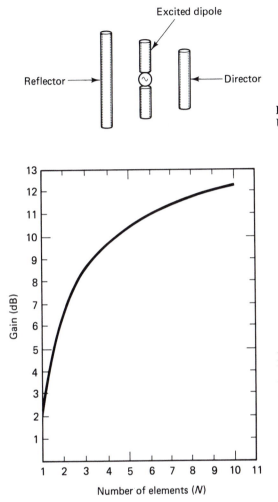

Figure 9.36 Three-element Yagi-Uda array.

Figure 9.37 Variation of gain of a Yagi-Uda array as a function of the total number of elements. The spacing between elements is 0.15λ, and diameters of conductors are 0.0025λ. (*From* H. E. Green, "Design data for short and medium length Yagi-Uda arrays," Institute of Engineers (Australia), Electrical Eng. Transactions, pp. 1–8, March 1966. Reprinted with permission)

below 12, with gains of approximately 10 to 15 dB, depending on the relative spacing and length of the array elements, as well as the number of the array elements. Figure 9.37 shows the variation in the array gain as a function of the number of elements.[¶]

Input impedance.   For a closely spaced two-element array, the induced current on the parasitic is approximately equal in magnitude and opposite in phase to the current on the excited element. The input impedance of the driven element is, hence,

$$\hat{Z}_{in} = \frac{\hat{V}_{in}}{\hat{I}_{in}} = \hat{Z}_{11} + \hat{Z}_{12}\frac{\hat{I}_{2}}{\hat{I}_{in}} = \hat{Z}_{11} - \hat{Z}_{12}$$

[¶]W. L. Stutzman and G. A. Thiele, *Antenna Theory and Design*. (New York: John Wiley & Sons, 1981), pp. 220–228.

For closely spaced elements of equal length, $\hat{Z}_{12} \approx \hat{Z}_{11}$ and $\hat{Z}_{in}$ is expected to be exceedingly small. This brings up a serious impedance matching problem between the excited element and the feed transmission line. Such a consideration requires the placement of the parasitic elements as far away from the excited antenna as possible without seriously deteriorating the gain. To minimize this impedance-matching problem further, a folded dipole is often used as the driven element. The folded dipole has an input (self-) impedance approximately four times that of a linear wire of the same length.** The Yagi-Uda array is a very narrow band antenna with a band width typically 1 percent.

Simple design procedure of Yagi-Uda array.   A simple design procedure of a Yagi-Uda array may be developed by employing the computer codes used in exercises 9.1 to 9.3. Assuming the second element in the array to be the driven one, we obtain

$$0 = \hat{Z}_{11} \hat{I}_1 + \hat{Z}_{12} \hat{I}_2 + \cdots + \hat{Z}_{1N} \hat{I}_N$$
$$\hat{V}_{in} = \hat{Z}_{21} \hat{I}_1 + \hat{Z}_{22} \hat{I}_2 + \cdots + \hat{Z}_{2N} \hat{I}_N$$
$$0 = \cdots$$
$$0 = \cdots$$
$$0 = \hat{Z}_{N1} \hat{I}_1 + \hat{Z}_{N2} \hat{I}_2 + \cdots + \hat{Z}_{NN} \hat{I}_N$$

(9.58)

In equation 9.58, the driving-point voltage in all the array elements was assumed zero (short circuited) except at the input port of the driving element. Hence, by specifying the number of the array elements, the lengths of all elements, and the spacing between them, the self- and mutual impedances in equation 9.58 can be calculated. The impedance values may either be obtained from Figures 9.31 to 9.33 or alternatively using formulas available elsewhere.†† Substituting these impedance values in equation 9.58, together with $\hat{V}_{in}$ (which may be $\hat{V}_{in} = 1\underline{/0°}$ for reference value), the currents in the various elements $\hat{I}_i, i = 1, \ldots, N$ may be calculated. The radiation pattern of the Yagi-Uda array may then be calculated using these currents in conjunction with equation 9.56 and the computer code developed in exercise 9.1.

9.6.6 Effect of Ground on Radiation Characteristics of Antennas

Another important effect that may be analyzed based on some of the array ideas developed in the previous sections is the effect of ground on the radiation characteristics of antennas. In the absence of any interfering object, reflections from the ground may have a significant impact on the input impedances and the radiation patterns of antennas. To analyze such an effect, let us consider a vertical electric dipole that is placed a distance d above an assumed perfectly conducting ground plane, as shown in Figure 9.38. In calculating the received radiation at the observation point P, the reflection

** W. L. Weeks, *Antenna Engineering* (New York: McGraw-Hill Book Company, 1968), pp. 180–184.
†† C. A. Balanis, *Antenna Theory: Analysis and Design* (New York: Harper & Row, 1982), chapter 7.

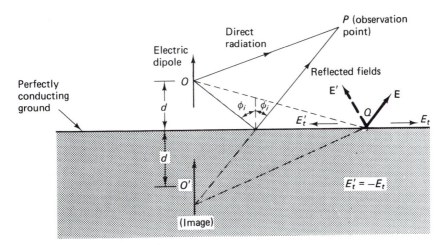

Figure 9.38   Short electric dipole radiating in the presence of ground plane.

from the ground must be considered. At the perfectly conducting surface, the angle of reflection is equal to the angle of incidence, and some of the reflected fields will be received at the observation point P. A rather simple way to account for the reflection from the ground is to introduce an image source beneath the surface of the conducting ground. The image sources are not real, but instead they are virtual (or imaginary) ones. They are introduced to provide a simple procedure for accounting for the reflections from the ground plane in the radiation zone above ground. The strength (magnitude) and direction of the image source, as well as its location beneath the ground plane, are adjusted such that it accounts exactly for the reflected fields. The total fields from the source and the image must also satisfy the boundary condition on the surface of the perfectly conducting ground. The vector electric field at Q is perpendicular to OQ in the absence of ground. The introduced image source at O' should be oriented and located such that the total tangential electric field is zero at Q. From Figure 9.38, it may be seen that locating the image source at an equal distance d beneath the conducting surface, and orienting the image source in the same direction as the original dipole, would result in an electric field $\mathbf{E}'$ that is equal in magnitude to $\mathbf{E}$ and is directed such that its tangential component exactly cancels the one from the original source. In other words, the magnitude, direction, and the location of the image were adjusted such that the total fields satisfy the boundary condition at the surface of the perfect conductor. To this end, the ground plane may be removed, and the reflection effects in the radiation zone above ground may be accounted for by adding the fields from the image source, such as the case at point P of Figure 9.38. In other words, the reflected field at P may be equivalently accounted for by considering the radiation from the image source. It should be emphasized, however, that the equivalence between reflections from the ground and the radiation from the image source are valid in the radiation zone above ground. Fields from either the original or the image source beneath the surface of the ground surface are incorrect and should not be used. In other words, the fields from the image source were introduced to provide a simple procedure for accounting for reflection in the radiation zone above ground and should be used for this purpose only.

If the electric dipole source is oriented parallel to the perfectly conducting ground plane, as shown in Figure 9.39a, the virtual (image) source should be opposite to that of the source so that the boundary condition would be satisfied. Combining the orientation of the image source ideas in the preceding two special cases of vertical and horizontal dipoles, it is possible to show that the image of an arbitrarily oriented source should be placed as shown in Figure 9.39b. This image theory, although developed for the special case of a perfectly conducting ground, has many applications in the antenna practice. Following are some examples.

Electric monopole above ground.   In many applications, including the commercial broadcasting AM stations and the mobile-communication services, a monopole stub mounted on a conducting plane is often used as an antenna. Such an antenna structure is particularly useful because of its simplicity and efficiency. The idealistic geometry of a monopole over a ground plane is shown in Figure 9.40. Based on the image theory described earlier, it may be shown that the image source is of equal length to the original source, and that its current is in the same direction as shown in Figure 9.40b. Hence, analysis of the radiation characteristics of a monopole of length ℓ above a perfectly conducting ground is equivalent to that of an electric dipole of length 2ℓ. This equivalence is clearly valid only in the radiation zone above the ground plane. Analysis of a dipole antenna was presented in detail in previous sections. Specifically, the electric and magnetic field expressions are given by equations 9.30 and 9.31, and the time-average power by equation 9.32. The total radiated power from the monopole, however, will be only in the half-space above the ground. Hence, for an equal current in both the dipole and monopole antennas, the monopole radiates half the amount of power radiated by the dipole. In other words, the monopole of length ℓ has a radiation resistance equal to half of that of the dipole of length 2ℓ. In fact, the input impedance

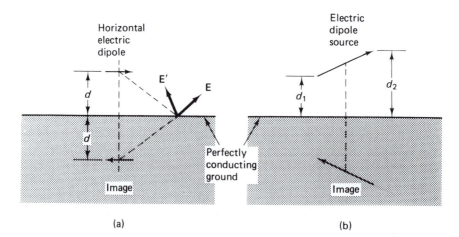

Figure 9.39   Orientation of images of electric dipole sources.

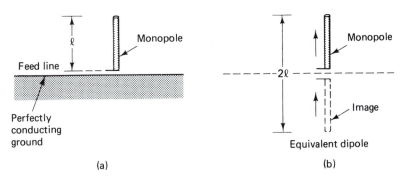

Figure 9.40   Vertical monopole over ground and its equivalent dipole.

of the monopole is half that of the dipole. Hence, for a quarter-wavelength monopole, the input impedance Z_{in} is given by

$$\hat{Z}_{in} = \frac{1}{2}(73 + j42) = 36.5 + j21 \ \Omega$$

The real value of the input impedance may be alternatively determined based on the calculation of the total radiated power. For a half-wavelength dipole, the total radiated power from equations 9.35 and 9.36 is given by

$$P_{tot} = 36.565 \, |\hat{I}_o|^2$$

This value was calculated by integrating the power density expression over the entire space surrounding the dipole—that is, $0 < \theta < \pi, 0 < \phi < 2\pi$. For the monopole case, the radiation space is limited to the upper half $0 < \theta < \pi/2, 0 < \phi < 2\pi$, which means that the total radiated power is given by

$$P_{\text{monopole}} = 18.283 \, |\hat{I}_o|^2$$

The radiation resistance in this case is, hence,

$$\frac{1}{2} R_a |\hat{I}_o|^2 = P_{\text{monopole}}$$

or

$$R_a = 36.565 \ \Omega$$

which is the same as the real part of $\hat{Z}_{in}$. Other radiation characteristics of the monopole, including the radiation pattern and the directivity, are the same as that of a dipole of double the length.

In practice and in particular for AM broadcasting stations (0.50 to 1.50 MHz), the construction of a $\lambda/4$ monopole involves building towers such as that shown in Figure 9.41. The monopole is approximately $\lambda/4$ high, with supporting wires that are made of small sections to minimize the induced currents on them. The tower is insulated from the ground, and is fed by coaxial or parallel-wire transmission lines. It may also be worth mentioning that some efforts should be made to improve the conductivity of the ground

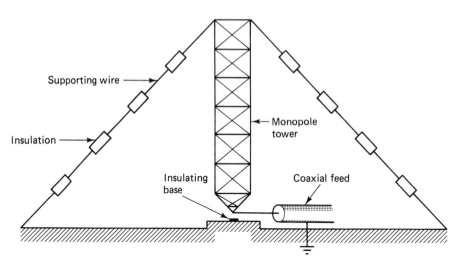

Figure 9.41 One possible practical realization of a monopole construction over for AM broadcast station.

(earth) so that the configuration in Figure 9.41 would resemble that of Figure 9.40b as closely as possible. Typically, the ground conductivity is improved by inserting a net of radial wires a few times longer than the height of the antenna. Salt might also be added to the ground to improve its conductivity. A vertical monopole antenna with an improved ground conductivity is shown in Figure 9.42. A closely spaced net of radial wires is clearly preferable for better ground conductivity. For economical reasons, however, significant reduction in the number of wires is often implemented. As few as four radial wires are sometimes used in portable dipole designs.

Calculation of radiation characteristics of antennas above ground.   The image theory developed earlier may also be used to calculate the radiation characteristics of antennas above ground. For example, let us consider the vertical linear antenna of Figure 9.43, which is placed at a distance d above ground. Assuming a perfectly conducting ground, ground reflections may be accounted for by introducing the image

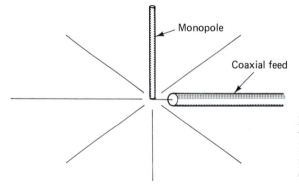

Figure 9.42 Monopole antenna with a net of radial wires to simulate the ground conductor or improve the conductivity of earth under the antenna.

shown in Figure 9.43. The far zone electric field from the two antennas (antenna and its image) is given from equation 9.31 by

$$E_\theta = \frac{j60\,I_o}{\sin\beta_o\,\ell} F(\theta) \left[\frac{e^{-j\beta_o r_1}}{r_1} + \frac{e^{-j\beta_o r_2}}{r_2} \right]$$

where the lengths of the two antennas were assumed to be the same as well as the pattern factor $F(\theta)$ and the maximum current I_o. Making the usual far field approximations—that is,

$$\frac{1}{r_1} = \frac{1}{r_2} = \frac{1}{r_o}, \qquad r_{\frac{1}{2}} = r_o \mp d\,\cos\theta$$

we obtain

$$E_\theta = \frac{j60\,I_o}{\sin\beta_o\,\ell} F(\theta) \frac{e^{-j\beta_o r_o}}{r_o} \left[e^{j\beta_o d\,\cos\theta} + e^{-j\beta_o d\,\cos\theta} \right]$$

or

$$E_\theta = \frac{j60\,I_o}{\sin\beta_o\,\ell} F(\theta) \frac{e^{-j\beta_o r_o}}{r_o} \left[2\,\cos(\beta_o d\,\cos\theta) \right] \tag{9.59}$$

The electric field expression in equation 9.59 may be considered as the multiplication of two terms; the first is the same expression as the electric field radiated from a single element, whereas the second $[2\,\cos(\beta_o d\,\cos\theta)]$ is an AF that accounts for the interaction between the two elements. In other words,

$$E_\theta = \text{electric field from a single element} \times \text{AF}$$

The same solution procedure outlined earlier, therefore, may be used to analyze the ground effects on the radiation characteristics of antennas or antenna arrays. For

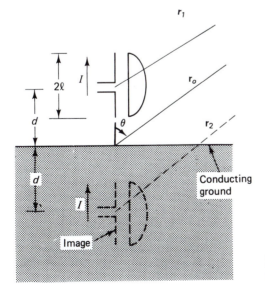

Figure 9.43 Vertical dipole antenna.

example, Figure 9.44 shows the radiation pattern of a horizontal 1/2λ dipole placed over a perfectly conducting plane. From Figure 9.44, it may be seen that the ground plane significantly altered or completely changed the originally circular radiation pattern of the λ/2 dipole. Such an effect is also a strong function of the separation distance between the antenna and the ground plane.

Clearly, the ground plane does not only impact the radiation pattern but also the input impedance of the antenna. As a result of the mutual coupling, the input impedance of the antenna is also expected to change. Assuming an equal and in-phase current in both the antenna and its image (perfectly conducting ground plane and small separation distances), the input impedance of the driven element is given by

$$\hat{Z}_{in} = \hat{Z}_{11} + \hat{Z}_{12}\frac{\hat{I}_2}{\hat{I}_1} = \hat{Z}_{11} + \hat{Z}_{12}$$

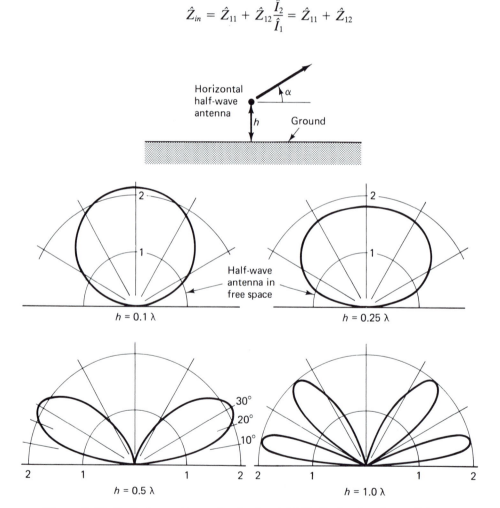

Figure 9.44   Radiation pattern of a horizontal λ/2 dipole placed parallel to a perfectly conducting ground plane. (*From* J. D. Kraus, *Antennas*, McGraw-Hill, New York, 1950, p. 306. Reprinted with permission).

From the input impedance curves of Figure 9.31 and the mutual impedance of collinear antennas of Figure 9.32, the variation of $\hat{Z}_{in}$ with the separation distance between the antenna and the ground plane shown in Figure 9.45 may be obtained. For the case of an electric dipole placed parallel to the ground plane, the input impedance is given by

$$\hat{Z}_{in} = \hat{Z}_{11} - \hat{Z}_{12}$$

where the image current in this case was assumed equal in magnitude, but opposite in phase to that of the driven antenna. The variation of $\hat{Z}_{in}$ versus the separation distance between the antenna and the ground plane is shown in Figure 9.46.

From both Figures 9.45 and 9.46, it may be seen that placing the antenna parallel to ground has a much larger impact on the input impedance values. Although in both cases, $\hat{Z}_{in}$ approaches its self-impedance value $\hat{Z}_{in} = \hat{Z}_{11} = 73 + j42$ asymptotically at larger values of h/λ, the rates of convergence of both curves to their asymptotic values are significantly different. For the vertical dipole, the asymptotic value is approached at $h/\lambda \approx 0.3$, whereas for the horizontal dipole values of $h/\lambda > 1$ are required for such a convergence. The large oscillations of $\hat{Z}_{in}$ around its asymptotic value for the parallel antenna case of Figure 9.46 should also be noticed.

Exercise 9.4. Given the two-element array shown in Figure 9.47, which is placed at distances d_1 and d_2 from a perfectly conducting ground plane, assume equal currents in the array elements, and that each antenna is a $\lambda/2$ dipole.

1. Construct the appropriate image diagram to account for the effect of ground.

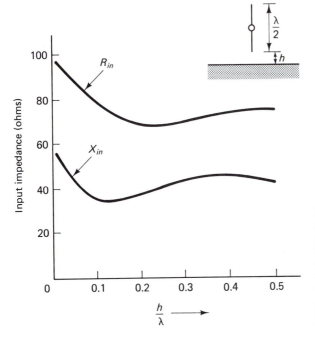

Figure 9.45 Real R_{in} and imaginary X_{in} parts of the input impedance of a $\lambda/2$ dipole placed at a height h above a perfectly conducting ground plane. (*From* W. L. Weeks, *Antenna Engineering* [New York: McGraw-Hill Book Company, 1968], p. 195. Reprinted with permission.)

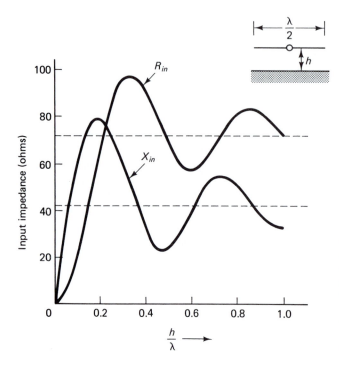

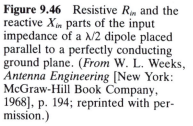

Figure 9.46 Resistive R_{in} and the reactive X_{in} parts of the input impedance of a $\lambda/2$ dipole placed parallel to a perfectly conducting ground plane. (*From* W. L. Weeks, *Antenna Engineering* [New York: McGraw-Hill Book Company, 1968], p. 194; reprinted with permission.)

2. Use the computer program developed in exercise 9.1 to obtain the radiation pattern of the four-element array obtained in part 1. Examine the variation of the radiation pattern for $0.1 < d_1 < 0.5$, $0.1 < d_2 < 0.5$, and $0.05 < d < 0.25$.

Corner reflector antenna.   The corner reflector antenna is yet another application that may be analyzed using the image theory described earlier. The basic structure of the antenna is shown in Figure 9.48.

It consists of two large ground planes intersecting at an angle γ. A center-fed dipole (or an array of parallel dipoles) is used as the driven antenna. For an appropriate operation of the corner reflector antenna, the width W_1 should be larger than the length of the antenna 2ℓ, and the length of each reflecting sheet W_2 should be larger than $2S$, where S is the distance from the driven element to the corner. The corner reflector is an effective directional antenna. Analysis of this antenna may be based on the image

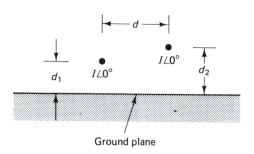

Figure 9.47 Two-element array above ground plane.

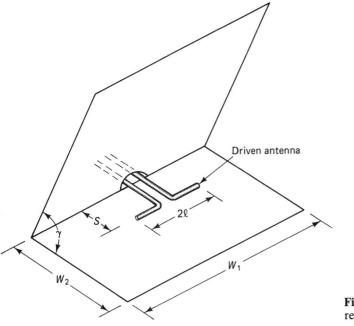

Figure 9.48   Geometry of a corner reflector.

theory, as shown in Figure 9.49. It may be shown that if the corner angle γ is an integer fraction of 180°—that is, $\gamma = 180°/n$, where n is an integer—there will be a finite number of images $(2n - 1)$ that may be used to account for the reflected radiation from the corner reflector. In the example shown in Figure 9.49, a 90° corner reflector is analyzed and three images in addition, of course, to the driven element were used to calculate the radiation pattern. Although all sources have currents of equal magnitudes, the directions of currents in the image sources 2 and 4 are opposite to those of the driven element and the image source 3.

The far field radiation resulting from the four sources is given by

$$E_\theta = \frac{j60\,I_o}{\sin\beta_o\,\ell}F(\theta)\left[\frac{e^{-j\beta_o r_1}}{r_1} - \frac{e^{-j\beta_o r_2}}{r_2} + \frac{e^{-j\beta_o r_3}}{r_3} - \frac{e^{-j\beta_o r_4}}{r_4}\right]$$

The negative sign in front of the second and fourth terms are due to the opposite direction of currents in these elements. Carrying out the usual far field approximation, we obtain

$$E_\theta = \frac{j60\,I_o}{\sin\beta_o\,\ell}F(\theta)\frac{e^{-j\beta_o r_o}}{r_o}\left[e^{j\beta_o S \cos\phi} - e^{-j\beta_o S \sin\phi} + e^{-j\beta_o S \cos\phi} - e^{j\beta_o S \sin\phi}\right]$$

$$= \text{radiation from each individual element} \times \text{AF}$$

where the AF is given by

$$\text{AF} = 2\cos(\beta_o S \cos\phi) - 2\cos(\beta_o S \sin\phi)$$

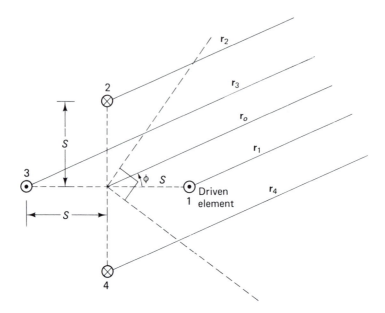

Figure 9.49 Image distribution in a 90° corner reflector. The direction of the current in sources 1 and 3 are out of the plane of the page, whereas sources 2 and 4 are directed into the plane.

Three radiation patterns of a 90° corner reflector for various values of S are shown in Figure 9.50. From Figure 9.50 it may be seen that, although higher gains may be achieved by further spacing the driven element away from the corner, care should be taken to avoid (unless desirable) the beam splitting shown in Figure 9.50b. The gain in field intensity relative to the driven $\lambda/2$ dipole alone is also shown in Figure 9.50.

A more detailed study of corner reflectors includes the calculation of the input impedance of the driven element taking into account the mutual coupling with its various images. For the 90° corner reflector, the four-element array is shown in Figure 9.49. The input impedance of the driven element may be calculated from

$$\hat{V}_1 = 1\,\underline{/0^\circ} = \hat{Z}_{11}\hat{I}_1 + \hat{Z}_{12}\hat{I}_2 + \hat{Z}_{13}\hat{I}_3 + \hat{Z}_{14}\hat{I}_4$$

$$\hat{Z}_{in} = \frac{\hat{V}_1}{\hat{I}_1} = \hat{Z}_{11} + \hat{Z}_{12}\frac{\hat{I}_2}{\hat{I}_1} + \hat{Z}_{13} + \frac{\hat{I}_3}{\hat{I}_1}\hat{Z}_{14}\frac{\hat{I}_4}{\hat{I}_1}$$

where $\hat{V}_1 = 1\underline{/0^\circ}$ is the reference voltage assumed to be applied to the input port of the driven element. Knowledge of the length of the array elements 2ℓ and the spacing S allow the calculation of the input and mutual impedances from Figures 9.31 to 9.33 or available equations in literature, as described in earlier sections. Substituting

$$\frac{\hat{I}_2}{\hat{I}_1} = \frac{\hat{I}_4}{\hat{I}_1} = -1$$

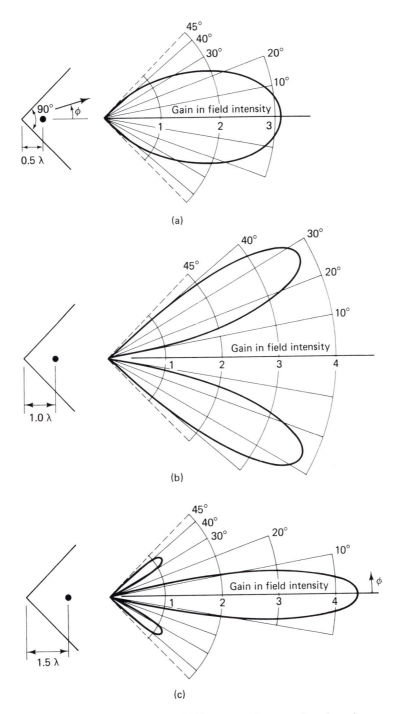

Figure 9.50 Radiation pattern of 90° corner reflector as function of spacing S. (a) $S = 0.5\lambda$. (b) $S = 1.0\lambda$. (c) $S = 1.5\lambda$. (*From* J. D. Kraus, *Antennas* [New York: McGraw-Hill Book Company, 1950], p. 332. Reprinted with permission.)

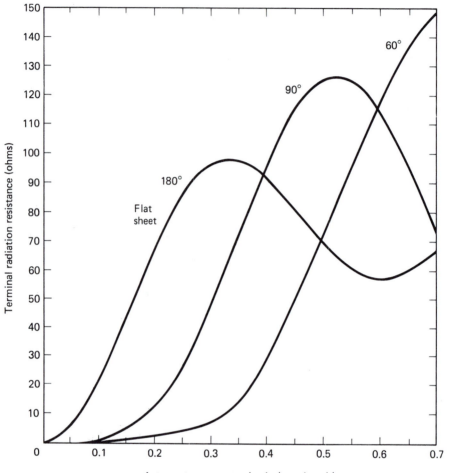

Figure 9.51 Input resistance of a λ/2 dipole placed in a corner (radiation) reflector of angles 180° (flat plate), 90°, and 60°. The radiation resistance is shown as a function of the element to corner spacing in wavelength. (*From* J. D. Kraus, *Antennas* [New York: McGraw-Hill Book Company, 1950], p. 333; reprinted with permission.)

and

$$\frac{\hat{I}_3}{\hat{I}_1} = 1$$

we obtain

$$\hat{Z}_{in} = \hat{Z}_{11} - \hat{Z}_{12} + \hat{Z}_{13} - \hat{Z}_{14} \qquad (9.60)$$

Figure 9.51 shows the variation of the terminal resistance

$$R_{in} = Re(\hat{Z}_{in})$$

as a function of the antenna-to-corner spacing in wavelength. The results in Figure 9.51 are shown for a 90° corner reflector, as well as for others, including a 60° corner reflector and a flat ground plate (i.e., 180° reflector). From Figure 9.51, it can be seen that the antenna-corner spacing is not only important in determining the radiation pattern and the gain of the corner reflector, but it also has a significant impact on the input impedance of the driven element. For a 50-Ω coaxial-fed driven element in a 90° corner reflector, $S/\lambda \approx 0.3$ is adequate for providing good impedance matching between the antenna and the coaxial transmission line.

SUMMARY

This chapter started with a description of physical aspects of radiation, mechanisms by which the electric field lines are detached from the physical structure of an antenna. It is shown that closed loops of electric field lines are separated from the antenna structure every half a period. Mathematical basis for calculating the radiation fields and engineering parameters that are often used to describe the characteristic properties of these radiation fields are then introduced based on studying the simple case of an electrically short current element. The following are some useful definitions and characteristics of radiation fields:

Near Fields: Near fields are particularly intense in the region near the antenna. They are associated with the electrostatic and magnetostatic fields because of the charge and current distributions on the antenna. Time-average power density associated with these fields is zero.

Far Fields: They are the dominant fields at a far distance from an antenna. The electric E_θ and magnetic H_ϕ fields are orthogonal and perpendicular to the direction of propagation $\mathbf{a}_r$. The ratio $E_\theta/H_\phi = \eta_o$, the intrinsic wave impedance. $\eta_o = \sqrt{\mu_o/\epsilon_o}$ in vacuum or air.

Radiation Pattern: Graphical representation of the relative distribution of the radiated power (at large distance r) as a function of θ and ϕ. To characterize a radiation pattern further, parameters such as half-power beam width, and E-plane and H-plane patterns may be used. The radiation pattern of a linear wire antenna is given by

$$|F(\theta)|^2 = \left| \frac{\cos(\beta_o \ell \cos \theta) - \cos(\beta_o \ell)}{\sin \theta} \right|^2$$

where 2ℓ is the length of the antenna and $\beta_o = 2\pi/\lambda$.

Directivity and Directive Gain $D(\theta, \phi)$: Directive gain $D(\theta, \phi)$ is the ratio of the radiation intensity in a given direction (θ, ϕ) to the radiation intensity of an isotropic source. Directivity (D_o) is the maximum value of directive gain. The directive gain of a half-wavelength linear wire antenna is

$$D(\theta, \phi) = 1.64 \left| \frac{\cos\left(\dfrac{\pi}{2} \cos \theta\right)}{\sin \theta} \right|^2$$

Radiation Resistance R_a: Value of an equivalent resistance that dissipates the same amount of power radiated by an antenna when the same current flows in both. R_a for a half-wavelength linear wire antenna is 73.13 Ω.

Antenna Array: Specific desirable characteristics of a radiation pattern may be achieved using an antenna array. The idea is based on capitalizing on the constructive and destructive interference effects between the radiation fields from individual elements. Specific design parameters of an array include the magnitude and phase of the input voltage to each antenna and the spacing as well as the geometric arrangements of these elements. For an N-element array of equal length 2ℓ and fed by currents of equal magnitudes (uniform array), the array factor $|AF|$ is given by

$$|AF| = \frac{\sin\frac{N}{2}(\beta_o d \cos\phi - \psi)}{\sin\frac{1}{2}(\beta_o d \cos\phi - \psi)}$$

where N is the number of elements, ψ is the progressive phase difference between the elements, and d is the separation distance between them. A graphical procedure for plotting the array factor versus the azimuthal angle ϕ is also described and demonstrated by examples 9.6 to 9.8.

The self- and mutual impedances of elements of an antenna array were then discussed, and Figures 9.31 to 9.33 were provided to determine these parameters for side-by-side and collinear arrangements of $\lambda/2$ dipoles. Arrays with parasitic (passive) elements were also analyzed, and a practical example of Yagi-Uda array (often used in television receivers) was described. The chapter was concluded with a discussion of the effect of ground on radiation characteristics of antennas. Image theory was introduced to account for the effect of a perfectly conducting ground plane, and a procedure to calculate the radiation characteristics of antennas above ground was described. It is shown that with the construction of the appropriate number of images with correct polarity, accounting for ground effects is identical to the analysis of an antenna array. The presented analysis of a corner reflector antenna is a good example of such a solution procedure.

PROBLEMS

1. Consider a short alternating current element of length $d\ell = 0.1$ m excited with a uniform (hypothetical) current distribution $\hat{I} = 1\underline{/0°}$. The operating frequency is $f = 10$ MHz.
   (a) Calculate the radiated electric and magnetic field components at $\theta = \frac{\pi}{2}$ and distances:
      (i) $r = 1$ m
      (ii) $r = 5$ m
      (iii) $r = 10$ m
   (b) Show that although the field components are dominated by the near nonradiated fields at $r = 1$ m, the calculations at $r = 5$ m represent a borderline case in which the near and far field components are approximately equal.

(c) Show that the calculations at $r = 10$ m are dominated by the far field components where E and H field components are mutually orthogonal and perpendicular to the direction of propagation, and their ratio is equal to the intrinsic wave impedance in free space.

2. Use the electric and magnetic field components in equations 9.14 and 9.15 to show that the θ component of the time-average density $\mathbf{P}_{av}$ is zero, whereas the radial component is given by equation 9.21.

3. The far field power density radiated from an antenna is given by

$$\mathbf{P} = K\frac{\sin\theta\,\cos\phi}{r^2}\,\mathbf{a}_r\ \text{W/m}^2 \qquad 0 \le \theta \le \pi, \qquad -\frac{\pi}{2} \le \phi \le \frac{\pi}{2}$$

where K is a constant. Determine the following:
(a) Azimuthal and elevation planes half-power beam widths.
(b) Azimuthal and elevation planes first null beam widths.
(c) Directivity and directive gain.

4. Repeat problem 3 for antennas having the following far field radiated power densities:
(a) $\mathbf{P} = K_a\dfrac{\sin^2\theta}{r^2}\,\mathbf{a}_r \qquad 0 \le \theta \le \pi,\, 0 \le \phi \le 2\pi$

(b) $\mathbf{P} = K_b\dfrac{\sin\theta\,\sin^2\phi}{r^2}\,\mathbf{a}_r \qquad 0 \le \theta \le \pi,\, 0 \le \phi \le 2\pi$

5. Given a short electric dipole of total length $\ell/\lambda_o = 0.056$ in air;
(a) Calculate the half-power beam width and the radiation resistance.
(b) If the same dipole is immersed in water $\epsilon = 81\epsilon_o$, calculate the new electrical length, its radiation resistance, and the new half-power beam width.

6. Consider a linear wire antenna of total length ℓ and carrying a sinusoidal current of amplitude $\hat{I}_o = 5\underline{/0°}\,A$ at a frequency of 500 MHz. Calculate the radiation electrical and magnetic field at a distance of 6 m. Make calculations at angles $\theta = 0°$, $30°$, $60°$, and $90°$, and for the following values of ℓ/λ:

$$\frac{\ell}{\lambda} = 0.5,\ 0.9,\ 1.25,\ \text{and}\ 1.5$$

Also calculate the time-average power density for all values of ℓ/λ and at all values of θ. Comment on values of radiation fields at $\theta = 0°$, and on the impact of the antenna length on the values of the power density at $\theta = 90°$.

7. Consider an array of two isotropic sources. Obtain an expression for the AF and sketch the radiation pattern for the following conditions:
(a) The separation distance is $d = \lambda/2$, and the two elements were fed equal currents but of $180°$ phase difference.
(b) The separation distance is $d = \lambda/4$, and the two elements were fed equal currents but of $90°$ phase difference.
(c) The separation distance is $d = \lambda$, and the two elements were fed equal currents but of $180°$ phase difference.

8. Use the graphical procedure described in section 9.6.2 to calculate the radiation pattern of the following N-element arrays:
(a) A five-element broadside array with $d = \lambda/2$.
(b) A five-element broadside array with $d = \lambda$.
(c) An end-fire five-element array with $d/\lambda = 2$, $d/\lambda = 5$, and $d/\lambda = 0.8$.

9. Consider a four-element array of uniformly excited, equally spaced elements. Sketch the array factor for the following operations:

(a) Broadside with $d = \lambda/2$.

(b) $d = \lambda/4$ and $\psi = \pi/2$.

10. Consider a uniform linear array of ten elements with an interspacing $d = \lambda/2$. Sketch the normalized array pattern and determine the half-power and first null beam widths for the following operations:

   (a) Broadside.

   (b) End fire.

11. Design a four-element, uniformly excited, equally spaced, linear array so that the main beam maximum occurs at $\phi = 45°$. Optimize your design so that the beam width is as small as possible without introducing additional main lobes in the radiation pattern.

12. Consider the uniformly excited, unequally spaced, linear array of four isotropic sources shown in Figure P9.12. The phase differences between the elements are also shown in Figure P9.12. Obtain an expression for the array factor and sketch the resulting pattern.

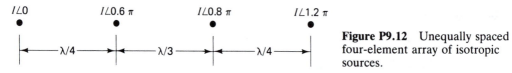

$1\angle 0$ $1\angle 0.6\,\pi$ $1\angle 0.8\,\pi$ $1\angle 1.2\,\pi$

|← $\lambda/4$ →|← $\lambda/3$ →|← $\lambda/4$ →|

Figure P9.12 Unequally spaced four-element array of isotropic sources.

13. Consider the array factor for the five-element nonuniform array shown in Figure 9.28a. Use this figure to sketch the radiation pattern of the following equispaced array elements:

   (a) $d = \lambda/2, \psi = 0$.

   (b) $d = \lambda/3, \psi = \pi/3$.

   (c) $d = \lambda/3, \psi = \pi/2$.

   In all cases, determine the half-power beam width.

14. A center-fed dipole of total length ℓ and diameter d is fed by a transmission line of 50-Ω characteristic impedance. Calculate the input reflection coefficient and the VSWR, assuming the following:

   (a) $\ell/\lambda = 0.2, \ell/d = 100$.

   (b) $\ell/\lambda = 0.2, \ell/d = 400$.

   (c) $\ell/\lambda = 0.2, \ell/d = 20$.

   (d) $\ell/\lambda = 0.5, \ell/d = 100$.

   (e) $\ell/\lambda = 0.7, \ell/d = 100$.

   (f) $\ell/\lambda = 0.25, 20 \le \ell/d \le 400$.

   Comment on the effect of length-to-diameter ratio at the first antenna resonance when $\ell/\lambda = 0.25$.

15. Consider a side-by-side two-element array. Each antenna is a dipole of length $\ell = \lambda/2$. Assuming equal currents, calculate the input impedance of each antenna for the following separation distances:

$$d = 0.3\lambda$$

$$d = 0.8\lambda$$

$$d = 1.2\lambda$$

16. Repeat the calculations of problem 15 for the case of a collinear arrangement with values of s the same as those of d.

17. Consider a three-element array of side-by-side arranged half-wavelength dipole antennas. The input voltages to the three elements are given by

$$\hat{V}_1 = 2\underline{/0°}$$

$$\hat{V}_2 = 2\underline{/-\frac{\pi}{2}}$$

$$\hat{V}_3 = 5\underline{/-\frac{\pi}{2}}$$

The elements are equally spaced with a separation distance $d = 0.3\lambda$.

(a) Calculate the currents in these elements, taking into account the mutual coupling between them.

(b) Based on the current values obtained in part (a) and the array geometry, determine the radiation pattern of the array.

(c) Compare the radiation pattern of part b with that based on neglecting the mutual coupling between the array elements.

18. This problem is intended to illustrate the design procedure of a Yagi-Uda array. Consider a three-element array placed in a side-by-side arrangement. The middle element is a center-fed dipole of total length $\ell = \lambda/2$. The reflector's length is $\ell/\lambda = 0.55$, whereas the director's length is $\ell/\lambda = 0.4$. The elements were spaced at distance $d = \lambda/4$.

(a) If the input voltage to the center-fed element is $\hat{V}_{in} = 1\underline{/0°}$, calculate the induced current on each element, taking into account the mutual coupling (using the mutual impedance value of Figure 9.33 as an approximation of the desired values).

(b) Use the obtained current values in part (a) and the given array geometry to calculate the radiation pattern of the Yagi-Uda array.

19. An infinitesimal dipole of length $\ell = \lambda/20$ is placed vertically above an infinite perfectly conducting ground plane. Calculate the angles at which nulls in the radiation pattern occur for the following cases:

(a) Dipole center is placed at a height of 0.5λ.

(b) Dipole center is placed at a height of 2λ.

20. The placement of antennas above ground has two main effects on their radiation characteristics. First, the mutual coupling between the antenna and its image alters the current distribution and, hence, the input impedance of the antennas. The radiation pattern also changes to account for the array factor of the two-element array consisting of the original antenna and its image.

(a) For a linear antenna of length $\ell = \lambda/2$ placed vertically above perfectly conducting ground, calculate the variation of the input impedance with the increase of the antenna height above ground. Choose $0.3 \leq d/\lambda \leq 1.5$, where d is the distance from the center of the antenna to the ground plane.

(b) Repeat part a if the antenna is placed horizontally above ground.

(c) Calculate the radiation pattern for each case as a function of d/λ. Make calculations for at least three values of d/λ.

21. Consider a corner reflector of an angle $\gamma = 60°$.

(a) Construct the appropriate number and polarity of the set of images that may be used to describe the radiation characteristics of the reflector.

(b) Obtain an expression for the array factor.

(c) Sketch the variation of the array factor with the separation distance S between the driven antenna and the vertex of the reflector.

APPENDIX A

VECTOR IDENTITIES AND OPERATIONS

A.1 VECTOR IDENTITIES

The following vector identities may be helpful in manipulating Maxwell's equations and in solving for the electromagnetic field quantities. Instead of listing these identities in terms of $\mathbf{E}$ and $\mathbf{H}$ fields, the following table is prepared in terms of the general vectors $\mathbf{A}$, $\mathbf{B}$, and $\mathbf{C}$.

$$(\mathbf{A} \times \mathbf{B}) \cdot \mathbf{C} = (\mathbf{B} \times \mathbf{C}) \cdot \mathbf{A} = (\mathbf{C} \times \mathbf{A}) \cdot \mathbf{B}$$

$$\mathbf{A} \times \mathbf{B} \times \mathbf{C} = (\mathbf{A} \cdot \mathbf{C})\mathbf{B} - (\mathbf{A} \cdot \mathbf{B})\mathbf{C}$$

$$\nabla \cdot (\Phi \mathbf{A}) = \mathbf{A} \cdot \nabla \Phi + \Phi \nabla \cdot \mathbf{A}, \qquad \Phi \text{ (Scalar)}$$

$$\nabla \cdot (\mathbf{A} + \mathbf{B}) = \nabla \cdot \mathbf{A} + \nabla \cdot \mathbf{B}$$

$$\nabla \cdot (\mathbf{A} \times \mathbf{B}) = \mathbf{B} \cdot \nabla \times \mathbf{A} - \mathbf{A} \cdot \nabla \times \mathbf{B}$$

$$\nabla \cdot \nabla \Phi = \nabla^2 \Phi$$

$$\nabla \cdot \nabla \times \mathbf{A} = 0$$

$$\nabla \times (\Phi \mathbf{A}) = \nabla \Phi \times \mathbf{A} + \Phi \nabla \times \mathbf{A}$$

$$\nabla \times (\mathbf{A} + \mathbf{B}) = \nabla \times \mathbf{A} + \nabla \times \mathbf{B}$$

$$\nabla \times (\mathbf{A} \times \mathbf{B}) = \mathbf{A} \nabla \cdot \mathbf{B} - \mathbf{B} \nabla \cdot \mathbf{A} + (\mathbf{B} \cdot \nabla)\mathbf{A} - (\mathbf{A} \cdot \nabla)\mathbf{B}$$

$$\nabla \times \nabla \Phi = 0$$

$$\nabla \times \nabla \times \mathbf{A} = \nabla(\nabla \cdot \mathbf{A}) - \nabla^2 \mathbf{A}$$

$$\int_v \nabla \cdot \mathbf{A} \, dv = \oint_s \mathbf{A} \cdot d\mathbf{s} \qquad \text{(Divergence theorem)}$$

$$\int_s \nabla \times \mathbf{A} \cdot d\mathbf{s} = \oint_c \mathbf{A} \cdot d\mathbf{c} \qquad \text{(Stokes's theorem)}$$

A.2 VECTOR OPERATIONS

Divergence of a vector A.   In the generalized curvilinear coordinate system, the divergence of the vector

$$\mathbf{A} = A_1 \mathbf{a}_1 + A_2 \mathbf{a}_2 + A_3 \mathbf{a}_3$$

is given by

$$\text{div } \mathbf{A} = \frac{1}{h_1 h_2 h_3} \left[\frac{\partial(A_1 h_2 h_3)}{\partial u_1} + \frac{\partial(A_2 h_1 h_3)}{\partial u_2} + \frac{\partial(A_3 h_1 h_2)}{\partial u_3} \right]$$

The various components of $\mathbf{A}$, the unit vectors $\mathbf{a}_1$, $\mathbf{a}_2$, $\mathbf{a}_3$, the independent variables u_1, u_2, and u_3, and the metric coefficients h_1, h_2, and h_3 in the Cartesian, cylindrical, and spherical coordinates are given by

	Cartesian	Cylindrical	Spherical
Independent variables u_1, u_2, u_3	x, y, z	ρ, ϕ, z	r, θ, ϕ
Vector components	A_x, A_y, A_z	A_ρ, A_ϕ, A_z	A_r, A_θ, A_ϕ
Unit vectors $\mathbf{a}_1, \mathbf{a}_2, \mathbf{a}_3$	$\mathbf{a}_x, \mathbf{a}_y, \mathbf{a}_z$	$\mathbf{a}_\rho, \mathbf{a}_\phi, \mathbf{a}_z$	$\mathbf{a}_r, \mathbf{a}_\theta, \mathbf{a}_\phi$
Metric coefficients h_1, h_2, h_3	$1, 1, 1$	$1, \rho, 1$	$1, r, r \sin\theta$

and explicit expressions for the divergence of **A** in these coordinate systems are then

$$\nabla \cdot \mathbf{A} = \frac{\partial A_x}{\partial x} + \frac{\partial A_y}{\partial y} + \frac{\partial A_z}{\partial z} \qquad \text{(Cartesian)}$$

$$\nabla \cdot \mathbf{A} = \frac{1}{\rho}\frac{\partial}{\partial \rho}(\rho A_\rho) + \frac{1}{\rho}\frac{\partial A_\phi}{\partial \phi} + \frac{\partial A_z}{\partial z} \qquad \text{(Cylindrical)}$$

$$\nabla \cdot \mathbf{A} = \frac{1}{r^2}\frac{\partial}{\partial r}(r^2 A_r) + \frac{1}{r\sin\theta}\frac{\partial}{\partial \theta}(A_\theta \sin\theta) + \frac{1}{r\sin\theta}\frac{\partial A_\phi}{\partial \phi} \qquad \text{(Spherical)}$$

Gradient of a scalar function Φ.   The gradient of a scalar function Φ in the generalized curvilinear coordinate system is given by

$$\text{grad }\Phi = \nabla\Phi = \frac{1}{h_1}\frac{\partial \Phi}{\partial u_1}\mathbf{a}_1 + \frac{1}{h_2}\frac{\partial \Phi}{\partial u_2}\mathbf{a}_2 + \frac{1}{h_3}\frac{\partial \Phi}{\partial u_3}\mathbf{a}_3$$

Substituting the independent variables (u_1, u_2, u_3), the base vectors $(\mathbf{a}_1, \mathbf{a}_2, \mathbf{a}_3)$, and the metric coefficients (h_1, h_2, h_3) in the various coordinate systems (as described in the divergence section), we obtain the following explicit expressions of the gradient in the Cartesian, cylindrical, and spherical coordinates:

$$\nabla\Phi = \frac{\partial \Phi}{\partial x}\mathbf{a}_x + \frac{\partial \Phi}{\partial y}\mathbf{a}_y + \frac{\partial \Phi}{\partial z}\mathbf{a}_z \qquad \text{(Cartesian)}$$

$$\nabla\Phi = \frac{\partial \Phi}{\partial \rho}\mathbf{a}_\rho + \frac{1}{\rho}\frac{\partial \Phi}{\partial \phi}\mathbf{a}_\phi + \frac{\partial \Phi}{\partial z}\mathbf{a}_z \qquad \text{(Cylindrical)}$$

$$\nabla\Phi = \frac{\partial \Phi}{\partial r}\mathbf{a}_r + \frac{1}{r}\frac{\partial \Phi}{\partial \theta}\mathbf{a}_\theta + \frac{1}{r\sin\theta}\frac{\phi\Phi}{\partial \phi}\mathbf{a}_\phi \qquad \text{(Spherical)}$$

Curl of vector A.   The curl of the vector $\mathbf{A} = A_1\mathbf{a}_1 + A_2\mathbf{a}_2 + A_3\mathbf{a}_3$ in the generalized curvilinear coordinate system is given by

$$\text{curl }\mathbf{A} = \begin{vmatrix} \dfrac{\mathbf{a}_1}{h_2 h_3} & \dfrac{\mathbf{a}_2}{h_1 h_3} & \dfrac{\mathbf{a}_3}{h_1 h_2} \\[2mm] \dfrac{\partial}{\partial u_1} & \dfrac{\partial}{\partial u_2} & \dfrac{\partial}{\partial u_3} \\[2mm] h_1 A_1 & h_2 A_2 & h_3 A_3 \end{vmatrix}$$

Substituting the components of **A**, independent variables, base vectors, and the metric coefficients in the various coordinate systems (as described in the divergence section), we obtain

$$\nabla \times \mathbf{A} = \left(\frac{\partial A_z}{\partial y} - \frac{\partial A_y}{\partial z}\right)\mathbf{a}_x + \left(\frac{\partial A_x}{\partial z} - \frac{\partial A_z}{\partial x}\right)\mathbf{a}_y + \left(\frac{\partial A_y}{\partial x} - \frac{\partial A_x}{\partial y}\right)\mathbf{a}_z \qquad \text{(Cartesian)}$$

$$\nabla \times \mathbf{A} = \left(\frac{1}{\rho}\frac{\partial A_z}{\partial \phi} - \frac{\partial A_\phi}{\partial z}\right)\mathbf{a}_\rho + \left(\frac{\partial A_\rho}{\partial z} - \frac{\partial A_z}{\partial \rho}\right)\mathbf{a}_\phi + \frac{1}{\rho}\left(\frac{\partial(\rho A_\phi)}{\partial \rho} - \frac{\partial A_\rho}{\partial \phi}\right)\mathbf{a}_z \qquad \text{(Cylindrical)}$$

$$\mathbf{\nabla} \times \mathbf{A} = \frac{1}{r \sin \theta} \left[\frac{\partial (A_\phi \sin \theta)}{\partial \theta} - \frac{\partial A_\theta}{\partial \phi} \right] \mathbf{a}_r$$

$$+ \frac{1}{r} \left[\frac{1}{\sin \theta} \frac{\partial A_r}{\partial \phi} - \frac{\partial (rA_\phi)}{\partial r} \right] \mathbf{a}_\theta$$

$$+ \frac{1}{r} \left[\frac{\partial (rA_\theta)}{\partial r} - \frac{\partial A_r}{\partial \theta} \right] \mathbf{a}_\phi \qquad \text{(Spherical)}$$

Laplacian of a Scalar Function Φ.   The Laplacian of a scalar function Φ is defined as

$$\nabla^2 \Phi \equiv \mathbf{\nabla} \cdot \mathbf{\nabla} \Phi$$

Hence, an expression for the Laplacian in the generalized curvilinear coordinate system may be obtained from those of the gradient and divergence described earlier. $\nabla^2 \Phi$ is given by

$$\nabla^2 \Phi = \frac{1}{h_1 h_2 h_3} \left[\frac{\partial}{\partial u_1} \left(\frac{h_2 h_3}{h_1} \frac{\partial \Phi}{\partial u_1} \right) + \frac{\partial}{\partial u_2} \left(\frac{h_1 h_3}{h_2} \frac{\partial \Phi}{\partial u_2} \right) + \frac{\partial}{\partial u_3} \left(\frac{h_1 h_2}{h_3} \frac{\partial \Phi}{\partial u_3} \right) \right]$$

Substituting the independent variables and the metric coefficients in the various coordinate systems as described earlier gives

$$\nabla^2 \Phi = \frac{\partial^2 \Phi}{\partial x^2} + \frac{\partial^2 \Phi}{\partial y^2} + \frac{\partial^2 \Phi}{\partial z^2} \qquad \text{(Cartesian)}$$

$$\nabla^2 \Phi = \frac{1}{\rho} \frac{\partial}{\partial \rho} \left(\rho \frac{\partial \Phi}{\partial \rho} \right) + \frac{1}{\rho^2} \frac{\partial^2 \Phi}{\partial \phi^2} + \frac{\partial^2 \Phi}{\partial z^2} \qquad \text{(Cylindrical)}$$

$$\nabla^2 \Phi = \frac{1}{r^2} \frac{\partial}{\partial r} \left(r^2 \frac{\partial \Phi}{\partial r} \right) + \frac{1}{r^2 \sin \theta} \frac{\partial}{\partial \theta} \left(\sin \theta \frac{\partial \Phi}{\partial \theta} \right) + \frac{1}{r^2 \sin^2 \theta} \frac{\partial^2 \Phi}{\partial \phi^2} \qquad \text{(Spherical)}$$

APPENDIX *B*

UNITS, MULTIPLES, AND SUBMULTIPLES

TABLE B.1 NAMES AND SI UNITS OF QUANTITIES COMMONLY USED IN ELECTROMAGNETICS

Quantity	Description	Units
Admittance (Y)	1/impedance	ohm^{-1} or Siemens (S)
Area (s)	length $\times$ length	meter^2 (m^2)
Attenuation constant (α)	rate at which amplitude attenuates per unit length	Neper/meter (Np/m)
Capacitance (C)	charge/potential	Coulomb/volt or Farad (F)

TABLE B.1 NAMES AND SI UNITS OF QUANTITIES COMMONLY USED
IN ELECTROMAGNETICS (Continued)

Quantity	Description	Units
Charge density (ρ)	either linear charge density $\rho_l (= Q/\text{length})$, surface charge density ρ_s $(= Q/\text{area})$ or volume charge density ρ_v $(= Q/\text{volume})$	ρ_l in $\dfrac{\text{Coulomb}}{\text{meter}} = \dfrac{\text{C}}{\text{m}}$ ρ_s in $\dfrac{\text{Coulomb}}{\text{meter}^2} = \dfrac{\text{C}}{\text{m}^2}$ ρ_v in $\dfrac{\text{Coulomb}}{\text{meter}^3} = \dfrac{\text{C}}{\text{m}^3}$
Charge (q, Q)	current × time	Coulomb (C)
Conductance	1/resistance	Siemens (S) or Ω^{-1}
Conductivity (σ)	conductance/length	Siemens/meter (S/m)
Current (I)	charge/time	Ampere (A)
Current density	surface sheet current density $\mathbf{J}_s$ volume current density $\mathbf{J}$	Ampere/meter (A/m) Ampere/meter^2 (A/m^2)
Electric dipole ($\mathbf{p}$)	$\mathbf{p} = q\mathbf{d}$, charge × distance	Coulomb × meter = C · m
Electric field intensity ($\mathbf{E}$)	force/charge or potential/length	Volt/meter
Electric flux (ψ_e)	$\psi_e = \oint_s \mathbf{D} \cdot d\mathbf{s} = Q$	Coulomb (C)
Electric flux density ($\mathbf{D}$)	electric flux/area	Coulomb/meter^2 (C/m^2)
Electric susceptibility (χ_e)	polarization $\mathbf{P} = \epsilon_o \chi_e \mathbf{E}$	χ_e is dimensionless
Force ($\mathbf{F}$)	mass × acceleration	Newton (N)
Frequency (f)	cycles/second	hertz (Hz) (s^{-1})
Impedance (Z_o, η)	characteristic impedance of transmission line Z_o intrinsic impedance of free-space vacuum η_o	ohms (Ω) ohms (Ω)
Inductance (L)	magnetic flux/current	Wb/A or Henry (H)
Length (ℓ,d)		meter (m)
Magnetic dipole ($\mathbf{m}$)	$\mathbf{m} = I\,d\mathbf{s}$, current × area	Ampere × meter^2 = A · m^2

TABLE B.1   NAMES AND SI UNITS OF QUANTITIES COMMONLY USED
IN ELECTROMAGNETICS (Continued)

Quantity	Description	Units
Magnetic field intensity (**H**)	$\oint_c \mathbf{H} \cdot d\ell = I$ or **H** is $\dfrac{\text{current}}{\text{distance}}$	Ampere/meter (A/m)
Magnetic flux (ψ_m)	$\psi_m = \displaystyle\int_s \mathbf{B} \cdot d\mathbf{s}$	Weber (Wb)
Magnetic flux density (**B**)	magnetic flux/area	Weber/m^2 (Wb/m^2)
Magnetic flux linkage	flux $\times$ number of turns	Weber $\cdot$ turn (Wb $\cdot$ turn)
Magnetic susceptibility (χ_m)	magnetization $\mathbf{M} = \chi_m \mathbf{H}$	χ_m is dimensionless
Magnetic vector potential (**A**)	$\mathbf{B} = \nabla \times \mathbf{A}$	Weber/meter (Wb/m)
Magnetization (**M**)	$\mathbf{M} \equiv$ magnetic dipole/volume	$\dfrac{\text{A} \cdot \text{m}^2}{\text{m}^3} = $ A/m
Mass (m)		kilogram (kg)
Period (T)	1/frequency	second (s)
Permeability (μ_o)	for vacuum	Henry/meter (H/m)
Permittivity (ϵ_o) (dielectric constant)	for vacuum or air at atmospheric pressure	Farad/meter (F/m)
Phase constant (β)	change in phase of wave per unit length	radian/meter (rad/m)
Polarization (**P**)	electric dipole/volume	Coulomb/meter^2 = C/m^2
Potential Φ or voltage (V)	work/charge	Volt (V)
Power	energy/time	Watt (W)
Power density (**P**) vector (Poynting vector)	power/area	Watt/meter^2 (W/m^2)
Propagation constant ($\hat{\gamma}$)	$e^{\pm \hat{\gamma} z}$ for a wave propagating along ($\pm z$) axis	meter^{-1} (m^{-1})
Radian frequency (ω)	$2\pi \times$ frequency	radian/second (rad/s)
Reactance	voltage/current	ohms (Ω)

TABLE B.1 NAMES AND SI UNITS OF QUANTITIES COMMONLY USED
IN ELECTROMAGNETICS (Continued)

Quantity	Description	Units
Reflection coefficient ($\hat{\Gamma}$)	reflected electric field or voltage/incident electric field or voltage	dimensionless
Relative permittivity (ϵ_r)	$\dfrac{\text{permittivity of material } \epsilon}{\text{permittivity of vacuum } \epsilon_o}$	$\dfrac{\epsilon}{\epsilon_o}$ (dimensionless)
Relative permittivity (μ_r)	$\dfrac{\text{permeability of material } \mu}{\text{permeability of vacuum } \mu_o}$	dimensionless
Reluctance ($\mathcal{R}$)	magnetic flux/current	Weber/meter (Wb/m) or Henry (H)
Resistance (R)	resistance = potential (voltage)/current	Volt/Ampere or ohm (Ω)
Resistivity	resistance $\times$ length	ohm $\cdot$ meter ($\Omega \cdot$ m)
Skin depth (δ)	distance at which amplitude reaches e^{-1} of its original value	meter (m)
Susceptance (B)	$\dfrac{1}{\text{reactance}}$	ohm^{-1} or Siemens (S)
Time (t)		second (s)
Transmission coefficient ($\hat{T}$)	transmitted field or voltage/ incident field or voltage	dimensionless
Velocity (v)	distance/time	m/s
Wavelength (λ)	velocity of propagation/ frequency	meter (m)
Work (energy)	force $\times$ distance	Joule (J)

TABLE B.2   MULTIPLES AND SUBMULTIPLES OF UNITS

Prefix	Symbol	Factor
exa (eksa)	E	10^{18}
peta	p	10^{15}
tera	T	10^{12}
giga	G	10^{9}
mega	M	10^{6}
kilo	k	10^{3}
milli	m	10^{-3}
micro	μ	10^{-6}
nano	n	10^{-9}
pico	p	10^{-12}
femto	f	10^{-15}
atto	a	10^{-18}
Angstrom	Å	10^{-10} m

TRIGONOMETRIC, HYPERBOLIC, AND LOGARITHMIC RELATIONS

C.1 TRIGONOMETRIC RELATIONS

$$e^{jx} = \cos x + j \sin x$$

$$\sin x = \frac{e^{jx} - e^{-jx}}{2j}$$

$$\cos x = \frac{e^{jx} + e^{-jx}}{2}$$

$$\sin(x \pm y) = \sin x \cos y \pm \cos x \sin y$$

$$\cos(x \pm y) = \cos x \cos y \mp \sin x \sin y$$

$$\tan(x \pm y) = \frac{\tan x \pm \tan y}{1 \mp \tan x \tan y}$$

$$\sin 2x = 2 \sin x \cos x$$

$$\cos 2x = \cos^2 x - \sin^2 x = 2 \cos^2 x - 1 = 1 - 2 \sin^2 x$$

$$\sin^2 x + \cos^2 x = 1$$

C.2 HYPERBOLIC RELATIONS

$$\sinh x = \frac{e^x - e^{-x}}{2}$$

$$\cosh x = \frac{e^x + e^{-x}}{2}$$

$$\tanh x = \frac{\sinh x}{\cosh x} = \frac{e^x + e^{-x}}{e^x - e^{-x}}$$

$$\sinh(jx) = j \sin x, \qquad j \sinh(x) = \sin(jx)$$

$$\cosh(jx) = \cos x, \qquad \cosh(x) = \cos(jx)$$

$$\sinh(x \pm y) = \sinh x \cosh y \pm \cosh x \sinh y$$

$$\cosh(x \pm y) = \cosh x \cosh y \pm \sinh x \sinh y$$

$$\cosh x + \sinh x = e^x$$

$$\cosh x - \sinh x = e^{-x}$$

C.3 LOGARITHMIC RELATIONS

$$\log xy = \log x + \log y$$

$$\log \frac{x}{y} = \log x - \log y$$

$$\log x^n = n \log x$$

$$\log x = \log_{10} x \qquad \text{(Common logarithm)}$$

$$\ln x = \log_e x, \qquad e = 2.71828 \qquad \text{(Natural logarithm)}$$

$$\log_{10} x = \log_e x \, \log_{10} e = 0.4343 \log_e x = 0.4343 \ln x$$

$$\ln x = \log_{10} x \, \log_e 10 = 2.3026 \log_{10} x$$

$$dB = 10 \log_{10} \text{(power ratio)} = 20 \log_{10} \text{(voltage ratio)}$$

$$1 \text{ Np} = \frac{1}{e} = 0.368 = 20 \log_{10} e \, 8.686 \text{ dB}$$

FREE-SPACE, ATOMIC, AND MATERIAL CONSTANTS

In this appendix we summarize some of the free-space (vacuum) constants, physical constants of electrons and protons, and some of the more important dielectric, magnetic, and conductive properties of materials. Table D.1 includes the free-space constants, Table D.2 provides the physical constants of electrons and protons, and Tables D.3 to D.5 list the dielectric, magnetic, and conductive properties of materials.

TABLE D.1 FREE-SPACE (VACUUM) CONSTANTS

Quantity	Symbol	Value
Velocity of light	c	2.997925×10^8 (m/s) $\sim 3 \times 10^8$ (m/s)
Dielectric constant (permittivity)	ϵ_o	8.854×10^{-12} (F/m) $\approx \dfrac{1}{36\pi} \times 10^{-9}$ (F/m)
Permeability	μ_o	$4\pi \times 10^{-7}$ (H/m)
Intrinsic impedance	$\eta_o = \sqrt{\dfrac{\mu_o}{\epsilon_o}}$	$120\pi \sim 377\ \Omega$
Wavelength λ_o	c/f (f is frequency)	meter (m)

TABLE D.2 ELECTRON, PROTON, AND OTHER PHYSICAL CONSTANTS

Quantity	Symbol	Value
Mass of electron	m	9.107×10^{-31} (kg)
Charge of electron	e	-1.602×10^{-19} (C)
Mass of proton	m_{pro}	1.673×10^{-27} (kg)
Boltzmann's constant	k	1.38045×10^{-23} Joule/degree Kelvin (J/K°)
Planck's constant	h	6.63×10^{-34} Js
Gravitational constant	G	6.67×10^{-11} Newton m^2/kg^2 (N m^2/kg^2)
Avogardo's constant	N_A = number of molecules per kilogram-mole	6.0225×10^{26} (kgmol)$^{-1}$

TABLE D.3 DIELECTRIC CONSTANT (ϵ_r) OF MATERIALS

Material	ϵ_r
Air	1.0006
Alcohol, ethyl	25
Asbestos fiber	4.8
Barium titanate	1200
Earth (dry)	7
Earth (moist)	15
Earth (wet)	30
Glass	4–10
Ice	4.2
Mica	5.4
Nylon	4
Paper	2–4
Polystyrene	2.56
Porcelain	6
Pyrex glass	5
Quartz	3.8
Rubber	2.5–3
Silica	3.8
Snow	3.3
Styrofoam	1.03
Teflon	2.1
Water (distilled)	81
Water (sea)	70

TABLE D.4 CONDUCTIVITIES

Material	σ (S/m)
Good conductors	
Silver	6.17×10^7
Copper	5.8×10^7
Gold	4.1×10^7
Aluminum	3.82×10^7
Tungsten	1.82×10^7
Brass	1.5×10^7
Bronze	1×10^7
Iron	1×10^7
Graphite	7×10^4
Poor conductors	
Sea water	4
Fresh water	10^{-3}
Earth (dry)	10^{-3}
Earth (moist)	1.2×10^{-2}
Earth (wet)	3×10^{-2}
Clay	10^{-4}
Insulators ($\sigma < 10^{-6}$)	
Glass	10^{-12}
Porcelain	10^{-10}
Diamond	2×10^{-13}
Polystyrene	10^{-16}
Quartz	10^{-17}
Rubber	10^{-15}

TABLE D.5 MAGNETIC PERMEABILITY (μ_r) OF MATERIALS

Material	μ_r
Diamagnetic	
Paraffin	0.99999942
Silver	0.9999976
Bismuth	0.999982
Water	0.9999901
Copper	0.99999
Gold	0.99996
Paramagnetic	
Aluminum	1.000021
Magnesium	1.000012
Platinum	1.0003
Palladium	1.00082
Tungsten	1.00008
Ferrimagnetic (Ferrites) Ferrites such as nickel ferrite ($NiO \cdot Fe_2O_3$) and magnesium ferrite ($MgO \cdot Fe_2O_3$) are nonmetallic (insulators) magnetic materials	
Ferrite (typical)	1000
Ferroxcube 3 ($Mn - Zn$ ferrite powder)	1500
Ferromagnetic (Nonlinear)	
Cobalt	250
Nickel	600
Mild steel (0.2 C)	2000
Iron (0.2 impurity)	5000
Silicon iron (4 Si)	7000
Mumetal	1×10^5
Purified iron (0.05 impurity)	2×10^5
Supermalloy	as high as 10^6

COSINE $C_i(x)$ AND SINE $S_i(x)$ INTEGRALS

In calculating the total power radiated and the radiation resistance and directivity of a linear wire antenna (Chapter 9), we encountered cosine $C_i(x)$ and sine $S_i(x)$ integrals of the form

$$C_i(x) = -\int_x^\infty \frac{\cos y}{y} dy = \int_\infty^x \frac{\cos y}{y} dy$$

$$S_i(x) = \int_0^x \frac{\sin y}{y} dy$$

To help students solve exercise problems, numerical values of these integrals are tabulated as a function of x.

x	$S_i(x)$	$C_i(x)$	x	$S_i(x)$	$C_i(x)$
0.0	0.0	$-\infty$	5.1	1.53125	-0.18348
0.1	0.09994	-1.72787	5.2	1.51367	-0.17525
0.2	0.19956	-1.04220	5.3	1.49731	-0.16551
0.3	0.29850	-0.64917	5.4	1.48230	-0.15439
0.4	0.39646	-0.37881	5.5	1.46872	-0.14205
0.5	0.49311	-0.17778	5.6	1.45667	-0.12867
0.6	0.58813	-0.02227	5.7	1.44620	-0.11441
0.7	0.68122	0.10051	5.8	1.43736	-0.09944
0.8	0.77209	0.19828	5.9	1.43018	-0.08393
0.9	0.86047	0.27607	6.0	1.42469	-0.06806
1.0	0.94608	0.33740	6.1	1.42087	-0.05198
1.1	1.02868	0.38487	6.2	1.41871	-0.03587
1.2	1.10805	0.42046	6.3	1.41817	-0.01989
1.3	1.18396	0.44574	6.4	1.41922	-0.00418
1.4	1.25622	0.46201	6.5	1.42179	0.01110
1.5	1.32468	0.47036	6.6	1.42582	0.02582
1.6	1.38918	0.47173	6.7	1.43120	0.03985
1.7	1.44959	0.46697	6.8	1.43787	0.05308
1.8	1.50581	0.45681	6.9	1.44570	0.06539
1.9	1.55777	0.44194	7.0	1.45460	0.07669
2.0	1.60541	0.42298	7.1	1.46443	0.08691
2.1	1.64870	0.40051	7.2	1.47509	0.09596
2.2	1.68762	0.37508	7.3	1.48644	0.10379
2.3	1.72221	0.34718	7.4	1.49834	0.11036
2.4	1.75248	0.31729	7.5	1.51068	0.11563
2.5	1.77852	0.28587	7.6	1.52331	0.11960
2.6	1.80039	0.25334	7.7	1.53611	0.12225
2.7	1.81821	0.22008	7.8	1.54894	0.12359
2.8	1.83210	0.18649	7.9	1.56167	0.12364
2.9	1.84219	0.15290	8.0	1.57419	0.12243
3.0	1.84865	0.11963	8.1	1.58637	0.12002
3.1	1.85166	0.08699	8.2	1.59810	0.11644
3.2	1.85140	0.05526	8.3	1.60928	0.11177
3.3	1.84808	0.02468	8.4	1.61981	0.10607
3.4	1.84191	-0.00452	8.5	1.62960	0.09943
3.5	1.83312	-0.03213	8.6	1.63857	0.09194
3.6	1.82195	-0.05797	8.7	1.64665	0.08368
3.7	1.80862	-0.08190	8.8	1.65379	0.07476
3.8	1.79339	-0.10378	8.9	1.65993	0.06528
3.9	1.77650	-0.12350	9.0	1.66504	0.05535
4.0	1.75820	-0.14098	9.1	1.66908	0.04507
4.1	1.73874	-0.15617	9.2	1.67205	0.03456
4.2	1.71837	-0.16901	9.3	1.67393	0.02391
4.3	1.69732	-0.17951	9.4	1.67473	0.01325
4.4	1.67583	-0.18766	9.5	1.67446	0.00268
4.5	1.65414	-0.19349	9.6	1.67316	-0.00771
4.6	1.63246	-0.19705	9.7	1.67084	-0.01780
4.7	1.61100	-0.19839	9.8	1.66757	-0.02752
4.8	1.58997	-0.19760	9.9	1.66338	-0.03676
4.9	1.56956	-0.19478	10.0	1.65835	-0.04546
5.0	1.54993	-0.19003	10.1	1.65253	-0.05352

x	$S_i(x)$	$C_i(x)$	x	$S_i(x)$	$C_i(x)$
10.2	1.64600	-0.06089	15.2	1.62575	0.03543
10.3	1.63883	-0.06751	15.3	1.62865	0.02955
10.4	1.63112	-0.07332	15.4	1.63093	0.02345
10.5	1.62294	-0.07828	15.5	1.63258	0.01719
10.6	1.61439	-0.08237	15.6	1.63359	0.01085
10.7	1.60556	-0.08555	15.7	1.63396	0.00447
10.8	1.59654	-0.08781	15.8	1.63370	-0.00187
10.9	1.58743	-0.08915	15.9	1.63280	-0.00812
11.0	1.57831	-0.08956	16.0	1.63130	-0.01420
11.1	1.56927	-0.08907	16.1	1.62921	-0.02007
11.2	1.56042	-0.08769	16.2	1.62657	-0.02566
11.3	1.55182	-0.08546	16.3	1.62339	-0.03093
11.4	1.54356	-0.08240	16.4	1.61973	-0.03583
11.5	1.53571	-0.07857	16.5	1.61563	-0.04031
11.6	1.52835	-0.07401	16.6	1.61112	-0.04433
11.7	1.52155	-0.06879	16.7	1.60627	-0.04786
11.8	1.51535	-0.06297	16.8	1.60111	-0.05087
11.9	1.50981	-0.05661	16.9	1.59572	-0.05334
12.0	1.50497	-0.04978	17.0	1.59014	-0.05524
12.1	1.50087	-0.04257	17.1	1.58443	-0.05657
12.2	1.49755	-0.03504	17.2	1.57865	-0.05732
12.3	1.49501	-0.02729	17.3	1.57285	-0.05749
12.4	1.49327	-0.01938	17.4	1.56711	-0.05708
12.5	1.49234	-0.01141	17.5	1.56146	-0.05610
12.6	1.49221	-0.00344	17.6	1.55597	-0.05458
12.7	1.49286	0.00443	17.7	1.55070	-0.05252
12.8	1.49430	0.01214	17.8	1.54568	-0.04997
12.9	1.49647	0.01961	17.9	1.54097	-0.04694
13.0	1.49936	0.02676	18.0	1.53661	-0.04348
13.1	1.50292	0.03355	18.1	1.53264	-0.03962
13.2	1.50711	0.03989	18.2	1.52909	-0.03540
13.3	1.51188	0.04574	18.3	1.52600	-0.03088
13.4	1.51716	0.05104	18.4	1.52339	-0.02610
13.5	1.52290	0.05576	18.5	1.52128	-0.02111
13.6	1.52905	0.05984	18.6	1.51969	-0.01596
13.7	1.53552	0.06327	18.7	1.51863	-0.01071
13.8	1.54225	0.06602	18.8	1.51810	-0.00540
13.9	1.54917	0.06806	18.9	1.51810	-0.00010
14.0	1.55621	0.06940	19.0	1.51863	0.00515
14.1	1.56330	0.07002	19.1	1.51967	0.01029
14.2	1.57036	0.06993	19.2	1.52122	0.01528
14.3	1.57733	0.06914	19.3	1.52324	0.02006
14.4	1.58414	0.06767	19.4	1.52572	0.02459
14.5	1.59072	0.06554	19.5	1.52862	0.02883
14.6	1.59701	0.06278	19.6	1.53192	0.03274
14.7	1.60296	0.05943	19.7	1.53557	0.03628
14.8	1.60850	0.05554	19.8	1.53954	0.03943
14.9	1.61360	0.05113	19.9	1.54377	0.04215
15.0	1.61819	0.04628	20.0	1.54824	0.04442
15.1	1.62226	0.04102			

The cosine $C_i(x)$ and sine $S_i(x)$ are also plotted as a function of x in Figure E-1.

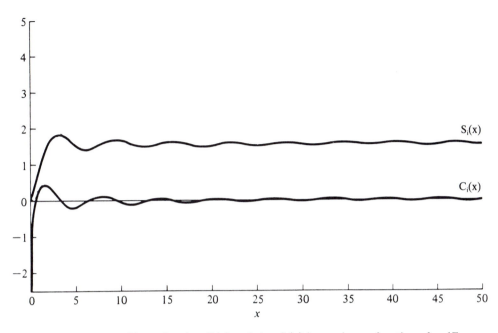

Figure E-1   Plots of cosine $C_i(x)$ and sine $S_i(x)$ integrals as a function of x. (*From* C. A. Balanis, *Antenna Theory*, *Analysis*, *and Design* [New York: Harper and Row, 1982], p. 747. Reprinted with permission by John Wiley & Sons.)

ANSWERS TO SELECTED PROBLEMS

CHAPTER 1

1.1

1. $a = |OP| \cos \alpha$, $b = |OP| \cos \beta$, $c = |OP| \cos \gamma$
2. $\mathbf{a}_p = \cos \alpha \, \mathbf{a}_x + \cos \beta \, \mathbf{a}_y + \cos \gamma \, \mathbf{a}_z$
3. Unit vector along $\mathbf{OQ} = \mathbf{a}_Q = \cos \alpha_1 \mathbf{a}_x + \cos \beta_1 \mathbf{a}_y + \cos \gamma_1 \mathbf{a}_z$
$\mathbf{a}_P \cdot \mathbf{a}_Q = \cos \theta = \cos \alpha \cos \alpha_1 + \cos \beta \cos \beta_1 + \cos \gamma \cos \gamma_1$

1.2

1. $\mathbf{B} = -2 \, \mathbf{a}_x - 3 \, \mathbf{a}_y - \mathbf{a}_z$
2. $5/\sqrt{14} = 1.3363$
3. $\theta = 110.93°$
4. $-0.841 \, \mathbf{a}_x + 0.535 \, \mathbf{a}_y + 0.0765 \, \mathbf{a}_z$

1.3

$0.0788 \, \mathbf{a}_x + 0.315 \, \mathbf{a}_y + 0.946 \, \mathbf{a}_z$

1.4

1. 20
2. $\theta = 41.08°$
3. 4.264

1.5

$B_x = 14.5$, $B_z = -25.5$, $C_y = -9$

1.6

1. $-0.9487\,\mathbf{a}_y + 0.3162\,\mathbf{a}_z$
2. area $= \sqrt{160}$

1.7

1. $\beta = -2$, $\alpha = -2$
2. $\mathbf{A}(x,y,z) = \dfrac{x-y}{\sqrt{x^2+y^2}}\mathbf{a}_x + \dfrac{x+y}{\sqrt{x^2+y^2}}\mathbf{a}_y + 3\,\mathbf{a}_z$

1.9

With respect to origin of **A**

$$\mathbf{C} = \mathbf{A} - \mathbf{B}' = 5\,\mathbf{a}_r - 3\,\mathbf{a}_\theta + 4\,\mathbf{a}_\phi$$

With respect to origin of **B**

$$\mathbf{C}' = \mathbf{A}' - \mathbf{B} = 4\,\mathbf{a}_r - 3\,\mathbf{a}_\theta - 5\,\mathbf{a}_\phi$$

1.10

1. $\mathbf{A} = 0.707\,\mathbf{a}_x + 0.707\,\mathbf{a}_y$
$\mathbf{B} = 1.73\,\mathbf{a}_y + 1.0\,\mathbf{a}_z$
2. $\theta = 52.24°$
3. $\mathbf{n} = 0.4475\,\mathbf{a}_x - 0.4475\,\mathbf{a}_y + 0.774\,\mathbf{a}_z$
4. $\theta_1 = 39.27°$

1.11

1. $c = 8$, $b = -3$
2. $b = \frac{8}{3}c + \frac{1}{3}$

1.13

$\mathbf{A}' = -\mathbf{a}_r + 2\,\mathbf{a}_\theta - 3\,\mathbf{a}_\phi$ (With respect to origin of **B**)
$\mathbf{A}' \times \mathbf{B} = 13\,\mathbf{a}_r + 5\,\mathbf{a}_\theta - \mathbf{a}_\phi$

1.14

Assume a unit positive test charge T at $(3,5,5)$

$$\mathbf{Q}_1\mathbf{T} = 2\,\mathbf{a}_x + 4\,\mathbf{a}_y + 5\,\mathbf{a}_z$$
$$\mathbf{Q}_2\mathbf{T} = 3\,\mathbf{a}_x + 3\,\mathbf{a}_y + 4\,\mathbf{a}_z$$
$$\mathbf{E}_T = 0.468\,\mathbf{a}_x + 0.527\,\mathbf{a}_y + 0.693\,\mathbf{a}_z \text{ V/m}$$

1.16

Place a unit positive test charge at P

$$\mathbf{Q}_A\mathbf{P} = 1.5\,\mathbf{a}_y + 2\,\mathbf{a}_z$$
$$\mathbf{Q}_B\mathbf{P} = -1.5\,\mathbf{a}_y + 2\,\mathbf{a}_z$$
$$\mathbf{E}_P = 86.33\,\mathbf{a}_y + 345.13\,\mathbf{a}_z$$

1.19

$\mathbf{F}_1 = 1/4\pi\epsilon_o(Q_1 Q_2/d^2)\,\mathbf{a}_{12}$
$\mathbf{F}_2 = 1/4\pi\epsilon_o[(Q_1 + Q_2)/2]^2/d^2\,\mathbf{a}_{12}$

1.21

1. $Q = 229nC$
2. $\alpha = 75.2°$

1.23

Force $\mathbf{F} = Qv_y B_o\,\mathbf{a}_x - Qv_x B_o\,\mathbf{a}_y = m\,\mathbf{a}$ (**a** is acceleration)
Therefore, acceleration is in the x-y plane.

1.24

$\mathbf{E} = v_o B_o[-14\,\mathbf{a}_y - 7\,\mathbf{a}_z]$

1.25

$\mathbf{B} = +E_x/v_y\,\mathbf{a}_z$

1.26

1. $\int_c \mathbf{F}\cdot d\ell = -25 + \rho_o^2$
2. $\int_{c_1} \mathbf{F}\cdot d\ell = 25 - \rho_o^2$
3. Yes

1.27

$W = 35 \times 10^{-6}\,J$

1.28

$\mathbf{A} = 4\,\mathbf{a}_y$

1.29

Element of length $d\ell$ along contour $= \rho d\phi\,\mathbf{a}_\phi$
The line integral $= \rho^2 = 9$

1.30

$W = 4J$

1.31

$\oint_s \mathbf{F}\cdot d\mathbf{s} = \pi^2/2 + \pi$

1.32

16π

1.33

1. 12
2. 0

1.34

1. 0
2. $\mathbf{E} = a\,\rho_s/\epsilon_o\,\rho\,\mathbf{a}_\rho$
3. $\mathbf{E} = (a - b)\,\rho_s/\epsilon_o\,\rho\,\mathbf{a}_\rho$

1.35

1. 0
2. $\mathbf{E} = a^2 \rho_s/\epsilon_o r^2 \, \mathbf{a}_r$
3. $\mathbf{E} = (a^2 - b^2) \rho_s/\epsilon_o r^2 \, \mathbf{a}_r$

1.36

1. $E_r = a^2 \rho_s/\epsilon_o r^2$ V/m $a < r < b$
   $= 0$ $r > b$
2. $\mathbf{J}_D = \partial(\epsilon_o \mathbf{E})/\partial t = -2 \times 10^{-4} a^2 \sin(10^5 t)/r^2 \, \mathbf{a}_r$ A/m

1.37

$E_r = 2\rho_o a^3/15 \epsilon_o r^2$ $r > a$
   $= (\rho_o r/3 \epsilon_o - \rho_o r^3/5 \epsilon_o a^2)$ $r < a$

1.39

$\psi_m = 2\pi B_o[\rho^2/2 - \rho^4/0.09] \sin \omega t$
$E_\phi = -\omega B_o[\rho/2 - \rho^3/0.09] \cos \omega t$

1.40

$i = -4 \cos 10^3 t$ A

1.42

emf $= 0.714 \times 10^4[\cos(10^4 t - 1.26) - \cos 10^4 t]$ V
$V = 4.868 \times 10^3[\cos(10^4 t - 1.26) - \cos(10^4 t)]$ V

1.43

$\mathbf{B} = B_\phi \mathbf{a}_\phi = 0.297 \, \mu_o/\rho \, \mathbf{a}_\phi$ $\rho > 0.5$
$\mathbf{B} = B_\phi \mathbf{a}_\phi = 1.25 \, \mu_o/\rho \, [1 - e^{-2\rho} - 2\rho e^{-2\rho}] \, \mathbf{a}_\phi$ $\rho < 0.5$

1.45

1. $B_\phi = \mu_o I_o \cos \omega t/2\pi\rho$ Wb/m^2
2. $\psi_m = \mu_o I_o b \cos \omega t/2\pi[\ell n(d + a)/d]$ Wb
   emf $= \mu_o I_o b \, \omega[\ell n(d + a)/d] \sin \omega t/2\pi$

1.46

$\psi_m = 3.84 \times 10^{-2} \mu_o I \cos \omega t$ Wb
emf $= 3.84 \times 10^{-2} \omega \mu_o I \sin \omega t$ V

1.47

1. Measure the vector magnetic field in terms of its three components. Each component is measured by placing a conducting loop with a unit vector (perpendicular to the plane of the loop) in the direction of the desired component of the magnetic field. The induced emf is proportional to the magnetic field component perpendicular to the loop. Three mutually perpendicular loops or one loop placed in three perpendicular positions measure the three components of the magnetic field.
2. emf $= -a \omega K_1 \cos \omega t[-1/(\rho_1 + b) + 1/\rho_1 - K^2 \ell n(\rho_1 + b)/\rho_1]$ V

CHAPTER 2

2.1

$$\nabla\psi = 3 E_o a^3 z \cos \phi/\rho^4 \mathbf{a}_\rho - E_o\left[1 - \left(\frac{a}{\rho}\right)^3\right] z \sin \phi/\rho \mathbf{a}_\phi + E_o\left[1 - \left(\frac{a}{\rho}\right)^3\right] \cos \phi \mathbf{a}_z$$

2.2

$\nabla \times \mathbf{A} = 2\,\mathbf{a}_z$
Vector $\mathbf{A}$ in cylindrical coordinates is $\mathbf{A} = \rho\,\mathbf{a}_\phi$
$\oint_c \mathbf{A} \cdot d\boldsymbol{\ell} = \int_s \nabla \times \mathbf{A} \cdot d\mathbf{s} = 2\pi$

2.3

$\oint_s \mathbf{F} \cdot d\mathbf{s} = \int \nabla \cdot \mathbf{F}\, dv = 72\pi$

2.5

1. $\nabla \times \mathbf{A} = 0$
2. $\nabla \times \mathbf{B} = 5x\,\mathbf{a}_y$
3. $\nabla \times \mathbf{C} = 0$
4. $\nabla \times \mathbf{D} = -2\rho\,\mathbf{a}_\phi$
5. $\nabla \times \mathbf{E} = -3y\,\mathbf{a}_x + x\,\mathbf{a}_y$
6. $\nabla \times \mathbf{F} = 0$

2.6

$\oint_c \mathbf{F} \cdot d\boldsymbol{\ell} = 1,\ \nabla \times \mathbf{F} = 2z\,\mathbf{a}_y,\ \int_{s_1} \nabla \times \mathbf{F} \cdot d\mathbf{s} = 1$

2.7

$\oint_c \mathbf{F} \cdot d\boldsymbol{\ell} = \int \nabla \times \mathbf{F} \cdot d\mathbf{s} = 0$

2.8

$\oint_c \mathbf{F} \cdot d\boldsymbol{\ell} = 2\pi,\ \nabla \times \mathbf{F} = 2\,\mathbf{a}_z$
$\int_{s_1} \nabla \times \mathbf{F} \cdot d\mathbf{s} = 2\pi,\ \int_{s_2 + s_3} \nabla \times \mathbf{F} \cdot d\mathbf{s} = 2\pi$

2.9

1. $\nabla \cdot \mathbf{A} = 0$
2. $\nabla \cdot \mathbf{B} = 0$
3. $\nabla \cdot \mathbf{C} = 3$
4. $\nabla \cdot \mathbf{D} = 24$
5. $\nabla \cdot \mathbf{E} = 3$

2.10

$\nabla \cdot \mathbf{F} = 3$
$\int_v \nabla \cdot \mathbf{F}\, dv = \oint_s \mathbf{F} \cdot d\mathbf{s} = 3\pi a^2 h/4$

2.11

$\nabla \times \mathbf{A} = -2\rho^2 \sin^2 \phi\,\mathbf{a}_\rho + 6\rho(z + 1)\sin^2\phi\,\mathbf{a}_z$
$\int_s \nabla \times \mathbf{A} \cdot d\mathbf{s} = 32\pi$

2.12

1. $\nabla \cdot \mathbf{A} = 0$
2. $\nabla \cdot \mathbf{B} = \phi/\rho$
3. $\nabla \cdot \mathbf{C} = 0$

4. $\nabla \cdot \mathbf{D} = 0$
5. $\nabla \cdot \mathbf{E} = 3$
6. $\nabla \cdot \mathbf{F} = z/\rho$

2.13

1. $\nabla \cdot \mathbf{A} = K, \ \nabla \times \mathbf{A} = 0$
2. $\nabla \cdot \mathbf{B} = 0, \ \nabla \times \mathbf{B} = -K \, \mathbf{a}_z$
3. $\nabla \cdot \mathbf{C} = 0, \ \nabla \times \mathbf{C} = 2K \, \mathbf{a}_z$
4. $\nabla \cdot \mathbf{D} = K/\rho, \ \nabla \times \mathbf{D} = 0$

2.14

$\nabla \times \mathbf{A} = -K/r \, \mathbf{a}_r - K \cot \theta/r \, \mathbf{a}_\theta$
$\int_s \nabla \times \mathbf{A} \cdot d\mathbf{s} = -\pi \, a \, K \, \cos \theta_1/2$

2.15

$\nabla \times \mathbf{E} = 0, \ \nabla \cdot (\epsilon_o \, \mathbf{E}) = \rho_v$

2.16

1. Satisfy Faraday's and Ampere's laws
2. $\mathbf{E}(z,t) = Re(\hat{\mathbf{E}}(z) \, e^{j\omega t})$
$\qquad\quad = 2 E_o \sin \beta z \, \sin \omega t \, \mathbf{a}_x$
$\quad \mathbf{B}(z,t) = 2 E_o/c \, \cos \beta z \, \cos \omega t \, \mathbf{a}_y$

2.17

$\nabla \cdot \mathbf{B} = 0, \ \mathbf{J} = [(\cos \phi \, \cos^2 \theta/\sin \theta) \, \mathbf{a}_\theta + (\sin \phi \, \sin 2\theta) \, \mathbf{a}_\phi]/\mu_o \, r^3$

2.18

$\mathbf{J}_D = \partial \epsilon_o \, \mathbf{E}/\partial t = -10 \, \epsilon_o \, \omega \sin \omega t \, \mathbf{a}_z \ \text{A/m}^2$
$I_D = \mathbf{J}_D \cdot \mathbf{S} = -0.1 \, \epsilon_o \, \omega \, \sin \omega t \ \text{A}$

2.20

$\nabla \cdot \mathbf{B} = 0, \ B_r = -\cos \theta \cos \phi - r \cos \phi/(3 \sin \theta)$

2.21

1 is not, whereas 2 is

2.22

$B_y = -\mu_o \, E_o(z^2/2) \cos \omega t$

2.25

$B_y = -(2 E_o z/\omega) \sin \omega t$

2.26

1. $\rho_v = 3.63 \, \epsilon_o$
2. $B_r = -\cos \phi(\cos \theta + r/3 \sin \theta)$

2.28

2. $f = 300$ MHz, $\lambda = 1$ m, $v_p = 3 \times 10^8$ m/s, negative z direction,
$\mathbf{H}(z,t) = -0.1 \cos (6\pi \times 10^8 t + 2\pi z) \, \mathbf{a}_y$ A/m

2.29

$f = 1{,}432$ MHz, $\lambda = \pi/15$ m,
$\mathbf{E}(z,t) = 40 \cos(9 \times 10^9 t - 30 z)\,\mathbf{a}_x$ V/m
$\mathbf{H}(z,t) = (1/3\pi) \cos(9 \times 10^9 t - 30 z)\,\mathbf{a}_y$ A/m

2.30

1. $\mathbf{a}_x$
2. positive z
3. $\lambda = 6$ m
4. 50 MHz, 20 ns
5. $\mathbf{H}(z,t) = 0.133 \cos(\omega t - \pi/3\, z + \pi/4)\,\mathbf{a}_y$
6. $\mathbf{E}(z,t) = 50 \cos(\omega t - \pi/3\, z + \pi/4)\,\mathbf{a}_x$, $\omega = 100\,\pi \times 10^6$ rad/s

2.31

1. negative z direction
2. $H_m = -0.25$ A/m, $\beta_o = 0.33$ rad/m, $\lambda = 6\pi$ m

2.34

1. $\mathbf{a}_y$
2. negative z direction
3. $f = 50$ MHz, $\lambda = 6$ m
4. $\mathbf{H}(z,t) = -0.0398 \cos(\pi \times 10^8 t + \pi/32)\mathbf{a}_x$

2.36

2. $f = 10$ GHz, $\beta = 209.4$ rad/m
$\mathbf{E}(z,t) = 200 \cos(\omega t - \beta z + \pi/4)\,\mathbf{a}_x$
$\hat{\mathbf{H}}(z) = (5/3\pi)\, e^{j\pi/4}\, e^{-j209.4\,z}\,\mathbf{a}_y$

CHAPTER 3

3.1

1. $\mathbf{P} = \epsilon_o \chi_e \mathbf{E} = 4.68\, \epsilon_o z^2 y\, \cos(10^8 t)\,\mathbf{a}_x$
2. $\rho_P = 0$
3. $\mathbf{J}_P = -4.68 \times 10^8\, \epsilon_o z^2 y\, \sin(10^8 t)\,\mathbf{a}_x$

3.2

1. $a < \rho < r_1$: $\mathbf{D} = \rho_\ell/2\pi\rho\,\mathbf{a}_\rho$
                           $\mathbf{E} = \rho_\ell/2\pi\epsilon_1\rho\,\mathbf{a}_\rho$
                           $\mathbf{P} = \rho_\ell/6\pi\rho\,\mathbf{a}_\rho$
   $r_1 < \rho < r_2$: $\mathbf{D} = \rho_\ell/2\pi\rho\,\mathbf{a}_\rho$
                           $\mathbf{E} = \rho_\ell/2\pi\epsilon_2\rho\,\mathbf{a}_\rho$
                           $\mathbf{P} = 7\rho_\ell/18\pi\rho\,\mathbf{a}_\rho$
   $\rho > r_2$: $\mathbf{D} = \mathbf{E} = \mathbf{P} = 0$
2. $\rho = a$, $\rho_{Ps} = -\rho_\ell/6\pi a$
   $\rho = r_1$, $\rho_{Ps} = -2\rho_\ell/9\pi r_1$
3. $r_1 < \rho < r_2$: $-\nabla \cdot \mathbf{P} = \rho_P = 0$

3.3

1. $\mathbf{H} = NI/2\pi\rho\,\mathbf{a}_\phi$ (All regions)
$\mathbf{B}_1 = 1500\,\mu_o\, NI/\pi\rho\,\mathbf{a}_\phi$, $\mathbf{M}_1 = 2999\, NI/2\pi\rho\,\mathbf{a}_\phi$ for $a < \rho < b$;
$\mathbf{B}_2 = (1 + 2/\rho)\,\mu_o\, NI/2\pi\rho\,\mathbf{a}_\phi$, $\mathbf{M}_2 = NI/\pi\rho^2\,\mathbf{a}_\phi$ for $\rho_1 < \rho < \rho_2$;

$\mathbf{B}_3 = \mu_o NI/2\pi\rho\, \mathbf{a}_\phi$, $\mathbf{M}_3 = 0$ for $c < \rho < d$

2. $\mathbf{J}_{m\Pi} = -NI/\pi\rho^3\, \mathbf{a}_z$

3. $\mathbf{J}_{ms} = -NI/c^2\, \mathbf{a}_z$ $\rho = c$,
$\mathbf{J}_{ms} = NI(1/b - 1499.5)/\pi b\, \mathbf{a}_z$ $\rho = b$

3.5

1. $H_{\phi 1} = \rho/4$, $0 < \rho < a$; $H_{\phi 2} = [a^2/4\rho + (a^3 - \rho^3)/6a\rho]$, $a < \rho < b$

4. At $\rho = a$, $H_{\phi 1} = a/4 = H_{\phi 2}$

3.6

1. $\mathbf{P} = 0.418\, \mathbf{a}_\rho - 2.508\, \mathbf{a}_\phi + 1.672\, \mathbf{a}_z$
$\rho_P = -0.418/\rho$

2. $\mathbf{D}_2 = (0.5 + 0.2 \times 10^{-6})\, \mathbf{a}_\rho - 0.492\, \mathbf{a}_\phi + 0.328\, \mathbf{a}_z$

3.8

1. $\mathbf{D} = Q/4\pi r^2\, \mathbf{a}_r$ (All regions)

$a < r < r_1$: $\mathbf{E}_1 = Q/4\pi\epsilon_o\epsilon_{r_1} r^2\, \mathbf{a}_r$,
                   $\mathbf{P}_1 = (\epsilon_{r_1} - 1)\, Q/\epsilon_{r_1} 4\pi r^2\, \mathbf{a}_r$
                   $\rho_{P_1} = 0$

$r_1 < r < r_2$: $\mathbf{E}_2 = Q/4\pi\epsilon_o\epsilon_{r_2} r^2\, \mathbf{a}_r$,
                   $\mathbf{P}_2 = (\epsilon_{r_2} - 1)\, Q/\epsilon_{r_2} 4\pi r^2\, \mathbf{a}_r$
                   $\rho_{P_2} = 0$

$r > r_2$: $\mathbf{E}_3 = Q/4\pi\epsilon_o r^2\, \mathbf{a}_r$,
                   $\mathbf{P}_3 = 0$

2. At $r = r_1$, $\rho_{Ps} = Q/4\pi r^2(1/\epsilon_{r_2} - 1/\epsilon_{r_1})$

3.9

$\mathbf{D}_2 = 6x\, \mathbf{a}_x + 8\sqrt{y}\, \mathbf{a}_y + 2.8\, \mathbf{a}_z$
$\mathbf{H}_1 = 4\, \mathbf{a}_x + 3y^2\, \mathbf{a}_y + 15.5\, \mathbf{a}_z$

3.10

At $y = 0$, $\rho_s = -j\omega\epsilon_o\mu_o a H_o \sin(\pi x/a)/\pi$
    $y = b$, $\rho_s = j\omega\epsilon_o\mu_o a H_o \sin(\pi x/a)/\pi$
    $x = 0$, $x = a$, $\rho_s = 0$
At $x = 0$, $\mathbf{J}_s = -H_o\, \mathbf{a}_y$
    $x = a$, $\mathbf{J}_s = -H_o\, \mathbf{a}_y$
    $y = 0$, $\mathbf{J}_s = -j\beta(a/\pi) H_o \sin(\pi x/a)\, \mathbf{a}_z + H_o \cos(\pi x/a)\, \mathbf{a}_x$

3.11

1. $\lambda = 0.0849$ m, $\beta = 74.01$ rad/m, $v_p = 84.9 \times 10^6$ m/s

2. $211.35\ \Omega$

3. $\mathbf{H}(z,t) = 0.47 \cos(2\pi \times 10^9 t - 74.01 z)\, \mathbf{a}_y$

3.12

$f \le 8.88$ MHz

3.14

$\alpha = 9.69$ Np/m
$\beta = 9.775$ rad/m

1. $\hat{E}_x = 200\, e^{-9.69z}\, e^{-j9.775z}$ V/m

2. $\hat{\eta} = 12.22 + j12.12\ \Omega$
$\hat{H}_y = 11.62\, e^{-9.69z}\, e^{-j(9.775z - 0.7812)}$ A/m $(0.7812$ rad $= 44.76°)$

3.15

2. At 20 GHz, $\alpha = 63.58$ Np/m
$z = 0.217$ m
At 20 kHz, $\alpha = 0.487$ Np/m
$z = 28.37$ m

3.16

1. $\mathbf{J}_D = -10\,\omega\,\epsilon_o\,\sin\omega t\,\mathbf{a}_z$
$I_D = -0.1\,\omega\,\epsilon_o\,\sin\omega t$
2. $I_{tot} = 0.4\cos\omega t - 0.044\sin\omega t$

3.18

$\mathbf{P}_{av} = E_o^2\,\sin^2(\pi x/a)/Z\,\mathbf{a}_z$ W/m^2
$P_{tot} = E_o^2\,ab/2Z$ W

3.19

1. f = 800 MHz, $\epsilon_r = 2.25$
2. $\mathbf{E}(z,t) = 50\cos(1.6\pi\times10^9 t - 25.13x)\,\mathbf{a}_z$
$\eta = 251.38\ \Omega$
$\mathbf{H}(z,t) = -0.1989\cos(1.6\pi\times10^9 t - 25.13x)\,\mathbf{a}_y$

3.20

2. 18.6

3.21

For nickel $\mu_r = 600$,

$\mathbf{J}_m = 599[(2r\cos\theta + (1/r^2)\cot\theta\sin\phi)\,\mathbf{a}_r - 3r\sin\theta\,\mathbf{a}_\theta - r\cos\theta\,\mathbf{a}_\phi]$

3.24

1. $\beta = 1$ rad/m, $\alpha = 0.693$, $\lambda = 6.28$ m,
$v_p = 1.508\times10^8$ m/s, $\delta = 1.45$ m
2. $\epsilon_r = 324$, f = 11.1 MHz,
$\mathbf{H}(z,t) = 6\cos(6.98\times10^7 t - (4\pi/3)z + 2\pi/3)\,\mathbf{a}_y$

CHAPTER 4

4.2

$\Phi = Q/4\pi\epsilon_o\sqrt{z^2 + a^2}$ V
$\mathbf{E} = (Q/4\pi\epsilon_o)[z/(z^2 + a^2)^{3/2}]\,\mathbf{a}_z$ V/m

4.3

$E_\rho = \rho_\ell/2\pi\epsilon_o\,\rho$ V/m
If reference of zero potential is assumed at ρ_o,

$\qquad \Phi = \rho_\ell/2\pi\epsilon_o\,\ell n(\rho_o/\rho)$

4.4

$\Phi = 2Q/4\pi\epsilon_o r - 2Qr/4\pi\epsilon_o r_1 r_2$ V
$\quad \approx 0 \qquad$ at $r \approx r_1 \approx r_2$

4.5

$$C = 1/[(1/C_1) + (1/C_2) + (1/C_3)]$$

$$C_1 = \frac{4\pi\epsilon_o}{\dfrac{1}{c+d} - \dfrac{1}{b}}, \quad C_2 = \frac{4\pi\epsilon_o\,\epsilon_r}{\dfrac{1}{c} - \dfrac{1}{c+d}}, \quad C_3 = \frac{4\pi\epsilon_o}{\dfrac{1}{a} - \dfrac{1}{c}}$$

4.6

$$\mathbf{E} = Q/4\pi\epsilon_o r^2\, \mathbf{a}_r, \qquad r > R$$
$$\mathbf{E} = r\,\rho_v/3\,\epsilon_o\, \mathbf{a}_r, \qquad r < R$$

4.24

$$\mathbf{B}_2 = 9\,\mu_o\,\mathbf{a}_x + 27\,\mu_o\,\mathbf{a}_z$$
$$\mathbf{B}_1 = 15.5\,\mu_o\,\mathbf{a}_x + 27\,\mu_o\,\mathbf{a}_z$$

4.29

$$L_{12} \approx \frac{\mu_o\,\pi\,a^2\,b^2}{2d^2}\,\mathrm{H}$$

4.30

$$L_{12} = \mu_o\,b\,\ell n((a+d)/d)/2\pi$$

CHAPTER 5

5.1

f = 150 MHz

5.2

1. $\mathbf{E}^{tot}(z,t) = 2E_m^+ \sin\beta z\, \sin\omega t\, \mathbf{a}_x$
$\mathbf{H}^{tot}(z,t) = 2(E_m^+/\eta) \cos\beta z\, \cos\omega t\, \mathbf{a}_y$
2. $\mathbf{P} = 4(E_m^{+2}/\eta) \sin\beta z\, \cos\beta z\, \sin\omega t\, \cos\omega t\, \mathbf{a}_z$
3. $\mathbf{P}_{ave} = 0$

5.3

1. $z = -21.23$ cm
2. $z = -10.63$ cm
3. $H_y(z,t) = 2.5 \cos\beta_1 z\, \cos\omega t$, $H_y(0,t) = 2.5 \cos\omega t$ A/m

5.4

1. $\epsilon_r = 9$
2. $z = -3.125$ cm
3. $\mathbf{J}_s = 3.5 \cos\omega t\, \mathbf{a}_x$

5.5

1. $\ell_1 = 7.5 \times 10^3$ m, $\ell_2 = 5.59$ m
2. $d = 4.184$ m

5.6

1. $\hat{\Gamma} = 0.33$
2. 82.5 V/m
3. 333.4 V/m
4. 0.884 A/m

5.7

1. $\mathbf{E}^T(z, t) = 96.5\, e^{-187.7\,z} \cos(2\pi \times 10^8 t - 189.3\, z + 0.691)\, \mathbf{a}_x$
2. $\mathbf{P}_{ave} = 62\, e^{-375.4\,z}\, \mathbf{a}_z$

5.9

1. $\mathbf{E}^{tot}(0) = 66.6\, e^{j2.8 \times 10^{-3}}\, \mathbf{a}_x$ V/m
2. $\hat{E}_{x2}^{tot}(O_2) = \hat{E}_{m3}^{+} = 68.38\, e^{-j92.4°}$ V/m

5.11

$\hat{\Gamma}(-0.1\lambda) = 0.5927\, e^{-j118.9°}$

5.12

$\hat{\Gamma}(-0.1\lambda) = 0.4\, e^{j46°}$

5.13

1. $\hat{\Gamma}(0_1) = 0.998\, e^{j178.1°}$
2. $\hat{E}_{m1}^{-} = 9.98\, e^{j178.1°}$ V/m

5.14

$\epsilon_r = 1.66,\ d = 58.2$ m

5.15

1. $\hat{E}_{m2}^{+} = 35.17\, e^{j125.81°}$ V/m

5.16

2. $z = -0.15$ m
3. $\mathbf{J}_s = 1/30\pi \cos \omega t\, \mathbf{a}_x$

5.17

2. Standing wave, $\epsilon_r = 2.25$; first null of **E** field is at $z = -0.628$ m, and of **H** field is at $z = -0.314$ m.

CHAPTER 6

6.1

1. $\lambda = 1.65$ m, $\mathbf{n}_\beta = -0.6\, \mathbf{a}_x + 0.8\, \mathbf{a}_y$
2. $\mathbf{E} = -\eta\, \mathbf{n}_\beta \times \mathbf{H} = -120\, \pi[(4 + 0.8j)\, \mathbf{a}_x + (3 + 0.6j)\, \mathbf{a}_y - 5.06\, \mathbf{a}_z]\, e^{-j3.8(-0.6x + 0.8y)}$

6.2

1. $\beta_x = 4.8$, $\mathbf{n}_\beta = 0.866\, \mathbf{a}_x + 0.3249\, \mathbf{a}_y - 0.379\, \mathbf{a}_z$
2. $\lambda = 1.13$ m
3. $\hat{\mathbf{H}} = (1/120\,\pi)(1.814\, \mathbf{a}_x - 5.09\, \mathbf{a}_y - 0.217\, \mathbf{a}_z)\, e^{-j(4.8x + 1.8y - 2.1z)}$

6.3

1. f = 30 MHz, $|\boldsymbol{\beta}| = 0.628$, $\mathbf{n}_\beta = 0.75\, \mathbf{a}_x - 0.433\, \mathbf{a}_y + 0.5\, \mathbf{a}_z$
2. $\mathbf{H}(z, t) = (1/96\,\pi)(-2\sqrt{3}\, \mathbf{a}_x + 2\, \mathbf{a}_y + 4\sqrt{3}\, \mathbf{a}_z) \cos[6\pi \times 10^7 t - 0.05\pi(3x - \sqrt{3}y + 2z)]$ A/m

6.4

1. f = 8 MHz

2. $\mathbf{n}_\beta = 0.433\,\mathbf{a}_x - 0.5\,\mathbf{a}_y - 0.75\,\mathbf{a}_z$
3. $\lambda = 12.5$ m

6.5

1. $\mathbf{n}_\beta = 0.5\,\mathbf{a}_y + \sqrt{3}/2\,\mathbf{a}_z$, $\mathbf{E}\cdot\mathbf{n}_\beta = 0$
2. $\mathbf{H}(z,t) = (1/120\,\pi)(5\sqrt{3}\,\mathbf{a}_y - 5\,\mathbf{a}_z)\cos[6\pi \times 10^7 t - 0.1\pi(y + \sqrt{3}z)]$

6.6

2. $\mu_r = 1$, $\epsilon_r = 4$

6.7

1. $\mathbf{P}_{ave} = (1/2)\,E_{x_o}H_{y_o}\cos^2(\beta_x x)\,\mathbf{a}_z$
2. Each of the electric field components show a traveling wave in the positive z direction and a standing wave along x. The real-time forms of these fields show this effect, and the fact that $\mathbf{P}_{ave}$ is in the z direction confirms it.

6.8

1. $\mathbf{P}_{ave} = 1.7\,E_o H_o \sin^2(0.5\beta z)\,\mathbf{a}_x$
2. Standing wave along z, traveling wave along x.

6.9

1. f = 238.7 MHz, $\mathbf{E}^i(z,t) = 25(0.87\,\mathbf{a}_x - 0.5\,\mathbf{a}_y)\cos(\omega t - (5x + 8.7z))$
2. $\theta_i = 30°$, $\theta_t = 18.43°$, $\hat{\tau}_\parallel = 0.7472$
$\hat{\mathbf{E}}^+ = (15.35\,\mathbf{a}_x - 2.954\,\mathbf{a}_z)\,e^{-j\beta_t \cdot \mathbf{r}}$, where
      $\boldsymbol{\beta}_t = 15.81(0.316\,\mathbf{a}_x + 0.949\,\mathbf{a}_z)$

6.11

1. $\theta_B = 67.79°$
2. $\hat{\mathbf{E}}^i = 5(0.378\,\mathbf{a}_x - 0.9258\,\mathbf{a}_z)\,e^{-j\beta(0.9258x + 0.378z)}$, $\beta = 9.93 \times 10^6$ rad/m
3. $\theta_t = 22.21°$, $\hat{\tau}_\parallel = 1.0$, $\boldsymbol{\beta}_t = 24.32 \times 10^6(0.378\,\mathbf{a}_x + 0.9258\,\mathbf{a}_z)$
4. $\theta_c = 24.09°$

6.12

1. f = 477.5 MHz, $\lambda = 0.628$ m
2. $\theta_i = 36.87°$, $\theta_t = 17.46°$
3. $\hat{\Gamma}_\parallel = -0.253$, $\hat{\Gamma}_\perp = -0.409$
$\hat{\mathbf{E}}^r(z) = (-0.7589\,\mathbf{a}_x - 0.818\,\mathbf{a}_y - 1.012\,\mathbf{a}_z)\,e^{-j(6x - 8z)}$

6.17

1. $\theta_t = 45°$, $\hat{\Gamma}_\parallel = 0.0718$,
$\hat{\mathbf{E}}^r = 0.0718\,E_m(\sqrt{3}/2\,\mathbf{a}_x + 1/2\,\mathbf{a}_z)\,e^{-j10\,\pi(-x + \sqrt{3}z)}$
2. $\boldsymbol{\beta}_t = 17.32\pi(\mathbf{a}_x + \mathbf{a}_z)$
$\hat{\mathbf{E}}^t = 0.536E_m(\mathbf{a}_x - \mathbf{a}_z)\,e^{-j17.32\pi(x + z)}$

CHAPTER 7

7.3

At $t = 0^+$, $v^+ = v_G Z_o/(Z_o + Z_G)$, $i^+ = v_G/(Z_o + Z_G)$
At $t \to \infty$, $v_L = v_G Z_L/(Z_L + Z_G)$

7.4

$v^+ = 3$ V, $\Gamma_G = 1/3$, $\Gamma_L = -1/5$
$v_R = v^+(1 + \Gamma_L)\,U(t - T) + v^+\Gamma_L\,\Gamma_G(1 + \Gamma_L)\,U(t - 3T) + v^+\Gamma_G^2\,\Gamma_L^2(1 + \Gamma_L)\,U(t - 5T)$
$+ \cdots U(t - T)$ is a unit step function at $t = T$.

7.5

$i_L = 2\,V_o/Z_o(1 - e^{-Z_o\,t/L})$, $t = 0$ (Inductor open circuit),
                                            $t \rightarrow \infty$ (Inductor short circuit)

7.6

$v_s = 1 + 4/9\ U(t - 4 \times 10^{-6}) + 4/45\ U(t - 6 \times 10^{-6}) + \ldots$ V
$v = 1$ V $0 \leq z \leq 150$ m, $v = 4/3$ 150 m $\leq z \leq 600$ m,
$v = 1/3$ 600 m $\leq z \leq 750$ m, $v = 8/15$ 750 m $\leq z \leq 900$ m

7.10

$v^+ = 6.67$ V, $\Gamma_G = -1/3$, $\Gamma_{11} = 0.201$, $\tau_{12} = 1.04$
1. $v_L = \tau_{12}v^+\,U(t - 5 \times 10^{-6}) + \Gamma_G\,\Gamma_{11}\tau_{12}v^+\,U(t - 9 \times 10^{-6}) +$
$\tau_{12}\Gamma_G^2\,\Gamma_{11}^2\,v^+\,U(t - 13 \times 10^{-6})$, t in seconds
3. $v_L(t \rightarrow \infty) = 6.5$ V
$i_L(t \rightarrow \infty) = 86.8$ mA

7.12

1. $\Gamma_G = 0.2$, $\Gamma_L = 0.5$
$v_s = 0.4$ V $0 \leq t \leq 1$ μs,
$v_s = 0.24$ V 6 μs $\leq t \leq 7$ μs,
$v_s = 0.024$ V 12 μs $\leq t \leq 13$ μs

7.14

1. $v = v^+\{(1 + (R - Z_o)/(R + Z_o)) + (1 - (R - Z_o)/(R + Z_o))\,e^{-t/\tau}\}$, $\tau = L/(R + Z_o)$
2. $v = v^+\{(1 + (R - Z_o)/(R + Z_o))\,e^{-t/\tau}\}$, $\tau = (R + Z_o)L/R\,Z_o$
3. $v = v^+\{2 - (1 - (R - Z_o)/(R + Z_o))\,e^{-t/\tau}\}$

7.16

1. $v_{tot} = 2v^+/R_{eq1} + 2v^+(R_2/R_{eq2} - 1/R_{eq1})\,e^{-R_{eq1}\,t/C\,R_{eq2}}$,
$R_{eq1} = 1 + Z_o/R_1$, $R_{eq2} = Z_o + R_2(1 + Z_o/R_1)$

7.17

1. $\hat{\gamma} = 0.6066 + j139.48$, $\hat{Z}_o = 126.13 - j0.548$ Ω
2. $\hat{Z}_{in} = 65.1 - j120.1$ Ω

7.19

1. $\hat{Z}_{in} = 195 + j180$ Ω
2. $\hat{V}_{in} = 79.8\,e^{j7.90}$ V, $P_{in} = 8.82$ W

7.20

$d/\lambda = 0.098$, $\ell_s/\lambda = 0.153$

7.22

1. $\hat{Z}_L = 30 - j40$
2. $d/\lambda = 0.0415$, $\ell_s/\lambda = 0.114$

7.23

$d_1/\lambda = 0.345\lambda$, $d_2/\lambda = 0.1$

7.25

$d_1/\lambda = 0.049\lambda$, $d_2/\lambda = 0.445$

7.30

1. f = 1.27 GHz, $\hat{Z}_{in} = 60 - j55$ Ω
3. $\ell_{min} = 0.15\lambda$, $\ell_{s1}/\lambda = 0.4$, $\ell_{s2}/\lambda = 0.303$

CHAPTER 8

8.1

1. $f_{c10} = 6.52$ GHz (TE), $f_{c01} = 12.5$ GHz (TE),
$f_{c20} = 13.04$ GHz (TE), $f_{c11} = 14.1$ GHz (TE, TM)
2. $f_{c10} = 3.26$ GHz, $f_{c01} = 6.25$ GHz,
$f_{c20} = 6.52$ GHz, $f_{c11} = 7.05$ GHz
3. $f_{c10} = 724.6$ MHz, $f_{c01} = 1.39$ GHz,
$f_{c20} = 1.45$ GHz, $f_{c11} = 1.57$ GHz

8.6

1. The mode may either be TM_{12} or TE_{12}.
2. f = 35 GHz
3. β = 494.5 rad/m
4. $f_{c12} = 25.84$ GHz, $\eta_{TE_{12}} = 558.8$ Ω, $\eta_{TM_{12}} = 254.3$ Ω

8.7

1. TM_{31} mode
2. $f_{c31} = 23.2$ GHz
3. f = 29.0 GHz
4. β = 364.7 rad/m

8.8

1. $f_{c21} = 3.61$ GHz, $f = 1.27 f_{c21} = 4.59$ GHz
2. β = 59.2 rad/m, η = 232.8 Ω

8.10

1. $a = 0.75$ cm, $b = 0.375$ cm
2. f_{c10} (air) = 20 GHz
3. Six TE modes—TE_{10}, TE_{01}, TE_{20}, TE_{11}, TE_{21}, and TE_{30}—and two TM modes—
TM_{11}, TM_{21}.

8.13

1. $f_{c10} = 1.97$ GHz, $f = 1.3 f_{c10} = 2.56$ GHz, $\eta_{TE_{10}} = 590$ Ω
2. Using equation (8.54), $H_o = 1.966$ A/m, amplitude of E_y from (8.44b) is 963.9 V/m.

8.14

1. $\lambda_g = 18$ cm, $f_c = 1.97$ GHz, f = 2.58 GHz
2. $\beta_{10} = 34.9$ rad/m, $\eta_{TE_{10}} = 583.8$ Ω

8.15

$Z_{in} = 660 - j467 \ \Omega$

8.18

Total power = 263 kW

8.19

For air and at f = 1.3 f$_c$,

$v = 3 \times 10^8$ m/s, $v_p = 4.7 \times 10^8$ m/s, $v_g = 1.92 \times 10^8$ m/s

For glass-filled wave guide at f = 1.1 f$_c$,

$v = 1.5 \times 10^8$ m/s, $v_p = 3.6 \times 10^8$ m/s, $v_g = 0.625 \times 10^8$ m/s

CHAPTER 9

9.1

1. Amplitude of magnetic field intensity is $8.13 \times 10^{-3} \underline{/-0.17°}$ at $r = 1$ m, $4.61 \times 10^{-4} \underline{/-13.7°}$ at $r = 5$ m, and $1.85 \times 10^{-4} \underline{/-55.6°}$ at $r = 10$ m.
2. At $r = 1$ m, the $1/r^2$ term is 1, whereas the $1/r$ term (i.e., $j\beta_o/r$) = 0.2096. The near-field term is approximately 5 times the radiating term. At $r = 5$ m, the $1/r^2$ term is 0.04 and the far-field term is 0.0419. A distance of $r = 5$ m, then, represents a borderline between near- and far-field radiation.

9.3

1. $\theta_E = 120°$, $\phi_H = 120°$
2. $\theta_N = 180°$, $\phi_N = 180°$
3. $D_o = 4$, $D(\theta, \phi) = 4 \sin\theta \cos\phi$

9.4

1. $\theta_E = 90°$, $\theta_N = 180°$, $D_o = 3/2$
2. $\theta_E = 120°$, $\phi_H = 90°$, $\theta_N = 180°$, $\phi_N = 180°$, $D_o = 8/\pi$

9.5

1. $\theta_E = 89.4°$, $R_a = 2.47 \ \Omega$
2. $\theta_E = 47.3°$

9.8

1. $\psi = 0$, $\beta_o d = \pi$
2. $\psi = 0$, $\beta_o d = 2\pi$
3. $\psi = \pm\beta_o d$, $\beta_o d = 4\pi$, 10π, and 1.6π

9.14

1. $Z_{in} = 19 - j70$, $\hat{\Gamma} = 0.78 \underline{/-68.5°}$
6. $\ell/\lambda = 0.25$, $\ell/d = 20$, $Z_{in} = 73 + j15$, $\hat{\Gamma} = 0.22 \underline{/26°}$

9.15

Assuming $Z_{11} = 73 + j42 \ \Omega$,

Z_{12} (from Figure 9.32a) = $44 - j33 \ \Omega$,

$Z_{in} = 117 + j9 \ \Omega$ ($d/\lambda = 0.3$)

9.16

$Z_{11} = 73 + j42 \ \Omega$,
Z_{12} (from Figure 9.32b) $= 0 - j7 \ \Omega$,
$Z_{in} = 73 + j35 \ \Omega$ ($d/\lambda = 0.3$)

9.18

Select $\ell/d = 100$
Z_{22} (driven) $= 50 + j15$, Z_{11} (reflector) $= 58 + j25$,
Z_{33} (director) $= 20 - j50$, $Z_{21} = Z_{12} = 60 - j30$,
$Z_{31} = Z_{13} = -10 - j25$, $Z_{21} = Z_{12} = 35 - j20$
$\hat{I}_1 = 0.02 \underline{/44°}$, $\hat{I}_2 = 0.026 \underline{/-109°}$, $\hat{I}_3 = 0.025 \underline{/132°}$

9.19

For $d = \lambda/2$, null is at $\theta = 60°$
For $d = 2\lambda$, nulls are at $\theta = 82.8°$, $68°$, $51.3°$, and $29°$

INDEX